Methods in Computational Molecular Physics

NATO ASI Series

Advanced Science Institutes Series

A series presenting the results of activities sponsored by the NATO Science Committee, which aims at the dissemination of advanced scientific and technological knowledge, with a view to strengthening links between scientific communities.

The series is published by an international board of publishers in conjunction with the NATO Scientific Affairs Division

A	**Life Sciences**	Plenum Publishing Corporation
B	**Physics**	New York and London
C	**Mathematical and Physical Sciences**	Kluwer Academic Publishers
D	**Behavioral and Social Sciences**	Dordrecht, Boston, and London
E	**Applied Sciences**	
F	**Computer and Systems Sciences**	Springer-Verlag
G	**Ecological Sciences**	Berlin, Heidelberg, New York, London,
H	**Cell Biology**	Paris, Tokyo, Hong Kong, and Barcelona
I	**Global Environmental Change**	

Recent Volumes in this Series

Volume 287—Coherence Phenomena in Atoms and Molecules in Laser Fields
edited by André D. Bandrauk and Stephen C. Wallace

Volume 288—Intersubband Transitions in Quantum Wells
edited by Emmanuel Rosencher, Børge Vinter, and Barry Levine

Volume 289—Nuclear Shapes and Nuclear Structure at Low Excitation Energies
edited by Michel Vergnes, Jocelyne Sauvage, Paul-Henri Heenen, and Hong Tuan Duong

Volume 290—Phase Transitions in Liquid Crystals
edited by S. Martellucci and A. N. Chester

Volume 291—Proton Transfer in Hydrogen-Bonded Systems
edited by T. Bountis

Volume 292—Microscopic Simulations of Complex Hydrodynamic Phenomena
edited by Michel Mareschal and Brad L. Holian

Volume 293—Methods in Computational Molecular Physics
edited by Stephen Wilson and Geerd H. F. Diercksen

Volume 294—Single Charge Tunneling: Coulomb Blockade Phenomena in Nanostructures
edited by Hermann Grabert and Michel H. Devoret

Series B: Physics

Methods in Computational Molecular Physics

Edited by

Stephen Wilson

Rutherford Appleton Laboratory
Oxfordshire, United Kingdom

and

Geerd H. F. Diercksen

Max-Planck-Institut für Astrophysik
Garching bei München, Germany

Springer Science+Business Media, LLC

Proceedings of a NATO Advanced Study Institute on
Methods in Computational Molecular Physics,
held July 22–August 2, 1991,
in Bad Windsheim, Germany

NATO-PCO-DATA BASE

The electronic index to the NATO ASI Series provides full bibliographical references (with keywords and/or abstracts) to more than 30,000 contributions from international scientists published in all sections of the NATO ASI Series. Access to the NATO-PCO-DATA BASE is possible in two ways:

—via online FILE 128 (NATO-PCO-DATA BASE) hosted by ESRIN, Via Galileo Galilei, I-00044 Frascati, Italy.

—via CD-ROM "NATO-PCO-DATA BASE" with user-friendly retrieval software in English, French, and German (© WTV GmbH and DATAWARE Technologies, Inc. 1989)

The CD-ROM can be ordered through any member of the Board of Publishers or through NATO-PCO, Overijse, Belgium.

Library of Congress Cataloging-in-Publication Data

Methods in computational molecular physics / edited by Stephen Wilson and Geerd H.F. Diercksen.
p. cm. -- (NATO ASI series. Series B, Physics ; v. 293)
"Proceedings of a NATO Advanced Study Institute on Methods in Computational Molecular Physics, held July 22-August 2, 1991 in Bad Windsheim, Germany"--T.p. verso.
"Published in cooperation with NATO Scientific Affairs Division."
Includes bibliographical references and index.

1. Molecules--Measurement--Congresses. 2. Molecular structure--Congresses. 3. Mathematical physics--Congresses. I. Wilson, S. (Stephen), 1950- . II. Diercksen, G. H. F. III. NATO Advanced Study Institute on Methods in Computational Molecular Physics (1991 : Bad Windsheim, Germany) IV. North Atlantic Treaty Organization. Scientific Affairs Division. V. Series.
QC170.M48 1992
539'.6'015118--dc20 92-13628
CIP

ISBN 978-1-4615-7421-7 ISBN 978-1-4615-7419-4 (eBook)
DOI 10.1007/978-1-4615-7419-4

Preface

This volume records the lectures given at a NATO Advanced Study Institute on *Methods in Computational Molecular Physics* held in Bad Windsheim, Germany, from 22nd July until 2nd. August, 1991.

This NATO Advanced Study Institute sought to bridge the quite considerable gap which exist between the presentation of molecular electronic structure theory found in contemporary monographs such as, for example, McWeeny's *Methods of Molecular Quantum Mechanics* (Academic Press, London, 1989) or Wilson's *Electron correlation in molecules* (Clarendon Press, Oxford, 1984) and the realization of the sophisticated computational algorithms required for their practical application. It sought to underline the relation between the electronic structure problem and the study of nuclear motion.

Software for performing molecular electronic structure calculations is now being applied in an increasingly wide range of fields in both the academic and the commercial sectors. Numerous applications are reported in areas as diverse as catalysis and interstellar chemistry, drug design and environmental studies, molecular biology and solid state physics. The range of applications continues to increase as scientists recognize the importance of molecular structure studies to their research activities. Recent years have seen a growing dependence of these applications on program packages, which are often not in the public domain and which may have a somewhat limited range of applicability dictated by the particular interests and prejudices of the program author. For example, the most widely used molecular electronic packages offer 'direct' SCF capabilities, energy gradients for molecular orbital functions (and correlation treatments developed with respect to a molecular orbital reference), and perturbation treatment of correlation effects again with respect to a molecular orbital model. Few can handle valence bond theories, which combine a simple, yet accurate, picture of molecular bonding in terms which are familiar to every chemist, with an ability to provide a rather accurate description of complete potential energy curves and surfaces, or relativistic effects in any well-founded fashion, in spite of their obvious importance to the chemistry of the heavy (and indeed not so heavy) elements. On the one hand, these packages have had a beneficial effect making molecular electronic structure calculations available to the non-specialist. Indeed, the majority of contemporary *ab initio* electronic structure calculations are carried out by non-specialists. On the other hand, the availability of such program packages might be said to have had a tendency to inhibit progress in molecular electronic structure theory and, more particularly, its computational realization. A major aim of this Advanced Study Institute and, therefore, this volume, was to stimulate research into the development of new theoretical computational techniques. The capabilities of contemporary techniques and algorithms is undoubtedly impressive, both in their accuracy and applicability, when compare with, say, what was available just ten years ago. However, it is clear that a great deal still remains to be done if the computational route to molecular information is to realize its full potential.

The introductory article by McWeeny gives a broad overview of molecular quantum mechanics and provides the background for the contributions which follow. In the article on the Born-Oppenheimer approximation, which lies at the heart of almost every application of quantum mechanics to molecular systems, Sutcliffe discusses the separation of electronic and nuclear motion in more detail than is usually given in text books and lays solid foundations for the later articles addressing the nuclear motion problem.

Most chemical processes involve quite small changes in the total energy of a molecular system and it is, therefore, necessary to consider corrections to independent electron models. The correlation problem is of central importance in molecular electronic structure theory. McWeeny gives a review of the techniques available for describing electron correlation and sets the scene for articles by Karwowski, McWeeny, Oddershede, Paldus and Wilson describing different aspects of a problem which has commanded the attention molecular physicists since the dawn of the computer age. Paldus gives a masterly overview of the group theoretical aspects of the many-body correlation problem. The "traditional" approach to the correlation problem, the configuration interaction expansion, is described in detail by Karwowski. A thorough discussion on the use of cluster expansions in the study of correlation effects is given by Paldus. Articles on the perturbation theory of electron correlation by Wilson deal with: basic Rayleigh-Schrödinger perturbation theory, many-body perturbation theory and the algebraic approximation, relativistic many-body perturbation theory and *concurrent computation* many-body perturbation theory. In his article, Oddershede describes response and propagator methods addressing first exact and then approximate polarization propagators. Two articles by McWeeny conclude the discussion of electron correlation. He provides one article on valence bond approaches, which have enjoyed a recent revival of interest, and a second on the the calculation of molecular properties.

The contribution by Helgaker and Jørgensen on the calculation of geometrical derivatives provides a link between articles on the electronic structure problem and those directed towards the study of nuclear motion. Jensen gives a timely and thorough review of the calculation of molecular rotation-vibration energies directly from the potential energy function, whilst Hubač and Svrček describes the many-body perturbation theory for the vibrational-electronic molecular hamiltonian.

The calculation of the electronic structure of periodic systems is a field of increasing importance and was addressed in lectures by Saunders describing the CRYSTAL program. These lectures are not reproduced in this volume since details of the latest version of the CRYSTAL program are to be published elsewhere (R. Dovesi, V.R. Saunders and C. Roetti *CRYSTAL92. An ab initio LCAO Hartree-Fock program for periodic systems*, University of Torino, 1992). The growing awareness of the effects of relativity in electronic structure theory is reflected in the contribution from Grant. The use of basis set expansions in quantum mechanical scattering theory for chemical reactions has lead to dramatic progress in recent years and, as Miller shows in his article, has a number of similarities with the electronic structure problem.

We are most grateful to the lecturers at the Institute not only for the high standard that they achieved in their lectures and accompanying notes but also for the stimulating atmosphere which they helped to create. It is our duty and privilege to thank the management and staff of the Kurhotel Residenz, Bad Windsheim, for their warm hospitality and attention to detail. We thank (in alphabetical order) Norbert Flocke, Peter Graf, Jan von Milczewski and Erich Schreiner for taking care of much of the local organization of the Institute. Finally, we are most grateful to NATO Division of Scientific Affairs for the generous support afforded to the Institute.

October, 1991

Stephen Wilson
Geerd H.F. Diercksen

Contents

INTRODUCTORY REMARKS

Stephen Wilson

Rutherford Appleton Laboratory
Chilton, Oxfordshire, OX11 0QX, England.

It is my pleasure, on behalf of the organizing committee†, to welcome you to this NATO Advanced Study Institute on *Methods in Computational Molecular Physics.*

It is a pleasure to welcome participants not only from ten of the member states of NATO (Belgium, Canada, Denmark, France, Germany, Italy, The Netherlands, Spain, Turkey, U.K., U.S.A.) and Sweden but also from South America (Brazil and Venezuela) and from Eastern Europe. Indeed, it is a particular pleasure to welcome participants from Czechoslovakia, Hungary, Lithuania, Poland, and what was formerly East Germany.

Some nine years ago, a similar Advanced Study Institute was organized in Bad Windsheim. Since that time there have been considerable advances in both theory and practice. On the theoretical side, breakthroughs have been made, for example, in the treatment of periodic systems, in the development of *ab initio* relativistic electronic structure theory, and the study of nuclear motion and the separation of the electronic and nuclear parts of the molecular structure problem. Dramatic progress has also been seen in computer technology and in computational techniques. A decade ago, *state-of-the-art* computers offered performance levels of tens of Mega-flops (10^6 floating point operations per second), today we measure performance in Giga-flops (10^9 floating point operations per second), expect tens of Gflops within a year, and are promised Tera-flop (10^{12} floating point operations per second) performance by the middle of the present decade. In 1982, the fruits of vector processing were beginning to be fully realized. Today, parallel processing is seen to be an increasingly important aspect of molecular electronic structure calculations.

The objective of all NATO Advanced Study Institutes is "To desseminate advanced knowledge not yet in university curricula and foster international scientific contacts through high-level teaching courses". I believe that the lecture courses at this Institute will provide a broad ranging, yet thorough, analysis of the most important aspects of contemporary research into the methods of computational molecular physics.

† G.H.F. Diercksen *(Max-Planck Institut fur Physik und Astrophysik, Garching)*
R. McWeeny *(Dipartimento di Chimica e Chimica Industriale, Universit'a di Pisa)*
S. Wilson *(Rutherford Appleton Laboratory)*

Methods in Computational Molecular Physics, Edited by Stephen Wilson and
Geerd H.F. Diercksen, Plenum Press, New York, 1992

AN OVERVIEW OF MOLECULAR QUANTUM MECHANICS

R. McWeeny

Dipartimento di Chimica e Chimica Industriale
Universitá di Pisa, Via Risorgimento 35, 56100 Pisa, Italy

1) Introductory Remarks

The fundamentals of molecular quantum mechanics have changed remarkably little since the 1930's: the most commonly used methods of constructing molecular wave functions are still dominated by expansions in terms of Slater determinants, in which individual electrons are described by spin-orbitals; the spin-orbitals (one-electron wavefunctions) still usually refer to an independent-particle model (IPM) in which each electron 'sees' an average field due to the nuclei and to all its companion electrons - in the spirit of Hartree and Fock; and simple first approximations, in which the wavefunction is built up from a few configurations of 'occupied' orbitals, are still systematically refined by adding more and more configurational functions (CFs) in a typical configuration interaction (CI) calculation. It is true that the techniques have changed: Fock space, for example, is now more familiar to quantum chemists than it used to be; the simple self-consistent field (SCF) of Hartree and Fock is nowadays often replaced by its multi-configuration counterpart; and CI may be admitted by diagrammatic perturbation theory or by unitary group methods. But, beautiful as some of these techniques may be, they are simply sophisticated and clever ways of doing the same old things!

What *has* changed, almost entirely as a result of vastly enhanced computing power, is our ability to employ an ever-increasing range of techniques in actual calculations on molecules. To use such computing power it is essential to have a good understanding of the theoretical techniques available: so let us make a global survey of the underlying fabric of the subject, starting from first principles and working our way forward to the present frontiers. This is an ambitious enough enterprise in itself, without saying a word about computers, and the intention in this chapter is therefore simply to lay some foundations: all the areas touched upon will be developed in full detail in later chapters – and some of the magnificent advances towards the goal of 'chemistry by computer' will become apparent.

2) The Schrödinger Equation, Hamiltonians and Wavefunctions

We recall that in Schrödinger's formulation of quantum mechanics the state of a many-particle system is described by a wavefunction

$$\Psi(\mathbf{q};t) = \Psi(q_1, q_2, ...q_n; t),$$

where $q_1, ...q_n$ are generalized coordinates of the particles involved (collectively denoted by $\mathbf{q}$) and t is the time. The time evolution of the system is then determined by solving the Schrödinger equation

$$i\hbar\partial\Psi/\partial t = \hat{H}\Psi, \tag{1}$$

where $\hat{H}$ is the Hamiltonian operator of the system, containing the coordinates q_i and the differential operators $\hat{p}_i$ associated with the conjugate momenta.

Methods in Computational Molecular Physics, Edited by Stephen Wilson and
Geerd H.F. Diercksen, Plenum Press, New York, 1992

In general, the full set of variables is much too large to deal with - even for small molecules - since it contains those of all the electrons and all the nuclei, including their spins ('non-classical' variables); and even at the outset we must therefore pass to a simplified model. The model most commonly used in molecular quantum mechanics is the one in which nuclear and electronic motions are 'separated', in first approximation, and in which all relativistic effects are neglected. The nature of this separation, and the methods of introducing nuclear-motion and relativistic effects, are taken up in later chapters. Here we assume the validity, at least as a good first approximation, of a non-relativistic Born-Oppenheimer Hamiltonian in which the nuclei occupy 'fixed' positions and only electronic variables are present. At this level, the spin does not appear in the Hamiltonian; but its presence must be recognized in constructing the wavefunction – which must be antisymmetric under an interchange of variables (space *and* spin) for any two electrons. To summarize, we start from (1) with

$$\hat{H} = \sum_i \hat{h}(i) + \frac{1}{2}\sum_{i,j}{}' g(i,j), \tag{2}$$

where (using atomic units)

$$\hat{h}(i) = -\tfrac{1}{2}\nabla^2(i) + V(i), \qquad g(i,j) = 1/r_{ij}; \tag{3}$$

and $\Psi = \Psi(\mathbf{x};t) = \Psi(\mathbf{x}_1, \mathbf{x}_2, ...\mathbf{x}_N; t)$. We also require (Pauli principle) that

$$\hat{P}\Psi = \epsilon_P \Psi, \tag{3}$$

$\epsilon_P(=\pm 1)$ being a parity factor for permutation P. We use a common convention, namely that $\mathbf{x}_i$ stands for the space-spin variables, collectively, of electron i, with $\mathbf{r}_i$ and s_i for space and spin separately. A particle permutation is thus equivalent to an *index* permutation, affecting both $\mathbf{x}_i$ and s_i simultaneously.

The calculation of molecular wavefunctions is usually concentrated on the *stationary states*, namely those particular solutions of (1) that have the form* $\Psi(\mathbf{x};t) = \Psi(\mathbf{x})f(t)$. It follows at once that $f(t) = \exp(-iEt/\hbar)$ and that $\Psi(\mathbf{x})$ must in this case satisfy the *time-independent* Schrödinger equation

$$\hat{H}\Psi = E\Psi. \tag{4}$$

Acceptable solutions exist only for specific values $E_1, E_2, ..E_K, ...$ of the separation parameter E and these *eigenvalues* of the operator $\hat{H}$ are in fact the *energies* of the posssible stationary states. The corresponding *eigenfunctions* $\{\Psi_K\}$ form a *complete set* in the usual sense, namely that an arbitrary function $\Phi(\mathbf{x}_1, \mathbf{x}_2, ...)$ may be developed in the form

$$\Phi(\mathbf{x}_1, \mathbf{x}_2, ...) = \sum_K c_K \Psi_K(\mathbf{x}_1, \mathbf{x}_2, ...) \tag{5}$$

with arbitrarily high accuracy. The summation is in principle *infinite* and if the spectrum of eigenvalues includes a continuum that part of the summation becomes an integral. In practice such expansions are always truncated; and wherever possible one tries to make use of sets that do not contain a continuum.

The property of completeness of eigenfunction sets is usually taken as axiomatic, owing to the obvious difficulty of finding rigorous proofs for sets of functions which are generally not even expressible in closed form**. Nevertheless this axiom lies at the root of almost all methods of calculating atomic and molecular wavefunctions.

3) Orbital Expansions of the Wavefunction

It is an elementary matter to show that, with neglect of the electron interaction terms $g(i,j)$ in

---

* No confusion should arise from the double usage of Ψ, the time t occurring as a parameter.

(3), the resultant Hamiltonian ($\hat{H}_0$, say) has exact eigenfunctions of *product form:*

$$\Phi^0_{ij...p} = \phi_i(\mathbf{r}_1)\phi_j(\mathbf{r}_2)...\phi_p(\mathbf{r}_N), \tag{6}$$

where the composite index $ij...p$ indicates any choice of the individual factors, called *orbitals*, provided these factors are drawn from the set $\{\phi_i\}$ of eigenfunctions of the *one*-electron operator $\hat{h}$ i.e. provided

$$\hat{h}\phi_i = \epsilon_i\phi_i \tag{7}$$

– a one-particle Schrödinger equation which is relatively easy to deal with numerically, even for a polyatomic system, in which case every ϕ_i is a *molecular orbital* (MO). The eigenvalue of $\hat{H}_0$ associated with the product function (6) is simply the sum of the *one*-electron energies, called 'orbital energies', of the 'occupied' orbitals

$$E^0_{ij...p} = \epsilon_i + \epsilon_j + ... + \epsilon_p. \tag{8}$$

The use of $\hat{H}_0$, or of any other 'model' Hamiltonian which contains only a sum of one-electron terms and *no interactions*, forms the basis of the *independent-particle model* IPM which has dominated molecular quantum mechanics for sixty years. The IPM has enormous interpretive value: a change of electronic state, for example, may be associated with a 'jump' of an electron from its original orbital (MO) ϕ_i to a new MO $\phi_{i'}$ and the change in the *total* electronic energy (8) will then be represented by $\epsilon'_i - \epsilon_i$ – easily visualized in terms of an energy level diagram. But the IPM also provides, in principle, complete sets which can be used in the expansion of *exact* wavefunctions for the *full* Hamiltonian (2); for an arbitrary function of the variables $\mathbf{r}_1, \mathbf{r}_2, ...\mathbf{r}_N$ can be represented as

$$\Phi = \sum_{ij...p} c_{ij...p}\Phi_{ij...p},$$

the summation being over the totality of all possible products (i.e. index choices). The superscript zero is from now on omitted, the use of $\hat{H}_0$ merely providing a particular expansion set.

So far, there is no spin! And indeed any spinless Hamiltonian will possess eigenfunctions which contain only spatial variables. It is the physics – not the mathematics – that tells us that a spin variable, and two spin eigenfunctions (α, β), must be admitted for every electron. To do so, it is sufficient to replace orbitals by *spin-orbitals*, writing $\psi_{i\sigma}(\mathbf{x}) = \phi_i(\mathbf{r})\eta_i(s)$ for the one-electron wavefunction for an electron in orbital ϕ_i with spin described by the factor η_i describing the spin state – which is either 'spin up' ($\eta_i = \alpha$) or 'spin down' ($\eta_i = \beta$). The discussion is then formally unchanged: with the single indices $i, j, ...$ now running over all possible *spin*-orbitals, the expansion of an acceptable wavefunction Ψ, with spin included, becomes

$$\Psi(\mathbf{x}_1, \mathbf{x}_2, ...\mathbf{x}_N) = \sum_{ij...p} c_{ij...p}\Phi_{ij...p}(\mathbf{x}_1, \mathbf{x}_2, ...\mathbf{x}_N) \tag{9}$$

where the typical expansion function is now

$$\Phi_{ij...p}(\mathbf{x}_1, \mathbf{x}_2, ...\mathbf{x}_N) = \psi_i(\mathbf{x}_1)\psi_j(\mathbf{x}_2)...\psi_p(\mathbf{x}_N) \tag{10}$$

– a *spin*-orbital product.

In the expansion (9), the antisymmetry requirement has not been insisted upon: but it is a trivial matter to show that the coefficients $c_{ij...p}$ cannot be independent if $\Phi_{ij...p}(\mathbf{x}_1, \mathbf{x}_2, ...\mathbf{x}_N)$ is to be antisymmetric under permutations. Thus, with the usual Dirac notation (and remembering the orthonormality of the spin-orbitals),

$$c_{ij...p} = < \Phi_{ij...p}|\Phi >$$

** In simple cases (e.g. for expansions of Fourier type) proofs of completeness *can* be given and for the coefficient of a product function in which the factors have merely been permuted we readily find

$$c_{\hat{P}[ij...p]} = \epsilon_P c_{ij...p}.$$

Consequently, using a standard ordering of the spin-orbitals (usually, in the IPM, ascending energy order) such that $i \leq j \leq ... \leq p$, the terms under the summation in (9) fall into sets of $N!$ whose coefficients differ only by a parity factor $\epsilon_P = \pm 1$. Such sets form *antisymmetrized products* or *Slater determinants* and the expansion of the wavefunction becomes, briefly,

$$\Psi = \sum_K c_K \Phi_K \tag{10}$$

where K now indicates an *ordered* index set $ij...p$ (in general $k_1k_2...k_N$, where the indices run over all menbers of the complete set) – or a spin-orbital *configuration*. The determinant Φ_K may be denoted by showing only the diagonal elements,

$$\Phi_K = (N!)^{-\frac{1}{2}} \det|\psi_i \psi_j ... \psi_p|, \tag{11}$$

and when the spin-orbitals are orthonormal this is a *normalized Slater determinant*. The systematic use of expansions such as (10) is usually referred to as the 'Slater method'. In order to determine the expansion coefficients, (11) may be substituted in the Schrödinger equation (4) and (after multiplying from the left by $\Phi^*_{ij...p}$ and integrating over all variables) the *operator* equation reduces to an infinite set of linear *algebraic* equations:

$$\mathbf{Hc} = E\mathbf{c} \tag{12}$$

where the infinite matrix $\mathbf{H}$ has elements

$$H_{KL} = <\Phi_K|\hat{H}|\Phi_L> = \int \Phi^*_K \hat{H} \Phi_L d\mathbf{x}_1 d\mathbf{x}_2 ... d\mathbf{x}_N, \tag{13}$$

whose evaluation forms the first step in calculating the wavefunction. The expansion is of course normally truncated and the solution of the linear equations proceeds by standard methods: it leads of necessity to an *approximate* solution of the operator equation – commonly referred to as the *algebraic approximation*. In many applications Φ_K may be a linear combination of determinants (e.g. exploiting some kind of symmetry) but the procedure is formally unchanged.
All numerical applications of the algebraic approximation have certain common features: in particular, they involve the computation of large numbers of 1- and 2-electron integrals over a set of *basis functions*, from which the orbitals are constructed (another algebraic approximation); and usually they involve some kind of *optimization* of the truncated wavefunction with respect to the parameters it contains. These aspects of a molecular calculation will be developed in due course.

The approach outlined above is not the only way of calculating a wavefunction. Although expansions of the type (10), in which the spin-orbitals are IPM wavefunctions for an electron in some kind of molecular 'effective field' (and are thus *molecular* orbitals), are basic to most present-day molecular computations, it must not be forgotten that the nature of the orbitals is arbitrary (as long as they form a complete 1-electron set) and that even the use of determinants is not obligatory. In pre-Slater days, for example, it was more common to employ *atomic* orbitals (AOs) - characteristic of the valence-bond (VB) approach - and the construction of antisymmetric wavefunctions followed an entirely different path, to which we now turn.

4) Permutation Symmetry: Alternative Expansions

Since the non-relativistic Hamiltonian we are using does not contain spin operators, it is perfectly possible to obtain, in principle, exact solutions of the Schrödinger equation which do not contain any spin variables! Whether or not such solutions are acceptable for describing a system of fermions then depends on how they behave under permutations of the 'orbital' (i.e. spatial) variables alone.

The possibility of using such functions rests upon group theoretical arguments which run briefly as follows.

We suppose that the g functions $\Phi_1, \Phi_2, ...\Phi_g$ comprise a g-fold degenerate set of exact eigenfunctions of the full Hamiltonian $\hat{H}$ (*not* an IPM Hamiltonian) with common energy E. Because $\hat{H}$ is invariant under electron permutations, any solution with the variables permuted is still a solution of the same equation and must therefore be a linear combination of the full set of solutions for the given energy E. We write

$$\hat{P}\Phi_\kappa = \sum_{\kappa'} \Phi_{\kappa'} D_{\text{orb}}(\hat{P})_{\kappa'\kappa}, \tag{14}$$

in which D_{orb} is the name of a *representation* of the permutation group S_N 'carried' by the spatial functions. In the absence of any 'accidental' degeneracy (the usual case) this representation is *irreducible* – it is an 'irrep'.

To recognize the existence of spin, we should attach spin factors to all the Φ_κ and seek linear combinations which (whilst remaining eigenfunctions with the same exact energy E) will conform to the Pauli principle. When we use spin *eigenfunctions* for this purpose, a similar argument applies: the operators $\hat{S}^2, \hat{S}_z$ are likewise invariant under electron permutations and will possess a common set of simultaneous eigenfunctions $\Theta_1, \Theta_2, ...\Theta_{f_S^N}$ with quantum numbers S, M, their number f_S^N being 'Wigner's number'

$$f_S^N = \binom{N}{\frac{1}{2}N - S} - \binom{N}{\frac{1}{2}N - S - 1}. \tag{15}$$

These functions will carry an irrep D_S, labelled by the spin quantum number* S and any product $\Phi_\kappa\Theta_\lambda$ will be a simultaneous eigenfunction of $\hat{H}, \hat{S}^2, \hat{S}_z$. What we need is a linear combination of all such products that will be totally antisymmetric. In fact, there is one such function and one only; it occurs when the two irreps are 'associate' (also called 'dual'), their corresponding matrices being related (in the real orthogonal case) by

$$\mathbf{D}_{\text{orb}}(\hat{P}) = \epsilon_P \mathbf{D}_S(\hat{P}) \tag{16}$$

for every permutation $\hat{P}$. This antisymmetric wavefunction (exact, for a non-relativistic Hamiltonian) is

$$\Psi = \sum_\kappa \Phi_\kappa \Theta_\kappa. \tag{17}$$

The Pauli principle now appears in a new light. For a given number of electrons N and given total spin S, the exact spatial eigenfunctions of $\hat{H}$, used in constructing the complete space-spin wavefunction, must comprise a degenerate set of the same dimensionality as the spin basis and carrying the same irrep (apart from the parity factor ϵ_P of the permutation group: *otherwise it is not possible to construct a physically acceptable wavefunction.* The necessary degeneracy of the spatial eigenfunctions is determined, through Wigner's formula (15), purely by N and S. Thus the singlet ground state of a 10-electron system cannot be constructed using a non-degenerate eigenfunction of the differential equation (4) – however low the corresponding energy may be. It can only be constructed from the lowest-energy 42-fold degenerate set! This, of course, is the real reason why the ground state of the neon atom cannot even roughly be represented by 10 electrons in a 1s orbital: the corresponding, non-degenerate, totally symmetric spatial function is not of acceptable symmetry.

The expectation values, in state Ψ of all spinless operators (including the energy), can be obtained from any *one* of the spatial functions Φ_κ: for the formula $< \hat{A} >=< \Psi|\hat{A}|\Psi > / < \Psi|\Psi >$ reduces at once to

$$< A >= \frac{< \Phi_\kappa|\hat{A}|\Phi_\kappa >}{< \Phi_\kappa|\Phi_\kappa >} \quad (\text{any } \kappa), \tag{18}$$

* The M value is irrelevant: the *same* irrep is repeated $2S + 1$ times, with one set of functions for each M.

and consequently we need seek only *one* spatial eigenfunction, so long as we insist on its being of pure symmetry species (S,κ) i.e. behaving under permutations like the κth basis vector of the irrep associate to D_S.

So far, all is exact. But normally, in computations, we must make an orbital expansion of Φ_κ. If we make a particular choice of κ, using the 'first' basis vector of the irrep (in some conventional sense), the symmetry label may be discarded and the spatial eigenfunction may be expanded in the form (cf. (10))

$$\Phi = \sum_K c_K F_K, \tag{19}$$

where all functions are of common symmetry species but those on the right are constructed from *orbital products* $\phi_{k_1}\phi_{k_2}...\phi_{k_N}$. The expansion is again infinite and the label K runs over all index sets $k_1k_2...k_N$ consistent with linear independence and the approach to completeness. Such expansions are of great generality but so far have been little used in computational work. They were first proposed by Matsen, as the basis for a 'spin-free quantum chemistry', but only in recent years have they been applied in actual molecular calculations.

5) Fock Space

Before passing to the optimization of approximate wavefunctions, it will be useful first to refer to the 'streamlined' versions of the Slater method which have become popular in recent years: they depend on the techniques of 'second quantization'.

The Slater determinants in the expansion (10) may be regarded as a particular realization of a set of basis vectors in an abstract linear space: when the space is enlarged, to include basis vectors for all possible numbers of electrons, it becomes 'Fock space'. The one-to-one correspondence between vectors and functions (i.e. spin-orbital determinants) may be indicated as

$$|ij...p> \Leftrightarrow \Phi_{ij...p}(\mathbf{x}_1,\mathbf{x}_2,...\mathbf{x}_N), \tag{20}$$

in which the N labels $i,j,...p$ indicate the spin-orbitals, normally assumed orthonormal, to which the N electrons are assigned. The same vector, or 'ket', may also be indicated by giving the *occupation numbers* (0 or 1) of the orbitals available; so the ket $|2\ 4\ 5\ ...\ >$ in the notation above could also be represented by $|0\ 1\ 0\ 1\ 1\ ...\ >$, showing that $\psi_2,\psi_4,\psi_5,...$ have been selected from the full set $\{\psi_i\}$. The occupation number notation is often convenient for a computer since, with M orbitals in the truncated set, M indices serve to specify any one of $\binom{M}{N}$ determinants.

What Fock-space operator corresponds to the Schrödinger Hamiltonian? A 1-electron operator connects kets which differ in just one index: how are such kets related? We need an operator which changes ψ_j, say, into ψ_i; it must 'destroy' (or 'annihilate') the electron in ψ_j and 'create' one in ψ_i. A 'substitution operator' is thus represented by $\hat{a}_i^+\hat{a}_j^-$, where $\hat{a}_i^+,\hat{a}_j^-$ are 'creation' and 'destruction' operators.

In order to work wih this new representation we have to construct an *isomorphism* between the Fock and Schrödinger spaces, so that operations in the one are 'echoed' by those in the other: more precisely, we seek those combinations of creation/annihilation products (with appropriate numerical coefficients) that will have the same matrix elements between Fock-space kets as the Schrödinger operators have between the corresponding Slater determinants. This isomorphism, for the kets and determinants, is indicated in (20); and antisymmetry of the determinant implies at once that

$$|ji...p> = -|ij...p>, \qquad |ii...p> = 0$$

– antisymmetry for exchange of indices and no spin-orbitals doubly occupied.

The properties of creation/annihilation products are discussed in terms of those of the *individual* operators - seldom used in the Schrödinger representation and providing an important generalization. By definition,

$$\hat{a}_q^+|ij...p> = |ij...pq> \; (\psi_q \textit{ not} \text{ initially occupied}), \quad = 0 \; (\text{otherwise})$$

- producing an $(N+1)$th electron in ψ_q unless there is one there already; while

$$\hat{a}_q^-|ij...pq> = |ij...p>$$

will indicate its subsequent destruction. The position of any index in the list must obviously be respected owing to antisymmetry: if q in the last equation had been n places from the end, a factor $(-1)^n$ would have appeared on the right.

It is also clear that *all* the basis vectors of Fock space can be produced by successive application of the creation operators to a 'vacuum state' $|0>$ corresponding to the total absence of electrons: thus

$$\begin{aligned} \hat{a}_{k_1}^+|0> &= |k_1> \Leftrightarrow \psi_{k_1}(\mathbf{x}_1) \\ \hat{a}_{k_2}^+\hat{a}_{k_1}^+|0> &= \hat{a}_{k_2}|k_1> = |k_1k_2> \Leftrightarrow (2!)^{-\frac{1}{2}}\det|\psi_{k_1}(\mathbf{x}_1)\psi_{k_2}(\mathbf{x}_2)| \end{aligned} \tag{21}$$

and so on. It is to be stressed that kets with two equal indices are identically zero and that operator products $\hat{a}_{k_1}^+\hat{a}_{k_2}^+...$ with repeated indices will therefore destroy any ket.

A characteristic of Schrödinger space is that it is a *metric* space; for any two elements there is a (Hermitian) *scalar product* defined as an integral over all electonic variables. The normalization integral for example is $<\Phi_{ij...p}|\Phi_{ij...p}> = \int \Phi_{ij...p}^*(\mathbf{x}_1,\mathbf{x}_2,...)\Phi_{ij...p}(\mathbf{x}_1,\mathbf{x}_2,...)d\mathbf{x}_1d\mathbf{x}_2...$. The corresponding scalar product in Fock space (using for a moment the same notation) will thus be

$$<\Phi_{ij...kp}|\Phi_{ij...kp}> = <\hat{a}_p^+\Phi_{ij...k}|\hat{a}_p^+\Phi_{ij...k}> = <\Phi_{ij...k}|(\hat{a}_p^+)^\dagger\hat{a}_p^+\Phi_{ij...k}> = 1,$$

defining the adjoint of an operator in the usual way. Thus $(\hat{a}_p^+)^\dagger$ *undoes* the work of $\hat{a}_p^+$. There is only one operator with this property – and it is $\hat{a}_p^-$. Similarly, $(\hat{a}_p^-)^\dagger = \hat{a}_p^+$ and it is therefore unnecessary to introduce the it two operators $\hat{a}_p^-$, $\hat{a}_p^+$; we simply write $\hat{a}_p$ for the destruction operator and $\hat{a}_p^\dagger$ for its adjoint, the creation operator.

With the new notation, the fundamental properties of the Fock-space operators are easily inferred: they are

$$\begin{aligned} \hat{a}_r^\dagger\hat{a}_s^\dagger + \hat{a}_s^\dagger\hat{a}_r^\dagger &= 0, \\ \hat{a}_r\hat{a}_s + \hat{a}_s\hat{a}_r &= 0, \\ \hat{a}_r\hat{a}_s^\dagger + \hat{a}_s^\dagger\hat{a}_r &= \delta_{rs}. \end{aligned} \tag{22}$$

These *anticommutation relations* are often written as $[\hat{a}_r^\dagger,\hat{a}_s^\dagger]_+ = 0$, etc., the plus sign indicating the *anti*-commutator.

The required operator isomorphism is now easily established. For a 1-electron operator such as $\hat{h}(i)$, the Schrödinger Hamiltonian contains $\sum_i \hat{h}(i)$ and we find

$$\sum_{i=1}^{N}\hat{h}(i) \Longleftrightarrow \sum_{r,s} <\psi_r|\hat{h}|\psi_s> \hat{a}_r^\dagger\hat{a}_s, \tag{23}$$

while for a typical 2-electron operator, such as $g(i,j)$ we find

$$\sum_{i,j=1}^{N} g(i,j) \Longleftrightarrow \sum_{r,s,t,u} <\psi_r\psi_s|g|\psi_t\psi_u>^\dagger \hat{a}_r^\dagger\hat{a}_s\hat{a}_u\hat{a}_t. \tag{24}$$

It is to be noted that in the Fock-space equivalents *there is no reference to the number of particles*: the operators are *number conserving* and connect kets only within any given N-electron subspace of Fock space (N arbitrary). Moreover, the summations on the right in (23) and (24) are over *all* the spin-orbitals of a complete set and the usual Schrödinger integrals such as $<\psi_r|\hat{h}|\psi_s>$ occur simply as *numerical parameters*. Truncation of the basis set and semi-empirical approximation of the integrals that remain makes no difference to the underlying theory: it merely leads to a *model system*, of which the Fock-space Hamiltonian provides a formally exact description. In this way the formalism provides a natural language for dealing with model Hamiltonians.

The generality of Fock space allows for the use of *single* creation or annihilation operators, to pass from states of an N-electron system to those of the *ion* with $N+1$ or $N-1$ electrons. The formalism is thus well adapted, as will be apparent in later chapters, to the discussion of electron removal and attachment processes. At this point, however, we use only number-conserving products, noting that all the familiar results of the traditional Slater approach can be derived without difficulty. Thus Slater's matrix element rules follow on writing $|1\ 2\ ...\ i\ ...\ N>= \hat{a}_N^\dagger\ ...\ \hat{a}_i^\dagger\ ...\ \hat{a}_1^\dagger|\mathrm{vac}>$ for the Fock-space equivalent of an N-electron determinant and using the Hamiltonian

$$\hat{H} = \sum_{r,s} <\psi_r|\hat{h}|\psi_s> \hat{a}_r^\dagger \hat{a}_s + \frac{1}{2}\sum_{r,s,t,u} <\psi_r\psi_s|g|\psi_t\psi_u>^\dagger \hat{a}_r^\dagger \hat{a}_s \hat{a}_u \hat{a}_t, \tag{25}$$

along with the anticommutation rules (22) to re-order and reduce the operator products.

We recall the results, namely, using $\Phi_{K'}$ for a determinant differing from Φ_K in *one* spin-orbital ($\psi_r' \neq \psi_r$) and $\Phi_{K''}$ for one with *two* spin-orbital differences ($\psi_r' \neq \psi_r$, $\psi_s' \neq \psi_s$),

$$\begin{aligned}
<\Phi_K|\hat{H}|\Phi_K> &= \sum_r <\psi_r|\hat{h}|\psi_{r'}> + \frac{1}{2}\sum_{r,s}{}'(<\psi_r\psi_s|g|\psi_r\psi_s> - <\psi_r\psi_s|g|\psi_s\psi_r>)\\
<\Phi_{K'}|\hat{H}|\Phi_K> &= <\psi_r'|\hat{h}|\psi_r> + \sum_{s(\neq r)}(<\psi_r'\psi_s|g|\psi_r\psi_s> - <\psi_r'\psi_s|g|\psi_s\psi_r>)\\
<\Phi_{K''}|\hat{H}|\Phi_K> &= (<\psi_r'\psi_s'|g|\psi_r\psi_s> - <\psi_r'\psi_s'|g|\psi_s\psi_r>)
\end{aligned} \tag{26}$$

and remember that these rules have been at the root of nearly all computations of electronic structure and properties for the last sixty years. They allow us to calculate all the matrix elements (13) and to proceed to the solution of the eigenvalue equation (12) which determines the approximate wavefunctions. It is because the expansion (10) usually converges very slowly (giving an eigenvalue equation of enormous dimensions) that further theoretical considerations are essential: in particular, how can we *optimize* the choice of the orbitals employed in order to obtain decent wavefunctions even with a severely truncated expansion? That this is a supremely important goal is well known – for the whole of Hartree-Fock theory and the SCF method is based on the use of just *one* term. A powerful generalization of the approach, embodied in multi-configuration (MC) SCF theory is now widely used, in a variety of forms. The Fock-space formalism often provides an elegant derivation of the 'master equations' on which the newer computational techniques are based, as will be seen presently; and the formalism is equally valuable for the inclusion of large-scale CI (e.g. by perturbation methods) in which the remaining terms in the expansion (10) are systematically admitted.

6) Optimization of the Wavefunction

At this point it is useful to analyse in a very general way the problem of optimizing the orbitals in a wavefunction of given form, as in SCF and MCSCF theory. We remember that, as follows from the Hylleraas-MacDonald theorem, an m-term approximation of the type (10), with coefficients determined by solution of the secular equations (12) leads to *upper bonds* to the first m energy eigenvalues (in ascending order); and that for a *single* term this implies the usual variation theorem

$$E_{\mathrm{stat}} = \frac{<\Psi|\hat{H}|\Psi>}{<\Psi|\Psi>} \geq E_1, \tag{27}$$

where E_{stat} is the stationary minimum value of the variation 'functional' (with respect to any parameters which Ψ may contain) and E_1 is the first exact eigenvalue.

The standard approach in all variational methods of optimizing the wavefunction is to consider the quantity (27), the expectation value of the Hamiltonian, namely

$$H(p_1, p_2, \dots) = \frac{< \Psi(p_1, p_2 \dots)|\hat{H}|\Psi(p_1, p_2, \dots) >}{< \Psi(p_1, p_2 \dots)|\Psi(p_1, p_2, \dots) >}, \tag{28}$$

as a function in 'parameter space' and to find the lowest stationary minimum by means of some suitable algorithm. The nature of the parameters is immaterial, so long as they are discrete and finite in number; the equations for the stationary point are always the same.

Let us introduce the following notation for the derivatives

$$\Psi^i = \partial\Psi/\partial p_i \qquad \Psi^{ij} = \partial^2\Psi/\partial p_i \partial p_j \Psi$$

and consider the variation of H around the point $\mathbf{p}_0 = (p_1^0 p_2^0 \dots)$, writing $\mathbf{p} = \mathbf{p}_0 + \mathbf{d}$ where $\mathbf{d} = (d_1 d_2 \dots)$ is a set of 'parameter displacements'. It is then not difficult to show (MMQM* Sect. 2.4), that up to second order in the displacements

$$H = H_0 + \begin{pmatrix} \mathbf{d} \\ \mathbf{d}^* \end{pmatrix}^\dagger \begin{pmatrix} \mathbf{g} \\ \mathbf{g}^* \end{pmatrix} + \frac{1}{2} \begin{pmatrix} \mathbf{d} \\ \mathbf{d}^* \end{pmatrix}^\dagger \begin{pmatrix} \mathbf{M} & \mathbf{Q} \\ \mathbf{Q}^* & \mathbf{M}^* \end{pmatrix} \begin{pmatrix} \mathbf{d} \\ \mathbf{d}^* \end{pmatrix} + \dots \tag{29}$$

In the first-order term, g contains components of the *gradient* vector, namely $g_i = \partial H/\partial p_i^* = \nabla_i H$ where the del on the left of the H indicates parameter variation in the *bra* factor; while $\mathbf{g}^*$ contains components $H\nabla_i = \partial H/\partial p_i$, for variation in the ket factor. Explicitly,

$$H\nabla_i = < \Psi|\Psi >^{-1} [< \Psi|\hat{H}|\Psi^i > - H < \Psi|\Psi^i >]. \tag{30}$$

In the second-order term of (29), $\mathbf{M}$ and $\mathbf{Q}$ are matrices with elements that may be denoted by

$$M_{ij} = (\nabla H \nabla)_{ij}, \qquad Q_{ij} = (\nabla\nabla H)_{ij}, \tag{31}$$

where Q_{ij}, for example, results when both variations are in the bra factor. Again there are symmetries –

$$M_{ji} = M_{ij}^*, \qquad Q_{ji} = Q_{ij} = (\nabla\nabla H)_{ij},$$

ensuring that the full partitioned matrix is Hermitian.

In terms of indices and summations, (29) may be written (remembering the dagger indicates transposition and starring)

$$H = H_0 + \sum_i d_i^* g_i + \sum_i d_i g_i^* + \sum_{i,j} d_i^* M_{ij} d_j + \frac{1}{2} \sum_{i,j} d_i^* d_j^* Q_{ij} + \frac{1}{2} \sum_{i,j} Q_{ij}^* d_i d_j + \dots . \tag{32}$$

It is assumed that all derivatives are evaluated at the 'initial point' $\mathbf{p}_0$ and it is evident that the second derivatives in $\mathbf{M}$ and $\mathbf{Q}$ are elements of a Hessian matrix.

The problem of optimization is how to reach the stationary point, given that the arbitrary starting point $\mathbf{p}_0$ is *not* such a point. To obtain a solution we assume the surface is roughly quadratic around $\mathbf{p}_0$ and seek a new point $\mathbf{p} = \mathbf{p}_0 + \mathbf{d}$, at which $H(\mathbf{p})$ will be stationary against a further small variation $\mathbf{d} \to \mathbf{d} + \delta\mathbf{d}$. With an obvious brief notation for the partitioned matrices, (29) becomes

---

* References shown in this way are to "*Methods of Molecular Quamtum Mechanics*" (see Bibliography)

$$H = H_0 + \mathbf{D}^\dagger\mathbf{G} + \tfrac{1}{2}\mathbf{D}^\dagger\mathbf{B}\mathbf{D}$$

and with $\mathbf{D} \rightarrow \mathbf{D} + \delta\mathbf{D}$ the energy variation around the *new* point $\mathbf{p}$ will be $\delta H = \delta\mathbf{D}^\dagger\mathbf{G} + \frac{1}{2}\delta\mathbf{D}^\dagger\mathbf{B}\mathbf{D} + \frac{1}{2}\mathbf{D}^\dagger\mathbf{B}\delta\mathbf{D}$. on adding the complex conjugate (noting that $\mathbf{B}^\dagger = \mathbf{B}$ and that the energy is *real*) we obtain the stationary condition

$$2\delta E = \delta\mathbf{D}(\mathbf{G} + \mathbf{D}\mathbf{D}) + (\mathbf{G} + \mathbf{B}\mathbf{D})^\dagger\delta\mathbf{D} = 0.$$

The two terms must separately vanish, for arbitrary $\delta\mathbf{D}$, and the resultant condition allows us to calculate* $\mathbf{D}$ $(= (\mathbf{d}/\mathbf{d}^*))$ which will carry us from the initial point $\mathbf{p}_0$ to the approximate stationary point $\mathbf{p}$.

The formal solution of the last equation, which provides the basis of all quadratic optimization algorithms, is clearly (in full)

$$\begin{pmatrix} \mathbf{d} \\ \mathbf{d}^* \end{pmatrix} = -\begin{pmatrix} \mathbf{M} & \mathbf{Q} \\ \mathbf{Q}^* & \mathbf{M}^* \end{pmatrix}^{-1} \begin{pmatrix} \mathbf{g} \\ \mathbf{g}^* \end{pmatrix}. \tag{33}$$

Of course this simplifies when all quantities are real (the usual case), reducing to

$$\mathbf{d} = -(\mathbf{M} + \mathbf{Q})^{-1}\mathbf{g}; \tag{34}$$

and in practice one solves directly the system of linear equations (e.g.by Gaussian elimination) instead of computing an inverse. But the main point is that, within a roughly quadratic 'basin', there is no difficulty of principle in proceeding straight to the stationary point *however general the form of the wavefunction.* Whenever it is feasible to calculate the second derivatives in (31), this type of *second-order* optimization process is to be preferred, convergence generally being much more rapid than in first-order processes which utilize only the *first* derivatives of the energy (e.g. as in the 'steepest descent' method).

An Example: Orbital Optimization

Let us make just one application, to illustrate the techniques introduced in these last two sections, one which leads directly to some of the most general and powerful methods at present available for the optimization of many-electron wavefunctions (see MMQM Ch.8).

Let Ψ be some orbital** approximation to the wavefunction, usually a truncated expansion of the form (10) with orbitals constructed (algebraic approximation) from some finite set $\{\psi_i\}$ of basis functions (initially including spin factors). When Ψ is normalized the variational energy is

$$E = H(\mathbf{p}) = <\Psi|\hat{H}|\Psi>, \tag{35}$$

in which $|\Psi>$ will be a vector of Fock space and $\hat{H}$ will assume the form (25). A variation of the wavefunction is then represented by

$$|\Psi> \rightarrow |\tilde{\Psi}> = \hat{U}|\Psi>, \tag{36}$$

where $\hat{U}$ must be a *unitary* operator ($\hat{U}^\dagger\hat{U} = 1$) in order to conserve normalization, $<\tilde{\Psi}|\tilde{\Psi}> = <\Psi|\hat{U}^\dagger\hat{U}|\Psi> = 1$. It is then convenient to write

$$\hat{U} = \exp(\hat{R}) = 1 + \hat{R} + \tfrac{1}{2}\hat{R}^2 + ... \qquad (\hat{R}^\dagger = -\hat{R}), \tag{37}$$

* We use $(\mathbf{d}/\mathbf{d}^*)$, for brevity, to indicate the column matrices in (29)

** When no confusion is likely, 'orbital' is taken to imply also 'spin-orbital'.

where $\hat{R}$ describes an 'infinitesimal rotation' and is *anti-Hermitian* to enforce $\hat{U}^\dagger\hat{U} = 1$.

The first-order energy variation will be

$$\delta E = <\Psi|\hat{U}^\dagger\hat{H}\hat{U}|\Psi> - <\Psi|\hat{H}|\Psi> = <\Psi|[\hat{H},\hat{R}]|\Psi>$$

(as follows from (37)) and the condition for a stationary point is thus

$$\delta E = <\Psi|[\hat{H},\hat{R}]|\Psi> = 0. \tag{38}$$

This 'generalized Brillouin condition', which corresponds to the well known Brillouin condition in simple Hartree-Fock (SCF) theory, leads quickly to the formulation of multi-configuration (MC-SCF) equations – studied in detail in later chapters.

To obtain one formulation of MCSCF theory we consider

$$\hat{R} = \sum_{st} \Delta_{st}\hat{a}_s^\dagger\hat{a}_t, \tag{39}$$

which from (23) is the Fock-space form of a general 1-electron operator, in this case containing 'rotation parameters' Δ_{st}. From (37), the parameter *matrix* must also be anti-Hermitian, $\mathbf{\Delta}^\dagger = -\mathbf{\Delta}$. It is clear that, in the 1-electron subspace spanned by the kets $\{\ |\psi_r>\ \}$,

$$\hat{R}|\psi_r> = \sum_{s,t}\Delta_{st}\hat{a}_s^\dagger\hat{a}_t|\psi_r> = \sum_s \Delta_{sr}\hat{a}_s^\dagger|vac> = \sum_s |\psi_s>\Delta_{sr}.$$

In other words, Δ provides a *matrix representation* of the operator $\hat{R}$; and consequently

$$\mathbf{V} = \exp(\Delta) \tag{40}$$

will provide a matrix representation of the full rotation operator $\hat{U}$ – which rotates *all* vectors of Fock space, for any number of electrons.

On using (39) in (38), the first-order energy variation becomes

$$\delta E = \sum_{s,t}\Delta_{st}(\hat{H}\hat{a}_s^\dagger\hat{a}_t - \hat{a}_s^\dagger\hat{a}_t\hat{H})$$

and (inserting (25) and using the anticommutation rules (22)) it is easy to reduce this expression to

$$\delta E = -\sum_{r,s}\Delta_{sr}(\epsilon_{rs} - \epsilon_{sr}^*), \tag{41}$$

where

$$\epsilon_{rs} = \sum_t <r|\hat{h}|t><\hat{a}_s^\dagger\hat{a}_t> + \sum_{t,u,v} <rv|g|tu><\hat{a}_s^\dagger\hat{a}_v^\dagger\hat{a}_u\hat{a}_t>. \tag{42}$$

This latter quantity takes a more familiar form on noting that the two Fock-space expectation values are, respectively, the 1- and 2-electron density matrix elements* ρ_{sr} and $\pi_{su,rt}$. Thus

$$\epsilon_{rs} = \sum_t <r|\hat{h}|t>\rho_{ts} + \sum_{t,u,v} <rv|g|tu>\pi_{tu,sv} \tag{43}$$

– first introduced (McWeeny,1955) in the traditional (Schrödinger-space) form of the MCSCF equations as a Lagrangian multiplier in an orthogonality-constrained orbital variation. The stationary value condition to be satisfied by the optimized orbitals will thus be, from (41),

$$\epsilon^\dagger = \epsilon, \tag{44}$$

which is probably the most succinct form of all MCSCF equations.

In order to to actually proceed from an arbitrary point p_0 in parameter space (the parameters here being the independent elements of the anti-Hermitian matrix Δ), one can make use of the second-order development (29). All quantities (integrals etc.) refer to the starting orbitals, with $\mathbf{d} = 0$; the gradient components in $\mathbf{g}$ are identified by reference to (41); the elements of $\mathbf{M}, \mathbf{Q}$ are similarly identified by going back to (38) and carrying second-order terms; and finally the required rotation parameters $(\Delta_{rs}, \Delta^*_{rs}, (r < s))$, contained in $\mathbf{d}, \mathbf{d}^*$, are obtained from (33). In this way one is lead (see, for example, MMQM Sect. 8.4) to the powerful MCSCF algorithms first formulated by Dalgaard and Jörgensen and further developed by Werner and Meyer.

7) The CI Problem: Methodology

The approximate wavefunction considered so far, consisting of a modest number of terms in the expansion (10) with orbitals and expansion coefficients fully** or partially optimized, is often called the 'reference (or root) function' and may in itself yield rather accurate values of the energy and other properties. But to obtain 'chemical' accuracy (remembering that to break a chemical bond may cost only 10^{-4} or less of the *total* electronic energy) it is necessary to include further configurations in which electrons have been promoted into the orbitals *not* used in the reference function, which form the complementary part of the complete set. In many applications (e.g. in any IPM approach) these are the 'virtual orbitals', which arise from the same eigenvalue equations as the occupied set but would describe physically the states of a 'ghost' electron in the field of the whole (i.e. neutral) molecule. In general, however, it is sufficient that they form a complementary set, so that completeness is approached as their number increases.

Clearly, even with a severely truncated basis set, the ensuing CI calculation rapidly reaches gigantic dimensions and nowadays it is not uncommon to include *millions* ot terms. It is therefore important to conduct the calculation as efficiently as possible. There are two main ways of doing this. Both will be fully developed later and here it is sufficient to explain the essentials of the procedures.

In the *direct* approach, we actually try to solve the full matrix eigenvalue equation (12), no matter how large it may be, by computing all matrix elements (not necessarily all at the same time) and using them to systematically improve the expansion coefficients c_κ. To do this (assuming that all 1- and 2-electron integrals are available) one needs to know how every integral contributes to every matrix element and how to use this information in constructing, using some suitable algorithm, an improved set of c_κ from a preceding approximate set. This procedure, in which the coefficients are iteratively refined without having to hold the full matrices $\mathbf{H}, \mathbf{M}$ in the computer (they would be much too big!) is usually known as 'direct CI': the knowhow that makes its most efficient implementation posssible comes from the theory of the *unitary group*.

The second main approach to the solution of (12) is based on refinement of the expansion coefficients by going to higher and higher orders of *perturbation theory*; and this is achieved most neatly (though it is not obligatory) by using the creation and annihilation operators already introduced and by the application of a *diagrammatic* classification of the correction terms in any given order.

Obviously, in a 'global survey', it is not possible to do more than drop hints as to how these methods proceed. Perturbation theory, in one form or another, will be familiar to everyone and the details will soon be filled in: the unitary group may be less familiar. So the rest of this introduction will be devoted to explaining briefly how the unitary group comes in and why it provides a powerful tool for large-scale CI (a fuller account is given in MMQM Ch.10).

First of all, why the *unitary* group? When we make a transformation from one basis to another

---

*At this point it is sufficient to define ρ_{sr} and $\pi_{su,rt}$ as the coefficients of $< r|\hat{h}|s >$ and $< rs|g|tu >$, respectively, in the expectation value of $\hat{H}$.

**Full optimization is difficult to achieve: ideally, orbitals should be allowed to 'float' away from the nuclei on which they are centred; and orbital exponents should also be optimized.

the corresponding 'rotation operator' ($\hat{U}$ in (37), for example) produces new basis functions which are linear combinations of the old. Thus, for a set of orbitals $\{\phi_i\}$,

$$(\phi_1' \ \phi_2' \ ...) = \hat{U}(\phi_1 \ \phi_2 \ ...) = (\phi_1 \ \phi_2 \ ...)\mathrm{U}, \tag{45}$$

where the columns of the matrix U contain the coefficients of $\phi_1, \phi_2, ...$ in each of the rotated basis functions $\phi_1', \phi_2', ...$. And when the basis sets are both orthonormal the matrix U must be *unitary*, $\mathrm{U}^\dagger\mathrm{U} = \mathrm{U}\mathrm{U}^\dagger = 1$. Moreover, in the N-electron space spanned by all possible products of N orbitals at a time, taken from $\{\phi_i\}$, the same rotation (45) will *induce* a corresponding rotation in which – in place of the matrix U – there stands a matrix $\mathrm{U} \times \mathrm{U} \times \ ...$ (N factors), where the x denotes the 'Krönecker product'. This transformation too is unitary.

In performing a full-CI calculation, to obtain approximate wavefunctions of given spin multiplicity, the totality of orbital products is used to span the 'CI space'; and consequently any transformation of orbitals (e.g. from MOs to localized linear combinations) will induce a unitary transformation in the CI space. Usually, spin factors are attached and Slater determinants are set up; but we know already (Section 4) that this is unnecessary, provided the orbital products are grouped into linear combinations which satisfy certain permutation symmetry reqirements. Thus, for a 2-electron system with a basis $\{\phi_i\}$ of m orbitals, we could expand the wavefunction in the form

$$\Phi(\mathbf{r}_1, \mathbf{r}_2) = \sum_{ij} c_{ij}\Omega_{ij}(\mathbf{r}_1, \mathbf{r}_2), \tag{46}$$

in which there would be m^2 products $\Omega_{ij} = \phi_i\phi_j$; and by solving a set of secular equations to determine the mixing coefficients we should find wavefunctions and energies at the 'full-CI' or 'basis set' limit. The same limit is reached, irrespective of the particular choice of basis (i.e. within a unitary transformation) in either the 1-electron space or the 2-electron space. In the *unitary group approach* (UGA) the aim is to calculate the matrix elements of the Hamiltonian *in the full-CI space* by exploiting the representation theory of the unitary group.

Very briefly, the Ω_{ij} products in (46) carry representations of *two groups*: under an orbital rotation $\hat{U}$ (transformation matrix U) each product turns into a new linear combination of products; and under an electron permutation each product turns into either itself or another product. Schematically

$$\begin{array}{ll} \{\Omega_{ij}\} \longrightarrow \{\hat{U}\Omega_{ij}\} & \text{(representation of U}(m)) \\ \downarrow & \\ \{\hat{P}\Omega_{ij}\} & \\ \text{(representation of S}_N) & \end{array}$$

The behaviour in the 'vertical' direction has been considered already in Section 4, where permuted products were combined in such a way as to provide *irreps* of S_N. In the present example the $m(m+1)$ combinations $\Omega_{ij} + \Omega_{ji} \quad (i \geq j)$ provide one irrep (appropriate to singlet states), while the $m(m-1)$ combinations $\Omega_{ij} - \Omega_{ji}$ provide another (appropriate to the triplet states). Similar considerations apply to the 'horizontal' behaviour: quite generally, the functions may be combined so as to provide bases for the irreps of $\mathrm{U}(m)$, thus reducing $\mathrm{U}(m)\times\mathrm{U}(m) \times ...$ into its irreducible components. In this way the possible combinations of the orbital products, which carry irreps of *both* groups, may be set out in a partitioned rectangular array in which the functions in any given block carry irreps of $\mathrm{U}(m)$, reading along a row; and irreps of S_m, reading down a column.

The value of reducing a representation is well known. In the present case, moving in the vertical direction, matrix elements between functions in different blocks (e.g. singlet and triplet) are zero, while those between functions within a column of the same block are determined *group theoretically* (from the irreps of the symmetric group S_N). Similar considerations apply in the horizontal direction: matrix elements between functions in different blocks are zero, while those between functions in a particular row of a given block are determined once we know the irreps of the *unitary* group.

To actually find the irreps, and subsequently the matrix elements of the Hamiltonian, it is sufficient to consider the *generators* of the two groups (those elements in terms of which all others may be expressed). For the symmetric group, these are the adjacent transpositions $\hat{P}_{n-1,n}$ which interchange electron indices $n-1$ and n. The situation is slightly more complicated for the unitary group because the number of elements is infinite. To see how the generators may be defined we recall the rotation operators used in Fock space (Section 5), where $\hat{U}|\phi_r>$ is a linear combination $\sum_r |\phi_s> U_{sr}$ and $|\phi_s>$ may in turn be expressed as

$$|\phi_s>= \hat{E}_{sr}|\phi_r>, \tag{47}$$

$\hat{E}_{sr}$ being the substitution operator $\hat{a}_s^\dagger \hat{a}_r$. An elementary 'two-by-two' rotation has a matrix $\mathbf{U}$ with only two non-zero elements, Δ_{rs} and Δ_{sr} $(= -\Delta_{rs}^*)$; and any finite transformation can clearly be expressed as a sequence of such rotations, each with its own (continuously variable) rotation parameter Δ_{rs}. The substitution operators $\hat{E}_{rs}$ thus play the part of the generators of a many-parameter continuous group.

As the discussion has been based on a spin-free approach, it is more appropriate to present the generators of the unitary transformations as ket-bra products: thus $\hat{E}_{sr} = |s><r|$ has the required property $\hat{E}_{sr}|p>= |s><r|p>= |s>$ $(p = r)$, $= 0$, otherwise. In Schrödinger language $\hat{E}_{sr}$ appears as an *integral operator* with kernel

$$E_{sr}(\mathbf{r};\mathbf{r}') = \phi(\mathbf{r})\phi^*(\mathbf{r}'), \tag{48}$$

so that, for an arbitrary function $f(\mathbf{r})$,

$$\hat{E}_{sr}f(\mathbf{r}) = \int E_{sr}(\mathbf{r};\mathbf{r}')f(\mathbf{r}')d\mathbf{r}' = \phi_s(\mathbf{r}) < \phi_r|f>, \tag{49}$$

this being the analogue of the symbolic statement preceding equation (48).

The last step is to make the connection with the CI problem and this must be, of course, through the Hamiltonian. In fact, *the Hamiltonian may be expressed in terms of the unitary group generators.* To see how this comes about we first note that $\hat{E}_{ss}(i)$, an operator in 1-electron space for electron i, is a *projection* operator: for any function $f(\mathbf{r}_i)$,

$$\hat{E}_{ss}(i)f(\mathbf{r}_i) = \phi_s(\mathbf{r}_i) < \phi_s|f>,$$

the component of f in 'direction' ϕ_s. If, then, we add together *all* the projected components in the (truncated) space $\{\phi_i\}$, we shall reproduce the original f: and this means that $\sum_s \hat{E}_{ss}(i)$ is equivalent to the *unit operator* in the 1-electron space. Consequently, as long as we work in 'full-CI' space (as determined by a basis set $\{\phi_i\}$), such unit operators may be inserted in the Hamiltonian to suggest alternative forms.

The 1-electron term in $\hat{H}$ may evidently be written as

$$\begin{aligned}\sum_{i=1}^{N} &= \sum_{i=1}^{N}[\sum_r \hat{E}_{rr}(i)]\hat{h}(i)[\sum_s \hat{E}_{ss}(i)] \\ &= \sum_{i=1}^{N}\sum_{r,s} |\phi_r(\mathbf{r}_i) < \phi_r|\hat{h}|\phi_s> \phi_s(\mathbf{r}_i)| = \sum_{r,s} < \phi_r|\hat{h}|\phi_s> \hat{E}_{rs}\end{aligned} \tag{50}$$

in which

$$\hat{E}_{rs} = \sum_{i=1}^{N} \hat{E}_{rs}(i) \tag{51}$$

is the usual symmetric sum of 1-electron operators (now the *generators* of orbital rotations) while

$\hat{h}$ itself appears only within the numerical coefficient $< \phi_r|\hat{h}|\phi_s >$. The situation is analogous to that obtaining in Fock space and (51) is closely parallel to (23).

The 2-electron terms may be dealt with in a similar way and the final result is

$$\hat{H} = \sum_{r,s} < \phi_r|\hat{h}|\phi_s > \hat{E}_{rs} + (1/2) \sum_{r,s,t,u} < \phi_r\phi_s|g|\phi_t\phi_u > (\hat{E}_{rt}\hat{E}_{su} - \delta_{ts}\hat{E}_{ru}). \tag{52}$$

This Hamiltonian contains only the *unitary group generators* (and their products) along with numerical parameters (1- and 2-electron integrals) which relate to the molecule studied and the choice of basis set: it provides a well-defined *model* of a chemical system and its exact eigenfunctions are attainable – corresponding to the full-CI or basis set limit.

The formulation of (52) brings us to some of the frontiers of current research. Full-CI calculations provide 'benchmarks' against which the quality of more severely truncated expansions can be measured: they may be completed by using (i) a graphical unitary group approach (GUGA) to obtain the matrix elements of $\hat{H}$ in terms of those of the generators $\hat{E}_{rs}$; and (ii) efficient methods of iteratively improving the CI expansion coefficients without actually holding all the elements of H (there may be hundreds of millions) in the computer. These are among the themes that are now to be developed in detail.

REFERENCES

Most of the topics touched upon in these notes are covered in more detail in the following books:

P. Jorgensen, J. Simons "Second Quantization-Based Methods in Quantum Chemistry" Academic Press, New York, 1981

R. McWeeny "Methods of Molecular Quantum Mechanics. 2nd Ed." Academic Press, London, 1989; Reprinted (paperback) with corrections, 1992 (referred to in the text as MMQM)

N. Szabo, S. Ostlund "Modern Quantum Chemistry" MacMillan, New York, 1982

S. Wilson "Electron Correlation in Molecules" Oxford University Press, Oxford, 1984

THE BORN-OPPENHEIMER APPROXIMATION

Brian T Sutcliffe

Deparment of Chemistry, University of York
York YO1 5DD, England

1. Introduction

Modern methods of electronic structure calculation can provide, at least for small molecules, accurate electronic energies at arbitary nuclear geometries. It is thus possible from a suitable number of electronic structure calculations to produce a fitted function that can be regarded as representative of the potential energy hypersurface or, to put it shortly, the potential. It is generally agreed that the potential plays a fundamental role in describing the relative motions of the nuclei. The potential for a single electronic state is thought to be able largely to account for the vibration-rotation spectra of the molecule in that state and also to account for its behaviour in elastic and inelastic scattering. It is thought too, that by considering potentials for a number of electronic states, accounts can be given of electronic-vibration- rotation spectra and indeed of reactive scattering.

In these lectures, I shall try to put these beliefs in context by considering the role of the potential in solutions of the full quantum mechanical problem for the molecule. In the first part of my talks I shall consider the clamped nucleus hamiltonian and some of the formal properties of its solutions. I shall then try to relate these solutions to the solutions of the full problem and consider how vibration-rotation spectra might be interpreted in terms of a single potential. Finally, I shall say a little about the many potential problem.

2. The Clamped Nucleus Problem

For the purposes of these talks it will be assumed that a molecule can be described by the ordinary Schrödinger form of the Hamiltonian for the appropriate number of electrons and nuclei and that the dominant interactions between the particles is the coloumb one. Spin will be considered only in so far as it determines the statistics of the particles.

Consider a collection of N particles labelled in the laboratory-fixed frame as $\underline{x}_i; i = 1, 2...N$ with masses m_i and charges $Z_i e$. The charge-numbers Z_i are positive

Methods in Computational Molecular Physics, Edited by Stephen Wilson and Geerd H.F. Diercksen, Plenum Press, New York, 1992

for a nucleus and minus one for an electron. In a neutral system the charge-numbers sum to zero. It will be convenient to think of the $\underline{x}_i$ as a column matrix of three cartesian components $x_{\alpha i}$, $\alpha = x, y, z$ and to think of the $\underline{x}_i$ collectively as the 3 by N matrix $\underline{x}$.

The separation between particles is then defined by :

$$x_{ij}^2 = \sum_{\alpha}(x_{\alpha j} - x_{\alpha i})^2 \tag{1}$$

where the α - sum runs over x y and z.

The laboratory-fixed form of the Schrödinger Hamiltonian describing this system of N particles is then:

$$\hat{H}(\underline{x}) = -\frac{\hbar^2}{2}\sum_{i=1}^{N} m_i^{-1}\nabla^2(\underline{x}_i) + \frac{e^2}{8\pi\epsilon_o}\sum_{i,j=1}^{N}{}' \frac{Z_i Z_j}{x_{ij}} \tag{2}$$

where the notation is standard. This will be taken as the full molecule Hamiltonian.

Since the aim initially is to explicate the standard clamped-nucleus electronic Hamiltonian, a notational distinction will be made between electronic and nuclear variables. Let the number of electrons in the problem be L and the number of nuclei, H so that $N = H + L$. Let the set of laboratory fixed coordinates be split into an electronic set, $\underline{x}_i^e, i = 1, 2, ...L$ and a nuclear set $\underline{x}_i^n, i = 1, 2, ...H$. The electronic charge-numbers will be taken explicitly as -1 and the electronic mass will be written as m. The charge-numbers Z_i and the masses m_i will, from now on, be taken as referring to the nuclei only.

The clamped-nucleus electronic Hamiltonian is then specified as:

$$\hat{H}^{cn}(\underline{a}, \underline{x}^e) = -\frac{\hbar^2}{2m}\sum_{i=1}^{L}\nabla^2(\underline{x}_i^e) - \frac{e^2}{4\pi\epsilon_0}\sum_{i=1}^{H}\sum_{j=1}^{L}\frac{Z_i}{|\underline{x}_j^e - \underline{a}_i|} + \frac{e^2}{8\pi\epsilon_0}\sum_{i,j=1}^{L}{}'\frac{1}{|\underline{x}_i^e - \underline{x}_j^e|} \tag{3}$$

This Hamiltonian is obtained from the laboratory-fixed one simply by assigning the values $\underline{a}_i$ to the nuclear variables $\underline{x}_i^n$, hence the designation *clamped-nucleus* for this form.

The Hamiltonian (3) specifies a perfectly well posed problem and arguments such as are given in [1] show that for any neutral or positively charged species the Hamiltonian has an infinite number of bound states (in the sense of being square integrable with negative eigenvalues) for arbitrary choice of a set of the $\underline{a}_i$. The eigenvalue for a particular choice of the $\underline{a}_i$ is called the *electronic energy* at the specified geometry.

Although the $\underline{a}_i$ in (3) are regarded simply as possible values of the $\underline{x}_i^n$ and hence chooseable at will, in practice in any sequence of calculations the sets are chosen so that no set in the sequence can be mapped on to any other set in the sequence merely by means of a uniform translation or a rigid rotation or rotation-inversion. Thus the $\underline{a}_i$ are regarded as specifying definite nuclear *geometries* rather than absolute positions in space. The electronic energies calculated with these constraints on the $\underline{a}_i$ can therefore be regarded as representative values of a single valued function $E(\underline{q})$ where the $\underline{q}$ are $3H - 6$ internal coordinates which are invariant to translations and rotations of the system as a whole. Since the nuclear repulsion energy is also invariant under these translations and rotations, that too can be written as a function of the chosen internal coordinates and hence the potential, as the sum of the electronic energy and the nuclear repulsion energy, has the required invariance. The choice of the $\underline{q}$ is completely open but in practical calculations it is not unusual to satisfy the translational invariance requirement by putting one nucleus in a fixed position (often at the origin of coordinates) and then specifying the positions of the remaining nuclei relative to this fixed one. The ways of satisfying the rotational invariance requirement are more varied but it is quite common to satisfy them choosing a pair of bonds to define a fixed plane. However this may be, it can be shown that, in principle, any set of translation-rotation invariant internal coordinates can be expressed in terms of any other set. (Since the relations between such sets are usually non-linear, the process of expressing one in terms of the other is at best messy and difficult and can be impossible in closed form.) Thus, if it is possible by choosing any set of internal coordinates to manipulate the full problem to yield a form like (3), then the position of the clamped nucleus Hamiltonian may be regarded as established.

It is widely believed that the decoupling of electronic and nuclear motions in such a way as to yield the clamped nucleus electronic Hamiltonian can be justified by reference to work presented in a paper [2] by Born and Oppenheimer published in 1927. Nowadays, however, it is more usual to attempt a justification in terms of an approach that was presented first by Born in 1951 and written up in a generally available form in Born and Huang [3] in 1954. This approach is usually assumed to supersede and subsume the older one and most popular expositions use this approach and call it the Born-Oppenheimer approximation. (It is interesting to notice that the same idea had actually been proposed by Slater [4] somewhat earlier than the original Born and Oppenheimer paper.) In the Born and Huang approach, the wave function for the full problem is expressed in terms of products of eigenstates of an electronic Hamiltonian with functions of the nuclear variables only. The functions of the nuclear variables are determined by solving a problem posed in terms of the potential energy surfaces associated with electronic eigenstates. Because the potentials are translation-rotation invariant it is clear that they must arise from the solution of a body-fixed problem from which translations and rotations have been separated off and the way in which such a body-fixed problem should be constructed is the subject of the next section. In the section that follows that, the nature of the product function separation is considered and an attempt is made there to place the clamped nucleus Hamiltonian. However to anticipate the outcome of that process here, it will be seen that it is not possible to assign that Hamiltonian a *unique* position.

3. The Full Molecule Problem

For ease of exposition the full problem will be considered first, without distinguishing between electrons and nuclei. In this way too, it will be possible to show what is generally the case irrespective of any particular coordinate choice. Again for ease of exposition, the move to fully internal coordinates will be accomplished in two stages. In the first, translationally invariant coordinates will be introduced to construct a space-fixed Hamiltonian and then rotationally invariant coordinates will be constructed to yield a body-fixed Hamiltonian.

First consider the space-fixed Hamiltonian. This may be constructed by the coordinate transformation symbolised by:

$$(\underline{t}\ \underline{X}_T) = \underline{x}\ \underline{V} \tag{4}$$

In (4) $\underline{t}$ is a 3 by $N-1$ matrix and $\underline{X}_T$ is a 3 by 1 matrix. $\underline{V}$ is an N by

N matrix which, from the structure of the left hand side of (4), has a special last column whose elements are :

$$V_{iN} = M_T{}^{-1} m_i, \qquad M_T = \sum_{i=1}^{N} m_i \tag{5}$$

so that $\underline{X}_T$ is the standard centre-of-mass coordinate.

$$\underline{X}_T = M_T{}^{-1} \sum_{i=1}^{N} m_i \underline{x}_i \tag{6}$$

The coordinates $\underline{t}_j, j = 1,2,....N-1$ are to be translationally invariant, so it is required on each of the remaining columns of $\underline{V}$ that:

$$\sum_{i=1}^{N} V_{ij} = 0, \qquad j = 1,2,....N-1 \tag{7}$$

and it is easy to see that (7) forces $\underline{t}_j \to \underline{t}_j$ as $\underline{x}_i \to \underline{x}_i + \underline{a}$, all i.

The $\underline{t}_i$ are independent if the inverse transformation

$$\underline{x} = (\underline{t}\,\underline{X}_T)\underline{V}^{-1} \tag{8}$$

exists. The structure of the right hand side of (8) shows that the bottom row of $\underline{V}^{-1}$ is special and it is easy to see that, without loss of generality, we may require its elements to be:

$$(\underline{V}^{-1})_{Ni} = 1 \qquad i = 1,2.....N \tag{9}$$

The inverse requirement on the remainder of $\underline{V}^{-1}$ implies that:

$$\sum_{i=1}^{N} (\underline{V}^{-1})_{ji} m_i = 0 \qquad j = 1,2,....N-1 \tag{10}$$

Writing the column matrix of the cartesian components of the partial derivative operator as $\partial/\partial\underline{x}_i$ then the coordinates change (4) gives:

$$\frac{\partial}{\partial \underline{x}_i} = m_i M_T{}^{-1} \frac{\partial}{\partial \underline{X}_T} + \sum_{j=1}^{N-1} V_{ij} \frac{\partial}{\partial \underline{t}_j} \tag{11}$$

and hence the Hamiltonian (2) in the new coordinates becomes:

$$\hat{H}(\underline{t},\, \underline{X}_T) = -\frac{\hbar^2}{2M_T}\nabla^2(\underline{X}_T) - \frac{\hbar^2}{2}\sum_{i,j=1}^{N-1} \mu_{ij}^{-1} \vec{\nabla}(\underline{t}_i).\vec{\nabla}(\underline{t}_j) + \frac{e^2}{8\pi\epsilon_o}\sum_{i,j=1}^{N}{}' \frac{Z_i Z_j}{f_{ij}(\underline{t})}. \tag{12}$$

Here

$$\mu_{ij}^{-1} = \sum_{k=1}^{N} m_k{}^{-1} V_{ki} V_{kj} \qquad i,j = 1,2,...N-1 \tag{13}$$

and f_{ij} is just x_{ij} as given by (1) but expressed as a function of the $\underline{t}_i$. Thus:

$$f_{ij}(\underline{t}) = (\sum_{\alpha}(\sum_{k=1}^{N-1}((\underline{V}^{-1})_{kj} - (\underline{V}^{-1})_{ki})t_{\alpha k})^2)^{1/2} \tag{14}$$

In (12) the $\vec{\nabla}(\underline{t}_i)$ are the usual grad operators expressed in the cartesian components of $\underline{t}_i$ and the first term represents the centre-of-mass kinetic energy. Since the centre-of-mass variable does not enter the potential term, the centre-of-mass problem may be separated off completely so that the full solution is of the form:

$$T(\underline{X}_T)\Psi(\underline{t}) \tag{15}$$

where:

$$T(\underline{X}_T) = \exp(i\underline{k}\underline{X}_T), \qquad \underline{k} \equiv (k_x, k_y, k_z) \tag{16}$$

and where the associated translational energy is:

$$E_T = \frac{|\underline{k}|^2}{2M_T} \tag{17}$$

It should be noticed that the translational wave function is not square integrable and that the translational energy is continuous. This is exactly what is to be expected given that the group of translations in three dimensions is a non-compact continuous group and has no finite dimension irreducible representations. It should also be noted that it is absolutely essential that the translational motion be separated from the problem before any approximate solution is attempted. This is in order to avoid the continuous spectrum, which cannot be approximated. Thus attention will be confined to the remaining terms in (12) which will be denoted collectively by $\hat{H}(\underline{t})$ and referred to as the *space-fixed* Hamiltonian. In the space-fixed Hamiltonian the inverse effective mass matrix $\underline{\mu}^{-1}$ and the form of the potential functions f_{ij} depend intimately on the choice of $\underline{V}$ and the choice of this is essentially arbitrary. In particular it should be observed that because there are only $N-1$ space fixed variables they cannot, except in the most conventional of senses, be thought of as particle coordinates and that the non-diagonal nature of $\underline{\mu}^{-1}$ and the peculiar form of the f_{ij} also militate against any simple particle interpretation of the space-fixed Hamiltonian.

For later purposes it is convenient to have available the angular momentum operator in terms of X_T and the $\underline{t}_i$. The total angular momentum operator may be written as:

$$\underline{\hat{L}}(\underline{x}) = \frac{\hbar}{i}\sum_{i=1}^{N}\underline{\hat{x}}_i\frac{\partial}{\partial\underline{x}_i} \tag{18}$$

where $\underline{\hat{L}}(\underline{x})$ are column matrices of cartesian components and the skew-symmetric matrix $\underline{\hat{x}}_i$ is:

$$\underline{\hat{x}}_i = \begin{pmatrix} 0 & -x_{zi} & x_{yi} \\ x_{zi} & 0 & -x_{xi} \\ -x_{yi} & x_{xi} & 0 \end{pmatrix} \tag{19}$$

The matrix $\underline{\hat{x}}_i$ can also be written in terms of the infintesimal rotation generators:

$$\underline{M}^x = \begin{pmatrix} 0 & 0 & 0 \\ 0 & 0 & 1 \\ 0 & -1 & 0 \end{pmatrix} \quad \underline{M}^y = \begin{pmatrix} 0 & 0 & -1 \\ 0 & 0 & 0 \\ 1 & 0 & 0 \end{pmatrix} \quad \underline{M}^z = \begin{pmatrix} 0 & 1 & 0 \\ -1 & 0 & 0 \\ 0 & 0 & 0 \end{pmatrix} \tag{20}$$

so that:

$$\underline{\hat{x}}_i = \sum_\alpha x_{\alpha i}\underline{M}^{\alpha T} \tag{21}$$

A variable symbol with a caret over it will, in future, be used to denote a skew-symmetric matrix as defined by (21).

Transforming to the coordinates $\underline{X}_T, \underline{t}_i$ gives:

$$\hat{\underline{L}}(\underline{x}) \rightarrow \frac{\hbar}{i}\hat{\underline{X}}_T\frac{\partial}{\partial \underline{X}_T} + \frac{\hbar}{i}\sum_{i=1}^{N-1}\hat{\underline{t}}_i\frac{\partial}{\partial \underline{t}_i} \tag{22}$$

and in future the second term will be denoted as $\hat{\underline{L}}(\underline{t})$ and called the space-fixed angular momentum.

To move from the space-fixed system to a body-fixed system, three variables are introduced which define the orientation of the system, together with $3N-6$ variables which describe the internal motions of the system. The internal variables are chosen to be invariant to orthogonal transformations. To construct the body-fixed system it is supposed that the three orientation variables are specified by means of an orthogonal matrix $\underline{C}$ which can be parameterized by the three Euler angles $\phi_m, m = 1,2,3$ as orientation variables. (It should be noted that for $N = 2$ only two orientation variables are required and this rather special case will be ignored in what follows.) Thus the space-fixed cartesian coordinates $\underline{t}$ can be thought of as being related to a body-fixed set $\underline{z}$ by

$$\underline{t} = \underline{C}\underline{z} \tag{23}$$

The above equation *defines* the variables $\underline{z}_i$ so that

$$\underline{z}_i = \underline{C}^T\underline{t}_i \tag{24}$$

and thus any orthogonal transformation of the $\underline{t}_i$ leaves the $\underline{z}_i$ invariant. However not all the $3N-3$ components of the $\underline{z}_i$ are independent for there must be three relations between them. What this means is that the components of $\underline{z}_i$ must be writeable in terms of $3N-6$ independent internal coordinates $q_i, i = 1,2....3N-6$. Of course it is possible that some of the q_i are some of the components of the $\underline{z}_i$ but generally speaking the q_i will be expressible in terms of scalar products of the $\underline{t}_i$ (and equally of the $\underline{z}_i$) since scalar products are the most general constructions that are invariant under orthogonal transformations of their consitutent vectors.[1]

It should be made explicit here that a particular choice of $\underline{C}$ does not force a particular choice of $\underline{q}$ and that different choices are possible for any given $\underline{C}$. Conversely, different choices of $\underline{C}$ are possible for any choice of $\underline{q}$. Some concrete examples of this sort of freedom of choice are given in [5].

In order to express the space-fixed differential operators in body-fixed terms it is necessary to obtain expressions for the partial derivatives of the body-fixed variables with respect to the space-fixed ones. To deal with the angular part first it should be noted that:

$$\frac{\partial}{\partial t_{\alpha i}}(\underline{C}^T\underline{C}) = \underline{0}_3 \tag{25}$$

because $\underline{C}$ is an orthogonal matrix and hence $\underline{C}^T\underline{C} = \underline{E}_3$.

From this it follows at once that:

$$\frac{\partial \underline{C}^T}{\partial t_{\alpha i}}\underline{C} = \hat{\underline{\omega}}^{\alpha i} \tag{26}$$

where $\underline{\hat{\omega}}^{\alpha i}$ is a skew-symmetric matrix of the same form as (19), containing three independent elements $\omega_\gamma^{\alpha i}$. Using the form (21), (26) can be rewritten as:

$$\frac{\partial \underline{C}}{\partial t_{\alpha i}} = \sum_\gamma \omega_\gamma^{\alpha i} \underline{CM^\gamma} \tag{27}$$

It is also convenient to introduce the matrix with elements $\Omega^i_{\beta\gamma}$ such that:

$$\omega_\gamma^{\alpha i} = \sum_\beta C_{\alpha\beta} \Omega^i_{\beta\gamma} \tag{28}$$

This ensures that the elements of the matrix $\underline{\Omega}^i$ are functions of the internal coordinates only and thus (27) becomes:

$$\frac{\partial \underline{C}}{\partial t_{\alpha i}} = \sum_\beta (\underline{CM^\beta})(\underline{C\Omega^i})_{\alpha\beta} \tag{29}$$

Recognising that $\underline{C}$ is a function of the ϕ_m only it follows that:

$$\frac{\partial \underline{C}}{\partial t_{\alpha i}} = \sum_{m=1}^{3} \frac{\partial \underline{C}}{\partial \phi_m} \frac{\partial \phi_m}{\partial t_{\alpha i}} \tag{30}$$

By analogy with (25) to (27) it follows that:

$$\frac{\partial \underline{C}}{\partial \phi_m} = \sum_\gamma (\underline{D}^{-1})_{m\gamma} \underline{CM^\gamma} \tag{31}$$

where $(\underline{D}^{-1})_{m\gamma}$ plays the same role in (31) that $\omega_\gamma^{\alpha i}$ does in (27).

From (29) to (31) it follows that:

$$\sum_{m=1}^{3} (\underline{D}^{-1})_{m\gamma} \frac{\partial \phi_m}{\partial t_{\alpha i}} = (\underline{C\Omega^i})_{\alpha\gamma}$$

or

$$\frac{\partial \phi_m}{\partial t_{\alpha i}} = (\underline{C\Omega^i D})_{\alpha m} \tag{32}$$

The process so far is, of course, purely formal since $\underline{C}$ has not been specified in terms of the $\underline{t}_i$ or the ϕ_m. However the process is not vacuous as will shortly be seen. A similar formal process enables it to be established that

$$\frac{\partial q_k}{\partial t_{\alpha i}} = (\underline{CQ^i})_{\alpha k} \tag{33}$$

---

[1]If only proper orthogonal transformations are considered the scalar triple products are also invariants but they change sign under improper operations. This and related matters are discussed in [6].

where the elements of $\underset{\sim}{Q}^i$ can be shown to be dependent on internal variables only using the fact that the q_k can be written as functions of the scalar products of the $\underline{t}_i$. Thus it is possible to write:

$$\frac{\partial}{\partial \underline{t}_i} = \underline{C}(\underline{\Omega}^i \underline{D} \frac{\partial}{\partial \underline{\phi}} + \underline{Q}^i \frac{\partial}{\partial \underline{q}}) \tag{34}$$

where $\partial/\partial\underline{\phi}$ and $\partial/\partial\underline{q}$ are column matrices of 3 and $3N-6$ partial derivatives respectively.

It is clear that (32) and (33) are expressions of the jacobian matrix elements for the transformation from the $(\underline{\phi}\,\underline{q})$ to the $\underline{t}$. It is useful to have expressions for the inverse transformations and these may, again formally, be obtained from (23). Thus because the $\underline{z}$ cannot be functions of the $\underline{\phi}$ then:

$$\frac{\partial t_{\alpha i}}{\partial \phi_m} = \left(\frac{\partial \underline{C}}{\partial \phi_m}\underline{z}_i\right)_\alpha \tag{35}$$

and using (31) it follows that:

$$\frac{\partial t_{\alpha i}}{\partial \phi_m} = (\underline{C}\hat{\underline{z}}_i \underline{D}^{-T})_{\alpha m} \tag{36}$$

Again using (23) it follows in a way analogous to (33) that:

$$\frac{\partial t_{\alpha i}}{\partial q_k} = (\underline{C}\tilde{\underline{Q}}^i)_{\alpha k} \tag{37}$$

and it can be shown that $\tilde{\underline{Q}}^i$ has elements that depend only on the q_k.

The relationship between the jacobian matrix and its inverse leads to the following expressions:

$$\sum_{i=1}^{N-1} \hat{\underline{z}}_i{}^T \underline{\Omega}^i = \underline{E}_3 \qquad \sum_{i=1}^{N-1} \tilde{\underline{Q}}^{i\,T} \underline{Q}^i = \underline{E}_{3N-6}$$

$$\sum_{i=1}^{N-1} \hat{\underline{z}}_i{}^T \underline{Q}^i = \underline{0}_{3,3N-6} \qquad \sum_{i=1}^{N-1} \tilde{\underline{Q}}^{i\,T} \underline{\Omega}^i = \underline{0}_{3N-6,3} \tag{38}$$

and

$$\underline{\Omega}^i \hat{\underline{z}}_j^T + \underline{Q}^i \tilde{\underline{Q}}^{jT} = \delta_{ij}\underline{E}_3 \tag{39}$$

These expressions are helpful in the formal manipulations that lead to body-fixed forms for the operators. They are also the origin of the "sum-rules" which constitute such a part of the manipulation of the Eckart Hamiltonian (see *e.g.* [7])

Finally it should be noted that on the formal variable change (23) it follows that:

$$\hat{\underline{t}}_i = |\underline{C}|\underline{C}\hat{\underline{z}}_i\underline{C}^T \tag{40}$$

where $|\underline{C}|$ is either plus or minus one according to whether $\underline{C}$ corresponds to a proper rotation or to an improper rotation.

It is convenient first to derive a body-fixed expression for the space-fixed angular momentum operator as given by the second term on the right in (21). Using this, (34) and (40), together with (38) yields:

$$\hat{\underline{L}}(\underline{t}) = -\frac{\hbar}{i}|\underline{C}|\underline{C}\underline{D}\frac{\partial}{\partial\underline{\phi}} \tag{41}$$

There is at this stage an element of choice for the definition of the body-fixed angular momentum. In this work it will be chosen as

$$\hat{\underline{L}}(\underline{\phi}) = \frac{\hbar}{i}\underline{D}\frac{\partial}{\partial\underline{\phi}} \tag{42}$$

and with the aid of (31) it is possible to show that the components of the operators with this choice obey the ***standard*** commutation conditions. It is somewhat more usual to chose the negative of (42) as the definition, in which case its components obey the celebrated anomalous commutation conditions. Whatever the choice, however the components of the body-fixed angular momentum operator are functions of the ϕ_m alone. With the present choice it may be imagined that there are angular momentum eigenfunctions $|JMk>$, just as in Brink and Satchler [8] or in Biedenharn and Louck [9] such that:

$$\hat{L}^2(\underline{t})|JMk> = \hat{L}^2(\underline{\phi})|JMk> = \hbar^2 J(J+1)|JMk>$$

$$\hat{L}_z(\underline{t})|JMk> = \hbar M|JMk>$$

$$\hat{L}_z(\underline{\phi})|JMk> = \hbar k|JMk> \tag{43}$$

Furthermore defining the step-up and -down operators in the usual way as $L_\pm = L_x \pm iL_y$ then:

$$\hat{L}_\pm(\underline{t})|JMk> = \hbar C^\pm_{JM}|JM \pm 1k>$$

$$\hat{L}_\pm(\underline{\phi})|JMk> = \hbar C^\pm_{Jk}|JMk \pm 1> \tag{44}$$

where the phase conventions are chosen analogously to the standard Condon and Shortley ones (see *eg* [8]) so that:

$$C^\pm_{Jj} = [J(J+1) - j(j \pm 1)]^{\frac{1}{2}} \tag{45}$$

If $\underline{C}$ is parameterized by the standard Euler angle choice made in ref. [8] or [9] then, it is easy to show that:

$$|JMk> = \left(\frac{2J+1}{8\pi^2}\right)^{\frac{1}{2}}(-1)^k \mathcal{D}^{J*}_{M-k}(\underline{\phi}) \tag{46}$$

where $\underline{\mathcal{D}}^J$ is the standard Wigner matrix as defined in ref. [8]] or [9]. The functions $|JMk>$ are often called *symmetric-top* eigenfunctions. Had the more usual choice of the negative of (42) been made for the body-fixed angular momentum operator the symmetric-top functions would have involved $\mathcal{D}^{J*}_{Mk}$ but in that case in order to get the second expression in (44) it is necessary to redefine the step-up and -down operators as $\hat{L}_\pm(\underline{\phi}) = \hat{L}_x(\underline{\phi}) \mp i\hat{L}_y(\underline{\phi})$. A more extended discussion of these matters can be found in Section (3.8) of ref [9] and also in [10] and [11].

In fact it can be shown [9], [12] that, whatever the parameterization of $\underline{C}$, the appropriate Wigner $\underline{\mathcal{D}}^1$ matrix can be written as:

$$\underline{\mathcal{D}}^1 = \underline{X}^{\dagger}\underline{C}\underline{X} \qquad (47)$$

with

$$\underline{X} = \begin{pmatrix} -1/\sqrt{2} & 0 & 1/\sqrt{2} \\ -i/\sqrt{2} & 0 & -i/\sqrt{2} \\ 0 & 1 & 0 \end{pmatrix} \qquad (48)$$

provided that $C_{\alpha\beta}$ is ordered $\alpha, \beta = x, y, z$ and the indices on $\underline{\mathcal{D}}^1$ run $+1, 0, -1$ across each row and down each column.

The elements of the general matrix $\underline{\mathcal{D}}^J$ can then be obtained (see Appendix 2 of ref. [12]) by repeated vector-couping of the elements of $\underline{\mathcal{D}}^1$.

It is thus perfectly possible to get expressions for the angular momentum eigenfunctions directly in terms of the elements of $\underline{C}$. (For details, see Section 6.19 of ref. [9].) This possibility is extensively exploited in Louck's derivation [13] of the Eckart Hamiltonian.

The transformation of the space-fixed kinetic energy operator into body-fixed form is a rather lengthy and tedious business to give in detail. However the final result can be stated reasonably straightforwardly and since the derivation itself is quite mechanical, simply involving letting (34) operate on itself and summing over i and j as in (12), there is no need to go into details. The resulting body-fixed operator is:

$$\hat{K}(\underline{\phi}, \underline{q}) = \frac{1}{2}\left(\sum_{\alpha\beta} M_{\alpha\beta}\hat{L}_\alpha\hat{L}_\beta + \hbar\sum_\alpha \lambda_\alpha \hat{L}_\alpha\right) - \frac{\hbar^2}{2}\left(\sum_{k,l=1}^{3N-6} G_{kl}\frac{\partial^2}{\partial q_k \partial q_l} + \sum_{k=1}^{3N-6} \tau_k \frac{\partial}{\partial q_k}\right) \qquad (49)$$

while the potential energy operator is built just as in (12) by means of (14) but now with $\underline{z}_k$ replacing $\underline{t}_k$ in (14) to give $f_{ij}(\underline{q})$.

The matrix $\underline{M}$ is a generalised inverse inertia tensor defined as:

$$\underline{M} = \sum_{i,j=1}^{N-1} \mu_{ij}^{-1}\underline{\Omega}^{i^T}\underline{\Omega}^j \qquad (50)$$

while $\underline{G}$ is given by:

$$\underline{G} = \sum_{i,j=1}^{N-1} \mu_{ij}^{-1}\underline{Q}^{i^T}\underline{Q}^j \qquad (51)$$

In the term linear in the angular momentum:

$$\lambda_\alpha = \frac{1}{i}\left(\nu_\alpha + 2\sum_{k=1}^{3N-6} W_{k\alpha}\frac{\partial}{\partial q_k}\right) \qquad (52)$$

with

$$\underline{W} = \sum_{i,j=1}^{N-1} \mu_{ij}^{-1}\underline{Q}^{i^T}\underline{\Omega}^j \qquad (53)$$

and

$$\nu_\alpha = \sum_{i,j=1}^{N-1} \mu_{ij}^{-1}\left(\sum_\beta\left((\underline{\Omega}^{i^T}\underline{M}^\beta\underline{\Omega}^j)_{\beta\alpha} + \sum_{l=1}^{3N-6} Q^i_{\beta l}\frac{\partial}{\partial q_l}\Omega^j_{\beta\alpha}\right)\right) \qquad (54)$$

The term (52) is associated with the Coriolis coupling and so no coordinate system can be found in which it will vanish.

In the term linear in the derivatives of the q_k :

$$\tau_k = \sum_{i,j=1}^{N-1} \mu_{ij}^{-1} (\sum_{\beta} ((\underline{\Omega}^{i^T} \underline{M}^{\beta} \underline{Q}^j)_{\beta k} + \sum_{l=1}^{3N-6} Q^i_{\beta l} \frac{\partial}{\partial q_l} Q^j_{\beta k})) \tag{55}$$

It is possible to choose a coordinate system in which this term vanishes.

It is clear from what has gone before and indeed from general group theoretical arguments that the eigenfunctions $\Psi(\underline{t})$ from (15) can be written as eigenfunctions of the body-fixed Hamiltonian in the form:

$$\Psi(\underline{t}) \rightarrow \Psi^{J,M}(\underline{\phi}, \underline{q}) = \sum_{k=-J}^{+J} \Phi^J_k(\underline{q}) |JMk> \tag{56}$$

where the internal coordinate function on the right hand side cannot depend on M because, in the absence of a field, the energy of the system does not depend on M, as will be seen.

It is thus possible to eliminate the angular motion from the problem and to write down on effective body-fixed Hamiltonian within any (J, M, k) rotational manifold, that depends only on the internal coordinates. Before showing how this is to be done it is appropriate to say something about the jacobian that is to be used in performing the integrals. Clearly the volume element for space-fixed integration is just $d\underline{t}_1 d\underline{t}_2 ....d\underline{t}_{N-1}$ because the transformation from the laboratory-fixed to the space-fixed frame is linear and so $|\underline{V}^{-1}| = |\underline{V}|^{-1}$ is simply a constant that can be ignored. The transformation from the space-fixed coordinates to the body-fixed ones is non-linear and so:

$$d\underline{t}_1 d\underline{t}_2 ....d\underline{t}_{N-1} = |\underline{J}|^{-1} d\phi_1 d\phi_2 d\phi_3 dq_1 dq_2 .....dq_{3N-6}$$

or:

$$d\underline{t} = |\underline{J}|^{-1} d\underline{\phi} d\underline{q} \tag{57}$$

where $|\underline{J}|$ is the determinant of the matrix constructed from (32) and (33). $|\underline{J}^{-1}|$ could, of course, equally well have been used where the derminant would have been constructed from (36) and (37) but in practice it will have been necessary to construct the $\underline{\Omega}^i$ and $\underline{Q}^i$ to get explicit forms for the operators, so it is easier to use these.

In many cases it is actually more convenient to construct the determinant of the metric derived from (32) and (33) which will be equal to $|\underline{J}|^{-2}$, to within a constant factor . The determinant is (see [6]):

$$|\underline{D}^{-1}|^2 \begin{vmatrix} \underline{M} & \underline{W}^T \\ \underline{W} & \underline{G} \end{vmatrix}^{-1} \tag{58}$$

where the matrices in the partitions are given by (50), (51) and (53). Simple algebra then shows that (again to within a constant factor):

$$|\underline{J}|^{-1} = |\underline{D}|^{-1} |\underline{M}|^{-\frac{1}{2}} |\underline{G} - \underline{W}\underline{M}^{-1}\underline{W}^T|^{-\frac{1}{2}} \tag{59}$$

It is the factor $|\underline{D}|^{-1}$ which is the angular part of the jacobian and it is easy to show that in the standard parameterization of [8], $|\underline{D}|^{-1} = \sin\phi_2$ as required for the usual interpretation of the matrix elements. The remaining terms in (59) are functions of the q_k alone. It is perhaps worthwhile noticing that in the Podolsky approach to the construction of the body-fixed Hamiltonian, which is the approach used by Watson [7] (and indeed many others) it is the form of the jacobian arising from (59) that is naturally used.

The choice is sometimes made to incorporate the internal coordinate part of the jacobian (or some of it) into the definition of the Hamiltonian. This is a fairly familiar process when working in spherical polars, for example, where the radial volume element r^2dr can be reduced to dr by writing the trial wavefunction $\psi(r)$ as $r^{-1}P(r)$ and modifying the Hamiltonian to refer to $P(r)$. This modification changes the derivative terms in the operator by $\partial/\partial r \rightarrow (\partial/\partial r - 1/r)$ and so on but alters none of the multiplicative or $\partial/\partial\theta$ terms. The resulting Hamiltonian is often said to be in *manifestly hermitian* form. Particular examples of this kind of construction can be found in [7] and in [13] while a general account is given in Section 35 of Kemble [14]. This process is often extremely useful in practice with specific coordinate choices, however it does not simplify matters at the level of formal exposition so it will not be considered further here.

To remove the rotational motion it is convenient to write $\hat{K}(\underline{\phi},\underline{q})$ given by (49) in two parts corresponding to the bracketing in that equation:

$$\hat{K}_R(\underline{\phi},\underline{q}) + \hat{K}_I(\underline{q}) \tag{60}$$

and to write the potential as $V(\underline{q})$. The matrix elements with respect to the angular functions of the operators that depend only on the q_k are quite trivial. Thus:

$$< J'M'k' \mid \hat{K}_I + V \mid JMk >= \delta_{J'J}\delta_{M'M}\delta_{k'k}(\hat{K}_I + V) \tag{61}$$

In what follows explicit allowance for the diagonal requirement on J and M will be assumed and the indices suppressed to save writing. Similarly the fact that the integration implied is over $\underline{\phi}$ only will be left implicit.

To deal with the first term in (60) is considerably more complicated and is best done by re-expressing the components of $\hat{\underline{L}}$ in terms of $\hat{L}_\pm(\underline{\phi})$ and $\hat{L}_z(\underline{\phi})$ and using (43) and (44). When this is done:

$$\begin{aligned}
&< JMk' \mid \hat{K}_R \mid JMk >= \\
&\frac{\hbar^2}{4}(b_{+2}C^+_{Jk+1}C^+_{Jk}\delta_{k'k+2} + b_{-2}C^-_{Jk-1}C^-_{Jk}\delta_{k'k-2}) \\
&+\frac{\hbar^2}{4}(C^+_{Jk}(b_{+1}(2k+1) + \lambda_+)\delta_{k'k+1} + C^-_{Jk}(b_{-1}(2k-1) + \lambda_-)\delta_{k'k-1}) \\
&+\frac{\hbar^2}{2}((J(J+1) - k^2)b + b_0k^2 + \lambda_0 k)\delta_{k'k}
\end{aligned} \tag{62}$$

In this expression:

$$\begin{aligned}
b_{\pm 2} &= (M_{xx} - M_{yy})/2 \pm M_{xy}/i \\
b_{\pm 1} &= M_{xz} \pm M_{yz}/i \\
b = (M_{xx} + M_{yy})/2 & \qquad b_0 = M_{zz}
\end{aligned} \tag{63}$$

and in terms of the λ_α in (52) λ_0 is λ_z and the $\lambda_\pm$ are :

$$\lambda_\pm = (\lambda_x \pm \lambda_y/i) \tag{64}$$

Thus within any rotational manifold it is the eigensolutions of the effective Hamilton given by (61) and (62) which are invariant to orthogonal transformations.

It should be noticed that it is not absolutely necessary to perform the passage to body-fixed coordinates in two stages and that one can pass directly to the body-fixed frame if it is desired. It is also the case that alternative forms to that given in (49) for the kinetic energy operator can be found and are sometimes useful (see e.g. [6]).

The underlying group structure implicit in the body-fixing as exhibited so far is $SO(3)$ rather than the full orthogonal group $O(3)$. The necessary extension involves an explicit consideration of parity and is somewhat involved. It has rather limited consequences but is discussed in more detail in [6].

It was remarked in explaining (57) that the transformation from the space-fixed to the body-fixed coordinates is non-linear and, in fact, it can be shown that this is a topological consequence of any transformation that allows rotational motion to be separated. Furthermore it can be shown that there is always some configuration of the particles that causes the jacobian to vanish. Clearly where the jacobian vanishes, the transformation is undefined. These and related matters are discussed in more detail in [15] but what it means in the present context is that not all possible $\Phi_k^J(\underline{q})$ are valid trial functions in (56) but only those which are strongly vanishing where the jacobian vanishes. If an illegitimate trial function is used in practice, divergent expectation values of the Hamiltonian result.

It should be stressed that the origin of these divergences is not physical, they arise simply as a consequence of coordinate choice. However a particular choice can obviously preclude the description of a possible physical state of a system. Thus suppose that a triatomic is described in the Eckart approach with the equilibrium geometry specified as bent. In this case the jacobian vanishes when the internal coordinates correspond to a linear geometry. The problem in this formulation then becomes ill-conditioned for states with large amplitude bending motions. Of course such large amplitude bending states are physically perfectly reasonable, it is just that they cannot be described in this formulation. A detailed discussion of these matters and an account of some proposals for coping with them in practice for triatomics is given in [5] [16] and [17]. The underlying problem is, however, well known and has long been studied.

From a physical point of view therefore it is clear that it is hopeless to attempt to describe all possible bound states $\Psi(\underline{t})$ in the form (56) by means of a *single* choice of internal and angular coordinates. In fact rather detailed topological arguments (which are summarised in [15]) show that this physical sense is mathematically sound and that more than one set of body-fixed coordinates is always required to describe the space-fixed set over its whole range. However in any particular case, a set of body-fixed coordinates can be found in which (56) is a valid expression.

It would be wrong to end this section without making it quite clear that although the transformation processes outlined above are formally impeccable their implementation in practice can be extremely difficult or even impossible. The nature of the difficulties will be apparent to students of [7] or [13] and an example of where things get impossible can be found in [18]. It has been suggested, [19], that computer-aided algebra systems might help to mitigate the difficulties but it is, of course, well known that it is not always possible to invert a non-linear transformation explicitly, so it is likely that there will always remain some desirable coordinate choices which are impossible in practice.

In the light of the discussion of this section it is now necessary to see how far it is possible to proceed in separating electronic and nuclear motions to yield an electronic problem that can be made to correspond to the usual clamped nucleus Hamiltonian and thus to yield a potential function that is invariant to translations and to orthogonal transformations of the laboratory-fixed coordinates.

4. Separating Electronic and Nuclear Motions

In the previous section it has been shown how the full problem can be reduced to a problem that, within a given rotational manifold, may be expressed entirely in terms of coordinates that are invariant under translations and rotations of the system as a whole. It is now necessary to decide how that reduced problem can be mapped onto some specific coordinate set so as to yield an electronic problem as much like (3) as possible and which yields a potential expressible entirely in terms of invariant coordinates derived from the laboratory-fixed nuclear variables. But from what has already been said it must be clear that the identification of electronic and nuclear variables among the space-fixed variables is purely conventional. No matter how this identification is made, the energy spectrum of the space-fixed Hamiltonian remains the same. Thus the choice can be made simply on grounds of convenience and utility while recognising that there can be disagreement about what is a reasonable choice. What follows in this section therefore cannot, by definition as it were, be regarded as definitive.

It seems reasonable to require that the space-fixed nuclear coordinates be expressible entirely in terms of the laboratory-fixed nuclear coordinates. Thus analogously to (4):

$$(\underline{t}^n \underline{X}) = \underline{x}^n \underline{V}^n \tag{65}$$

Here $\underline{t}^n$ is a 3 by $H-1$ matrix and $\underline{X}$ is a 3 by 1 matrix. $\underline{V}^n$ is an H by H matrix whose last column is special, with elements:

$$V^n_{iH} = M^{-1} m_i, \qquad M = \sum_{i=1}^{H} m_i \tag{66}$$

so that $\underline{X}$ is the centre-of-nuclear-mass coordinate. The elements in each of the first $H-1$ columns of $\underline{V}^n$ sum to zero, precisely as in (7), to ensure translational invariance.

If the $\underline{t}^n$ are independent then:

$$\underline{x}^n = (\underline{t}^n \underline{X})(\underline{V}^n)^{-1} \tag{67}$$

just as in (8) and, just as in (9), the bottom row of $(\underline{V}^n)^{-1}$ is special with elements:

$$((\underline{V}^n)^{-1})_{Hi} = 1 \qquad i = 1, 2, ...H \tag{68}$$

while like (10) the inverse requirement on the remaining rows gives:

$$\sum_{i=1}^{H}((\underline{V}^n)^{-1})_{ji}m_i = 0 \qquad\qquad j = 1,2,....H-1 \tag{69}$$

The space-fixed electronic coordinates will have to involve the laboratory-fixed nuclear coordinates so that (65) may be generalised as:

$$(\underline{t}^e\underline{t}^n\underline{X}) = (\underline{x}^e\underline{x}^n)\begin{pmatrix} \underline{V}^e & \underline{0} \\ \underline{V}^{ne} & \underline{V}^n \end{pmatrix} \tag{70}$$

where $\underline{t}^e$ is a 3 by L matrix. It is not possible to choose $\underline{V}^{ne}$ to be a null matrix and to satisfy simultaneously the translational invariance requirements as specified by (7) while leaving the whole matrix non-singular. Given that the inverse of (70) exists , however, then (67) may be generalised as :

$$(\underline{x}^e\underline{x}^n) = (\underline{t}^e\underline{t}^n\underline{X})\begin{pmatrix} (\underline{V}^e)^{-1} & \underline{0} \\ \underline{B} & (\underline{V}^n)^{-1} \end{pmatrix} \tag{71}$$

where

$$\underline{B} = -(\underline{V}^n)^{-1}\underline{V}^{ne}(\underline{V}^e)^{-1} \tag{72}$$

The bottom row of $\underline{B}$ is special in the same way as is the bottom row of $(\underline{V}^n)^{-1}$ and consists of the elements:

$$B_{Hi} = 1, \qquad\qquad i = 1,2,......H \tag{73}$$

Using (71) and the definitions of $\underline{X}$ and $\underline{X}_T$ it follows that

$$\underline{X}_T = \underline{X} + \sum_{i=1}^{L} s_i^e\underline{t}_i^e + \sum_{i=1}^{H-1} s_i^n\underline{t}_i^n \tag{74}$$

where

$$s_i^e = M_T{}^{-1}m\sum_{j=1}^{L}((V^e)^{-1})_{ij}, \qquad\qquad s_i^n = M_T{}^{-1}m\sum_{j=1}^{L} B_{ij} \tag{75}$$

so that using (74) $\underline{X}$ can be eliminated in favour of $\underline{X}_T$ whenever necessary.

In fact it is obvious from the definition (70) that a change from $\underline{X}$ to $\underline{X}_T$ has no effect on the expressions for $\underline{t}^e$ and $\underline{t}^n$ but from (71) it is seen that such a change does affect the expression for the inverse. Using (71) and (74) :

$$\underline{x}_i^e = \underline{X}_T + \sum_{j=1}^{L}\underline{t}_j^e((\underline{V}^e)_{ji}^{-1} - s_j^e) + \sum_{j=1}^{H-1}\underline{t}_j^n(B_{ji} - s_j^n) \tag{76}$$

and

$$\underline{x}_i^n = \underline{X}_T - \sum_{j=1}^{L}\underline{t}_j^e s_j^e + \sum_{j=1}^{H-1}\underline{t}_j^n((\underline{V}^n)_{ji}^{-1} - s_j^n) \tag{77}$$

If the expression (76) is substituted into $\underline{x}_{ij}^e = |\underline{x}_j^e - \underline{x}_i^e|$ then it is seen that the resulting form is not generally expressible in terms of the $\underline{t}_i^e$ alone and that furthermore, the form is not invariant under permutation of identical nuclei neither does it change in the ordinary way under the permutation of electrons. These features are obviously undesirable for they are not reflected in the standard expression for the clamped-nucleus Hamiltonian. It is clear therefore that it would be wise further to restrict the form of $\underline{V}^{ne}$ and of $\underline{V}^e$.

To construct a transformation such that the space-fixed electronic coordinates are invariant under any permutation of identical nuclei so that under such a permutation the t_i^e are unaffected and also such that they change in the usual manner under permutation of laboratory-fixed electronic variables, consider the general permutation of identical particles. This can be written as:

$$\mathcal{P}(\underline{x}^e\underline{x}^n) = (\underline{x}^e\underline{x}^n)\begin{pmatrix} \underline{P}^e & \underline{0} \\ \underline{0} & \underline{P}^n \end{pmatrix} \tag{78}$$

where $\underline{P}^e$ and $\underline{P}^n$ are standard permutation matrices.

Using (70) and (71), it follows that:

$$\mathcal{P}(\underline{t}^e\underline{t}^n\underline{X}) = (\underline{t}^e\underline{t}^n\underline{X})\begin{pmatrix} (\underline{V}^e)^{-1} & \underline{0} \\ \underline{B} & (\underline{V}^n)^{-1} \end{pmatrix}\begin{pmatrix} \underline{P}^e & \underline{0} \\ \underline{0} & \underline{P}^n \end{pmatrix}\begin{pmatrix} \underline{V}^e & \underline{0} \\ \underline{V}^{ne} & \underline{V}^n \end{pmatrix} \tag{79}$$

To achieve the required invariance, the matrix on the right hand side of (79) must be block diagonal and this occurs only if:

$$\underline{B}\underline{P}^e\underline{V}^e + (\underline{V}^n)^{-1}\underline{P}^n\underline{V}^{ne} = \underline{0}_{H,L} \tag{80}$$

The most general way in which this can be achieved is to require the following relations to hold:

$$\underline{P}^e\underline{V}^e = \underline{V}^e\underline{P}^e \tag{81}$$

$$\underline{P}^n\underline{V}^{ne} = \underline{V}^{ne} \tag{82}$$

$$\underline{V}^{ne}\underline{P}^e = \underline{V}^{ne} \tag{83}$$

If these relations hold then $\underline{P}^e$ can be taken as a common factor to the right and, using (72) for $\underline{B}$, $(\underline{V}^n)^{-1}$ may be taken as a common factor to the left in (80). The factored matrix then vanishes identically.

The physical content of the requirement (82) is simply that every member of a set of identical nuclei must enter into the definition of $\underline{t}^e$ in the same way. The physical content of the requirement (83) is that any electronic variable in the problem should have exactly the same relationship to the nuclear variables as does any other electronic variable. Thus (83) is satisfied by requiring all the columns of $\underline{V}^{ne}$ to be identical and from now on a typical column will be denoted $\underline{v}$.

The most general form for $\underline{V}^e$ that satisfies (81) is:

$$(\underline{V}^e)_{ij} = \delta_{ij} + a \tag{84}$$

where a is a constant. The inverse has elements:

$$((\underline{V}^e)^{-1})_{ij} = \delta_{ij} - a/(1 + La) \tag{85}$$

This shows that a can take any value (including 0) except $-1/L$. Using these results in (72) it follows that the columns of $\underline{B}$ are identical one with another and the typical column will from now on be written as $\underline{b}$. Using (76) with these restrictions it follows that x_{ij}^e becomes t_{ij}^e, as required. Using the results in (75) gives:

$$s_i^e = \frac{m}{M_T(1 + La)}, \qquad s_i^n = M_T^{-1}Lmb_i, \qquad \underline{b} = -\frac{(\underline{V}^n)^{-1}\underline{v}}{(1 + La)} \tag{86}$$

If the relations (81) to (83) are satisfied then it is easy to show that if in (79), X is replaced by X_T, the equation generalises to:

$$\mathcal{P}(\underline{t}^e\underline{t}^n\underline{X}_T) = (\underline{t}^e\underline{t}^n\underline{X}_T)\begin{pmatrix} \underline{P}^e & \underline{0} & \underline{0} \\ \underline{0} & \underline{H} & \underline{0} \\ \underline{0} & \underline{0} & 1 \end{pmatrix} \tag{87}$$

where

$$(\underline{H})_{ij} = ((\underline{V}^n)^{-1}\underline{P}^n\underline{V}^n)_{ij} \qquad i,j = 1,2,.....H-1 \tag{88}$$

The $H-1$ by $H-1$ matrix $\underline{H}$ is not in general in standard permutational form neither is it orthogonal even though it has determinant ± 1 according to the sign of $|\underline{P}^n|$.

In future it will be assumed that a coordinate system has been chosen according to (70) and (71) in which the conditions (81) to (83) are satisfied.

The above discussion gives an account of a reasonable and reasonably convenient way of partitioning $\underline{V}$ and hence its inverse, if the division of the problem into electronic and nuclear parts is to be recognized and made explicit. The form of the space-fixed operators remains unchanged from that given in the previous section but the partition made here does enable a more specific structure to be given to them with parts attributable to the types of particle.

Thus the derivative operator (11) can now be distinguished as consisting of two parts:

$$\frac{\partial}{\partial \underline{x}_i^e} = mM_T{}^{-1}\frac{\partial}{\partial \underline{X}_T} + \sum_{j=1}^{L}(\delta_{ij} + a)\frac{\partial}{\partial \underline{t}_j^e} \tag{89}$$

$$\frac{\partial}{\partial \underline{x}_i^n} = m_i M_T{}^{-1}\frac{\partial}{\partial \underline{X}_T} + v_i\sum_{j=1}^{L}\frac{\partial}{\partial \underline{t}_j^e} + \sum_{j=1}^{H-1} V_{ij}^n\frac{\partial}{\partial \underline{t}_j^n} \tag{90}$$

The space-fixed Hamiltonian arising from the last two terms in (12) expands into three parts:

$$\hat{H}(\underline{t}) \rightarrow \hat{H}^e(\underline{t}^e) + \hat{H}^n(\underline{t}^n) + \hat{H}^{en}(\underline{t}^n, \underline{t}^e) \tag{91}$$

Here

$$\hat{H}^e(\underline{t}^e) = -\frac{\hbar^2}{2\mu}\sum_{i=1}^{L}\nabla^2(\underline{t}_i^e) - \frac{\hbar^2}{2\mu'}\sum_{i,j=1}^{L}{}' \vec{\nabla}(\underline{t}_i^e).\vec{\nabla}(\underline{t}_j^e) + \frac{e^2}{8\pi\epsilon_0}\sum_{i,j=1}^{L}{}' \frac{1}{|\underline{t}_j^e - \underline{t}_i^e|} \tag{92}$$

with

$$\mu^{-1} = m^{-1} + \mu'^{-1} \tag{93}$$

$$\mu'^{-1} = m^{-1}a(2 + La) + \sum_{k=1}^{H} m_k^{-1}v_k{}^2 \tag{94}$$

and

$$\hat{H}^n(\underline{t}^n) = -\frac{\hbar^2}{2}\sum_{i,j=1}^{H-1}\mu_{ij}^{-1}\vec{\nabla}(\underline{t}_i^n).\vec{\nabla}(\underline{t}_j^n) + \frac{e^2}{8\pi\epsilon_0}\sum_{i,j=1}^{H}{}' \frac{Z_iZ_j}{f_{ij}(\underline{t}^n)} \tag{95}$$

where $\mu_{ij}{}^{-1}$ is defined just as in (13) but in terms of the nuclear masses only and using $\underline{V}^n$. Similarly $f_{ij}(\underline{t}^n)$ is defined just as in (14) but using the $\underline{t}_i^n$ only and $(\underline{V}^n)^{-1}$.

Finally:

$$\hat{H}^{en}(\underline{t}^n,\underline{t}^e) = -\hbar^2 \sum_{i=1}^{H-1} \mu_i^{-1}\vec{\nabla}(\underline{t}_i^n).\sum_{j=1}^{L}\vec{\nabla}(\underline{t}_j^e) - \frac{e^2}{4\pi\epsilon_0}\sum_{i=1}^{H-1}\sum_{j=1}^{L}\frac{Z_i}{f'_{ij}(\underline{t}^n,\underline{t}^e)} \tag{96}$$

with

$$\mu_i^{-1} = \sum_{k=1}^{H} m_k^{-1} v_k V_{ki}^n \tag{97}$$

while f'_{ij} is the electron-nucleus distance and so it is the modulus :

$$|\underline{x}_i^n - \underline{x}_j^e| = |\sum_{k=1}^{H-1} \underline{t}_k^n((\underline{V}^n)_{ki}^{-1} - b_k) + a/(1+La)\sum_{k=1}^{L}\underline{t}_k^e - \underline{t}_j^e| \tag{98}$$

The space-fixed angular momentum operator (the second term in (22)) can be written as:

$$\hat{\underline{L}}(\underline{t}^n,\underline{t}^e) = \frac{\hbar}{i}\sum_{i=1}^{H-1}\hat{\underline{t}}_i^n \frac{\partial}{\partial \underline{t}_i^n} + \frac{\hbar}{i}\sum_{i=1}^{L}\hat{\underline{t}}_i^e\frac{\partial}{\partial \underline{t}_i^e} \tag{99}$$

and is of this form, irrespective of the precise choices made for $\underline{t}^e$ and $\underline{t}^n$.

The expressions so far obtained are reasonably general, if somewhat cumbersome. However were it to be thought desirable to attempt to make *ab-initio* calculations directly with the space-fixed form or to do body-fixed calculations similarly then the rather awkward form of the Hamiltonian could have computational advantages. It is nevertheless possible to simplify (96) somewhat and if one is thinking of an approach to the problem through the standard potential energy surface then the simplification is actually advantageous.

Let the elements of $\underline{v}$ be chosen as

$$v_k = -\alpha M^{-1} m_k, \qquad \alpha = (1 + La) \tag{100}$$

where the choice for α is determined by the translational invariance requirements as explained in connection with (71). Thus the $\underline{t}_i^e$ are the electronic coordinates referred to the centre-of-nuclear-mass scaled by α. This choice satisfies the condition expressed in equation (82) and from (69) and (86) $\underline{b} = \underline{0}$. Because of this, s_i^n vanishes and the electron-nucleus attraction term (98) is somewhat simplified by the loss of the parts that depend on the elements of $\underline{b}$. Substituting (100) for v_k into (97) it follows at once that μ_i^{-1} vanishes so that the electron-nucleus coupling term in the kinetic energy part of (96) vanishes. For the time being, choices for a or, equivalently, α will be left open.

In moving to the body-fixed frame it would seem reasonable to require that the matrix $\underline{C}$ is specified entirely in terms of the space-fixed nuclear variables $\underline{t}^n$. This means that the case $H = 2$ must be excluded much as the case $N = 2$ had to be excluded in the previous section. The simplest system that can be considered, therefore is a triatomic.[2]

The three relations required to define the body-fixed frame arise therefore from three relations among the body-fixed nuclear cartesian variables given by (*c.f.* (26)):

$$\underline{z}_i^n = \underline{C}^T \underline{t}_i^n \qquad i = 1, 2, ...H-1 \tag{101}$$

There are therefore $3H - 6$ internal coordinates, which will be denoted q_k just as before, that are expressible as functions of the scalar products of the $\underline{t}_i^n$ in terms of which the $\underline{z}_i^n$ can be expressed completely. The remaining $3L$ internal coordinates are made up of the three cartesian components of each of the L body-fixed electronic cartesians given by:

$$\underline{z}_i = \underline{C}^T \underline{t}_i^e \qquad i = 1, 2, \ldots. L \tag{102}$$

where there is no explicit superscript on the electronic body-fixed variables.

The jacobian matrix elements (32) are now in two parts:

$$\frac{\partial \phi_m}{\partial t_{\alpha i}^n} = (\underline{C}\underline{\Omega}^i \underline{D})_{\alpha m} \qquad \frac{\partial \phi_m}{\partial t_{\alpha i}^e} = 0 \tag{103}$$

where the $\underline{\Omega}^i$ are functions of the q_k only.

The jacobian matrix elements (33) develop into four parts:

$$\frac{\partial q_k}{\partial t_{\alpha i}^n} = (\underline{C}\underline{Q}^i)_{\alpha k} \qquad \frac{\partial q_k}{\partial t_{\alpha i}^e} = 0 \tag{104}$$

where the $\underline{Q}^i$ are functions of the q_k only and

$$\frac{\partial z_{\gamma j}}{\partial t_{\alpha i}^n} = (\underline{C}\underline{\Omega}^i \hat{\underline{z}}_j)_{\alpha\gamma} \qquad \frac{\partial z_{\gamma j}}{\partial t_{\alpha i}^e} = \delta_{ij} C_{\alpha\gamma} \tag{105}$$

It follows therefore that (34) becomes an expression in two parts:

$$\frac{\partial}{\partial \underline{t}_i^n} = \underline{C}(\underline{\Omega}^i \underline{D}\frac{\partial}{\partial \underline{\phi}} + \underline{Q}^i \frac{\partial}{\partial \underline{q}} + \underline{\Omega}^i \sum_{j=1}^{L} \hat{\underline{z}}_j \frac{\partial}{\partial \underline{z}_j}) \tag{106}$$

and

$$\frac{\partial}{\partial \underline{t}_i^e} = \underline{C}\frac{\partial}{\partial \underline{z}_i} \tag{107}$$

There are similar developments in in the expressions associated with the inverse transformation so that (36) becomes:

$$\frac{\partial t_{\alpha i}^n}{\partial \phi_m} = (\underline{C}\hat{\underline{z}}_i^n \underline{D}^{-T})_{\alpha m} \qquad \frac{\partial t_{\alpha i}^e}{\partial \phi_m} = (\underline{C}\hat{\underline{z}}_i \underline{D}^{-T})_{\alpha m} \tag{108}$$

while (37) goes into four parts:

$$\frac{\partial t_{\alpha i}^n}{\partial q_k} = (\underline{C}\bar{\underline{Q}}^i)_{\alpha k} \qquad \frac{\partial t_{\alpha i}^e}{\partial q_k} = 0 \tag{109}$$

and

$$\frac{\partial t_{\alpha i}^n}{\partial z_{\gamma j}} = 0 \qquad \frac{\partial t_{\alpha i}^e}{\partial z_{\gamma j}} = \delta_{ij} C_{\alpha\gamma} \tag{110}$$

The relationship between the jacobian and its inverse leads to exactly the same form of equations as are given in (38) and (39) but now i and j are to be interpreted as referring to space-fixed nuclear coordinates and the sums are to run to $H - 1$ rather than to $N - 1$.

Although the above body-fixed relationships depend on the choices (70) and (101) they are not otherwise dependent on a particular way of choosing the electronic and nuclear coordinates. The expression for the angular momentum similarly remains

[2] In fact even the triatomic is somehat special because three points always define a plane and the presence of a plane allows rather special mappings under the inversion operation. This is discussed in [6].

general and is still carried entirely by the Euler angles as in (41) and (42).

The body-fixed Hamiltonian now arises by transforming the terms in (91) and here the special choices for the variables implicit in (81) to (83) will be assumed, together with a centre-of-nuclear mass choice (100). The change in $\hat{H}^e(\underline{t}^e)$ as given by (92) is trivial and is effected simply by replacing $\underline{t}_i^e$ by $\underline{z}_i$ everywhere it occurs. The result will be denoted as $\hat{H}(\underline{z})$, dropping the superscript as the provenance of the operator is clear from the variable name. Furthermore since the first term in (96) vanishes because of the centre-of-nuclear mass choice, it is sufficient to transform (98) and this too is done simply by replacing $\underline{t}_i^e$ by $\underline{z}_i$ and $\underline{t}_i^n$ by $\underline{z}_i^n(\underline{q})$ wherever they occur. The transformation to body-fixed form of the kinetic energy operator from (95) is a little more involved however, because of the presence in the last term in (106) of body-fixed electronic variables. The required form is a slight extension of (49), namely:

$$\hat{K}(\underline{\phi},\underline{q},\underline{z}) = \frac{1}{2}\left(\sum_{\alpha\beta} M_{\alpha\beta}\hat{L}_\alpha\hat{L}_\beta + \hbar\sum_{\alpha}(\lambda_\alpha + 2(\underline{M}\hat{\underline{l}})_\alpha)\hat{L}_\alpha\right)$$

$$-\frac{\hbar^2}{2}\left(\sum_{k,l=1}^{3H-6} G_{kl}\frac{\partial^2}{\partial q_k \partial q_l} + \sum_{k=1}^{3H-6}\tau_k\frac{\partial}{\partial q_k}\right) + \frac{\hbar^2}{2}\left(\sum_{\alpha\beta} M_{\alpha\beta}\hat{l}_\alpha\hat{l}_\beta + \sum_\alpha \lambda_\alpha \hat{l}_\alpha\right) \tag{111}$$

where $\hat{\underline{l}}$ is a 3 by 1 column matrix of cartesian components:

$$\hat{\underline{l}} = \frac{1}{i}\sum_{i=1}^{L}\hat{\underline{z}}_i\frac{\partial}{\partial \underline{z}_i} \tag{112}$$

The expressions for $\underline{M}$, $\underline{G}$ and $\underline{W}$ are of exactly the same form as given in (50) (51) and (53) as are the expressions for λ_α, ν_α and τ_k of the same form as given in (52), (54) and (55). However (see below (95)) the sums extend only to $H-1$ and μ_{ij}^{-1} is defined using $\underline{V}^n$ and the nuclear masses only.

The last term in (111), which couples the body-fixed electronic and nuclear variables, is composed of operators which define the Coriolis coupling and so cannot be transformed away by some particular coordinate choice. Had the kinetic energy operator in (96) not been transformed away by means of the centre-of-nuclear mass choice (100), then the coupling expression found at this stage would have been more extensive and would have involved mixed derivatives in the body-fixed electronic and nuclear variables. The nuclear repulsion term is found by expressing the $\underline{t}^n$ in terms of the q_k to yield $f_{ij}(\underline{q})$ as in the discussion of (95).

The overall structure of the body-fixed problem is essentially preserved under the identification of particle types so that the separation is much as before. Equation (60) can now be written as:

$$\hat{K}_R(\underline{\phi},\underline{q},\underline{z}) + \hat{K}_I(\underline{q},\underline{z}) \tag{113}$$

where the first term is just the term in brackets that begins (111) and second term consists of the remaining terms from that equation together with the kinetic energy operator from $\hat{H}(\underline{z})$. It is useful to recognize a division of V in (61) by writing it as

$$V \rightarrow V(\underline{q}) - V(\underline{q},\underline{z}) + V(\underline{z}) \tag{114}$$

The first term is the nuclear repulsion potential term from (95) realised as a function of the q_k, the second term is the electron-nucleus attraction potential term from (98) realised as a function of the q_k and the $\underline{z}_i$ and the last term is just the electronic repulsion term from (92) expressed in terms of the $\underline{z}_i$.

Similarly it is useful to expand the notation of (56) somewhat so that:

$$\Psi^{J,M}(\underline{\phi},\underline{q},\underline{z}) = \sum_{k=-J}^{+J} \Phi_k^J(\underline{q},\underline{z})|JMk> \tag{115}$$

and using functions like this as the basis in removing the rotational motion, the process yields equations exactly like (61) and (62) but making explicit the electronic variables in the second term in (111) means that the λ_α term in (62) has to be slightly extended:

$$\lambda_\alpha \rightarrow \lambda_\alpha + 2(\underline{M}\hat{\underline{l}})_\alpha \tag{116}$$

The volume element for integration is a simple development of (57) namely:

$$d\underline{t} = |J|^{-1} d\underline{\phi} d\underline{q} d\underline{z} \tag{117}$$

where $|J|$ is exactly as given by (59) but with the constituent matrices as functions of the nuclear variables alone (see below (95)) and thus the volume element for $\underline{z}$ is in standard cartesian form.

To effect the uncoupling of the nuclear ($\underline{q}$) part of the problem from the electronic ($\underline{z}$) part in the manner proposed by Born and Huang [3] is formally quite similar to the uncoupling of rotational ($\underline{\phi}$) part from the internal ($\underline{q}$) part as undertaken above. The internal motion function from (115) is expressed in terms of a sum of products of the form:

$$\Phi_{kp}^J(\underline{q})\psi_p(\underline{q},\underline{z}) \tag{118}$$

where p labels the electronic state and the sum is over p. The function $\psi_p(\underline{q},\underline{z})$ is assumed known, just as $|JMk>$ is assumed known and the effective nuclear motion Hamiltonian is obtained in terms of matrix elements of the effective internal motion Hamiltonian between the $\psi_p(\underline{q},\underline{z})$ with respect to the variables $\underline{z}$, just as the effective internal motion Hamiltonian itself is expressed in terms of matrix elements of the full body-fixed Hamiltonian between the $|JMk>$ with respect to the $\underline{\phi}$. The effective nuclear motion Hamiltonian then contains the electronic state labels p as parameters, in much the same way that the full effective Hamiltonian for internal motion contains the angular momentum labels k. Of course the analogy between the two derivations is simply a formal one. There is no underlying symmetry structure in the effective nuclear problem and neither is the sum over p of definite extent as is the sum over k.

In fact Hunter [20] has shown (at least for the case $J = 0$) that the *exact* wavefunction can be written as a single product of this form. However in Hunter's form ψ is not determined as the solution of any sort of electronic problem but is rather obtained as a conditional probability amplitude by a process of integration and is to be associated with a marginal probability amplitude Φ to constitute a complete probability amplitude. The work of Czub and Wolniewicz [21] would seem to indicate however, that it would be very difficult to use this scheme to define, *ab-initio*, a potential in terms of which nuclear motion functions could be calculated. Thus unless the full function is known it seems not possible to determine its parts factored in this way. For all practical purposes then we must use the full Born and Huang approach. In the original formulation it was stipulated that the set of known functions, $\psi_p(\underline{q}, \underline{z})$, were to be looked on as exact solutions of a problem like:

$$(\hat{H}(\underline{z}) - V(\underline{q}, \underline{z}))\psi_p(\underline{q}, \underline{z}) = E_p(\underline{q})\psi_p(\underline{q}, \underline{z})$$

that is:

$$\hat{H}^{elec}(\underline{q}, \underline{z})\psi_p(\underline{q}, \underline{z}) = E_p(\underline{q})\psi_p(\underline{q}, \underline{z}) \tag{119}$$

Because in this equation there are no terms which involve derivatives with respect to the q_k, there is no development with respect to $\underline{q}$ in $E_p(\underline{q})$ or $\psi_p(\underline{q}, \underline{z})$. Thus the $\underline{q}$ act here simply as parameters that can be chosen at will.

In fact it is not absolutely essential for what follows to require the ψ_p to be eigenfunctions of $\hat{H}^{elec}$. A reasonably concise and useful form can be obtained simply by requiring that:

$$\int \psi^*_{p'}(\underline{q}, \underline{z})\psi_p(\underline{q}, \underline{z})d\underline{z} \equiv < \psi_{p'}|\psi_p >_{\underline{z}} = \delta_{p'p} \tag{120}$$

and, using the above abbreviation to denote integration over all $\underline{z}$ only :

$$< \psi_{p'}|\hat{H}^{elec}|\psi_p >_{\underline{z}} = \delta_{p'p}E_p(\underline{q}) \tag{121}$$

The requirements (120) and (121) can be met in a simple and practical way by requiring the ψ_p to be solutions of a linear variation problem with matrix elements determined by integration over the $\underline{z}$ alone, for each and every value assigned to $\underline{q}$.

The effective nuclear motion Hamiltonian, depending only on the $\underline{q}$, is expressed in terms of matrix elements of the Hamiltonian (113) between pairs of functions like (115) but with internal coordinate parts like (118) integrated over the $\underline{z}$ as well as the angular factors. Doing this yields equations rather like (61) and (62) but, as explained above, coupled not only between different rotational states, labelled by k, but also between different electronic states, labelled by p. In deriving them it should be remembered that the product rule must be used when considering the effect of derivative operators with respect to the q_k because *both* terms in the product (118) depend on the $\underline{q}$ variables.

The term analogous to (61)becomes:

$$< JMk'p' \mid \hat{K}_I - V(\underline{q}, \underline{z}) + V(\underline{q}) \mid JMkp >_{\underline{z}} \quad =$$

$$\delta_{p'p}\delta_{k'k}(\hat{K}_H + E_p(\underline{q}) + V(\underline{q})) + \delta_{k'k}\gamma_{p'p}(\underline{q}) \tag{122}$$

where the designation of the angular integration variables has been left implicit as before, as have the diagonal requirement on J and M. The term $\hat{K}_H$ consists of the second group of terms from (111), namely the nuclear kinetic energy terms and the last term in (122) is given by:

$$\gamma_{p'p}(\underline{q}) = \frac{\hbar^2}{2}\Bigg(\sum_{\alpha\beta} < \psi_{p'}|\hat{l}_\alpha\hat{l}_\beta|\psi_p >_{\underline{z}} M_{\alpha\beta} + \sum_{\alpha} < \psi_{p'}|\hat{l}_\alpha|\psi_p >_{\underline{z}} \lambda_\alpha$$

$$- \sum_{k,l=1}^{3H-6} G_{kl}(< \psi_{p'}|\frac{\partial^2}{\partial q_k \partial q_l}|\psi_p >_{\underline{z}} + < \psi_{p'}|\frac{\partial}{\partial q_k}|\psi_p >_{\underline{z}} \frac{\partial}{\partial q_l} + < \psi_{p'}|\frac{\partial}{\partial q_l}|\psi_p >_{\underline{z}} \frac{\partial}{\partial q_k})$$

$$+ \sum_{k=1}^{3H-6}(\frac{2}{i} < \psi_{p'}|(\underline{W}\hat{\underline{l}})_k\frac{\partial}{\partial q_k}|\psi_p >_{\underline{z}} - \tau_k < \psi_{p'}|\frac{\partial}{\partial q_k}|\psi_p >_{\underline{z}})\Bigg) \tag{123}$$

In (122) the role of $E_p(\underline{q}) + V(\underline{q})$ as a potential in the nuclear variables is clear.

The expression (62) can be developed in a similar fashion now using $\hat{K}_R(\underline{\phi}, \underline{q}, \underline{z})$ and integrating out over the $\underline{z}$ as well as the $\underline{\phi}$. This gives:

$$< JMk'p' \mid \hat{K}_R \mid JMkp >_{\underline{z}} \quad =$$

$$\frac{\hbar^2}{4}(b_{+2}C^+_{Jk+1}C^+_{Jk}\delta_{k'k+2} + b_{-2}C^-_{Jk-1}C^-_{Jk}\delta_{k'k-2})\delta_{p'p}$$

$$+\frac{\hbar^2}{4}(C^+_{Jk}(b_{+1}(2k+1) + \lambda_+)\delta_{k'k+1} + C^-_{Jk}(b_{-1}(2k-1) + \lambda_-)\delta_{k'k-1})\delta_{p'p}$$

$$+\frac{\hbar^2}{4}(C^+_{Jk}\gamma^+_{p'p}(\underline{q})\delta_{k'k+1} + C^-_{Jk}\gamma^-_{p'p}(\underline{q})\delta_{k'k-1})$$

$$+\frac{\hbar^2}{2}((J(J+1) - k^2)b + b_0k^2 + \lambda_0 k)\delta_{k'k}\delta_{p'p} + \frac{\hbar^2}{2}\delta_{k'k}k\gamma^0_{p'p}(\underline{q}) \tag{124}$$

The meaning of the terms in this equation is exactly as in (116) but defined in relation to the nuclear variables alone. The γ terms simply extend the definition of the λ terms as in (116) and are specified by means of:

$$\gamma^\alpha_{p'p}(\underline{q}) = 2(< \psi_{p'}|(\underline{M\hat{l}})_\alpha|\psi_p >_{\underline{z}} + \frac{1}{i}\sum_{k=1}^{3H-6} W_{k\alpha} < \psi_{p'}|\frac{\partial}{\partial q_k}|\psi_p >_{\underline{z}}) \tag{125}$$

with γ^0 and $\gamma^\pm$ defined in terms of the γ^α in a manner analogous to the definition of the equivalent λ quantities in (64).

If H^{cn} could be identified with H^{elec} then a justifiable computational scheme would now be clear. A set of functions $\psi_p(\underline{q}, \underline{z})$ would be generated using the linear variation method on H^{elec} in a basis of trial functions chosen at a sequence of values of $\underline{q}$. In $\gamma_{p'p}$ and $\gamma^\alpha_{p'p}$ the coefficients that arise from integrating over the $\underline{z}$ variables would be calculated for each chosen value of $\underline{q}$ just as $E_p(\underline{q})$ is calculated and these, together with $E_p(\underline{q})$ values, would have to be extended into functions over all $\underline{q}$, perhaps by fitting to analytic forms. The effective nuclear motion problem specified by (122) and (124) could then be solved in the basis Φ^J_{nkp} (where n is used to label the components of the basis) to yield nuclear motion energies and states. It would be hoped that, at least in the electronic ground state, it would be sufficient to use a single electronic state in (118) to account for most of the nuclear motion states of interest. If that were the case then only a single potential would be needed and only diagonal elements of the γ matrices would arise. Then one could use a linear variation approach, for example, to the solution of the effective nuclear motion problem within any rotational manifold as specified by J.

However the relationship between (3) and (119) remains to be discovered and it is also necessary, in view of what was said at the end of the previous section about the possible vanishing of the jacobian, to investigate more closely the status of the product approximation (118). These matters will be discussed in the next section.

5. The Clamped Nucleus Hamiltonian

From what has been said in the previous section it is clearly reasonable to make the choice (100) for v_k if it is hoped to map the clamped-nucleus Hamiltonian (3) onto the body-fixed electronic Hamiltonian (119). Making that choice, (94) becomes:

$$\mu'^{-1} = m^{-1}a(2 + La) + \alpha^2 M^{-1} \tag{126}$$

and while the form of (92) remains unchanged, the form of the electron-nucleus attraction term becomes:

$$f'_{ij}(\underline{q},\underline{z}) = |\underline{x}_i^n - \underline{x}_j^e| = |\sum_{k=1}^{H-1} \underline{z}_k^n(\underline{q})(\underline{V}^n)_{ki}^{-1} - \underline{z}_j + a/(1+La)\sum_{k=1}^{L} \underline{z}_k| \tag{127}$$

Thus the form of body-fixed electronic Hamiltonian that can probably be best compared with the clamped-nucleus Hamiltonian (3) can be written:

$$\hat{H}^{elec}(\underline{q},\underline{z}) = -\frac{\hbar^2}{2\mu}\sum_{i=1}^{L}\nabla^2(\underline{z}_i) - \frac{e^2}{4\pi\epsilon_0}\sum_{i=1}^{H-1}\sum_{j=1}^{L}\frac{Z_i}{f'_{ij}(\underline{q},\underline{z})} + \frac{e^2}{8\pi\epsilon_0}\sum_{i,j=1}^{L}{}' \frac{1}{|\underline{z}_j - \underline{z}_i|}$$

$$-\frac{\hbar^2}{2\mu'}\sum_{i,j=1}^{L}{}' \vec{\nabla}(\underline{z}_i).\vec{\nabla}(\underline{z}_j) \tag{128}$$

for if the correspondence $\underline{x}_i^e \rightarrow \underline{z}_i$ is made in (3) then it matches (128) fairly closely. The mismatches arise from the form of (127) and with the last term in (128) and, unfortunately, it is not possible to eliminate both of these mismatches simultaneously. The choice $a = 0$ (and the consequent choice $\alpha = 1$) removes the last term in (127) and enables, as will be shown, the electron-nucleus attraction terms in (3) and in (128) to be matched. However with this choice μ' becomes M and μ becomes $mM/(m + M)$. Thus the last term in (128) persists and this has no matching term in (3) neither does μ match m, though this last is not important for it is simply a matter of scale. Alternatively the choice $\alpha = (M/M_T)^{\frac{1}{2}}$, with the consequent choice $a = (\alpha - 1)/L$, causes μ'^{-1} to vanish and makes $\mu = m$. Thus the last term in (128) now vanishes but the last term in (127) persists and prevents a matching with the electron-nucleus attraction term in (3). This last choice is analogous to the Radau choice of heliocentric coordinates (see *e.g* Smith [22]) but here the centre-of-nuclear mass plays the part of the distinguished coordinate.

How these unavoidable discrepancies should be managed is a matter of judgement but it is suggested that it is reasonable to make the choice $a = 0$ and simply to *redefine* $\hat{H}^{elec}$ as consisting of all but the last term in (128) and to extend the definition of $\gamma_{p'p}(\underline{q})$ in (123) so that the operator there includes the term neglected here. That is:

$$\gamma_{p'p}(\underline{q}) \Rightarrow \gamma_{p'p}(\underline{q}) - \frac{\hbar^2}{2M}\sum_{i,j=1}^{L}{}' < \psi_{p'}|\vec{\nabla}(\underline{z}_i).\vec{\nabla}(\underline{z}_j))|\psi_p >_{\underline{z}} \tag{129}$$

This seems a sensible course of action because all the operators relevant to the integration in (123) involve derivatives of the electronic coordinates and all the multiplicative factors involve reciprocals of the nuclear masses. Furthermore the added term might be hoped to be the smallest of the terms there, because it involves the reciprocal of the total nuclear mass. In any case, its inclusion in the diagonal terms will produce at most a nuclear-mass dependent constant energy shift for any electronic state. If (123) is so modified and (128) equivalently truncated then all that remains is to establish the matching of the electron-nucleus attraction terms in (3) with (127). To do that, let it be supposed that the choice $\underline{x}^n = \underline{a}$ has been made and a set of constant space-fixed coordinates $\underline{t}^n(\underline{a})$ has been defined according to (65). Then, using (67) it follows that:

$$\underline{a}_i = \sum_{k=1}^{H-1} \underline{t}_k^n(\underline{a})(\underline{V}^n)^{-1}_{ki} + \underline{X}(\underline{a}) \tag{130}$$

where $\underline{X}(\underline{a})$ is the centre-of-nuclear mass for the configuration chosen. This set of constant space-fixed coordinates will generate a constant matrix $\underline{C}(\underline{a})$ which will in turn, generate according to (101), a set of constant body-fixed cartesians, $\underline{z}^n(\underline{q}(\underline{a}))$. The constant internal coordinates, $\underline{q}(\underline{a})$, are generated in terms of scalar products of the constant space-fixed coordinates. It follows then that:

$$\sum_{k=1}^{H-1} \underline{z}_k^n(\underline{q}(\underline{a}))(V^n)^{-1}_{ki} = \underline{C}^T(\underline{a})(\underline{a}_i - \underline{X}(\underline{a})) = \underline{b}_i \tag{131}$$

so that (127) becomes

$$|\underline{a}_i - \underline{x}_i^e| = |\underline{b}_i - \underline{z}_i| \tag{132}$$

Now although the left hand side does not match the right on making the correspondence $\underline{x}_i^e \rightarrow \underline{z}_i$, at a deeper level a match can be made. This is because the set of all $\underline{b}_i$ define a geometrical object that differs from that defined by the set of all $\underline{a}_i$ at most by a uniform constant translation and a constant rigid rotation, so that the electron-nucleus attraction terms will yield identical expectation values with respect to integration over the appropriate free variable $\underline{x}_i^e$ or $\underline{z}_i$. Thus, without loss of generality $\underline{b}_i$ can be replaced by $\underline{a}_i$ on the right hand side of (132) if it is done for all i. To see this in another way, it is clear that it is perfectly possible to choose a set of $\underline{a}_i$ such that $\underline{X}(\underline{a}) = 0$ and such that $\underline{C}(\underline{a}) = \underline{E}_3$ and if this is done then the matching is apparent at once for then $\underline{b}_i = \underline{a}_i$. So if (129) is used in place of (123) in defining $\gamma_{p'p}(\underline{q})$ then the form of (128) that may be used in (119) is:

$$\hat{H}^{elec}(\underline{q}(\underline{a}), \underline{z}) = -\frac{\hbar^2}{2\mu}\sum_{i=1}^{L}\nabla^2(\underline{z}_i) - \frac{e^2}{4\pi\epsilon_0}\sum_{i=1}^{H-1}\sum_{j=1}^{L}\frac{Z_i}{|\underline{a}_i - \underline{z}_i|} + \frac{e^2}{8\pi\epsilon_0}\sum_{i,j=1}^{L}{}' \frac{1}{|\underline{z}_j - \underline{z}_i|} \tag{133}$$

The clamped nucleus Hamiltonian (3) may be mapped exactly onto this form simply by the correspondences $m \rightarrow \mu$ and $\underline{x}_i^e \rightarrow \underline{z}_i$ providing that care is taken to see that no two different sets of $\underline{a}_i$ map onto the same $\underline{q}(\underline{a})$. Thus the usual practice of clamped nucleus Hamiltonian calculation can be fitted precisely into a full calculational strategy if the approach outlined in this section is adopted. It is not claimed, of course, that this is the only way that it could be done but it is one well defined way.

As explained in Section (2), the clamped-nucleus problem is well defined in its own right and its solution can be undertaken without any thought for a particular body fixed coordinate system. However if a full calculation is to be attempted by utilising clamped nucleus solutions in the product form (118) then, to ensure that the calculations are meaningful, the functions $\Phi^J_{kp}(\underline{q})$ for the chosen body-fixed coordinate system, must be chosen so as to vanish when the jacobian vanishes because electronic functions alone will not generally vanish there. Thus, somewhat paradoxical though it might seem, it is formally possible to generate non-existent states in this approach unless a very careful watch is kept on the behaviour or the nuclear motion functions.

Of course even though the clamped-nucleus problem does not determine a particular choice of body-fixed coordinates, generally speaking the form of its solutions certainly influence such choices. Thus the classical Eckart approach to the description of molecular spectra (see *e.g* [7]) is developed from the view that, if the ground-state potential (assumed here to be generated from a clamped-nucleus calculation) has a very deep minimum, then that minimum can be regarded as defining the equilibrium nuclear geometry. The body-fixed coordinates are then defined so as to provide a first order description of the rotational motion in terms of a rigid top with the shape of the equilibrium molecular geometry. The internal motion is described by means of normal coordinates which are expressed in terms of displacements from the equilibrium geometry and, again to first order, describe the motions as small amplitude simple-harmonic vibrations. This approach generally works very well indeed and is the approach in which it is most usual to interpret the rotation-vibration spectra of molecules. However it will fail if the amplitude of the internal motions is sufficiently great as to force three nuclei to be co-linear for in this case the jacobian vanishes. Indeed sometimes it is quite difficult to spot this failure, as is evidenced by some (justified) criticism of work in which the author was involved (see [23]). To deal with such large amplitude motions it is necessary to use a different body-fixing prescription which, in this case, would be one that could deal with the co-linearity problem. The way in which this was done for the criticised work mentioned above can be found in [17] but the problem is a general one and solvable only on a case by case basis.

In the preceeding discussion it has beeen assumed that it is sufficient to use just one potential to describe the full problem. Suppose, however, that it is wished to describe electronic-rotation-vibration spectra, then at least two potentials will be necessary. Let it be supposed further that both potentials have deep minima so that the Eckart approach can be used on both. Because the minima will generally define different equilibrium geometries however, the coordinate systems will be different for each potential. The relevant jacobians will vanish in different regions and thus the design of adequate trial functions will be even trickier than it is in the single potential case. But even supposing that this is not a problem, the use of two distinct coordinate systems gives rise to difficulties in computation because in order to evaluate any matrix elements it is necessary to express the functions and operators involved in a common coordinate system. (This problem was identified first by Duchinsky [24] and his analysis of the problem is often used to account, at least qualitatively, for intensity effects in polyatomic spectra in much the same way as the Franck-Condon analysis is used in similar circumstances for diatomics.) The quantitative treatment of such problems is as yet very much underdeveloped.

Even though two potentials were necessary in the case just discussed it was still assumed that the potentials were well separated in the sense that they had no common or close values in the regions where the nuclear motion wavefunctions were non-vanishing. However there are a number of experimental situations of interest which seem to be not accountable for in terms of well separated potentials and to understand which, the coupling terms in (122) and (129) must be invoked. The chief of these experimental situations are labelled as the Jahn-Teller effect and the Renner-Teller effect, though it would be as well to be a little sceptical about the "effect" nomenclature. Here too quantitative treatment is rather underdeveloped but some interesting work has been begun by Carter *et al.* [25].

6. Conclusions

In these lectures I have tried to establish the position of the clamped-nucleus electronic Hamiltonian as it it is usually worked with, in the context of the full Schrödinger Hamiltonian for a molecule. To do this involved what perhaps seemed a large digression into the body-fixed form of the full problem. I hope, however, that it is clear that this was necessary in order to establish a context in which the separation of electronic and nuclear motions could be fruitfully considered. In actually making this separation, a number of choices were made which were matters of judgement and motivated by the desired end. However it should be recognised that other judgements could have been made and that those that actually were made, are defended solely on the grounds of utility and reasonableness.

Assuming that these choices are made then the position of the clamped-nucleus Hamiltonian has been elucidated. The usual practices in bound-state nuclear motion calculations on a single potential have been examined and been found to be justifiable provided that care is taken to use nuclear motion wavefunctions that vanish at geometries where the jacobian for the body-fixing transformation vanishes. It has also proved possible to provide a rational formal framework for calculations involving more than one potential but much remains to be discovered about the extent to which this formalism can be given successful computational substance.

There is much more that could be said about the role of permutational symmetry and of particle statistics but this would need another full course of lectures as would any attempt to extend the discussion to consider scattering states. Finally I recognise that the much of the discussion of body-fixing that has been given, is of a rather formal and abstract nature. It would have been nice to have been able to flesh it out by working an example or two but to do so would have involved pages of rather rebarbative algebra even to manage to show how to get the Eckart Hamiltonian. This is usually something of a turn-off when done in public and I feel that it is best done between consenting adults in private so I should be jolly happy to talk more to any interested persons.

References

[1] B. Simon, Quantum Mechanics for Hamiltonians defined as quadratic forms , Princeton U.P. (1971)

[2] M. Born and J.R. Oppenheimer, Ann.der Phys., 84, 457-484 (1927)

[3] M. Born and K. Huang, in Dynamical Theory of Crystal Lattices , Appendix 8, Oxford U.P. (1954)

[4] J.C. Slater, Proc. Nat. Acad. Sci., 13, 423-430, (1927)

[5] B.T. Sutcliffe and J. Tennyson, Int. J. Quant Chem., 39, 183-196, (1991)

[6] B. T. Sutcliffe, in Methods of Computational Chemistry Vol. 5, Ed. S. Wilson, Plenum Press, New York and London (1991).

[7] J.K.G. Watson, Mol. Phys., 15, 479-490 (1968)

[8] D.M. Brink and G.R. Satchler, Angular Momentum, 2nd ed. Clarendon Press, Oxford (1968)

[9] L.C. Biedenharn and J.C. Louck, Angular Momentum in Quantum Physics, Addison- Wesley, Reading ,Mass. (1982)

[10] J. M. Brown and B. J. Howard, Mol. Phys., 31, 1517-1525, (1976)

[11] R. N. Zare, Angular Momentum, Chap 3.4, Wiley, New York, (1988)

[12] G. Ezra, Symmetry properties of molecules, Lecture Notes in Chemistry 28, Springer-Verlag, Berlin (1982)

[13] J.C. Louck, J. Mol. Spec., 61, 107-137 (1976)

[14] E.C. Kemble, The fundamental principles of Quantum Mechanics, McGraw-Hill, New York, (1937)

[15] B.T. Sutcliffe, in Theoretical models of chemical bonding, Pt. 1 (Z Maksic, ed.), pp 1-28, Springer-Verlag Berlin (1990)

[16] R. Bartholomae, D. Martin and B.T. Sutcliffe, J. Mol. Spec., 87, 367-381 (1981)

[17] J. Tennyson and B. T. Sutcliffe, J. Mol. Spec., 101, 71-82 (1983)

[18] S. Carter and N. C. Handy, Mol. Phys. , 53, 1033-1039 (1984)

[19] N.C. Handy, Mol. Phys., 61., 207-233 (1987)

[20] G. Hunter, Int. J. Quant. Chem., 9, 237-242 (1975)

[21] J. Czub and L. Wolniewicz, Mol. Phys., 36, 1301-1308 (1978)

[22] F. T. Smith, Phys. Rev. Letts., 45, 1157-1160 (1980)

[23] S. Carter and N. C. Handy, J. Mol. Spec., 95, 9-19 (1982)

[24] F. Duchinsky, Acta. Phys. Chem. URSS, 7, 551-566 (1937)

[25] S. Carter, N. C. Handy, P. Rosmus and G. Chambaud, Mol. Phys., 71, 605-622 (1990)

THE ELECTRON CORRELATION PROBLEM

R.McWeeny

Dipartimento di Chimica e Chimica Industriale
Universitá di Pisa, Via Risorgimento 35, 56100 Pisa, Italy

1) Introduction

The so-called 'electron correlation problem' has been a dominant theme in molecular quantum mechanics for the last forty years; there is little new to say but it is of great importance to define the concept precisely and to investigate how correlation effects can be admitted in wavefunction calculations.

After making a large basis set calculation, typically achieving near Hartree-Fock accuracy, it is commonly stated that the remaining error in the energy (usually not more than 1%) is associated with 'electron correlation' - and that its elimination constitutes the 'electron correlation problem'. This problem has been recognized since the early days of quantum mechanics: as long ago as 1934, Wigner proposed as a measure of 'correlation' the *correlation energy* defined as

$$E_{\text{cor}} = E_{\text{exact}} - E_{\text{IPM}} \tag{1}$$

where E_{exact} is an 'exact' non-relativistic limit, while E_{IPM} refers to an independent-particle model. But *what* independent-particle model? Löwdin (1959) adopts this definition, taking E_{IPM} to be the *Hartree-Fock* energy obtained as the lowest eigenvalue of the (Fock) Hamiltonian

$$\hat{H}^{\text{F}} = \sum_i \hat{F}(i), \qquad \hat{F} = \hat{h} + \hat{G}, \tag{2}$$

where $\hat{h}$ is the usual 1-electron Hamiltonian while $\hat{G}$ contains an 'average' (Hartree-Fock) field due to the presence of other electrons. Consequently the electrons are regarded as 'pseudo-independent' rather than *strictly* independent, the meaning of 'IPM' being that the many-electron Hamiltonian of the *model* is simply a sum of *one*-electron operators. The lowest-energy (antisymmetric) eigenfunction of (2) is a single Slater determinant, in the absence of degeneracy; and this one-term approximation is usually improved by admitting CI. The inclusion of correlation effects has thus come to be associated with the admission of CI.

As Löwdin remarked, however, the definition (1) "should be modified somewhat if the state under consideration shows spin or orbital degeneracy"; and this situation is not uncommon, even when the ground state *appears* to be non-degenerate. Thus, for the hydrogen molecule *in non-equilibrium geometry*, the H-F wavefunction (with 2 electrons in the same MO) may be inappropriate: for large internuclear distance it fails to describe the true independent-particle limit in which there is one electron in each of two separate hydrogen atoms. The error can of course be corrected by admitting CI - but this correction has nothing to do with correlation in the sense of 'lack of independence' (for nothing could be more independent than 2 electrons at infinite separation!), being instead an artefact connected with the use of a variational wavefunction which 'does not dissociate correctly'. There are other complications: for arbitrary geometry, the HF IPM is frequently found to be unstable, each doubly-occupied orbital 'splitting' into one for each electron; and 'unrestricted' (UHF) functions, besides being physically unsatisfactory in violating spin requirements (mixing multiplicities), can lead to energy surfaces with discontinuous first derivatives (Mayer 1978) even if they give a correct dissociation limit. So the simplicity of the definition (1) is deceptive: at best, it is a 'rule of thumb' useful mainly for systems in equilibrium geometry.

2) Density Functions and Correlation

An alternative approach to the correlation concept is provided by the statistical interpretation of the Schrödinger wavefunction (McWeeny,1954):

$$\rho_{sys}(\mathbf{x}_1, \mathbf{x}_2, \ldots \mathbf{x}_N) = \Psi(\mathbf{x}_1, \mathbf{x}_2, \ldots \mathbf{x}_N)\Psi^*(\mathbf{x}_1, \mathbf{x}_2, \ldots \mathbf{x}_N) \quad (3)$$

is a *probabitity density*: multiplied by the volume element $d\mathbf{x}_1 d\mathbf{x}_2 \ldots d\mathbf{x}_N$ it gives the probability of finding electron 1 in $d\mathbf{x}_1$ at point $\mathbf{x}_1$, electron 2 simultaneously in $d\mathbf{x}_2$ at point $\mathbf{x}_2$, and so on. If we are interested not in *all* electrons simultaneously being present at* N given points, but rather for example that 1 and 2 are at $\mathbf{x}_1, \mathbf{x}_2$ and the others anywhere, then we simply integrate (3) over all variables except $\mathbf{x}_1$ and $\mathbf{x}_2$. But, because $\Psi\Psi^*$ is *symmetric* under any permutation of variables, the same result would follow for *any* pair i, j; and evidently the probability of finding $d\mathbf{x}_1$ and $d\mathbf{x}_2$ simultaneously occupied (by *any* electrons) will be $\pi(\mathbf{x}_1, \mathbf{x}_2)$ where

$$\pi(\mathbf{x}_1, \mathbf{x}_2) = N(N-1) \int \Psi(\mathbf{x}_1, \mathbf{x}_2, \ldots \mathbf{x}_N)\Psi^*(\mathbf{x}_1, \mathbf{x}_2, \ldots \mathbf{x}_N) d\mathbf{x}_3 \ldots d\mathbf{x}_N \quad (4)$$

is a *pair density* – as commonly used in the theory of liquids. The pair density in (4), however, also includes the spin variables: if we are not interested in the spins ('up' or 'down') of the two electrons we must make further integrations, over spins s_1, s_2, to obtain an 'ordinary' pair density

$$\Pi(\mathbf{r}_1, \mathbf{r}_2) = \int \pi(\mathbf{x}_1, \mathbf{x}_2) ds_1 ds_2, \quad (5)$$

which refers only to two points in ordinary 3-dimensional space.

The functions (4) and (5) are special cases of the *reduced density matrices* (Husimi 1940; McWeeny 1955,1960; Löwdin 1955; see also MMQM** Ch.5) which we define presently: they allow us to discuss correlation as a *dynamical* phenomenon. The probability of electrons being found close together is reduced by their mutual repulsion – but 'reduced' with respect to what? And here the IPM re-appears: for if the electrons were truly independent the probabiity of finding $d\mathbf{x}_1, d\mathbf{x}_2$ simultaneously occupied would be a *product* (the joint probability of two independent events occurring) –

$$\pi(\mathbf{x}_1, \mathbf{x}_2) = \rho(\mathbf{x}_1)\rho(\mathbf{x}_2) \qquad \text{(IPM)}, \quad (6)$$

where $\rho(\mathbf{x})$ is the probability for finding an electron 'at' $\mathbf{x}$ without reference to any others. The function $\rho(\mathbf{x})$ would also be obtained explicitly from (3), by integration over all the 'unwanted' variables, in the form

$$\rho(\mathbf{x}_1) = N \int \Psi(\mathbf{x}_1, \mathbf{x}_2, \ldots \mathbf{x}_N)\Psi^*(\mathbf{x}_1, \mathbf{x}_2, \ldots \mathbf{x}_N) d\mathbf{x}_2 \ldots d\mathbf{x}_N \quad (7)$$

and again, when spin is of no interest, a *spatial* density may be obtained by integration as in (5):

$$P(\mathbf{r}_1) = \int \rho(\mathbf{x}_1) ds_1. \quad (8)$$

With a spin independent Hamiltonian we might expect, at the IPM level, $\Pi(\mathbf{r}_1, \mathbf{r}_2) = P(\mathbf{r}_1)P(\mathbf{r}_2)$ as the analogue of (6): but we shall find that even for a single Slater determinant *neither result is true*.

---

* The term 'at' is used with the more precise meaning of the preceding sentence.

** See Ch.1 Bibiography

First, a slight generalization is needed: the density functions do indeed determine the expectation values of the 1-electron potential energy (as the potential energy of a distribution of charge of density $-eP(\mathbf{r})$ in the field of the nuclei), and of the 2-electron repulsion energy (a simple statistical average with pair function $\Pi(\mathbf{r}_1, \mathbf{r}_2)$. But they do not determine the expectation values of *operator* quantities such as the kinetic energy. For this purpose we need the density 'matrices'

$$\rho(\mathbf{x}_1;\mathbf{x}_1') = \int \Psi(\mathbf{x}_1,\mathbf{x}_2 \dots \mathbf{x}_N)\Psi^*(\mathbf{x}_1',\mathbf{x}_2 \dots \mathbf{x}_N)d\mathbf{x}_2 \dots d\mathbf{x}_N, \tag{9}$$

and

$$\pi(\mathbf{x}_1,\mathbf{x}_2;\mathbf{x}_1',\mathbf{x}_2') = \int \Psi(\mathbf{x}_1,\mathbf{x}_2 \dots \mathbf{x}_N)\Psi^*(\mathbf{x}_1',\mathbf{x}_2' \dots \mathbf{x}_N)d\mathbf{x}_3 \dots d\mathbf{x}_N, \tag{10}$$

where the primed variables are associated with Ψ^*. The probability densities (7) and (4) are seen to be 'diagonal elements' of the density matrices (9) and (10), in which the unprimed and primed variables (analogous to row and column indices) have been identified.

The functions (9) and (10) allow one to 'eliminate' the wavefunction in all expectation value expressions for 1- and 2-electron quantities. Thus, quite generally (i.e. for *any kind of wavefunction*)

$$< \Psi|\sum_i \hat{A}(i)|\Psi > = \int_{\mathbf{x}_1'=\mathbf{x}_1} [\hat{A}(1)\rho(\mathbf{x}_1;\mathbf{x}_1')]d\mathbf{x}_1 \rightarrow \int_{\mathbf{r}_1'=\mathbf{r}_1} [\hat{A}(1)P(\mathbf{r}_1;\mathbf{r}_1')]d\mathbf{r}_1 \qquad (11),$$

where the form on the extreme right applies to operators that do not work on spins. The corresponding result for 2-electron operators is

$$< \Psi|\sum_{i,j} \hat{B}(i,j)|\Psi > = \int_{\substack{\mathbf{x}_1'=\mathbf{x}_1 \\ \mathbf{x}_2'=\mathbf{x}_2}} [\hat{B}(1,2)\pi(\mathbf{x}_1,\mathbf{x}_2;\mathbf{x}_1',\mathbf{x}_2')]d\mathbf{x}_1 d\mathbf{x}_2. \tag{12}$$

The identification of primed and unprimed variables (*after* applying the operator), followed by integration, corresponds to taking the *trace* in a matrix representation. Thus, in a finite spin-orbital basis $\{\psi_i\}$,

$$\rho(\mathbf{x}_1;\mathbf{x}_1') = \sum_{r,s} \rho_{rs}\psi_r(\mathbf{x}_1)\psi_s^*(\mathbf{x}_1') \tag{13}$$

and (11) gives at once, collecting the quantities ρ_{rs} and $< \psi_s|\hat{A}|\psi_r >$ into square matrices,

$$< \Psi|\sum_i \hat{A}(i)|\Psi > = \sum_r (\mathbf{A}\rho)_{rr} = \mathrm{tr}(\mathbf{A}\rho), \tag{14}$$

with a similar result for (12).

For the Hamiltonian (1.2) we easily obtain

$$E = < \hat{H} > = \int [\hat{h}(1)\rho(\mathbf{x}_1;\mathbf{x}_1')]_{\mathbf{x}_1'=\mathbf{x}_1} + \frac{1}{2}\int g(1,2)\pi(\mathbf{x}_1,\mathbf{x}_2)d\mathbf{x}_1 d\mathbf{x}_2, \tag{15}$$

where primed variables are not needed in the 2-electron term (no differential operator) and the diagonal element $\pi(\mathbf{x}_1,\mathbf{x}_2) = \pi(\mathbf{x}_1,\mathbf{x}_2;\mathbf{x}_1,\mathbf{x}_2)$ is sufficient. In the algebraic approximation (for any spin-orbital basis) the last result becomes

$$E = \mathrm{tr}\mathbf{h}\rho + \tfrac{1}{2}\mathrm{tr}\mathbf{g}\pi \tag{16}$$

and the elements of the matrices ρ, π, referred to the finite basis, may thus be obtained from the energy expectation value expression – and vice versa. Thus

$$\begin{aligned} \rho_{sr} &= \text{coefficient of } < \psi_r|\hat{h}|\psi_s > \text{ in } < \hat{H} > \\ \pi_{tu,rs} &= \text{coefficient of } < \psi_r\psi_s|g|\psi_t\psi_u > \text{ in } < \hat{H} > \end{aligned} \tag{17}$$

where the matrix elements of the 2-electron quantities are labelled by the double indices. Needless to say, the elements of the *spinless* density matrices (i.e. P_{rs} and $\Pi_{rs,tu}$), referred to an *orbital* basis, are obtained in a precisely similar way as coefficients of the integrals $< \phi_s|\hat{h}|\phi_r >$ and $< \phi_t\phi_u|g|\phi_r\phi_s >$.

Equation (17) reveals an important link with the second-quantization approach. For when $\hat{H}$ is expressed in the form (1.25) the energy expression contains expectation values of creation/annihilation operator products: and by comparison it follows that

$$\rho_{sr} =< \hat{a}_r^\dagger \hat{a}_s >, \qquad \pi_{tu,rs} =< \hat{a}_r^\dagger \hat{a}_s^\dagger \hat{a}_u \hat{a}_t > \tag{18}$$

– confirming the correct usage of density matrix notation in (1.43).

It is now easy to examine the nature of the correlation, if any, implicit in the simple closed-shell HF approximation. From Slater's rules (1.26), ρ is *diagonal* with respect to the occupied spin-orbitals, with unit elements, all other elements being zero. When the spin-orbitals are written $\psi_r = \phi_r\alpha$, $\bar{\psi}_r = \phi_r\beta$, this means that

$$\rho(\mathbf{x};\mathbf{x}') = P_\alpha(\mathbf{r};\mathbf{r}')\alpha(s)\alpha(s') + P_\beta(\mathbf{r};\mathbf{r}')\beta(s)\beta(s'), \tag{19}$$

where the spatial functions on the right (sums of orbital products $\phi_r(\mathbf{r})\phi_r^*(\mathbf{r}')$) are labelled according to the spin factors they carry. On identifying primed and unprimed variables and integrating over spin, the electron density $P(\mathbf{r})$ is seen to be the sum

$$P(\mathbf{r}) = P_\alpha(\mathbf{r}) + P_\beta(\mathbf{r}) \tag{20}$$

of the densities of 'up-spin' and 'down-spin' electrons – equal in this case since the wavefunction describes a singlet state (zero total spin).

Examination of the 2-electron terms gives a similar result for $\pi(\mathbf{x}_1,\mathbf{x}_2;\mathbf{x}_1',\mathbf{x}_2')$, with 6 non-zero terms of which 2 disappear on integrating over spin. The pair function (diagonal element of the resultant $\Pi(\mathbf{r}_1,\mathbf{r}_2;\mathbf{r}_1',\mathbf{r}_2')$) may be expressed in terms of the functions P_α, P_β in (19) and is a sum of 4 terms:

$$\Pi_{\alpha\alpha}(\mathbf{r}_1,\mathbf{r}_2) = P_\alpha(\mathbf{r}_1)P_\alpha(\mathbf{r}_2) - P_\alpha(\mathbf{r}_1;\mathbf{r}_2)P_\alpha(\mathbf{r}_2;\mathbf{r}_1) \tag{21a}$$
$$\Pi_{\alpha\beta}(\mathbf{r}_1,\mathbf{r}_2) = P_\alpha(\mathbf{r}_1)P_\beta(\mathbf{r}_2) \tag{21b}$$
$$\Pi_{\beta\alpha}(\mathbf{r}_1,\mathbf{r}_2) = P_\beta(\mathbf{r}_1)P_\alpha(\mathbf{r}_2) \tag{21c}$$
$$\Pi_{\beta\beta}(\mathbf{r}_1,\mathbf{r}_2) = P_\beta(\mathbf{r}_1)P_\beta(\mathbf{r}_2) - P_\beta(\mathbf{r}_1;\mathbf{r}_2)P_\beta(\mathbf{r}_2;\mathbf{r}_1). \tag{21d}$$

Thus, *for electrons of opposite spin*, the motions of the electrons are completely uncorrelated; but, *for electrons of like spin*, the IPM probability (of product form) is reduced by a term arising from the off-diagonal part of the 1-electron density matrix. We may use (21a) to define a *conditional* probability density

$$\frac{\Pi_{\alpha\alpha}(\mathbf{r}_1,\mathbf{r}_2)}{P_\alpha(\mathbf{r}_2)} = P_\alpha(\mathbf{r}_1) - \frac{P_\alpha(\mathbf{r}_2,\mathbf{r}_1)P_\alpha(\mathbf{r}_1,\mathbf{r}_2)}{P_\alpha(\mathbf{r}_2)}, \tag{22}$$

which gives the probability of finding an up-spin electron at $\mathbf{r}_1$ *when another is known to be at* $\mathbf{r}_2$; and the second term on the right in (22) thus defines a 'correlation hole' i.e. the amount by which $P_\alpha(\mathbf{r}_1)$ is reduced by the presence of an up-spin 'reference electron' at point $\mathbf{r}_2$. If (22) is written in the form

$$\frac{\Pi_{\alpha\alpha}(\mathbf{r}_1,\mathbf{r}_2)}{P_\alpha(\mathbf{r}_2)} = P_\alpha(\mathbf{r}_1)[1 + f_{\alpha\alpha}(\mathbf{r}_1\mathbf{r}_2)] \tag{23}$$

then $f_{\alpha\alpha}(\mathbf{r}_1,\mathbf{r}_2)$ is a *correlation factor* for up-spin electrons at points $\mathbf{r}_1,\mathbf{r}_2$.

The correlation factor in (23) was first calculated (analytically) for a free-electron gas by Wigner and Seitz (1933); it has been extensively discussed by Slater (1951) and provides a basis for the well known 'X_α' method (see Slater 1974). The corresponding correlation hole is often called the 'Fermi hole' being characteristic of fermions and a consequence of the antisymmetry of the wavefunction. For $r_{12} \to 0, f_{\alpha\alpha}(\mathbf{r}_1,\mathbf{r}_2) \to -1$, and there is consequently zero probability of finding two like-spin electrons at the same point in space – a completely general statement of Pauli's exclusion principle. In fact $\Pi_{\alpha\alpha}(\mathbf{r}_1, \mathbf{r}_2)$ goes to zero like r_{12}^2 and the correlation factor therefore shows no cusp at the origin.

For electrons of *different* spin, on the other hand, the correlation factor (missing in a 1-determinant approximation) has a sharp negative peak for $r_{12} \to 0$ and it is this cusp that is extremely difficult to obtain without very extensive CI. The results of some representative calculations of the 'coulomb hole' (a reflection of the electrostatic repulsion between the electrons) have been discussed by Cooper and Pounder (1978, 1980, 1983) who show that, even with only a few thousand configurations, the cusp is fairly well defined.

Various properties of the two types of correlation hole (McWeeny, 1960) are well known and this statistical approach to the discussion of dynamical correlation has been elaborated by Kutzelnigg et al (1968). The immediate concern at this point is how to admit correlation effects in practice.

3) Linked Cluster Expansions

To include coulomb correlation effects by standard CI methods it is necessary to admit very long expansions in order to describe the correct cusp behaviour for $r_{ij} \to 0$. This may be done by perturbation methods; but first we consider the general nature of the expansion.

The basic idea is to start from a *reference function* Φ_0, usually (as will be assumed here) a 1-determinant wavefunction of HF type, and to add terms corresponding to single, double, triple,.. 'excitations' in which electrons are promoted from the occupied orbitals in Φ_0 to a complementary set (e.g. the 'virtual orbitals' which are the 'unused' solutions of the HF equations). Thus, with an obvious notation,

$$\Psi = C_0\Phi_0 + \sum_{i,m} C_i^m \Phi_i^m + \sum_{\substack{i<j \\ m<n}} C_{ij}^{mn}\Phi_{ij}^{mn} + ...$$

and this may be written in second-quantization language, without normalizing, as

$$|\Psi> = (1 + \sum_{i_1} \hat{u}_{i_1} + \sum_{i_1 i_2} \hat{u}_{i_1 i_2} + ...)|\Phi_0>, \qquad (24)$$

in which the operator $\hat{u}_{i_1} = \sum_{m_1} C_{i_1}^{m_1} \hat{a}_{m_1}^{\dagger} \hat{a}_{i_1}$ generates the appropriate linear combination of single-excitation functions, $\hat{u}_{i_1 i_2}$ generates a combination of doubly excited functions, and so on. Such terms are called 1-cluster, 2-cluster, ... n-cluster functions, referring to the excitation of 'clusters' of n electrons at a time from the reference function; and (24) is a 'cluster development' (in Fock space) of the wavefunction.

When the orbitals in an IPM reference function Φ_0 are varied, using for example the exponential operator defined in (1.37) and (1.39), Φ_0 becomes (in terms of the original orbitals) a function Φ_0' whose cluster development contains 1-cluster, 2-cluster and many-cluster terms. But the wavefunction still corresponds to an IPM and therefore does not admit correlation effects: this is because the clusters are of a very special kind, the n-cluster terms being *products* of the 1-cluster terms and thus 'reducible'. The orbital-variation operator $\exp(C_{i_1}^{m_1})^{\dagger}\hat{a}_{m_1}\hat{a}_{i_1}$ will contain products

$$C_{i_2}^{m_2} C_{i_1}^{m_1} \hat{a}_{m_2}^{\dagger} \hat{a}_{i_2} \hat{a}_{m_1}^{\dagger} \hat{a}_{i_1}$$

and the anticommutation rules allow the product to be re-ordered as $\hat{a}_{m_1}^{\dagger}\hat{a}_{m_2}^{\dagger}\hat{a}_{i_2}\hat{a}_{i_1}$ – the operator which generates the 2-cluster function $\Phi_{i_1 i_2}^{m_1 m_2}$ by removing electrons from ψ_{i_1}, ψ_{i_2} and putting them back in ψ_{m_2}, ψ_{m_1} (in that order). But this 2-cluster function occurs in the expansion (24) with the coefficient

$$C_{i_1 i_2}^{m_1 m_2} = C_{i_1}^{m_1} C_{i_2}^{m_2}$$

and it is in this sense that the 2-cluster terms are reducible to products of 1-cluster terms. A similar discussion applies to clusters of higher order: a 4-cluster term, for example, may be simply a product of two 2-cluster terms. In order to escape the limitations of the IPM, it is necessary to admit n-cluster terms which are *not* reducible to products of cluster functions for smaller numbers of particles. Every $\hat{u}$-operator in (24) should be written so as to clearly separate the reducible and irreducible terms; and this is done by using an exponential operator which generalizes (1.37), namely

$$\exp(\hat{T}) = \exp(\hat{T}_1 + \hat{T}_2 + \ldots), \tag{25}$$

in which

$$\hat{T}_1 = \sum_{i_1} \hat{v}_{i_1}, \qquad \hat{v}_{i_1} = \sum_{m_1} D_{i_1}^{m_1} \hat{a}_{m_1}^{\dagger} \hat{a}_{i_1} \tag{26}$$

and higher terms are defined in a similar way. The coefficients $D_{i_1}^{m_1}$, $D_{i_1 i_2}^{m_1 m_2}$, ... are then characteristic of the *irreducible* clusters and are not factorizable but completely independent. By expanding the exponential (25) and noting the properties of the operators it follows that

$$\exp(\hat{T}) = 1 + \sum_{i_1} \hat{v}_{i_1} + \sum_{i_1<i_2} (\hat{v}_{i_1 i_2} + \hat{v}_{i_1}\hat{v}_{i_2}) + \ldots, \tag{27}$$

where the irreducible 2-cluster terms are separated from the products of 1-cluster terms – and so on, throughout the development.

How can the terms in the cluster development be interpreted? It will be sufficient to consider one term only – that produced by the operator $\hat{v}_{1,2} = \sum_{m_1 m_2} D_{1,2}^{m_1 m_2} \hat{a}_{m_1}^{\dagger} \hat{a}_{m_2}^{\dagger} \hat{a}_2 \hat{a}_1$. If we write the reference function, corresponding to the Fock-space ket $|\Phi_0>$ as

$$\Phi_0(\mathbf{x}_1, \mathbf{x}_2, \ldots \mathbf{x}_N) = (N!)^{\frac{1}{2}} \hat{A}_N \Omega(\mathbf{x}_1, \mathbf{x}_2, \ldots \mathbf{x}_N),$$

where $\hat{A}_N$ is the usual N-electron antisymmetrizer and the product $\psi_1(\mathbf{x}_1)\psi_2(\mathbf{x}_2)\ldots \psi_N(\mathbf{x}_N)$ is denoted by Ω, then $\hat{v}_{1,2}$ produces the ket whose Schrödinger equivalent will be

$$\sum_{m_1 m_2} D_{1,2}^{m_1 m_2} \hat{A}_N[\psi_{m_1}(\mathbf{x}_1)\psi_{m_2}(\mathbf{x}_2)][\psi_3(\mathbf{x}_3)\ldots \psi_N(\mathbf{x}_N)].$$

On noting that $\hat{A}_N \hat{A}_2 = \hat{A}_N$ ($\hat{A}_2$ being a *two*-electron antisymmetrizer) and that $\sqrt{2}\hat{A}_2[\psi_1\psi_2]$ will be a normalized Slater determinant, the correspondence may be written

$$\hat{v}_{1,2}|\Phi_0> \Leftrightarrow (N!)^{\frac{1}{2}} \hat{A}_N \frac{1}{\sqrt{2}} [\psi_{1,2}(\mathbf{x}_1, \mathbf{x}_2)\Omega(\mathbf{x}_3, \ldots \mathbf{x}_N)_{(1,2)}], \tag{28}$$

where

$$\psi_{1,2}(\mathbf{x}_1, \mathbf{x}_2) = \sum_{m_1 m_2} D_{1,2}^{m_1 m_2} \sqrt{2}\hat{A}_2[\psi_{m_1}(\mathbf{x}_1)\psi_{m_2}(\mathbf{x}_2)] \tag{29}$$

and is thus an infinite expansion of a general 2-electron wavefunction in terms of normalized Slater determinants.

It is not difficult to verify that the correspondence just established is general: in Schrödinger language, a *p*-cluster term is one in which *p* spin-orbitals (with their electrons) have been removed from the Hartree-Fock 'sea' and replaced by a 'fully-correlated' (in principle 'exact') *p*-electron cluster function, the product then being antisymmetrized. It is also evident that by suitably choosing the orbitals of the reference function the 1-cluster terms may be eliminated: when Φ_0 is a 1-determinant function this is the case (in good approximation) with HF orbitals (the Brillouin theorem); more generally it characterizes the 'Brueckner orbitals'. The leading terms in the correlation correction then refer to the 2-clusters and their products; and in most current work it is assumed that the *irreducible* clusters for more than two particles can be safely neglected. At this level of approximation the cluster development takes the form, in Schrödinger language,

$$\Psi = (N!)^{\frac{1}{2}} \hat{A}_N[\Omega_0 + \sum_{(ij)} \frac{\Psi_{ij}}{\sqrt{2}}\Omega_{ij} + \sum_{(ij)(kl)} \frac{\Psi_{ij}}{\sqrt{2}}\frac{\Psi_{kl}}{\sqrt{2}}\Omega_{ijkl} + \ldots, \tag{30}$$

where the summations are over distinct pairs and combinations of pairs. Clearly, a CI calculation with all possible 'double-excitations' from the Hartree-Fock sea would yield only the first two terms in the expansion (3); but the reducible 4-cluster, 6-cluster, and higher terms all make an important contribution to the correlation energy - and they are so numerous that their inclusion by CI methods is not easy. It is for this reason that much attention has been given to 'correlated pair' approximations of various kinds.

4) Methods of Calculation

There are essentially two main approaches - short of performing full-CI - to the calculation of the numerical parameters (expansion coefficients) defining the various functions in a cluster development: the first concentrates on the direct calculation of the 2-clusters, neglecting all other (irreducible) terms; the second employs perturbation theory, generally in high order, to obtain approximate values of the expansion coefficients in clusters of arbitrary type - and of course the corresponding energy. Since both approaches are developed in great detail later, the presentation here is intended only to provide a preliminary orientation. Further details may be found in MMQM Ch.9 and in the other books cited.

Correlated-Pair Methods

There are many forms of 'correlated-pair theory' in which (30) provides the basic ansatz for the wavefunction. The earlier work has been reviewed by Hurley (1976): more recent developments were initiated largely by Kutzelnigg (1977), Meyer (1977), and Ahlrichs (1979,1983) and are usually presented in Fock-space form. The basic problem is how to optimize the pair functions (i.e. the coefficients in the operators $\hat{T}_2$ which generate them) and the various approaches in use differ mainly in the extent to which they recognize the interdependence of the pairs.

Formally, one could assume the full expansion (xx) to be exact, so that

$$(\hat{H} - E)|\Psi> = (\hat{H} - E)\exp(\hat{T})|\Phi_0> = 0, \tag{31}$$

and then take scalar products from the left with a number of independent vectors (such as Φ_i^m, Φ_j^n) sufficient to determine the coefficients of all the creation/annihilation products. The resultant equations, which are not of the usual secular-equation type, are of course non-linear and were first formulated (using a diagrammatic analysis) by Cizeck (1966). When the ansatz (30) is assumed, in place of the exact development, this procedure corresponds to an application of the 'method of moments' (Bangudu et al, 1974)). The value of E obtained is then no longer a variational upper bound to the energy and there is no monotonic convergence to the exact results; but the number of equations needed is relatively small and progress is possible. In this case, with $\hat{T}_1, \hat{T}_2$ denoting the sums of 1- and 2-cluster operators in (25), it is sufficient to consider the conditions

$$\left.\begin{aligned} <\Phi_0|(\hat{H} - E)|(1 + \hat{T}_2)\Phi_0> &= 0 \\ <\Phi_{ij}^{mn}|(\hat{H} - E)|(1 + \hat{T}_2 + \tfrac{1}{2}\hat{T}_2^2)\Phi_0> &= 0 \end{aligned}\right\} \quad (\text{all } im, jn). \tag{32}$$

Together, these conditions determine all the numerical coefficients in the operators. The explicit form of the equations, which are quadratic in the D_{ij}^{mn}, reveals the enormous computational difficulty of obtaining a solution. The most primitive approximation is to retain only the linear terms, obtaining an 'independent-pair' model in which each pair ij is 'decoupled' from the others. The next step is to introduce a coupling, through the quadratic terms, and this leads to various 'coupled electron pair approximations' (CEPA). Some of these possibilities were anticipated and explored in the early work of Sinanoglu (1964) and Nesbet (1965).

The techniques for implementing correlated-pair calculations are now at an advanced stage and are fully discussed in later chapters.

Perturbation Methods

The other main approach in calculating the expansion coefficients in the general cluster development is to regard the reference function Φ_0 as the exact wavefunction for a 'model' system (in this discussion the HF-IPM), treating the difference between the 'real' and 'model' Hamiltonians as a *perturbation* $\delta\hat{H}$, and then to use a suitable form of perturbation theory to estimate the effect of admitting $\delta\hat{H}$.

To simplify and systematize the calculation of high order perturbation corrections, and in some cases even to provide partial sums to infinite order, it is customary to employ diagrammatic techniques. The only point to be made here is that the diagrams, which simply indicate graphically the connected sequences of 'incoming' and 'outgoing' orbital indices in the products of matrix elements which appear in the numerators of perturbation terms, show very clearly the significance of the reducible and irreducible clusters: the reducible clusters are associated with diagrams that separate into 'unlinked' subdiagrams, while the irreducible clusters are associated with those which cannot be separated, being fully 'linked'. Since the *reducible* clusters are determined 'automatically' by the irreducible clusters of lower order, it is not surprising that the energy can be determined from the irreducible clusters alone and that, consequently, there exists a '*linked-cluster expansion*' giving the perturbation energy to arbitrarily high order.

The fact that a linked cluster expansion exists does not, of course, guarantee its convergence: but when the series *does* converge (and this is usually the case for the ground states of neutral molecules in near-equilibrium geometry) the application of diagrammatic perturbation theory provides a powerful method of obtaining results of very high precision.

5) Alternative Approaches: Correlation Factors

In view of the slow convergence of CI methods, as a means of approximating the correlation cusp, it is not surprising that many efforts have been made to follow the classic approaches of Hylleraas (1929) and James and Coolidge (1933), in which interelectronic distances ('r_{12} terms') are explicitly included in the variational wavefunction. This is the procedure followed, for example, in the work of Kolos and Wolniewicz (1968) on the hydrogen molecule. Unfortunately, the generalization to many-electron systems raises enormous computational problems which have not as yet been fully solved. The wavefunction in the general case may be represented by

$$\Psi(\mathbf{x}_1, \mathbf{x}_2, \dots \mathbf{x}_N) = \Phi_0(\mathbf{x}_1, \mathbf{x}_2, \dots \mathbf{x}_N) C(\mathbf{r}_1, \mathbf{r}_2, \dots r_N), \tag{33}$$

where the 'correlation factor' typically has the form

$$C(\mathbf{r}_1, \mathbf{r}_2, \dots r_N) = \Pi_{i>j} F(\mathbf{r}_i, \mathbf{r}_j), \tag{34}$$

and is symmetric under electron permutations. The function $F(\mathbf{r}_i, \mathbf{r}_j)$ might depend, for example, both on the mean position vector $\mathbf{R}_{ij}$ of pair ij and on the interparticle distance r_{ij}; but even the simple form $1 + \frac{1}{2} r_{ij}$ would apparently be satisfactory in the limit $r_{ij} \to 0$. In either case, however, the energy functional is bound to contain, in addition to the usual 1- and 2-electron integrals, formally 3- and 4-electron integrals (whose integrands depend on up to *four* spatial points): and it is the nature and number of these terms that leads to severe computational problems.

For gaussian orbitals and correlation factors that contain only *even* powers of r_{ij}, all the necessary integrals can be evaluated (Boys, 1960; Longstaff and Singer, 1960); but unfortunately even powers do not give the correct cusp behaviour. Nevertheless, progress has been made along such lines (Sims and Hagstrom, 1971; Karunkaren and Christofferson, 1975; Preiskorn and Woznicki, 1982,1984; Frye et al, 1990) and it is possible that valuable insight into correlation effects in few-electron systems will be obtained using this approach.

For correlation factors containing *odd* powers, in particular r_{ij}, Kutzelnigg (1985) has given a useful general analysis of the 'cusp problem' and has presented results of calculations on the He isoelectronic series which show considerable promise. In some very recent work (Kutzelnigg and Klopper 1991) this approach has been generalized and, with the inclusion of apparently well-founded approximations, provides a practical basis for inclusion of the cusp in CI calculations on few-electron systems. Preliminary results are impressive and further progress seems likely.

It must be admitted at this point, however, that the prospect of performing direct 'Hylleraas-type' calculations on *many*-electron molecules is not very encouraging; and it is therefore worth remarking in conclusion that the essential features of the coulomb hole *can* be incorporated in a semi-empirical way in wavefunction calculations. Semi-empirical rules for adding correlation corrections to the energy have been in use for many years (e.g. Clementi 1963); but a more satisfactory theoretical insight into the origin of such corrections is provided by an *approximate* treatment of the correlation hole associated with a wavefunction of the form (33). This approach, which cannot be described here, was proposed and implemented by Colle and Salvetti (1975) and leads to a 'correlation energy functional', involving the electron density and pair function calculated without correlation (e.g. in HF approximation). The correlation energies computed using this functional are often in good agreement with those obtained from experimental data, even for excited states (Colle and Salvetti,1979), and the same functional applies remarkably well (McWeeny, 1976) even to the free electron gas, giving a realistic dependence on the electron density.

In summary, the above observations suggest that the 'correlation problem' (except in the case of very simple systems) will remain a problem for many years to come, in spite of immense progress at the computational level, and that new theory and new approximation schemes are still needed.

References

Ahlrichs, R., 1979, Comput. Phys. Commun. **17**, 31

Ahlrichs, R., 1983, in "Methods in Computational Molecular Physics" (ed. G. H. F. Diercksen and S. Wilson, Reidel, Dordrecht)

Bangudu, E. A., Jankowski, K. and Dion, D. R., 1974, Chem. Phys. Lett, **19**, 418

Boys, S. F., 1960, Proc. Roy. Soc. (Lond.) **A258**, 402

Cizek, J., 1966, J. Chem. Phys. **45**, 4256

Clementi, E., 1963, J. Chem. Phys. **38**, 996, 1001, 2248

Colle, R. and Salvetti, O., 1975, Theor. Chim. Acta **39**, 329

Colle, R. and Salvetti, O., 1979, Theor. Chim. Acta **53**, 55

Cooper, I. L. and Pounder, C. M. N., 1978, Theor. Chim. Acta **47**, 51

Cooper, I. L. and Pounder, C. M. N., 1980, Int. J. Quantum Chem. **17**, 759

Cooper, I. L. and Pounder, C. M. N., 1983, Int. J. Quantum Chem. **23**, 257

Hurley, A. C., 1976, "Electron Correlation in Small Molecules" (Academic Press, London)

Husimi, K., 1940, Proc.Phys.-Math. Soc. Jpn. **22**,264

Hylleraas, E. A., 1929, Z. Phys. **54**, 347

James, H. and Coolidge, A. S., 1933, J. Chem. Phys. **1**, 825

Karunakaran, K. M. and Christoffersen, R. E., 1975, **62**, 1992

Kolos, W. and Wolniewicz, L., 1968, J. Chem. Phys. **49**, 404

Kutzelnigg, W., 1977, in "Modern Theoretical Chemistry Vol.3" (ed H. F. Schaeffer, Plenum, New York)

Kutzelnigg, W., 1985, Theor. Chim. Acta **68**, 445

Kutzelnigg, W., Del Re, G. and Berthier, G., 1968, Phys. Rev. **172**, 49

Kutzelnigg, W. and Klopper, W., 1991, J. Chem. Phys. **94**, 1985

Longstaff, J. V. L. and Singer, K., 1960, Proc. Roy. Soc. (Lond.) **A258**, 421

Löwdin, P.-O., 1955, Phys. Rev. **97**, 1474

Löwdin, P.-O., 1959, Adv. Chem. Phys. **2**, 207

McWeeny, R., 1954, Proc. Roy. Soc. (Lond.) **A223**, 63

McWeeny, R., 1955, Proc. Roy. Soc. (Lond.) **A232**, 114

McWeeny, R., 1960, Rev. Mod. Phys. **32**, 335

Mayer, I., 1978, Int. J. Quantum Chem. **14**, 29

Meyer, W., 1977, in "Modern Theoretical Chemistry Vol.3" (ed. H. F. Schaeffer, Plenum, New York)

Nesbet, R. K., 1965, Adv. Chem. Phys. **9**, 321

Preiskorn, A. and Woznicki, W., 1982, Chem. Phys. Lett. **86**, 369

Preiskorn, A. and Woznicki, W., 1984, Mol. Phys. **52**, 1291

Sims, J. S. and Hagstrom, 1971, Phys. Rev. **A4**, 908

Sinanoglu, O., 1964, Adv. Chem. Phys. **6**, 315

Slater, J. C., 1951, Phys. Rev. **81**, 305

Slater, J. C., 1974, "Quantum Theory of Molecules and Solids Vol.4" (McGraw-Hill, New York)

UNITARY GROUP APPROACH TO THE MANY-ELECTRON CORRELATION PROBLEM

Josef Paldus

Department of Applied Mathematics
Department of Chemistry and (GWC)[2]
University of Waterloo
Waterloo, Ontario, N2L 3G1, Canada

The fascination with symmetry and the multitude of its manifestations in nature can be traced back to the earliest times of human civilization. Its importance as a basic unifying principle for our understanding of the physical universe can hardly be overestimated. Initially it was of course the most conspicuous geometrical symmetry, with its unquestionable aesthetic appeal, that enabled scientists to systematically classify crystal structures and unravel the complexities of molecular spectra. More recently, especially with the advent of quantum mechanics, an entirely new universe of very subtle symmetries was opened up and the intrinsic elegance and mathematical beauty of fundamental physical laws of nature was revealed, manifesting itself in countless phenomena ranging from the discovery of antiparticles to our understanding of the periodic table and chemical bonding. Nowadays, these limits are further expanded by probing various dynamical and gauge symmetries, quasi-symmetries and so-called supersymmetries.

While classical geometrical symmetries deal with at most three-dimensional objects and are easily comprehended and appreciated by our senses, various symmetries that are relevant in the submicroscopic world characterize objects and concepts that can only be defined in highly-dimensional abstract spaces. Even here, however, there is a fundamental difference between present-day classical symmetries that are characterized by the so-called *invariance*, *symmetry* or *degeneracy group* of the Hamiltonian of the studied system (see, for example, Hamermesh, 1962; McWeeny, 1963; Chen, 1989; and references therein) and the rather recently introduced *dynamical groups* or *spectrum generating algebras* (Bohm *et al.*, 1988) that are required to contain all the bound states of the system within a single irreducible representation (e.g., SO(4,2) or SO(2,1) for the hydrogen atom (Barut, 1971; Adams *et al.*, 1988), U(4) or SO(4) for the molecular vibronic problems (Iachello, 1981; Iachello and Levine, 1982), SO(3,2) or SO(3,1) for scattering problems (Iachello, 1986), U(N) for molecular electronic states (Lemus and Frank, 1991), etc.).

An appropriate dynamical group for an n-orbital model of an N-electron system that is described by a spin-independent Hamiltonian is the unitary group U(2 n) (Paldus, 1988). This observation constitutes a basis for the so-called Clifford algebra unitary group approach (CAUGA) (Paldus and Sarma, 1985; Paldus *et al.*, 1987, Paldus and Jeziorski, 1988). However, it is often useful to relax the definition of the dynamical group given above and to consider as a dynamical group (in a weak sense) of the system any group

Methods in Computational Molecular Physics, Edited by Stephen Wilson and
Geerd H.F. Diercksen, Plenum Press, New York, 1992

(usually a Lie group), or rather its Lie algebra, if its Hamiltonian can be expressed in terms of the generators (or infinitesimal operators) of this algebra (or group) (Wormer, 1975). Thus, in the N-electron case, the various subgroups in the chain

$$U(2\,n) \supset Spin(m) \supset SO(m) \supset U(n), \qquad (1)$$

where $m = 2n+1$ or $m = 2n$, can equally be useful. In fact, the smallest group in this chain U(n), together with the Gel'fand-Tsetlin (GT) (1950) canonical chain

$$U(n) \supset U(n\text{-}1) \supset \dots \supset U(2) \supset U(1) \qquad (2)$$

represents the basis of the so-called *unitary group approach* (UGA) (Paldus, 1974, 1976; Paldus and Boyle, 1980*a*) as well as of the corresponding graphical representation (GUGA) (Shavitt, 1977, 1978).

Since the required N-particle spaces can be built as appropriate tensor powers of a chosen one-particle space, we can conveniently exploit the representation theory of a relevant dynamical Lie group to construct the symmetry-adapted states, as well as to compute the required matrix elements of the Hamiltonian or other operators. Most importantly, however, such a formulation of the problem often enables penetrating insight into the problem and enables both a compact representation of quantities involved and a construction of versatile and efficient algorithms for the actual computations.

The relevance of unitary groups in the quantum mechanical many-body problem was already recognized by Weyl (1931) and Jordan (1935) in the early thirties. Although the second edition of Hermann Weyl's (1931) book on "Group Theory and Quantum Mechanics" culminates with the branching rules for GL(n) or U(n), it was not until the early fifties when a convenient orthonormal basis based on the chain (2), as well as the explicit form of representations, were found (Gel'fand and Tsetlin, 1950). The potential usefulness of this formalism for the nuclear many-body problem was first recognized by Moshinsky (1966) and for the many-electron problem by Moshinsky and Seligman (1971).

In the meantime, the representation theory of unitary groups was greatly advanced thanks to the work by Biedenharn and Louck (for references see Biedenharn and Louck, 1981*a*,*b*), who extended Schwinger's (1965) boson calculus for SU(2) to the general case. It was not, however, until 1974 when it was realized that in the many-electron case the general formalism can be drastically simplified (Paldus, 1974) that the extensive practical exploitation of UGA began. The subsequent developments (Paldus, 1976; Shavitt, 1977,1978; Paldus and Boyle, 1980*a*) provided then a very versatile formalism that enabled numerous practical exploitations in actual molecular electronic structure calculations (see, e.g., Hinze, 1981, and references therein). In particular we must mention the possibility of an efficient generation and representation of a spin-adapted N-electron basis constructed from a given set of n orthonormal orbitals using the concept of step numbers and Shavitt graphs (Paldus, 1974, 1976; Shavitt, 1977, 1978), the latter representing essentially an expanded branching diagram (Boyle and Paldus, 1980) characterizing the Yamanouchi-Kotani coupling scheme (Yamanouchi, 1937; Kotani *et al.*, 1963), and the segment factorization of matrix elements for both single generators and generator products that arise in the two-electron part of the Hamiltonian (Shavitt, 1978, 1988; Drake and Schlesinger, 1977; Paldus and Boyle, 1980*a*; Boyle and Paldus, 1980). In fact, these concepts can be usefully extended to more than two-column irreducible representations (Kent and Schlesinger, 1989*a*,*b*, 1991, and references therein). The versatility of UGA is perhaps best characterized by the fact that it permits very diverse computational implementations, particularly in large scale configuration interaction (CI) program packages, ranging from the integral driven (Siegbahn, 1979, 1980*a*,*b*, 1981, 1984; Lischka *et al.*, 1981, Shepard *et al.*, 1988), loop-driven (Brooks and Schaefer, 1979), shape (Saxe *et al.*, 1982) or internal interaction block-driven (Saunders and van Lenthe, 1983) to matrix element driven approaches (Downward and Robb, 1977; Hegarty and Robb, 1979; Robb and Hegarty, 1981) based on harmonic level excitation diagrams (HLED) (Paldus, 1977). Moreover, this formalism is not only suitable for large scale CI calculations, but also in coupled cluster (CC) (see, e.g., Banerjee and Simons, 1981; Baker and Robb, 1983; Hoffmann and Simons, 1988, 1989; Paldus and Jeziorski, 1988; Jeziorski and Paldus, 1988, 1989, Mukhopadhyay and Mukherjee, 1989), Green's function (Born and Shavitt, 1982, Born, 1982, 1985) perturbation theory (PT) (Downward and Robb, 1977; Hegarty and Robb, 1979) and other approaches (cf., e.g., Siegbahn *et al.*, 1980) to the many-electron correlation problem. We also mention calculations of second order properties using CI methodology (Bendazzoli *et al.*, 1989).

It is also worth noting, that the basic formalism exploiting the U(n) generator concept was also essential for the development of partially spin-adapted (with respect to S_z but not S^2) algorithms that employ Slater determinantal basis (or, rather, Waller-Hartree (1929) double-determinantal basis) and that are particularly suitable for modern vector processors in view of their simplicity, as long as full CI is considered (cf., Knowles and Handy, 1989, and references therein). Even here, however, the insight afforded by the unitary group structure enabled further improvement of the codes, particularly by exploiting the CAUGA chain (1) (Zarrabian *et al.*, 1989; Harrison and Zarrabian, 1989).

One of the definite drawbacks of UGA is undoubtedly the fact that the Gel'fand-Tsetlin (GT) states, or corresponding chain (2), represent a rather unphysical coupling scheme and, above all, imply the lexical labeling that makes it difficult to introduce a physically meaningful truncation scheme for the N-electron basis. Such truncation is essential in actual applications, since the dimension of relevant CI spaces rapidly increases with the size of the one-electron orbital basis set employed. The above mentioned difficulties stem partially from the fact that the lexical labels of GT states that are associated with identical orbital occupancies are not contiguous and can in fact be more or less "randomly" distributed in the N-electron spin-adapted basis set. For similar reasons even the codes based on partially spin-adapted determinantal bases are restricted to full CI (for a possible extension, see Bajpai *et al.*, 1991). Moreover, GT basis is also unsuitable for valence bond (VB) applications, since most of the GT (or Yamanouchi-Kotani) states correspond to non-canonical (in Rumer's (1932) sense) VB structures.

All these problems can be effectively removed when the CAUGA formalism is relied upon (Paldus *et al.*, 1989). Indeed, very recently, this formalism was exploited in the VB context (Li and Paldus, 1991*a,b* and Paldus and Li, 1991) for the Pariser-Parr-Pople (PPP)-type semiempirical Hamiltonians (cf., Parr, 1963 and references therein). Initial applications yielded very encouraging results. Of course, the well known N! problems is partially avoided with the PPP-type Hamiltonians, although the PPP-VB scheme that we employ relies on overlap-enhanced atomic orbitals (OEAO's) (Gerratt and Raimondi, 1980; McWeeny, 1988). Nonetheless, a more efficient way of handling the orbital non-orthogonality in VB-type approaches would be most welcome in spite of an important progress that was recently achieved (Verbeek, 1990; Verbeek and van Lenthe, 1991).

Most of the developments that we briefly reviewed above, particularly those concerning the standard UGA to the large scale CI problem, have been extensively described in numerous review articles (Paldus, 1976, 1977, 1986, 1988; Shavitt, 1983, 1988; Robb and Niazi, 1984, 1990), conference proceedings on the subject (Hinze, 1981), chapters in books (Pauncz, 1971; Wilson, 1984; McWeeny, 1989) and even in a monograph (Matsen and Pauncz, 1986). It would thus be superfluous to repeat them here. In particular, we refer the reader to the most recent reviews by this author (Paldus, 1988), Shavitt (1988) and Robb and Niazi (1990).

Let us, finally, point out some recent developments in the UGA methodology. There has been continued effort to extend UGA to partitioned bases that would enable an effective handling of larger systems. An earlier work (Paldus and Boyle, 1980*b*; Boyle and Paldus, 1980; Gould and Paldus, 1986; Gould, 1986) on system partitioning was recently addressed from the viewpoint of a general Racah-Wigner calculus for U(n) irreducible representations that are pertinent to quantum chemical models of many-electron systems (Li and Paldus, 1990*a*). In this study the standard Clebsch-Gordan coefficients and isoscalar factors (or reduced Wigner coefficients) for both symmetric (S(N)) and unitary (U(n)) groups were extended to transformation coefficients and corresponding isoscalar factors relating canonical Young-Yamanouchi or GT bases to simple partitioned bases. For the two-column representation detailed explicit expressions for the relevant reduced Wigner coefficients were given. This study was now extended to generator products as well and will appear in the same series (Li and Paldus, 1991*c*). Another approach to this problem was pursued by Lin (1991). We also mention in this connection a general work on the relationship between the S(N) and U(n) isoscalar factors and higher order U(n) invariants, in particular U(n) Racah coefficients and 9*j* symbols (Li and Paldus, 1990*b*).

Several studies were also devoted to spin-dependent UGA. Since the total spin S represents usually a very good quantum number, it is reasonable to assume that one considers first a standard spin-independent Hamiltonian, in which case UGA can be employed, and only subsequently studies the spin-dependent (spin-orbit, spin-other-orbit, spin-spin, etc.) interactions. To do so in general, one needs the matrix elements of general spin-orbital U($2n$) generators in the GT U(n) bases that carry relevant irreducible representations interacting via spin-dependent components of the Hamiltonian. This problem was considered earlier by Gould and Chandler (1984). However, the resulting formalism was not directly amenable to computer implementation and, more importantly, the results for the zero-shift components of the U($2n$) generators were missing an important term. These shortcomings were recently rectified by Gould and Paldus (1990) and the matrix elements of the U($2n$) generators in the U(n) × U(2) adapted GT basis were given in terms of a single U(n) adjoint tensor operator $\mathbf{\Delta} = \mathbf{E}(\mathbf{E} + N/2 - n - 2)$, where $\mathbf{E} = \|E_{ij}\|$ is the matrix of U(n) generators E_{ij} and N designates the number of electrons. The relevant explicit expressions for segment values were worked out by Gould and Battle (1991). In the context of the atomic central field problem, Kent and Schlesinger (1990) employed the graphical methods of spin algebras (see Kent and Schlesinger, 1989*a,b*) to treat the rank-K, ($K = 0,1,2$) tensors arising from the coupling of single-particle spin operators. Currently, the actual implementation of these developments within the *ab initio* codes is being attempted (Battle, 1991).

Another recent development formulates a fully spin-adapted approach to many-electron density matrices within the context of UGA (Gould *et al.*, 1990). For spin-independent systems, the one-electron spin-density operator was expressed as a second degree polynomial in U(n) generators. Extension to spin-dependent systems then leads naturally to the appearance of total-spin transition densities possessing interesting algebraic properties which were investigated in some detail. A similar formalism for the two-electron density matrices was also developed and is open to further generalizations. It is remarkable that the same second degree polynomial $\mathbf{\Delta}$ that plays the key role in the above mentioned spin-dependent theory also prominently figures in the density matrix formalism. On the practical side, these developments should provide a versatile method for the direct calculation of one- and two-body density matrices in the UGA framework.

In closing, let us remark that the UGA to molecular electronic structure continues to provide a useful tool in investigating interesting chemical systems (e.g., Jeung, 1989, 1991, Jeung *et al.*, 1990), sometimes even very large ones (e.g., iron (II) porphyrin bis(water), Rohmer, 1989). We also see continuously new aspects of group theoretical and algebraic structure of quantum theoretical description of atomic and molecular structure to be discovered (see, e.g., Čížek *et al.*, 1983; Wen *et al.*, 1989, 1990; Zhou and Pulay, 1989; Klimko and Mestechkin, 1990; Koutecký and Scheuch, 1990; Nomura, 1990; Planelles and Karwowski, 1990; etc.). In view of the inherent beauty of the representation theory of semisimple Lie groups and algebras (see, e.g., Humphreys; 1972), and the potential wealth of new mathematics for hardly explored domain of representations of non-compact topological groups, there is to be expected the presence of a rich algebraic structure in the mathematical description of the submicroscopic world that is yet to be found.

ACKNOWLEDGEMENTS

The continued support by NSERC is greatly appreciated. The author is very much indebted to Mrs. Helen Warren for her immaculate typesetting of the camera-ready manuscript. Great appreciation and wholehearted thanks are due to Drs. Geerd Diercksen and Stephen Wilson for creating a most inducive atmosphere for learning and scientific exchange at the 1991 NATO School in Bad Windsheim as well as for an exemplary organization and enlightening program of the School.

References

Adams, B.G., Čížek, J., and Paldus, J., 1988, *Advan. Quantum Chem.*, **19**:1; also reprinted in Bohm *et al.*, 1988, pp. 103-207.
Bajpai, N., Sarma, C.R., and Paldus, J. 1991, *J. Mol. Struct. (Theochem)*, **227**:67.
Baker, H., and Robb, M.A., 1983, *Mol. Phys.*, **50**:1077.
Banerjee, A., and Simons, J., 1981, *Int. J. Quantum Chem.*, 19:207.
Barut, A.O., 1971, "Dynamical Groups and Generalized Symmetries in Quantum Theory", University of Canterbury, Christchurch, New Zealand.
Battle, J.S., 1991, private communication.
Bendazzoli, G.L., Palmieri, P., and Rettrup, S., 1989, *J. Chem. Phys.*, 91:5518.
Biedenharn, L.C., and Louck, J.D., 1981*a*, "Angular Momentum in Quantum Physics, Theory and Application", Chap. 5, Addison-Wesley, Reading, Mass.
Biedenharn, L.C., and Louck, J.D., 1981*b*, "The Racah-Wigner Algebra in Quantum Theory", Addison-Wesley, Reading, Mass.
Born, G., 1982, *Int. J. Quantum Chem.*, Symp. **16**:633.
Born, G., 1985, *Int. J. Quantum Chem.*, **28**:335.
Born, G., and Shavitt, I., 1982, J. Chem. Phys., **76**:558.
Bohm, A., Ne'eman, Y., and Barut, A.O., eds., 1988, "Dynamical Groups and Spectrum Generating Algebras", Vols. 1 and 2, World Scientific, Singapore.
Boyle, M., and Paldus, J., 1980, *Phys. Rev. A*, **22**:2316.
Brooks, B.R., and Schaefer, H.F., 1979, *J. Chem. Phys.*, 70:5092.
Chen, J.-Q., 1989, "Group Representation Theory for Physicists", World Scientific, Singapore.
Čížek, J., Pauncz, R., and Vrscay, E.R., 1983, *J. Chem. Phys.*, 78:2468.
Downward, M.J., and Robb, M.A., 1977, *Theor. Chim. Acta*, **46**:129.
Drake, G.W.F., and Schlesinger, M., 1977, *Phys. Rev. A*, **15**:1990.
Gel'fand, I.M., and Tsetlin, M.L., 1950, *Dokl. Akad. Nauk SSSR*, 71:825, 1070.
Gerratt, J., and Raimondi, M., 1980, *Proc. Roy. Soc. (London) A*, **371**:525.
Gould, M.D., 1986, *Int. J. Quantum Chem.*, **30**:364.
Gould, M.D., and Battle, J.S., 1991, unpublished results.
Gould, M.D., and Chandler, G.S., 1984, *Int. J. Quantum Chem.*, **26**:441.
Gould, M.D., and Paldus, J., 1986, *Int. J. Quantum Chem.*, **30**:327.
Gould, M.D., and Paldus, J., 1990, *J. Chem. Phys.*, 92:7394.
Gould, M.D., Paldus, J., and Chandler, G.S., 1990, *J. Chem. Phys.*, **93**:4142.
Hamermesh, M., 1962, "Group Theory and Its Application to Physical Problems", Addison-Wesley, Reading, Mass.
Harrison, R.J., and Zarrabian, S., 1989, *Chem. Phys. Lett.*, **158**:393.
Hegarty, D., and Robb, M.A., 1979, *Mol. Phys.*, **38**:1795.
Hinze, J., ed., 1981, "The Unitary Group for the Evaluation of Electronic Energy Matrix Elements", Lecture Notes in Chemistry, Vol. 22, Springer-Verlag, Berlin.
Hoffmann, M.R., and Simons, J., 1988, *J. Chem. Phys.*, **88**:993.
Hoffmann, M.R., and Simons, J., 1989, *J. Chem. Phys.*, 90:3671.
Humphreys, J.E., 1972, "Introduction to Lie Algebras and Representation Theory", Springer-Verlag, New York.
Iachello, F., 1981, *Chem. Phys. Lett.*, 78:581.
Iachello, F., 1986, *in*: "Symmetries in Science II", B. Gruber and R. Lenczewski, eds., Plenum Press, New York, pp. 247-254.
Iachello, F., and Levine, R.D., 1982, *J. Chem. Phys.*, **77**:3046.
Jeung, G.-H., 1989, *Mol. Phys.*, **67**:747.

Jeung, G.-H., 1991, *Chem. Phys. Lett.*, **176**:233.
Jeung, G.-H., Broyer, M., and Labastie, P., 1990, *Chem. Phys. Lett.*, **165**:494.
Jeziorski, B., and Paldus, J., 1988, *J. Chem. Phys.*, **88**:5673.
Jeziorski, B., and Paldus, J., 1989, *J. Chem. Phys.*, **90**:2714.
Jordan, P., 1935, *Z. Phys.*, **94**:531.
Kent, R.D., and Schlesinger, M., 1989*a*, *Phys. Rev. A*, **39**:19, 3260.
Kent, R.D., and Schlesinger, M., 1989*b*, *Phys. Rev. A*, **40**:536.
Kent, R.D., and Schlesinger, M., 1990, *Phys. Rev. A*, **42**:1155.
Kent, R.D., and Schlesinger, M., 1991, *J. Math. Phys.*, **32**:1102.
Klimko, G.T., and Mestechkin, M.M., 1990, *Int. J. Quantum Chem.*, **37**:753.
Knowles, P.J., and Handy, N.C., 1989, *Comput. Phys. Commun.*, **54**:75.
Kotani, M., Amemiya, A., Ishiguro, E., and Kimura, T., 1963, "Tables of Molecular Integrals", 2nd ed., Maruzen, Tokyo.
Koutecký, J., and Scheuch, C., 1990, *Int. J. Quantum Chem.*, **37**:373.
Lemus, R., and Frank, A., 1991, *Phys. Rev. Lett.*, **66**:2863.
Li, X., and Paldus, J., 1990*a*, *J. Math. Chem.*, **4**:295.
Li, X., and Paldus, J., 1990*b*, *J. Math. Phys.*, **31**:1589
Li, X., and Paldus, J., 1991*a*, *J. Mol. Struct.* (*Theochem*), **229**:249.
Li, X., and Paldus, J., 1991*b*, *Int. J. Quantum Chem.*, in press.
Li, X., and Paldus, J., 1991*c*, *J. Math. Chem.*, to appear.
Lin, H.-L., 1991, *Theor. Chim. Acta*, **79**:43,53.
Lischka, H., Shepard, R., Brown, F., and Shavitt, I., 1981, *Int. J. Quantum Chem.*, **15**:91.
Matsen, F.A., and Pauncz, R., 1986, "The Unitary Group in Quantum Chemistry", Elsevier, Amsterdam.
McWeeny, R., 1963, "Symmetry - An Introduction to Group Theory and Its Applications", Pergamon Press, New York.
McWeeny, R., 1988, *Int. J. Quantum Chem.*, **34**:25.
McWeeny, R., 1989, "Methods of Molecular Quantum Mechanics", 2nd ed., Academic Press, New York, Chap. 10.
Moshinsky, M., 1966, *in*: "Many-Body Problems and Other Selected Topics in Theoretical Physics", M. Moshinsky, T.A. Brody and G. Jacob, eds., Gordon and Breach, New York, pp. 289-377; reissued by Gordon and Breach in 1968 as a separate monograph entitled: "Group Theory and the Many-Body Problem".
Moshinsky, M., and Seligman, T.H., 1971, *Ann. Phys.* (*N.Y.*), **66**:311.
Mukhopadhyay Jr., D., and Mukherjee, D., 1989, *Chem. Phys. Lett.*, **163**:171.
Nomura, M., 1990, *J. Math. Phys.*, 31:351.
Paldus, J., 1974, *J. Chem. Phys.*, 61:5321.
Paldus, J., 1976, *in* "Theoretical Chemistry: Advances and Perspectives", Vol. 2, H. Eyring and D. Henderson, eds., Academic Press, New York.
Paldus, J., 1977, *in*: "Electrons in Finite and Infinite Structures", P. Phariseau and L. Scheire, eds., Plenum Press, New York, pp. 411-429.
Paldus, J., 1986, *in*: "Symmetries in Science II", B. Gruber and R. Lenczewski, eds., Plenum Press, New York, pp. 429-446.
Paldus, J., 1988, *in*: "Mathematical Frontiers in Computational Chemical Physics", D.G. Truhlar, ed., Springer-Verlag, New York, pp. 262-299.
Paldus, J., and Boyle, M.J., 1980*a*, *Phys. Scripta*, **21**:295.
Paldus, J., and Boyle, M.J., 1980*b*, *Phys. Rev. A*, **22**:2299.
Paldus, J., and Jeziorski, B., 1988, *Theor. Chem. Acta*, 73:81.
Paldus, J., and Li, X., 1991, *Israel J. Chem.*, in press.
Paldus, J., and Sarma, C.R., 1985, *J. Chem. Phys.*, 83:5135.
Paldus, J., Gao, M.-J., and Chen, J.-Q., 1987, *Phys. Rev. A*, 35:3197.
Paldus, J., Rettrup, S., and Sarma, C.R., 1989, *J. Mol. Struct.* (*Theochem*), **199**:85.
Parr, R.G., 1963, "The Quantum Theory of Molecular Electronic Structure", Benjamin, New York.
Pauncz, R., 1971, "Spin Eigenfunctions: Construction and Use", Plenum Press, New York.
Planelles, J., and Karwowski, J., 1990, *J. Phys. A: Math. Gen.*, 23:5083.

Robb, M.A., and Hegarty, D., 1981, *in*: "The Unitary Group for the Evaluation of Electronic Energy Matrix Elements", Lecture Notes in Chemistry, Vol. 22, J. Hinze, ed., Springer-Verlag, Berlin, pp. 100-107.
Robb, M.A., and Niazi, U., 1984, *Comput. Phys. Rep.*, 1:127.
Robb, M.A., and Niazi, U., 1990, *in*: "Reports in Molecular Theory", Vol. 1, pp. 23-55.
Rohmer, M.-M., 1989, *Inorg. Chem.*, 28:4574.
Rumer, G., 1932, *in*: "Göttingen Nachrichten, Ges. d. Wiss. Nachrichten, Math.-Phys. Klasse", Heft 4, pp. 337-341.
Saunders, V.R., and van Lenthe, J.H., 1983, *Mol. Phys.*, 48:923.
Saxe, P., Fox, D.J., Schaefer, H.F., and Handy, N.C., 1982, *J. Chem. Phys.*, 77:5584.
Schwinger, J., 1965, *in*: "Quantum Theory of Angular Momentum", L.C. Biedenharn and H. van Dam, eds., Academic Press, New York, pp. 229-279.
Shavitt, I., 1977, *Int. J. Quantum Chem.*, Symp. 11:131.
Shavitt, I., 1978, *Int. J. Quantum Chem,.*, Symp. 12:5.
Shavitt, I., 1983, *in*: "New Horizons in Quantum Chemistry", P.-O. Löwdin and B. Pullman, eds., Reidel, Dordrecht, The Netherlands, pp. 279-293.
Shavitt, I., 1988, *in*: "Mathematical Frontiers in Computational Chemical Physics", D.G. Truhlar, ed., Springer-Verlag, New York, pp. 300-349.
Shepard, R., Shavitt, I., Pitzer, R.M., Comeau, D.C., Pepper, M., Lischka, H., Szalay, P.G., Ahlrichs, R., Brown, F.B., and Zhao, J.-G., 1988, *Int. J. Quantum Chem.*, Symp. 22:149.
Siegbahn, P.E.M., 1979, *J. Chem. Phys.*, 70:5391.
Siegbahn, P.E.M., 1980*a*, *J. Chem. Phys.*, 72:1647.
Siegbahn, P.E.M., 1980*b*, *Int. J. Quantum Chem.*, 18:1229.
Siegbahn, P.E.M., 1981, *in*: "The Unitary Group for the Evaluation of Electronic Energy Matrix Elements", Lecture Notes in Chemistry, Vol. 22, J. Hinze, ed., Springer-Verlag, Berlin, pp. 119-135.
Siegbahn, P.E.M., 1984, *Chem. Phys. Lett.*, 109:417.
Siegbahn, P., Heiberg, A., Roos, B., and Levy, B., 1980, *Phys. Scripta*, 21:323.
Verbeek, J., 1990, Ph.D. Thesis, University of Utrecht, The Netherlands.
Verbeek, J., and van Lenthe, J.H., 1991*a*, *J. Mol. Struct.(Theochem)*, in press.
Verbeek, J., and van Lenthe, J.H., 1991*b*, *Int. J. Quantum Chem.*, 40:201.
Waller, I., and Hartree, R.D., 1929, *Proc. Roy. Soc. (London) A*, 124: 119.
Wen, Z., Wang, Y., Don, Y., and Liu, Z., 1989, *Theor. Chim. Acta*, 76:269.
Wen, Z., Wang, Y., and Zhang, Z., 1990, *Int. J. Quantum Chem.*, 37:631.
Weyl, H., 1931, "Gruppentheorie und Quantenmechanik", 2nd ed., Hirzel, Leipzig (English translation: "The Theory of Groups and Quantum Mechanics", Methuen, London, 1931; reissued by Dover, New York, 1949 and 1964).
Wilson, S., 1984, "Electron correlation in molecules", Clarendon Press, Oxford, Chap. 5.
Wormer, P.E.S., 1975, Ph.D. Thesis, University of Nijmegen, The Netherlands.
Yamanouchi, Y., 1937, *Proc. Phys.-Math. Soc. Jpn.*, 19:436.
Zarrabian, S., Sarma, C.R. and Paldus, J., 1989, *Chem. Phys. Lett.*, 155:183.
Zhou, X.-F., and Pulay, P., 1989, *J. Comp. Chem.*, 10:935.

THE CONFIGURATION INTERACTION APPROACH TO ELECTRON CORRELATION

Jacek Karwowski
Instytut Fizyki
Uniwersytet Mikołaja Kopernika
Grudziądzka 5, 87-100 Toruń, Poland

1. INTRODUCTION

Beyond-Hartree-Fock methods describing the electronic structure of molecules, develop in two main directions: (1) taking into account the electron correlation effects, and (2) extending the basic formulation of the theory to include the relativistic effects. Though the electron correlation correction to the total energy is smaller than that due to the relativistic effects already for Z=14 (silicon), the majority of theoretical contributions in the area of computational molecular physics is concerned with the first of these two subjects. From a formal point of view, the methods designed to describe the electron correlation range from simple implementations of the variational principle to highly sophisticated approaches based on most advanced mathematical theories. All main areas of this broad field have been reviewed during this meeting. The general aspects of the electron correlation problem have been discussed by McWeeny (1991). The perturbation theory of electron correlation has been reviewed by Wilson (1991) and the propagator methods by Oddershede (1991). Finally, the coupled cluster theory and its links with other methods have been described by Paldus (1991). The present contribution is concerned with the configuration interaction (CI) method. It emphasizes an approach based on the permutational symmetry of the wave function.

Among methods describing the electron correlation, the CI method is conceptually the simplest. It is a straightforward consequence of application of the variational principle to a trial wavefunction chosen as a linear combination of antisymmetrized products of orthonormal spinorbitals. The coefficients of the linear expansion are here the only variational parameters. The origin of the CI method may be traced back to the early

Methods in Computational Molecular Physics, Edited by Stephen Wilson and
Geerd H.F. Diercksen, Plenum Press, New York, 1992

works of Hylleraas (1928), Parr, Craig and Roos (1950), Boys (1953), and others. Probably the first rigorous treatment of the method has been given by Löwdin (1955). Its main difficulties are rooted in data handling associated with matrix eigenvalue equations of extremely large dimensions. Therefore, due to inadequate computing facilities, applications of the CI method in large scale molecular calculations were rather scarce before the end of the 1960's.

Apart of its simplicity, there are several other reasons that give the CI method a special status among other methods describing the electron correlation. In the limit of full CI (FCI), it offers an exact solution to the Hamiltonian eigenvalue problem in an antisymmetric finite-dimensional Hilbert space. If the set of orbitals approaches completeness, solutions to the resulting FCI approach the exact solutions to the corresponding Schrödinger equation. Besides, due to the McDonald theorem (McDonald 1933, Löwdin 1955), each CI energy is an upper bound to the exact eigenvalue of the Hamiltonian. Therefore the FCI results are very informative as benchmark tests for other methods (e.g. Handy 1980, 1984, Bartlett et al. 1983, Brown et al. 1984, Bauschlicher et al. 1986, 1987, 1989).

Since FCI expansions, even for small molecules and for moderate one-electron orbital basis sets, become nontractably large, developing selection schemes aimed at constructing limited CI expansions which contain the maximum information about the molecular system in the most compact CI wavefunction (see e.g. Burton 1983, Burton and Gray 1983, Knowles 1989, Knowles and Handy 1989, Duch 1990) is crucial for designing efficient CI algorithms. On the other hand, tendency towards maximization of the tractable CI expansions stimulates works on new formulations of the method. A multitude of new approaches to CI has been created during the last two decades. The largest impact upon the further development of the method has been caused by the following four concepts:

(1) Formulating the direct CI (DCI) approach (Roos 1972, Roos and Siegbahn 1977);

(2) Introducing techniques of the unitary groups to CI (Paldus 1976, 1981) resulting in what is called the unitary group approach (UGA);

(3) Introducing a graphical way of representing the CI basis (Shavitt 1977a, 1978, 1981) that gave the graphical UGA (GUGA);

(4) Defining in the GUGA graph its simple *external* part (Siegbahn 1979, 1981) opened a way to construct efficient GUGA algorithms in large orbital spaces.

Using the above concepts within a formulation that exploits consequences of the invariance of the Hamiltonian with respect to permutations of the electronic variables, resulted in another powerful technique known as the symmetric group approach (SGA). SGA originates from works of Kotani et al. (1955), Ruedenberg (1971), and Karwowski (1973, 1973a). A review of the mathematical foundations of the method has been given by Dinesha et al. (1981). Combining the original SGA formalism with the ideas of DCI and of GUGA Duch and Karwowski (1981, 1985) obtained the symmetric group graphical approach (SGGA) to the CI method. However, in spite of the "CI revolution" triggered by the recent developments in computer technology, high expectations concerning CI expressed in early works (Löwdin 1955, 1959) remained largely unfulfilled. The main source of the problem is very slow convergence of the CI expansion. Recent works aimed at emulation of the FCI results by relatively compact expansions (Knowles 1989, Duch 1990, Harrison 1991, Zarrabian et al. 1991) may mark another milestone in the process of building an optimum approach to the electron correlation problem.

Several review papers dealing with the CI method appeared during the last decade. Very good introduction may be found in the books by McWeeny (1989), Szabo and Ostlund (1982), Matsen and Paunez (1986). The classical review by Shavitt (1977) is well supplemented by more recent papers of Robb and Niazi on the GUGA (1984), of Duch and Karwowski on the SGGA (1985) and of McWeeny and Sutcliffe (1985) on general aspects of the method. A report on the large scale FCI benchmark calculations was recently published by Bauschlicher and Langhoff (1989). General review papers on the CI method have been published very recently by Duch (1991) and by Karwowski (1991).

2. THE SYMMETRIC GROUP AND THE HAMILTONIAN EIGENVALUE PROBLEM

When analyzing an eigenvalue problem of an N-electron operator it is most important to find those transformations, which leave the operator invariant. The pertinent transformations form a group known as the *invariance group* of the operator. It may be easily shown that eigenfunctions of the operator which belong to the same eigenvalue form a basis for an irreducible representation of the invariance group. This property not only allows us to determine transformation properties of the eigenfunctions but also, as we shall discuss later, essentially simplifies problem of the matrix element evaluation. An operator may have several invariance groups. For example, the nonrelativistic Hamiltonian of an N-electron atom is invariant with respect to rotations in the three-dimensional space of coordinates and with respect to permutations of electrons. Therefore to its invariance groups belong the *symmetric group* S_N consisting of $N!$ permutations $\hat{P}$ and the *rotation group* $O(3)$. There may be more invariance groups and finding all of them is, in general, a difficult task. Detailed discussion of this subject may be found in many textbooks as e.g. Hamermesh (1962) or Wybourne (1974). Its specific aspects, relevant for the molecular structure theory, have been presented, among others, by Sutcliffe (1982), McWeeny and Sutcliffe (1985), Paldus (1988).

Some symmetries are consequences of very general laws of the nature. For example, the invariance of all N-electron operators under the symmetric group S_N results from the indistiguishability of electrons. All spin-independent operators are invariant with respect to rotations in the spin space and all operators that depend exclusively on spin variables are invariant with respect to rotations in the coordinate space. Other symmetries are more specific. For example, the point group D_{6h} (consisting of the rotations around the 6-fold axis and of the appropriate reflections), is the invariance group of rather few molecules (e.g., the benzene molecule in its nuclear equilibrium configuration). Our further considerations are concerned with the eigenvalue problem of the Hamiltonian operator. Two of its invariance properties: relative to permutations of the electron coordinates and relative to rotations in the spin space are essential for this approach.

In the nonrelativistic and clamped-nuclei approximation the N-electron Hamiltonian may be expressed as

$$\hat{H} = \sum_{i=1}^{N} \hat{h}_1(i) + \sum_{i>j}^{N} \hat{h}_2(i,j), \tag{1}$$

where $\hat{h}_1$ includes the kinetic energy and interactions of the electron with an external field (usually with the electrostatic field of the nuclei), $\hat{h}_2(i,j) = 1/r_{ij}$ describes the electron-electron interactions and i, j stand for the electron coordinates. Since the Hamiltonian is spin-independent, rotations in the spin space leave it invariant. In consequence the Hamiltonian, the square of the total spin operator ($\hat{S}^2$) and its projection ($\hat{S}_z$) form a set of commuting operators:

$$[\hat{H}, \hat{S}^2] = [\hat{H}, \hat{S}_z] = [\hat{S}^2, \hat{S}_z] = 0. \tag{2}$$

The permutational invariance results in

$$[\hat{H}, \hat{P}] = [\hat{S}^2, \hat{P}] = [\hat{S}_z, \hat{P}] = 0. \tag{3}$$

But different permutations, in general, do not commute with each other.

In order to describe relativistic effects, in particular the spin-orbit coupling and transitions between states of different multiplicity, spin-dependent terms have to be added to the Hamiltonian. They may be treated as a perturbation within a nonrelativistic formalism. The spin-dependent problems are briefly discussed later in this work. However, the most rigorous treatment based on the relativistic Dirac-Coulomb Hamiltonian remains beyond the scope of these lectures. In this meeting it is discussed by Grant (1991). In the case of a spin-dependent Hamiltonian, the total spin is not a constant of motion and the commutation relations (2) are not fulfilled. However the permutational invariance remains valid independently of the form of the Hamiltonian.

Let $< \sigma|SM, l >$ be an eigenfunction of $\hat{S}^2$ and $\hat{S}_z$ which depends upon spin variables only (σ stands collectively for the spin variables $\sigma_1, \sigma_2, \ldots, \sigma_N$ of the N-electron system). The corresponding eigenvector is denoted $|SM, l >$. This means that

$$\hat{S}^2 |SM, l >= S(S+1)|SM, l >, \tag{4}$$

$$\hat{S}_z |SM, l >= M|SM, l >, \quad l = 1, 2, \ldots, f, \tag{5}$$

where the index $l = 1, 2, \ldots, f$ distinguishes independent spin functions belonging to the same S, M pair and (Kotani 1955, Pauncz 1979)

$$f \equiv f(S,N) = \frac{2S+1}{N+1}\binom{N+1}{N/2-S}. \tag{6}$$

The set of vectors $|SM,l>$, $l = 1,2,\ldots,f$ may be considered as a basis in a subspace of the N-electron spin space, which is called the (SM) space. Since S_N is an invariance group for these two operators, their eigenfunctions form bases for its irreducible representations, i.e.

$$\hat{P}|SM,l> = \sum_{j=1}^{f} V_S^N(P)_{jl}|SM,j>, \tag{7}$$

where $\mathbf{V_S^N}$ are the matrices of these representations. Index S labels the representation and f is its dimension. We assume that the basis in the (SM) space is chosen orthonormal. Then the matrices $\mathbf{V_S^N}$ are orthogonal. As one can show (see e.g. Pauncz 1979), the representation matrices depend upon N but are M-independent.

The basis $|SM,l>_{l=1}^{f}$ is complete in the (SM) space. Therefore, every normalized eigenfunction of $\hat{S}^2$ and $\hat{S}_z$ which depends upon both spin (σ) and space $(\mathbf{r})$ coordinates of the N electrons may be expanded as

$$\Psi_{SM}(\mathbf{r},\sigma) = \frac{1}{\sqrt{f}}\sum_{l=1}^{f} \Phi_{Sl}(\mathbf{r}) < \sigma|SM,l>, \tag{8}$$

where $\Phi_{Sl}(\mathbf{r})$ are normalized and M-independent functions of the spatial coordinates of the electrons. Let us assume that in equation (8) we expand a wavefunction $\Psi_{SM}^{\mathcal{E}}(\mathbf{r},\sigma)$ that is an eigenfunction of a spin free Hamiltonian (1)

$$\hat{H}\Psi_{SM}^{\mathcal{E}} = \mathcal{E}\Psi_{SM}^{\mathcal{E}} \tag{9}$$

and fulfills the Pauli principle

$$\hat{P}\Psi_{SM}^{\mathcal{E}} = (-1)^p\Psi_{SM}^{\mathcal{E}}, \tag{10}$$

where p is the parity of $\hat{P}$. Then the expansion coefficients, which are denoted now $\Phi_{Sl}^{\mathcal{E}}$, are also eigenfunctions of $\hat{H}$ corresponding to the same eigenvalue. Their transformation properties result directly from equations (8) and (10). Indeed, from equation (7) we have

$$\hat{P}\Psi_{SM}(\mathbf{r},\sigma) = \frac{1}{\sqrt{f}}\sum_{l=1}^{f}\sum_{j=1}^{f} V_S^N(P)_{jl} < \sigma|SM,j> \hat{P}\Phi_{Sl}^{\mathcal{E}}(\mathbf{r}) \tag{11}$$

Combining equations (8), (10) and (11) and remembering that

$$\sum_{m=1}^{f} V_S^N(P)_{ml}V_S^N(P)_{mj} = \delta_{lj}, \tag{12}$$

we finally get

$$\hat{P}\Phi^{\mathcal{E}}_{Sl} = (-1)^p \sum_{j=1}^{f} V^N_S(P)_{jl}\Phi^{\mathcal{E}}_{Sj}. \tag{13}$$

Equation (13) may be rewritten as

$$\hat{P}\Phi^{\mathcal{E}}_{s_l} = \sum_{j=1}^{f} U^N_S(P)_{jl}\Phi^{\mathcal{E}}_{s_j}, \tag{14}$$

where

$$\mathbf{U}^{\mathbf{N}}_{\mathbf{S}}(P) = (-1)^p \mathbf{V}^{\mathbf{N}}_{\mathbf{S}}(P). \tag{15}$$

Equation (14) says that the set of eigenfunctions of $\hat{H}$ associated with the same eigenvalue carries another representation of $\mathcal{S}_N$. Representations $\mathbf{V}^{\mathbf{N}}_{\mathbf{S}}(P)$ and $\mathbf{U}^{\mathbf{N}}_{\mathbf{S}}(P)$ are mutually dual. For those familiar with the labelling irreducible representations of $\mathcal{S}_N$ by means of the Young diagrams (see e.g. Hamermesh 1962), it may be useful to know that the representations carried by the spin functions are limited to those whose Young shapes consist of only two rows, while the representations carried by the space functions correspond to the two-column shapes.

3. THE CONFIGURATION INTERACTION SPACE

Since the Hamiltonian eigenvalue equation in a many-electron case cannot be solved exactly, we have to design methods of approximation $\Phi^{\mathcal{E}}_{Sj}$. In the most practical and commonly used approach, Φ is expanded into a linear combination of products of one-electron orbitals. If the set of the orbitals is complete, then the expansion is exact but it is infinite. If the expansion is finite and the coefficients are determined by the Ritz variational principle, the resulting approximation is referred to as the CI method.

The Basis Functions

To build a model space for the CI method, we replace Φ by a linear combination of the primitive configuration functions taken in the form

$$\Psi^0_\lambda(\mathbf{r}) = \prod_{i=1}^{N} \psi_{\lambda i}(\mathbf{r}_i), \tag{16}$$

where $\psi_{\lambda i}$ are chosen from an orthonormal set $\psi_1, \psi_2, \ldots, \psi_K$ of orbitals. In the product, which defines the *orbital configuration* λ, s_λ orbitals appear once (singly occupied orbitals or *singles*) and d_λ orbitals — twice (doubly occupied orbitals or *doubles*). The numbers of singles and doubles in λ are connected by the obvious relation

$$s_\lambda + 2d_\lambda = N. \tag{17}$$

We adopt from now on a convention, that in all products of orbitals the position index of an orbital in the product is equal to the designation index of the electron described by this orbital, i.e. if $\psi_{\lambda i}(\mathbf{r}_j)$ is contained in the product, then $i = j$. As a consequence of this convention we usually omit the electron designation indices in the products of orbitals. Besides, for simplicity, we shall write λ_i rather than $\psi_{\lambda i}$ and, whenever it does not lead to any confusion, s/d rather than s_λ/d_λ. We assume that in the orbital products the position indices of singles are always less than those of doubles and that both singles and doubles stand in an ascending order of their indices, i.e.

$$\lambda_1 < \lambda_2 < \ldots < \lambda_s; \qquad \lambda_{s+1} = \lambda_{s+2} < \ldots < \lambda_{N-1} = \lambda_N. \tag{18}$$

The primitive products (16) do not have the transformation properties (14) fulfilled by the exact space eigenfunctions. In order to obtain a *symmetric group adapted* basis of configurations one may use appropriate projection operators. If we define a projection operator

$$\mathcal{O}_{Sk}^{m} = \sqrt{\frac{f(S,N)}{N!2^d}} \sum_P U_S^N(P)_{km} \hat{P}, \tag{19}$$

then

$$\Phi_{Sk}^{(m)} = \mathcal{O}_{Sk}^{m} \Psi_\lambda^0(\mathbf{r}), \tag{20}$$

form an orthonormal set and fulfill equation (14) for $m \leq f(S,s)$ and vanish for $m > f(S,s)$ (Kotani et al. 1955, Hamermesh 1962). Substituting the above result into equation (8) we get, for each configuration λ, a set of $f(S,s)$ antisymmetric and spin adapted *configuration state functions* (CSFs)

$$\Psi_{SM,m}^{\lambda}(\mathbf{r},\sigma) = \frac{1}{\sqrt{N!2^d}} \sum_P \sum_{k=1}^{f(S,N)} U_S^N(P)_{km} < \sigma|SM,k> \hat{P}\Psi_\lambda^0(\mathbf{r}). \tag{21}$$

The last equation, due to relations (15) and (17), may be rewritten as

$$\Psi_{SM,m}^{\lambda}(\mathbf{r},\sigma) = \xi_\lambda \hat{A}[< \sigma|SM,m> \Psi_\lambda^0(\mathbf{r})], \tag{22}$$

where

$$\hat{A} = \frac{1}{N!} \sum_P (-1)^p \hat{P} \tag{23}$$

is the antisymmetrizer and

$$\xi_\lambda = \sqrt{N!2^{-d}} \tag{24}$$

is the normalization constant. In a more general form, equation (22) reads:

$$|\lambda; SM,m> = \xi_\lambda \hat{A}[|\lambda> |SM,m>], \quad m = 1,2,\ldots,f(S,s). \tag{25}$$

The number of CSFs corresponding to a given configuration $|\lambda>$ is equal to $f(S,s)$ rather than $f(S,N)$. This is because the antisymmetry requirement restricts the spin coupling schemes in this part of $<\sigma|SM,m>$ which corresponds to a double in Ψ_λ^0. Since Ψ_λ^0 is symmetric upon transpositions within doubles, the corresponding spin functions must be antisymmetric. Therefore the spins of those electron pairs that correspond to doubles must be coupled in $<\sigma|SM,m>$ to two-electron singlets. Therefore the N-electron spin function $<\sigma_1,\sigma_2,\ldots,\sigma_N|SM,m>$ consists of the s-electron spin function $<\sigma_1,\sigma_2,\ldots,\sigma_s|SM,m>$, $s=1,2,\ldots,f$ multiplied by a product of d singlet-coupled pairs. This kind of spin functions has been defined by Salmon and Ruedenberg (1972) as *geminally antisymmetric*. Though the coupling of the first s spins may be, in principle, arbitrary, we assume that they are coupled according to the Yamanouchi-Kotani scheme (Kotani et al. 1955, Pauncz 1979). In this case the spin space may be described by the van Vleck branching diagram or, more precisely, by the reversed branching diagram (Duch and Karwowski 1985).

Parallel to the permutation operators $\hat{P}$ acting on the coordinates of electrons, we define operators $\overline{P}$ acting on the orbital indices. The $\hat{P}$ and $\overline{P}$ operators, when acting on Ψ_λ^0, are connected by relation $\overline{P}=\hat{P}^{-1}$. We denote (st) for a transposition of the indices λ_s and λ_t and I for the unit operator. The subgroup of $\mathcal{S}_N$, for which Ψ_λ^0 is invariant, is denoted Π_λ. It consists of transpositions within doubles and its order is 2^d. Permutations of this subgroup acting on a spin function $|SM,m>$ either change its sign (if the permutation is odd) or leave it invariant (in the even case). Therefore, all permutations belonging to the $\hat{P}\Pi_\lambda$ coset of $\mathcal{S}_N$ give identical contributions to the CSF.

All CSFs that can be constructed from a given set of orbitals form an orthonormal basis in an antisymmetric and spin-adapted Hilbert space $\mathcal{H}^A(N,K,S,M)$ referred to as the *full CI space*. The full CI space is a model space in which the majority of methods used to describe the electronic structure of molecules has been formulated. Many different approaches to constructing a basis in $\mathcal{H}^A$ have been developed. The approach in which it is based on the permutational symmetry of the wavefunction is called the *symmetric group approach* (SGA).

The Space

In SGA constructing the full CI space may be described in the following way. We start from a given one-electron orbital space, $\mathcal{U}_K^o$, spanned by a set of K orbitals

$$\mathcal{U}_K^o=\{\psi_j\}_{j=1}^K, \tag{26}$$

and the two-dimensional spin space $\mathcal{U}_2^s$, spanned by two one-electron spin functions α and β. Then the N-electron orbital space spanned by our primitive functions $\Psi_\lambda^0(\mathbf{r})$

$$\mathcal{V}_\psi(N,K)=\{\Psi_\lambda^0(\mathbf{r})\}_{all\ \lambda} \tag{27}$$

is obtained as the N-fold product of the one-electron orbital space with appropriate restrictions concerning the orbital occupation numbers (cf. equations (17) and (18)) The N-electron spin space corresponding to the pair S, M of eigenvalues of $\hat{S}^2$ and S_z is spanned by the set of the total spin eigenfunctions

$$\mathcal{W}(N,S,M) = \{|SM,k>\}_{k=1}^{f(S,N)}. \tag{28}$$

It is a subspace of the 2^N- dimensional complete spin space $\mathcal{W}(N,S,M) \subset (U_2^s)^{\otimes N}$. Finally, the full CI space is taken as the antisymmetric part of the product of the two N-electron spaces — of the spin one and of the orbital one:

$$\mathcal{H}^A(N,K,S,M) = \hat{A}[\mathcal{W}(N,S,M) \otimes (\mathcal{V}_\psi(N,K)]. \tag{29}$$

The logical scheme of building the full CI space in SGA may be represented by the following diagram:

$$\left.\begin{array}{ccccccc} \mathcal{U}_K^o & \rightarrow & \{\otimes N\} & \rightarrow & \{\lambda\} & \rightarrow & \mathcal{V}_\psi(N,K) \\ \mathcal{U}_2^s & \rightarrow & \{\otimes N\} & \rightarrow & \{M,S\} & \rightarrow & \mathcal{W}(N,S,M) \end{array}\right\} \rightarrow \{\otimes\} \rightarrow \{\hat{A}\} \rightarrow \mathcal{H}^A(N,K,S,M).$$

The independent handling of the orbital and the spin spaces makes SGA particularly simple and useful in designing CI algorithms.

The full CI space and, in fact, the same basis in this space, may be constructed using a quite different approach. One can start combining the one-electron orbital and spin spaces to form the one-electron spin-orbital space:

$$\mathcal{U}_{2K}^{so} = \mathcal{U}_K^o \otimes \mathcal{U}_2^s. \tag{30}$$

Let us note that the concept of *spinorbital*, basic in this approach, in principle does not appear in SGA. The N-electron space is formed by taking the antisymmetric part of the N-fold product of $\mathcal{U}_{2K}^{so}$. The basis in this space are all N-electron Slater determinants which may be constructed from the set of $2K$ spinorbitals. Finally, taking the spin-adapted combinations of the Slater determinants we obtain the basis in $\mathcal{H}^A(N,K,S,M)$. In this approach the structure of the full CI space may be studied using its invariance properties under unitary transformations of the orbitals. The corresponding invariance group is the *unitary group in K dimensions*, $U(K)$. Exploiting this invariance results in the *unitary group approach* (UGA), another powerful group-theory based technique (Paldus 1976, 1981, 1988, Shavitt 1977a, 1978, 1981, Drake and Schlesinger (1977), Robb and Niazi 1984, Matsen and Pauncz 1986).

4. THE CONFIGURATION INTERACTION METHOD — GENERAL CONSIDERATIONS

The Hamiltonian (1) and its eigenvalue equation (9) may be projected onto the full CI space. Using the resolution of the identity in $\mathcal{H}^A(N,K,S,M)$

$$\hat{I}_{\mathcal{H}} = \sum_{\lambda} \sum_{m=1}^{f(S,s_\lambda)} |\lambda; SM, m >< \lambda; SM, m| \tag{31}$$

one can transform equation (9) into

$$\mathbf{H}^{(SM)}\mathbf{C}^{E(SM)} = E\mathbf{C}^{E(SM)} \tag{32}$$

where $\mathbf{H}^{(SM)}$ is the Hamiltonian matrix with the elements

$$H^{\lambda\mu}_{mn} =< \lambda; SM, m|\hat{H}|\mu; SM, n > . \tag{33}$$

The components of its eigenvector $\mathbf{C}^{E(SM)}$ are equal to the projections of the exact eigenfunction $|\Psi^{\mathcal{E}}_{SM} >$ to the pertinent basis functions in $\mathcal{H}^A$:

$$C^{E(SM)}_{\lambda m} =< \lambda; SM, m|\Psi^{\mathcal{E}}_{SM} > \tag{34}$$

Hereafter, the designation SM is, for simplicity, omitted.

Since $\mathcal{H}^A$ is a subspace of the infinite Hilbert space in which $\hat{H}$ is defined, the eigenvalues of the Hamiltonian matrix $\mathbf{H}$, except the case of a complete orbital space, are not equal to the eigenvalues of $\hat{H}$. If the consecutive eigenvalues are numbered so that

$$\mathcal{E}_1 \leq \mathcal{E}_2 \leq \mathcal{E}_3 \leq \ldots, \quad \text{and} \quad E_1 \leq E_2 \leq E_3 \leq \ldots \tag{35}$$

then

$$\mathcal{E}_1 \leq E_1, \ \mathcal{E}_2 \leq E_2, \ \mathcal{E}_3 \leq E_3, \ldots . \tag{36}$$

The last inequalities are also valid if the system of equations (32) is restricted to any subspace of $\mathcal{H}^A$. This is a consequence of the fact that equation (32) in an arbitrary subspace of $\mathcal{H}^A$ results directly from the Ritz variational principle applied to a trial function

$$|E, SM >= \sum_{\lambda m} C^E_{\lambda m}|\lambda; SM, m > . \tag{37}$$

They are referred to as the *CI equations*, the corresponding $\mathbf{C}^E$ as the *CI vectors* and the wavefunctions $|E, SM >$ with the coefficients determined by the CI equations — the *CI wavefunctions*. Equation (37) is frequently called the *CI expansion* and the set of $|\lambda; SM, m>$ functions - the *CI basis*. The CI energies fulfill another important set of inequalities, known as the McDonald theorem (McDonald 1933, Löwdin 1955)

$$E_{k-1}^{J} \leq E_{k}^{J+1} \leq E_{k}^{J}, \tag{38}$$

where the superscripts stand for the number of terms in the corresponding CI expansions.

Hamiltonian in the CI Space

Hamiltonian (1) in a space

$$\mathcal{F}_{\psi}(K) = \hat{A}[\mathcal{U}_{K}^{o} \otimes \mathcal{U}_{2}^{s}]^{\otimes \tilde{\mathcal{N}}}, \tag{39}$$

where $\tilde{\mathcal{N}}$ is an arbitrary integer, may be expressed in the second-quantization form:

$$\hat{H} = \sum_{pq}^{2K} a_{p}^{\dagger} a_{q} \{p(1)|q(1)\} + \frac{1}{2} \sum_{pqrs}^{2K} a_{p}^{\dagger} a_{r}^{\dagger} a_{q} a_{s} \{p(1)q(1)|r(2)s(2)\}, \tag{40}$$

where p, q, r, s are spinorbital indices, $a_{p}^{\dagger}/a_{p}$ are the fermion creation/annihilation operators and $\{p|q\}/\{pq|rs\}$ denote the one-/two-electron integrals in the spinorbital basis. The Hamiltonians defined in equations (1) and (40) have the same matrix elements in subspaces of $\mathcal{F}_{\psi}(K)$ and in this sense they are equivalent. It should be stressed that this equivalence is limited to the subspaces of $\mathcal{F}_{\psi}(K)$ only. In particular, in all subspaces of $\mathcal{H}^{A}(N, K, S, M)$ the second quantization Hamiltonian (40) have the same matrix elements as the "exact" one, defined in equation (1). This equivalence remains valid if the Hamiltonian depends upon spin variables. The only modifications concern the one- and two-electron integrals which contain the appropriate spin-dependent operators (cf. section 5 of these notes).

If $\hat{h}_1$ and $\hat{h}_2$ are spin-independent, equation (40) may be subject of an important simplification. First, we can use orbitals rather than spinorbitals. Then, writing $a_{p\alpha}^{\dagger}$ for spinorbital $\psi_p \alpha$ and $a_{p\beta}^{\dagger}$ for $\psi_p \beta$ and introducing a spin-traced quantity

$$E_{pq} = a_{p\alpha}^{\dagger} a_{q\alpha} + a_{p\beta}^{\dagger} a_{q\beta}, \tag{41}$$

we may integrate over the spins. In effect we obtain

$$\hat{H} = \sum_{ij}^{K} (i|j) E_{ij} + \frac{1}{2} \sum_{ijkl}^{K} (ij|kl)(E_{ij} E_{kl} - \delta_{jk} E_{il}), \tag{42}$$

where i, j, k, l refer to orbitals and

$$(i|j) = < \psi_i(1)|\hat{h}(1)|\psi_j(1) >, \tag{43}$$

$$(ij|kl) = < \psi_i(1)| < \psi_k(2)|\hat{h}_2(1,2)|\psi_l(2) > |\psi_j(1) > \tag{44}$$

are one- and two-electron integrals. The operators E_{ij} are spin independent. When they act on a product of orbitals they replace ψ_j by ψ_i. Therefore they are called *replacement operators* (Duch and Karwowski 1985). They are also referred to as the *shift operators* (Matsen 1978), *excitation operators* (Kutzelnigg 1985) and appear to be the *generators of the unitary group* $U(K)$ Paldus (1976). The replacement operators act in the orbital space only and (Matsen 1978)

$$E_{ij} = \sum_{l=1}^{N} |\psi_i(l) >< \psi_j(l)|. \tag{45}$$

If $i = j$ then

$$E_{jj}|\lambda >= n_j^\lambda|\lambda >, \tag{46}$$

where $n_j^\lambda = 0, 1, 2$ is the occupation number of ψ_j in λ. Then, E_{jj} is the occupation number operator.

The replacement operators and their generalizations are most useful in compact and transparent formulations of various many-electron theories, including perturbation theory (e.g. Duch 1989), theory of spin-adapted reduced Hamiltonians (Karwowski et al. 1986), theory of density operators (Kutzelnigg 1985), Clifford algebra formalism (Paldus and Jeziorski 1988), and others.

The Orbitals

As it has already been mentioned, the full CI space is invariant with respect to linear transformations of the orbitals. Therefore any linearly independent mixture of the original spinorbitals will lead to the same results. There are many examples which illustrate this equivalence. The best known one is that the molecular orbital and the corresponding valence bond approach converge to the same FCI limit (see e.g. McWeeny 1989). Another example, very useful in constructing model Hamiltonians for large organic molecules, is equivalence of the FCI expansions based on molecular and atomic orbitals, provided they span the same one-electron space (Paldus 1976, Mazumdar and Soos 1981). But even with the largest computers and the most efficient matrix-eigenvalue techniques, the full CI space is usually much too large. Its dimension, in the case of $\hat{S}_z$ and $\hat{S}^2$ adaptation, is given by the Weyl-Paldus dimension formula (Paldus 1974)

$$D(N, K, S) = \frac{2S+1}{K+1}\binom{K+1}{N/2-S}\binom{K+1}{N/2+S+1}, \tag{47}$$

and reaches 10^6 already for $N = 10$, $K = 14$, $S = 0$. Therefore selection of the one-electron basis, in a given orbital space, which gives some reasonably fast convergence of the CI expansion belongs to the most important tasks in practical calculations.

There are many ways of determining orbitals for CI calculations, most of them used occasionally, frequently in test calculations only. Usually the molecular orbitals are taken as linear combinations of some localized orbitals $\{\phi_c\}_{c=1}^{K'}$, with $K' \geq K$, referred to as "atomic", though frequently they are centered in properly selected points rather than in the atomic nuclei. Then, in this case,

$$\psi_k = \sum_{c=1}^{K'} b_{ck}\phi_c. \tag{48}$$

Most commonly the coefficients b_{ck} are determined in a Roothaan-Hartree-Fock SCF procedure (Roothaan 1951, 1960). Another choice is to use the *natural orbitals* (Löwdin 1955, 1959). The natural orbitals are eigenfunctions of the first-order density matrix corresponding to a given wavefunction and assure the fastest convergence of the CI expansion corresponding to this wavefunction. The difficulty in using these orbitals is that the density matrix and, therefore, also the CI wavefunction is required to calculate the natural orbitals. Thus, they can be obtained only after the CI calculation is completed. Several methods allowing to construct approximate natural orbitals have been developed and successfully applied (e.g. Bender and Davidson 1969, Davidson 1975a, Meyer 1977). Finally one should mention procedures in which both coefficients in the CI expansion (equations (37)) and in the molecular orbitals (equation (48)) are simultaneously optimized. The resulting techniques are known as multiconfiguration self consistent field (SCF) methods. They usually lead to rather compact expansions of the wavefunctions, however frequently cause convergence problems. A review of the MCSCF techniques has been recently published by McWeeny and Sutcliffe (1985).

The Integrals

The one- and two-electron integrals in the basis (48) of molecular orbitals are constructed from the analogous integrals in the basis of atomic orbitals by the linear transformation

$$(i|j) = \sum_{cd}^{K'} b_{ci}^* b_{dj} < \phi_c(1)|\hat{h}_1(1)|\phi_d(1) >, \tag{49}$$

$$(ij|kl) = \sum_{cdef}^{K'} b_{ci}^* b_{dj} b_{ek}^* b_{fl} < \phi_c(1)| < \phi_e(2)|\hat{h}_2(1,2)|\phi_f(2) > |\phi_d(1) > . \tag{50}$$

The integral transformation, though formally very simple, for large basis sets requires extensive computing facilities. It is also one of the most time-consuming steps in a CI calculation. The number of terms in the two-electron integral transformation formula is proportional to K'^4 and the number of the molecular integrals is proportional to K^4. Therefore, the number of terms to be computed in a straightforward application of equation (50) is proportional to $(KK')^4 \approx K^8$. That gives $\sim 10^{16}$ multiplications for K=100 — a task not affordable even on the best computer installations. A number of methods aimed at efficient performing the transformation have been developed (cf. Diercksen (1974), Saunders and van Lenthe (1983)). The methods are based on breaking up the fourfold sum in equation (50) into a sequence of four partial summations. In effect the transformation becomes a K^5 process.

Some work has also been done in developing general CI formalism based on the four-component one-electron spinors (Esser 1984). Then the Hamiltonian is spin-dependent and the total spin operators are not constants of the motion. In consequence, the CSFs adapted to different $\hat{S}^2$ and $\hat{S}_z$ eigenvalues may appear in the same CI expansion. The problem is more complicated than the nonrelativistic one and no large-scale fully relativistic CI calculations were reported. A complete bibliography of the subject up to 1985 may be found in a compilation by Pyykkö (1985).

Common in atomic structure calculations (and very efficient) is using the numerical Hartree-Fock (HF) orbitals. For molecules, numerical solving of the HF equations became feasible quite recently (McCullough 1986, Laaksonen at al. 1986) and for diatomics only. Performing large scale molecular CI calculations based on numerical HF orbitals remains a tempting challenge.

The Configuration Interaction Expansion

Usually the orbital basis consists of the Roothaan-Hartree-Fock orbitals. Then the CI expansion limited to a single CSF chosen to be the Roothaan-Hartree-Fock N-electron wavefunction gives the Roothaan-Hartree-Fock energy (E_{HF}). The CI calculation is, in this case, aimed at obtaining the correlation energy, i.e. the difference between the exact nonrelativistic energy of the system and E_{HF}. However one should remember, that the Roothaan-Hartree-Fock energy is equal to the Hartree-Fock energy only in the limit of a complete basis of atomic orbitals. Therefore the names *self-consistent-field (SCF) wavefunction* and *SCF energy* (E_{SCF}) are more appropriate. The difference between a CI and the corresponding SCF energy is commonly though not quite precisely, referred to as the correlation energy obtained in the CI calculation.

With the above choice of the orbitals, the CI expansion is usually dominated by one or by a few CSFs — *the reference CSFs*. The orbitals used to construct the reference CSFs are referred to as the occupied or *internal orbitals* — as distinct from the virtual or *external orbitals,* which do not appear in the reference CSFs. The internal orbitals which are doubly occupied in all reference configurations are "frozen" in the space of the reference CSFs. The remaining internal orbitals, are distinguished as the *active orbitals*. Then, the orbital space is divided into two major subspaces — the *internal space* and the *external space*. The internal space consists of the *doubly occupied space* and the *active space*.

The most commonly used configuration selection scheme is based on the excitation multiplicity criterion. One starts with the SCF wavefunction as the reference configuration and adds all configurations in which no more than a predetermined number, say m, of the occupied orbitals have been replaced by the virtual ones ($m = N$ for the FCI). In the simplest and the most important version of this scheme only single and double excitations out of the reference configuration are included. This scheme is referred to as *singly and doubly excited CI* (SD CI). The SD CI is very frequently used because of its formal simplicity. The fact that the singly and doubly excited configurations are the only ones which interact directly (have non-zero matrix elements) with the reference configuration may justify this kind of truncating the CI expansion if the SCF wavefunction gives a good zeroth-order description, as it frequently happens for small molecules, close to their equilibrium geometries. However, with growing dimension of the molecule, SD CI becomes increasingly poor due to its inherent limitations known as the size-consistency error. An instructive discussion of this subject may be found in the book by Hurley (1976).

Another, much more efficient way to improve the CI expansion, is using several reference configurations with SD CI out of all of them. The resulting *multi-reference* (MR) SD CI expansion may reach any arbitrary accuracy by enlarging the number of the reference configurations. Several extrapolation schemes aimed at recovering the FCI energies from a series of MR SD CI calculations have been developed and claimed to be very efficient (e.g., nearly exact recovering the FCI energy for H_2O in double-zeta basis using only 5% of the FCI expansion was reported by Burton and Gray (1983).

A reasonable compromise between accuracy and compactness of the CI expansion is based on the idea of the *complete active space* (CAS) (Roos 1980, Siegbahn et al. 1980). CAS is the N-electron, spin- and symmetry-adapted space spanned by all CSFs which may be constructed using the active orbitals. In the CAS SCF method the wavefunction is expressed as a linear combination of the CAS basis functions. Both the mixing coefficients and the orbitals are subject to a multi-configuration SCF optimization procedure. A similar approach, developed independently by Ruedenberg et al. (1982) is known as the *fully optimized reaction space* (FORS) method.

The optimized CAS expansion may be supplemented by a set of configurations singly and doubly excited relatively to the ones in the active (internal) space. As it was shown by Siegbahn (1979, 1981) contributions from the doubly excited configurations forming the external space, may be included in a very simple way. This observation is of a fundamental importance in constructing very efficient CI programs and will be discussed later. An excellent study on this subject has been given by Saunders and van Lenthe (1983).

An inadequate treatment of the singularity in the electron-electron interaction operator is, to a large extent, responsible for the slow convergence of CI expansions (Kato 1957, Kutzelnigg 1985a). Therefore explicitly including the interelectronic distance to the CI wavefunction may considerably reduce the necessary length of the expansion. The resulting method, known as the *Hylleraas-CI*, leads to very high accuracy and compact wavefunctions. Yet, due to numerical difficulties, it has been applied to very small systems only (cf. Urdaneta et al. 1988 and references therein). Recently, much more efficient approaches to r_{12}-dependent CI have been reported (Klopper et al. 1991).

5. MATRIX ELEMENTS

Algorithms of evaluation of the Hamiltonian matrix elements are of basic importance not only for the CI method but also for designing many other approaches to the molecular electronic structure theory. Though the case of a spin independent Hamiltonian (equations (1) and (42)) is most important for this presentation, a spin-dependent scalar (as to distinguish from a spinor Coulomb-Dirac-Fock) Hamiltonian will be considered. This is because in SGA a spin-dependent case in nearly as simple as the spin independent one, while spin-dependent phenomena are frequently omitted in theoretical studies because of the lack of appropriate algorithms.

Let us assume that the operators $\hat{h}_1$ and $\hat{h}_2$ in equation (1) depend on both space and spin variables. To simplify the algebra, let both $\hat{h}_1$ and $\hat{h}_2$ be products of two scalar operators: one of them spin-independent and the other one depending upon spin only. This formalism, by only minor and rather obvious modification, can be extended to the most important case of scalar products of two vector operators, as it is, for example, in the case of the spin-orbit interaction. Then, the spin-dependent N-electron Hamiltonian $\hat{H}'$ reads:

$$\hat{H}' = \sum_{i=1}^{N} \hat{h}_1^{\sigma}(\sigma_i)\hat{h}_1^{u}(i) + \sum_{i>j}^{N} \hat{h}_2^{\sigma}(\sigma_i,\sigma_j)\hat{h}_2^{u}(i,j), \tag{51}$$

where the superscripts σ and u refer, respectively, to the spin- and the space-dependence.

Since the Hamiltonian commutes with all permutations, its matrix element (equation (33)), after using equations (23) - (25), is

$$H'^{\lambda\mu}_{mn} = 2^{-(d_\lambda+d_\mu)/2} \sum_{P}(-1)^p < \lambda| < SM,m|\hat{P}\hat{H}'|S'M',n > |\mu > . \tag{52}$$

Let us note that in the spin-dependent case the matrix elements between CSFs belonging to different pairs of SM eigenvalues do not necessarily vanish. It is easy to observe that permutations belonging to the double coset $\Pi_\lambda \hat{P} \Pi_\mu$ of S_N give identical contributions to (52). Hence, formula (52) may be simplified to

$$H'^{\lambda\mu}_{mn} = \sum_{q} \alpha^q_{\lambda\mu}(-1)^p < \lambda| < SM,m|\hat{P}_q\hat{H}'|S'M',n > |\mu > . \tag{53}$$

where the sum runs over all distinct double cosets $\Pi_\lambda \hat{P}_q \Pi_\mu$ and

$$\alpha^q_{\lambda\mu} = 2^{-(d_\lambda+d_\mu)/2} |\Pi_\lambda \hat{P}_q \Pi_\mu|, \tag{54}$$

with $|\Pi_\lambda \hat{P}_q \Pi_\mu|$ being the number of elements in the double coset generated by $\hat{P}_q$. If the Hamiltonian is spin-independent then the integration over the spin and over the orbital variables in equation (53) may be separated. Remembering equations (7) and (15) we get

$$H^{\lambda\mu}_{mn} = \sum_q \alpha^q_{\lambda\mu} U^N_S(\hat{P}_q)_{mn} < \overline{P}\lambda|\hat{H}|\mu > \delta(S,S')\delta(M,M'),$$

$$m = 1,2,\ldots,f(S,s_\lambda), \quad n = 1,2,\ldots,f(S,s_\mu). \tag{55}$$

Substituting the explicit form of the spin-dependent Hamiltonian (52) into the general matrix element expression we get

$$H'^{\lambda\mu}_{mn} = \sum_q (-1)^p \alpha^q_{\lambda\mu} [\sum_{i=1}^N < SM,m|\hat{P}_q \hat{h}^g_1(\sigma_i)|S'M',n >< \overline{P}_q\lambda|\hat{h}^u_1(i)|\mu > +$$

$$\sum_{i>j}^N < SM,m|\hat{P}_q \hat{h}^g_2(\sigma_i,\sigma_j)|S'M',n >< \overline{P}_q\lambda|\hat{h}^u_2(i,j)|\mu >]. \tag{56}$$

If in the integrals over the orbital variables $< \overline{P}\lambda|\hat{h}^u|\mu >$, $|\hat{P}\lambda >$ and $|\mu >$ differ by r orbitals (i.e. if there are more than r mismatches in the two sequences of the orbitals) and $\hat{h}^u$ depends on coordinates of less than r electrons then, due to orthogonality of the orbitals, these integrals vanish. The permutations which bring orbitals of λ into maximum coincidence with orbitals of μ are defined as the *line up permutations* (Harris 1967, Ruedenberg 1971, Karwowski 1973) and denoted $\overline{P}_o$. Then

$$< \overline{P}\lambda|\hat{h}^u_1(i)|\mu >= (\lambda_k|\mu_i), \tag{57}$$

if $r \leq 1$, $\overline{P} \in \Pi_\lambda \overline{P}_o \Pi_\mu$, and λ_k and μ_i may (or may not) be different while all the other orbitals in both configurations are the same.

The two-electron-integral contributions do not vanish only if $r \leq 2$ and either $\overline{P} \in \Pi_\lambda \overline{P}_o \Pi_\mu$ or $\overline{P} \in \Pi_\lambda (kl) \overline{P}_o \Pi_\mu$, where (kl) means transposition of λ_k and λ_l. In the first case

$$< \overline{P}\lambda|\hat{h}^u_2(i,j)|\mu >= (\lambda_k\mu_i|\lambda_l\mu_j), \tag{58}$$

where

$$\frac{< \mathbf{r}|\mu >}{< \mathbf{r}|\overline{P}\lambda >} = \frac{\psi_{\mu i}(\mathbf{r}_i)\psi_{\mu j}(\mathbf{r}_j)}{\psi_{\lambda k}(\mathbf{r}_i)\psi_{\lambda l}(\mathbf{r}_j)}. \tag{59}$$

Also here the pairs of indices (λ_k, λ_l) and (μ_i, μ_j) may (but do not have to) be different. Similarly, in the second case

$$< (kl)\overline{P}\lambda|\hat{h}^u_2(i,j)|\mu >= (\lambda_l\mu_i|\lambda_k\mu_j). \tag{60}$$

Otherwise the N-electron integrals over the orbital variables vanish.

As we see, the integration over the orbital variables selects, for each $h_2^u(i,j)$, *at most* two double cosets giving non-zero contributions to the matrix elements. Finally, $H'^{\lambda\mu}_{mn}$, if not equal to zero, is expressed as a linear combination of the one- (equation (57)) and two-electron (equations (58) and (60)) integrals. The coefficients in these combinations are expressed as matrix elements of $\hat{P}_q \hat{h}_t^\sigma$, $t = 1, 2$, between the pure spin eigenfunctions of $\hat{S}^2$ and $\hat{S}_z$. If $\hat{h}^\sigma = \hat{I}$ (spin-independent operators), the coefficients become matrix elements of the appropriate representation matrices of S_N (eqations (7), (15)). In most general terms, a matrix element of $\hat{H}'$ may be expressed as a linear combination of the one- and two-electron integrals:

$$H'^{(\lambda\mu)}_{mn} = \sum_{ij}^{K} G^{ij}_{\lambda m,\mu n}(i|j) + \frac{1}{2}\sum_{ijkl}^{K} G^{ijkl}_{\lambda m,\mu n}(ij|kl), \tag{61}$$

where $G^{ij}_{\lambda m,\mu n}$ and $G^{ijkl}_{\lambda m,\mu n}$ are referred to as the *coupling constants*.

The same results may be easily obtained by substitution the second-quantized form of the Hamiltonian (equation 40) into equation (53) and subsequent separation of the space and spin integrations. The procedure is particularly simple for a spin-independent Hamiltonian. Combining equations (42) and (55) one gets

$$H^{(\lambda\mu)}_{mn} = \sum_{ij} \Gamma^{ij}_{\lambda m,\mu n}(i|j) + \frac{1}{2}\sum_{ijkl} \Gamma^{ijkl}_{\lambda m,\mu n}(ij|kl), \tag{62}$$

where, for simplicity, the Kronecker deltas appearing in equation (55) have been omitted. In this case the coupling constants have also an alternate meaning. They are, respectively, the first- and the second-order reduced density matrices, if $(\lambda m) = (\mu n)$, or reduced transition matrices, otherwise.

After some algebra (Duch and Karwowski 1985) one may show that

$$\Gamma^{ij}_{\lambda m,\mu n} = \alpha_{\lambda\mu} < \hat{P}\lambda|E_{ij}|\mu > U^N_S(P)_{mn} \tag{63}$$

and

$$\Gamma^{ijkl}_{\lambda m,\mu n} = \alpha_{\lambda\mu} < \hat{P}\lambda|\hat{\Omega}_{ijkl}|\mu > U^N_S(P)_{mn}, \tag{64}$$

where

$$\hat{\Omega}_{ijkl} = E_{ij}E_{kl} - \delta_{jk}E_{il} \tag{65}$$

and

$$\alpha_{\lambda\mu} = 2^{q/2}, \quad q = 0,1,2,3,4, \tag{66}$$

with the value of q determined by relations between the orbital occupation numbers in configurations λ and μ. Due to orthogonality of the orbitals, the density/transition matrix elements (63) or (64) vanish, unless the result of action of E_{ij} or $\hat{\Omega}_{ijkl}$ on $|\mu>$ contains $\overline{P}|\lambda>$. This implies specific relations between the occupation numbers of ψ_i and ψ_j or ψ_i , ψ_j , ψ_k and ψ_l in both configurations. In consequence the sums in equation (62) reduce to a few terms only. Moreover, $<\overline{P}\lambda|E_{ij}|\mu>= 0,1,2$ and $<\overline{P}\lambda|\hat{\Omega}_{ijkl}|\mu>= 0,1,2,4$, depending upon the occupation numbers of the orbitals involved. A generalization of equations (63) and (64) for the case of spin-dependent operators, based on our previous considerations (equations (56) - (61)), gives

$$G^{ij}_{\lambda m,\mu n} = (-1)^p \alpha_{\lambda\mu} <\overline{P}\lambda|E_{ij}|\mu><SM,m|\hat{P}h^\sigma_1(\sigma_{j\mu})|S'M',n> \tag{67}$$

and

$$G^{ijkl}_{\lambda m,\mu n} = (-1)^p \alpha_{\lambda\mu} <\overline{P}\lambda|\hat{\Omega}_{ijkl}|\mu><SM,m|\hat{P}h^\sigma_2(\sigma_{j\mu},\sigma_{l\mu})|S'M',n>, \tag{68}$$

where the subscript $j\mu$ refers to the coordinates of the electron occupying ψ_j in μ. As one can see, apart from a very similar general structure, there are two differences between the matrix element equations in the spin-independent the spin-dependent formalism:
(1) Matrix elements of the permutation operators between the N-electron spin functions (i.e. the representation matrices of S_N) appearing in the spin-independent case, in the spin-dependent case are replaced by the corresponding matrix elements of the permutation operators multiplied by the one- or two-electron spin operators.
(2) The matrices representing spin-dependent operators, in general, are not diagonal in the S, M quantum numbers.

Complete sets of equations for the coupling constant matrices $\mathbf{\Gamma}^{ij}_{\lambda\mu}$ and $\mathbf{\Gamma}^{ijkl}_{\lambda\mu}$, including explicit forms of the line-up permutations, have been reported by Duch and Karwowski (1985). Analogous expressions derived within the UGA formalism may be found in a paper by Robb and Niazi (1984). The N-electron spin integrals appearing in equations (67) and (68) have been calculated by Karwowski (1975) and, within the unitary group approach, by Kent and Schlesinger (1990). As we see, within the symmetric-group based formalism, the way transition/density matrix elements (63) and (64) depend upon the spin coupling scheme is reflected by the form of the $\mathbf{U}^N_S$ matrices only. Therefore, for given λ, μ and i,j or i,j,k,l, a block of coupling constants corresponding to all values of m and n may be constructed at once. In consequence, matrix elements in SGA may be expressed (and coded) in a very compact way (Duch and Karwowski, 1987). Besides, the general formulation remains valid for any spin coupling scheme: changing the spin coupling corresponds to a similarity transformation performed on the $\mathbf{U}^N_S$ matrices. The last property opens another possibility of optimizing the CI algorithm — through selecting the most appropriate spin coupling scheme (cf. Ponnapalli, Schlesinger and Kent, 1991 and references therein).

6. STRATEGY OF CI CALCULATIONS

If the size of a CI problem is small, i.e. if the **H** matrix and all two-electron integrals can be stored in the computer core memory, designing a CI algorithm is quite straightforward. First the **H** matrix is constructed and stored. Next, its eigenvalues and eigenvectors are calculated using an arbitrary diagonalization method. However, in most case, realistic description of molecular structure may be expected only if the CI expansion is rather large. For example, only 75% of the valence correlation energy was obtained from a more than two million term CI expansion for Ne atom (Bauschlicher et al. 1985). Estimated 57 million CSFs are necessary to account for bound properties of Cr_2 (Walsch et al. 1983). Also very large orbital basis sets, and in consequence very large integral files, are frequently necessary. It has been recently shown that not only the f- and g-type atomic orbitals, but also the h-type ones should be included in expansions of the molecular orbitals in order to get the dissociation energies of such molecules as N_2, O_2 or F_2 accurate up to 1 *mhartree* in MR CI calculations (Jankowski et al. 1985, Ahlrichs et al. 1985a). The problem becomes much worse for the excited states. Even for such a small system like the H_3 molecule, a set of carefully selected 95 orbitals is necessary if one wishes to obtain reliable description of both ground and 8 lowest excited states close to their equilibrium geometry (Diercksen et al. 1990).

Diagonalization

If the CI matrices are sufficiently large (say larger than 10^4), only iterative methods are suitable for solving a matrix eigenvalue equation (32). The iterative methods suitable for solving CI eigenvalue problems should have one feature in common: they should be well adapted to calculating a small number of the lowest eigenvalues and of the corresponding eigenvectors of large, symmetric matrices. The method should be applicable also in cases of nearly degenerate or degenerate eigenvalues. Several methods fulfilling these conditions are available (see e.g. Parlett 1980, Shavitt 1970, Davidson 1975, 1989) Most commonly used is the Davidson method (1975) not only because it is numerically stable and fast convergent, but also because it can give simultaneously several lowest eigenvalues and eigenvectors.

The simplest iterative procedure may be obtained by moving everything but the terms associated with the diagonal elements of **H** in equations (32) to the right hand side and assuming that this side of the equations can be obtained from the CI vector of the previous iteration. Then, first, equations (32) are rewriten as

$$C_k = -[\sum_{l(\neq k)} H_{kl} C_l] \,/\, [H_{kk} - E], \tag{69}$$

where the superscript E in the eigenvector designation has been omitted. Next, they are transformed to a form suitable for a simple iterative scheme of solution:

$$C_k^{(p+1)} = C_k^{(p)} - [\sum_l H_{kl}C_l^{(p)} - E^{(p)}C_k^{(p)}] \; / \; [H_{kk} - E^{(p)}], \tag{70}$$

where (p), $(p+1)$ stand for the iteration numbers. The starting vector may be obtained by a diagonalization of a small Hamiltonian matrix defined in a basis of the most important CSFs. A straightforward application of this procedure is, in general, not very efficient. Its convergence may be slow and frequently it is divergent. In the Davidson (1975) method, a small Hamiltonian eigenvalue problem is set up in a space where the basis vectors are the $\mathbf{C}^{(p)}$ vectors. The Hamiltonian matrix in this space is obtained as a byproduct in constructing $\mathbf{C}^{(p+1)}$. Its maximum dimension is equal to the number of the calculated eigenvalues times the number of iterations. Convergence is normally achieved in several iterations, but if the dimension of the small space becomes too large, it may be reduced before the iteration process is finished.

Direct CI Method

The main part of all the iterative methods is multiplication of the Hamiltonian matrix by its approximate eigenvector, i.e. finding

$$\mathbf{B}^{(p+1)} = \mathbf{H}\mathbf{C}^{(p)}, \tag{71}$$

where the superscripts refer to the iteration numbers. At this step two different strategies have been proposed. In the first one, known as the conventional CI (Shavitt 1977), the Hamiltonian matrix elements are computed and stored. Sequential calculation of the Hamiltonian matrix elements requires random access to the molecular integrals. The method is feasible as long as the list of integrals can be stored in the central memory of the computer. However the number of two-electron integrals is proportional to K^4 and in typical molecular calculations $K \simeq 10^2$. Hence, one has to deal with about 10^8 integrals. Then, in calculations aimed at results comparable to the experimental ones, neither the Hamiltonian matrix nor the list of integrals can be stored in the central computer memory.

Overcoming these difficulties is offered by the second strategy, proposed by Roos (1972) and developed by Roos and Siegbahn (1977) and known as the *direct CI* (DCI) approach. In DCI solving matrix eigenvalue problem and evaluation of matrix elements is combined into a single algorithm. The product vector $\mathbf{B}^{(p+1)}$ is here constructed directly from a list of one- and two-electron integrals, without constructing $\mathbf{H}$. This avoids the auxiliary storage problems and dramatically reduces the data transferring tasks. The core of a DCI algorithm may be summarized as follows. Given the orbital indices i, j, k, l; find all pairs of configurations λ, μ for which the coupling constants $\Gamma^{ijkl}_{\lambda m,\mu n}$ do not vanish; determine values of the coupling constants; multiply them by the appropriate integrals and by the (μn) component of $\mathbf{C}$; increment the (λm) component of **B** by the resulting product. As one can see, a single coupling constant corresponding to a given i, j, k, l set of indices, may contribute to many components of **B** and **C**. Therefore a random access to these vectors is necessary, i.e they should be kept in the core memory.

Development of the DCI methods induced some fundamental changes in the strategy of designing CI algorithms. Sets of explicit formulas of different degrees of sophistication and efficiency, developed and implemented in the seventies (Harris 1967, Ruedenberg 1971, Karwowski 1973, 1973a, Sarma and Rettrup 1977, Wormer and Paldus 1979) had one feature in common — before a formula could be applied, the pair of CSFs had to be at least looked at (in fact, the two configurations had to be compared to each other in order to find relations between the orbital occupation numbers). Since an overwhelming majority of the CI matrix elements are equal to zero, in the cases of sufficiently large CI expansions the comparison itself consumed more than 90% of the total time used for construction of the CI matrix. In the DCI, by definition, only those pairs of configurations are considered which generate nonvanishing coupling constants, i.e. which give different from zero matrix elements. This considerably reduces computational effort but also makes the logical part of the algorithm much more complex.

Formulation of graphical approaches to CI became a key to the final success of the DCI methods. In these approaches, instead of individual pairs of configurations, the entire spaces are handled. In the graphical unitary group approach (GUGA) of Shavitt (1977a, 1978) the CI space spanned by a set of spin-adapted CSFs is described by a graph composed of a network of paths. Each of the paths describes a spin-adapted CSF. An adaptation of this idea to the SGA resulted in the symmetric group graphical approach (SGGA) to the CI method (Duch and Karwowski 1981, 1985). In SGA the CI basis is described by two independent graphs: the orbital graph and the spin graph, closely related to the van Vleck branching diagram. The orbital function $|\lambda>$ and the spin function $|SM,m>$ are represented in the appropriate graph by a path. Two different paths in a graph form a loop. The loop corresponds to a matrix element and is associated with the line-up permutation. Only some easily recognizable loops lead to non-zero matrix elements. The orbital indices i,j,k,l define shapes of the non-zero loops and, in consequence, allow to find all pairs of configurations coupled by $\Gamma^{ijkl}_{\lambda m,\mu n}$ and the corresponding permutations $\hat{P}_0$. Hence, the non-zero contributions to the CI matrix may be identified by recognizing specific patterns in the orbital graph. The orbital graph gives a symbolic form of the coupling constants. Their numerical values may be derived from the spin branching diagram. In effect not only calculating zeros is avoided but also, if several pairs of configurations lead to the same matrix element, then the element is calculated only once. If the Yamanouchi-Kotani spin functions (Kotani et al. 1955, Pauncz 1979) are used, the $\mathbf{U}^N_S$ matrices may be constructed directly from the spin graph (Duch and Karwowski 1985). A careful optimization of this algorithm allows the computation of the representation matrices with a speed similar to that of a reading them from a file (Duch 1986a). This structure of the algorithm removes one of the main bottlenecks of many CI programs — constructing and storing a file containing all information about the structure of the CI matrix, the so called *formula tape*. A compact and transparent representation of the orbital space allows for various kinds of optimization — from optimizing ways of dealing with the external part of the orbital space, as discussed by Saunders and van Lenthe (1983), to algorithms in which the graphs are cut into two parts processed independently and then combined together again (Olsen et al. 1988). A review of CI methods based on graphical representations of different model spaces has been published by Duch (1986).

A similar approach may be applied also to spin dependent Hamiltonians. However the degree of complication is here slightly higher. Instead of the representation matrices of $\mathcal{S}_N$, matrices that depend upon the form of the spin-dependent part of the Hamiltonian appear in the coupling constant expressions (cf. equations 67, 68). A complete set of spin-dependent Hamiltonian matrix elements met in atomic CI calculations have recently been derived by Fraga et al. (1986). A general UGA spin-dependent theory has been formulated by Gould and Chandler (1984). However, spin-dependent CI has never been implemented in large-scale calculations.

One major disadvantage of the graphical methods, resulting from their global character, is lack of sufficient flexibility in defining the basis of CSFs. To some extent a limited CI basis may be defined by modifying the shape of the graph, by adding or removing some of its elements (as certain vertices, in which different paths cross, or certain arcs of which the paths are composed). However one vertex or one arc is associated with many configurations. An easy way of defining CI basis is the main (and perhaps the only) advantage of the conventional CI. In graphical approaches an arbitrary basis may be defined using an *index vector*. In order to construct the index vector, to each path in a graph its *lexical index* (Shavitt 1977a, 1978) is attached. The lexical index numbers the CSFs in the case of a GUGA graph and the orbital configurations or pure spin functions in the case of SGGA graphs. The index vector facilitates a mapping of the path lexical indices to the components of the CI vectors (Siegbahn 1981, Saunders and van Lenthe 1983, Duch and Karwowski 1985). Using the index vector one can, at least in principle, define an entirely arbitrary CI expansion — the configurations which are in the graph but should not appear in the CI vector may be marked in the index vector and then skipped. Unfortunately, this approach reduces efficiency of the algorithm — the configurations not included to the expansion have to be partly processed when constructing the coupling constants.

An optimized CI expansion corresponding to an internal space or to a FCI CAS expansion may be supplemented by a set of configurations singly and doubly excited relatively to the ones in the active space. As shown by Siegbahn (1979, 1981), the coupling constants (63) and (64) may be factorized into a simple external part X and a complicated internal part G, respectively, as:

$$\Gamma^{ij}_{\lambda m,\mu n} = G^{ij}_{\lambda m,\mu n} X^{ij}_{\lambda m,\mu n}, \tag{72}$$

and

$$\Gamma^{ijkl}_{\lambda m,\mu n} = G^{ijkl}_{\lambda m,\mu n} X^{ijkl}_{\lambda m,\mu n}. \tag{73}$$

If the external orbitals are spin coupled first (within the Yamanouchi-Kotani spin coupling scheme), then the external parts of the coupling constants do not depend upon the reference configurations and may assume values ± 1 or $\pm\sqrt{2}$ and 2. This property of the external part of the orbital space is of fundamental importance in constructing efficient CI programs (Saunders and van Lenthe 1983).

Implementations

The best known conventional (sometimes called *formula driven*) CI programs are the codes of Buenker and Peyerimhoff (1974, 1975) and of Davidson et al. (1988). The direct CI programs are mainly based on either GUGA or on SGGA. Among the GUGA programs to the most frequently quoted belong: the *integral driven* code by Siegbahn (1979), the *loop driven* (1979) by Brooks and Schaefer, the *shape driven* (1982) by Saxe and coworkers, and the program by Lischka et al. (1981). Some of these programs are included in large, generally available systems (Dupuis et al. 1989). The SGA-based programs have been reported by Ruttink and van Schaik (1982), by Duch and Karwowski (1987) and by Guldberg et al. (1987). One should also mention a program by Liu and Yoshimine based on a symbolic matrix element method (1981) and the FCI program based on the determinant CI expansion by Handy (1980, 1984). Efficient CI codes are based on rather sophisticated algorithms (Saunders and van Lenthe 1983, Robb and Niazi 1984, Duch and Karwowski 1985). Therefore checking their correctness and localizing errors are not easy tasks. A solution to this problem is offered by a recently published general and simple in implementations method allowing to locate errors in programs calculating CI matrices (Diercksen and Karwowski 1987).

Efficiency of DCI codes is mainly a consequence of taking an advantage of the simple structure of the external space. Contributions from each of the external orbitals to the "internal part" of a given coupling constant are the same (equations (72) and (73)). Therefore, in a computer program, the corresponding increments to the appropriate components of **B** and **C** vectors may be added within a simple DO-loop over the external orbitals. Optimum strategies for dealing with the external space contributions have been discussed in detail by Saunders and van Lenthe (1983).

The available computer storage rather than the computation time is the main factor limiting tractable size of the CI expansions. The storage requirement in DCI calculations is proportional to the length of the CI vectors. Though only a very small fraction of the CI expansion coefficients is large enough to contribute to the final energy in a visible way, all must be stored. Excluding the unimportant configurations should result in compact, easily tractable expansions. Let us note, that due to the normalization condition, the maximum number of the expansion coefficients that are not smaller than ε is ε^{-2}, e.g. 10^6 for $\varepsilon = 0.001$. The very idea of eliminating all configurations contributing less than a certain threshold and its implementation within the classical CI approach may be traced back to early works by Buenker and Peyerimhoff (1974, 1975). Recent formulations due to Knowles (1989), Knowles and Handy (1989) and due to Duch (1990) opened a way to selecting the most important configurations during the process of solving the DCI eigenvalue problem. In an encouraging recent calculation on NH_3, FCI results have been obtained using only 0.1% – 0.3% of the FCI expansion vector (Knowles and Handy 1989). In another algorithm Harrison (1991) exploits the sparsity of the Hamiltonian matrix and the full-CI vector by combining a configuration selection scheme and the perturbation theory. The approach is aimed at approximating FCI results up to the "chemical accuracy", i.e. 0.1 kcal/mol or better. It has very small memory requirements and the only limitation is the available computation time. It seems that the algorithms, as mentioned above, opened a new generation of CI codes.

Another group of CI algorithms, called *intermediate projection CI* (Olsen et al. 1988, 1990, Zarrabian et al. 1989 Duch 1990) originated from an idea of Siegbahn (1984). The most time-consuming, complicated and difficult to vectorize part of the standard direct CI programs is calculation of the two-electron coupling constants. In the new method, the two-electron coupling constants are implicitly created in the diagonalization step. The CI vector is updated by taking the trace of a product of three matrices: the matrix of the one-electron coupling constants $\mathbf{\Gamma}$, another matrix which is constructed from $\mathbf{\Gamma}$ and from the trial CI vector $\mathbf{C}$, and the symmetry blocked two-electron integral matrix. In effect only the one-electron coupling constants have to be computed and, in some cases, stored. The resulting algorithm vectorizes very well and offers many possibilities of improvement. In particular, it may be formulated for the full (Siegbahn 1984, Olsen et al. 1990) as well as for the restricted CI (Olsen et al. 1988, Duch 1990) expansions. The intermediate states used to decompose the two-electron coupling constant matrix onto a product of matrices may be taken as N or as $N-2$ (Zarrabian et al. 1989) electron functions. Taking this last possibility, one gets a much shorter expansion and, hence, a more efficient algorithm. Finally, the method may take explicitly advantage of sparseness of the Hamiltonian matrix (Olsen et al. 1990). It seems, that combining the idea of flexible CI with the intermediate projection techniques will result in a most powerful CI method (Olsen et al. 1990, Duch 1990).

7. ODDS AND ENDS

Though the approaches aimed at emulation of the FCI in restricted spaces are promising and may entirely change our computational strategies very soon, for the time being various, sometimes drastic, approximations are unavoidable. SD CI and MR SD CI with very few references are the most commonly used versions of the method. Among many interesting methods which may reduce the numerical effort one should certainly mention the externally and internally contracted CI methods by Siegbahn (1983), offering approximations to MR SD CI obtained by freezing relative weights of certain groups of configurations.

Another method was recently formulated by Olsen et al. (1987). It is based on a spectral resolution of a symmetric matrix formed from the CI coefficients of the doubly excited CSFs. Several eigenvectors of this matrix became new variables in the variational procedure. The resulting energy expression is a quartic function of the new variables and therefore non-linear optimization procedures must be used. In return for the higher numerical complexity, the storage requirements are considerably reduced.

A *fast CI method* has been proposed by Bendazzoli et al. (1982, 1985) and, independently, under the name *superdirect CI* by Duch (1989a). The method needs nearly no computer memory but gives only energies. In the direct CI constructing and saving the Hamiltonian matrix is avoided. In the superdirect approach one avoids constructing and saving the eigenvectors as well (in a sense, the many-body perturbation theory is a superdirect approach). The energy values, possessing variational bound properties, are obtained by solving a low-order secular equation, which depends upon matrix elements of the Hamiltonian powers, as for example $det| < i|H|j > -E < i|j > | = 0$, with $|j >= \hat{H}^j|0 >$ and $|0 >$ — a fixed initial vector. Since calculation of the Hamiltonian power expectation values is not an easy task, no large scale computations using this method were performed, though some pilot calculations are encouraging.

An attractive answer to the excited state problem is the *state specific CI* developed by Nicolaides (1989). In this approach both the orbital space and the CI expansion are individually selected for each state. In effect the memory requirements are considerably reduced but the problem becomes nonorthogonal.

Size Consistency

All single-reference CI methods, based on expansions determined according to an excitation multiplicity criterion, behave incorrectly when the size of the molecule grows up. We say that the limited CI methods are not *size consistent.* A method is said to be size-consistent if the computed energy $E(A + B)$ of a composite system $A + B$, with A and B at infinite distance from each other, is the same as if the method was applied to A and B separately and the energies were added, i.e. $E(A + B) = E(A) + E(B)$.

It is easy to see that the SD CI is not size-consistent. The wavefunction for $(A + B)$

at the infinite separation should be a product of the wavefunctions for A and for B. However in the SD CI calculations for the separate systems there are local double excitations on both A and B. In consequence the product function will contain certain quadruple excitations and SD CI gives $E(A) + E(B) < E(A + B)$. In general, HF and FCI methods are size-consistent, but CI limited to any degree of excitation is not. Most discouraging is the following theorem, valid for a system of n spatially separated electron pairs (Hurley 1976): "Any truncation of the CI expansion to a finite order of excitation leads, as n tends to infinity, to zero overlap with the exact wavefunction and a zero value for the calculated correlation energy per localized electron pair". If we consider the case where each of the separated two-electron systems is a hydrogen molecule, then SD CI yields, 100%, 98%, 93%, 86% and 70% of the exact correlation energy if, respectively, n is equal to 1, 2, 5, 10 and 25. The SD CI deteriorates as the size of the system increases — it is a much worse approximation for large molecules than it is for small ones. For infinite systems, as e.g. for crystals, it is completely useless — the correlation energy given by SD CI is then exactly zero. Multiple excitations are of little importance when they first occur (quadruple for $n = 2$, sextuple for $n = 3$) but their contribution increases sharply with the number of electrons. For accuracy of $1 mhartree$ quadruple excitations are essential for $N > 4$ and also sextuple for $N > 14$.

In chemical applications the lack of size consistency is a very serious drawback. For example, the chemical reaction energy ΔE for the process $A + B \rightarrow C + D$ cannot be evaluated by simple using the truncated CI energies of the individual species A, B, C, D since the errors due to the truncation of CI in systems containing different number of electrons are different. One should rather evaluate $\Delta E = E(A + B) - E(C + D)$, where $E(A + B)$ and $E(C + D)$ are the truncated CI energies of the corresponding "supermolecules" at large $A - B$ and $C - D$ separations, expecting cancellation of the electron correlation errors, similar for $(A + B)$ and for $(C + D)$. An alternative might be running CI calculations for A, B, C and D capable to simulate FCI results, as discussed in the previous section. Both options are, however, costly and often not feasible.

A careful analysis of the origin of the size consistency error, based on the many-body perturbation theory, leads to approximate formulas correcting SD CI energies for this error. Most commonly known is the Davidson (1974) correction

$$\Delta E_{DC} = (1 - C_0^2)(E_{CI}^{SD} - E_{SCF}), \tag{74}$$

where E_{CI}^{SD} is the energy yielded by SD CI and C_0 is the coefficient of the reference function in the CI expansion. Since SD CI and FCI are equivalent for two-electron systems, there is no size consistency error in this case. Nevertheless $\Delta E_{DC} \neq 0$ also for $N = 2$. Though the N-dependence of the size-consistency corrections was considered already long ago (cf. Pople et al. 1977), only very recently it has been proved that ΔE_{DC} should be rescaled by the factor $(N - 2)(N - 3)/N(N - 1)$ (Meissner 1988).

The size consistency defect is not restricted to the energy only. The limited CI wavefunctions deteriorate with an increase of the number of electrons. It has important consequences in calculating properties, especially in large systems (Meunier et al. 1976, 1979). Introducing corrections to the wavefunctions is, however, much more difficult. A size-consistent modification of SD CI, known as the *coupled pair functional* (CPF) method has recently been formulated by Ahlrichs et al. (1985). It is based on the variation of a size-consistency corrected energy functional derived from the SD CI expressions. However a general problem of size-consistency in limited CI remains open.

Semiempirical Methods

There is little, if any, doubt about importance of electron correlation effects in *ab initio* electronic structure calculations. On the other hand, the role of CI in semiempirical methods is not always obvious. One may hope that the electron correlation effects can be taken into account through properly defined empirical parameters. Effective operators have been introduced both in atomic (Cowan 1981) and in molecular (Freed 1974, Iwata and Freed 1975) structure theory to describe in a rigorous way this process. Yet the parameters may compensate only some part of the electron correlation effects. Certainly the non-dynamic correlation (Urban et al. 1987) cannot be described in this way. In the case of atoms, mixing of states derived from the same electronic configuration must always be taken into account within parametric theories. In semiempirical theories of molecules mixing of singly excited configurations is necessary if we intend to describe excited electronic states. But, if we wish to treat correctly the non-dynamic correlation, CAS FCI in the space of the molecular orbitals used to describe the ground and the excited states should be performed.

Clearly, the way a theory is parametrized must be consistent with the level of approximation used, i.e. introducing CI within a parametrization scheme developed for a theory without CI would give, in general, worse results. If a theory is parametrized directly on experiment, then the optimum parameters are the ones, which minimize the standard deviation

$$\delta = \left[\frac{1}{n} \sum_{i=1}^{n} (F_i^{calc} - F_i^{exp})^2 \right]^{1/2} \tag{75}$$

between the calculated (F_i^{calc}) and the reference (F_i^{exp}) quantities. Usually F_i^{exp} are experimental energies of a molecule or of an atom. However the rapid increase in the accuracy of very high level *ab initio* calculations coupled with a shortage of good experimental data is likely to result in a change in emphasis in fitting models — parameters for semiempirical methods will be more frequently derived using also some *ab initio* data. If δ is minimized for several different CI expansions, then it may be taken as a measure of importance of CI — as long as the quality of the fit changes with adding more CI, the effect of CI is not compensable by the empirical parameters (Karwowski 1973b). Several semiempirical CI programs have been reported as e.g. Tavan and Schulten (1979), Zerner (1989).

Semiempirical methods are designed to study structure and properties of large molecules. In this context the correlation effects should be treated at the same level of accuracy for all, large and small systems belonging to the same class, i.e. described by the same set of parameters. But a limited CI is not size-extensive. To overcome this difficulty, several methods, which take an advantage of simplicity of semiempirical model Hamiltonians in constructing large scale CI algorithms, have been formulated. The basic simplification is due to neglect of the differential overlap between orthogonalized atomic orbitals. This assumption results in the well known ZDO (zero-differential overlap), CNDO (complete neglect of differential overlap), and other related approximations. Its consequence is that only Coulomb two-electron integrals, γ_{ik}, are different from zero. Thus, we have

$$\gamma_{ik} = \delta_{ij}\delta_{kl}(ij|kl), \tag{76}$$

and Hamiltonian (42) reads

$$\hat{H}^{SE} = \sum_{ik} \beta_{ik} E_{ik} + \frac{1}{2}\sum_{ijkl} \gamma_{ik}(E_{ii}E_{kk} - \delta_{ik}E_{ik}), \tag{77}$$

where $\beta_{ik} = (i|k)$. In most cases both β_{ik} and γ_{ik} are empirical parameters. Symmetry conditions and further simplifying assumptions (as e.g., the tight binding approximation) usually reduce their number to a few only. In particular, the extensively studied Hubbard Hamiltonian is obtained by setting $\gamma_{ik} = \gamma_{11}\delta_{ik}$. The two-electron part of $\hat{H}^{SE}$ is expressed in terms of the occupation number operators only and therefore it is diagonal and the coupling constants may only be 1, 2 or 4, in any CSF basis constructed from products of orthonormal atomic orbitals. The off-diagonal CI matrix elements depend upon the one-electron part only and therefore are very easy to calculate. A disadvantage of this Hamiltonian is necessity of using rather large CI. The one-electron space is here defined by atomic rather than by molecular orbitals. Therefore a one-term CI expansion corresponds to a single valence structure in an orthogonal orbital basis and usually it gives a very poor description of a delocalized molecular state. On the other hand, using this Hamiltonian, full CI or nearly full CI calculations are easy to perform. In effect it has been applied to studies on correlation effects in spectra of polyenes and polyacenes (Tavan and Schulten 1979) and in biologically important systems such as retinal dyes (Grossjean and Tavan 1988). An interesting study on electronic excitations in finite and infinite polyenes, demonstrating that properties of polyacetylene and of related conducting and semiconducting polymers are dominated by effects of electron correlation was recently reported by Tavan and Schulten (1987). A series of papers by Soos at al. (cf. Mazumdar and Soos 1981, Ramanesha and Soos 1984 and references therein) contributes wealth of information about spectra and properties of π-electron systems described by the FCI solutions of the semiempirical Hamiltonian (77). Finally,

$\hat{H}^{SE}$ is frequently used to test new approaches to CI and to other methods of molecular structure theory. In this area, the best known are works by Paldus et al. (Paldus 1976, Takahashi, Paldus and Čižek 1983) where references to other papers may be found.

ACKNOWLEDGEMENTS

I am grateful to G. H. F. Diercksen for his kind hospitality at the Max-Planck-Institut für Astrophysik in Garching, where a substantial part of these notes has been prepared and to B. Wybourne for his critical reading the manuscript.

REFERENCES

Ahlrichs, R., Scharf, P. and Ehrhardt, C., 1985, J. Chem. Phys. **82**, 890.

Ahlirichs, R., Scharf, P. and Jankowski, K., 1985a, Chem. Phys. **98**, 381.

Bartlett, R. J., Sekino, H. and Purvis III, G. D., 1983, Chem. Phys. Lett. **98**, 66.

Bauschlicher, C. W. and Langhoff, S. R., 1987, Chem. Phys. Lett. **135** 67.

Bauschlicher, C. W. and Langhoff, S. R., 1989, Comments At. Mol. Phys. **23**, 1.

Bauschlicher, C. W., Langhoff, S. R., Taylor, P. R. and Partridge, H., 1985, Chem. Phys. Lett. **126**, 436.

Bauschlicher, C. W. and Taylor, P. R., 1986, J. Chem. Phys. **85**, 6510.

Bendazzoli, G. L., Fano, G., Ortolani, G. and Palmieri, P., 1982, J. Chem. Phys. **76**, 2498.

Bendazzoli, G. L., Palmieri, P., Fano, G. and Ortolani, G., 1985, J. Chem. Phys. **83**, 674.

Bender, C. F. and Davidson, E. R., 1969, Phys. Rev. **138**, 23.

Boys, S. F., 1953, Proc. Roy. Soc. (London) **A217**, 136, 235.

Brooks, B. R. and Schaefer, H. F., 1979, J. Chem. Phys. **70**, 5092.

Brown, F. B., Shavitt, I. and Shepard, R., 1984, Chem. Phys. Lett. **105**, 363.

Buenker, R. J. and Peyerimhoff, S. D., 1974, Theoret. Chim. Acta (Berlin) **35**, 33.

Buenker, R. J. and Peyerimhoff, S. D., 1975, Theoret. Chim. Acta (Berlin) **39**, 217.

Burton, P. G., 1983, Intern. J. Quantum Chem. **23** 613.

Burton, P. G. and Gray, P. D., 1983, Chem. Phys. Lett. **96**, 453.

Cowan, R. D., 1981, *The Theory of Atomic Structure and Spectra*, University of California, Berkeley.

Davidson, E. R., 1974, in *The World of Quantum Chemistry*, ed. R. Daudel and B. Pullman, Reidel, Dordrecht.

Davidson, E. R., 1975, J. Comput. Phys. **17**, 87.

Davidson, E. R., 1975a, *Reduced Density Matrices in Quantum Chemistry*, Academic Press, New York.

Davidson, E. R., 1989, Comput. Phys. Commun. **53**, 49.

Davidson, E. R. and Feller, D., 1988, *MELDF*, Quantum Chemistry Group, Indiana University.

Diercksen, G. H. F., 1974, Theoret. Chim. Acta (Berlin) **33**, 1.

Diercksen, G. H. F., Duch, W. and Karwowski, J., 1990, Chem. Phys. Lett. **168**, 69.

Diercksen, G. H. F. and Karwowski, J., 1987, Comp. Phys. Commun. **47**, 83.

Dinesha, K. V., Sarma, C. R. and Rettrup, S., 1981, Advan. Quantum Chem **14**, 125.

Drake, G. W. F. and Schlesinger, M., 1977, Phys. Rev. A **15**, 1990.

Duch, W., 1986, *GRMS or Graphical Representation of Model Spaces* (Lecture Notes in Chemistry, Vol. 42), Springer, Berlin.

Duch, W., 1986a, Chem. Phys. Lett. **124**,442.

Duch, W., 1989, J. Chem. Phys. **94**, 2452.

Duch, W., 1989a, Chem. Phys. Lett. **162**, 56.

Duch, W., 1990, Intern. J. Quantum Chem. **S24**, 682.

Duch, W., 1991, J. Mol. Structure (Theochem) **xx**, xxx.

Duch, W. and Karwowski, J., 1981, in *The Unitary Group for the Evaluation of Electronic Energy Matrix Elements*, (Lecture Notes in Chemistry, vol. 22), ed. J. Hinze, Springer, Berlin, p. 260.

Duch, W. and Karwowski, J., 1985, Comp. Phys. Reports **2**, 94.

Duch, W. and Karwowski, J., 1987, Theoret. Chim. Acta (Berlin) **71**, 187.

Dupuis, M., Watts, J. D., Villar, H. O. and Hurst, G. J. B., 1989, Comput. Phys. Commun. **52**, 415.

Esser, M., 1984, Intern. J. Quantum Chem. **26**, 313. Fraga, S., Kłobukowski, M., Muszyńska, J., Saxena, K. M. S. and Sordo, J. A., 1986, Phys. Rev. A **34**, 23.

Freed, K. F., 1974, J. Chem. Phys. **60**, 1765.

Gould, M. D. and Chandler, G. S., 1984, Intern. J. Quantum Chem. **25**, 553; **25**, 603; **25**, 1089; *errata*, 1985, *ibid.* **27**, 787.

Grant, I. P., 1991, *Relativistic Electronic Structure Theory*, this volume.

Grossjean, M. F. and Tavan, P., 1988, J. Chem. Phys. **88**, 4884.

Guldberg, A., Rettrup, S., Bendazzoli, G. L. and Palmieri, P., 1987, Intern. J. Quantum Chem. **S21**, 513.

Hamermesh, M., 1962, *Group Theory and its Applications to Physical Problems*, Addison Wesley, Reading.

Handy, N. C., 1980, Chem. Phys. Lett. **74**, 280.

Handy, N. C., 1984, Faraday Symp. Chem. Soc. **19**, 17.

Harris, F., 1967, J. Chem. Phys. **46**, 2769; **47**, 1047.

Harrison, R. J., 1991, J. Chem. Phys. **94**, 5021.

Hurley, A. C., 1976, *Electron Correlation in Small Molecules*, Academic Press, New York.

Hylleraas, E. A., 1928, Z. Phys. **48**, 469.

Iwata, S. and Freed, K. F., 1975, Chem. Phys. **11**, 433.

Jankowski, K., Becherer, R., Scharf, P., Schiffer, H. and Ahlrichs, R., 1985, J. Chem. Phys. **82**, 1413.

Karwowski, J., 1973, Theoret. Chim. Acta (Berlin) **29**, 151.

Karwowski, J., 1973a, Chem. Phys. Lett. **19**, 279.

Karwowski, J., 1973b, J. Mol. Structure **19**, 143.

Karwowski, J., 1975, Acta Phys. Polon. **A48**, 553.

Karwowski, J., 1991, in: *Strucure, Interactions, Reactivity*, ed. S. Fraga, Elsevier, Amsterdam.

Karwowski, J. Duch, W. and Valdemoro, C., 1986, Phys. Rev. A **33**, 2254.

Kato, T., 1957, Commun. Pure Appl. Math. **10**, 151.

Kent, R. D. and Schlesinger, M., 1990, Phys. Rev. A **42**, 1155.

Klopper, W., Röhse, R. and Kutzelnigg, W., 1991, Chem. Phys. Lett. **178**, 455.

Knowles, P. J., 1989, Chem. Phys. Lett. **155**, 513.

Knowles, P. J. and Handy, N. C., 1989, J. Chem. Phys. **91**, 2396.

Kotani, M., Amemiya, A., Ishiguro, E. and Kimura, T., 1955, *Table of Molecular Integrals*, Maruzen, Tokyo.

Kutzelnigg, W., 1985, J. Chem. Phys. **82**, 4166.

Kutzelnigg, W., 1985a, Theoret. Chim. Acta (Berlin) **68**, 445.

Laaksonen, L., Pyykkö, P., and Sundholm, D., 1986, Comput. Phys. Reports **4**, 313.

Lischka, H., Schepard, R., Brown, F. B. and Shavitt, I., 1981, Intern. J. Quantum. Chem. **S15**, 91.

Liu, B. and Yoshimine, M., 1981, J. Chem. Phys. **74**, 612.

Löwdin, P. -O., 1955, Phys. Rev. **97**, 1509.

Löwdin, P. -O., 1959, Advan. Chem. Phys. **2**, 207.

McCullough, E. A., 1986, Comput. Phys. Reports **4**, 265.

McDonald, J. K. L., 1933, Phys. Rev. **43**, 830.

McWeeny, R., 1989, *Methods in Molecular Quantum Mechanics*, Academic Press, New York.

McWeeny, R., 1991, *The Electron Correlation Problem*, this volume.

McWeeny, R., and Sutcliffe, B. T., 1985, Comp. Phys. Reports **5**, 217.

Matsen, F. A., 1978, Advan. Quantum Chem. **11**, 223.

Matsen, F. A. and Pauncz, R., 1986, *The Unitary Group in Quantum Chemistry*, Elsevier, Amsterdam.

Mazumdar, S. and Soos, Z. G., 1981, Phys. Rev. **B 23**, 2810.

Meissner, L., 1988, Chem. Phys. Lett. **146**, 204.

Meunier, A. and Levy, B., 1979, Intern. J. Quantum Chem. **16**, 955.

Meunier, A., Levy, B. and Berthier, G., 1976, Intern. J. Quantum Chem. **10**, 1061.

Meyer, W., 1977 in: *Methods of Electronic Structure Theory*, ed. H. F. Schaefer III, Plenum, New York, p. 413 and references therein.

Nicolaides, C. A., 1989, in *Quantum Chemistry - Basic Aspects, Actual Trends*, ed. R. Carbo, Elsevier, Amsterdam, p. 343.

Oddershede, J., 1991, *Response and Propagator Methods*, this volume.
Olsen, J., Jørgensen, P. and Simons, J., 1990, Chem. Phys. Lett. **169**, 463.
Olsen, J., Malmqvist, P., Roos, B. O., Lindh, R. and Widmark, P. O., 1987, Chem. Phys. Lett. **133**, 91.
Olsen, J., Roos, B. O., Jørgensen, P. and Jensen, H. J. Aa., 1988, J. Chem. Phys. **89**, 2185.
Paldus, J., 1974, J. Chem. Phys. **61**, 5321.
Paldus, J., 1976, in: *Theoretical Chemistry. Advances and Perspectives*, ed. H. Eyring and D. Henderson, vol. 2, Academic Press, New York, p 131.
Paldus, J., 1981, in *The unitary Group for the Evaluation of Electronic Energy Matrix Elements*, (Lecture Notes in Chemistry, vol. 22), ed. J. Hinze, Springer, Berlin, p. 1.
Paldus, J., 1983, in *New Horizons of Quantum Chemistry*, ed. P. O. Löwdin and B. Pullman, Reidel, Dordrecht, p. 31.
Paldus, J., 1988, in *Mathematical Frontiers in Computational Chemical Physics*, ed. D. Truhlar, Springer, Berlin, p. 262.
Paldus, J., 1991, *Cluster Expansions for Electron Correlation Effects*, this volume.
Paldus, J. and Jeziorski, B., 1988, Theoret. Chim. Acta (Berlin) **73**, 81.
Parlett, B. N., 1980, *The Symmetric Eigenvalue Problem*, Prentice Hall, New York.
Parr, R. G., Craig, D. P. and Roos, I. G., 1950, J. Chem. Phys. **18**, 1561.
Pauncz, R., 1979, *Spin Eigenfunctions*, Plenum, New York.
Ponnapalli, P. S., Schlesinger, M. and Kent, R. D., 1991, J. Math. Phys. **32**, 1203.
Pople, J. A., Seeger, R. and Krishan, R., 1977, Intern. J. Quantum Chem. **S11**, 149.
Pyykkö, P., 1986, *Relativistic Theory of Atoms and Molecules. A Bibliography 1916-1985*, Lecture Notes in Chemistry, vol. 41. Springer, Berlin.
Ramanesha, S. and Soos, Z. G., 1984, J. Chem. Phys. **80**, 3278.
Robb, M. A. and Niazi, U., 1984, Comp. Phys. Reports **1**, 127.
Roos, B. O., 1972, Chem. Phys. Lett. **15**, 153.
Roos, B. O., 1980, Intern. J. Quantum Chem. **14**, 175.
Roos, B. O. and Siegbahn, P. E. M., 1977, in: *Methods of Electronic Structure Theory*, ed. H. F. Schaefer III, Plenum, New York, p. 277.
Roothaan, C. C. J., 1951, Rev. Mod. Phys. **23**, 69.
Roothaan, C. C. J., 1960, Rev. Mod. Phys. **32**, 179.
Ruedenberg, K., 1971, Phys. Rev. Lett. **27**, 1105.
Ruedenberg, K., Schmidt, M. W., Gilbert, M. M. and Elbert, S. T., 1982, J. Chem. Phys. **71**, 41.
Ruttink, P. J. A. and van Schaik, M. M. M., 1982, in *Proceedings of the 5-th Seminar on Computational Methods in Quantum Chemistry, Gröningen 1982*, Max-Planck-Institut für Physik und Astrophysik, Institut für Astrophysik, Garching, p. 273.
Salmon, W. I. and Ruedenberg, K., 1972, J. Chem. Phys. **57**, 2776.

Sarma, C. R. and Rettrup, S., 1977, Theoret. Chim. Acta (Berlin) **46**, 63.

Saunders, V. R. and van Lenthe, J. H., 1983, Mol. Phys. **48**, 923.

Saxe, P., Fox, D. J., Schaefer, H. F. and Handy, N. C., 1982, J. Chem. Phys. **77**, 5584.

Shavitt, I., 1970, J. Comput. Phys. **6**, 124.

Shavitt, I., 1977 in: *Methods of Electronic Structure Theory*, ed. H. F. Schaefer III, Plenum, New York, p. 189.

Shavitt, I. 1977a, Int. J. Quantum Chem. **S11**, 131.

Shavitt, I. 1978, Int. J. Quantum Chem. **S12**, 5.

Shavitt, I. 1981, in *The unitary Group for the Evaluation of Electronic Energy Matrix Elements*, (Lecture Notes in Chemistry, vol. 22), ed. J. Hinze, Springer, Berlin, p. 51.

Siegbahn, P. E. M., 1979, J. Chem. Phys. **70**, 5391.

Siegbahn, P. E. M., 1981, in *The unitary Group for the Evaluation of Electronic Energy Matrix Elements*, (Lecture Notes in Chemistry, vol. 22), ed. J. Hinze, Springer, Berlin, p. 119.

Siegbahn, P. E. M., 1983, Intern. J. Quantum Chem. **23**, 1869.

Siegbahn, P. E. M., 1984, Chem. Phys. Lett. **109**, 417.

Siegbahn, P. E. M., Heiberg, A., Roos, B. O. and Levy, B., 1980, Physica Scripta **21**, 323.

Sutcliffe, B. T., 1982, in *Methods in Computational Molecular Physics*, (NATO ASI Series C, Vol.113), ed. G. H. F. Diercksen and S. Wilson, D. Reidel, Dordrecht, p. 115.

Szabo, A. and Ostlund, N. S., 1982, *Modern Quantum Chemistry*, Mac Millian, New York.

Takahashi, M., Paldus, J. and Čižek, J., 1983, Intern. J. Quantum Chem. **24**, 707.

Tavan, P. and Schulten, K., 1979, J. Chem. Phys. **70**, 5407; 5414.

Tavan, P. and Schulten, K., 1987, Phys. Rev. B **36**, 4337.

Urban, M., Černušak, V, Kellö, V. and Noga, J., 1987, in *Methods in Computational Chemistry*, vol. 1, ed. S. Wilson, Plenum, New York, p. 117.

Urdaneta, C., Largo-Cabrerizo, A., Levin, J., Lie, G. C., and Clementi, E., 1988, J. Chem. Phys. **88**, 2091.

Walsch, S. P., Bauschlicher, C. W., Roos, B. O. and Nelin, C. H., 1983, Chem. Phys. Lett. **103**, 175.

Wilson, S., 1991, *The Perturbation Theory of Electron Correlation*, this volume.

Wormer, P. E. S. and Paldus, J., 1979, Intern. J. Quantum Chem. **16**, 1307.

Wybourne B. G. (1974), *Classical Groups for Physicists*, Wiley, New York.

Zarrabian, S., Kazempour, M. K. and Estévez, 1991, Chem. Phys. Lett. **178**, 55.

Zarrabian, S., Sarma, C. R. and Paldus, J., 1989, Chem. Phys. Lett. **155**, 183.

Zerner, M. C., 1989, *ZINDO, a General Semi-Empirical Program Package*, Department of Chemistry, University of Florida, Gainesville.

COUPLED CLUSTER THEORY

Josef Paldus

Department of Applied Mathematics
Department of Chemistry and (GWC)²
University of Waterloo
Waterloo, Ontario, N2L 3G1, Canada

1. INTRODUCTION

The necessity to account for many-electron correlation effects in atomic and molecular electronic structure calculations was realized in the sixties, when the first attempts to compute dissociation or activation energies of simple chemical processes (e.g., Wahl, 1964; Schaefer, 1971,1972) and various molecular properties (e.g., Huo, 1965) were made. The initial hope that the configuration interaction (CI) approach limited to doubly excited configurations, originating from a single reference state, will provide a satisfactory description of correlation effects soon had to be given up (see, e.g., Bender and Davidson, 1968,1969; Barr and Davidson, 1970; Davidson, 1974). By that time the basic structure of correlated wavefunctions (Hubbard, 1957,1958) and the problem of size extensivity (Brueckner, 1955; Goldstone, 1957; Primas, 1965) were well understood [although this term was employed only later (Bartlett and Purvis, 1978)], thanks to the developments in solid state (e.g., Gell-Mann and Brueckner, 1957; Quinn and Ferrell, 1958) and nuclear (e.g., deShalit and Feshbach, 1974; Eisenberg and Greiner, 1972) physics. In particular, the linked cluster theorem of the many-body perturbation theory (MBPT) and the connected cluster structure of the exact wavefunctions (Hubbard, 1958b) were firmly established. The exponential ansatz for the wave operator, implied by Hubbard's (1958b) work, was exploited in the context of the nuclear correlation problem by Coester (1958) and Coester and Kümmel (1960). Independently of these developments, the idea of cluster expansion, originally developed in statistical physics (e.g., Mayer and Mayer, 1940; Kahn and Uhlenbeck, 1938), and its approximate exploitation in molecular electronic structure calculations, were introduced by Sinanoğlu (1961, 1962). A systematic procedure yielding explicit equations for the connected cluster amplitudes in the single reference closed-shell (CS) case was then developed by Čížek (1966a, 1969, 1991), relying primarily on the field theoretical formulation of the MBPT by Tolmachev (1963). This formulation thus encompassed both Brueckner (1955) theory of nuclear matter and Gell-Mann and Brueckner (1957) theory of electron gas, as well as Sinanoğlu's (1961, 1962) electron-pair theory. The theory was also applied (Čížek, 1966a) to a semi-empirical

Methods in Computational Molecular Physics, Edited by Stephen Wilson and Geerd H.F. Diercksen, Plenum Press, New York, 1992

π-electron model of benzene. The first *ab initio* application of this theory (Paldus *et al.*, 1972) showed much promise and the single reference CC theory witnessed much development in the intervening years, primarily thanks to a continuous effort by Bartlett's group (for reviews, see Bartlett, 1981, 1989; Paldus, 1983; Bartlett *et al.*, 1984, Jankowski, 1987; Lindgren and Morrison, 1982; Wilson, 1984).

The successes and the desirable features of the CC method for CS ground states stimulated numerous efforts in trying to develop the generalized version of the theory that would be applicable to arbitrary open-shell (OS) situations, where single reference theory is often inadequate. In contrast to CI methodology, however, such an extension turned out to be most challenging, not only due to numerous conceptual difficulties, but also due to the sheer complexity of the resulting theory that makes practical implementation very demanding. Although impressive progress has been made in this direction and a number of controversies have been resolved, one must nonetheless admit that even now, three decades after the introduction of the CC ansatz, the basic formal structure of OS theory and our knowledge of the cluster structure of general OS wave functions can hardly be regarded as definitively established. Part of the difficulty stems from the fact that in contrast to the single reference case, several options are opened for OS theory. The existing OSCC formalisms may be roughly classified into the three basic categories. The oldest and best developed are the so-called *Fock space* or *valence universal* methods (Coester, 1969; Offerman *et al.*, 1976; Offerman, 1976; Mukherjee *et al.*, 1977, Lindgren, 1978; Ey, 1978; Kutzelnigg, 1984; Haque and Mukherjee, 1984; Stolarczyk and Monkhorst, 1985, 1988; Haque and Kaldor, 1985, 1986; Kaldor, 1987; Lindgren and Mukherjee, 1987; Kutzelnigg *et al.*, 1987; Koch and Mukherjee, 1988; Mukherjee and Pal, 1989; Lindgren, 1991; and references therein). The next group of methods, characterized as the *Hilbert space* or *state-universal* approaches, is based on Jeziorski and Monkhorst (1981) cluster ansatz and recently witnessed its first concrete implementations at various levels of sophistication (Paldus, 1983; Laidig and Bartlett, 1984; Laidig *et al.*, 1987; Jeziorski and Paldus, 1988; Meissner *et al.*, 1988, 1989; Paldus *et al.*, 1989b; Balková *et al.*, 1991; Paldus *et al*, 1991; Piecuch and Paldus, 1991b). Finally, the last group can be labeled as consisting of *one-state* or *state-selective* approaches. This is the most "heterogeneous" group, since it may involve approaches based on the equations of motion or linear response theory (e.g., Monkhorst, 1977; Mukherjee and Mukherjee, 1979; Emrich, 1981; Ghosh *et al.*, 1981, 1982; Dalgaard and Monkhorst, 1983; Ghosh and Mukherjee, 1984; Sekino and Bartlett, 1984; Prasad *et al.*, 1985; Takahashi and Paldus, 1986; Salter *et al.*, 1987; Geertsen and Oddershede, 1986), as well as various single state (but multi-configuration) reference approaches (e.g., Harris, 1977; Paldus *et al.*, 1978; Saute *et al.*, 1979; Nakatsuji and Hirao, 1978; Nakatsuji, 1978, 1979, 1983; Banerjee and Simons, 1981, 1982; Baker and Robb, 1983; Tanaka and Terashima, 1984; Hoffmann and Simons, 1988; Janssen and Schaefer, 1991; Li and Paldus, 1991). These approaches are perhaps least satisfactory from a purely theoretical viewpoint but, on the other hand, are preferable to multi-reference approaches in practical applications, since they avoid simultaneous consideration of a quasi-degenerate manifold of states. An extensive review of various developments in OSCC formalisms was recently given by Mukherjee and Pal (1989).

Unfortunately, it is well beyond the scope of these lecture notes to give justice to all the interesting and useful developments in CC methodology that took place during the past three decades. The CC approach is nowadays being exploited well beyond the atomic and molecular domain, particularly in various problems of condensed matter physics (e.g., Arponen *et al.*, 1987, 1988; Bishop and Arponen, 1990), nuclear physics (Kümmel *et al.*, 1982; Kümmel, 1984), theory of homogeneous electron-liquid (Bishop and Lührmann, 1978, 1982; Arponen and Pajanne, 1982; Emrich and Zabolitzky, 1984), relativistic field theory (e.g., Hsue *et al.*, 1985; Kaulfuss, 1985), and elsewhere. The reader may get some idea about these developments by consulting the proceedings of last year's workshop on CC theory (Bartlett, 1991; in particular, Bishop, 1991) as well as a popular overview by Bishop and Kümmel (1987). We shall restrict ourselves to the basics of CC theory as exploited today in molecular electronic structure calculations, trying to establish the required theoretical foundations of this approach and its relationship to CI and MBPT methods. We shall start with the well understood CS case and only later will try to open a window to the plethora of OS theories. Our primary goal will be to expound theoretical foundations and interrelationships of various approaches rather than to provide an exhaustive overview of existing theories and applications. We thus begin with a brief summary of the notation and formalism and an exploration of the MBPT foundations of CC theory. In subsequent chapters we then touch upon various aspects of both CS and OSCC theories.

2. BASIC NOTATION AND FORMALISM

Since our exposition will employ the second quantization formalism and in view of a close relationship between the CC approaches and the MBPT, we first briefly summarize the required basic facts and establish the notation employed. Nowadays, there exist numerous sources where the reader can familiarize himself with these topics. We shall in particular follow our earlier exposition employing a time-independent approach (Paldus and Čížek, 1975; Paldus, 1981).

2.1. Second quantization formalism

The second quantization formalism associates with a given one-electron orthonormal basis of spin-orbitals $|A\rangle$ a set of annihilation (X_A) and creation ($X_A^\dagger$) operators that satisfy the anticommutation relations

$$\{X_A, X_B\} = \{X_A^\dagger, X_B^\dagger\} = 0, \quad \{X_A, X_B^\dagger\} = \langle A|B\rangle = \delta_{AB}\,, \tag{1}$$

where $\{A, B\} \equiv AB + BA$ and $A^\dagger$ designates the Hermitian conjugate of A. These operators act in the so-called *Fock space* $\mathcal{F}$, given by the direct sum of N-particle spaces $\mathcal{H}_N$,

$$\mathcal{F} = \bigoplus_{N=0}^{\infty} \mathcal{H}_N\,, \tag{2}$$

including the one-dimensional zero-particle space $\mathcal{H}_0$, spanned by the vacuum state $|0\rangle$, with properties

$$X_A|0\rangle = 0\,, \qquad \langle 0|0\rangle = 1\,. \tag{3}$$

The N-electron subspace $\mathcal{H}_N$ of $\mathcal{F}$ can be viewed as an antisymmetric component of the N-th tensor power of the defining single-particle space $\mathcal{H}_1 = \text{Span}\{|A\rangle\}$, i.e. $\mathcal{H}_N = \mathcal{A}(\mathcal{H}_1^{\otimes N})$, and may thus be represented as a linear span of N-electron configuration states $|\{A_1 A_2 \cdots A_N\}\rangle$,

$$|\{A_1 A_2 \cdots A_N\}\rangle = \sum_{P\in S_N} (-1)^p |A_{p1} A_{p2} \cdots A_{pN}\rangle\,, \tag{4}$$

where

$$|A_{i_1}A_{i_2}\cdots A_{i_N}\rangle = |A_{i_1}\rangle|A_{i_2}\rangle\cdots|A_{i_N}\rangle \equiv |A_{i_1}\rangle\otimes|A_{i_2}\rangle\otimes\cdots\otimes|A_{i_N}\rangle\,, \qquad (4')$$

and the sum extends over all $N!$ permutations $P : i \mapsto p_i$ of the symmetric group S_N. The parity of P is designated by $(-1)^p$. Each such configuration is related to the vacuum state $|0\rangle$ as follows

$$|\{A_1A_2\cdots A_N\}\rangle = X^\dagger_{A_1}X^\dagger_{A_2}\cdots X^\dagger_{A_N}|0\rangle\,, \qquad (5)$$

so that any N-electron state can be represented in terms of creation and annihilation operators and of the vacuum state $|0\rangle$ (or, for bra states, its dual $\langle 0|$).

Moreover, any particle-number preserving*) k-body operator $\hat{\Theta}_k$ on $\mathcal{H}_N$,

$$\hat{\Theta}_k = \sum_{i_1<i_2<\cdots<i_k=1}^{N} \hat{\theta}_k(\mathbf{x}_{i_1},\mathbf{x}_{i_2},\cdots,\mathbf{x}_{i_k})\,, \qquad (6)$$

can be extended to $\mathcal{F}$ and expressed in the form

$$\begin{aligned}\hat{\Theta}_k &= (k!)^{-2}\sum_{\substack{A_1\cdots A_k\\B_1\cdots B_k}}\langle A_1\cdots A_k|\hat{\theta}_k|B_1\cdots B_k\rangle_{\mathcal{A}}\,X^\dagger_{A_1}\cdots X^\dagger_{A_k}X_{B_k}\cdots X_{B_1}\\ &= (k!)^{-1}\sum_{\substack{A_1\cdots A_k\\B_1\cdots B_k}}\langle A_1\cdots A_k|\hat{\theta}_k|B_1\cdots B_k\rangle\,X^\dagger_{A_1}\cdots X^\dagger_{A_k}X_{B_k}\cdots X_{B_1}\,,\end{aligned} \qquad (6')$$

where the antisymmetrized and standard matrix elements are related as follows

$$\langle A_1\cdots A_k|\hat{\theta}_k|B_1\cdots B_k\rangle_{\mathcal{A}} = \sum_{P\in S_k}(-1)^p\langle A_1\cdots A_k|\hat{\theta}_k|B_{p_1}\cdots B_{p_k}\rangle\,, \qquad (7)$$

and

$$\begin{aligned}&\langle A_1\cdots A_k|\hat{\theta}_k|B_1\cdots B_k\rangle =\\ &= \int d^4x_1\cdots\int d^4x_k\,\phi^*_{A_1}(\mathbf{x}_1)\cdots\phi^*_{A_k}(\mathbf{x}_k)\hat{\theta}_k(\mathbf{x}_1,\cdots,\mathbf{x}_k)\phi_{B_1}(\mathbf{x}_1)\cdots\phi_{B_k}(\mathbf{x}_k)\,,\end{aligned} \qquad (8)$$

where

$$\phi_A(\mathbf{x}) = \langle\mathbf{x}|A\rangle, \qquad \mathbf{x} = (\mathbf{r}, s)\,, \qquad (9)$$

$\mathbf{r}$ and s designating the position and spin coordinates, respectively and $\int d^4x$ indicates integration over spatial coordinates and summation over spin. Thus, for example, for a standard atomic or molecular electronic Hamiltonian, involving at most two-body interactions, we have

$$H = Z + V = \sum_{A,B}\langle A|\hat{z}|B\rangle X^\dagger_A X_B + \tfrac{1}{2}\sum_{ABCD}\langle AB|\hat{v}|CD\rangle X^\dagger_A X^\dagger_B X_D X_C\,. \qquad (10)$$

*) In fact, the second quantization formalism was primarily designed to handle the systems in which the particle number may not be conserved, particularly the problems arising in quantum field theory and quantum statistics. However, this formalism is also very useful for particle-number conserving systems, such as considered in the following text.

Any expression involving N-particle wave functions and particle-number conserving operators can thus be written in the second quantized form in terms of creation and annihilation operators and the vacuum state $|0\rangle$ and/or its dual $\langle 0|$, in addition to, of course, scalar factors involving operator matrix elements (7) or (8) and coefficients arising from the expansion of wave functions in terms of configurations (4) or (5). The antisymmetry of the wave functions involved is reflected in the anticommutation relations (1). Consequently, all the combinatorial problems that are associated with the antisymmetry of the wave function or configuration states (4) are now automatically taken care of when the fermionic algebra of anticommuting creation and annihilation operators is employed.

The actual algebraic manipulations are greatly simplified by the property (3) (and its Hermitian conjugate $\langle 0|X_A^\dagger = 0$) and by the so-called time-independent Wick's theorem. To state this theorem one introduces the normal ordered product, in which all the creators are to the left of all the annihilators,

$$N[M_1 \cdots M_k] = (-1)^p X_{B_1}^\dagger \cdots X_{B_r}^\dagger X_{B_{r+1}} \cdots X_{B_k}\,, \tag{11}$$

where $M_i = X_{A_i}$ or $M_i = X_{A_i}^\dagger$ and $(-1)^p$ is again the parity of the permutation $P : i \equiv A_i \mapsto B_i$, bringing the product $(M_1 \cdots M_k)$ to its N-product form. Thus, the N-product operation rearranges the operators into the desired order assuming that all operators exactly anticommute. The advantage of this concept stems from the fact that, for example, the vacuum mean value of such a product must vanish as long as $k \geq 1$. For completeness, we also define the N-product of an empty set of operators as $N[\emptyset] = 1$. We next define the contraction of two operators as

$$\underbrace{M_1 M_2} = M_1 M_2 - N[M_1 M_2]\,. \tag{12}$$

Clearly, all contractions vanish except for $\underbrace{X_A X_A^\dagger} = 1$ (that is precisely the case when the operators do not exactly anticommute, see Eq. (1)). Defining, finally, the N-product with contractions (N-product of operators some of which are contracted) as a product of these contractions and the N-product of uncontracted operators (where again we assume exact anticommutation except that the order of contracted operators must not change), we can state Wick's theorem as follows

$$M_1 \cdots M_k = N[M_1 \cdots M_k] + \sum N[M_1 \underbrace{\cdots\cdots\cdots} M_k]\,, \tag{13}$$

where the sum extends over all possible N-products with contractions. Clearly, when we evaluate the vacuum mean value $\langle 0|M_1 \cdots M_k|0\rangle$, only fully contracted terms may contribute.

Wick's theorem can be generalized to products in which some of the operators are already in the N-product form. In such cases, one must drop on the right hand side of (13) all the terms involving contractions between operators that appear in the same N-product on the left hand side. Thus, using the N-product form of operators, we can decrease the number of contraction schemes that must be taken into account.

2.2. Particle-hole formalism

When referring all the configurations to the true vacuum state $|0\rangle$, we need as many creators as there are particles. This can become very cumbersome for large N and unmanageable for extended systems when $N \to \infty$. It is thus advantageous to employ a more appropriate reference state, which involves the same number of particles as the studied system (or an average number of particles when particle number is not conserved). From a practical viewpoint we should of course choose for a reference the configuration that best describes the state we consider. We refer to the spin-orbitals that constitute such a reference configuration as forming a Fermi sea and divide the entire spin-orbital set into the occupied (or hole) and unoccupied (or virtual, or particle) spin-orbitals. Let us designate at this stage occupied and virtual spin-orbitals by subscripted (A_i) and superscripted (A^i) labels, respectively, while the generic spin-orbitals will be labeled by unsubscripted letters ($A, B, C, \ldots$).

Assume, thus, that $|\Phi_0\rangle$ is a new reference state,

$$|\Phi_0\rangle = |\{A_1 \cdots A_N\}\rangle \tag{14}$$

and re-adapt accordingly the second quantization formalism. Designating the particle-hole (p-h) formalism operators by the letter Y, we reverse the role of creation and annihilation operators for the hole states and preserve it for the particle states, i.e.

$$\begin{aligned} Y_{A_i} &= X_{A_i}^{\dagger}, \quad Y_{A_i}^{\dagger} = X_{A_i}, \\ Y_{A^i} &= X_{A^i}, \quad Y_{A^i}^{\dagger} = X_{A^i}^{\dagger}. \end{aligned} \tag{15}$$

The p-h operators Y_A, $Y_A^{\dagger}$ satisfy the same anticommutation relations as the operators $X_A, X_A^{\dagger}$, and the annihilation condition, Eq. (3), now takes the form

$$Y_A|\Phi_0\rangle = 0 \qquad \text{or} \qquad \langle\Phi_0|Y_A^{\dagger} = 0\,, \tag{16}$$

for any A. We can thus follow the same development as in the preceding section, redefining accordingly the relevant concepts.

Thus, we define the normal product relative to $|\Phi_0\rangle$ as in (11),

$$n[M_1 \cdots M_k] = (-1)^p\, Y_{B_1}^{\dagger} \cdots Y_{B_r}^{\dagger}\, Y_{B_{r+1}} \cdots Y_{B_k}\,, \tag{17}$$

and designate it by the lower case letter n in order to distinguish it from the pure vacuum case (11). Similarly, we define p-h contractions (placing now the contraction symbol above the operators)

$$\overbrace{M_1 M_2} = M_1 M_2 - n[M_1 M_2]\,, \tag{18}$$

and n-products with contractions, $n[M_1 \cdots\cdots\cdots M_k]$. In terms of these p-h concepts, the Wick theorem, or its generalized form, take the same form as in Section 2.1.

It is straightforward to carry out the transition (15) from the true vacuum operators X_A, $X_A^{\dagger}$ to the p-h ones, Y_A, $Y_A^{\dagger}$, as long as we know the character of the spin-orbital A involved. This is not the case in general, as for example in the operator representation (6′) or (10). We could, of course, split each spin-orbital sum into separate summations over the hole and particle labels and carry out the transition for each specific case. It is, however, much easier, to simply keep the true vacuum operators X_A, $X_A^\dagger$, and work out the corresponding p-h contractions (18). For example, we easily find that

$$\begin{aligned} \overbracket{X_A X_B} &= \overbracket{X_A^\dagger X_B^\dagger} = 0 \,, \\ \overbracket{X_A^\dagger X_B} &= h(A)\langle A|B\rangle \,, \\ \overbracket{X_A X_B^\dagger} &= p(A)\langle A|B\rangle \,, \end{aligned} \tag{19}$$

where we introduced the p-h "step-functions"

$$\begin{aligned} h(A_i) &= 1,\ h(A^i) = 0 \,, \\ p(A_i) &= 0,\ p(A^i) = 1 \,. \end{aligned} \tag{20}$$

The reader can easily find the p-h contractions for mixed cases involving both operators X and Y.

We note, finally, that in order to benefit from the generalized Wick theorem, we bring the operators involved to their n-product form. We recall that for the Hamiltonian (10) we get (cf., Paldus and Čížek, 1975)

$$\begin{aligned} \hat{H} &= \hat{H}_N + \langle \hat{H} \rangle \,, \qquad \langle \hat{H} \rangle \equiv \langle \Phi_0 | \hat{H} | \Phi_0 \rangle \\ \hat{H}_N &= \hat{F}_N + \hat{V}_N \,, \qquad \hat{F}_N = \hat{Z}_N + \hat{G}_N \end{aligned} \tag{21}$$

$$\hat{\Xi}_N = \sum_{AB} \langle A|\hat{\xi}|B\rangle\, n[X_A^\dagger X_B] \,, \qquad \hat{\Xi} = \hat{F}, \hat{Z}, \hat{G};\ \hat{\xi} = \hat{f}, \hat{z}, \hat{g} \tag{21'}$$

$$\hat{V}_N = \tfrac{1}{2} \sum_{ABCD} \langle AB|\hat{v}|CD\rangle\, n[X_A^\dagger X_B^\dagger X_D X_C] \,, \tag{21''}$$

$$\langle A|\hat{f}|B\rangle = \langle A|\hat{z}|B\rangle + \langle A|\hat{g}|B\rangle \,, \tag{22}$$

$$\langle A|\hat{g}|B\rangle = \sum_{C_1} \langle AC_1|\hat{v}|BC_1\rangle_{\mathcal{A}} \,, \tag{22'}$$

$$\langle AB|\hat{v}|CD\rangle_{\mathcal{A}} = \langle AB|\hat{v}|CD\rangle - \langle AB|\hat{v}|DC\rangle \,, \tag{7'}$$

the last equation being a special case of the definition (7).

2.3. Diagrammatic representation

As the relationships (16) and (19) imply, only very few contraction schemes actually yield a nonvanishing contribution when we employ Wick's theorem in any particular case. For both computational and conceptual reasons, it is very convenient to visualize these contraction schemes graphically (Čížek, 1966a, 1969; Paldus and Čížek, 1975; Paldus, 1981). To this effect, we represent creators and annihilators $X_A^\dagger$ and X_A, in terms of which our operators are expressed, by oriented lines that, respectively, leave and enter various vertices, which in turn indicate their origin, and with which we associate appropriate scalar factors. In this exposition we need to represent certain one- and two-body operators, stemming from the Hamiltonian (10) or (21), along with various k-body excitation operators, or closely related cluster operators, of the general form

$$\begin{aligned}\hat{\Gamma}_k &= (k!)^{-1} \sum_{\substack{A^1\cdots A^k\\ B_1\cdots B_k}} \langle A^1\cdots A^k|\hat{\gamma}|B_1\cdots B_k\rangle \prod_{i=1}^{k}(X^\dagger_{A^i}X_{B_i})\\ &= (k!)^{-2} \sum_{\substack{A^1\cdots A^k\\ B_1\cdots B_k}} \langle A^1\cdots A^k|\hat{\gamma}|B_1\cdots B_k\rangle_{\mathcal{A}} \prod_{i=1}^{k}(Y^\dagger_{A^i}Y^\dagger_{B_i}),\end{aligned} \tag{23}$$

each operator pair $Y^\dagger_{A^i}Y^\dagger_{B_i} = X^\dagger_{A^i}X_{B_i}$ exciting a particle from the spin-orbital $|B_i\rangle$ to $|A^i\rangle$ or, equivalently, creating a particle-hole pair (A^i, B_i). The Goldstone and Hugenholtz form of the diagrams that we associate with these quantities is shown in Fig. 1. Then, in order to evaluate the desired expressions of the type $\hat{\Xi}_i\cdots\hat{\Xi}_j\hat{\Gamma}_k\cdots\hat{\Gamma}_\ell|\Phi_0\rangle$, or the matrix elements of the form $\langle\Phi_0|\hat{\Gamma}^\dagger_i\hat{\Xi}_j\cdots\hat{\Xi}_k\hat{\Gamma}_\ell\cdots\hat{\Gamma}_m|\Phi_0\rangle$, we simply represent each operator $\hat{\Xi}_r$ or $\hat{\Gamma}_s$ by an appropriate diagram (Fig. 1) and apply Wick's theorem, depicting individual contractions by interconnecting corresponding lines representing the individual creation and annihilation operators. It is easily realized that only those

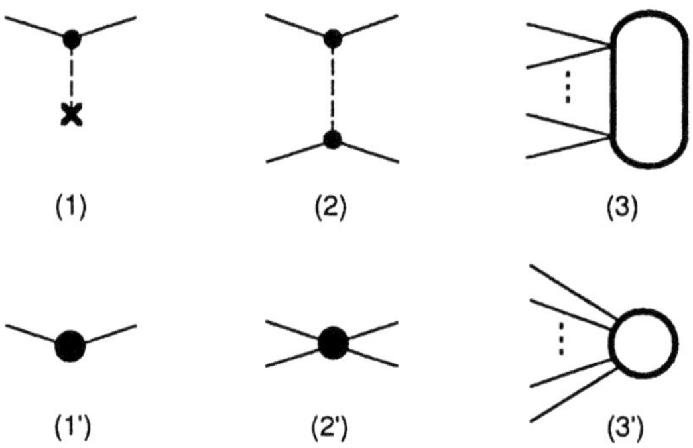

Fig. 1. Goldstone (1-3) and Hugenholtz (1′-3′) (unlabeled and non-oriented) vertices representing one-body (1,1′) and two-body (2,2′) terms of the Hamiltonian (10) or (21) [in the latter case, F-vertices are represented by full circles rather than by crosses], also referred to as *interaction vertices*, and a k-body (or kp-kh) excitation operator $\hat{\Gamma}_k$, Eq. (23), or $\hat{C}^{(k)}_{K_j} \equiv C_j$, Eq. (27); or corresponding configuration $|\Phi_j\rangle = \hat{C}_j|\Phi_0\rangle$, or a linear combination of such configurations given by $\hat{\Gamma}_k|\Phi_0\rangle$, which are referred to as *excitation vertices* (3, 3′). The same vertices are also used to represent k-body cluster operators (cf., Section 4.5).

contractions need be considered that preserve the orientation of lines. Each contraction is thus represented by an oriented *internal line* leaving and entering some vertex (when the operators are in the n-product form, no internal lines interconnecting their vertices are allowed), while the uncontracted operators are represented by *external lines.* The spin-orbital labels that we attach to oriented lines are referred to as *free* if they represent summation indices and as *fixed* otherwise. Stripping the diagram of free labels we obtain a *skeleton*, whose group of automorphisms G determines the *weight* w of the diagram, $w = |G|^{-1}$. We also see from Eqs. (19) that the *hole lines* are oriented from left to right, while the *particle lines* extend in the opposite direction. Any line issuing and ending on the vertices representing the same operator must be a hole line. Moreover, in view of Eq. (16), no external line can extend in the direction of the vacuum state $|\Phi_0\rangle$. Thus, all the *resulting diagrams* (*R-diagrams*) representing nonvanishing contraction schemes that result when the Wick theorem is applied, may only have internal lines when we calculate Fermi vacuum mean values, $\langle\Phi_0|\Gamma\cdots\Xi\cdots\Gamma'|\Phi_0\rangle$, and are referred to as *vacuum diagrams*. Similarly, when calculating expressions of the type $\Gamma\cdots\Xi\cdots\Gamma'|\Phi_0\rangle$, no external line can extend to the right.

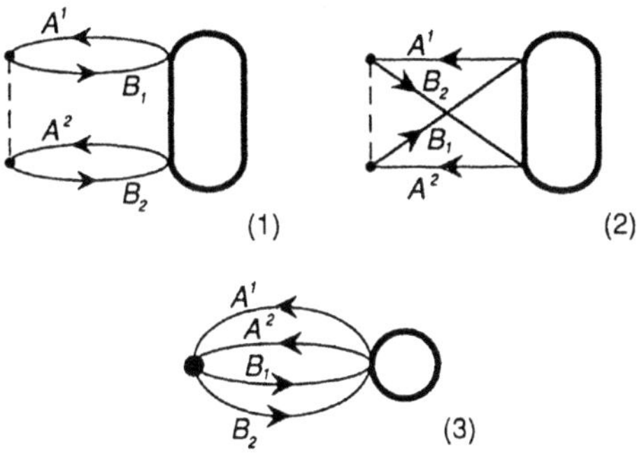

Fig. 2. Resulting Goldstone (1,2) and Hugenholtz (3) diagrams representing the mean value $\Delta E = \langle\Phi_0|\hat{H}_N\hat{T}_2|\Phi_0\rangle \equiv \langle H_N T_2\rangle$, Eq. (24).

Thus, a general procedure for evaluation of expressions consisting of various operators $\hat{\Omega}_i$ and Fermi vacuum state(s) $|\Phi_0\rangle$ (or $\langle\Phi_0|$) can be summarized as follows:

(i) Represent the operators $\hat{\Omega}_i$ by appropriate skeletons (cf. Fig. 1), placing them in the same left-right order as found in the product $\hat{\Omega}_1\hat{\Omega}_2\cdots$.

(ii) Form all permissible R-skeletons and label their oriented lines with spin-orbital indices to form all distinct R-diagrams. (Clearly, only one spin-orbital label per each oriented line is required; only when both labels on a contracted line are fixed do we have an internal line labeled by two indices. With each such doubly indexed line we associate the Kronecker delta involving these indices. The spin-orbital labels must reflect the hole or particle character of the oriented lines.)

(iii) Determine the weight w of the resulting skeleton, associate the scalar factor with each operator vertex (outgoing lines label the bra state and ingoing lines the ket state) and particle-hole operator pair $X_A^\dagger X_B$ with each pair of external lines, labeled by A and B, that terminate and initiate an open path of oriented lines, sum over all free labels and attach the phase factor $(-1)^{h+\ell}$, where h designates the number of internal hole lines and ℓ the number of closed loops of oriented lines.

Thus, for example, with the Goldstone type diagrams (1) and (2) of Fig. 2 that represent possible R-diagrams when evaluating the energy mean value $\Delta E = \langle\Phi_0|\hat{H}_N\hat{T}_2|\Phi_0\rangle$, where $\hat{V}_N$ is given by Eq. (21″) and represented by diagram (2) of Fig. 1, while $\hat{T}_2$ is given by Eq. (23) for $k = 2$, $\hat{\Gamma} = \hat{T}$ and is represented by the excitation type diagram (3) of Fig. 1, we get

$$\Delta E = \tfrac{1}{2} \sum_{A^1,A^2,B_1,B_2} \left(\langle B_1B_2|\hat{v}|A^1A^2\rangle - \langle B_1B_2|\hat{v}|A^2A^1\rangle\right)\langle A^1A^2|\hat{t}_2|B_1B_2\rangle\,, \tag{24}$$

since $w = \frac{1}{2}$, $h = 2$, and $\ell = 2$ and 1 for the first and the second R-diagram, Fig. 2, respectively.

To reduce the number of R-diagrams, we can employ Hugenholtz (or degenerate) diagram representation. For example, in the example just considered, we obtain only one R-diagram, Fig. 2(3). With Hugenholtz vertices we must associate antisymmetrized matrix elements. Unfortunately, Hugenholtz diagrams do not determine uniquely the phase of the resulting expressions. This drawback can be easily avoided when we represent each Hugenholtz diagram by one of its Goldstone representatives (any such diagram can be used), in which case we usually refer to them as Brandow diagrams. Thus, considering diagram (1) of Fig. 2 as a Brandow diagram (i.e., a Goldstone representative of the Hugenholtz diagram (3)), we get

$$\Delta E = \tfrac{1}{4} \sum_{A^1,A^2,B_1,B_2} \langle B_1B_2|\hat{v}|A^1A^2\rangle_A\langle A^1A^2|\hat{t}_2|B_1B_2\rangle_A\,, \tag{24′}$$

since now $w = \frac{1}{4}$. It is easily seen that this expression is equivalent to the former one, Eq. (24). The reader should verify that the same result is also obtained when considering the second Goldstone representative, Fig. 2(2), as a Brandow diagram.

3. MBPT ORIGINS OF THE CLUSTER ANSATZ

Using the diagrammatic technique just described, we can obtain an efficient and perspicuous formulation of the many-body perturbation theory (MBPT). For details the reader is referred to our earlier expositions (Paldus and Čížek, 1975; Paldus, 1981).

3.1. Time independent MBPT

Taking the Hamiltonian (10) in the n-product form, Eq. (21), we choose the separable one-electron part $\hat{F}_N$ as an unperturbed Hamiltonian $\hat{H}_0$, and the two-electron part $\hat{V}_N$ as a perturbation $\hat{W}$, so that (for details, see Paldus and Čížek, 1975 or Paldus, 1981)

$$\begin{aligned} \hat{H}_N &= \hat{H} - \langle\hat{H}\rangle = (\hat{H}_0)_N + \hat{W}\,, \\ (\hat{H}_0)_N &= \hat{H}_0 - \langle\hat{H}_0\rangle = \hat{F}_N = \hat{Z}_N + \hat{G}_N; \qquad \hat{W} = \hat{V}_N\,. \end{aligned} \tag{25}$$

The unperturbed eigenvalue problem

$$\hat{H}_0|\Phi_j\rangle = E_j^{(0)}|\Phi_j\rangle\,, \tag{26}$$

reduces to the one-electron problem (in fact the Hartree-Fock problem)

$$\hat{f}|B\rangle = \omega_B|B\rangle \tag{26′}$$

in view of the separability of $\hat{H}_0$. Here we assume that we have chosen $|\Phi_0\rangle = |\{A_1 \cdots A_N\}\rangle$ as a reference, with the energy $E_0^{(0)} = \sum_{i=1}^{N} \omega_{A_i}$. Thus, any unperturbed eigenstate $|\Phi_j\rangle$, $j \neq 0$, Eq. (26), represents an excited state configuration that may be expressed in terms of this reference state, namely

$$|\Phi_j\rangle \equiv |\Phi_{K_j}^{(k)}\rangle = \left| \begin{matrix} B^1 \cdots B^k \\ B_1 \cdots B_k \end{matrix} \right\rangle = \left(\prod_{i=1}^{k} X_{B^i}^{\dagger} X_{B_i} \right) |\Phi_0\rangle = \hat{C}_{K_j}^{(k)} |\Phi_0\rangle \,, \tag{27}$$

where $K_j \equiv \{B^1 \cdots B^k; B_1 \cdots B_k\}$ designates the index set defining the excitation operator $\hat{C}_{K_j}^{(k)} = \prod_{i=1}^{k}(Y_{B^i}^{\dagger} Y_{B_i}^{\dagger})$ and the corresponding configuration $|\Phi_{K_j}^{(k)}\rangle$, k being the excitation order, while the corrresponding eigenvalue $E_j^{(0)}$ is given by

$$E_j^{(0)} - E_0^{(0)} \equiv -\Delta_j = \sum_{i=1}^{k} (\omega_{B^i} - \omega_{B_i}) \,. \tag{28}$$

Expanding then the exact eigenvalues ΔE_i and eigenstates $|\Psi_i\rangle$ of $\hat{H}_N$ in the Rayleigh-Schrödinger perturbation series (for simplicity we consider the ground state case, $i = 0$, drop the subscript 0 at E's and the circumflex over the operator symbols) we get

$$\Delta E = \sum_{n=0}^{\infty} \Delta E^{(n)} \,, \tag{29}$$

$$|\Psi_0\rangle = \sum_{n=0}^{\infty} |\Psi_0^{(n)}\rangle \,, \tag{30}$$

the superscript (n) indicating the order of perturbation theory. The actual expressions for $\Delta E^{(n)}$ and $|\Psi_0^{(n)}\rangle$ are most easily found using the bracketing technique: thus, $\Delta E^{(n)}$ is given by the so-called principal term $\langle W(PW)^{n-1}\rangle$, plus the renormalization term $R^{(n)}$ that is given by the sum of all possible "bracketings" of the principal term, taken with the sign $(-1)^k$, k being the number of bracketings. Each "bracketing" means taking the Fermi vacuum mean value of any contiguous subset of operators $W(PW)^i$ in the principal term, excluding the first and the last W. Here,

$$P = \frac{1 - |\Phi_0\rangle\langle\Phi_0|}{E^{(0)} - H_0} = \sum_{j \neq 0} \Delta_j^{-1} |\Phi_j\rangle\langle\Phi_j| \,, \tag{31}$$

so that $P|\Phi\rangle \equiv P_0|\Phi_0\rangle = 0$, with the denominator given by Eq. (28), and bracketings within the bracketings are allowed. Thus, generally

$$\Delta E^{(n)} = \langle W(PW)^{n-1}\rangle + R^{(n)} \,. \tag{32}$$

The corresponding eigenstate correction $|\Psi_0^{(n)}\rangle$ is then obtained by taking $\Delta E^{(n+1)}$ and removing the leftmost W together with the bra $\langle\Phi_0|$. Thus, for the first four orders we have (clearly $\Delta E^{(0)} = 0$)

$$\Delta E^{(1)} = \langle W\rangle = 0, \qquad |\Psi_0^{(0)}\rangle = |\Phi_0\rangle \,, \tag{33a}$$

$$\Delta E^{(2)} = \langle WPW\rangle, \qquad |\Psi_0^{(1)}\rangle = PW|\Phi_0\rangle \,, \tag{33b}$$

$$\Delta E^{(3)} = \langle W(PW)^2\rangle, \qquad |\Psi_0^{(2)}\rangle = (PW)^2|\Phi_0\rangle \,, \tag{33c}$$

$$\Delta E^{(4)} = \langle W(PW)^3\rangle - \langle WPW\rangle\langle WP^2W\rangle,\ |\Psi_0^{(3)}\rangle = \Big[(PW)^3 - \langle WPW\rangle P^2W\Big]|\Phi_0\rangle, \tag{33d}$$

and similarly for higher orders. It is useful to note that the total number of bracketings (including the principal term) in the n-th order energy term (32) equals $(2n-2)!/[n!(n-1)!]$ (Tong, 1962).

3.2. Diagrammatic MBPT

Representing expressions (33) diagramatically, we find that the nth order principal energy term is given by all possible vacuum diagrams having n two-body interaction vertices representing $W = V_N$ [Fig. 1(2) or 1(2′)]. These diagrams may be either connected or disconnected. However, in view of the presence of the operator P, in every permissible R-diagram there must be at least one (in fact two) oriented line(s) passing in the space between any two interaction vertices, and the evaluation rules given in Section 2.3 are supplemented by the denominator rule, assigning to every pair of neighboring interaction vertices a denominator Δ_j^{-1}, Eq. (28), where the spin-orbital energies ω_{B^i} and ω_{B_i} are determined by the labels carried by the particle and hole fermion lines passing through the interval separating these vertices. Since our perturbation W is in the n-product form, all internal lines must interconnect distinct interaction vertices (in this way we have eliminated all the so-called Hartree-Fock diagrams as well as $\Delta E^{(1)}$).

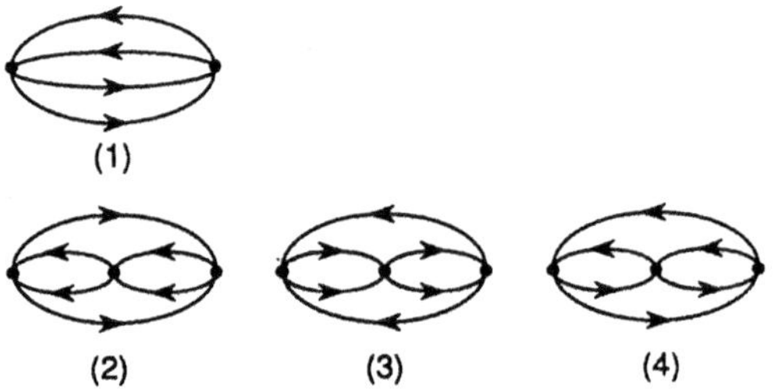

Fig. 3. Hugenholtz skeletons representing the second (1) and the third (2-4) order MBPT energy contributions $\Delta E^{(2)}$, Eq. (33b), and $\Delta E^{(3)}$, Eq. (33c), respectively.

Clearly, for $n \leq 3$, only connected diagrams can appear. The possible Hugenholtz skeletons, representing the second and third order energy, are shown in Fig. 3. The disconnected diagrams that can occur in the fourth order are shown in Fig. 4. It is not difficult to see that, in fact, their contribution exactly cancels the contribution from the renormalization term $\langle WPW\rangle\langle WP^2W\rangle$, which equals the product of the second order term $\Delta E^{(2)}$ times a similar term, in which all the denominators are squared. Now, since the numerators of both diagrams in Fig. 4 are identical, we only examine their denominator part. Let us designate the denominator associated with the top disconnected part by a and that with the bottom part by b. Taking the sum of the contributions from both diagrams, we clearly get (deleting the common numerator factor)

$$\frac{1}{b}\cdot\frac{1}{a+b}\cdot\frac{1}{b}+\frac{1}{a}\cdot\frac{1}{a+b}\cdot\frac{1}{b}=\left(\frac{1}{b}+\frac{1}{a}\right)\frac{1}{b(a+b)}=\frac{1}{ab^2}\,. \tag{34}$$

This is precisely the expression that we get in the renormalization part. We thus see that the *disconnected* diagrams arising from the principal term are exactly cancelled by the renormalization terms. This result will be seen to hold in general, so that we can write

$$\Delta E^{(n)} = \langle W(PW)^{n-1}\rangle_L , \tag{35}$$

where the subscript L designates the connected or linked component. A diagram is referred to as *unlinked*, if it contains a disconnected vacuum component. For energy diagrams, *linked* is thus synonymous with *connected.*

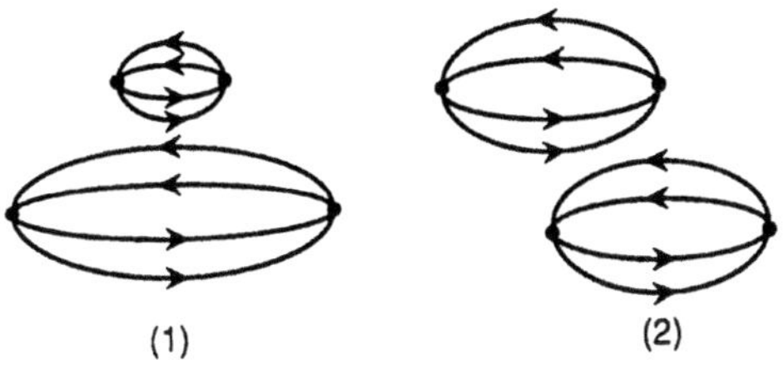

Fig. 4. Hugenholtz skeletons representing the fourth order MBPT disconnected or unlinked energy contributions that are cancelled by the renormalization term $R^{(4)}$, Eq. (33d). See text for details.

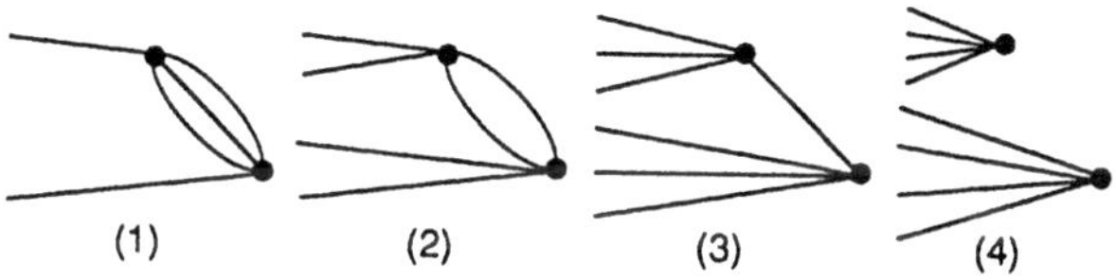

Fig. 5 Hugenholtz skeletons representing the second order MBPT component of the wave function $|\Psi_0^{(2)}\rangle$, Eq. (33c).

Considering, next, the wave function contributions, Eqs. (33), we find that the principal term is represented by all possible diagrams with no external lines extending to the right but at least one line extending to the left. We then also attach the denominator, originating from the external oriented lines extending beyond the leftmost vertex, when evaluating these diagrams. For example, in the second order, we get the skeletons that are obtained from those of Fig. 5 by orienting the lines in all permissible ways. The number of external lines determines the excitation order of the contribution ($2k$ external lines for the k-times excited configurations). Although the last diagram (4) of Fig. 5 is disconnected, it is linked and must be included when evaluating $|\Psi_0^{(2)}\rangle$. In the third order, however, we also get unlinked contributions, which may be simply obtained from those of Fig. 4 by removing the leftmost interaction vertex. An exactly the same reasoning as given above, Eq. (34), shows that these unlinked terms will again exactly cancel the renormalization term in Eq. (33d). Thus, the n-th order contribution to the wave function $|\Psi_0^{(n)}\rangle$ is given by all linked diagrams. In fact, a similar factorization, as illustrated by Eq. (34), can also be applied to the disconnected but linked wave function diagrams, thus revealing the connected cluster structure of the exact wave function, as will be shown in Section 3.3. However, before concluding this section, the following remark should be made.

When evaluating the MBPT diagrams, we sum independently over all spin-orbital labels (since all labels are free labels in this case). We thus include terms that correspond to states in which a given spin-orbital is occupied more than ones. Clearly, all such terms must vanish in view of the Pauli principle (i.e., the fact that $(X_A^\dagger)^2 = X_A^2 = 0$) and we could eliminate them by appropriately restricting the summation labels. We refer to these terms as the *exclusion principle violating* (EPV) *terms.* In the diagrammatic representation these terms are easily recognized as having two identically labeled particle or hole lines at some point. It may be shown that, generally, the EPV contributions cancel out. However, it may happen, that the cancelling terms are of a different type, i.e. one linked and the other unlinked, as illustrated in Fig. 6. Individually, these diagrams do not vanish, but cancel out one another [diagram (i) has one more closed loop than diagram (ii)]. Thus, once we eliminate all the unlinked diagrams due to the cancellation by renormalization terms, we must keep the EPV terms in the linked component as well. This will be automatically the case if we consider all summations over the spin-orbital labels to be unrestricted, which is in fact very convenient from the computational point of view.

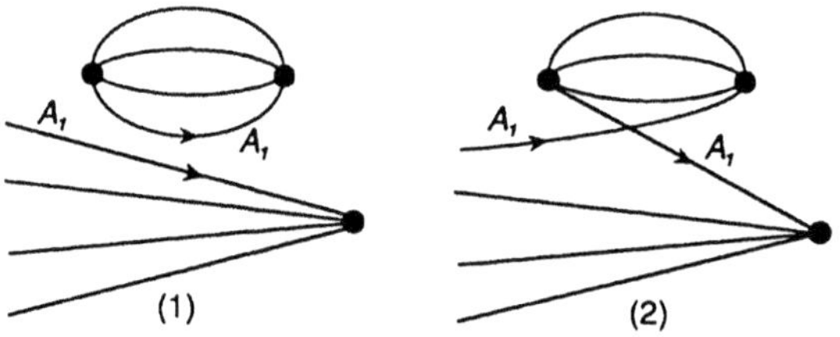

Fig. 6. An illustration of the EPV terms. See text for details.

3.3. Factorization Lemma

The factorization of unlinked diagrams that leads to the cancellation of renormalization terms is but a special case of a general result that holds for any disconnected diagram, linked or unlinked. Before we state this general result, let us introduce some useful concepts and consider a simple example.

Diagrams that only differ in the left-right ordering of their vertices are referred to as different *time-versions* of the same diagram (this terminology originates in the time-dependent version of the MBPT). If a disconnected vacuum part is involved, the diagram with a vanishing denominator (called eufemistically a "dangerous denominator") is disallowed. Moreover, if in the process that permutes the left-right order of vertices no fermion line changes its particle or hole character, we get a *time-version of the first kind.*

As an example, consider a wave-function-type 3rd order linked diagram consisting of two disconnected parts (1a) and (1b) of Fig. 7, whose denominators a, b and c are indicated by vertical dashed lines (for simplicity we suppress line orientation). In this case, three distinct time versions of the first kind [Fig. 7(2-4)] can be formed. The matrix elements, weights, operators associated with external lines and phase factors being identical in each case, we consider only the denominator part, obtaining

$$\frac{1}{(a+b)bc} + \frac{1}{(a+b)(a+c)c} + \frac{1}{(a+b)(a+c)a} = \frac{1}{a} \cdot \frac{1}{bc} . \qquad (36)$$

Thus, again, the total contribution from these three time-versions factorizes into a product of contributions that are associated with its disconnected parts, Fig. 7(1a,1b).

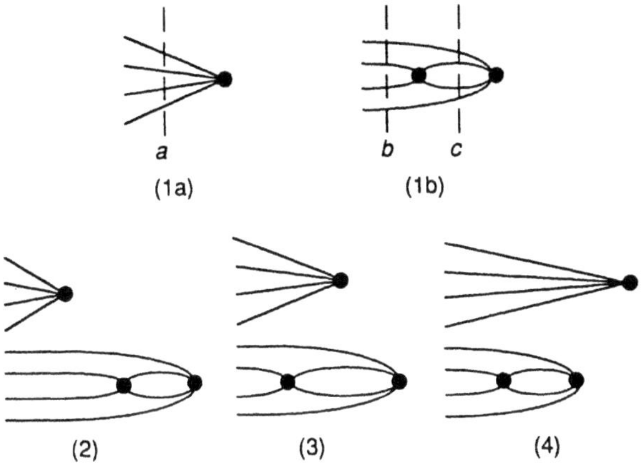

Fig. 7. An illustration of the factorization lemma using the 3rd order MBPT wave function diagrams of the type shown (see text for details). The disconnected parts of these 3rd order linked diagrams (2-4) are represented by diagrams (1a) and (1b). The denominators of disconnected parts, designated as a, b and c, are indicated by vertical dashed lines.

To prove the factorization in general, we apply mathematical induction (Frantz and Mills, 1960). We consider all possible time versions (of the first kind) for a linked diagram consisting of two disconnected parts a and b, which themselves, for the sake of generality, may be disconnected. We designate their energy denominators by Δ_i^a, $i = 1, \ldots, m$ and Δ_i^b, $i = 1, \ldots, n$, respectively, numbering them along the time axis (i.e., the rightmost denominators are Δ_1^a and Δ_1^b). The denominator contribution from possible time-versions of the first kind, corresponding to all available orderings of interaction vertices in both disconnected parts relative to one another (the numerator part being identical in all cases) can be written as

$$D_{mn}^{ab} = \sum_{\{\alpha,\beta\}} \prod_{p=1}^{m+n} (\Delta_{\alpha(p)}^a + \Delta_{\beta(p)}^b)^{-1} , \qquad (37)$$

where the summation extends over all sets of $(m+n)$ integer pairs $\Gamma_p := (\alpha(p), \beta(p))$ defined as follows:

(i) either $\Gamma_1 = (1,0)$ or $\Gamma_1 = (0,1)$,

(ii) either $\Gamma_{p+1} = (\alpha(p)+1, \beta(p))$ or $\Gamma_{p+1} = (\alpha(p), \beta(p)+1)$,

(iii) $\Gamma_{m+n} = (m,n)$,

assuming that we set

$$\Delta_0^a \equiv \Delta_0^b \equiv 0\,. \tag{38}$$

Condition (i) characterizes the two possible time orderings for the rightmost vertex, while (iii) expresses the fact that the leftmost denominator is always $(\Delta_m^a + \Delta_n^b)$. Condition (ii) then indicates the fact that after going through the first p vertices in the whole diagram, the next vertex can be encountered either in the part a or in the part b. Thus, for example, the three time versions (2-4) in Fig. 7 are characterized by the following integer pairs: (0,1), (0,2), (1,2) for diagram (2), (0,1), (1,1), (1,2) for diagram (3) and (1,0), (1,1), (1,2) for diagram (4).

For the separate disconnected parts, the denominators are given by the products of Δ_i^a and Δ_i^b, respectively, so that

$$D_m^a \equiv D_{m0}^{ab} = \prod_{i=1}^{m}(\Delta_i^a)^{-1}\,, \qquad D_n^b \equiv D_{0n}^{ab} = \prod_{i=1}^{n}(\Delta_i^b)^{-1}\,. \tag{39}$$

It is also useful to define

$$D_0^a = D_0^b = D_{00}^{ab} \equiv 1\,. \tag{40}$$

We can then simply state the desired *factorization lemma* as follows

$$D_{mn}^{ab} = D_m^a D_n^b\,. \tag{41}$$

This is easily proved using mathematical induction. Clearly, the lemma holds when $m = 0$ or $n = 0$, since

$$D_{m0}^{ab} = D_m^a D_0^b = D_m^a \qquad \text{or} \qquad D_{0n}^{ab} = D_0^a D_n^b = D_n^b\,, \tag{42}$$

in agreement with (39). Assume, thus, that the lemma holds for $M = m-1$, $N = n$ and $M = m$ and $N = n-1$; $m, n \geq 1$. Clearly, all the terms in D_{mn}^{ab} can be divided into two disjoint classes according to whether the leftmost interaction occurs in the a or b subdiagram, respectively, as schematically shown in Fig. 8. The last denominator being always the same as required by (iii), namely $(\Delta_m^a + \Delta_n^b)$, we can write

$$D_{mn}^{ab} = (\Delta_m^a + \Delta_n^b)^{-1}(D_{m-1,n}^{ab} + D_{m,n-1}^{ab})\,, \tag{43}$$

since all the remaining factors are identical to those characterizing the disconnected diagrams that result when the leftmost vertex is deleted. This result holds even when m or n equals 1. Thus, since the lemma (41) holds for $M = m-1$, $N = n$ and $M = m$, $N = n-1$ by assumption, we can write (43) as

$$\begin{aligned} D_{mn}^{ab} &= (\Delta_m^a + \Delta_n^b)^{-1}(D_{m-1}^a D_n^b + D_m^a D_{n-1}^b) \\ &= (\Delta_m^a + \Delta_n^b)^{-1}(\Delta_m^a D_m^a D_n^b + \Delta_n^b D_m^a D_n^b) = D_m^a D_n^b\,, \end{aligned} \tag{44}$$

since

$$D_m^a = D_{m-1}^a(\Delta_m^a)^{-1}\,, \tag{45}$$

or

$$D^a_{m-1} = \Delta^a_m D^a_m\,, \tag{45'}$$

in view of (39), and similarly for the b part.

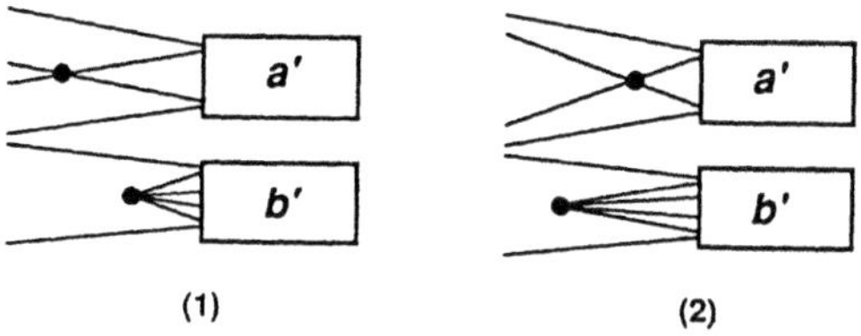

Fig. 8. A schematic representation of two distinct classes of linked wave-function-type MBPT diagrams that differ in the ordering of their two leftmost vertices. See text for details.

3.4. Linked Cluster Theorem

As indicated in Section 3.2, all the renormalization terms are cancelled by the unlinked terms originating from the principal one, so that the net result in each order is given by the linked diagrams (including the EPV ones). We now conjecture this result for an arbitrary order, writing (we drop the subscript 0 whenever confusion cannot arise)

$$|\Psi\rangle = \sum_{n=0}^{\infty}{}^{L}(PW)^n|\Phi\rangle \equiv \sum_{n=0}^{\infty}\left\{(PW)^n|\Phi\rangle\right\}_L\,, \tag{46}$$

where L indicates that the sum extends over linked diagrams only. This implies that the exact energy ΔE, Eq. (29), is given by

$$\Delta E = \sum_{n=0}^{\infty}\langle\Phi|W|\Psi^{(n)}\rangle = \sum_{n=0}^{\infty}\langle W(PW)^n\rangle_L\,, \tag{47}$$

i.e. by the sum of linked vacuum diagrams. To prove the *linked cluster theorem*, Eq. (46), we apply the operator $(E^{(0)} - H_0)$ to $|\Psi\rangle$, obtaining

$$\begin{aligned}(E^{(0)} - H_0)|\Psi\rangle &= (E^{(0)} - H_0)\Big[|\Phi\rangle + \sum_{n=1}^{\infty}{}^{L}(PW)^n|\Phi\rangle\Big]\\ &= (E^{(0)} - H_0)P\sum_{n=1}^{\infty}{}^{L}W(PW)^{n-1}|\Phi\rangle\\ &= (1 - |\Phi\rangle\langle\Phi|)\sum_{n=1}^{\infty}{}^{L}W(PW)^{n-1}|\Phi\rangle\\ &= \sum_{n=1}^{\infty}{}^{L}W(PW)^{n-1}|\Phi\rangle - |\Phi\rangle\sum_{n=1}^{\infty}{}^{L}\langle W(PW)^{n-1}\rangle\,,\end{aligned} \tag{48}$$

where we used Eq. (31) in the third step. Since in each term $(PW)^n|\Phi\rangle$ with $n \geq 1$ there is at least one external fermion line extending to the left, the last term in

(48) must vanish (note that the action of $(E^{(0)} - H_0)$ only removes the leftmost denominator, leaving the operator part unchanged), so that

$$\begin{aligned}(E^{(0)} - H_0)|\Psi\rangle &= \sum_{n=0}^{\infty}{}^{L} W(PW)^n|\Phi\rangle \\ &= W\sum_{n=0}^{\infty}{}^{L}(PW)^n|\Phi\rangle - \Big\{W\sum_{n=0}^{\infty}{}^{L}(PW)^n|\Phi\rangle\Big\}_U,\end{aligned} \tag{49}$$

where in the last equation we took the operator W outside the summation symbol and included the term compensating for the possible unlinked terms (indicated by the subscript U) that may thus arise. These terms must be represented by diagrams that are unlinked due to the leftmost W interaction, and linked when this leftmost interaction vertex is removed, as shown schematically in Fig. 9a. Removing this last

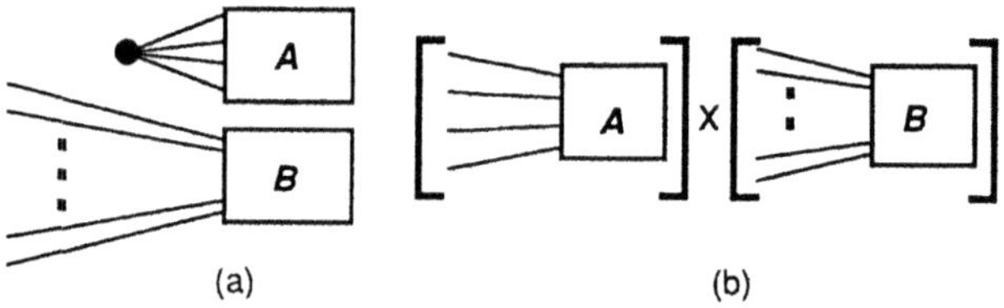

Fig. 9. Schematic representation of the unlinked terms given by the last term in Eq. (49), diagram (a), and their factorization when the leftmost vertex is removed, diagram (b). See text for details.

vertex (recall that the last denominator in all terms of (48) is absent), we can apply factorization lemma obtaining the product of contributions from the diagrams A and B, as indicated in Fig. 9b. The addition of the last interaction vertex will not change this factorization and will only replace the operators associated with the external lines of A by the corresponding matrix element associated with the added W vertex. Thus, the last term of (49) must equal $\Delta E|\Psi\rangle$, since the vacuum part in Fig. 9a represents an arbitrary linked energy diagram, while the remaining part may contain any linked wave function diagram. This gives that

$$(E^{(0)} - H_0)|\Psi\rangle = W|\Psi\rangle - \Delta E|\Psi\rangle \tag{50}$$

or, equivalently,

$$H_N|\Psi\rangle = \Delta E|\Psi\rangle\,, \tag{50'}$$

thus proving our hypothesis of the linked cluster theorem as expressed by Eqs. (46) and (47).

3.5. The Connected Cluster Structure of the Exact Wave Function

We now rewrite the linked cluster form of the wave function $|\Psi\rangle$, Eq. (46), in yet simpler form thus revealing its connected cluster structure that is of basic importance for coupled cluster approaches. We achieve this by classifying the linked diagrams, representing various terms in (46), by the number r of disconnected parts they contain.

The class $r = 1$ thus contains all possible *connected diagrams* of the wave function type and we represent it by some operator T defined as follows

$$T|\Phi\rangle = \sum_{n=1}^{\infty}{}^{C} (PW)^n |\Phi\rangle . \tag{51}$$

The connected diagrams appearing in (51) may be classified by PT order n as well as by the excitation order j, giving the number of external lines (equal to $2j$) they contain. We can thus write

$$T = \sum_{j=1}^{N} T_j , \tag{52}$$

since we can restrict ourselves to at most N-times excited states when considering an N-electron system, and

$$T_j = \sum_{n=1}^{\infty} T_j^{(n)} , \tag{53}$$

where

$$T_j^{(n)}|\Phi\rangle = \{(PW)^n|\Phi\rangle\}_{C;j} \tag{53'}$$

represents all n-th order connected diagrams with $2j$ external lines. We can also write

$$\begin{aligned} T_j &= (j!)^{-1} \sum_{\{B^i,B_i\}} \langle B^1 \cdots B^j|t_j|B_1 \cdots B_j\rangle \, X^\dagger_{B^1} \cdots X^\dagger_{B^j} X_{B_j} \cdots X_{B_1} \\ &= (j!)^{-2} \sum_{\{B^i,B_i\}} \langle B^1 \cdots B^j|t_j|B_1 \cdots B_j\rangle_{\mathcal{A}} \, X^\dagger_{B^1} \cdots X^\dagger_{B^j} X_{B_j} \cdots X_{B_1} , \end{aligned} \tag{54}$$

where $\{B^i, B_i\}$ designates independent summations over all hole and particle labels involved and the subscript $\mathcal{A}$ indicates an antisymmetrized matrix element, Eq. (7).

Now, all possible linked diagrams containing two disconnected parts ($r = 2$) can be expressed in view of the factorization lemma as $\frac{1}{2}T^2|\Phi\rangle$, since

$$\tfrac{1}{2}T^2 = \tfrac{1}{2}(\sum_j T_j)^2 = \tfrac{1}{2}\sum_i T_i^2 + \sum_{i<j} T_i T_j , \tag{55}$$

and contributions from disconnected parts commute,

$$[T_i, T_j] = 0 . \tag{56}$$

Thus, factor $\frac{1}{2}$ is needed either to correct the double counting of non-equivalent parts or as an appropriate topological factor when both parts are equivalent.

In general, a component with r disconnected parts can be represented by the term

$$(r!)^{-1}T^r|\Phi\rangle ,$$

so that we can express $|\Psi\rangle$, Eq. (46), as

$$\begin{aligned} |\Psi\rangle &= |\Phi\rangle + T|\Phi\rangle + \tfrac{1}{2}T^2|\Phi\rangle + \cdots + (r!)^{-1}T^r|\Phi\rangle + \cdots \\ &= (1 + T + \tfrac{1}{2}T^2 + \cdots + (r!)^{-1}T^r + \cdots)|\Phi\rangle \\ &= e^T|\Phi\rangle . \end{aligned} \tag{57}$$

We can thus write the exact wave function $|\Psi\rangle$ in terms of only connected diagrams that define the operator T, Eq. (51). This implies that we can express the *wave operator* W that transforms the unperturbed independent particle model (IPM) wave function $|\Phi\rangle$ into the exact wave function $|\Psi\rangle$,

$$|\Psi\rangle = W|\Phi\rangle\,, \tag{58}$$

as an exponential of the so-called *cluster operator* T,

$$W = e^T\,, \tag{59}$$

which in turn is defined in terms of connected wave function diagrams.

The expansion (57) or (59) in terms of connected diagram contributions is usually referred to as the *cluster expansion* in view of its similarity with an analogous exponential expansion exploited in statistical mechanics (Mayer and Mayer, 1940). The operator T is called the *cluster operator*, and its j-particle component T_j, Eq. (54), as the *j-times excited* or *j-body cluster.* The form of the wave operator given by Eq. (59) is also referred to as the *cluster expansion ansatz.*

One of the important implications of this ansatz stems from the basic property of the exponential mapping that transforms additive structures into multiplicative ones. Thus, considering a system consisting of two independent, non-interacting subsystems A and B,

$$\begin{aligned} |\Psi^X\rangle &= e^{T^{(X)}}|\Phi^{(X)}\rangle, \qquad X = A, B \\ \left[T^{(A)}, T^{(B)}\right] &= 0\,, \end{aligned} \tag{60}$$

we find that

$$|\Psi^{(AB)}\rangle = e^{T(A)+T(B)}|\Phi^{(AB)}\rangle\,, \tag{61}$$

since

$$|\Phi^{(AB)}\rangle = |\Phi^{(A)}\rangle|\Phi^{(B)}\rangle\,, \tag{62}$$

so that the energy is given by the sum of subsystem energies

$$E^{(AB)} = E^{(A)} + E^{(B)}\,, \tag{63}$$

which is the essence of the size extensivity.

4. CLOSED SHELL COUPLED CLUSTER THEORY

We have just seen (Section 3.5) that the exact nondegenerate wave function $|\Psi\rangle$ can be written in terms of the cluster expansion, assuming that a satisfactory single reference IPM wave function $|\Phi\rangle$ is available. This is indeed the case for closed shell ground states of most molecular systems, at least in the vicinity of their equilibrium geometries. It is both convenient and effective to exploit the cluster structure of the exact wave function and to devise a formalism that is based on additive extensive quantities, characterizing connected clusters, as indicated schematically by Eqs. (60)-(63).

Ideally, one would exploit the cluster ansatz, Eq. (57), in the variational approach and derive equations that determine cluster components T_i variationally. In a certain sense, this is done in the Hartree-Fock theory, where the cluster operator T is truncated to its one-body component T_1, whose matrix elements or cluster amplitudes $\langle B^1|t_1|B_1\rangle$ can be regarded as infinitesimal variational parameters which give the admixture of unoccupied spin-orbitals $|B^1\rangle$ into the occupied ones $|B_1\rangle$. Thus, varying the one-particle spin-orbital basis we can achieve that the first order variation of the mean energy functional will vanish. Clearly, this cannot be done once at least two body cluster components are present. Since, generally, the energy mean value contains all positive powers of the cluster components considered, we cannot simply apply the variational principle. In fact, there seems to be a basic incompatibility between the variational and size-extensive approaches, since the terms that must be included in the energy in order to ensure its upper bound property may be shown to possess the unlinked character, thus making the energy size-nonextensive. Although this incompatability decreases as we approach the exact, or more precisely, full CI or full CC limit, it plays a crucial role at the energetically most important and computationally manageable two-body, or at most three-body, level. We thus directly employ the cluster ansatz in the Schrödinger equation and determine the cluster components by requiring that this equation be satisfied in a certain subspace spanned by the corresponding low-order excited configurations. We shall see that such an approach generally yields an energy independent, non-linear system of algebraic equations for the cluster components. In fact, these equations can be regarded as recursion formulas for the generation of higher order MBPT diagrams for the connected components of the wave function of a certain type, depending on the approximation employed. Thus, by solving these equations, we not only implicitly construct all such diagrams to infinite order of PT, but evaluate them at the same time. Once the solution is found, we compute the energy, which only depends on the one- and two-body components.

We shall next derive the CC equation for the most important case when the cluster operator T is approximated through its two-body component T_2,

$$T \approx T_2\,, \tag{64}$$

exploiting the diagrammatic procedure developed by Čížek (1966a). This method can be easily extended to an arbitrary approximation of the T operator, although the computational requirements become quickly formidable when higher than two-body cluster components are considered. We shall also examine the relationship of CC and other approaches to the many-electron correlation problem, particularly CI and MBPT. Finally, we shall discuss the calculation of other properties than the energy and indicate the limitations of single reference approaches.

4.1. Cluster ansatz and its diagrammatic representation

We consider an N-electron system, characterized by the Hamiltonian (10), with a non-degenerate ground state. Choosing an appropriate IPM description $|\Phi\rangle$ for this state (usually the Hartree-Fock approximation) as a reference, we invoke the cluster ansatz, Eq. (57), for the exact wave function $|\Psi\rangle$, with T given by Eqs. (52) and (54) in terms of its cluster amplitudes. We recall that the k-body cluster operators T_k mutually commute,

$$[T_i, T_j] = 0\,, \tag{65}$$

and are automatically in the n-product form. They represent, in fact, excitation operators, since

$$\begin{aligned} T_j|\Phi\rangle &= (j!)^{-1} \sum_{\{B^i,B_i\}} \langle B^1 \cdots B^j|t_j|B_1 \cdots B_j\rangle|\Phi_{K_i}^{(j)}\rangle \\ &= \sum_{\substack{B^1<B^2<\cdots<B^j \\ B_1<B_2<\cdots<B_j}} \langle B^1 \cdots B^j|t_j|B_1 \cdots B_j\rangle_{\mathcal{A}}\,|\Phi_{K_i}^{(j)}\rangle\,, \end{aligned} \tag{66}$$

where $|\Phi_{K_i}^{(j)}\rangle = \left| \begin{matrix} B^1 \cdots B^j \\ B_1 \cdots B_j \end{matrix} \right\rangle$ represents a j-times excited configuration, Eq. (27), defined by the index set $K_i \equiv \{B^1, \cdots B^j; B_1, \cdots, B_j\}$.

Diagrammatically, the one- and two-body components of the Hamiltonian (10), or their n-product forms (21), as well as the cluster opreators T_j, Eq. (54), will be represented in a standard way (Fig. 1). We shall refer to these diagrams as the F, V or T diagrams, respectively. It thus remains to discsuss the diagrammatic representation of the cluster expansion, Eq. (57).

We can represent a general product $\prod_{j=1}^{k}(n_j!)^{-1}T_j^{n_j}$ by k disconnected groups, each containing n_j disconnected T_j diagrams. Since the cluster components T_j mutually commute, Eq. (65), the order of individual T_j diagrams is immaterial and we shall place their diagrammatic representatives along a hypothetical vertical line to indicate this fact. Since these disconnected parts may be arbitrarily permuted, the weight of such a *multiple* (or M) diagram is

$$w_M^X = \prod_{j=1}^{k}(n_j!)^{-1}[w_{T_j}^X]^{n_j}, \; X = G, H \tag{67}$$

the superscript distinguishing Goldstone and Hugenholtz weights,

$$w_{T_j}^G = (j!)^{-1}, \qquad w_{T_j}^H = (j!)^{-2}. \tag{68}$$

The operator associated with an M diagram is clearly a product of operators corresponding to its disconnected parts,

$$D_M^X = \prod_j D_{T_j}^X(\rho_j)\,, \tag{69}$$

where the argument $\rho_j \equiv \{B^1 \cdots B^j; B_1 \cdots B_j\}$ designates the index set labeling external lines. For Goldstone diagrams, where individual paths of oriented fermion lines are easily distinguished, we assume that the pair of indices B^i, B_i labels the terminal lines of a given open path, as schematically indicated in Fig. 10.

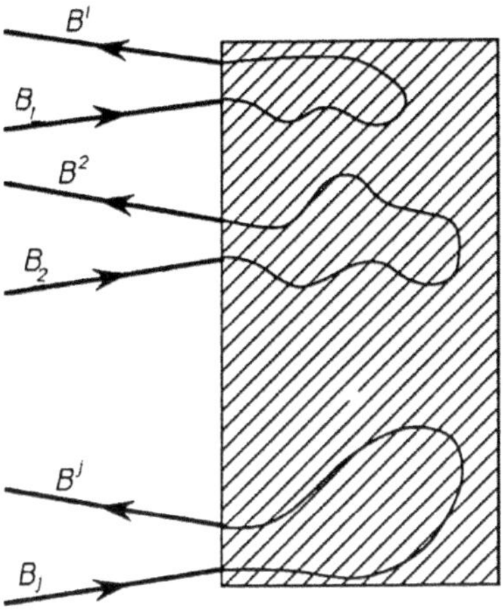

Fig. 10. Schematic representation of open path labeling in Goldstone or Brandow diagrams.

Thus, formally

$$e^T = \sum_M w_M^X D_M^X \,, \tag{70}$$

with the sum extending over all posssible M diagrams, including an empty M diagram representing 1 (diagrammatically represented as [1]). Schematically, we thus represent the cluster expansion as

$$e^T = [1] + -\boxed{\mathrm{T}} + \begin{array}{c} -\boxed{\mathrm{T}} \\ -\boxed{\mathrm{T}} \end{array} + \begin{array}{c} -\boxed{\mathrm{T}} \\ -\boxed{\mathrm{T}} \\ -\boxed{\mathrm{T}} \end{array} + \cdots \,, \tag{70$'$}$$

where each box $\boxed{\mathrm{T}}$ represents an arbitrary T_j diagram (the horizontal line on the left indicates that all fermion lines, of which there is at least one, extend in that direction), while an M-diagram having n connected components is represented by n such vertically stacked boxes.

4.2. Connected and disconnected clusters

To illustrate the significance of connected and disconnected cluster components, we compare the cluster expansion, Eq. (57), with an ordinary CI expansion of the exact wave function $|\Psi\rangle$. Writing the latter in terms of excitation operators $C_k \equiv \Gamma_k$, Eq. (23), so that

$$|\Psi\rangle = \sum_{k=0}^{N} C_k |\Phi\rangle \,, \tag{71}$$

and choosing the intermediate normalization ($C_0 = \hat{1}$), we obtain when comparing both expansions that

$$C_1 = T_1\,, \tag{72a}$$

$$C_2 = T_2 + \tfrac{1}{2}T_1^2\,, \tag{72b}$$

$$C_3 = T_3 + T_1T_2 + \tfrac{1}{3!}T_1^3\,, \tag{72c}$$

$$C_4 = T_4 + \tfrac{1}{2}T_2^2 + T_1T_3 + \tfrac{1}{2}T_1^2T_2 + \tfrac{1}{4!}T_1^4, \text{ etc.} \tag{72d}$$

or, generally

$$C_k = T_k + \sum_{\mathcal{P}_k} \prod_{j=1}^{p} (n_j!)^{-1} T_j^{n_j}\,, \tag{73}$$

with the sum extending over all nontrivial partitions $\mathcal{P}_k$ of k, $k = \sum_{j=1}^{p} j\,n_j$, $1 \leq n_j \leq k$, $2 \leq p < r$. The first term on the right hand side, T_k, represents the k-body *connected* cluster component, while the remaining terms represent *disconnected* k-body clusters.

Should we choose Brueckner (maximum overlap) orbitals as our one-electron states, then

$$T_1 = 0\,, \tag{74}$$

and formulas (72) considerably simplify, namely

$$C_1 = T_1 = 0\,, \tag{75a}$$

$$C_2 = T_2\,, \tag{75b}$$

$$C_3 = T_3\,, \tag{75c}$$

$$C_4 = T_4 + \tfrac{1}{2}T_2^2\,, \text{ etc.} \tag{75d}$$

Thus, in this case, the first disconnected cluster contribution occurs in the tetraexcited component.

For molecular systems with a non-metallic closed-shell ground state, the T_1 component is usually small, even when Hartree-Fock orbitals are employed and by far the largest contribution to the correlation energy arises from the doubly excited component, $C_2 \approx T_2$. This is also true even for highly delocalized systems (e.g., benzene) if the number of electrons is not very large. The next important contribution to the energy comes from the tetraexcited component C_4. It was shown by Sinanoğlu (1961,1962) that in this case the major contribution arises from the disconnected component $\frac{1}{2}T_2^2$. This fact is usually indicated by the symbolic inequality

$$T_4 \ll \tfrac{1}{2}T_2^2\,. \tag{76}$$

The approximation $C_4 \approx \frac{1}{2}T_2^2$ is well satisfied for nondegenerate ground states of medium-sized nonmetallic systems, and reasonably well even for small metallic ones (in the benzene π-electron model, for example, the T_4 component represents less than 3% of the total quadruply excited component when measured by the corresponding probability amplitudes; Čížek *et al.*, 1969; Paldus and Boyle, 1982). Consequently, setting

$$T \approx T_2\,, \tag{77}$$

provides an excellent approximation for medium sized systems, referred to as coupled pair many-electron theory (CPMET) or CCD (coupled clusters with doubles) for short. When also T_1 clusters are considered, one obtains CCSD approach. In view of smallness of one-body clusters, one can usually neglect higher order than linear terms in T_1 amplitudes. For example, so-called quadratic CI (QCISD) method (Pople *et al.*, 1987, 1989) represents in fact an approximation to the CCSD, which considers only the T_1T_2 term in monoexcited CC equations in addition to the linear terms T_1 considered throughout (Paldus *et al.*, 1989a, 1990).

Next in importance are the T_3 clusters, whose inclusion leads to the CCSDT approach, which is computationally very demanding in view of a large number of T_3 amplitudes (see, e.g., Bartlett, 1989). We shall discuss these various approaches and approximations in Sections 4.4 and 4.5 after we derive the basic CC formalism.

4.3. Connected cluster theorem

4.3.1. *Diagrammatic derivation*

We shall first demonstrate that all disconnected diagram contributions cancel out when we employ the cluster ansatz (57) in the Schrödinger equation for the non-degenerate ground state,

$$H_N|\Psi\rangle = \Delta E|\Psi\rangle\,. \tag{78}$$

Here $H_N = F_N + V_N$, Eq. (21), and $\Delta E = E - \langle H\rangle = E - E^{HF}$ represents the correlation energy, assuming that we employ the restricted Hartree-Fock solution $|\Phi\rangle$ as the Fermi vacuum. Using the cluster ansatz (57) in Eq. (78), we get

$$H_N e^T|\Phi\rangle = \Delta E\, e^T|\Phi\rangle\,. \tag{79}$$

We can now employ the diagrammatic technique to evaluate the left hand side of this equation to any desired order in T, relying on generalized Wick's theorem, since both H_N and e^T (or any of their components) are in the n-product form. Thus, no resulting (or R) diagrams with internal lines interconnecting vertices of the same component (i.e., F, V or T vertex) are permitted. Moreover, since all the lines of any connected T component extend to the left, no internal lines interconnecting distinct T vertices are allowed.

We shall carry out the evaluation of the left hand side of (79) in a general, schematic fashion in order to determine its overall structure. To this purpose we shall represent schematically the F or V vertices as $\boxed{\mathrm{H}}$, with attached fermion lines (suppressing arrows for simplicity) as needed. Clearly, at most four fermion lines can be incident with this vertex. We shall similarly represent the T vertices and the cluster expansion (57) as indicated in Eq. (70′). We make a convention that each explicitly shown line indicates at least one fermion line extending in that direction (when no lines are shown, they may extend in either direction). Similarly, a line interconnecting the two vertices indicates that there is at least one internal fermion line between these components.

Using these conventions, we can schematically represent the left hand side of (79) as follows

$$H_N e^T|\Phi\rangle = \boxed{\mathrm{H}}\left\{[1] + -\boxed{\mathrm{T}} + \begin{matrix}-\boxed{\mathrm{T}}\\ -\boxed{\mathrm{T}}\end{matrix} + \cdots\right\}|\Phi\rangle\,, \tag{80}$$

and applying generalized Wick's theorem we get

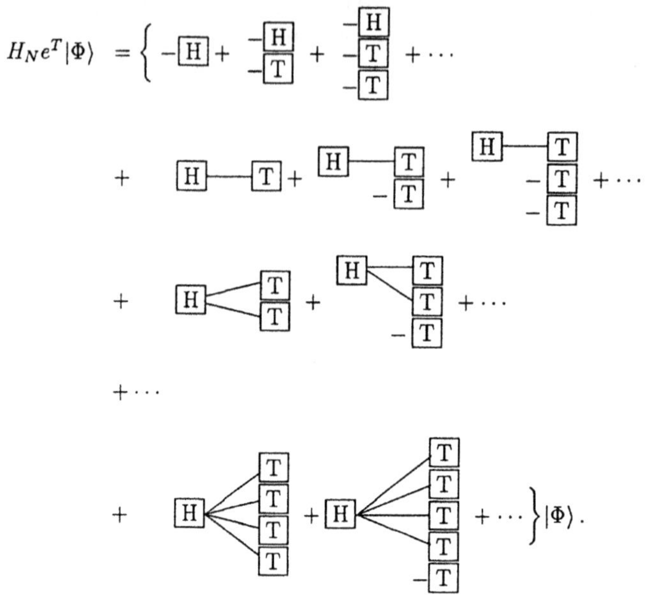

(81)

Since the individual disconnected components in each term on the right hand side of (81) mutually commute (show why!), we can rewrite (81) as follows

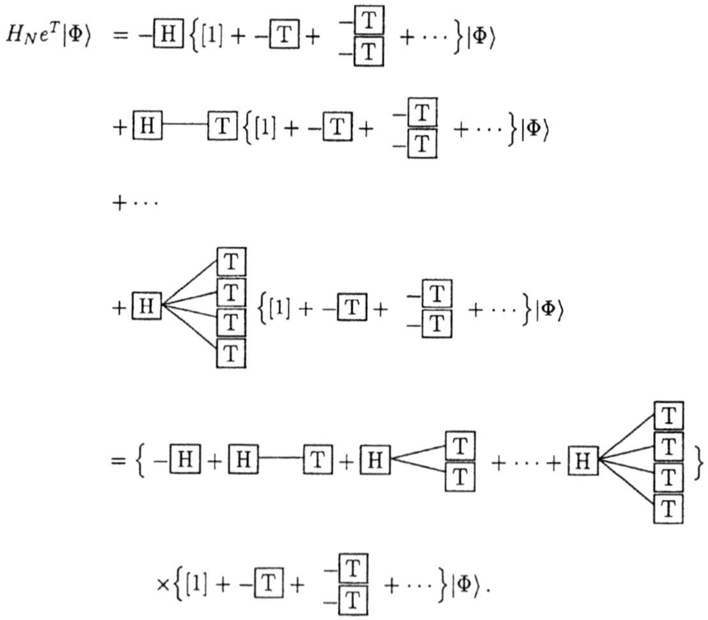

(82)

Note that the first line in Eq. (81) corresponds to the terms with no contractions, the second line to the terms with at least one contraction between the operator part and one connected T component, etc., until the last row contains the terms in which each fermion line of the V vertex is joined with one connected T component. Further, the last factor in (82) is obviously the cluster expansion of $|\Psi\rangle$, Eqs. (57) and (70), while the first factor on the right hand side represents all possible connected R-diagrams that can be formed from one F or one V diagram and one M diagram, i.e., zero, one, two, three or four T diagrams. Moreover, these two factors commute, since all disconnected components mutually commute.

In algebraic form we can thus write this result as follows

$$H_N e^T|\Phi\rangle = (H_N e^T)_C\, e^T|\Phi\rangle = e^T(H_N e^T)_C|\Phi\rangle\,, \tag{83}$$

where the subscript C indicates the connected component. This factorization property, expressed by Eq. (83), is sometimes referred to as the *connected cluster theorem.* The connected component $(H_N e^T)_C$ contains at most four T vertices, so that we can also write

$$(H_N e^T)_C = \sum_{n=0}^{4}(n!)^{-1}\{H_N T^n\}_C\,. \tag{84}$$

Making use of the connected cluster theorem, Eq. (83), in the Schrödinger equation (79), and multiplying each side from the left by the inverse e^{-T} of the wave operator $W = e^T$, we find the desired relationship

$$(H_N e^T)_C|\Phi\rangle = \Delta E|\Phi\rangle\,, \tag{85}$$

which can be regarded as the connected cluster form of the Schrödinger equation (78). We can also regard the operator $(H_N e^T)_C$, acting in the one-dimensional "model space" $\mathcal{M}_0$ that is spanned by the chosen reference configuration $|\Phi\rangle$, $\mathcal{M}_0 = \mathrm{Span}\{|\Phi\rangle\}$, as the so-called "model Hamiltonian" $H^{\mathrm{eff}} = (H_N e^T)_C$, since it yields the exact energy ΔE when acting on the model space wave function. This is an important concept which will enable us to generalize the CC theory to the multi-reference open shell case.

Projecting now Eq. (85) onto the reference configuration $|\Phi\rangle$, we get

$$\Delta E = \langle (H_N e^T)_C\rangle \equiv \langle H_N e^T\rangle_C\,, \tag{86}$$

while projecting onto the excited configuration states $|\Phi_K^{(k)}\rangle$, Eq. (27), that are orthogonal to $|\Phi\rangle$, we obtain the energy independent set of equations for cluster components, namely

$$\langle\Phi_K^{(k)}|(H_N e^T)_C|\Phi\rangle = 0\,. \tag{87}$$

Since the matrix elements of the connected cluster components (also called cluster amplitudes) $\langle B^1\cdots B^k|\hat{t}_k|B_1\cdots B_k\rangle$ are coefficients of the linearly independent configurations $|\Phi_K^{(k)}\rangle$ in an appropriate k-times excited subspace, Eq. (66), we obtain the same number of equations of the type (87) as there are unknown cluster amplitudes. We shall also see that by truncating the cluster operator T at the k-body level, $T \approx \sum_{j=1}^{k} T_j$, will automatically decouple the chain of equations (87) without any additional approximations. However, the CC equations (87) are generally nonlinear and may possess multiple solutions. There is also a possibility to use a larger projection space than corresponds to the cluster amplitudes considered, obtaining an overcomplete set of CC equations that can be solved in the least square sense (Jankowski *et al.*, 1991).

4.3.2. *Algebraic derivation*

Before deriving an explicit form of the CC equations for some important approximations (i.e., CPMET or CCD), we wish to briefly indicate an alternative way of deducing the connected cluster form of the Schrödinger equation, Eq . (85). We can pre-multiply Eq. (79) by e^{-T} before the factorization, obtaining

$$(e^{-T}H_N e^T)|\Phi\rangle = \Delta E|\Phi\rangle\,, \tag{79'}$$

and evaluate the operator product on the left-hand side using the well-known expansion*)

$$e^{-A}B\,e^{A} = \sum_{n=0}^{\infty}(n!)^{-1}[\cdots[[B,\underbrace{A],A],\cdots,A]}_{n}\,. \tag{88}$$

In our case this expansion terminates,

$$\begin{aligned} e^{-T}H_N e^T &= H_N + [H_N,T] + \tfrac{1}{2}[[H_N,T],T] \\ &\quad + \tfrac{1}{6}[[[H_N,T],T],T] + \tfrac{1}{24}[[[[H_N,T],T],T],T]\,, \end{aligned} \tag{89}$$

since each commutator eliminates one creation or annihilation operator in H_N. Each commutator involving term on the right hand side of (89) contains both connected and disconnected terms, but the latter ones cancel out. To see the connection with the previous derivation, let us evaluate these commutators diagrammatically.

Consider, thus, a commutator

$$[H_N,T] = H_N T - T H_N\,. \tag{90}$$

The last term TH_N is automatically in the N-product form since T contains only particle-hole creators. This is also apparent from the diagrammatic representation, since all the fermion lines issuing from a T-vertex must extend to the left and cannot be contracted with those issuing from H_N. We can thus represent this term as

$$TH_N = \begin{array}{c}\boxed{\mathrm{H}}\\ -\boxed{\mathrm{T}}\end{array}\,. \tag{91a}$$

The first term $H_N T$ in Eq. (90), on the other hand, is identical with that evaluated earlier [cf. Eqs. (80)-(82)], so that

$$H_N T = \begin{array}{c}\boxed{\mathrm{H}}\\ -\boxed{\mathrm{T}}\end{array} + \boxed{\mathrm{H}}-\boxed{\mathrm{T}}\,. \tag{91b}$$

Thus, the disconnected terms cancel out and we can write

$$[H_N,T] = \boxed{\mathrm{H}}-\boxed{\mathrm{T}} = (H_N T)_C\,. \tag{90'}$$

---

*) This result is often referred to as the Baker-Campbell-Hausdorff (BCH) formula. In fact, it represents only one lemma needed in the proof of the BCH formula.

In a similar way we obtain

$$\tfrac{1}{2}[[H_N, T], T] = \tfrac{1}{2}\left[\boxed{\mathrm{H}}\!-\!\boxed{\mathrm{T}}, -\boxed{\mathrm{T}}\right]$$

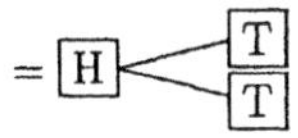

$$= \tfrac{1}{2}(H_N T^2)_C \,, \tag{92}$$

and, generally

$$e^{-T} H_N e^T = \sum_{n=0}^{4} (n!)^{-1} \{H_N T^n\}_C \,, \tag{93}$$

which is equivalent with Eq. (84).

In this connection it is worthwhile to mention that the algebraic derivation based on Eq. (90) is greatly facilitated if we express the operators in terms of U(n) generators (see, e.g., Paldus, 1976). Since the latter can be realized in terms of creation and annihilation operators as follows

$$e^A_B = X^\dagger_A X_B \,, \tag{94}$$

we can write

$$H = \sum_{AB} \langle A|\hat{f}|B\rangle N\{e^A_B\} + \tfrac{1}{2} \sum_{ABCD} \langle AB|\hat{v}|CD\rangle N\{e^A_C e^B_D\} \,, \tag{95}$$

$$T_j = (j!)^{-2} \sum_{\{B^i, B_i\}} \langle B^1 \cdots B^j|\hat{t}_j|B_1 \cdots B_j\rangle_A \prod_{i=1}^{j} e^{B^i}_{B_i} \,, \tag{96}$$

with the generators satisfying the U(n) commutation relations

$$[e^A_B, e^C_D] = \delta^C_B e^A_D - \delta^A_D e^C_B \tag{97}$$

and the Hermitian property

$$(e^A_B)^\dagger = e^B_A \,. \tag{98}$$

It is also convenient to introduce higher order operators, referred to as *replacement* (or *excitation*) operators (Kutzelnigg, 1985; Paldus and Jeziorski, 1988)

$$\begin{aligned} e^{ABC\cdots}_{IJK\cdots} &= X^\dagger_A X^\dagger_B X^\dagger_C \cdots X_K X_J X_I \\ &= N\{e^A_I e^B_J e^C_K \cdots\} \,. \end{aligned} \tag{99}$$

They can also be defined recursively, i.e.

$$e^{AB}_{IJ} = e^A_I e^B_J - \delta^B_I e^A_J \,, \tag{100a}$$

$$e^{ABC}_{IJK} = e^{AB}_{IJ} e^C_K - \delta^C_I e^{AB}_{KJ} - \delta^C_J e^{AB}_{IK} \,, \text{ etc.} \tag{100b}$$

A similar rule applies when multiplying higher than the first order replacement operators, e.g.

$$\begin{aligned} e^{AB}_{IJ} e^{CD}_{KL} &= e^{ABCD}_{IJKL} \\ &+ \delta^C_I e^{ABD}_{KJL} + \delta^C_J e^{ABD}_{IKL} + \delta^D_I e^{ABC}_{LJK} + \delta^D_J e^{ABC}_{ILK} \\ &+ \delta^C_I \delta^D_J e^{AB}_{KL} + \delta^C_J \delta^D_I e^{AB}_{LK} \,. \end{aligned} \tag{101}$$

In fact, this rule represents an algebraic realization of Wick's theorem, each Kronecker delta representing a contraction. Thus, in the above example, the first term on the right hand side gives the normal product, the terms in the second row correspond to those involving a single contraction and, finally, those in the last row represent doubly contracted terms. Consequently, the N-product terms on the right hand side of (95) can be simply written as $N\{e^A_C e^B_D\} = e^{AB}_{CD}$. This Wick-type multiplication rule also easily yields all the commutators that we need to evaluate. For example, we have that

$$[e^{AB}_{IJ}, e^{C}_{K}] = \delta^C_I e^{AB}_{KJ} + \delta^C_J e^{AB}_{IK} - \delta^A_K e^{CB}_{IJ} - \delta^B_K e^{AC}_{IJ}\,, \tag{102a}$$

or

$$\begin{aligned}[e^{AB}_{IJ}, e^{CD}_{KL}] &= \delta^C_I e^{ABD}_{KJL} + \delta^C_J e^{ABD}_{IKL} + \delta^D_I e^{ABC}_{LJK} + \delta^D_J e^{ABC}_{ILK} \\ &\quad -\delta^A_K e^{BCD}_{JIL} - \delta^A_L e^{BCD}_{JKI} - \delta^B_K e^{ACD}_{IJL} - \delta^B_L e^{ACD}_{IKJ} \\ &\quad +\delta^C_I \delta^D_J e^{AB}_{KL} + \delta^C_J \delta^D_I e^{AB}_{LK} - \delta^A_K \delta^B_L e^{CD}_{IJ} - \delta^A_L \delta^B_K e^{CD}_{JI}\,, \text{ etc.,}\end{aligned} \tag{102b}$$

the basic commutation relationship for U(n) generators, Eq. (97), representing a special case. Clearly,

$$e^{A(1)A(2)\cdots A(n)}_{B(1)B(2)\cdots B(n)} = (-1)^p e^{A(p_1)A(p_2)\cdots A(p_n)}_{B(1)B(2)\cdots B(n)} = (-1)^p e^{A(1)A(2)\cdots A(n)}_{B(p_1)B(p_2)\cdots B(p_n)}\,, \tag{103}$$

for any permutation $P : i \to p_i$ of S_n.

The replacement operators and the above given algebraic rules can be extended to a spin-adapted (or, in fact, orthogonally spin-adapted) form (see, e.g., Paldus and Jeziorski, 1988; Jeziorski and Paldus, 1988, 1989 [Appendix]).

4.4. CC formalism: basic considerations

We shall first illustrate how to find an explicit form for the CC equations using, as an example, the most important case when only pair-cluster amplitudes T_2 are considered. An extension of this procedure to cases where singly- and/or triply-excited clusters are also considered is straightforward.

The correlation energy ΔE is always given by Eq. (86), and involves at most doubly-excited clusters. Thus, the relevant Brandow-type diagrams are immediately found,

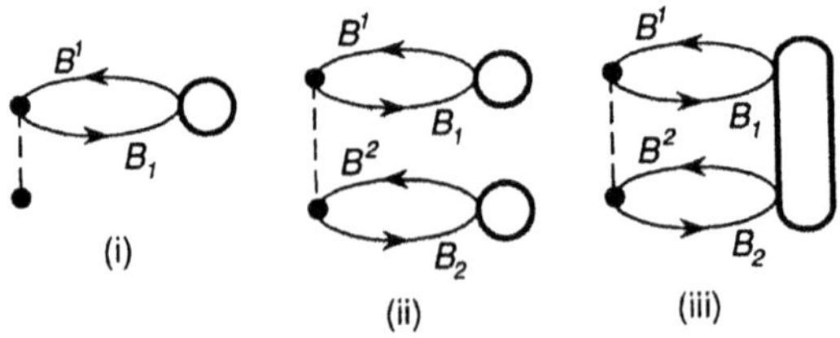

(104)

and give the explicit result

$$\Delta E = \sum_{B_1,B^1} \langle B_1|\hat{f}|B^1\rangle\langle B^1|\hat{t}_1|B_1\rangle + \tfrac{1}{2} \sum_{B_1,B_2,B^1,B^2} \langle B_1B_2|\hat{v}|B^1B^2\rangle_A \Big(\langle B^1|\hat{t}_1|B_1\rangle\langle B^2|\hat{t}_1|B_2\rangle + \tfrac{1}{2}\langle B^1B^2|\hat{t}_2|B_1B_2\rangle_A\Big) . \qquad (105)$$

Consequently, at most doubly-excited clusters contribute *directly* to the energy, the higher-excited ones contributing only indirectly via their interaction with mono- and bi-excited clusters. In fact even the singly-excited clusters contribute primarily indirectly (if Brueckner orbitals are employed, their direct contribution vanishes entirely), since their amplitudes $\langle B^1|\hat{t}_1|B_1\rangle$ are usually small and appear quadratically with the two-electron integrals, Eq. (105). Further, when HF orbitals are used, we have that

$$\langle B_1|\hat{f}|B^1\rangle = 0 , \qquad (106)$$

so that the one-particle contribution, diagram (104i), vanishes. Thus, for most practical purposes, the correlation energy is given by the expression

$$\begin{aligned}\Delta E &\approx \tfrac{1}{4} \sum_{B_1,B_2,B^1,B^2} \langle B_1B_2|\hat{v}|B^1B^2\rangle_A\langle B^1B^2|\hat{t}_2|B_1B_2\rangle_A \\ &= \tfrac{1}{2} \sum_{B_1,B_2,B^1,B^2} \langle B_1B_2|\hat{v}|B^1B^2\rangle\Big(\langle B^1B^2|\hat{t}_2|B_1B_2\rangle - \langle B^1B^2|\hat{t}_2|B_2B_1\rangle\Big) ,\end{aligned} \qquad (107)$$

assuming that the HF (or Brueckner) orbitals are employed. The second form for ΔE, Eq. (107), corresponds to the Goldstone type R-diagrams.

Let us next briefly examine the relative importance of the various cluster components in the exact wave function, assuming that the HF reference is employed, so that Eq. (106) holds. We see immediately that under this assumption the T_2 clusters appear already in the first order of PT, while the T_1 and T_3 clusters will appear for the first time in the second order PT, as the following typical diagrams illustrate (for simplicity, the line orientation is suppressed)

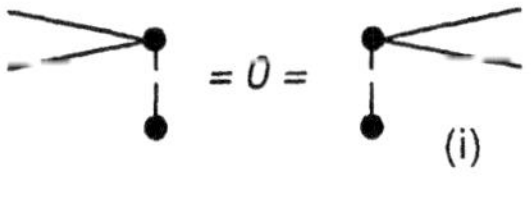

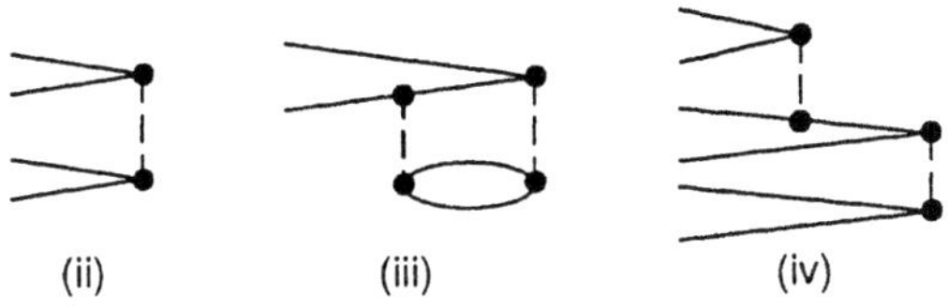

(108)

For quadruply-excited clusters the connected component contributes for the first time in the third order of PT, while the disconnected component $\frac{1}{2}T_2^2$, Eq. (72d), contributes in the second order, as the following typical diagrams indicate:

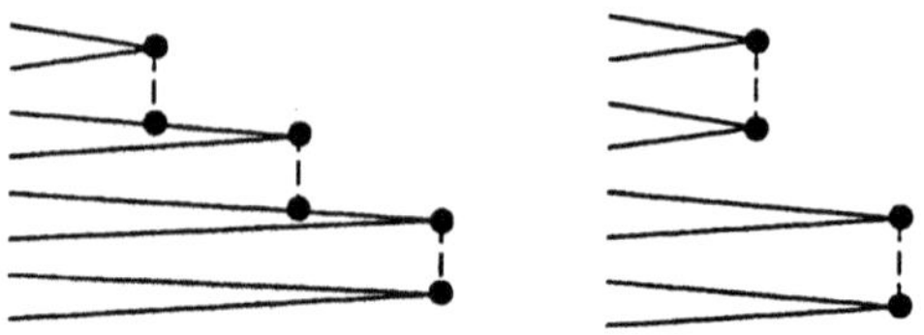

(i) (ii)

(109)

We have summarized the lowest order of PT, in which a given wave function contribution occurs for the first time, in Table I.

Table I. The lowest order of PT in which a given cluster component contributes for the first time in the exact wave function, assuming that Eq. (106) holds.

Cluster:	T_1	T_2	T_3	T_4	$\frac{1}{2}T_2^2$	T_1T_2
PT order:	2	1	2	3	2	3

These simple considerations clearly show the importance of pair clusters: their connected contribution, corresponding to the following doubly-excited component (cf. Eq. (66))

$$T_2|\Phi\rangle = \sum_{B^1<B^2} \sum_{B^1<B^2} \langle B^1B^2|\hat{t}_2|B_1B_2\rangle_A \left| \begin{matrix} B^1 & B^2 \\ B_1 & B_2 \end{matrix} \right\rangle, \qquad (110)$$

usually yields well over 90% of the correlation energy for small systems (say, with less than 10 electrons) and over 80% even for medium-size systems, unless their ground state $|\Phi\rangle$ is quasidegenerate with some of the doubly-excited configuration(s). However, even for the Be atom, in which the well known $2s^2 - 2p^2$ quasidegeneracy is very strong, the pair clusters alone yield well over 90% of the correlation energy.

The next important contribution, according to the classification given in Table I, comes from the T_1, T_3 and $\frac{1}{2}T_2^2$ clusters. The T_1 contribution depends very much on the orbitals employed [it exactly vanishes for the Brueckner orbitals, Eq. (74)]. For the RHF orbitals, this contribution is usually non-negligible, though rather small. The T_3 contribution has usually been heuristically neglected, since it requires a considerable computational effort in comparison with the pair contribution. Only relatively recently has its effect been studied in detail (Lee *et al*, 1984; Lee and Bartlett, 1984, Raghavachari, 1985; Urban *et al.*, 1985, Noga and Bartlett, 1987; Noga *et al.*, 1987, Scuseria and Schaefer, 1988; Gauss and Cremer, 1989; Scuseria and Lee, 1990; Lee *et al.*, 1990; Piecuch *et al.*, 1990; Lee and Rendell, 1991; He and Cremer, 1991). Full CCSDT results proved to give highly accurate correlation energies, particularly for geometries that are in the vicinity of the equilibrium ones. For singly bonded systems this is also the case when the bonds have been significantly stretched (twice their equilibrium values, Noga and Bartlett, 1987). An excellent approximation of the

T_3 contribution can often be obtained using a perturbative estimate (see references above).

Undoubtedly, by far the most important rôle from amongst the terms that first contribute to the wave function in the second order of PT is played by the disconnected tetraexcited clusters $\frac{1}{2}T_2^2$. These terms distinguish linear-CCD (L-CCD) and full CCD methods, and play a particularly crutial role when the RHF reference configuration becomes quasidegenerate with some low-lying biexcited configuration (Jankowski and Paldus, 1980; Paldus *et al.*, 1988), even though in nondegenerate situations the L-CCD method often provides an excellent approximation to the exact FCI result (since it overestimates the correlation energy by roughly the same amount that is missed due to the neglect of triexcitations, see e.g. Paldus, 1983; Bartlett *et al.*, 1979,1984). We can thus conclude that the importance of the T_1, T_3 and $\frac{1}{2}T_2^2$ terms, which first contribute to the wave function in the second order of PT, is roughly proportional to their number and, though essential for an accurate description, particularly in quasidegenerate situations, is much less significant than the T_2 clusters that appear in the first order. It is also worthwhile to note, that while in the tetraexcited case it is the disconnected component $\frac{1}{2}T_2^2$ that plays the decisive role, the reverse is true for the triexcited clusters, where the connected component T_3 is the most important one. This fact immediatley follows from a simple PT argument (cf. Table I), considering the lowest order of the PT in which these clusters contribute for the first time (Paldus *et al.*, 1972). The prevailing importance of disconnected four-body clusters $\frac{1}{2}T_2^2$ as compared with the connected ones, T_4, Eq. (109), is also easy to understand in terms of a simple physical picture: The disconnected terms are clearly associated with correlations between two "independent" electron pairs, while the connected ones describe the true four-body correlations. This was in fact the key idea that led to earlier independent-pair theories (Sinanoğlu, 1961,1962; Nesbet, 1958, 1965, 1969) and to various forms of the approximate coupled-pair theories (Ahlrichs, 1979; Ahlrichs *et al.*, 1975; Ahlrichs and Scharf, 1983; Hurley, 1976; Koch and Kutzelnigg, 1981; Kutzelnigg, 1977; Meyer, 1971,1974,1977).

4.5. Explicit CC equations

4.5.1. *CCD approximation (spin-orbital form)*

Let us now illustate the derivation of explicit CC equations for the most important case of pair clusters T_2, Eq. (77). Considering the pair-cluster approximation, Eq. (77), we must determine the cluster components $\langle B^1 B^2|\hat{t}_2|B_1 B_2\rangle_{\mathcal{A}}$, Eq. (110), by solving the pertinent CCD equations. A compact form of these equations is given by Eq. (87) with $k = 2$ and $T \approx T_2$. Their explicit form may be easily found by employing the diagrammatic technique.

We can represent the bra-state, associated with the doubly-excited configuration

$|\Phi_{K_C}^{(2)}\rangle \equiv \left| \begin{matrix} C^1 & C^2 \\ C_1 & C_2 \end{matrix} \right\rangle$, $K_C \equiv \{C^1, C^2; C_1, C_2\}$, by the excitation diagram $\bar{C}$ that is

conjugate to the standard excitation diagram C, Fig. 1(3), as indicated in Fig. 11(1). Here the spin-orbital labels defining the configuration K_C are fixed, so that the weight w_C equals 1 as does the scalar factor d_C that we associate with this diagram, since only projection onto $|\Phi_{K_C}^{(2)}\rangle$ is involved.

We thus must form all possible connected R-diagrams (using the Brandow representation) that involve one $\bar{C}$ and one H vertex, as well as zero, one or two T_2 vertices, as shown schematically in Fig. 11(2), (it is easily seen that no such R-diagrams can be formed with three or four T_2 vertices). Since $w_{\bar{C}} = d_{\bar{C}} = 1$, it is customary to drop the $\bar{C}$-vertex and form corresponding open diagrams, Fig. 11(3).

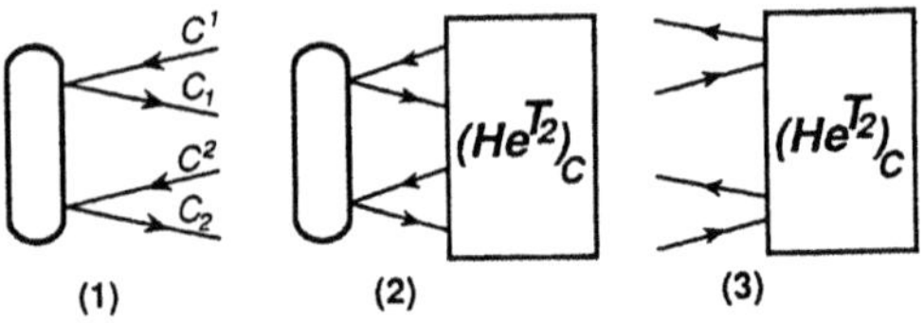

Fig. 11. Congugate diagram $\bar{C}$ representing the projection onto the biexcited state $\left| \begin{matrix} C^1C^2 \\ C_1C_2 \end{matrix} \right\rangle$, (diagram 1) and a schematic representation of CCD diagrams (2,3). A standard representation drops the $\bar{C}$ vertex as illustrated by diagram (3).

To proceed in a systematic way, we first construct corresponding Hugenholtz R-skeletons that contain relevant vertices, but ignore the orientation of fermion lines (Fig. 12).

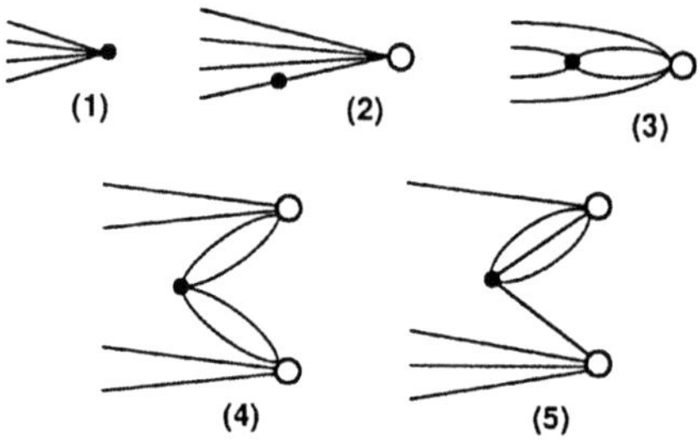

Fig. 12. Nonoriented Hugenholtz skeletons characterizing the CCD method.

Next, we introduce the orientation of fermion lines and represent each Hugenholtz skeleton by one Goldstone representative. We finally label all fermion lines in all permissible ways, using first the fixed labels $C^i, C_i (i = 1,2)$ for the external lines and B^i, B_i free (i.e., summation) labels, as necessary, for the internal lines (Fig. 13). To avoid unnecessary repetition, we introduce the particle and hole symmetrization operators $\mathcal{S}_x (x = p, h)$,

$$\mathcal{S}_p = \hat{1} + (C^1, C^2) \text{ and } \mathcal{S}_h = \hat{1} + (C_1, C_2)\,, \tag{111}$$

where (A, B) designates the transposition of the spin-orbital labels A and B, and employ them whenever the particle or hole external lines are nonequivalent (in case of diagram 7, Fig. 13, either S_h or S_p can be used). Clearly, the action of the transposition (C_1, C_2) or (C^1, C^2) will change the number of closed loops by one [cf. Figs. 10 and 11(2)] and thus will introduce an additional phase factor (-1). Thus, when evaluating the diagrams in Fig. 13, we can simply associate with each symmetrizer $\mathcal{S}_x(x = p, h)$ the corresponding antisymmetrizer $\mathcal{A}_x(x = p, h)$,

$$\mathcal{A}_p = \hat{1} - (C^1, C^2) \text{ and } \mathcal{A}_h = \hat{1} - (C_1, C_2), \tag{111'}$$

(cf., Piecuch and Paldus, 1989,1990).

It thus remains to determine the Hugenholtz weights w_H and phase factors for each diagram, which we collected in Table II. Designating the contributions from the absolute ($j = 0$, diagram 1 of Fig. 13), linear ($j = 1$, diagrams 2-6 of Fig. 13) and quadratic ($j = 2$, diagrams 7-10 of Fig. 13) terms in the T_2 cluster components by $\Lambda^{(j)}(C^1C^2; C_1C_2)$, we can write the CCD equations as

$$\sum_{j=0}^{2} \Lambda^{(j)}(C^1C^2; C_1C_2) = 0, \tag{112}$$

where (we drop circumflexes over f, v and t_2 for simplicity)

$$\Lambda^{(0)}(C^1C^2; C_1C_2) = \langle C^1C^2|v|C_1C_2\rangle_A, \tag{113}$$

$$\begin{aligned}\Lambda^{(1)}(C^1C^2; C_1C_2) &= \sum_{B^1} \mathcal{A}_p \langle C^1|f|B^1\rangle\langle B^1C^2|t_2|C_1C_2\rangle_A \\ &- \sum_{B_1} \mathcal{A}_h \langle B_1|f|C_1\rangle\langle C^1C^2|t_2|B_1C_2\rangle_A \\ &+ \tfrac{1}{2} \sum_{B^1B^2} \langle C^1C^2|v|B^1B^2\rangle_A \langle B^1B^2|t_2|C_1C_2\rangle_A \\ &+ \tfrac{1}{2} \sum_{B_1B_2} \langle B_1B_2|v|C_1C_2\rangle_A \langle C^1C^2|t_2|B_1B_2\rangle_A \\ &- \sum_{B^1B_1} \mathcal{A}_p\mathcal{A}_h \langle B_1C^1|v|C_2B^1\rangle_A \langle B^1C^2|t_2|C_1B_1\rangle_A\end{aligned} \tag{114}$$

and

$$\begin{aligned}\Lambda^{(2)}(C^1C^2; C_1C_2) &= \sum_{B^1B^2B_1B_2} \langle B_1B_2|v|B^1B^2\rangle_A \\ &\times (\mathcal{A}_h \langle C^1B^1|t_2|C_1B_1\rangle_A \langle C^2B^2|t_2|C_2B_2\rangle_A \\ &+ \tfrac{1}{4} \langle C^1C^2|t_2|B_1B_2\rangle_A \langle B^1B^2|t_2|C_1C_2\rangle_A \\ &+ \tfrac{1}{2} \mathcal{A}_p \langle C^1B^1|t_2|B_1B_2\rangle_A \langle B^2C^2|t_2|C_1C_2\rangle_A \\ &+ \tfrac{1}{2} \mathcal{A}_h \langle B^1B^2|t_2|C_1B_1\rangle_A \langle C^1C^2|t_2|B_2C_2\rangle_A).\end{aligned} \tag{115}$$

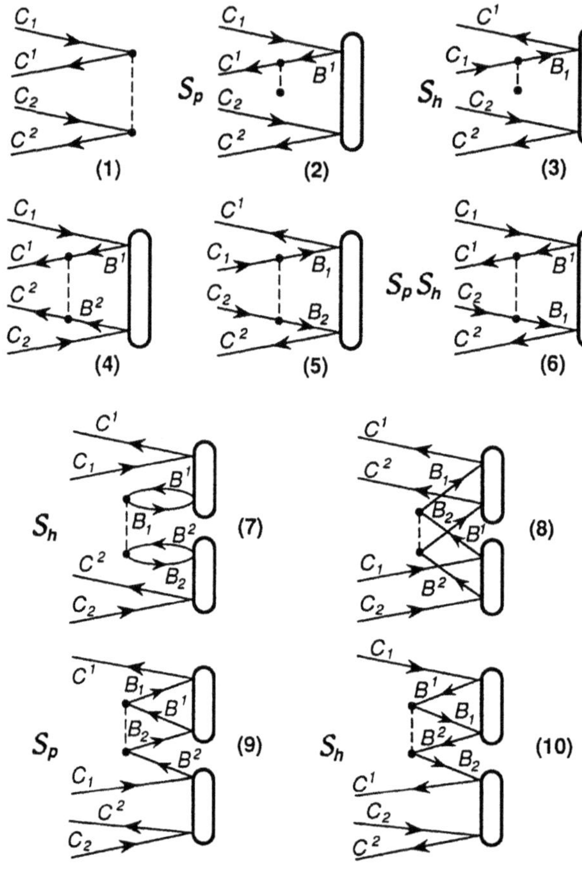

Fig. 13. Brandow diagrams representing the CCD equations. These diagrams are easily obtained from the skeletons of Fig. 12.

Table II. Hugenholtz weights w_H, number of internal hole lines (h) and closed loops (ℓ), the phase factor $\phi = (-1)^{h+\ell}$ and the overall factor f associated with the CCD Brandow-type R-diagrams of Fig. 13. Number of T_2-vertices involved is designated by j.

Diagram (Fig. 13)	j	w_H	$(h-2)$	$(\ell-2)$	ϕ	f
1	0	1	0	0	1	1
2	1	1	0	0	1	$\mathcal{A}_p$
3	1	1	1	0	-1	$-\mathcal{A}_h$
4	1	$\frac{1}{2}$	0	0	1	$\frac{1}{2}$
5	1	$\frac{1}{2}$	2	0	1	$\frac{1}{2}$
6	1	1	1	0	-1	$-\mathcal{A}_p\mathcal{A}_h$
7	2	1	2	2	1	$\mathcal{A}_h$
8	2	$\frac{1}{4}$	2	0	1	$\frac{1}{4}$
9	2	$\frac{1}{2}$	2	0	1	$\frac{1}{2}\mathcal{A}_p$
10	2	$\frac{1}{2}$	2	0	1	$\frac{1}{2}\mathcal{A}_h$

The CCD equations (112) that determine the T_2 cluster amplitudes are thus energy independent non-linear algebraic equations. Arranging the t_2 matrix elements $\langle D^1D^2|t_2|D_1D_2\rangle_A$ in some arbitrary but fixed order and designating them simply as the unknowns by x_j, the system (112) takes the form

$$a_i + \sum_j b_{ij}x_j + \sum_{j\leq k} d_{ijk}x_jx_k = 0\,, \tag{116}$$

where the absolute, linear and quadratic coefficients a_i, b_{ij} and c_{ijk}, respectively, are given by the matrix elements of the Hamiltonian H_N,

$$\begin{aligned} a_i &= \langle i|H_N|\Phi_0\rangle\,, \\ b_{ij} &= \langle i|H_N|j\rangle\,, \\ d_{ijk} &= \langle i|H_N|j,k\rangle_C\,, \end{aligned} \tag{117}$$

where $|i\rangle$ or $|j\rangle$ designate the appropriate biexcited configurations $\left|\begin{matrix} D^1D^2 \\ D_1D_2 \end{matrix}\right\rangle$ and $|j,k\rangle = |j\rangle|k\rangle$. Neglecting completely the inter-pair interaction terms, Eq. (115), we obtain a linear system of equations. This approximation has been termed the linear CPMET (L-CPMET) or L-CCD method. We shall examine its relationship with the MBPT and CI approaches in Section 4.6.

4.5.2. *Higher than CCD approximations*

We now very briefly outline an extension of the CCD approach in which T_1 and T_3 clusters are also explicitly considered. Here we emphasize that the desirable property of size extensivity will hold for any approximation, in which any subset of connected CC diagrams is considered. Following the same procedure as in the preceding section, it is straightforward to derive CC equtions for any chosen approximation scheme. Of course, once higher than two-body clusters are involved, rather formidable equation systems may arise.

The simplest extension of the basic CCD approach, which is also relatively easy to implement computationally, considers both one- and two-body clusters and results in the so-called CCSD approach. Here

$$T \approx T_1 + T_2 \,, \tag{118}$$

so that two coupled sets of non-linear equations for the t_1 and t_2 amplitudes must be simultaneously considered. Projecting onto the monoexcited, $|\Phi_{K_C}^{(1)}\rangle = \left| \begin{matrix} D^1 \\ D_1 \end{matrix} \right\rangle$, as well as biexcited, $|\Phi_{K_C}^{(2)}\rangle = \left| \begin{matrix} D^1 D^2 \\ D_1 D_2 \end{matrix} \right\rangle$, configurations in Eq. (87) and using the approximation (118), we can schematically write

$$\langle \Phi_{K_C}^{(1)} | H_N (1 + T_1 + T_2 + \tfrac{1}{2} T_1^2 + T_1 T_2 + \tfrac{1}{3!} T_1^3) | \Phi_0 \rangle_C = 0 \,, \tag{119a}$$

$$\begin{aligned} \langle \Phi_{K_C}^{(2)} | H_N (1 + T_1 + T_2 + \tfrac{1}{2} T_2^2 + \tfrac{1}{2} T_1^2 + T_1 T_2 + \tfrac{1}{3!} T_1^3 \\ + \tfrac{1}{2} T_1^2 T_2 + \tfrac{1}{4!} T_1^4) | \Phi_0 \rangle_C = 0 \,, \end{aligned} \tag{119b}$$

where again the subscript C indicates the connected part. Thus, the CCSD equations will also contain quartic terms in mono-excited cluster amplitudes. Although these higher order than quadratic terms (i.e., $\frac{1}{2}T_1^2 T_2$, $\frac{1}{3!}T_1^3$ and $\frac{1}{4!}T_1^4$) will only contribute in rather high orders of the MBPT (in fact, they contribute for the first time in the 5th, 6th and 8th order in the wave function and in the 7th, 8th and 10th order in the energy, assuming that we employ RHF orbitals; when non-HF orbital basis is used, as may be useful in handling of large systems using localized orbital bases, these terms may already contribute to the wave function in the 3rd and 4th order), they can be easily handled in numerical computations, since their number is small (Bartlett, 1989 and *loc. cit.*). As an illustration, we list in Fig. 14 the diagrams characterizing the first equation (119a), resulting from projection with monoexcited configurations. For simplicity, we list only nonoriented Hugenholtz skeletons and assume that HF orbitals are employed. These skeletons can be easily expanded into the relevant Brandow diagrams and corresponding CC equations derived.

When also triexcited cluster components are considered, many more terms will arise and the computational effort becomes very demanding, so that only relatively small problems can be handled at the present time. The first account of triexcited clusters was carried out in the early seventies (Paldus *et al.*, 1972) and showed rather clearly that very accurate correlation energies can be obtained in this approximation. The full CCSDT at the *ab initio* level (Noga and Barlett, 1987; Scuseria and Schaefer, 1988b) confirmed these expectations. However, as already pointed out earlier, simpler techniques were sought to account for a relatively small effect of three-body clusters and several, both practical and quite reliable schemes, are now to our disposal (see the references and discussion in Section 4.4).

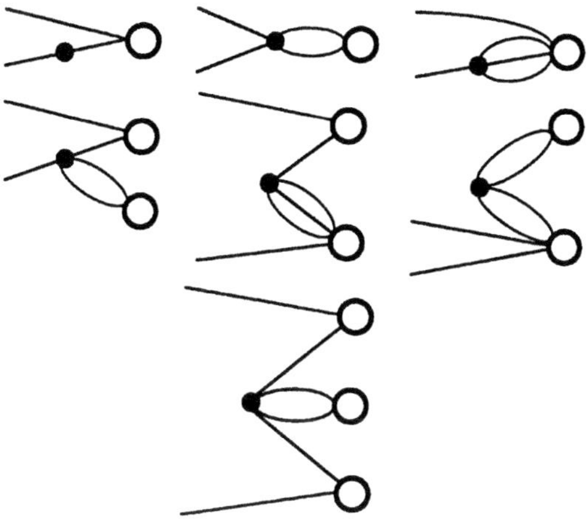

Fig. 14. Nonoriented Hugenholtz skeletons characterizing the monoexcited component of CCSD equations, Eq. (119a).

It is also worth remarking that in the case of 3-body clusters (and, similarly, triexcited configurations) we shall encounter for the first time linear dependencies when considering all distinct configurations. As was shown earlier (Paldus *et al.*, 1972), one can ignore these dependencies and consider larger system of CC equations, while introducing appropriate constraints. However, one can remove these dependencies, when properly spin-adapted CC equations are used (see Section 4.7). We thus end this section with an observation that the above outlined diagrammatic technique enables us to easily derive the explicit system of CC equations for any approximation considered and a reminder, that a very general list of relevant diagrams, involving even highly excited cluster components, was presented by Hoffmann and Schaefer (1986).

4.6. Relationship with other approaches

Let us now explore the relationship between the CC approaches, particularly the basic CCD method and its linear version L-CCD, and both the variational (i.e., CI) and perturbative (i.e., MBPT) approaches. This inter-relationship between these various methodologies will provide us with a better understanding of the approximations involved and the limitations that ensue.

4.6.1. *Variational approaches*

Consider, first, the truncation schemes that lead to the limited CI and corresponding CC approaches. We shall concentrate on the simplest, yet most basic, case involving the CCD method. Corresponding variational schemes may thus involve, in addition to the biexcited configuration manifold, also the higher even-number of times excited configurations. In particular, we shall deal with the D-CI (CI limited to doubly excited configurations) and DQ-CI (CI limited to doubly and quadruply excited configurations) approaches.

We thus write the CI expansion in terms of excitation operators C_k, Eq., (71), and assume the intermediate normalization $C_0 = \hat{1}$. Though it is irrelevant at this stage whether we consider the excitation operators in their simple spin-orbital form, as we have done so far, or in fact employ a spin-adapted form of these operators (see Section 4.7), it is nonetheless essential for the following derivation to assume that the relevant configurations $|\Phi_K^{(k)}\rangle = G_K^{(k)}|\Phi\rangle$ are orthonormal. Thus, writing

$$C_k = \sum_K c_K^{(k)} G_K^{(k)}, \tag{120}$$

we easily arrive at the well-known chain of CI equations

$$\sum_{\ell=k-2}^{k+2} \sum_L \left(H_{KL}^{(k\ell)} - \Delta E\, \delta_{KL}^{(k\ell)}\right) c_L^{(\ell)} = 0\,, \tag{121}$$

with $k = 0, 1, \ldots, N$ and K ranging over all orthonormal configurations of the given excitation order k. In the special case considered by us, we neglect the monoexcited terms, $C_1 = T_1 = 0$, as well as the higher odd-number of times excited configurations.

We can thus write the chain (121) in an explicit form as

$$\begin{gathered}
\sum_L H_{0,L}^{(0,2)} c_L^{(2)} = \Delta E \\
H_{K,0}^{(2,0)} + \sum_L H_{K,L}^{(2,2)} c_L^{(2)} + \sum_L H_{K,L}^{(2,4)} c_L^{(4)} = \Delta E\, c_K^{(2)}\,, \\
\sum_L H_{K,L}^{(4,2)} c_L^{(2)} + \sum_L H_{K,L}^{(4,4)} c_L^{(4)} + \sum_L H_{K,L}^{(4,6)} c_L^{(6)} = \Delta E\, c_K^{(4)}\,, \text{etc.}
\end{gathered} \tag{122}$$

Each Eq. (121) for $k > 0$ represents a system of equations for all possible ordered configurations K of the given excitation order k. When all equations for all $k = 0, 1, \ldots, N$ are considered, we obtain the full CI (FCI) approximation (representing the exact solution in a chosen model space). Various limited CI (L-CI) approximations then result by decoupling the FCI chain (121) or (122). For a given excitation order (or level) k, this is achieved by neglecting all $H_{KL}^{(k\ell)}$ matrix elements with $\ell > k$ (in our case $\ell = k + 2$ terms). Thus, the D-CI approximation results by neglecting all the $H_{KL}^{(2,4)}$ matrix elements in the second equation (122), yielding

$$\begin{aligned}
\sum_L a_L^* c_L &= \Delta E^{(\mathrm{D-CI})}\,, \\
a_K + \sum_L b_{KL} c_L &= \Delta E^{(\mathrm{D-CI})}\, c_K\,,
\end{aligned} \tag{123}$$

where

$$a_K = H^{(2,0)}_{K,0},\ b_{KL} = H^{(2,2)}_{K,L},\ c_K = c^{(2)}_K,\ K = 1,\ldots,M \tag{124}$$

with M designating the number of biexcited configurations. More compactly, we write (123) in the matrix form

$$\begin{bmatrix} 0 & \mathbf{a}^\dagger \\ \mathbf{a} & \mathbf{b} \end{bmatrix} \begin{bmatrix} 1 \\ \mathbf{c} \end{bmatrix} = \Delta E^{(\mathrm{D-CI})} \begin{bmatrix} 1 \\ \mathbf{c} \end{bmatrix}, \tag{125}$$

where $\mathbf{a} = [a_K]_{M\times 1}$, $\mathbf{b} = [b_{KL}]_{M\times M}$ and $\mathbf{c} = [c_K]_{M\times 1}$.

In a similar way we can truncate at the tetraexcited ($k = 4$) level by neglecting all $H^{(4,6)}_{K,L}$ matrix elements, obtaining DQ-CI equations, which will have the following general form

$$\begin{bmatrix} 0 & \mathbf{a}^\dagger & 0 \\ \mathbf{a} & \mathbf{b} & \mathbf{p}^\dagger \\ 0 & \mathbf{p} & \mathbf{q} \end{bmatrix} \begin{bmatrix} 1 \\ \mathbf{c}^{(2)} \\ \mathbf{c}^{(4)} \end{bmatrix} = \Delta E^{(\mathrm{DQ-CI})} \begin{bmatrix} 1 \\ \mathbf{c}^{(2)} \\ \mathbf{c}^{(4)} \end{bmatrix}, \tag{126}$$

where $\mathbf{p}$ and $\mathbf{q}$ are matrices of tetra-bi and tetra-tetra-excited CI matrix elements.

Let us now compare this decoupling with that characterizing the CC approaches. For the $T \approx T_2$ approximation, we can start from the same chain of equations (122) as in the CI case. However, the truncation of this chain is now achieved very differently, since all higher than biexcited configurations arise as powers of T_2, i.e.

$$C_{2j} \approx (j!)^{-1} T_2^j\,. \tag{127}$$

Consider, thus, DQ-CI equations (126) and employ the approximation (127). We note that in this case $C_2 = T_2$, since $T_1 = 0$, so that

$$c^{(2)}_K = t^{(2)}_K\,, \tag{128}$$

and the matrices $\mathbf{a}$ and $\mathbf{b}$ in (126) are the same as in (125).

It thus remains to consider the matrix elements $H^{(2,4)}_{K,L}$, which had to be neglected in order to achieve the decoupling in the D-CI case. In a diagrammatic language, these terms are given by the R-diagrams that can be formed from one $\bar{C}$ vertex, Fig. 11(1), representing the bra configuration, one H vertex, Fig. 1(1), 1(2) (or 1(1′), 1(2′)), representing the Hamiltonian, and two C-type biexcited vertices (represented by the same diagrams as T vertices, except that the corresponding scalar factors now equal 1) as shown schematically in Fig. 15(1). Clearly, in all these R-diagrams (skeletons), the fermion lines issuing from the bra $\bar{C}$-vertex and the interaction V-vertex must be joined with the lines issuing from the ket C-vertices (which in turn implies that in a nonvanishing $H^{(2,4)}_{KL}$ matrix element, the (spin)-orbital labels characterizing the configuration K must form a subset of those characterizing L), so that these R-diagrams can be of two basic types, namely connected or disconnected, as illustrated by skeletons (2) and (3) of Fig. 15. Evidently, the connected diagrams, Fig. 15(2), are precisely those arising from skeletons 4 and 5 of Fig. 12, while the disconnected ones factorize and equal

$$\Delta E \cdot c_K^{(2)} . \tag{129}$$

Thus, the disconnected terms exactly cancel the right hand side term in the second equation (122), decoupling at the same time the CI chain and yielding the energy independent CCD equation

$$a_K + \sum_L b_{KL} t_L^{(2)} + \sum_{L \leq M} d_{KLM} t_L^{(2)} t_M^{(2)} = 0 , \tag{130}$$

the last term originating from connected diagrams of the kind shown in Fig. 15(2),

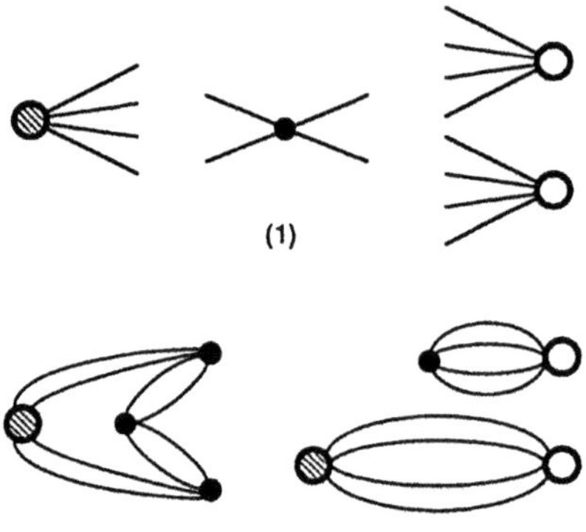

Fig. 15. Schematic representation of the CCD decoupling of the FCI chain of equations (122). Diagram 1 shows the relevant vertices for the construction of R-diagrams representing bi-tetraexcited CI matrix elements $H_{KL}^{(2,4)}$ using the pair-cluster approximation $C_4 \approx \frac{1}{2} T_2^2$. Hugenholtz skeletons 2 and 3 are typical representatives of linked and unlinked components of $H_{KL}^{(2,4)}$.

and thus being identical with the last term in Eq. (116). The energy is then given by the first equation (122) so that

$$\Delta E^{(\mathrm{CCD})} = \sum_K a_K^* t_K^{(2)} = \mathbf{a}^\dagger \mathbf{t}^{(2)} . \tag{131}$$

We have thus recovered the CCD equations directly from the CI chain (Čížek and Paldus, 1971b; Paldus *et al.*, 1989a). Although in this case the same number of amplitudes $t_K^{(2)}$ is involved as there are CI biexcited coefficients $c_K^{(2)}$ (in fact, Eq. (128) holds exactly in this approximation), the CCD approach does not simply neglect the important effect of tetraexcitations as the D-CI does, but approximates them by introducing the pair-pair interaction nonlinear terms, since now $C_4 \approx \frac{1}{2} T_2^2$. The CC decoupling procedure is thus more satisfactory from a physical viewpoint, even though the upper bound property of the variational approach is lost. More importantly, the CC approach is size extensive as warranted by the exponential ansatz for the wave operator, Eq. (59).

A few remarks are now in place. First of all, we note that higher order CC approaches may be similarly derived by appropriately decoupling the CI chain, as the CCD approach above (for the CCSD case, see Paldus *et al.*, 1989a). However, to obtain an equivalent set of CC equations that results directly from Eq. (87) one has to eliminate lower-excitation order equations from the higher-order ones (see Paldus *et al.*, 1989a). This may turn out to be quite laborious for the higher order cluster equations, while it is automatically achieved in Eq. (87), thanks to the exponential ansatz and factorization lemma.

Further, the CCD equations (130) are equivalent, though not necessarily identical, to the CCD system (116) (particularly when we employ the spin-adapted version of the theory, Section 4.7). In fact, in most CCD applications the matrix $[b_{ij}]$, Eq. (117), is not Hermitian, while $\mathbf{b}$, Eq. (125), being the biexcited state submatrix of the CI matrix, always will be. This difference arises from different normalizations of t-matrix elements that is often convenient to employ. However, the algebraic system of CCD equations (116) is not changed when we multiply its equations by arbitrary, nonzero constants. In this way we can always achieve that $\tilde{\mathbf{b}} = [b_{ij}]$ is Hermitian (or symmetric in the real case) and (116) is identical with the system (130). When this is the case, the CCD energy is given by a simple scalar product (131). We can thus assume, without restricting generality in any way, that the $\mathbf{a}$ and $\mathbf{b}$ matrices characterizing the CCD problem, Eq. (116), are identical with those forming the D-CI matrix, Eq. (125). This identification is important for revealing a deeper connection between the CI and CC approaches.

When solving the CCD system (116) or (130), it is often convenient to start with the solution of the corresponding linear system

$$a_K + \sum_L b_{KL}\tilde{t}_L^{(2)} = 0\,, \tag{132}$$

that defines the L-CCD approximation with the energy

$$\Delta E_L^{(\mathrm{CCD})} \equiv \Delta E_L = \mathbf{a}^\dagger \tilde{\mathbf{t}}\,. \tag{133}$$

The L-CCD energies, as already noted, are usually very close to the FCI ones (to within a few percent of the correlation energy) unless the HF reference is quasidegenerate with some low-lying biexcited configuration.

It is worth mentioning that within the scope of approximate coupled electron pair approximation (CEPA) methods, which may be regarded as simplified versions of the CCD or CCSD approaches (see, e.g., Jankowski and Paldus, 1980; Adams *et al.*, 1979,1981b), the L-CCD approximation is referred to as the CEPA(0) approach. We shall also give a simple MBPT interpretation of this approximation. Next, however, let us examine its relationship with the closely related D-CI approach.

Both the D-CI and L-CCD methods employ exactly the same information about the Hamiltonian of the system considered, namely that given by the submatrices $\mathbf{a}$ and $\mathbf{b}$ of the D-CI problem, Eq. (125). However, while the variational correlation energy $\Delta E^{(\text{D-CI})}$ is given by the lowest eigenvalue of the D-CI matrix, Eq. (125), the L-CCD energy $\Delta E_L \equiv \Delta E_L^{(\text{CCD})}$ is obtained by first solving a linear system of energy independent inhomogeneous algebraic equations (132) and, second, by evaluating the scalar product (133), i.e.

$$\mathbf{a} + \mathbf{b}\tilde{\mathbf{t}} = 0, \qquad \Delta E_L \equiv \Delta E_L^{(\text{CCD})} = \mathbf{a}^\dagger \tilde{\mathbf{t}} . \tag{134}$$

Obviously, the L-CCD approach breaks down if $\mathbf{b}$ is singular. This will be the case if one or more eigenvalues of $\mathbf{b}$ vanish. However, in such case there exists a state in the doubly excited subspace which is degenerate with the reference state $|\Phi\rangle \equiv |\Phi_0\rangle$, so that our basic assumption that $|\Phi_0\rangle$ is nondegenerate breaks down and the degenerate or multi-reference version of the theory (Chapter 5) should be used.

In view of this similarity between the L-CCD and D-CI approaches, it is not surprising that both energies $\Delta E^{(\text{D-CI})}$ and $\Delta E_L^{(\text{CCD})}$ can be obtained as an extremal value of the same general functional (Ahlrichs, 1979; Paldus *et al.*, 1982)

$$G(\mathbf{X}) = \mathbf{X}^\dagger \mathbf{H}^{(\text{D})} \mathbf{X} , \tag{135}$$

where

$$\mathbf{H}^{(\text{D})} = \begin{bmatrix} 0 & \mathbf{a}^\dagger \\ \mathbf{a} & \mathbf{b} \end{bmatrix} , \tag{136}$$

with $\mathbf{a}$, $\mathbf{b}$ given by (125). In the D-CI case, $\mathbf{X}$ represents a unit normalized vector

$$\mathbf{X}^{(\text{D-CI})} \equiv \mathbf{Y} / \parallel \mathbf{Y} \parallel = \mathbf{Y}(\mathbf{Y}^\dagger \mathbf{Y})^{-\frac{1}{2}} , \tag{137}$$

with $\mathbf{Y}$ arbitrary, while in the L-CCD case, $\mathbf{X}$ is intermediately normalized,

$$\mathbf{X}^{(\text{L-CCD})} = \begin{bmatrix} 1 \\ \mathbf{x} \end{bmatrix} . \tag{138}$$

Indeed, this gives

$$G(\mathbf{X}^{(\text{D-CI})}) = \mathbf{Y}^\dagger \mathbf{H} \mathbf{Y} / (\mathbf{Y}^\dagger \mathbf{Y}) , \tag{139}$$

a standard quantum mechanical variational ansatz whose minimization leads to the DCI eigenvalue problem

$$\mathbf{H}^{(\text{D})} \mathbf{Y}_i = \lambda_i \mathbf{Y}_i , \tag{140}$$

that is equivalent to (125), so that $\Delta E^{(\text{D-CI})} = \min\{\lambda_i\} \equiv \lambda_0 = \min G(\mathbf{X}^{(\text{D-CI})})$, or

$$G(\mathbf{X}^{(\text{L-CCD})}) = \mathbf{x}^\dagger \mathbf{a} + \mathbf{a}^\dagger \mathbf{x} + \mathbf{x}^\dagger \mathbf{b} \mathbf{x} , \tag{141}$$

in which case the condition for an extremum becomes

$$\mathbf{a} + \mathbf{b}\mathbf{x} = 0 , \tag{142}$$

that is identical with (132) or (134) when we identify the unknowns $\mathbf{x}$ and $\tilde{\mathbf{t}}$, and

$$\Delta E_L = \mathbf{a}^\dagger \mathbf{x}_e = -\mathbf{a}^\dagger \mathbf{b}^{-1} \mathbf{a} = G\left(\begin{bmatrix} 1 \\ \mathbf{x}_e \end{bmatrix} \right) , \tag{143}$$

with $\mathbf{x}_e$ designating the solution of (142).

This formulation provides us with an interesting geometrical interpretation. Both the D-CI and L-CCD energies can be found as an extremum of the same functional $G(\mathbf{X})$ but with different constraints: in the D-CI case, the trial vectors $\mathbf{X}$ are restricted to the unit sphere $\| \mathbf{X} \| = 1$, while in the L-CCD case they must belong to the tangent plane $x_0 = 1$ to this unit sphere at the point corresponding to the reference state $|\Phi_0\rangle$. This interpretation also nicely explains a very different behavior of both correlation energies: the D-CI is proportional to the square root of the particle number N when a system of weakly interacting subsystems is considered with $N \to \infty$, while the L-CCD energy is size extensive and thus proportional to N. Indeed, in the D-CI case, the variational manifold is compact, namely a unit sphere in the $(M+1)$-dimensional Euclidean space, while in the L-CCD case, the variational manifold is a linear space given by the tangent hyperplane to the unit sphere just mentioned, at the point characterizing the reference state $|\Phi_0\rangle$. Thus, if we would regard the D-CI variational manifold as a Lie group, then the corresponding L-CCD manifold is given by the corresponding Lie algebra. Consequently, the noninteracting systems behave additively in the linear L-CCD space, so that the size-extensivity holds, while this is not the case in the D-CI manifold, which has a multiplicative, rather than additive, structure. In view of the exponential ansatz characterizing the CC approaches, the size extensivity is preserved in the CCD and higher approaches as well, even though the respective variational functional may not exist.

To establish an explicit relationship between the D-CI and L-CCD energies (Paldus *et. al.*, 1982), we observe that the L-CCD equations (134) or (142) are closely related with matrix inverse of $\mathbf{H}^{(\mathrm{D})}$, Eq. (136). Blocking $[\mathbf{H}^{(\mathrm{D})}]^{-1}$ in the same way as $\mathbf{H}^{(\mathrm{D})}$,

$$[\mathbf{H}^{(\mathrm{D})}]^{-1} = \begin{bmatrix} p & \mathbf{q}^{\dagger} \\ \mathbf{q} & \mathbf{r} \end{bmatrix}, \tag{144}$$

with $\mathbf{r}$ Hermitian, we find immediatley that $\mathbf{H}^{(\mathrm{D})}[\mathbf{H}^{(\mathrm{D})}]^{-1} = \mathbf{1}$ implies

$$\begin{aligned} \mathbf{a}^{\dagger}\mathbf{q} &= 1\,, \\ \mathbf{a}p + \mathbf{b}\mathbf{q} &= \mathbf{0}\,. \end{aligned} \tag{145}$$

Since $p \neq 0$, this last condition becomes identical with the L-CCD system (142) when we set

$$\mathbf{x}_e \equiv \tilde{\mathbf{t}} = p^{-1}\mathbf{q}\,, \tag{146}$$

so that

$$\Delta E_L^{(\mathrm{CCD})} = p^{-1} = \left([\mathbf{H}^{(\mathrm{D})}]^{-1}\right)_{00}^{-1}, \tag{147}$$

i.e.

$$\Delta E_L^{(\mathrm{CCD})} = \left[\sum_k |y_{0k}|^2 \lambda_k^{-1}\right]^{-1}, \tag{148}$$

where y_{0k} are components of the eigenvector $\mathbf{Y}_0$, Eq. (140), corresponding to the lowest eigenvalue λ_0, so that

$$\Delta E^{(D-CI)} = \lambda_0\,. \tag{149}$$

We also note that the relationship (148) leads immediately to the Davidson (1974) correction and related estimates of the correlation energy, since to a very good approximation we get from (148) that (Paldus *et al.*, 1982)

$$\Delta E_L^{(\mathrm{CCD})} \approx \lambda_0 |y_{00}|^{-2} . \tag{150}$$

We have already seen above that the source of the size inconsistency of the D-CI approach is the presence of unlinked terms, Eq. (129), which in the CCD or L-CCD approach are cancelled by the unlinked terms $\frac{1}{2}\langle\Phi_K^{(2)}|H_N T_2^2|\Phi_0\rangle_U$ arising from the disconnected $\frac{1}{2}T_2^2$ component (note that this cancellation is carried out before we truncate the CCD approach to L-CCD). It must be noted, however, that this cancellation involves both EPV and non-EPV diagrams (cf. Section 3.2), since the summations over the (spin)-orbital labels in each factor of (129) are independent. As a consequence, L-CCD "overcorrects" the D-CI result and the EPV terms are then again corrected for through the non-linear EPV terms in the CCD approach. The most important terms in this respect are those represented by diagrams that may be separated over one or two hole lines (diagrams 8 and 10 of Fig. 13). These are precisely the terms that form the basis of various CEPA(k), $k \neq 0$ approaches. In fact, it can be shown that the contributions from the remaining non-linear terms (diagrams 7 and 9 of Fig. 13) almost precisely cancel out, at least when no quasidegeneracy is present (Jankowski and Paldus, 1980; Adams *et al.*, 1981b; Chiles and Dykstra, 1981; Bachrach *et al.*, 1981). The methods neglecting these terms (so called ACP or ACCD or ACPQ approaches) can be shown to correct, under certain circumstances, for the effect of connected quadruply excited clusters that become important in quasidegenerate situations (see, e.g., references above and Paldus *et al.*, 1984; Takahashi and Paldus, 1985; Piecuch *et al.*, 1990), and invariably yield better correlation energies than the full CCD (or CCSD when also one-body clusters are considered) approach.

In this regard it is interesting to compare D-CI and L-CCD approaches for the case of two-electron systems (see also Pople *et al.*, 1987, 1989 and Paldus *et al.*, 1989a,1990). In this special case, D-CI represents the FCI and will yield an exact result within the model space considered, while the L-CCD result will only give an approximate energy. This is due to the fact that the latter approach includes a correction for the unlinked tetraexcited clusters. However, all such components are unphysical in this case or, in other words, are of the EPV type. These terms will then be subsequently cancelled by the non-linear CCD (or CCSD) terms, thus restoring the exact D-CI result, when performing the CC(S)D calculation. Obviously, all the non-linear CCD terms in the two-electron case considered are also of the EPV type. In fact, in the two-electron case, all the terms that distinguish between the D-CI, CCD and L-CCD approximations are of this type.

Clearly, the CCD (or higher CC approaches) or its simple linear version L-CCD, were designed to study the many-electron correlation effects in many-electron, rather than two-electron, systems. For these systems, the number of non-EPV unlinked contributions will be much larger than the number of corresponding EPV terms. Consequently, the corrections for the unlinked tetraexcited terms, restoring the size extensivity that is invalidated in the D-CI approach (except for the 2-electron systems), are bound to be much more important than the "overcorrection" due to the inclusion of the EPV terms that are necessary to achieve the factorization and that are corrected for when the proper "coupling" of pair clusters is taken into account in the full CC(S)D approach.

4.6.2. *Perturbative approaches*

The relationship between the CC approaches and the MBPT is most obvious in the diagrammatic representation. For this purpose, we first isolate the diagonal terms b_{ii}, Eq. (116) [or b_{KK}, Eq. (130)] from the remaining linear CCD terms, since they involve the one-particle orbital energies ω_A forming the denominators of the PT contributions, i.e. we write equations (116) in the form

$$-b_{ii}x_i = a_i + \sum_j{}' b_{ij}x_j + \sum_{j\leq k} d_{ijk}x_jx_k \,, \tag{151}$$

where the prime on the first summation symbol indicates the omission of the diagonal term. Further, let us separate the one-particle (spin)-orbital energy contribution to the diagonal term (arising from diagrams 2 and 3 of Fig. 13), designating it by Ω_i, so that

$$b_{ii} = \Omega_i + \Delta_i \,,$$

where

$$\Omega_i = \omega_{C^1} + \omega_{C^2} - \omega_{C_1} - \omega_{C_2} \,, \tag{152}$$

assuming that $K_i \equiv \{C^1C^2; C_1C_2\}$. Thus, Δ_i designates the two-electron part of b_{ii} given by diagrams 4-6 of Fig. 13. In order to achieve some flexibility in the choice of PT denominators (or, equivalently, in the separation of the Hamiltonian into the unperturbed component and the perturbation), we define

$$b_{ii} = b_{ii}^{(0)} + b_{ii}^{(1)} \,, \tag{153}$$

with

$$\begin{aligned} b_{ii}^{(0)} &= \Omega_i + (1-\lambda)\Delta_i \,, \\ b_{ii}^{(1)} &= \lambda\Delta_i \,, \qquad \lambda \in [0,1] \,. \end{aligned} \tag{154}$$

The $\lambda = 1$ limit then corresponds to the Rayleigh-Schrödinger PT (RSPT), briefly described in Chapter 3, that is also referred to in this context as the Møller-Plesset PT, assuming that the HF reference is used. On the other hand, the $\lambda = 0$ limit corresponds to the Epstein-Nesbet PT that results when we start from the CI matrix, use its diagonal as the unperturbed Hamiltonian and treat the off-diagonal part as the perturbation. Thus, splitting the diagonal matrix elements b_{ii} as indicated by Eqs. (153) and (154), we can go continuously from the Møller-Plesset form ($\lambda = 1$) to the Epstein-Nesbet one ($\lambda = 0$).

Using now the partitioning (153) in Eq. (151) we get

$$-b_{ii}^{(0)}x_i = a_i + b_{ii}^{(1)}x_i + \sum_j{}' b_{ij}x_j + \sum_{j\leq k} d_{ijk}x_jx_k \,. \tag{155}$$

Similarly as in the Jacobi algorithm for an iterative solution of a system of linear algebraic equations, we can write Eq. (155) in a recursive form

$$x_i^{[n+1]} = -(b_{ii}^{(0)})^{-1}\left[a_i + b_{ii}^{(1)}x_i^{[n]} + \sum_j{}' b_{ij}x_j^{[n]} + \sum_{j\leq k} d_{ijk}x_j^{[n]}x_k^{[n]}\right], \tag{156}$$

where the superscript $[n]$ designates the n-th iterate.

In order to start with the chosen reference state, we take vanishing components for the initial approximation,

$$x_i^{[0]} := 0\,, \tag{157}$$

so that

$$x_i^{[1]} = -a_i/b_{ii}^{(0)} \tag{158}$$

and

$$\Delta E_L^{[1]} = -\sum_i |a_i|^2/b_{ii}^{(0)}\,, \tag{159}$$

obtaining the well-known second order PT result. In subsequent iterations we shall be obtaining higher and higher orders of the PT. It must be noted, however, that subsequent iterates do not simply correspond to the next PT order but, due to the presence of the non-linear terms in Eq. (156), various orders of PT theory become intermixed.

To obtain a better idea about the MBPT structure of the CCD (or higher level) theory, it is instructive to carry out the above outlined iterative procedure diagrammatically. When we choose $\lambda = 1$, we can simply ignore the one-particle diagrams 2 and 3 of Fig. 13 (by incorporating them via Eq. (156) into the denominator rule of the MBPT) and interpret the remaining diagrams as PT wave function diagrams (cf. Chapter 3). In this way Eq. (156) can be written in the following symbolic manner

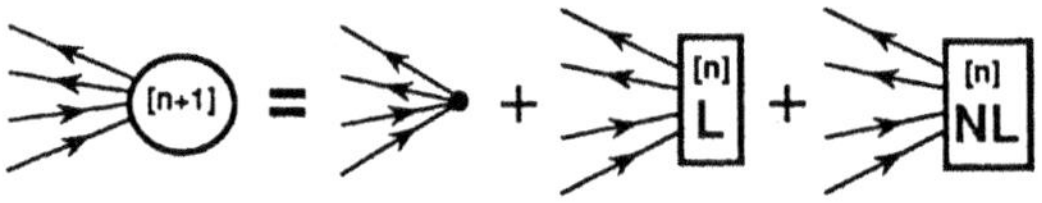

(156′)

where the factor $\left(b_{ii}^{(0)}\right)^{-1}$ is now implicitly included in the right hand side diagrams as the leftmost denominator factor and the last two terms are obtained from the linear (L) and nonlinear (NL) CCD diagrams of Fig. 13 by replacing their t_2 vertices with the n-th iterate pair cluster contribution to the exact wave function, represented by the diagram shown [for the $(n+1)$-st order] on the left hand side of Eq. (156′). Thus, using again (157) as the starting approximation, we obtain for the first order iterate from Eq. (156′)

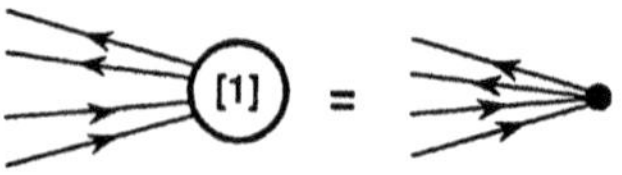

(158′)

which yields the energy $\Delta E_L^{[1]}$ when "substituted" into the energy diagram 3 of Fig. 2,

$$\Delta E_L^{[1]} =$$  (159′)

We thus recover the 2nd order MBPT energy assuming that we are using HF reference.

Consider now, at least very schematically, the next iteration. From the linear terms in Eq. (156′) we get the following contributions

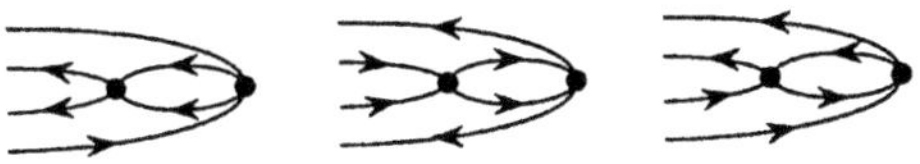 (160)

that yield the 3rd order MBPT energy (cf. Fig. 3) when substituted into the diagram 3 of Fig. 2. Thus, with the HF reference, the first non-trivial Jacobi iteration of the L-CCD system yields the energy correct up to the 3rd order of the MBPT.

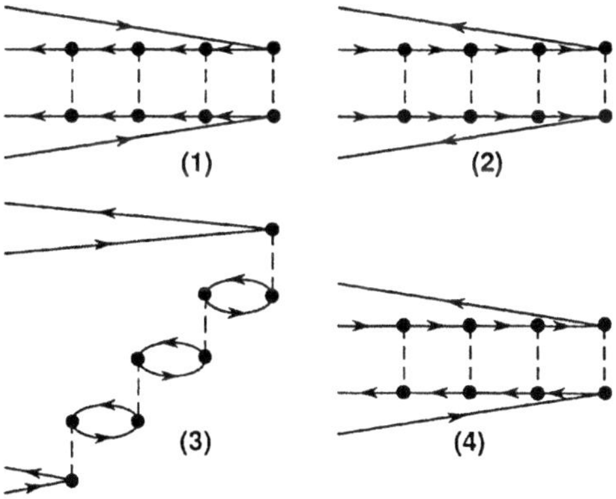

Fig. 16. Examples of the Goldstone-type wave function diagrams that are accounted for in all orders of PT in the L-CCD, or equivalent CEPA(0) or DMBPT(∞), approaches.

In higher orders of PT these linear terms yield all possible ladder-type diagrams (1 and 2 of Fig. 16), ring diagrams (3 of Fig. 16) and ring diagrams with exchange (4 of Fig. 16), as well as their various combinations. We note, that both ladders and rings (with or without exchange) can be summed to infinite order of PT by solving a

Dyson-like equation of the general propagator or Green's function theory and yield, respectively, Brueckner and the random phase approximation (RPA). It is noteworthy that both these approximations represent special cases of the L-CPMET approach, which includes not only all ladders and rings but also their various combinations. It must be added, however, that the above mentioned approximations have been developed to treat various *extended* systems (nuclear matter, electron gas, etc.) and that it is very difficult to solve the CC equations for these models, since they become a system of integro-differential equations (see, e.g., Freeman, 1977; Bishop and Lührmann, 1978,1982; Bishop, 1991).

We also see immediately from the diagrams (160) and of Fig. 16 that the subsequent iterations of the linear part always yield doubly excited wave function diagrams with at most doubly excited intermediate states. Consequently, a systematic iteration of the linear part of (156′) yields all possible PT diagrams, to all orders, in which all intermediate states are doubly excited configurations. Such an approximation was independently carried out by Bartlett and Shavitt (1977) from the standpoint of the MBPT, and was termed the DMBPT(∞) approximation. We thus find that both CEPA(0) and DMBPT(∞) approaches are equivalent with the L-CCD approximation. In this regard it is interesting to compare L-CCD and D-CI approaches within the "canonical" representation of the biexcited subspace that diagonalizes $\mathbf{b}$, Eq. (125),

$$\mathbf{b}\mathbf{v} = \mathbf{v}\boldsymbol{\beta}, \qquad \mathbf{v}^{\dagger}\mathbf{v} = \mathbf{v}\mathbf{v}^{\dagger} = \mathbf{1}, \qquad \boldsymbol{\beta} = [\beta_i \delta_{ij}], \tag{161}$$

i.e., by transforming the doubly-excited subspace to its "principal axes". Transforming, correspondingly, the D-CI matrix, we get an equivalent matrix

$$\begin{bmatrix} 1 & 0 \\ 0 & \mathbf{v} \end{bmatrix}^{\dagger} \begin{bmatrix} 0 & \mathbf{a}^{\dagger} \\ \mathbf{a} & \mathbf{b} \end{bmatrix} \begin{bmatrix} 1 & 0 \\ 0 & \mathbf{v} \end{bmatrix} = \begin{bmatrix} 0 & \boldsymbol{\alpha}^{\dagger} \\ \boldsymbol{\alpha} & \boldsymbol{\beta} \end{bmatrix}, \tag{162}$$

where

$$\boldsymbol{\alpha} = \mathbf{v}^{\dagger}\mathbf{a}, \tag{163}$$

having the same eigenvalues as the original D-CI matrix, while the L-CCD problem becomes

$$\boldsymbol{\alpha} + \boldsymbol{\beta}\boldsymbol{\tau} = 0, \qquad \Delta E_L^{(\mathrm{CCD})} = \mathbf{a}^{\dagger}\tilde{\mathbf{t}} = \boldsymbol{\alpha}^{\dagger}\boldsymbol{\tau}. \tag{164}$$

Thus

$$\Delta E_L^{(\mathrm{CCD})} = -\boldsymbol{\alpha}^{\dagger}\boldsymbol{\beta}^{-1}\boldsymbol{\alpha} = -\sum_{i=1}^{M} |\alpha_i|^2/\beta_i, \tag{165}$$

so that the whole L-CCD energy is obtained in the 2nd order of the MBPT. This expression also reveals directly the singular behavior of the L-CCD approximation as $\beta_i \rightarrow 0$. This is in fact the simplest case of an intruder state problem, since the singularity occurs when one of the "canonical" (in practice the lowest) biexcitations becomes degenerate with the reference state. Finally, we note that the corresponding variational D-CI problem is now characterized by the following secular equation

$$\det \begin{bmatrix} -\Delta E^{(\mathrm{D-CI})} & \boldsymbol{\alpha}^{\dagger} \\ \boldsymbol{\alpha} & \boldsymbol{\beta} - \Delta E^{(\mathrm{D-CI})}\mathbf{1} \end{bmatrix} = 0, \tag{166}$$

which can be expressed in the form

$$\Delta E^{(\mathrm{D-CI})} = -\sum_{i=1}^{M} |\alpha_i|^2/(\beta_i - \Delta E^{(\mathrm{D-CI})}). \tag{167}$$

Thus, the D-CI energy $\Delta E^{(\mathrm{D-CI})}$ is given by the 2nd order Brillouin-Wigner PT, rather than by the RSPT which yields the L-CCD energy $\Delta E_L^{(\mathrm{CCD})}$, Eq. (165). This is also consistent with the fact that the latter is size extensive, while the former one is not.

Consider, finally, the MBPT contributions resulting from the nonlinear CCD terms. Using the first non-trivial approximation for the t_2 clusters, Eq. (158′), on the right hand side of Eq. (156′), the nonlinear terms yield the 3rd order contribution to the wave function (the 4th order contribution to the energy). For example, the contribution from the diagram 9 or 10 of Fig. 13 (ignoring line orientation for simplicity, i.e. from diagram 5 of Fig. 12) gives the diagram

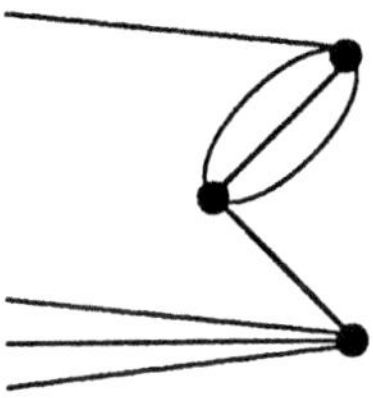

. (168)

If we recall that the time ordering of the T_i vertices was immaterial (since they mutually commute), we immediately realize that the above diagram (168) represents in fact the sum of all possible time versions of the first kind, namely,

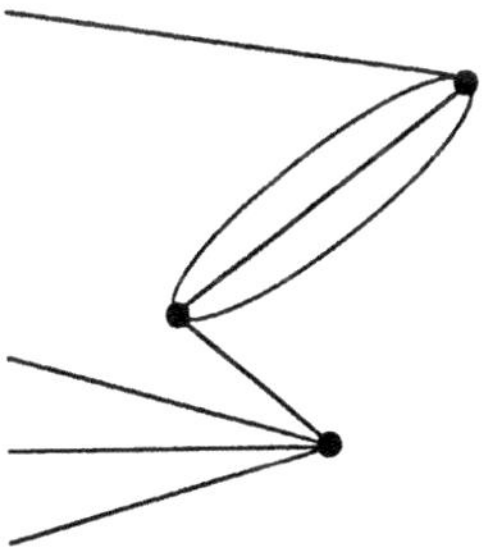

and

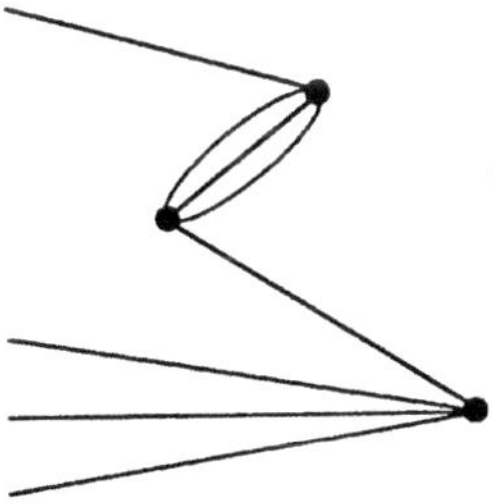

, (169)

as may be easily verified using the factorization lemma (Section 3.3). It is then easy to draw all the fourth order contributions to the energy that arise from the CCD nonlinear terms in the second iteration: they are easily obtained from the Hugenholtz skeletons of Fig. 17 by introducing all possible line orientations.

In the subsequent iteration we use all the diagrams (160) and all possible diagrams of the type (169) and replace with them the t_2-vertices on the right hand side of (156′). The linear terms simply increase the PT order by one in each iteration. However, using the k-th and ℓ-th order diagrams in the nonlinear part, we obtain the $(k+\ell+1)$-th order contribution.

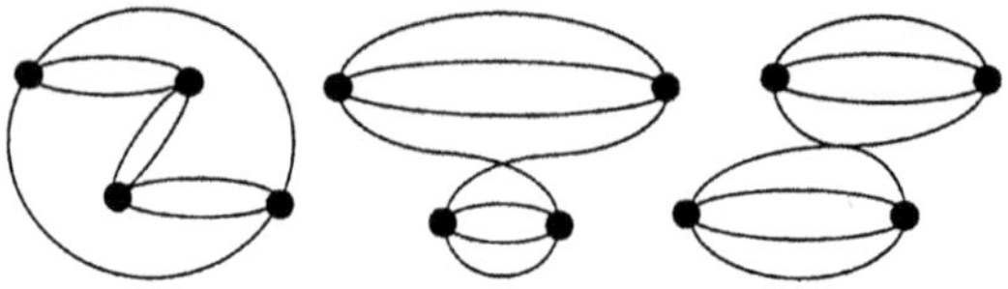

(1) (2) (3)

Fig. 17. Nonoriented Hugenholtz skeletons characterizing the fourth order MBPT energy contributions obtained with the CCD method.

We can thus regard the CCD (or, in general, CCX, X = D, SD, SDT, etc.) equations as a convenient recursive scheme for a systematic generation of higher and higher order MBPT diagrams. By solving these equations, we not only implicitly construct all these diagrams, but we also simultaneously evaluate them and sum them to infinite order of PT. In fact, we can employ the general CC approach as a simple way to formulate the MBPT and to obtain the linked cluster expansion for the energy. When all cluster components are accounted for, the result is completely equivalent to the MBPT. When approximations are introduced (such as the pair approximation $T \approx T_2$, for example), certain diagrams are omitted from the MBPT expansion. In the CCD approach, for example, the following 4th order diagrams are missing

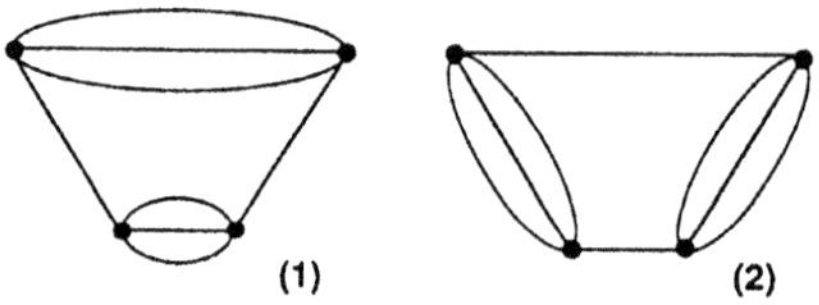

(1) (2) (170)

since they involve the connected triexcited (diagram 1) and monoexcited (diagram 2) intermediate states. Thus, in order to include these diagrams, we have to use CCSDT or CCSD approximations, respectively. However, except for the diagrams of the type (170), all the 4th order diagrams are accounted for by the CCD method. The reader can easily analyze the MBPT content of any CC method in any order of PT (see also, Bartlett and Kucharski, 1986; He and Cremer, 1991).

4.7 Spin-adapted CC formalism

An important invariance property of non-relativistic atomic and molecular electronic Hamiltonians is their spin independence. Generally, we can express a $2n$-dimensional spin-orbital one-electron space $\mathcal{H}_1$ as a tensor product of an n-dimensional orbital space $\mathcal{H}_1^{(\mathrm{orb})}$, $\mathcal{H}_1^{(\mathrm{orb})} = \mathrm{Span}\{|a\rangle\}$, and a two-dimensional spin space $\mathcal{H}_1^{(\mathrm{spin})}$, $\mathcal{H}_1^{(\mathrm{spin})} = \mathrm{Span}\{|\frac{1}{2}\rangle \equiv |\frac{1}{2},\frac{1}{2}\rangle, |-\frac{1}{2}\rangle \equiv |\frac{1}{2}-\frac{1}{2}\rangle\}$. As a spin-orbital basis for $\mathcal{H}_1$ it is then convenient to choose a standard tensor product basis, namely

$$|A\rangle = |a\rangle \otimes |\alpha\rangle \equiv |a\rangle|\alpha\rangle\,, \tag{171}$$

designating the corresponding orbital and spin functions by the lower case letters of the Latin and Greek alphabets, respectively.

In view of the spin-independence of the forces involved, the one- and two-body matrix elements simplify as follows

$$\langle A|\hat{z}|B\rangle = \langle a|\hat{z}|b\rangle\langle\alpha|\beta\rangle\,, \tag{172a}$$

$$\langle AB|\hat{v}|CD\rangle = \langle ab|\hat{v}|cd\rangle\langle\alpha|\gamma\rangle\langle\beta|\delta\rangle\,, \tag{172b}$$

so that the electronic Hamiltonian (10) or (95) takes the form

$$\hat{H} = \sum_{a,b}\langle a|\hat{z}|b\rangle E_{ab} + \tfrac{1}{2}\sum_{a,b,c,d}\langle ab|\hat{v}|cd\rangle N[E_{ac}E_{bd}]\,, \tag{173}$$

where E_{ab} designates a U(n) generator that represents a partial trace of the corresponding spin-orbital operator E_{AB} over the spin space, i.e.

$$E_{ab} = \sum_{\alpha=-1/2}^{1/2} E_{a\alpha,b\alpha} = \sum_{\alpha} X^{\dagger}_{a\alpha}X_{b\alpha}\,. \tag{174}$$

Clearly, the Hamiltonian (173) commutes with the total spin operators $\hat{S}_z$ and $\hat{S}^2$. Thus, the explicit dependence on spin-variables can be eliminated from our formulas and spin-free formalism obtained. In this way the number of cluster amplitudes can be significantly reduced (by about a factor of two for pair cluster components) and the repetitions in the one- and two-electron integral lists avoided.

Elimination of the explicit spin-dependence can be achieved in various ways. The simplest procedure relies on the fact that in view of the spin-independence of the interactions involved, the spin-orientation must be preserved along each fermion path. Thus, similarly as in the MBPT case, we can simply assign a factor of two to each closed loop of oriented lines, assuming that we employ Goldstone diagrams (Čížek, 1966a; Paldus and Čížek, 1975). It can be shown, however, that this simple spin-adaptation procedure leads to the cluster amplitudes that correspond to a nonorthogonal set of configuration states and, in the case when higher than biexcited cluster components are considered, even to a linearly dependent set (Paldus *et al.*, 1972). Although this linear dependence only marginally increases the size of the system of CC equations, and the linear dependencies can be eliminated if desired, this formulation lacks the symmetry characteristics that we normally employ in variational approaches, thus making the comparison with spin-adapted limited CI methods more complicated. It is thus desirable to carry out a spin-adaptation that is based on orthogonal configuration states, similarly as in the CI approaches. We then refer to such a procedure as an *orthogonal* spin-adaptation. Such an adaptation can be achieved in a similar way as in the CI methods. However, since the CC formulation is usually brought to the spin-orbital level using the diagrammatic approach, as we have seen above, it is most straightforward to employ graphical methods of spin algebras (Jucys *et al.*, 1960; Jucys and Bandzaitis, 1977; Brink and Satchler, 1968; El Baz and Castel, 1972; Lindgren and Morrison, 1982; Wilson, 1984). A simple introduction that is sufficient for the present purposes can be found in the Appendix of the paper by Paldus *et al.* (1977). This procedure can be extended to account for other symmetries as well, in particular to obtain the angular momentum adapted CC formalism for atoms (Adams and Paldus, 1981; Adams *et al.*, 1981; see also Stedman, 1990).

Since it is beyond the scope of this review to go into details of spin-algebra based spin-adaptation procedures, we only briefly mention the underlying principles, relying on the basic CCD method, and refer the reader to the original literature. We first have to choose a suitable coupling scheme leading to spin-adapted configurations (clearly, for closed-shell ground states, only singlet coupled states are required). It can be easily seen that the most appropriate is the particle-particle–hole-hole (*pp-hh*) coupling scheme (Čížek, 1966b; Paldus *et al.*, 1977), which leads to the following spin-adapted biexcited configurations*)

$$\left| \begin{matrix} r & s \\ a & b \end{matrix} \right\rangle_{S_i} = N_{ab}^{rs}\,(2S_i+1)^{-1} \sum_{\rho,\sigma,\alpha,\beta} \sum_{\tau} \langle \tfrac{1}{2}\alpha, \tfrac{1}{2}\beta | S_i \tau \rangle \times \langle \tfrac{1}{2}\rho, \tfrac{1}{2}\sigma | S_i \tau \rangle X_{r\rho}^{\dagger} X_{a\alpha} X_{s\sigma}^{\dagger} X_{b\beta} |\Phi\rangle\,, \tag{175}$$

where $\langle s_1\sigma_1,\, s_2\sigma_2 | s_3\sigma_3\rangle$ are SU(2) Clebsch-Gordan (CG) coefficients, and the normalization factor N_{ab}^{rs},

$$N_{ab}^{rs} = [(1+\delta_{rs})(1+\delta_{ab})]^{-\frac{1}{2}}\,, \tag{176}$$

assures normalization in case of double occupancies. For simplicity, the virtual orbitals are labeled by r, s and the occupied ones by a, b. The intermediate spin coupling quantum number S_i assumes only two values: $S_i = 0$ or 1 for singlet and triplet intermediate coupling, respectively. The resulting states (175) possess the following symmetry properties

$$\left| \begin{matrix} r & s \\ a & b \end{matrix} \right\rangle_{S_i} = \left| \begin{matrix} s & r \\ b & a \end{matrix} \right\rangle_{S_i} = (-1)^{S_i} \left| \begin{matrix} r & s \\ b & a \end{matrix} \right\rangle_{S_i} = (-1)^{S_i} \left| \begin{matrix} s & r \\ a & b \end{matrix} \right\rangle_{S_i}, \tag{177}$$

so that the triplet coupled configuration ($S_i = 1$) automatically vanishes when $a = b$ and/or $r = s$. Defining the pair-cluster amplitudes in the same way (Paldus, 1977) we can write the expansion (110) for the doubly excited component $T_2|\Phi\rangle$ of the exact wave function (or, more precisely, its singlet component) as

$$\hat{T}_2|\Phi\rangle = \sum_{\substack{a\leq b \\ r\leq s}} \sum_{S_i} \langle rs|\hat{t}_2|ab\rangle_{S_i} \left| \begin{matrix} r & s \\ a & b \end{matrix} \right\rangle_{S_i}, \tag{178}$$

*) Note that configurations (175) differ by the phase factor $(-1)^{S_i+1}$ from those employed earlier (Čížek, 1966b; Paldus *et al.*, 1977, Paldus, 1977; Adams and Paldus, 1979,1981; Takahashi and Paldus, 1986).

with the cluster components having the same symmetry properties as the corresponding configurations, i.e.

$$\langle rs|\hat{t}_2|ab\rangle_{S_i} = \langle sr|\hat{t}_2|ba\rangle_{S_i} = (-1)^{S_i}\langle rs|\hat{t}_2|ba\rangle_{S_i} = (-1)^{S_i}\langle sr|\hat{t}_2|ab\rangle_{S_i} . \tag{179}$$

In this way we are guaranteed that the number of cluster coefficients is the same as the number of spin-adapted biexcited configurations.

The coupling scheme defining the spin-adapted states (175) is then represented by an appropriate spin graph, which consists of two CG vertices (each having three incident lines representing coupled spin angular momenta), whose lines, labeled with the intermediate spin quantum number τ, are interconnected, representing the sum over τ [cf. Eq. (175)], leading to the *pp-hh* coupled states. We then associate with each spin-orbital R-diagram corresponding *orbital diagram* and *spin graph*. The former is simply obtained by replacing spin-orbital labels (capitals) with orbital ones [lower case letters, see Eq. (171)]. The spin graph is then obtained by interconnecting the free lines of spin graphs representing spin-adapted cluster components involved and a bra-state that projects onto a given spin-adapted configuration (note that interaction vertices may be ignored since they are spin independent): this is easily achieved by interconnecting those lines which correspond to the internal fermion lines or paths of the orbital R-diagram. This graph is then "evaluated" by breaking it, with the help of simple separation rules, into basic SU(2) invariants represented by various $3n$-j coefficients. It is then sufficient to multiply the contribution of each orbital R-diagram by the value of the associated spin graph. The reader is referred to original papers for details (D-CI: Paldus *et al.*, 1977; CCD: Paldus, 1977; approximate version of CCSDT: Adams and Paldus, 1979; full CCSD: Piecuch and Paldus, 1989; CCSDT-1 and ACPTQ: Piecuch and Paldus, 1990).

The resulting orthogonally spin-adapted formalism not only involves a minimal number of unknown cluster amplitudes, and thus equations, and is simply related with the corresponding CI formalism (just as is the simple spin-orbital version, Section 4.6.1), but also enables a simple formulation of various approximate approaches (e.g., Paldus *et al.*, 1984a, Piecuch and Paldus, 1990) and leads to much sparser matrices of CC nonlinear coefficients than the nonorthogonally spin-adapted approaches (e.g., Adams *et al.*, 1981a).

4.8. Calculation of properties

Once the energy and the wave function is determined by solving the CC equations, it is often desirable to calculate various static and dynamic properties of the system. For static properties, this may be achieved by evaluating mean values of appropriate (usually one-body) operators or, equivalently, the corresponding one- or two-body density matrices. This is not as straightforward as it may seem at first sight (not only for the CC approaches, but also for some standard ones) and for this reason one often employs so-called finite field method to achieve this goal. A direct evaluation of properties may then be achieved basically in two ways: first, by computing the expectation value of the operator involved or, second, by relying on the linear response theory or one of its variants.

Calculation of the expectation value of some one- or two-body operator W, i.e.,

$\langle W \rangle \equiv \langle \Psi | W | \Psi \rangle / \langle \Psi | \Psi \rangle$, was already discussed by Čížek (1969). With the wave function $|\Psi\rangle$ written in the cluster expansion form, Eq. (57), we thus have to evaluate the expression

$$\langle W \rangle = \langle \Phi | e^{T^\dagger} W e^T | \Phi \rangle / \langle \Phi | e^{T^\dagger} e^T | \Phi \rangle \,. \tag{180}$$

Although one can easily show the factorization of the numerator into the connected component involving W and the remainder that exactly cancels the denominator, so that

$$\langle W \rangle = \langle \Phi | e^{T^\dagger} W e^T | \Phi \rangle_C \,, \tag{181}$$

the resulting expression (even when $T \approx T_2$) involves infinitely many terms since, in contrast with $e^{-T} W e^T$, the expansion of $e^{T^\dagger} W e^T$ never terminates. Thus, even if the cluster amplitudes were small, so that a numerical truncation would be possible (say, with quadratic terms), the resulting method would be rather awkward to perform. It is thus desirable to explore other possible avenues, as was first done by Monkhorst (1977) [see also, Dalgaard and Monkhorst, 1983].

Similarly as in the finite field method (or linear response theory), let us consider a perturbed Hamiltonian with the one-body operator W, whose mean value we wish to evaluate, as a perturbation

$$H(\lambda) = H + \lambda W = Z + V + \lambda W \,, \tag{182}$$

λ being a real parameter. Taking the normal product form (Section 2.2),

$$H_N(\lambda) = H_N + \lambda W_N = F_N + V_N + \lambda W_N \,, \tag{183}$$

where

$$W_N = W - \langle \Phi | W | \Phi \rangle \,, \tag{183'}$$

we consider the time-independent eigenvalue problem

$$H_N(\lambda) | \Psi(\lambda) \rangle = \Delta E(\lambda) | \Psi(\lambda) \rangle \,. \tag{184}$$

Let

$$\Delta E(\lambda) = E(\lambda) - \langle \Phi | H(\lambda) | \Phi \rangle = \sum_{n=0}^{\infty} \lambda^n \Delta E^{(n)} \,, \tag{185}$$

and

$$E(\lambda) = \sum_{n=0}^{\infty} \lambda^n E^{(n)} \,, \tag{186}$$

so that

$$\Delta E^{(0)} = E^{(0)} - \langle \Phi | H | \Phi \rangle \,, \tag{187a}$$

$$\Delta E^{(1)} = E^{(1)} - \langle \Phi | W | \Phi \rangle \,, \tag{187b}$$

and

$$\Delta E^{(k)} = E^{(k)}, \; k > 1 \,. \tag{187c}$$

Recalling Hellman-Feynman theorem, we have that in fact

$$\frac{\langle \Psi | W | \Psi \rangle}{\langle \Psi | \Psi \rangle} = \left. \frac{\partial \Delta E(\lambda)}{\partial \lambda} \right|_{\lambda=0} = \Delta E^{(1)} \,, \tag{188}$$

where $|\Psi\rangle \equiv |\Psi(0)\rangle$.

We now write the cluster expansion for $|\Psi(\lambda)\rangle$ as

$$|\Psi(\lambda)\rangle = e^{T(\lambda)}|\Phi\rangle\,, \tag{189}$$

with $T(\lambda)$ also expanded in powers of the coupling constant λ,

$$T(\lambda) = \sum_{n=0}^{\infty} \lambda^n T^{(n)}\,. \tag{190}$$

Since all $T^{(n)}$ are automatically in the n-product form and contain an even number of operators, we have that

$$[T^{(i)}, T^{(j)}] = 0, \qquad \forall i, j \tag{191}$$

so that $e^{T(\lambda)}$ is well defined. Substituting now (189) into the Schrödinger equation (184) and applying the connected cluster theorem (cf. Section 4.3), we get

$$\{(H_N + \lambda W_N)e^{T(\lambda)}|\Phi\rangle\}_C = \Delta E(\lambda)|\Phi\rangle\,. \tag{192}$$

Evidently, in the zeroth order ($n = 0$) we recover the original unperturbed problem, while in the general k-th order we get, schematically

$$\begin{aligned}(H_N e^{T^{(0)}} T^{(k)}|\Phi\rangle)_C &+ \sum_{k_1+k_2+\cdots+k_j=k} (H_N e^{T^{(0)}} T^{(k_1)}\cdots T^{(k_j)}|\Phi\rangle)_C \\ &+ \sum_{k_1+k_2+\cdots+k_j=k-1} (W_N e^{T^{(0)}} T^{(k_1)}\cdots T^{(k_j)}|\Phi\rangle)_C = \Delta E^{(k)}|\Phi\rangle\,.\end{aligned} \tag{193}$$

Again, $\Delta E^{(k)}$ is given by vacuum diagrams, while the projection on corresponding excited configurations yields the CC equations for the $T^{(i)}$ amplitudes. Obviously, we assume the unperturbed problem to be solved, so that the cluster amplitudes defining $T^{(0)}$ are known.

Now, we can calculate $T^{(k)}$ amplitudes order by order. Generally, we only have to solve linear algebraic equations, since $T^{(k)}$ can only appear linearly in (193). Further, only the absolute terms depend on W, the coefficient matrix being W-independent and thus the same for any property considered. Moreover, in view of (188), we only need to calculate the first order term $\Delta E^{(1)}$. In this case, (193) simplifies to

$$(H_N e^{T^{(0)}} T^{(1)}|\Phi\rangle)_C + (W_N e^{T^{(0)}}|\Phi\rangle)_C = \Delta E^{(1)}|\Phi\rangle\,. \tag{194}$$

Let us briefly indicate the structure of the resulting equations. Similarly to the expansion for $T \equiv T^{(0)}$, Eq. (52), we write

$$T^{(1)} = \sum_{j=1}^{N} T_j^{(1)}\,, \tag{195}$$

and we represent these cluster components by the same diagrams as before (diagram 3 or 3′ of Fig. 1), except that we now enclose the integer n ($n = 0$ or 1 in our case) in the vertex symbol.

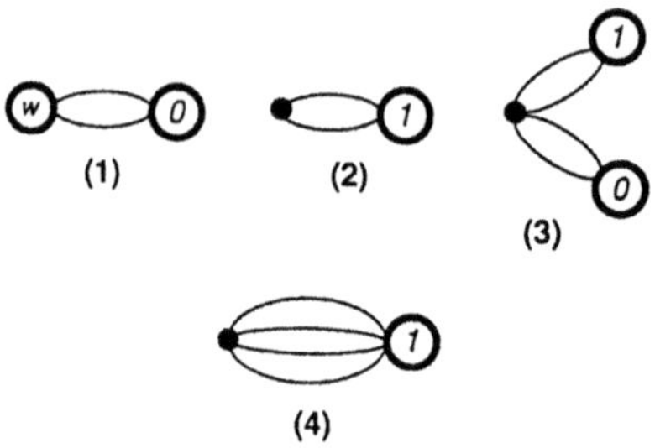

Fig. 18. Nonoriented Hugenholtz skeletons yielding the mean value $\langle\Psi|H|\Psi\rangle/\langle\Psi|\Psi\rangle$ of the one-body operator W in the state $|\Psi\rangle$ as described by the SR-CC methods. The standard one- and two-body interaction vertices are designated by small full circles, while the W-vertices are labeled by w. One- and two-body cluster vertices (large open circles) are labeled with the PT order n, i.e. vertex labeled n ($n = 0, 1$ in the case considered) designates the cluster component of $T^{(n)}$, Eq. (190).

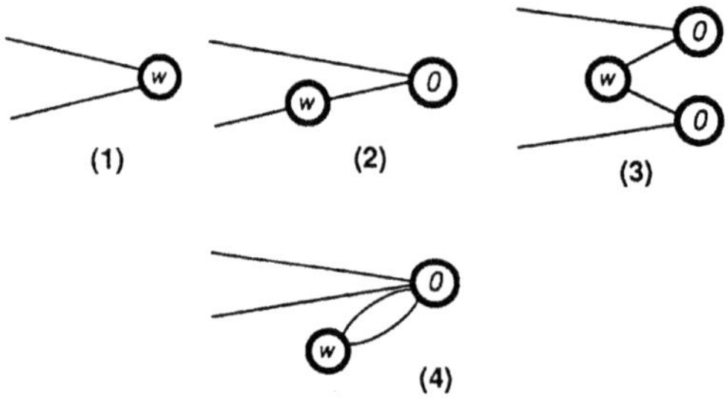

Fig. 19. Nonoriented Hugenholtz skeletons characterizing the absolute terms $A_i = \langle\Phi_i|W_N e^{T^{(0)}}|\Phi\rangle_C$ in Eq. (197). The same notation as in Fig. 18 is employed.

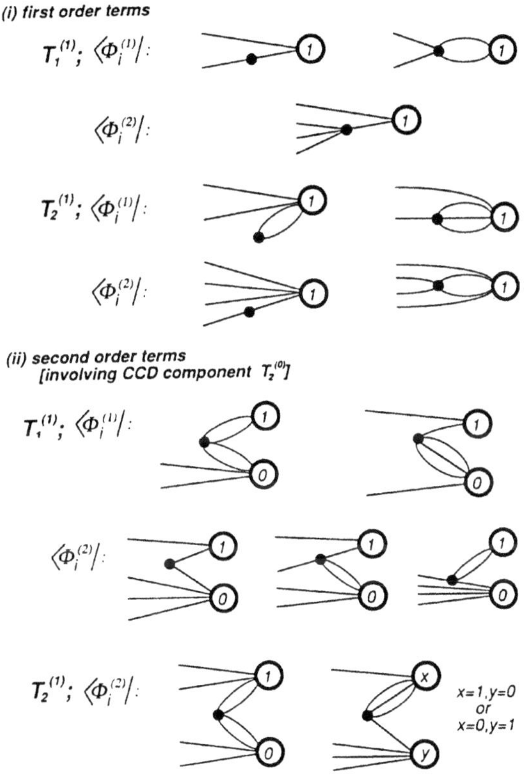

Fig. 20. First and second order nonoriented Hugenholtz skeletons characterizing the coefficients $B_{ij} = \langle\Phi_i|H_N\, e^{T^{(0)}}T^{(1)}|\Phi\rangle_C$ of the linear system (197). For vertex designation see caption to Fig. 18.

The desired mean value $\Delta E^{(1)}$ is thus given by the vacuum terms

$$\Delta E^{(1)} = \langle\Phi|(H_N T^{(1)} + W)e^{T^{(0)}}|\Phi\rangle_C\,, \tag{196}$$

represented by the diagrams of Fig. 18. The linear system determining the cluster amplitudes of $T^{(1)}$ then has the form

$$A_i + B_{ij}t_j^{(1)} = 0\,, \tag{197}$$

where the coefficients A_i and B_{ij} are given by the diagrams representing the terms $\langle\Phi_i|W_N e^{T^{(0)}}|\Phi\rangle_C$ and $\langle\Phi_i|H_N e^{T^{(0)}}T^{(1)}|\Phi\rangle_C$, respectively. Their typical representatives, assuming the approximation $T^{(1)} \approx T_1^{(1)} + T_2^{(1)}$, are shown in Figs. 19 and 20, respectively. In a similar way we could handle the two-body operators.

For the dynamical properties, we can proceed in an analogous manner as above, employing a time-dependent perturbation theory. In this way we can also calculate excitation energies and corresponding oscillator strengths (Monkhorst, 1977; Sekino and Bartlett, 1984; Takahashi and Paldus, 1986). In fact, this approach is rather closely related to the Green function perturbative approach (Paldus and Čížek, 1974) and the one-state Hermitized (or variational) CC approach to open-shell systems with linear ansatz for the excitation operator (Paldus *et al.*, 1978; Saute *et al.*, 1979).

4.9. Computational implementation

We cannot possibly do justice to this important topic in the available space and will only mention a few basic concepts and algorithms, and direct the reader to the original literature on the subject. Clearly, the computational strategy and algorithms used will very much depend on the type of applications intended and models employed: obviously, we must employ very different computational algorithms and strategy in methodological studies intended to test various approximation schemes and in programs designed for production runs in more or less routine applications, or in handling extended system and finite ones. Since the latter will be of primary interest to most readers, we shall restrict ourselves to pointing out two basic aspects on which the programs intended for large scale molecular electronic structure applications are usually based, namely, (i) the algorithm used for the solution of large scale algebraic systems of CC equations and, (ii) the recursive formulation used for the evaluation of multi-linear (bilinear in the CCD case) forms appearing in CC equations. Let us first mention that both these aspects have been first utilized by the Florida (Bartlett and Purvis, 1978,1980) and Carnegie-Mellon (Pople *et al.*, 1978) groups, and were thoroughly reviewed, together with their applications, by Bartlett (1981). In the meantime, similar programs have also been developed by other groups (e.g., Taylor *et al.*, 1976; Scuseria *et al.*, 1987; Lee and Rice, 1988), often including an analytical evaluation of energy gradients (Fitzgerald *et al.*, 1985; Scheiner *et al.*, 1987; Scuseria and Schaefer, 1988; Gauss and Cremer, 1988; Koch *et al.*, 1990; Rendell and Lee, 1991; Lee and Rendell, 1991; see also conceptual developments: Jorgensen and Simons, 1983; Adamowicz *et al.*, 1984; Bartlett, 1986; Fitzgerald *et al.*, 1986; Salter *et al.*, 1989). It is important to note that most of these implementations that are primarily intended for large scale *ab-initio* applications are based on spin-nonadapted, spin-orbital version of CCX (X = D, SD, SDT, etc.) methods. This not only simplifies recursive formulation that facilitates code vectorization and thus very efficient utilization of modern vector processors, but also permits an exploitation of unrestricted HF orbitals, which in turn enables applications to various open-shell systems and guarantees proper dissociation with only a single determinantal reference. We hasten to say, however, that this can hardly be regarded as an ultimate solution of the open-shell problem. Not only the corresponding wave functions are contaminated by higher spin multiplets, but most importantly, the resulting potential energy curves (or surfaces) show a non-analytic behavior due to the onset of various triplet (or non-singlet, non-doublet, etc.) instabilities (Čížek and Paldus, 1967,1971; Paldus and Veillard, 1977, 1978; Paldus, 1990).

Relatively small systems of CCD (or, generally, CC) equations can be efficiently solved using Gauss' elimination for the linear system and Newton-Raphson procedure for the non-linear ones. For larger systems, the Gauss elimination is too core demanding and must be replaced by an iterative procedure (Jacobi, Gauss-Seidl, etc.). For large implementations only iterative methods can be used. Using a simple, Jacobi-type iterative scheme (that we employed in Section 4.6.2 when inter-relating CCD and MBPT approaches) is usually not satisfactory: generally, the convergence is rather slow and often the method does not converge at all (which is not surprising since the corresponding PT expansion also often diverges). For this reason, Purvis and Barlett (1981) introduced a new algorithm, referred to as *reduced linear equation* (RLE) method, that employs a similar strategy in building up a reduced linear system as do the efficient diagonalization algorithms. Conceptually, this method is very similar to that proposed by Pople *et al.* (1979) that was formulated for coupled HF theory and successfully exploited in MCSCF (Lengsfield and Liu, 1981) and CI-PT (Daborn *et al.*, 1980), but leads to a much faster convergence. It can be shown (Wormer *et al.*, 1982) that both methods represent different versions of the conjugate gradient method (cf., e.g., Hestenes, 1980), which is very appropriate for the CC-type equations (Paldus *et al.*, 1982). The faster convergence of the RLE method is due to the fact that it corresponds to the preconditioned conjugate gradient method. In essence, the coefficient matrix b is split into its diagonal ($\mathbf{D}$) and off-diagonal ($\mathbf{\Delta}$) parts,

$$\mathrm{b} = \mathbf{D} + \mathbf{\Delta}\,, \tag{198}$$

and the system (142) is transformed with $\mathbf{D}^{-1/2}$ so that the new system arises

$$\mathbf{A} + \mathbf{B}\,\mathrm{y} = 0\,, \tag{199}$$

where

$$\begin{aligned} \mathbf{A} &= \mathbf{D}^{-1/2}\mathrm{a}\,, \\ \mathbf{B} &= \mathbf{D}^{-1/2}\mathrm{b}\mathbf{D}^{-1/2} = \mathbf{1} + \mathbf{D}^{-1/2}\mathbf{\Delta}\mathbf{D}^{-1/2}\,, \end{aligned} \tag{200}$$

so that

$$\mathrm{x} = \mathbf{D}^{-1/2}\mathrm{y}\,. \tag{201}$$

Since b is normally a diagonally dominant matrix, this preconditioning transforms the coefficient matrix to the one with 1's along the main diagonal and rather small off-diagonal entries, so that its eigenvalues are clustered around 1. As is well known, this results in a good convergence of the conjugate gradient procedure. Of course, when solving the nonlinear CC systems, the nonlinear terms are handled in a similar way as the linear ones.

Let us, finally, discuss the evaluation of nonlinear terms that must be carried out in each iteration. As already mentioned, this is ideally done by presorting the integrals and/or the cluster amplitudes, so that this evaluation can be carried out as consecutive scalar products of appropriate arrays. In designing this algorithm the diagrammatic representation can be again helpful, since it enables us to conveniently split a given diagram into an "effective interaction" and a cluster amplitude. For example, consider the CCD nonlinear term associated with the diagram 9 of Fig. 13 (given by the third term on the right hand side of Eq. (115) and the first term of $\mathcal{A}_p$, Eq. (111′), i.e. the identity operator $\hat{1}$)

$$\Lambda_3^{(2)}(C^1C^2;C_1C_2) = \tfrac{1}{2} \sum_{B^1,B^2,B_1,B_2} \langle B_1B_2|\hat{v}|B^1B^2\rangle_{\mathcal{A}} \langle C^1B^1|\hat{t}_2|B_1B_2\rangle_{\mathcal{A}} \langle B^2C^2|\hat{t}_2|C_1C_2\rangle_{\mathcal{A}} \,. \tag{202}$$

We can factorize it as follows

$$\Lambda_3^{(2)}(C^1C^2;C_1C_2) = \tfrac{1}{2} \sum_{B^2} \langle C^1|\hat{\xi}|B^2\rangle \langle B^2C^2|\hat{t}_2|C_1C_2\rangle_{\mathcal{A}} \,, \tag{203}$$

with

$$\langle C^1|\hat{\xi}|B^2\rangle = \sum_{B^1,B_1,B_2} \langle C^1B^1|\hat{t}_2|B_1B_2\rangle_{\mathcal{A}} \langle B_1B_2|\hat{v}|B^1B^2\rangle_{\mathcal{A}} \,, \tag{204}$$

i.e., over the particle line B^2. Presorting suitably the two-electron integrals, each right hand side of (203) and (204) can be evaluated as a simple scalar product (cf., e.g., Purvis and Bartlett, 1982). In this way, one can handle all the terms. Clearly, the diagrammatic representation can be very useful both to find most suitable factorizations to group terms possessing a similar structure (e.g., T_2 and corresponding $\frac{1}{2}T_1^2$ terms, or T_3, T_1T_2 and $\frac{1}{3!}T_1^3$ terms, etc.). A systematic factorization of this kind, for a general CCSDT procedure, was worked out by Kucharski and Bartlett (1991).

5. MULTI-REFERENCE CC THEORY

So far we have concentrated our attention on closed-shell nondegenerate ground states (or the lowest energy state of a given symmetry species), where the CC theory is best developed and continues to provide valuable information (see, e.g., reviews by Bartlett, 1981,1989). It often provides very good results even in highly quasidegenerate situations (e.g., Jankowski and Paldus, 1980; Paldus *et al.*, 1988), particularly when higher than pair clusters are taken into account (Bartlett *et al.*, 1989,1990; Kucharski and Bartlett, 1989; Noga *et al.*, 1989; Lee *et al.*, 1990; Lee and Scuseria, 1990; Rendell *et al.*, 1990, Scuseria and Lee, 1990; Watts *et al.*, 1990; Oliphant and Adamowicz, 1991; Lee and Rendell, 1991; Lee and Rice, 1991; Rendell *et al.*, 1991, Rice *et al.*, 1991). Lee and Taylor (1989) even developed a diagnostic for determining the quality of SR correlation methods. However, consideration of higher than two-body cluster amplitudes can be rather costly and has its inherent limitations. Particularly when we wish to explore entire potential energy surfaces for strongly interacting states of the same symmetry, we can hardly avoid a genuine multi-reference (MR) approach.

An extension of the cluster ansatz (57) or (59) to the open-shell (OS) or MR cases is by no means an unambiguous and easy task, as already indicated in the Introduction. However, the underlying principles and basic equations of the MR formalism are well established. We shall thus first concentrate on this formal development and only subsequently explore different cluster ansätze that generalize (59) to OS situations.

Generally, just as in the SR case, we have to establish an appropriate *reference* or *model space*, which is now multidimensional, find an appropriate wave operator U, which is determined in this case by the so-called *Bloch equation* and, finally, calculate the *effective* or *model Hamiltonian* representation on the model space and find its eigenvalues and eigenstates, just as we evaluated the SR correlation energy as the mean value of the "effective Hamiltonian" HU in the reference state $|\Phi\rangle$, Eq. (86). However, in constrast to the SR case, where the cluster ansatz, Eq. (59), for the wave operator $U \equiv W$ is unambiguous in view of the connected cluster theorem, several avenues are opened to us in the MR case, thanks to a greater freedom in defining the cluster amplitudes in this case. Nonetheless, irrespective of the choice of the ansatz for U, the basic equations that determine it (Bloch equation), and the effective Hamiltonian, are well defined. We thus first concentrate on this general aspect of the OS or MR-CC theory (or, in fact, MR-MBPT) and only subsequently discuss various cluster ansätze for the wave operator U. The derivation of the requisite formalism is not difficult, although certain care must be exercised if one is to avoid pitfalls that often plagued the literature on this subject. We shall thus proceed quite slowly, giving enough attention to detail.

5.1. Model space

For nondegenerate closed-shell states we are able to achieve a reasonable description with a single configuration $|\Phi\rangle$, thus employing one-dimensional "model space". Moreover, this configuration can be employed as a Fermi vacuum and unambiguously partitions the spin-orbitals into the occupied and virtual ones. This type of description must be, of course, fundamentally changed when considering general OS systems or, in fact, even highly quasidegenerate closed-shell systems. In this case we have to simultaneously consider a group of states that arises through a strong interaction of a number of (i.e., at least two) configurations and cannot be reasonably described by a single configuration over a range of relevant nuclear framework geometries. We thus must find a suitable set of configurations $\{|\Phi_i\rangle\}$ that are capable of providing a reasonable zeroth-order description of the states considered. Typically, these configurations will only differ in the occupancies of a few spin-orbitals, whose energies are lying in the vicinity of a hypothetical Fermi level for the states involved. We thus subdivide (cf. Fig. 21) the orthonormal spin-orbitals $|A\rangle$ of the system*) into three disjoint classes, namely the *hole* or *core* (C), *valence* or *open-shell* or *active* (A or V), and *particle* or *virtual* or *excited* (E) spin-orbitals. The core spin-orbitals are occupied in all reference configurations $|\Phi_i\rangle$, while the virtual ones are always unoccupied. Thus, the reference configurations $|\Phi_i\rangle$ differ by the occupancies of valence spin-orbitals.**)

When all possible occupations are considered for the valence spin-orbitals, we obtain a *complete reference space*. Although this complete space may be unnecessarily large

*) These can be RHF, MCSCF or other (spin)-orbitals.

**) The same terminology is also employed for spin-independent Hamiltonians when the spin-orbital representation, Eq. (171), is employed. In this case, all the core orbitals are doubly occupied in all reference configurations.

for the description of states that are of interest to us, it is essential for the theoretical developments that follow. Only later on we shall also consider *incomplete* (i.e., truncated) *model spaces.*

For definiteness, we shall assume that there are K core and L valence spin-orbitals. The number of excited spin-orbitals may, in principle, be infinitely large, even though in practical applications we always deal with finite-dimensional (spin)-orbital spaces. Analogously as in the SR case, we shall also require that our theory is invariant with respect to separate unitary transformations of C, A and E spin-orbitals. The invariance for the C and E spin-orbitals follows in the same way as in the SR case, while the invariance with respect to transformations of A spin-orbitals clearly implies the necessity to consider a complete model space. The invariance with respect to C and E

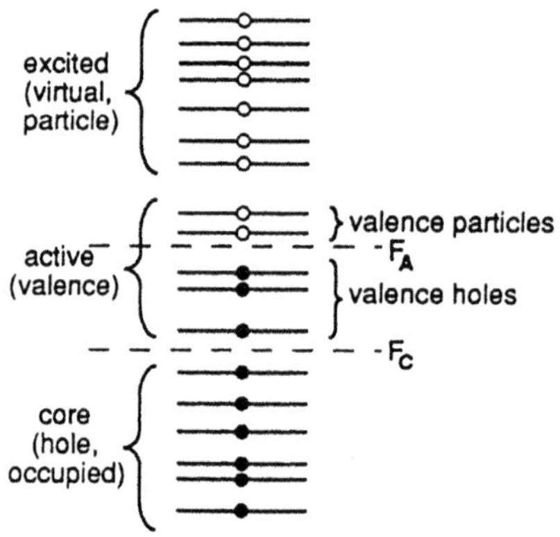

Fig. 21. A schematic representation of (spin)-orbital energy levels and the (spin)-orbital classification employed in the MR approaches. The horizontal dashed levels indicate two possible choices for the Fermi level: the first (F_C) employs the core state vacuum and leads to particle-only formalism (no valence holes are possible), while the second possibility (F_A) characterizes a general case involving both valence particles and holes. A "variable" (defined by the bra-state) Fermi level is used in Hilbert space approaches.

spin-orbitals then implies the basis set independence of the resulting formalism, which is essential when we wish to use localized spin-orbitals (for C) (e.g., Takahashi and Paldus, 1985) or numerical (Lindgren and Salomonson, 1980; Lindgren and Morrison, 1982) or geminal (Jeziorski *et al.*, 1984) techniques.

Consider now a model space $\mathcal{M}_0$ that is spanned by all possible, orthonormal, N-electron configurations $|\Phi_i\rangle$, $i = 1,\ldots,M$, in which K core spin-orbitals are fully occupied and remaining (N-K) electrons occupy valence spin-orbitals, leaving virtual orbitals empty.*) Thus, the model space $\mathcal{M}_0$,

$$\mathcal{M}_0 = \text{Span}\{|\Phi_i\rangle | 1 \leq i \leq M\}, \tag{205}$$

is an M-dimensional subspace of a general N-electron space $\mathcal{H}_N$ (and thus of a Fock space $\mathcal{F}$, Eq. (2)), where

$$M \equiv \dim \mathcal{M}_0 = \begin{pmatrix} L \\ N-K \end{pmatrix}, \tag{206}$$

and represents a complete reference space for a chosen spin-orbital subdivision into the C, A and E-types. We shall associate with this subspace the projection operator P, $P : \mathcal{H}_N \rightarrow \mathcal{M}_0$,

$$P = \sum_{i=1}^{M} P_i, \qquad P_i = |\Phi_i\rangle\langle\Phi_i|, \tag{207a}$$

$$P_i^2 = P_i,\ P_i^\dagger = P_i,\ P^2 = P,\ P^\dagger = P \tag{207b}$$

as well as the projection operator Q onto the orthogonal complement $\mathcal{M}_0^\perp$ of $\mathcal{M}_0$, $Q : \mathcal{H}_N \rightarrow \mathcal{M}_0^\perp$,

$$Q = 1 - P. \tag{207'}$$

Assume, next, that we can construct appropriate (linearly independent) linear combinations $|\Phi_i'\rangle$ of $|\Phi_i\rangle$,

$$|\Phi_i'\rangle = \sum_{j=1}^{M} a_{ij}|\Phi_j\rangle,\ (i = 1, \ldots, M) \tag{208}$$

that represent a reasonable zeroth order approximation to the lowest M exact states $|\Psi_i\rangle$ of the Hamiltonian H of the system,

$$H|\Psi_i\rangle = E_i|\Psi_i\rangle. \tag{209}$$

This requirement can be more precisely stated in the PT language. Suppose that $|\Phi_i'\rangle$'s are eigenstates of some approximate Hamiltonian H_0 (not necessarily Hermitian) and consider a general Hamiltonian $H(\lambda) = H_0 + \lambda H'$, such that $H(1) = H$. Designating the eigenstates of $H(\lambda)$ as $|\Psi_i(\lambda)\rangle$, so that $|\Psi_i\rangle = |\Psi_i(1)\rangle$, we thus require that

$$|\Phi_i'\rangle = \lim_{\lambda \rightarrow 0} |\Psi_i(\lambda)\rangle, \tag{210}$$

which will be the case if H_0 is a reasonable approximation to H.

The M mutually orthogonal eigenstates $|\Psi_i\rangle$, $i = 1, \ldots M$ of H, Eq. (209), which

*) Of course, in the presence of symmetry (space, spin, etc.) $\mathcal{M}_0$ may be broken down into a direct sum of invariant subspaces, labeled by the irreducible representations of the invariance groups involved, and each invariant subspace can be considered as an independent model space. However, for the sake of simplicity, we consider a simple spin-orbital formalism (for spin-adapted formalism, see Jeziorski and Paldus, 1988; Paldus *et al.*, 1989b; Piecuch and Paldus, 1991a).

thus correspond to configurations $|\Phi_i'\rangle$ or $|\Phi_i\rangle$, spanning $\mathcal{M}_0$, span an M-dimensional subspace of $\mathcal{H}_N$, namely*)

$$\mathcal{M} = \text{Span}\{|\Psi_i\rangle | 1 \leq i \leq M\}, \quad \dim \mathcal{M} = M. \tag{211}$$

Since $|\Phi_i'\rangle$'s represent a reasonable approximation to $|\Psi_i\rangle$'s, we can further assume that no function from $\mathcal{M}$ is orthogonal to all of $\mathcal{M}_0$, since any state from $\mathcal{M}$ should have a reasonably large component contained in $\mathcal{M}_0$, so that certainly $P(\sum_i a_i|\Psi_i\rangle) \neq 0$ for any choice of a_i's that do not simultaneously vanish. This in turn implies that the projections of $|\Psi_i\rangle$'s onto $\mathcal{M}_0$

$$|\tilde{\Phi}_i\rangle = P|\Psi_i\rangle, \qquad (i = 1, \ldots M) \tag{212}$$

are linearly independent (though no longer orthogonal), so that also

$$\mathcal{M}_0 = \text{Span}\{|\tilde{\Phi}_i\rangle | 1 \leq i \leq M\}. \tag{213}$$

Indeed, assuming that $|\tilde{\Phi}_i\rangle$'s are linearly dependent, so that there exist scalars a_i, not all zero, such that $\sum_i a_i|\tilde{\Phi}_i\rangle = 0$, would imply that $\sum_i a_i|\Psi_i\rangle$ is orthogonal to $\mathcal{M}_0$, since

$$P(\sum_i a_i|\Psi_i\rangle) = \sum_i a_i|\tilde{\Phi}_i\rangle = 0. \tag{214}$$

Since both sets $\{|\Phi_i\rangle | 1 \leq i \leq M\}$ and $\{|\tilde{\Phi}_i\rangle | 1 \leq i \leq M\}$ are linearly independent and span $\mathcal{M}_0$, there must exist a nonsingular transformation relating these two basis sets of $\mathcal{M}_0$, so that we can write

$$P|\Psi_i\rangle \equiv |\tilde{\Phi}_i\rangle = \sum_j c_{ji}|\Phi_j\rangle, \qquad \det \mathbf{C} = |c_{ij}| \neq 0. \tag{215}$$

Thus, the restriction $\tilde{P}$ of P to $\mathcal{M}$, i.e. $\tilde{P} = P \downarrow \mathcal{M}$, represents a bijection (i.e., one-to-one and onto mapping) of $\mathcal{M}$ and $\mathcal{M}_0$,

$$\tilde{P} : \mathcal{M} \to \mathcal{M}_0. \tag{216}$$

Consequently, there must exist the inverse of $\tilde{P}$, which we designate by $\tilde{U}$, so that

$$\tilde{P}^{-1} \equiv \tilde{U} : \mathcal{M}_0 \to \mathcal{M} \tag{217}$$

is also one-to-one and onto. We next extend $\tilde{U}$ to the whole of $\mathcal{H}_N$ (or, in fact, to the whole Fock space $\mathcal{F}$)

$$U = \tilde{U} \uparrow \mathcal{H}_N \qquad (\text{or } U = \tilde{U} \uparrow \mathcal{F}), \tag{218}$$

by defining

$$U : \mathcal{H}_N \to \mathcal{M}, \qquad (\text{or } \mathcal{F} \to \mathcal{M}), \tag{219a}$$

---

*) Note that at this stage we could simply define the subspaces $\mathcal{M}_0$ and $\mathcal{M}$ of $\mathcal{H}_N$ or $\mathcal{F}$ by selecting M exact eigenstates $|\Psi_i\rangle$, $i = 1, 2, \ldots M$ of H that we are interested in and that define $\mathcal{M}$, Eq. (211), and then consider M linearly independent zero-order functions $|\Phi_i\rangle$, $i = 1, \ldots, M$ that have pairwise nonvanishing overlap with $|\Psi_i\rangle$'s, namely $\langle\Psi_i|\Phi_i\rangle \neq 0$, $i = 1, \ldots, M$, to define $\mathcal{M}_0$, Eq. (205). We prefer, however, a "constructive" approach, starting with an appropriate one-electron spin-orbital basis and yielding a concrete model space, since it better reflects the actual procedure that one employs in applications.

with

$$U \downarrow \mathcal{M}_0 \equiv \tilde{U} \qquad \text{and} \qquad U \downarrow \mathcal{M}_0^{\perp} \equiv 0 \,. \tag{219b}$$

In view of (212) we thus have that

$$U|\tilde{\Phi}_i\rangle = |\Psi_i\rangle, \qquad (i = 1, \ldots, M) \tag{220}$$

while (219) implies that

$$UQ = U(1 - P) = 0 \,, \tag{221}$$

or

$$UP = U \,. \tag{222}$$

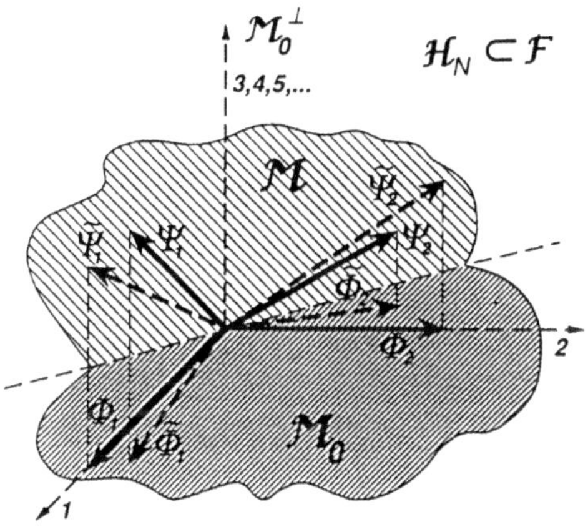

Fig. 22. An illustration of the relationship between the model ($\mathcal{M}_0$) and exact ($\mathcal{M}$) subspaces and their basis sets, Eq. (225), for the two-dimensional case ($M = 2$).

Thus, UP acts as the identity on $\mathcal{M}$, since clearly

$$|\Psi_i\rangle = U|\tilde{\Phi}_i\rangle = UP|\Psi_i\rangle \,. \tag{223}$$

We also define "projections" of $|\Phi_i\rangle$ by U, similarly to (212), as

$$|\tilde{\Psi}_i\rangle = U|\Phi_i\rangle, \qquad (i = 1, \ldots, M) \,. \tag{224}$$

Again, it easily follows that $|\tilde{\Psi}_i\rangle \in \mathcal{M}$ and $|\tilde{\Psi}_i\rangle, i = 1, \ldots, M$ are linearly independent; i.e., span the subspace $\mathcal{M}$. This is immediately obvious since $\tilde{U} \equiv U \downarrow \mathcal{M}_0$ is one-to-one and onto linear mapping and $|\Phi_i\rangle$ span $\mathcal{M}_0$.

We have thus established two basis sets for each of the M-dimensional spaces considered: the model space $\mathcal{M}_0$, defined as a span of approximate N-electron states or configurations $|\Phi_i\rangle, i = 1, \ldots, M$ (represented by Slater determinants in the spin-orbital formalism) and the hypothetical "exact" space $\mathcal{M}$, spanned by exact eigenstates $|\Psi_i\rangle, i = 1, \ldots, M$, of the considered Hamiltonian H, Eq. (209). Thus, in summary

$$\begin{aligned} \mathcal{M}_0 &= \mathrm{Span}\{|\Phi_i\rangle | 1 \le i \le M\} = \mathrm{Span}\{|\tilde{\Phi}_i\rangle | 1 \le i \le M\}, \\ \mathcal{M} &= \mathrm{Span}\{|\Psi_i\rangle | 1 \le i \le M\} = \mathrm{Span}\{|\tilde{\Psi}_i\rangle | 1 \le i \le M\}, \end{aligned} \tag{225}$$

and

$$\begin{aligned} U|\tilde{\Phi}_i\rangle &= |\Psi_i\rangle, \quad P|\Psi_i\rangle = |\tilde{\Phi}_i\rangle, \\ U|\Phi_i\rangle &= |\tilde{\Psi}_i\rangle, \quad P|\tilde{\Psi}_i\rangle = |\Phi_i\rangle. \end{aligned} \tag{226}$$

The relationship between these various states and the subspaces $\mathcal{M}_0$ and $\mathcal{M}$ is schematically illustrated for the two-dimensional case in Fig. 22. While the untilded bases $\{|\Phi_i\rangle\}$ and $\{|\Psi_i\rangle\}$ are orthogonal, the tilded bases, in general, are not. However, the relationship between the tilded and untilded bases in the model space $\mathcal{M}_0$, Eq. (215), also applies (although in a dual manner) to the corresponding bases for the "exact subspace" $\mathcal{M}$, namely

$$|\Psi_i\rangle = \sum_j c_{ji} |\tilde{\Psi}_j\rangle, \tag{227}$$

as may be easily verified. Thus,

$$\langle \Psi_i | \tilde{\Psi}_j \rangle = (\mathbf{C}^{-1})_{ij}, \tag{227'}$$

while

$$c_{ij} = \langle \Phi_i | \tilde{\Phi}_j \rangle. \tag{227''}$$

Similarly to the expression (207a) for P, we can write U in the form

$$U = \sum_i |\tilde{\Psi}_i\rangle\langle\Phi_i|, \tag{228}$$

since [cf. Eqs. (207a), (222) and (224)]

$$\begin{aligned} U = UP &= U \sum_i |\Phi_i\rangle\langle\Phi_i| = \sum_i U|\Phi_i\rangle\langle\Phi_i| \\ &= \sum_i |\tilde{\Psi}_i\rangle\langle\Phi_i|. \end{aligned} \tag{229}$$

Clearly, the expression (228) holds on the whole $\mathcal{H}_N$ (or $\mathcal{F}$), since $U|\Xi\rangle = 0$ for any $|\Xi\rangle \in \mathcal{M}_0^{\perp}$. For the product PU, we find that on $\mathcal{M}_0$

$$PU \downarrow \mathcal{M}_0 = P\tilde{U} = \tilde{P}\tilde{U} = \hat{\mathbf{1}} \downarrow \mathcal{M}_0 = P \downarrow \mathcal{M}_0, \tag{230}$$

or, equivalently,

$$PU|\Phi_i\rangle = |\Phi_i\rangle = P|\Phi_i\rangle, \tag{231}$$

as well as

$$PU|\tilde{\Phi}_i\rangle = |\tilde{\Phi}_i\rangle = P|\tilde{\Phi}_i\rangle\,, \tag{231'}$$

so that $PU = P$ on $\mathcal{M}_0$. Since on $\mathcal{M}_0^\perp$ both sides vanish, we have that generally

$$PU = P\,. \tag{232}$$

In fact, this relationship is equivalent to the assumption of the *intermediate normalization*, which in the multireference case requires that

$$\boxed{\langle\Phi_i|\tilde{\Psi}_j\rangle = \delta_{ij}\,.} \tag{233}$$

Indeed, with this condition we have that

$$PU = \sum_{ij}|\Phi_i\rangle\langle\Phi_i|\tilde{\Psi}_j\rangle\langle\Phi_j| = \sum_i|\Phi_i\rangle\langle\Phi_i| = P\,. \tag{234}$$

Finally, note that U, similarly as P, is idempotent, since

$$U^2 = (UP)^2 = U(PU)P = UP^2 = UP = U\,, \tag{235}$$

or directly from (228)

$$U^2 = \sum_{i,j}|\tilde{\Psi}_i\rangle\langle\Phi_i|\tilde{\Psi}_j\rangle\langle\Phi_j| = \sum_i|\tilde{\Psi}_i\rangle\langle\Phi_i| = U\,. \tag{236}$$

However, in contrast to P, U is not Hermitian, as may be easily seen from (228). Of course, it acts as the identity on $\mathcal{M}$,

$$U|\Psi_i\rangle = |\Psi_i\rangle \qquad \text{and} \qquad U|\tilde{\Psi}_i\rangle = |\tilde{\Psi}_i\rangle\,, \tag{237}$$

as may be easily verified. U may thus be regarded as a non-Hermitian projection onto the exact subspace $\mathcal{M}$. In summary,

$$\begin{aligned} PU &= P, & UP &= U\,,\\ PUP &= P, & UPU &= U\,,\\ P^2 &= P, & U^2 &= U\,,\\ (1-P)^2 &= 1-P, & (1-U)^2 &= 1-U\,. \end{aligned} \tag{238}$$

In view of the intermediate normalization (233), we have for each $|\tilde{\Psi}_i\rangle$ that

$$|\tilde{\Psi}_i\rangle = |\Phi_i\rangle + \ell.c.(\text{determinants from } \mathcal{M}_0^\perp)\,, \tag{239}$$

where $\ell.c.$ stands for a linear combination. Equivalently, $|\Psi_i\rangle$ has the form

$$|\Psi_i\rangle = \sum_j c_{ji}|\Phi_j\rangle + \ell.c.(\text{determinants from} \mathcal{M}_0^\perp)\,, \tag{239'}$$

since

$$c_{ji} = \langle\Phi_j|\tilde{\Phi}_i\rangle = \langle\Phi_j|P|\Psi_i\rangle = \langle\Phi_j|P^\dagger|\Psi_i\rangle$$
$$= \langle P\Phi_j|\Psi_i\rangle = \langle\Phi_j|\Psi_i\rangle . \qquad (227''')$$

The states $|\tilde{\Psi}_i\rangle$ are obviously more convenient for our purposes since, analogously as in the SR case, we can write

$$|\tilde{\Psi}_i\rangle = (1 + \hat{C}(i))|\Phi_i\rangle , \qquad (240)$$

where $\hat{C}(i)$ designates an excitation operator into the $\mathcal{M}_0^\perp$ relative to the reference $|\Phi_i\rangle$.

5.2 Effective Hamiltonian and Bloch equation

In the preceding section we established two equidimensional subspaces of a general N-electron space $\mathcal{H}_N$. The first one, $\mathcal{M}_0$, is determined by our choice of one-electron functions (spin-orbitals) and their partitioning into the C, A and E spin-orbitals. The second one, $\mathcal{M}$, is a hypothetical space spanned by the corresponding exact eigenstates of the actual Hamiltonian. We have established two basis sets in each space $\mathcal{M}_0$ and $\mathcal{M}$ as well as an isomorphism $\mathcal{M}_0 \sim \mathcal{M}$. The basis sets in $\mathcal{M}_0$ and $\mathcal{M}$ are interrelated through the projection-like maps P and U, and the transformations relating the two bases (one orthogonal and the other not) in each space are related by duality, Eqs. (215) and (227). Thus, if we know $|\tilde{\Phi}_i\rangle$ in terms of $|\Phi_i\rangle$, we also know the relationship between $|\Psi_i\rangle$ and $|\tilde{\Psi}_i\rangle$.

Now, the key idea that enables us to "translate" the exact problem, Eq. (209), into the problem formulated entirely within our model space $\mathcal{M}_0$, is that of an *effective* or *model Hamiltonian*. This is a Hamiltonian that is defined on $\mathcal{M}_0$, yet yields the exact energies that are associated with the corresponding exact eigenstates $|\Psi_i\rangle$ from $\mathcal{M}$. On the first glance, this may sound too good to be realistic. We must keep in mind, however, that we are not trying to determine *all* the eigenvalues of H but only M of them, whose eigenstates have non-negligible components in $\mathcal{M}_0$. Indeed, if we project the Schrödinger equation onto $\mathcal{M}_0$, using the operator P of the preceding section, we get

$$PH|\Psi_i\rangle = E_i P|\Psi_i\rangle = E_i|\tilde{\Phi}_i\rangle . \qquad (241)$$

Thus, using (220) or (223) on the left hand side, we find that

$$PHU|\tilde{\Phi}_i\rangle = E_i|\tilde{\Phi}_i\rangle , \qquad (242)$$

which can be formally written as

$$H^{(\text{eff})}|\tilde{\Phi}_i\rangle = E_i|\tilde{\Phi}_i\rangle , \qquad (243)$$

with the effective or model Hamiltonian $H^{(\text{eff})}$ given by the expression

$$\boxed{H^{(\text{eff})} = PHU = PHUP .} \qquad (244)$$

Clearly, $H^{(\text{eff})}$ is defined on $\mathcal{M}_0$, so that we can simply determine its matrix representative using the known basis $\{|\Phi_i\rangle\}$ of $\mathcal{M}_0$ and obtain $|\tilde{\Phi}_i\rangle$ by diagonalization. Acting with Q on (243) or (244) shows that

$$\boxed{Q\,H^{(\text{eff})}P = 0\,,} \tag{245}$$

which formally expresses the fact that the action of the effective Hamiltonian on the model space does not take us outside this space.

This generalizes a similar procedure that we employed in the SR case. Notice that the form of the effective Hamiltonian is in fact the same in both cases (in the SR case we designated the wave operator by W) as Eqs. (59) and (86) imply. The key factor is, of course, the knowledge of the wave operator U (or W). Again, as in the SR case, we start from the original Schrödinger equation, Eq. (209), to find the energy independent equation(s) determining U. Using the fact that U acts as the identity on $\mathcal{M}$, Eq. (237), we can write

$$UH|\Psi_i\rangle = E_i U|\Psi_i\rangle = E_i|\Psi_i\rangle = H|\Psi_i\rangle\,, \tag{246}$$

so that using Eq. (220) we get

$$UHU|\tilde{\Phi}_i\rangle = HU|\tilde{\Phi}_i\rangle\,. \tag{247}$$

In view of (219b) we conclude that

$$\boxed{UHU = HU} \tag{248}$$

holds on the entire space $\mathcal{H}_N$. This is the well-known *Bloch equation* that determines the wave operator U in the MR case (Bloch, 1958; Lindgren, 1974,1978; Kvasnička, 1974,1977). This operator U is referred to as the *Bloch wave operator* (in contrast to Kato's (1966) projectors that are always Hermitian or orthogonal).

To see that (248) reduces to the SR-CC equations in the one-dimensional case, $\mathcal{M}_0 = \text{Span}\{|\Phi_0\rangle\}$, we simply set

$$U = WP_0 = e^T P_0\,, \tag{249}$$

since U must vanish on $\mathcal{M}_0^\perp$. Substituting (249) into (248) and letting it act on $|\Phi_0\rangle \equiv |\Phi\rangle$, we get after multiplying on the left with e^{-T}

$$P_0 He^T|\Phi\rangle = (e^{-T}He^T)|\Phi\rangle\,. \tag{250}$$

We thus get for the energy, relying on Eqs. (243) and (244), that

$$\Delta E = \langle He^T\rangle = \langle e^{-T}He^T\rangle\,, \tag{251}$$

while the projection with configurations $|\Phi_K^{(i)}\rangle$ from $\mathcal{M}_0^\perp$ gives the SR-CC equations in the form

$$\langle\Phi_K^{(i)}|e^{-T}He^T|\Phi\rangle = 0\,, \tag{252}$$

that are equivalent with those given earlier, cf., Eqs. (84), (87) and (93). Thus, in this case, we see that

$$H^{(\text{eff})} = P_0 He^T P_0 \qquad \text{and} \qquad \Delta E = \langle H^{(\text{eff})}\rangle\,, \tag{253}$$

as in the general case.

The Bloch equation (248) is often expressed in different forms that easily follow from the above given derivation. Thus, for example, using the definition of the effective Hamiltonian, Eq. (244), we can also write (248) in one of the following forms

$$HUP = UH^{(\text{eff})}P \qquad \text{or} \qquad HU = UH^{(\text{eff})}, \tag{254}$$

$$HUP = UPH^{(\text{eff})}P, \tag{255}$$

or even

$$HUP = UPHUP. \tag{255'}$$

Partitioning the Hamiltonian H into an unperturbed part H_0 that commutes with P,

$$[H_0, P] = 0, \tag{256}$$

and the perturbation H',

$$H = H_0 + H', \tag{257}$$

we can also write the Bloch equation (or the generalized Bloch equation, Lindgren, 1974,1978; Lindgren and Morrison, 1986; Kvasnička, 1974,1981) in the form

$$[U, H_0]P = (H'U - UPH'U)P, \tag{258}$$

which has often been used as a starting point in both the MBPT and CC approaches.

5.3. Cluster ansätze for the wave operator

We now come to a crucial point in the development of the OS-CC theories, where different avenues can be followed. This nonuniqueness, leading to different viable formulations, was often a source of confusion in the past. Although we shall essentially follow the historical path, we shall try to avoid as much as possible various misconceptions and pitfalls that confounded the often fallacious development of this subject, lest the reader get confused. Thus, nowadays, we can roughly classify all the OS-CC approaches into the three categories mentioned in the Introduction, namely the Fock space, Hilbert space and state-selective approaches. We shall primarily concentrate on the first two, that represent genuine MR theories, while the approaches of the latter category concentrate on a single state at a time (although using a MR configuration as a reference state) and in fact encompass a wide range of different methodologies that go beyond the scope of these notes.

The principal difficulty that we are facing here is easy to comprehend when we realize that unlike in the SR case, where the reference configuration $|\Phi\rangle$ unambiguously defines the hole and particle states, so that various cluster operators T_i mutually commute, Eqs. (56) and (65), since they only involve particle creation and hole annihilation operators (i.e., all operators are creation operators in the p-h formalism), the OS formalism involves active spin-orbitals and the cluster amplitudes can contain both creation and annihilation valence operators and thus no longer commute. In diagrammatic language, there can be no internal lines interconnecting various T vertices in the CS case, while such lines can appear in the OS theory.

Let us start by examining various alternatives that offer themselves when we try to generalize the SR formalism to the MR case. First, if we wish to retain the *p-h* second quantization formalism, together with a useful diagrammatic representation it provides, we have to define a suitable Fermi vacuum. This may be done in several ways:

(i) We can choose for the Fermi vacuum a configuration, in which all the core states are occupied, while all the active and virtual spin-orbitals are unoccupied (cf. Fig. 21, Fermi level F_C). The reference configurations of an N-electron model space are then obtained by acting with $(N-C)$ valence creation operators on this vacuum. The main advantage of this choice is that only valence creation operators are involved or, in other words, that only *valence particles* are present in the model space, while the fact that a relatively large number of such particles may be needed is certainly not welcome.

(ii) We can choose the Fermi level to lie within the manifold of active states (Fig. 21, Fermi level F_A), in which case both valence particles and valence holes are present in the reference space. Consequently, the formalism becomes more complex, though the number of operators involved (and thus the number of fermion lines in our diagrams) is smaller than in case (i) above.

(iii) We can also work with different Fermi vacua for different reference configurations. This idea was first exploited by Hose and Kaldor (1979,1980,1981,1982) in their MR-MBPT. This choice is particularly suitable for Hilbert space CC approaches. Its weak point is the necessity to inter-relate various vacua in evaluating matrix elements, especially when we wish to employ orthogonally spin-adapted formalism.

Second, and most important, is the generalization of the cluster ansatz for the wave operator. We shall address this problem more rigorously later on. Here we only wish to mention basic options that arise. We must keep in mind that we wish to uniquely characterize all possible excitations outside the model space and express a general wave function in the factorizable (exponential) cluster form. There are basically two possibilities for such a cluster ansatz:

(i) We consider a single cluster operator T that is sufficiently general so that an appropriate cluster expansion in terms of this operator gives a unique wave operator U (or Ω) that produces appropriate exact states $|\tilde{\Psi}_i\rangle$ when acting on model functions $|\Phi_i\rangle$, Eq. (224). We shall see that this is only possible when we consider simultaneously all n-valence electron model spaces that "interconnect" the chosen Fermi vacuum and the N electron states considered, i.e. $0 \leq n \leq (N-C)$. Unique operators T and U (or Ω) that encompass the action of various $T^{(n)}$ and U_n for all $0 \leq n \leq (N-C)$ are then referred to as *valence universal* operators, since they apply to any valence space occupancy that is required.

(ii) We use a number of cluster operators and corresponding exponential ansätze, one for each distinct reference configuration, writing the total wave operator as an appropriate linear combination (Jeziorski and Monkhorst, 1981). Since each such cluster operator $T^{(i)}$ is defined relative to one reference configuration $|\Phi_i\rangle$ and produces a corresponding linear combination of exact states $|\tilde{\Psi}_i\rangle$, $|\tilde{\Psi}_i\rangle = \exp(T^{(i)})|\Phi_i\rangle$, $i = 1, \ldots, M$, this ansatz and the corresponding formalism are referred to as *state universal* approaches.

(iii) For the sake of completeness we mention other numerous possibilities, usually representing special cases of the above two basic ansätze. The simplest and earliest attempts, designed to handle various excited and ionized states, used a simple modification of the standard closed-shell ansatz, namely

$$|\Psi\rangle = W \exp(T^{(0)})|\Phi_0\rangle \,, \tag{259}$$

with W producing the desired state (Čížek, 1966a,1969; Paldus *et al.*, 1978; Saute *et al.*, 1979), which enabled Hermitian variational formulation. In fact, this ansatz can be regarded as a special case of valence universal ansätze used by Offermann *et al.* (1976) and Ey (1978). Various other state-selective approaches usually involve a simple, heuristic, exponential ansatz for a given state. Although this type of ansätze can seldom give a fully size-extensive formalism, they certainly are of considerable practical value in view of their simplicity (similarly as the standard SR closed-shell CC approach) and certainly deserve further exploration (Paldus *et al.*, 1978; Saute *et al.*, 1979; Nakatsuji and Hirao, 1978; Nakatsuji, 1979; Hirao and Nakatsuji, 1981; Banerjee and Simons, 1981, 1982; Baker and Robb, 1983; Tanaka and Terashima, 1984; Hoffmann and Simons, 1988, 1989). In particular, we draw the reader's attention to the possibility of exploiting presently available symbol manipulation software packages (such as MAPLE, MATHEMATICA or MACSYMA) or in fact specially designed software to handle complex algebraic manipulations that may arise in specific OS applications, as recently developed by Janssen and Schaefer (1991).

In order to give a more precise formulation of the valence and state universal cluster ansätze, we first present some useful formal developments (Paldus and Jeziorski, 1988; Jeziorski and Paldus, 1989).

5.4 Model algebras

Let us consider the algebra of replacement operators $e^{RST\cdots}_{ABC\cdots}$ that we introduced in Sec. 4.3.2. To simplify our notation, we use the following labelling of C, A and E spin-orbitals: $a, b, c, d \in C; i, j, k, \ell \in A; r, s, t, u \in E; \alpha, \beta, \gamma, \delta \in C \cup A; \rho, \sigma, \tau, \omega \in A \cup E$ and $\lambda, \kappa, \mu, \nu \in C \cup A \cup E$. Then the definition (99) takes the form

$$e^{\lambda\kappa\cdots}_{\mu\nu\cdots} = X^\dagger_\lambda X^\dagger_\kappa \cdots X_\nu X_\mu \,. \tag{260}$$

Designating, next, the core orbital Fermi vacuum as $|\Phi\rangle$,

$$|\Phi\rangle = (\prod_{a \in C} X^\dagger_a)|0\rangle \,, \tag{261}$$

we define model spaces $\mathcal{M}_0^{(n)}$ for systems with $n = 0, 1, \ldots, L$ valence electrons as

$$\begin{aligned}
\mathcal{M}_0^{(0)} &= \mathrm{Span}\{|\Phi\rangle\}\,, \\
\mathcal{M}_0^{(1)} &= \mathrm{Span}\{X_i^\dagger|\Phi\rangle | 1 \le i \le L\}\,, \\
\mathcal{M}_0^{(2)} &= \mathrm{Span}\{X_i^\dagger X_j^\dagger|\Phi\rangle | 1 \le i,j \le L\}\,, \\
&\vdots \\
\mathcal{M}_0^{(L)} &= \mathrm{Span}\{(\prod_{i \in A} X_i^\dagger)|\Phi\rangle\}\,,
\end{aligned} \tag{262}$$

assuming a definite ordering of C, A and E spin-orbitals and using a product symbol for an ordered product. Corresponding "exact" spaces into which these model spaces "evolve" when all inter-electronic correlations are accounted for are designed as $\mathcal{M}^{(n)}, n = 0, 1, \ldots, L$. They are linear spans of corresponding exact solutions $|\Psi_q^{(n)}\rangle$, $H_n|\Psi_q^{(n)}\rangle = E_q^{(n)}|\Psi_q^{(n)}\rangle$, where H_n designates the $(K+n)$-electron Hamiltonian. We shall assume that for each pair of spaces $\mathcal{M}_0^{(n)}$ and $\mathcal{M}^{(n)}$, the same relationships hold as derived in Section 5.1, including the intermediate normalization, and we shall designate the corresponding quantities by the same symbol with an attached superscript $(n), n = 0, 1, \ldots, L$, indicating the n-valence electron subshell. Thus, $\mathcal{H}_N$ becomes $\mathcal{H}_{(K+n)}$, or $\mathcal{H}^{(n)}$ for short, and we have that

$$\begin{aligned}
(P_n)^2 &= P_n,\ (U_n)^2 = U_n,\ P_n U_n = P_n,\ U_n P_n = U_n\,, \\
|\tilde{\Psi}_q^{(n)}\rangle &= U_n|\Phi_q^{(n)}\rangle,\ |\Psi_q^{(n)}\rangle = U_n|\tilde{\Phi}_q^{(n)}\rangle\,, \text{etc.},
\end{aligned} \tag{263}$$

with $q \equiv I_n = \{i_1 i_2 \cdots i_n\}$ labeling the individual states of $\mathcal{M}_0^{(n)}$ or $\mathcal{M}^{(n)}$.

We next define a *model algebra* $\mathcal{A}$ as an associative operator algebra generated by all the excitation operators e_α^ρ (recall that $\alpha \in C \cup A$, $\rho \in A \cup E$) and by the identity, with the product being an ordinary product of operators. Thus, as a vector space, $\mathcal{A}$ has a basis

$$1,\ e_\alpha^\rho,\ e_{\alpha\beta}^{\rho\sigma},\ e_{\alpha\beta\gamma}^{\rho\sigma\tau}, \cdots\,. \tag{264}$$

Clearly, $\mathcal{A}$ is not Abelian, since valence spin-orbitals can be both created and annihilated, so that the operators $e_{\alpha\beta\gamma\cdots}^{\rho\sigma\tau\cdots}$ do not generally commute. All operators from $\mathcal{A}$ conserve particle number and thus leave each of the $\mathcal{H}^{(n)}$ spaces invariant. Moreover, $\mathcal{A}$ consists of two subalgebras: an *internal algebra* $\mathcal{A}_{\mathrm{int}}$, spanned by the identity and all replacement operators having only valence indices, $\mathcal{A}_{\mathrm{int}} = \mathrm{Span}\{1, e_j^i, e_{k\ell}^{ij}, \cdots\}$, and an *external algebra* $\mathcal{A}_{\mathrm{ext}}$, spanned by the remaining basis elements, i.e. $\mathcal{A}_{\mathrm{ext}} = \mathrm{Span}\{e_\alpha^\rho, e_{\alpha\beta}^{\rho\sigma}, \cdots\}$, where each replacement operator carries at least one non-valence label, so that

$$\mathcal{A} = \mathcal{A}_{\mathrm{int}} \oplus \mathcal{A}_{\mathrm{ext}}\,. \tag{265}$$

In fact, $\mathcal{A}_{\mathrm{ext}}$ is a two-sided ideal of $\mathcal{A}$ and maps each model space $\mathcal{M}_0^{(n)}$ into its orthogonal complement $(\mathcal{M}_0^{(n)})^\perp$, while $\mathcal{M}_0^{(n)}$ are invariant under the action of $\mathcal{A}_{\mathrm{int}}$. The algebras $\mathcal{A}$, $\mathcal{A}_{\mathrm{int}}$ and $\mathcal{A}_{\mathrm{ext}}$ can also be considered as Lie algebras with commutators defining the Lie product (Paldus and Jeziorski, 1988; Jeziorski and Paldus,

1989). We have seen in Section 4.3.2 that the commutators play an important role in many-body theory, since they preserve connectedness of operators. The same holds for the so-called cumulants (Jeziorski and Paldus, 1989). It is also useful to introduce a commutative product in $\mathcal{A}$, designated by an asterix, by defining

$$\begin{aligned} e^{\rho}_{\alpha} * e^{\sigma}_{\beta} &= e^{\rho\sigma}_{\alpha\beta}, \\ e^{\rho\sigma}_{\alpha\beta} * e^{\tau}_{\gamma} &= e^{\rho\sigma\tau}_{\alpha\beta\gamma}, \text{ etc.} \end{aligned} \tag{266}$$

Note that this asterix product is essentially the normal product $N[\cdots]$ defined earlier. Using this new product, we define "normal ordered" exponential function, denoted by the bold face symbol **e**, as follows

$$\mathbf{e}^{X} = 1 + X + \frac{1}{2!}X * X + \frac{1}{3!}X * X * X + \cdots. \tag{267}$$

This is equivalent to the normal ordered exponential ansatz first introduced by Lindgren (1978). Clearly, the series (267) terminates if $X \in \mathcal{A}_{\text{ext}}$ since $\mathcal{A}_{\text{ext}}$ is nilpotent.

5.5. Valence-universal operators

Recall that for Bloch wave operators U_n we have [cf. Eqs. (224) and (240)]

$$|\tilde{\Psi}^{(n)}_q\rangle = U_n|\Phi^{(n)}_q\rangle = (1 + \hat{C}_n(q))|\Phi^{(n)}_q\rangle, \tag{268}$$

where the excitation operator $\hat{C}_n(q) \equiv \hat{C}(q)$ can be expanded in $\mathcal{A}_{\text{ext}}$,

$$\hat{C}(q) = C^{\alpha}_{\rho}(q)e^{\rho}_{\alpha} + \left(\frac{1}{2!}\right)^2 C^{\alpha\beta}_{\rho\sigma}(q)e^{\rho\sigma}_{\alpha\beta} + \cdots. \tag{269}$$

Here we used Einstein summation convention and the summations extend over spin-orbitals $\alpha, \beta, \cdots$ $(\rho, \sigma, \cdots)$ that are occupied (unoccupied) in $|\Phi^{(n)}_q\rangle$. Consequently, the superscript and subscript sets in $C^{\alpha\beta\cdots}_{\rho\sigma\cdots}(q)$ are distinct, $\{\alpha, \beta, \cdots\} \cap \{\rho, \sigma, \cdots\} = \emptyset$. These coefficients represent *renormalized CI coefficients* and are antisymmetric in both superscripts and subscripts. We can also introduce the coefficient corresponding to $\{\alpha\} = \{\rho\} = \emptyset$, which in view of the intermediate normalization equals unity, $C(q) \equiv C(I_n) = 1$.

Recalling now that U_n is a nonorthogonal projector onto $\mathcal{M}^{(n)}$, Eq. (238), we see immediately that we cannot represent it by a simple exponential, e.g. e^{T_n}, since its idempotency would imply that $U_n = 1$. We can resolve this dilemma by employing an exponential ansatz for an operator Ω, defined on $\mathcal{F}$, and acting on each $\mathcal{M}^{(n)}_0$ in exactly the same way as U_n, so that $\Omega P_n = U_n$, but differing from U_n in its action on $(M^{(n)}_0)^{\perp}$. Such an operator, however, cannot be defined within a single n-valence space $\mathcal{H}^{(n)}$, since this would amount to inverting P_n. This basic idea was already implied in Coester's (1969) work and was later fully elaborated by Mukherjee *et al.* (1977) [see also Offermann *et al.*, 1976, and Ey, 1978], who required that all wave operators $U_n, n = 0, 1, \ldots, L$ be universally represented on corresponding $\mathcal{M}^{(n)}_0$ by $\Omega \in \mathcal{A}$. Expanding this operator using the basis (264) we can thus write

$$\Omega = \omega + \omega^{\alpha}_{\rho}e^{\rho}_{\alpha} + \left(\frac{1}{2!}\right)^2 \omega^{\alpha\beta}_{\rho\sigma}e^{\rho\sigma}_{\alpha\beta} + \cdots, \tag{270}$$

where $\omega^{\alpha\beta\cdots}_{\rho\sigma\cdots}$ are again antisymmetric in both sub- and super-scripts. However, in contrast to CI coefficients, the sets of upper and lower indices may now overlap, $\emptyset \neq \{\alpha, \beta, \cdots\} \cap \{\rho, \sigma, \ldots\} \in A$. Through this additional degree of freedom we can achieve that the amplitudes $\omega^{\alpha\beta\cdots}_{\rho\sigma\cdots}$ contain the same amount of information as all the q (or I_n) dependent CI coefficients $C^{\alpha\beta\cdots}_{\rho\sigma\cdots}(I_n)$. In fact, by projecting (268) with a general function from $(\mathcal{M}_0^{(n)})^{\perp}$, we get

$$C^{\alpha_1\cdots\alpha_m}_{\rho_1\cdots\rho_m}(I_n) = \langle \Phi^{\rho_1\cdots\rho_m}_{\alpha_1\cdots\alpha_m}(I_n)|U_n|\Phi(I_n)\rangle \,, \tag{271}$$

where

$$|\Phi^{\rho_1\cdots\rho_m}_{\alpha_1\cdots\alpha_m}(I_n)\rangle = e^{\rho_1\cdots\rho_m}_{\alpha_1\cdots\alpha_m}|\Phi(I_n)\rangle \tag{272}$$

and

$$|\Phi(I_n)\rangle \equiv |\Phi_q^{(n)}\rangle \,. \tag{272'}$$

Using the expansion (270) on the right hand side of (271), we can then express the renormalzied CI coefficients $C^{\alpha\beta\cdots}_{\rho\sigma\cdots}(I_n)$ in terms of $\omega^{\alpha\beta\cdots}_{\rho\sigma\cdots}$ amplitudes. Explicitly, we find (Jeziorski and Paldus, 1989)

$$\begin{aligned}
C^{\alpha_1\cdots\alpha_m}_{\rho_1\cdots\rho_m} &= \omega^{\alpha_1\cdots\alpha_m}_{\rho_1\cdots\rho_m}\,, && \text{for } n=0\,,\\
C^{\alpha_1\cdots\alpha_m}_{\rho_1\cdots\rho_m}(i) &= \omega^{\alpha_1\cdots\alpha_m}_{\rho_1\cdots\rho_m} + \omega^{\alpha_1\cdots\alpha_m i}_{\rho_1\cdots\rho_m i}\,, && \text{for } n=1\,,\\
C^{\alpha_1\cdots\alpha_m}_{\rho_1\cdots\rho_m}(ij) &= \omega^{\alpha_1\cdots\alpha_m}_{\rho_1\cdots\rho_m} + \omega^{\alpha_1\cdots\alpha_m i}_{\rho_1\cdots\rho_m i} \\
&\quad + \omega^{\alpha_1\cdots\alpha_m j}_{\rho_1\cdots\rho_m j} + \omega^{\alpha_1\cdots\alpha_m ij}_{\rho_1\cdots\rho_m ij}\,, && \text{for } n=2\,,\\
&\text{etc.,}
\end{aligned} \tag{273}$$

assuming that $\{\alpha_1, \ldots, \alpha_m\} \cap \{\rho_1, \ldots, \rho_m\} = \emptyset$ and $\{\alpha_1, \ldots, \alpha_m\} \in C \cup I_n$ (including the case $\{\alpha\} = \{\rho\} = \emptyset$ when $C(I_n) = \omega = 1$). Inverting these relationships, we obtain ω amplitudes in terms of renormalized MR-CI coefficients. In this way the family of wave operators U_n, $n = 0, 1, \ldots, L$ can be uniquely represented by a valence universal operator Ω from $\mathcal{A}$.

The same argument can be given for any set of operators $A_n : \mathcal{H}^{(n)} \to \mathcal{H}^{(n)}$, $n = 0, 1, \ldots, L$. The corresponding valence universal operator is then simply designated as A or, for greater emphasis, as $\mathcal{P}(\mathbf{A})$ with $\mathbf{A} \equiv \{A_0, A_1, \ldots, A_L\}$. In case that a family $\mathbf{D} = \{D_0, D_1, \ldots D_L\}$ can be obtained by restricting a Fock space operator D to $\mathcal{H}^{(n)}$, $D_n = D \downarrow \mathcal{H}^{(n)}$, we write simply $\mathcal{P}(D)$. Clearly, in general $\mathcal{P}(D) \neq D$, since D does not belong to $\mathcal{A}$ (since it may involve arbitrary generators $e^{\mu\nu\cdots}_{\lambda\kappa\cdots}$).

Similarly, operators acting in $\mathcal{M}_0^{(n)}$, for example $H_n^{(\text{eff})}$, $H_n^{(\text{eff})} = P_n H U_n$, can be represented by a single operator G from $\mathcal{A}_{\text{int}}$, so that $H_n^{(\text{eff})} = G P_n$. An explicit prescription for the construction of G was also given by Jeziorski and Paldus (1989).

Finally, Bloch equations for U_n, $HU_n = U_n H_n^{(\text{eff})}$, can be represented as

$$\mathcal{P}(H\Omega) = \Omega G \,, \tag{274}$$

since $\mathcal{P}(\Omega G) = \Omega G$, but $\mathcal{P}(H\Omega) \neq H\Omega$ (although, misleadingly, these equations are sometimes written as $H\Omega = \Omega G$.

Since $\Omega - 1 \equiv \hat{\Omega} \in \mathcal{A}_{ext}$ (see Eq. (270) and recall that $\omega = 1$ assuming the intermediate normalization) is nilpotent, the inverse Ω^{-1} must exist. Thus, premultiplying (274) with Ω^{-1} and projecting on $\mathcal{A}_{ext}$ we can eliminate G, obtaining

$$\mathcal{P}_{ext}(\Omega^{-1}H\Omega) = 0\,, \tag{275}$$

where $\mathcal{P}_{ext}(A)$ designates the exterior part of $A \in \mathcal{A}$. This equation is often used to derive explicit CC equations for OS cluster amplitudes. Its main advantage stems from the fact that it does not involve the effective interaction operator G so that no disconnected terms can creep into the effective Hamiltonian when truncating the expansion for the wave operator Ω.

5.6. Basic OS cluster ansätze

We can use (274) to generate PT expansion for G and to show that the resulting correlation energies contain only connected terms, thus assuring the size extensivity. However, when we truncate the Ω expansion, as we always have to do in practical applications, a rather delicate cancellation of disconnected terms may no longer be complete and the size extensive property sacrificed. It is thus essential to formnulate the theory in terms of connected quantities only, as we have seen in detail in the SR case. This may be achieved in different ways, as already alluded to above. We now briefly describe two basic approaches that enable such a formulation.

5.6.1. *Fock space approaches*

Fock space approaches employ the cluster ansatz for a single, valence universal, wave operator Ω, representing wave operators for all $n = 0, 1, \ldots, L$ valence levels. Writing simply (Mukherjee, 1979; Haque and Mukherjee, 1984)

$$\Omega = e^{W}\,, \tag{276}$$

leads, however, to a formalism in which also the contractions between various W-clusters must be considered. Classifying various terms constituting W by their *valence rank* r, given by the number of valence annihilators they contain, we write

$$W = \sum_{r=0}^{L} W^{(r)}\,. \tag{277}$$

Using then the ansatz (276) in (275) leads to a formalism in which different valence rank components of W are strongly coupled. Representing W-vertices by empty circles, one thus needs, in addition to typical CC diagrams arising in the SR case, also diagrams of the type

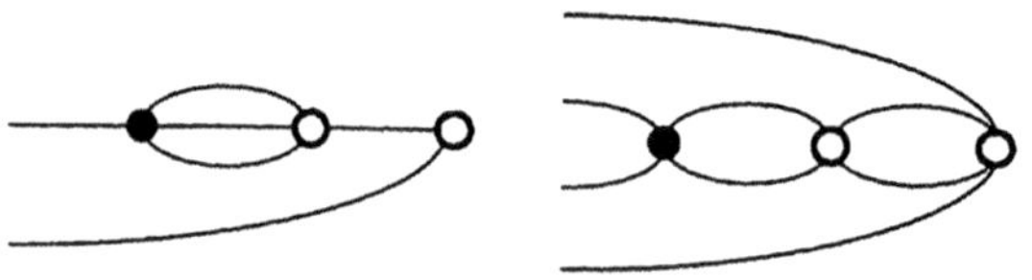

(278)

Of course, the lines interconnecting W-vertices always carry valence labels (usually indicated by double arrows when the orientation of fermion lines is introduced). Consequently, the relationship between the Ω and W amplitudes is rather complicated (see also Jeziorski and Paldus, 1989).

These complications can be avoided by employing a normal ordered exponential ansatz of Lindgren (1978),

$$\Omega = \mathrm{e}^{S} \equiv N[\exp(S)], \tag{279}$$

with S designating the connected part of Ω. In fact, Lindgren initially employed this ansatz for a single Bloch equation with a fixed valence electron number n (i.e., $n = L$). In general, this would lead to an underdetermined system of CC equations, since the number of cluster amplitudes exceeds that of equations (we know from Section 5.5 that Bloch equations for all $n = 0, 1, \ldots, L$ are needed for a complete specification of cluster amplitudes). However, Lindgren replaced these remaining equations by certain sufficiency conditions that guaranteed the connectivity of cluster amplitudes (these conditions are particularly easy to implement in the diagrammatic formulation by considering only connected diagrams). As shown later (Haque and Mukherjee, 1984; Lindgren and Mukherjee, 1987; Jeziorski and Paldus, 1989), these sufficiency conditions in fact imply the adoption of the valence universality assumption. Moreover, when we use the normal product ansatz (279) to calculate Bloch wave operators $U_n = \Omega P_n$, we get (Jeziorski and Paldus, 1989)

$$\begin{aligned} U_0 &= e^{S^{(0)}} P_0 \,, \\ U_1 &= [e^{S^{(0)}} * (1 + S^{(1)})] P_1 \,, \\ U_2 &= [e^{S^{(0)}} * (1 + S^{(1)} + \tfrac{1}{2} S^{(1)} * S^{(1)} + S^{(2)})] P_2, \text{ etc.,} \end{aligned} \tag{280}$$

where $S^{(r)}$ designates the valence rank r part of S. This leads to a cluster ansatz used earlier by Coester (1969) and the Bochum group (Offermann *et al.*, 1976, Ey, 1978) which decouples different valence rank components and enables their recursive calculation, starting with $S^{(0)}$. Moreover, $S^{(n)}$ component is linearly related with the wave operator U_n.

Evidently, all the above approaches are equivalent when carried out in their entirety. However, when introducing various approximations by truncating the cluster expansions employed, important differences arise. It is beyond the scope of these introductory notes to analyze the advantages and shortcomings of various methods and approximations, which are of course very important for practical implementations and which continue to be explored in actual applications. For a more comprehensive exposition of these and related problems we refer the reader to a recent review by Mukherjee and Pal (1989).

Similar diversity arises when deriving the explicit form of the OS-CC equations that determine cluster amplitudes of S or W. Thus, one can either use Eq. (274) or (275). The latter one can be written in the form

$$\langle \Phi^{\rho\sigma\cdots}_{\alpha\beta\cdots} | \Omega^{-1} H \Omega | \Phi^{(n)}_q \rangle = 0, \qquad \forall n, q, \Phi^{\rho\sigma\cdots}_{\alpha\beta\cdots}\,, \tag{281}$$

and the corresponding energies may be found using the effective Hamiltonian with matrix elements

$$H^{(\mathrm{eff})}_{pq} = \langle \Phi^{(n)}_p | \Omega^{-1} H \Omega | \Phi^{(n)}_q \rangle \,. \tag{282}$$

Similarly, the former one leads to the equations obtained by projecting the Fock space Bloch equation onto the $(\mathcal{M}_0^{(n)})^\perp$ with Q_n, yielding

$$Q_n H \Omega P_n = Q_n \Omega P_n H_n^{(\text{eff})} P_n \,, \qquad \forall n \tag{283}$$

with the effective Hamiltonian evaluated by computing matrix elements

$$H_{pq}^{(\text{eff})} = \langle \Phi_p^{(n)} | H \Omega | \Phi_q^{(n)} \rangle \,. \tag{284}$$

Mukherjee and Pal (1989) call the former ones, Eqs. (281) and (282), "similarity-transformation-based" methods and the latter ones "Bloch-equation-based" methods. We refer the reader to their review and the original papers for more detail on the explicit form of these equations and related matters (elimination of spectator lines, subsystem embedding conditions, etc.).

5.6.2. *Hilbert space approaches*

The connected CC formalism that does not involve valence universality assumption and thus relies only on the quantities and concepts associated with a fixed valence electron number sector of the Fock space, namely $\mathcal{H}^{(n)}$ (with $n = L$, of course), may be achieved by assuming a separate cluster ansatz for every reference configuration (Jeziorski and Monkhorst, 1981; see also Silverstone and Sinanoğlu, 1966), so that

$$U = \sum_q \exp(T^{(q)}) P_q \,, \tag{285}$$

where $P_q = |\Phi_q\rangle\langle\Phi_q|$, Eq. (207a), and $T^{(q)}$ represents a connected cluster operator associated with the reference configuration $|\Phi_q\rangle \equiv |\Phi_q^{(n)}\rangle$. Thus, the exact eigenstates are given as

$$|\Psi_p\rangle = \sum_q C_{qp} \exp(T^{(q)}) |\Phi_q\rangle \,, \tag{286}$$

where the coefficients C_{qp} are given by the components of the eigenvectors of the effective Hamiltonian with elements

$$H_{qp}^{(\text{eff})} = \langle \Phi_q | e^{-T^{(p)}} H e^{T^{(p)}} | \Phi_p \rangle = \langle \Phi_q | (H_{N_p} e^{T^{(p)}})_C | \Phi_p \rangle \,. \tag{287}$$

The CC equations that determine the cluster amplitudes of various $T^{(p)}$ operators are again derived by substituting the ansatz (285) into Bloch equation acting on a reference state $|\Phi_p\rangle$, premultiplying with $\exp(-T^{(p)})$, and projecting onto the excited configurations relative to $|\Phi_p\rangle$, obtaining

$$\langle G_i \Phi_p | e^{-T^{(p)}} H e^{T^{(p)}} | \Phi_p \rangle = \sum_{q(\neq p)} \langle G_i \Phi_p | e^{-T^{(p)}} e^{T^{(q)}} | \Phi_q \rangle H_{qp}^{(\text{eff})} \,, \tag{288}$$

since

$$(T^{(p)})^\dagger |\Phi_q\rangle = 0 \,. \tag{289}$$

Here G_i designates a general excitation operator from $\mathcal{A}_{\text{ext}}$. The connectedness of the left-hand side is easily established as in the SR case, while for the right hand side a special PT consideration, as given by Jeziorski and Monkhorst (1981), must be invoked. It should also be noted that the same excited configuration can arise from different reference states. One of the advantages of this formulation is the fact that the cluster amplitudes do not carry any spectator labels. On the other hand, this formalism is not as easily amenable to symmetry adaptation as is the Fock space formalism.

To get a better idea of a general structure of Hilbert space CC equqations, we expand the exponentials in (288) and neglect cubic and higher order terms, obtaining

$$\sum_{r=0}^{2} \Lambda_p^{(r)}(G_i) = \sum_{r=1}^{2} \Xi_p^{(r)}(G_i)\,, \tag{290}$$

where

$$\begin{aligned} \Lambda_p^{(0)}(G_i) &= \langle G_i\Phi_p|H|\Phi_p\rangle\,, \\ \Lambda_p^{(1)}(G_i) &= \langle G_i\Phi_p|[H,T^{(p)}]|\Phi_p\rangle\,, \\ \Lambda_p^{(2)}(G_i) &= \tfrac{1}{2}\langle G_i\Phi_p|[[H,T^{(p)}],T^{(p)}]|\Phi_p\rangle\,, \end{aligned} \tag{291}$$

and

$$\begin{aligned} \Xi_p^{(1)}(G_i) &= \sum_{q(\neq p)} \langle G_i\Phi_p|T^{(q)} - T^{(p)}|\Phi_q\rangle H_{qp}^{(\mathrm{eff})}\,, \\ \Xi_p^{(2)}(G_i) &= \tfrac{1}{2} \sum_{q(\neq p)} \langle G_i\Phi_p|(T^{(q)} - T^{(p)})^2 + [T^{(q)},T^{(p)}]|\Phi_q\rangle H_{qp}^{(\mathrm{eff})}\,. \end{aligned} \tag{292}$$

Thus, for each reference $|\Phi_p\rangle$, the left-hand side has exactly the same form as in the SR case considered in Chapter 4, involving the absolute, linear and quadratic terms in cluster amplitudes $\Lambda_p^{(0)}$, $\Lambda_p^{(1)}$ and $\Lambda_p^{(2)}$, respectively. Specifically, for the two-body amplitudes $T_2^{(p)}$ and $G_i \in \mathcal{A}_{\mathrm{ext}}$ representing a double excitation relative to $|\Phi_p\rangle$, $G_i \equiv G_i^{(p)} = e_{C_1C_2}^{C^1C^2}$, the explicit form of $\Lambda^{(0)}$, $\Lambda^{(1)}$ and $\Lambda^{(2)}$ is precisely the same as in Eqs. (113)-(115), except that the summations over the hole and particle states are restricted in such a way that no cluster amplitudes involving only valence indices arise. However, in contrast to the SR case, the equations for different reference states $|\Phi_p\rangle$ are now mutually coupled by the coupling terms $\Xi^{(r)}$.

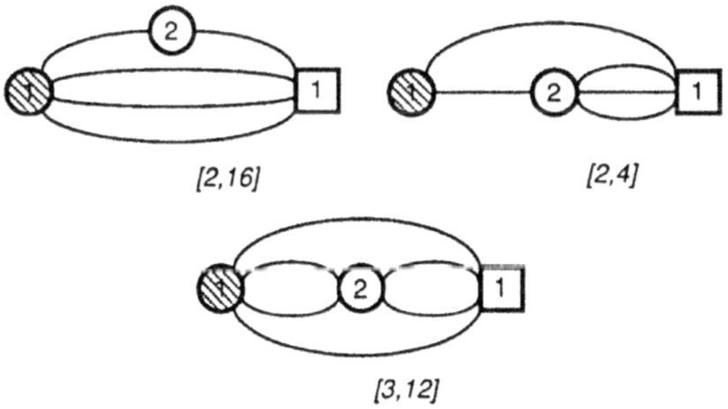

Fig. 23. Nonoriented Hugenholtz-type skeletons representing the linear coupling terms $\Xi^{(1)}$, Eq. (292). The first integer in the square bracket gives the number of possible oriented skeletons, while the second integer indicates the number of distinct Hugenholtz or Brandow diagrams that arise when fixed labels are introduced in all possible ways. The latter are easily handled using the symmetrizers (antisymmetrizers), Eq. (111) [Eq. (111′)]. See text for details.

The general structure of these terms is best illustrated using a suitable diagrammatic representation. To this effect we write the effective Hamiltonian relative to the reference $|\Phi_p\rangle$ as

$$^{(p)}H^{(\text{eff})} = H_{pp}^{(\text{eff})} P_p + \tilde{T}^{(p)} P \,, \tag{293}$$

with

$$\tilde{T}^{(p)} = \sum_{q(\neq p)} H_{qp}^{(\text{eff})} \Omega_p^q \,, \tag{294}$$

where $\Omega_p^q \in \mathcal{A}_{\text{int}}$ transforms $|\Phi_p\rangle$ into $|\Phi_q\rangle$, $|\Phi_q\rangle = \Omega_p^q|\Phi_p\rangle$, or $\Omega_p^q = |\Phi_q\rangle\langle\Phi_p|$, while $P_p = |\Phi_p\rangle\langle\Phi_p|$. Representing then $T^{(p)}$ vertices by open circles labeled by p, $\tilde{T}^{(p)}$ vertex by an open rectangle labeled by p (clearly, all the fermion lines attached to this vertex carry valence labels) and the excited state $|G_i\Phi_p\rangle$, $G_i \in \mathcal{A}_{\text{ext}}$, onto which we project, by a dashed circle labeled by p, we can represent the linear ($\Xi^{(1)}$) and quadratic ($\Xi^{(2)}$) coupling terms by the diagrams shown in Figs. 23 and 24, respectively.

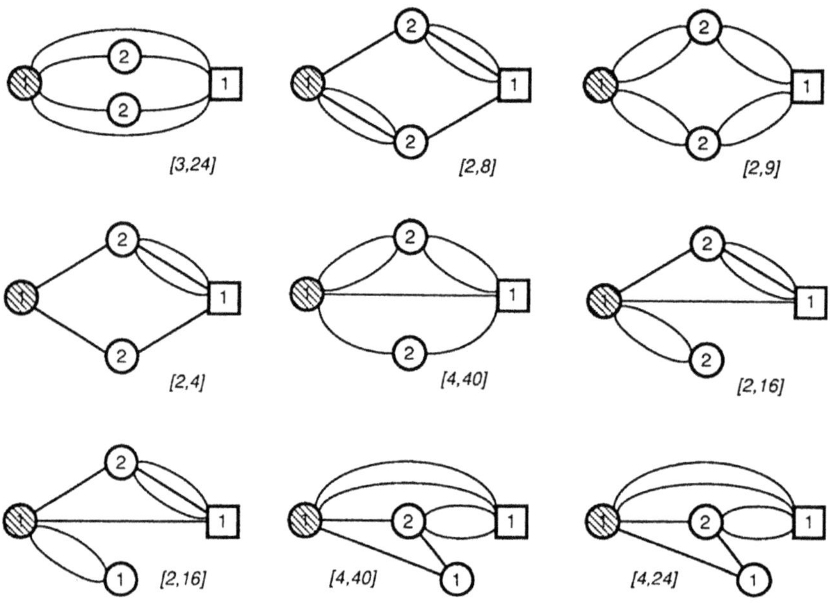

Fig. 24. Nonoriented Hugenholtz-type skeletons representing the quadratic coupling terms $\Xi^{(2)}$, Eq. (292). See caption to Fig. 23 and text for details.

Note that the diagrammatic representation employs the pertinent ket state in the evaluated expression as a vacuum. Similar derivation of the Hilbert space CC equations was given by Meissner *et al.* (1988) for both spin-orbital and spin-adapted closed-shell-type reference states and, for the special case of two closed-shell-type references (corresponding to two active orbitals of different symmetry) by Jeziorski and Paldus (1988) and Paldus *et al.* (1989) for the orthogonally spin-adapted case. While the former author employed diagrammatic representation, the latter work was carried out using algebraic formalism (using spin-adapted replacement operators). Recently, the orthogonally spin-adapted coupling terms were rederived using diagrammatic approach and graphical methods of spin algebras (Piecuch and Paldus, 1991). Initial applications to simple model systems (Paldus *et al.*, 1989,1991; Balková *et al.*, 1991) indicate that in the absence of intruder states, already the linear approximation with one- and two-body clusters relative to each reference works extremely well, while the quadratic approximation works satisfactorily even when intruders are present. The full details of these results will be given elsewhere (Pylypow, 1991; Paldus *et al.*, 1991; Piecuch *et al.*, 1991).

5.7. Cluster conditions

One of the basic assumptions for a successful and efficient exploitation of CC approaches is their rapid convergence, so that higher than pair-cluster amplitudes are either negligible or can be reliably accounted for in an approximate manner as products of lower order amplitudes. In the closed-shell case, for example, we require that four-body clusters T_4 are negligible, so that the tetraexcited states are well approximated by the disconnected clusters, schematically

$$\hat{C}_4 \approx \tfrac{1}{2}\hat{C}_2^2 , \tag{295}$$

where $\hat{C}_n$ operator creates the n-fold excited part of the CI wave function when acting on the reference configuration $|\Phi\rangle$. The validity of this so-called *cluster condition* was investigated for several systems, and its breakdown in quasidegenerate situations is well established (Čížek *et al.*, 1969; Jankowski and Paldus, 1980; Paldus and Boyle, 1982; Wilson *et al.*, 1983, Paldus *et al.*, 1984b). This and similar cluster conditions follow easily once the relationship between the CI and connected CC amplitudes is known. In the single reference case this relationship, as expressed by Eqs. (72), is easily established (Primas, 1965). Unfortunately, the relationship between various MR-CI and OS-CC ansätze is much more involved and thus more difficult to establish (cf. Section 5.5).

Basic cluster conditions for both the Fock and Hilbert space CC ansätze were recently derived by Jeziorski and Paldus (1989). For the latter case, the resulting conditions are very similar to those characterizing the SR formalism, as may be expected in view of the form of the corresponding cluster ansatz, Eqs. (285) or (286). However, for the Fock space formalism, the relationships involved have a more complex structure, primarily due to the presence of spectator or passive orbital labels and the necessity to interrelate different valence sectors of the Fock space.

Thus, while the valence universal counterpart of (295) has the same form, namely $\Omega_4 \approx \frac{1}{2}\Omega_2^2$ (neglecting again Ω_1 for simplicity), the actual implied relationships between the MR-CI (renormalized) amplitudes may differ substantially from the SR or Hilbert space cases. For example, assuming that one of the upper indices in CI amplitudes is of valence type (i.e., $\alpha = a$, $\beta = b$, $\gamma = c$ and $\delta = i$; valence rank one) and none of the lower indices equals to i, we get [cf. Eq. (7.10) of Jeziorski and Paldus, 1989]

$$C^{abci}_{\rho\sigma\tau\omega}(i) \cong \mathcal{A}_{[\rho\sigma][\tau\omega]}\left[C^{ab}_{\rho\sigma}C^{ci}_{\tau\omega}(i) + C^{ac}_{\rho\sigma}C^{ib}_{\tau\omega}(i) + C^{ai}_{\rho\sigma}(i)C^{bc}_{\tau\omega}\right] , \tag{296}$$

where the antisymmetrizer $\mathcal{A}_{[\rho\sigma][\tau\omega]}$ permutes indices between the square brackets,

$$\mathcal{A}_{[\rho\sigma][\tau\omega]} = 1 - (\rho\tau) - (\rho\omega) - (\sigma\tau) - (\sigma\omega) + (\rho\tau)(\sigma\omega) , \tag{297}$$

$(\mu\nu)$ designating the transposition of μ and ν. Notice that this cluster condition represents a homogeneous *linear* relationship between CI amplitudes corresponding to states with one valence electron with the coefficients given by biexcited CI amplitudes. Other cluster conditions, requiring that the four-body cluster amplitudes be small, can be found in the above mentioned paper (Jeziorski and Paldus, 1989).

Very recently the validity of these cluster conditions, for both valence universal and nonuniversal theories, was tested for the minimum basis set H_4 and H_8 model systems (Jankowski *et al.*, 1991). It was found that these conditions are well satisfied for the geometries that correspond to degenerate or highly quasidegenerate situations, but may be severely violated in a nondegenerate region (where the SR ansatz holds well) and, generally, in the presence of intruder states. It would be certainly highly desirable to explore the cluster structure of a great variety of OS systems in order to establish meaningful truncation schemes and the limits of their validity in diverse OS situations, and possibly find suitable criteria for a classification of these systems.

5.8 Incomplete model spaces

As already mentioned, the requirement of using a complete model space, in order to warrant the size-extensivity of the resulting energies, may be both inconvenient and costly in practical implementations, regardless whether the valence universal or nonuniversal cluster ansatz is employed. Further, the larger the model space is, the more likely we encounter intruder states. Finally, when calculating various "differential" properties with Fock space theories, for example excitation energies of systems with closed-shell ground state, it should be sufficient to rely on a model space that is spanned by corresponding singly excited (i.e., $1p$-$1h$) configurations rather than on the full model space (consisting of all np-nh configurations).

The incomplete model spaces have thus been discussed (Brandow 1967,1977; Hose and Kaldor, 1979,1981,1982; Jeziorski and Monkhorst, 1981; Lindgren, 1985) and exploited (e.g., Haque and Kaldor, 1986; Kaldor, 1987; Pal *et al.*, 1988; Ben-Shlomo and Kaldor, 1988; Koch and Mukherjee, 1988) in both the MBPT and CC approaches for some time. Recently, new light has been shed on our understnading of the conditions that guarantee size-extensivity even when truncated model spaces are employed (see, Lindgren and Mukherjee, 1987; Mukherjee and Pal, 1989). It seems that the principal reason why disconnected contributions show up in the effective Hamiltonian when incomplete model spaces are used, and thus invalidate the size extensive character of resulting energies (see Chaudhuri *et al.*, 1989), is a presupposition of an intermediate normalization for the wave operator, $P\Omega P = P$ [cf. Eqs. (232), (238) and (263)], assuming that Ω is valence universal (Mukherjee, 1986). Thus, replacing an intermediate normalization with the so-called size-extensive normalization (Mukherjee, 1986), size-extensive CC approaches based on an incomplete model space can be formulated.

An important step in restricting a general model space was the introduction of *quasi-complete model spaces* by Lindgren (1985a). In this case, valence orbitals are subdivided into distinct subgroups and the occupancy within each subgroup is preserved. For example, considering a closed shell system and choosing a finite number of spin-orbitals below and above the Fermi level as the two subgroups in the active space, the set of monoexcited configurations involving these levels spans such a quasi-complete model space. Another example are so-called *isolated incomplete model spaces* of Kutzelnigg *et al.* (1987).

Let us note, finally, that the necessary and sufficient conditions for the connectivity of $H^{(\mathrm{eff})}$ when an arbitrary model space is employed were recently formulated by Lindgren and Mukherjee (1987).

6. CONCLUSIONS

During the past decade, particularly following the realization of the importance of size-extensivity in molecular electronic structure calculations, the CC methods witnessed a steady growth in methodological development as well as in their computational implementation and exploitation. Today, their relevance and usefulness extends far beyond the domain of quantum chemical many-electron correlation problem (e.g., Bishop and Kümmel, 1987; Bishop, 1991).

In these notes we have tried to address two aspects of the CC formalism that are of special significance for the molecular many-electron correlation problem. First, an elucidation of fundamental concepts and origins of the CC approach and an outline of this formalism for the SR case, presently firmly established and widely exploited. Second, an overview of the present status and an elementary unified exposition of various MR extensions of the CC formalism. The development of the OS-CC methodology followed a rather tortuous path, but lately began to "crystalize" into a few standardized and well defined approaches. We have thus concentrated on basic methodological developments, practically ignoring the vast field of specific implementations and applications that would certainly deserve an up-to-date systematic and exhaustive review. But even on the methodological side we had to restrict outselves to only the most rudimentary aspects of the theory, ignoring advances made in the design of actual algorithms and codes. Nevertheless, it is hoped that these notes will provide a leisurely introduction to the subject, will help to clarify its basic concepts and guide the reader through an often perplexing maze of OS generalizations and developments.

For the future, it is hoped that promising and useful exploitations of the Fock space formalism, that seems to be particularly suitable for the calculation of various "differential" properties (i.e. excitation energies and energetics of various attachment and detachment processes, yielding ionization potentials, electron affinities, double ionization energies, etc.), will be explored on a yet larger scale and the currently emerging applications of the Hilbert space methodology will enable a global exploration of potential energy surfaces, so that eventually both approaches will enable routine applications to be carried out, as is nowadays the case with SR calculations of closed-shell ground states. Undoubtedly, this development would be facilitated were a better grasp and understanding of the cluster structure of general OS wave functions available to us. Then suitable approximation and truncation schemes could be designed a priori, thus obliterating much of the trial and error searching for them. These developments could also be facilitated by the design and study of suitable model systems that typify, interrelate and possibly unify various OS situations.

In view of the genuine complexity of OS-CC approaches, be they of valence universal type or not, it is important to search for simplifications and reliable approximations. Here the exploration of incomplete model spaces certainly comes to the fore. The possibilities offered by both Fock and Hilbert space approaches should be investigated in parallel, since it is too early to abandon one for the other. Clearly, the Fock space formalism seems to be particularly suitable for a study of "differential" properties as already noted. The fact that one has to simultaneously consider the whole sequence of systems differing in electron number (i.e., in most cases the parent molecule or radical together with several of its ions) seems to be both a blessing and a curse. The usual argument in its favour, namely that one obtains information about the system and its ions at the same time, is sound in principle, although not particularly useful in actual computations, since it is well known that in order to obtain reliable information about widely different electronic states (even of the same system) one has to employ different basis sets. This is not possible, however, in the existing versions of the Fock space formalism. The development and exploitation of SR multi-configurational approaches should also be actively pursued, since they are most satisfactory from a computational viewpoint, assuming that size-extensivity is not significantly violated.

Finally, let us emphasize the importance of obtaining an intimate understanding of the relationship between the MR-CC and MR-CI methods, similarly as in the SR case (Paldus *et al.*, 1982). This would enable not only the design of useful approximate schemes (à la Davidson correction) but also an exploitation of efficient algorithms and computer codes, developed for large scale variational calculations, in the CC packages.

ACKNOWLEDGEMENTS

The continued support by NSERC is greatly appreciated. The author is very much indebted to Mrs. Helen Warren for her immaculate typesetting and patience when preparing the camera-ready manuscript. Sincere thanks are also due to Dr. Piotr Piecuch for reading parts of the manuscript and useful comments. Great appreciation and wholehearted thanks are also due to Drs. Geerd Diercksen and Steve Wilson for creating a most inducive atmosphere for learning and scientific exchange at the 1991 NATO School in Bad Windsheim as well as for an exemplary organization and enlightening program of the School.

References

Adamowicz, L., Laidig, W.D., and Bartlett, R.J., 1984, *Int. J. Quantum Chem.*, Symp. **18**:245.

Adams, B.G., and Paldus, J., 1979, *Phys. Rev. A*, **20**:1.

Adams, B.G., and Paldus, J., 1981, *Phys. Rev. A*, **24**:2302.

Adams, B.G., Jankowski, K., and Paldus, J., 1979, *Chem. Phys. Lett.*, **67**:144.

Adams, B.G., Jankowski, K., and Paldus, J., 1981a, *Phys. Rev. A*, **24**:2316.

Adams, B.G., Jankowski, K., and Paldus, J., 1981b, *Phys. Rev. A*, **24**:2330.

Ahlrichs, R., 1979, *Comp. Phys. Commun.*, **17**:31.

Ahlrichs, R., and Scharf, P., 1983, *in*: "Ab Initio Methods in Quantum Chemistry I," K. P. Lawley, ed., John Wiley and Sons, New York, pp. 501-537.

Ahlrichs, R., Lischka, H., Staemmler, V., and Kutzelnigg, W., 1975, *J. Chem. Phys.*, **62**:1225.

Arponen, J., and Pajanne, E., 1982, *J. Phys. C*, **15**:2665,2683.

Arponen, J.S., Bishop, R.F., and Pajanne, E., 1987, *Phys. Rev. A*, **36**:2519,2539.

Arponen, J., Bishop, R.F., Pajanne, E., and Robinson, N.I., 1988, *Phys. Rev. A*, **37**:1065.

Bachrach, S.M., Chiles, R.A., and Dykstra, C.E., 1981, *J. Chem. Phys.*, **75**:2270.

Baker, H., and Robb, M.A., 1983, *Mol. Phys.*, **50**:1077.

Balková, A., Kucharski, S.A., Meissner, L., and Bartlett, R.J., 1991, *Theor. Chim. Acta*, **80**:335.

Banerjee, A., and Simons, J., 1981, *Int. J. Quantum Chem.*, **19**:207.

Banerjee, A., and Simons, J., 1982, *J. Chem. Phys.*, **76**:4548.

Barr, T.L., and Davidson, E.R., 1970, *Phys. Rev. A*, **1**:644.

Bartlett, R.J., 1981, *Annu. Rev. Phys. Chem.*, **32**:359.

Bartlett, R.J., 1986, *in*: "Geometrical Derivatives of Energy Surfaces and Molecular Properties," P. Jørgensen and J. Simons, ed., Reidel, Dordrecht, pp. 35-61.

Bartlett, R.J., 1989, *J. Phys. Chem.*, **93**:1697.

Bartlett, R.J., ed., 1991, Special issues of *Theor. Chim. Acta*, Vol. **80**; Nos. 2-6, with Proceedings of the Workshop on CC Theory at the Interface of Atomic Physics and Quantum Chemistry, Institute for Theoretical and Molecular Physics at the Harvard-Smithsonian Center for Astrophysics, Cambridge, MA.

Bartlett, R.J., and Kucharski, S., 1986, *Adv. Quantum Chem.*, **18**:281.

Bartlett, R.J., and Purvis, G.D., 1978, *Int. J. Quantum Chem.*, **14**:561.

Bartlett, R.J., and Purvis, G.D., 1980, *Phys. Scripta*, **21**:255.

Bartlett, R.J., and Shavitt, I., 1977, *Chem. Phys. Lett.*, **50**:190; Erratum (1978), **57**:157.

Bartlett, R.J., Shavitt, I., and Purvis, G.D., 1979, *J. Chem. Phys.*, **71**:281.

Bartlett, R.J., Dykstra, C.E., and Paldus, J., 1984, *in*: "Advanced Theories and Computational Approaches to the Electronic Structure of Molecules," C.E. Dykstra, ed., Reidel, Dordrecht, pp. 127-159.

Bartlett, R.J., Kucharski, S.A., Noga, J., Watts, J.D., and Trucks, G.W., 1989, *in*: "Many-Body Methods in Quantum Chemistry," Lecture Notes in Chemistry, Vol. 52, U. Kaldor, ed., Springer-Verlag, Berlin, Heidelberg, New York, pp. 125-149.

Bartlett, R.J., Watts, J.D., Kucharski, S.A., and Noga, J., 1990, *Chem. Phys. Lett.*, **165**:513.

Bender, C.F., and Davidson, E.R., 1968, *J. Chem. Phys.*, **49**:4222.

Bender, C.F., and Davidson, E.R., 1969, *Phys. Rev.*, **183**:23.

Ben-Shlomo, S., and Kaldor, U., 1988, *J. Chem. Phys.*, **89**:956.

Bishop, R.F., 1991, *Theor. Chim. Acta*, **80**:95, and references therein.

Bishop, R.F., and Arponen, J.S., 1990, *Int. J. Quantum Chem.*, Symp. **24**:197.

Bishop, R.F., and Kümmel, H.G., 1987, *Phys. Today*, **40**:52.

Bishop, R.F., and Lührmann, K.H., 1978, *Phys. Rev. B*, **17**:3757.

Bishop, R.F., and Lührmann, K.H., 1982, *Phys. Rev. B*, **26**:5523, and references therein.

Bloch, C., 1958, *Nucl. Phys.*, **6**:329.

Brandow, B., 1967, *Rev. Mod. Phys.*, **39**:771.

Brandow, B., 1977, *Adv. Quantum Chem.*, **10**:187.

Brink, D.M., and Satchler, G.R., 1968, "Angular Momentum," 2nd ed., Clarendon, Oxford.

Brueckner, K.A., 1955, *Phys. Rev.*, **100**:36.

Chiles, R.A., and Dykstra, C.E., 1981, *Chem. Phys. Lett.*, **80**:69.

Chaudhuri, R., Guha, S., Sinha, D., and Mukherjee, D., 1989, *in*: "Many-Body Methods in Quantum Chemistry," U. Kaldor, ed., Lecture Notes in Chemistry, Vol. 52, Springer-Verlag, Berlin, Heidelberg, New York, pp. 171-198.

Čížek, J., 1966a, *J. Chem. Phys.*, **45**:4256.

Čížek, J., 1966b, *Theor. Chim. Acta*, **6**:292.

Čížek, J., 1969, *Advan. Chem. Phys.*, **14**:35.

Čížek, J., 1991, *Theor. Chim. Acta.*, **80**:91.

Čížek, J., and Paldus, J., 1967, *J. Chem. Phys.*, **47**:3976.

Čížek, J., and Paldus, J., 1971a, *Phys. Rev. A*, **3**:525.

Čížek, J., and Paldus, J., 1971b, *Int. J. Quantum Chem.*, **5**:359.

Čížek, J., Paldus, J., and Šroubková, L., 1969, *Int. J. Quantum Chem.*, **3**:149.

Coester, F., 1958, *Nucl. Phys.*, **7**:421.

Coester, F., 1969, *in*: "Lectures in Theoretical Physics," Vol. 11B, K.T. Mahanthappa and W.E. Brittin, eds., Gordon and Breach, New York, pp. 157-186.

Coester, F., and Kümmel, H., 1960, *Nucl. Phys.*, **17**:477.

Daborn, G.T., Ferguson, W.I., and Handy, N.C., 1980, *Chem. Phys.*, **50**:255.

Dalgaard, E., and Monkhorst, H.J., 1983, *Phys. Rev. A*, **28**:1217.

Davidson, E.R., 1974, *in:* "The World of Quantum Chemistry," R. Daudel and B. Pullman, eds., Reidel, Dordrecht, pp. 17-30.

deShalit, A., and Feschbach, H., 1974, "Theoretical Nuclear Physics," Vol. 1, Nuclear Structure, Chap. 3, John Wiley and Sons, New York.

Eisenberg, J.M., and Greiner, W., 1972, "Nuclear Theory", Vol. 3, Microscopic Theory of Nucleus, Chap. 4, North-Holland Publishing Co., Amsterdam

El Baz, E., and Castel, B., 1972, "Graphical Methods of Spin Algebras in Atomic, Nuclear and Particle Physics", M. Dekker, New York.

Emrich, K., 1981, *Nucl. Phys. A*, **351**:379.

Emrich, K., and Zabolitzky, J.G., 1984, *Phys. Rev. B*, **30**:2049.

Ey, W., 1978, *Nucl. Phys. A*, **296**:189.

Fitzgerald, G., Harrison, R.J., Laidig, W.D., and Bartlett, R.J., 1985, *Chem. Phys. Lett.*, **117**:433.

Fitzgerald, G., Harrison, R.J., and Bartlett, R.J., 1986, *J. Chem. Phys.*, **85**:5143.

Frantz, L.M., and Mills, R.L., 1960, *Nucl. Phys.*, **15**:16.

Freeman, D.L., 1977, *Phys. Rev. B*, **15**, 5512.

Gauss, J., and Cremer, D., 1988, *Chem. Phys. Lett.*, **150**:280.

Gauss, J., and Cremer, D., 1989, *Chem. Phys. Lett.*, **163**:549.

Geertsen, J., and Oddershede, J., 1986, *J. Chem. Phys.*, **85**:2112.

Gell-Mann, M., and Brueckner, K., 1957, *Phys. Rev.*, **106**:364.

Ghosh, S., and Mukherjee, D., 1984, *Proc. Ind. Acad. Sci.*, **93**:947.

Ghosh, S., Mukherjee, D., and Bhattacharyay, S.N., 1981, *Mol. Phys.*, **43**:173.

Ghosh, S., Mukherjee, D., and Bhattacharyay, S.N., 1982, *Chem. Phys.*, **72**:961.

Goldstone, J., 1957, *Proc. Roy. Soc. (London) A*, **239**:267.

Haque, A., and Kaldor, U., 1985, *Chem. Phys. Lett.*, **117**:347; **120**:261.

Haque, A., and Kaldor, U., 1986, *Int. J. Quantum Chem.*, **29**:425.

Haque, A., and Mukherjee, D., 1984, *J. Chem. Phys.*, **80**:5058.

Harris, F., 1977, *Int. J. Quantum Chem.*, Symp. **11**:403.

He, Z., and Cremer, D., 1991, "Analysis of CC and QCI Theory in Terms of 6th Order PT," preprint.

Hestenes, M., 1980, "Conjugate Direction Methods in Optimization,", Springer-Verlag, Berlin, Heidelberg, New York.

Hirao, K., and Nakatsuji, H., 1981, *Chem. Phys. Lett.*, **79**:292.

Hoffmann, M.R., and Schaefer III, H.F., 1986, *Adv. Quantum Chem.*, **18**:207.

Hoffmann, M.R., and Simons, J., 1988, *J. Chem. Phys.*, **88**:993.

Hoffmann, M.R., and Simons, J., 1989, *J. Chem. Phys.*, **90**:3671.

Hose, G., and Kaldor, U., 1979, *J. Phys. B*, **12**:3827.

Hose, G., and Kaldor, U., 1980, *Phys. Scripta*, **21**:357.

Hose, G., and Kaldor, U., 1981, *Chem. Phys.*, **62**:419.

Hose, G., and Kaldor, U., 1982, *J. Phys. Chem.*, **86**:2133.

Hsue, C.S., Kümmel, H., and Ueberholz, P., 1985, *Phys. Rev. D*, **32**:1435.

Hubbard, J., 1957, *Proc. Roy. Soc. (London) A*, **240**:539.

Hubbard, J., 1958a, *Proc. Roy. Soc. (London) A*, **243**:336.

Hubbard, J., 1958b, *Proc. Roy. Soc. (London) A*, **244**:199.

Huo, W., 1965, *J. Chem. Phys.*, **43**:624.

Hurley, A.C., 1976, "Electron Correlation in Small Molecules," Academic Press, New York.

Jankowski, K., 1987, *in*: "Methods in Computational Chemistry," Vol. 1., S. Wilson, ed., Plenum Press, New York, pp. 1-116.

Jankowski, K., and Paldus, J., 1980, *Int. J. Quantum Chem.*, **18**:1243.

Jankowski, K., Paldus, J., and Piecuch, P., 1991a, *Theor. Chim. Acta*, **80**:223.

Jankowski, K., Paldus, J., and Wasilewski, J., 1991b, *J. Chem. Phys.*, **95**:3549.

Janssen, C.L., and Schaefer III, H.F., 1991, *Theor. Chim. Acta*, **79**:1.

Jeziorski, B., and Monkhorst, H.J., 1981, *Phys. Rev. A*, **24**:1668.

Jeziorski, B., and Paldus, J., 1988, *J. Chem. Phys.*, **88**:5673.

Jeziorski, B., and Paldus, J., 1989, *J. Chem. Phys.*, **90**:2714.

Jeziorski, B., Monkhorst, H.J., Szalewicz, K., and Zabolitzky, J.G., 1984, *J. Chem. Phys.*, **81**:386.

Jørgensen, P., and Simons, J., 1983, *J. Chem. Phys.*, **79**:334.

Jucys, A.P., and Bandzaitis, A.A., 1977, "The Theory of Angular Momentum in Quantum Mechanics," 2nd ed., Mokslas, Vilnius (in Russian).

Jucys, A.P., Levinson, I.B., and Vanagas, V.V., 1960, "Mathematical Apparatus of the Theory of Angular Momentum," Institute of Physics and Mathematics of the Academy of Sciences of the Lithuanian S.S.R., Mintis, Vilnius, (in Russian); English translations: Israel Program for Scientific Translations, Jerusalem, 1962, and Gordon and Breach, New York, 1964.

Kahn, B., and Uhlenbeck, G.E., 1938, *Physica (Utrecht)*, **5**:399.

Kaldor, U., 1987, *J. Chem. Phys.*, **87**:467.

Kaldor, U., ed., 1989, "Many-Body Methods in Quantum Chemistry," Lecture Notes in Chemistry, Vol. 52, Springer-Verlag, Berlin, Heidelberg, New York.

Kato, T., 1966, "Perturbation Theory for Linear Operators," Springer-Verlag, Berlin, Heidelberg, New York.

Kaulfuss, U., 1985, *Phys. Rev. D*, **32**:1421.

Koch, H., Jensen, H.J.Aa, Jørgensen, P., Helgaker, T., Scuseria, G.E., and Schaefer III, H.F., 1990, *J. Chem. Phys.*, **92**:4924.

Koch, S., and Kutzelnigg, W., 1981, *Theor. Chim. Acta*, **59**:387.

Koch, S., and Mukherjee, D., 1988, *Chem. Phys. Lett.*, **145**:321.

Kucharski, S.A., and Bartlett, R.J., 1989, *Chem. Phys. Lett.*, **158**:550.

Kucharski, S.A., and Bartlett, R.J., 1991, Complete Linearization of the Coupled-Cluster Single, Double, Triple, and Quadruple Excitation Equations, *Theor. Chim. Acta*, in press.

Kümmel, H., 1984, *in*: "Nucleon-Nucleon Interaction and Nuclear Many-Body Problem," S.S. Wu and T.T.S. Kuo, eds., World Scientific, Singapore, p. 46.

Kümmel, H., Lührmann, K.H., and Zabolitzky, J.G., 1982, *Phys. Rep. C*, **90**:160.

Kutzelnigg, W., 1977, *in*: "Methods of Electronic Structure Theory," H.F. Schaefer III, ed., Plenum Press, New York, pp. 129-188.

Kutzelnigg, W., 1984, *J. Chem. Phys.*, **80**:822.

Kutzelnigg, W., 1985, *J. Chem. Phys.*, **82**:4166, and references therein.

Kutzelnigg, W., 1989, *in*: "Aspects of Many-Body Effects in Molecules and Extended Systems," O. Mukherjee, ed., Lecture Notes in Chemistry, Vol. 50, Springer-Verlag, Berlin, Heidelberg, New York, pp. 35-67.

Kutzelnigg, W., Mukherjee, D., and Koch, S., 1987, *Chem. Phys.*, **87**:5902.

Kvasnička, V., 1974, *Czech. J. Phys. B*, **24**:605.

Kvasnička, V., 1977a, *Czech. J. Phys. B*, **27**:599.

Kvasnička, V., 1977b, *Adv. Chem. Phys.*, **36**:345.

Kvasnička, V., 1981, *Chem. Phys. Lett.*, **79**:89.

Laidig, W.D., and Bartlett, R.J., 1984, *Chem. Phys. Lett.*, **104**:424.

Laidig, W.D., Saxe, P., and Bartlett, R.J., 1987, *J. Chem. Phys.*, **86**:887.

Lee, T.J., and Rendell, A.P., 1991a, *Chem. Phys. Lett.*, **177**:491.

Lee, T.J., and Rendell, A.P., 1991b, *J. Chem. Phys.*, **94**:6229.

Lee, T.J., and Rice, J.E., 1988, *Chem. Phys. Lett.*, **150**:406.

Lee, T.J., and Rice, J.E., 1991, *J. Chem. Phys.*, **94**:1215.

Lee, T.J., and Scuseria, G.E., 1990, *J. Chem. Phys.*, **93**:489.

Lee, T.J., and Taylor, P.R., 1989, *Int. J. Quantum Chem.*, Symp. **23**:199.

Lee, T.J., Rendell, A.P., and Taylor, P.R., 1990a, *J. Chem. Phys.*, **93**:6636.

Lee, T.J., Rendell, A.P., and Taylor, P.R., 1990b, *J. Phys. Chem.*, **94**:5463.

Lee, Y.S., and Bartlett, R.J., 1984, *J. Chem. Phys.*, **80**:4371.

Lee, Y.S., Kucharski, S.A., and Bartlett, R.J., 1984, *J. Chem. Phys.*, **81**:6906; Erratum (1985) **82**:5761.

Lengsfield, B.H., and Liu, B., 1981, *J. Chem. Phys.*, **75**:478.

Lindgren, I., 1974, *J. Phys. B*, **7**:2441.

Lindgren, I., 1978, *Int. J. Quantum Chem.*, Symp. **12**:33.

Lindgren, I., 1985a, *Phys. Scripta*, **32**:291.

Lindgren, I., 1985b, *Phys. Scripta*, **32**:611.

Lindgren, I., 1991, *J. Phys. B*, **24**:1143.

Lindgren, I., and Morrison, J., 1982, "Atomic Many-Body Theory," Springer-Verlag, Berlin, Heidelberg, New York.

Lindgren, I., and Mukherjee, D., 1987, *Phys. Rep.*, **151**:93.

Lindgren, I., and Salomonson, S., 1980, *Phys. Scripta*, **21**:335.

Li, X., and Paldus, J., 1991, unpublished results.

Mayer, J.E., and Mayer, M.G., 1940, "Statistical Mechanics," Chap. 13, John Wiley and Sons, New York.

Meissner, L., Jankowski, K., and Wasilewski, J., 1988, *Int. J. Quantum Chem.*, **34**:535.

Meissner, L., Kucharski, S.A., and Bartlett, R.J., 1989, *J. Chem. Phys.*, **91**:6187.

Meyer, W., 1971, *Int. J. Quantum Chem.*, **5**:341.

Meyer, W., 1974, *Theor. Chim. Acta*, **35**:277.

Meyer, W., 1977, *in*: "Methods of Electronic Structure Theory," H.F. Schaefer III, ed., Plenum Press, New York, pp. 413-446.

Monkhorst, H.J., 1977, *Int. J. Quantum Chem.*, Symp. **11**:421.

Mukherjee, D., 1979, *Pramana*, **12**:203.

Mukherjee, D., 1986a, *Chem. Phys. Lett.*, **125**:207.

Mukherjee, D., 1986b, *Int. J. Quantum Chem.*, Symp. **20**:409.

Mukherjee, D., ed., 1989, "Aspects of Many-Body Effects in Molecules and Extended Systems," Lecture Notes in Chemistry, Vol. 50, Springer-Verlag, Berlin, Heidelberg, New York.

Mukherjee, D., and Mukherjee, P.K., 1979, *Chem. Phys.*, **37**:325.

Mukherjee, D., and Pal, S., 1989, *Adv. Quantum Chem.*, **20**:292.

Mukherjee, D., Moitra, R.K., and Mukhopadhyay, A., 1977, *Mol. Phys.*, **33**:955.

Mukherjee, D., Kutzelnigg, W., and Koch, S., 1987, *J. Chem. Phys.*, **87**:5902.

Nakatsuji, H., 1978, *Chem. Phys. Lett.*, **59**:362.

Nakatsuji, H., 1979, *Chem. Phys. Lett.*, **67**:329.

Nakatsuji, H., 1983, *Int. J. Quantum Chem.*, Symp. **17**:241 and references therein.

Nakatsuji, M., and Hirao, K., 1978, *J. Chem. Phys.*, **68**:2053,4279.

Nesbet, R.K., 1958, *Phys. Rev.*, **109**:1632.

Nesbet, R.K., 1965, *Adv. Chem. Phys.*, **9**:321.

Nesbet, R.K., 1969, *Adv. Chem. Phys.*, **14**:1.

Noga, J., and Bartlett, R.J., 1987, *J. Chem. Phys.*, **86**:7041; Erratum (1988) **89**:3401.

Noga, J., Bartlett, R.J., and Urban, M., 1987, *Chem. Phys. Lett.*, **134**:126.

Noga, J., Kucharski, S.A., and Bartlett, R.J., 1989, *J. Chem. Phys.*, **90**:3399.

Offerman, R., 1976, *Nucl. Phys. A*, **273**:368.

Offerman, R., Ey, W., and Kümmel, H., 1976, *Nucl. Phys. A*, **273**:349.

Oliphant, N., and Adamowicz, L., 1991, *J. Chem. Phys.*, **94**:1229.

Pal, S., Rittby, M., Bartlett, R.J., Sinha, D., and Mukherjee, D., 1988, *J. Chem. Phys.*, **88**:4357.

Paldus, J., 1976, *in*: "Theoretical Chemistry: Advances and Perspectives," Vol. 2, D. Henderson and H. Eyring, eds., Academic Press, New York, pp. 131-290.

Paldus, J., 1977, *J. Chem. Phys.*, **67**:303.

Paldus, J., 1981, "Diagrammatic Methods for Many-Fermion Systems," (Lecture Notes) University of Nijmegen, Nijmegen.

Paldus, J., 1983, *in:* "New Horizons of Quantum Chemistry," P.O. Löwdin and B. Pullman, eds., Reidel, Dordrecht, pp. 31-60.

Paldus, J., 1990, *in:* "Self-Consistent Field: Theory and Applications," R. Carbó and M. Klobukowski, eds., Elsevier, Amsterdam, pp. 1-45.

Paldus, J., and Boyle, M.J., 1982, *Int. J. Quantum Chem.*, **22**:1281.

Paldus, J., and Čížek, J., 1974, *J. Chem. Phys.*, **60**:149.

Paldus, J., and Čížek, J., 1975, *Adv. Quantum Chem.*, **9**:105.

Paldus, J., and Jeziorski, B., 1988, *Theor. Chim. Acta*, **73**:81.

Paldus, J., and Veillard, A., 1977, *Chem. Phys. Lett.*, **50**:6.

Paldus, J., and Veillard, A., 1978, *Mol. Phys.*, **35**:445.

Paldus, J., Čížek, J., and Shavitt, I., 1972, *Phys. Rev. A*, **5**:50.

Paldus, J., Adams, B.G., and Čížek, J., 1977, *Int. J. Quantum Chem.*, **11**:813.

Paldus, J., Čížek, J., Saute, M., and Laforgue, A., 1978, *Phys. Rev. A*, **17**:805.

Paldus, J., Wormer, P.E.S., Visser, F., and van der Avoird, A., 1982a, *J. Chem. Phys.*, **76**:2458.

Paldus, J., Wormer, P.E.S., Visser, F., and van der Avoird, A., 1982b, *in:* "Proceedings of the 5th Seminar on Computerational Methods in Quantum Chemistry," P. Th. van Duijnen and W.C. Nieuwpoort, eds., Max-Planck-Institut für Physik und Astrophysik, Munich, Germany, pp. 13-30.

Paldus, J., Čížek, J., and Takahashi, M., 1984a, *Phys. Rev. A*, **30**:2193.

Paldus, J., Takahashi, M., and Cho, R.W.H., 1984b, *Phys. Rev. B*, **30**:4267.

Paldus, J., Wormer, P.E.S., and Bénard, M., 1988, *Coll. Czech. Chem. Commun.*, **53**:1919.

Paldus, J., Čížek, J., and Jeziorski, B., 1989a, *J. Chem. Phys.*, **90**:4356.

Paldus, J., Pylypow, L., and Jeziorski, B., 1989b, *in:* "Many-Body Methods in Quantum Chemistry," U. Kaldor, ed., Lecture Notes in Chemistry, Vol. 52, Springer-Verlag, Berlin, Heidelberg, New York, pp. 151-170.

Paldus, J., Čížek, J., and Jeziorski, B., 1990, *J. Chem. Phys.*, **93**:1485.

Paldus, J., Pylypow, L., Piecuch, P., and Jeziorski, B., 1991, unpublished results.

Piecuch, P., and Paldus, J., 1989, *Int. J. Quantum Chem.*, **36**:429.

Piecuch, P., and Paldus, J., 1990, *Theor. Chim. Acta*, **78**:65.

Piecuch, P., and Paldus, J., 1991a, Orthogonally Spin-Adapted Multi-Reference Hilbert Space Coupled-Cluster Formalism: Diagrammatic Formulation, *Theor. Chim. Acta*, in press.

Piecuch, P., and Paldus, J., 1991b, unpublished results.

Piecuch, P., Zarrabian, S., Paldus, J., and Čížek, J., 1990, *Phys. Rev. B*, **42**:3351.

Pople, J.A., Krishnan, R., Schlegel, H.B., and Binkley, J.S., 1978, *Int. J. Quantum Chem.*, **14**:545.

Pople, J.A., Krishnan, R., Schlegel, H.B., and Binkley, J.S., 1979, *Int. J. Quantum Chem.*, Symp. **13**:225.

Pople, J.A., Head-Gordon, M., and Raghavachari, K., 1987, *J. Chem. Phys.*, **87**:5968.

Pople, J.A., Head-Gordon, M., and Raghavachari, K., 1989, *J. Chem. Phys.*, **90**:4635.

Prasad, M.D., Pal, S., and Mukherjee, D., 1985, *Phys. Rev. A*, **31**:1287.

Primas, H., 1965, *in*: "Modern Quantum Chemistry," Vol. 2, O. Sinanoğlu, ed., Academic Press, New York, pp. 45-74.

Purvis III, G.D., and Bartlett, R.J., 1981, *J. Chem. Phys.*, **75**:1284.

Purvis III, G.D., and Bartlett, R.J., 1982, *J. Chem. Phys.*, **76**:1910.

Pylypow, L., 1991, Ph.D. Thesis, University of Waterloo, in preparation.

Quinn, J.J., and Ferrell, R.A., 1958, *Phys. Rev.*, **112**:812.

Raghavachari, K., 1985, *J. Chem. Phys.*, **82**:4607.

Rendell, A.P., and Lee, T.J., 1991, *J. Chem. Phys.*, **94**:6219.

Rendell, A.P., Lee, T.J., and Taylor, P.R., 1990, *J. Chem. Phys.*, **92**:7050.

Rendell, A.P., Lee, T.J., and Komornicki, A., 1991, *Chem. Phys. Lett.*, **178**:462.

Rice, J.E., Taylor, P.R., Lee, T.J., and Almlöf, J., 1991, *J. Chem. Phys.*, **94**:4972.

Salter, E.A., Sekino, H., and Bartlett, R.J., 1987, *J. Chem. Phys.*, **87**:502.

Salter, E.A., Trucks, G.W., and Bartlett, R.J., 1989, *J. Chem. Phys.*, **90**:1752.

Saute, M., Paldus, J., and Čížek, J., 1979, *Int. J. Quantum Chem.*, **15**:463.

Schaefer III, H.F., 1971, *J. Chem. Phys.*, **54**:2207.

Schaefer III, H.F., 1972, "The Electronic Structure of Atoms and Molecules: A Survey of Rigorous Quantum Mechanical Results," Addison-Wesley, Reading, Mass.

Scheiner, A.C., Scuseria, G.E., Rice, J.E., Lee, T.J., and Schaefer III, H.F., 1987, *J. Chem. Phys.*, **87**:5361.

Scuseria, G.E., and Lee, T.J., 1990, *J. Chem. Phys.*, **93**:5851.

Scuseria, G.E., and Schaefer III, H.F., 1988a, *Chem. Phys. Lett.*, **146**:23.

Scuseria, G.E., and Schaefer III, H.F., 1988b, *Chem. Phys. Lett.*, **152**:382.

Scuseria, G.E., Scheiner, A.C., Lee, T.J., Rice, J.E., and Schaefer III, H.F., 1987, *J. Chem. Phys.*, **86**:2881.

Sekino, H., and Bartlett, R.J., 1984, *Int. J. Quantum Chem.*, Symp. **18**:255.

Shepard, M., 1984, *J. Chem. Phys.*, **80**:1225.

Silverstone, H., and Sinanoğlu, O., 1966, *J. Chem. Phys.*, **44**:1899,3608.

Sinanoğlu, O., 1961, *Proc. Roy. Soc. (London) A*, **260**:379.

Sinanoğlu, O., 1962, *J. Chem. Phys.*, **36**:207.

Stedman, G.E., 1990, "Diagram Techniques in Group Theory," Cambridge University Press, Cambridge.

Stolarczyk, L., and Monkhorst, H.J., 1985, *Phys. Rev. A*, **32**:725,743.

Stolarczyk,, L., and Monkhorst, H.J., 1988, *Phys. Rev. A*, **37**:1908,1926.

Takahashi, M., and Paldus, J., 1985a, *Phys. Rev. B*, **31**:5121.

Takahashi, M., and Paldus, J., 1985b, *Int. J. Quantum Chem.*, **28**:459.

Takahashi, M., and Paldus, J., 1986, *J. Chem. Phys.*, **85**:1486.

Tanaka, R., and Terashima, H., 1984, *Chem. Phys. Lett.*, **106**:558.

Taylor, P.R., Bacskay, G.B., Hush, N.S., and Hurley, A.C., 1976a, *J. Chem. Phys.*, **69**:1971.

Taylor, P.R., Bacskay, G.B., Hush, N.S., and Hurley, A.C., 1976b, *Chem. Phys. Lett.*, **41**:444.

Tolmachev, V.V., 1963, "The Field Theoretical Form of Perturbation Theory Applied to Many-Electron Problems in Atoms and Molecules," University of Tartu, Tartu, Estonia, (in Russian).

Tong, B.Y., 1962, *Proc. Phys. Soc. (London)*, **80**:1101.

Urban, M., Noga, J., Cole, S.J., and Bartlett, R.J., 1985, *J. Chem. Phys.*, **83**:4041.

Wahl, A.C., 1964, *J. Chem. Phys.*, **41**:2600.

Watts, J.D., Černušák, I., Noga, J., Bartlett, R.J., Bauschlicher, Jr., C.W., Lee, T.J., Rendell, A.P., and Taylor, P.R., 1990, *J. Chem. Phys.*, **93**:8875.

Wilson, S., 1984, "Electron Correlation in Molecules," Clarendon Press, Oxford.

Wilson, S., Jankowski, K., and Paldus, J., 1983, *Int. J. Quantum Chem.*, **23**:1781.

Wormer, P.E.S., Visser, F., and Paldus, J., 1982, *J. Comp. Phys.*, **48**:23.

THE PERTURBATION THEORY OF ELECTRON CORRELATION. I. BASIC RAYLEIGH-SCHRÖDINGER PERTURBATION THEORY

Stephen Wilson

Rutherford Appleton Laboratory
Chilton, Oxfordshire, OX11 0QX, England

1. Introduction

This is the first of four articles describing the application of perturbation theory to the electron correlation problem in atoms and molecules. In this article, the basic formalism of Rayleigh-Schrödinger perturbation theory is presented in some detail. In the second article we describe the many-body perturbation theory paying particular attention to the formulation within the algebraic approximation, *i.e.* by using finite analytic basis sets. The use of relativistic many-body perturbation theory to describe both relativistic and correlation effects in atoms and molecules is described in the third article. In the final article, recent work on the *concurrent computation* many-body perturbation theory (*cc*MBPT) computer program, which is designed to exploit the power of multiprocessor supercomputers, is described. Throughout this work, references have been chosen to amplify the text and to indicate key papers and review articles. No attempt has been made to provide a comprehensive catalogue of the large and expanding literature of the perturbation approach to the problem of electron correlation.

The exact solution of the Schrödinger equation is, in general, associated with considerable mathematical difficulties. Perturbation theory is the most general and systematic technique for the approximate solution of quantum mechanical eigenvalue problems. The term "perturbation" and the associated concept of expansion in some small parameter were introduced into quantum mechanics by analogy with the perturbation method of classical mechanics which had played such an important rôle in the solution of problems in celestial mechanics.

We shall confine our attention to problems which do not involve time. It is assumed that the eigenproblem for some model problem which affords a reasonable zero-order approximate solution to the problem of interest can be solved exactly. A perturbation, $\mathfrak{H}_1$, is added to the zero-order hamiltonian, $\mathfrak{H}_0$, associated with the model problem, to recover the full hamiltonian for the system, $\mathfrak{H}$. A parameter λ is introduced in the hamiltonian, $\mathfrak{H} = \mathfrak{H}_0 + \lambda\, \mathfrak{H}_1$, so as to interpolate between the zero-order hamiltonian, λ=0, and the full hamiltonian, λ=1. By making expansion in powers of λ for the exact energy and the exact wave function and then equating the coefficients for each power of λ, a set of equations can be obtained which are then employed to obtain corrections to the zero-order energy and wave function.

The outline of the present article is as follows:- The basic equations of Rayleigh-Schrödinger perturbation theory for a single perturbation are given in section 2.1. In section 2.2, the application of these equations to non-degenerate states is

Methods in Computational Molecular Physics, Edited by Stephen Wilson and Geerd H.F. Diercksen, Plenum Press, New York, 1992

considered. Perturbed wave functions and the coefficients in the expansion for the energy are discussed in sections 2.3 and 2.4, respectively. Wigner's 2n+1 rule is presented in section 2.5; this simplifies substantially the calculation of the energy coefficients in each order of the perturbation expansion. In section 3, the generalization of perturbation theory to cases which involve two perturbations is discussed. In section 3.1, the basic equations of double Rayleigh-Schrödinger perturbation theory are described, whilst in sections 3.2 and 3.3 the perturbed wave functions and energy coefficients, respectively, are presented. Section 4 is devoted to the sum-over-states formulae which are frequently employed in practical applications. Sum-over-states formulae are presented for single perturbation theory and for double perturbation theory in section 4.1 and 4.2, respectively. In section 5, we turn our attention to the case in which the reference function is degenerate. The basic equations of degenerate Rayleigh-Schrödinger perturbation theory are presented in section 5.1. The zero order, first order and higher order terms in the degenerate Rayleigh-Schrödinger perturbation theory are discussed in sections 5.2, 5.3 and 5.4, respectively. When the reference function displays near degeneracy then the so-called quasi-degenerate Rayleigh-Schrödinger perturbation theory is often thought to be appropriate and, therefore, some details of this approach are given in section 5.5. We examine the use of perturbation expansions based on non-orthogonal functions in section 6 giving the basic formalism in section 6.1, details of the perturbed wave function in section 6.2 and the energy coefficients in section 6.3. The Lennard-Jones Brillouin Wigner expansion is introduced in section 7 with the basic formalism being presented in section 7.1 and the relation with the Rayleigh-Schrödinger perturbation theory being described in section 7.2.

2. Perturbation theory

2.1 Rayleigh-Schrödinger perturbation theory

Let the quantum mechanical eigenvalue problem whose solutions are required be

$$\mathfrak{H}\,\Psi_i = E_i\,\Psi_i,\ \ i=0,1,2,... \qquad (1)$$

where $\mathfrak{H}$ is a self-adjoint hamiltonian operator. In the present work, equation (1) is, of course, the quantum eigenproblem for the electronic structure of an atom or molecule. Suppose that the eigenvalue problem corresponding to some model description of the system

$$\mathfrak{H}_0\,\Phi_i = \mathrm{E}_i\,\Phi_i,\ i=0,1,2,... \qquad (2)$$

where $\mathfrak{H}_0$ is some self-adjoint model hamiltonian, has been solved. Equation (2) may be referred to as the zero-order problem or reference problem. Typically, in studies of atoms and molecules, equation (2) may correspond to the Hartree-Fock approximation or some other independent electron model.

In order to develop a perturbation series for the systematic correction of the solutions of the zero-order problem, put

$$\mathfrak{H} = \mathfrak{H}_0 + \lambda\,\mathfrak{H}_1 \qquad (3)$$

where

$$\mathfrak{H}_1 = \mathfrak{H} - \mathfrak{H}_0 \qquad (4)$$

$\mathfrak{H}_1$ is termed the perturbation operator and λ the perturbation parameter. As λ is increased from zero to unity, equation (3) interpolates between the zero-order hamiltonian appearing in equation (2) and the perturbed hamiltonian appearing in equation (1).

Substituting equation (3) into equation (1) yields

$$(\mathcal{H}_0 + \lambda\, \mathcal{H}_1)\, \Psi_i(\lambda) = E_i(\lambda)\, \Psi_i(\lambda), \quad i=0,1,2,... \tag{5}$$

where the dependence of the eigenvalues and eigenfunctions on the perturbation parameter has been indicated. It will be assumed that $E_i(\lambda)$ and $\Psi_i(\lambda)$ are continuous functions of λ. This assumption may be invalidated if the perturbation operator, $\mathcal{H}_1$, introduces new singularities or if it changes the nature of existing singularities. Clearly, $E_i = E_i(0)$ and $\Phi_i = \Psi_i(\lambda)$.

Often an atomic or molecular system may be subjected to several perturbations simultaneously. For example, the zero-order model may be the Hartree-Fock description of a free molecule and perturbation theory may be employed to take account of electron correlation effects together with the response to an external electric field. In such cases, multiple perturbation theory is used. The total hamiltonian for the perturbed system is written as

$$\mathcal{H} = \mathcal{H}_0 + \sum_{m=1}^{m=n} \lambda_m\, \mathcal{H}_m \tag{6}$$

and the eigenproblem whose solutions are required takes the form

$$(\mathcal{H}_0 + \sum_{m=1}^{m=n} \lambda_m\, \mathcal{H}_m)\, \Psi_i(\lambda_1,\lambda_2,...,\lambda_n) = E_i(\lambda_1,\lambda_2,...,\lambda_n)\, \Psi_i(\lambda_1,\lambda_2,...,\lambda_n) \tag{7}$$

Clearly, $E_i = E_i(0,0,...,0)$ and $\Phi_i = \Psi_i(0,0,...,0)$.

2.2 Non-degenerate Rayleigh-Schrödinger perturbation theory

Consider the case of a single perturbation operator, $\mathcal{H}_1$, and a non-degenerate reference function, Φ_i.

The eigenvalue, $E_i(\lambda)$, in equation (5) is assumed to be a continuous function of λ and thus a power series expansion can be made

$$E_i(\lambda) = E_i + \sum_{p=1}^{\infty} \lambda^p\, \varepsilon_{ip} \tag{8}$$

ε_{ip} is the p th. order energy, or energy coefficient, in the perturbation series for E_i, the exact energy of the i th. state.

$$\varepsilon_{ip} = (p!)^{-1} \left[\partial^p E_i(\lambda) / \partial \lambda^p \right]_{\lambda=0} \tag{9}$$

Similarly, for the wave function, $\Psi_i(\lambda)$, an expansion is also made

$$\Psi_i(\lambda) = \Phi_i + \sum_{p=1}^{\infty} \lambda^p\, \chi_{ip} \tag{10}$$

in which χ_{ip} is the p th. order wave function

$$\chi_{ip} = (p!)^{-1} \left[\partial^p \Psi_i(\lambda)/\partial \lambda^p \right]_{\lambda=0} \tag{11}$$

Substituting the expansion for the eigenvalue and the eigenfunction, equations (8) and (10), respectively, into the Schrödinger equation, equation (5), gives

$$(\mathfrak{H}_0 + \lambda \mathfrak{H}_1)(\Phi_i + \sum_{p=1}^{\infty} \lambda^p \chi_{ip}) =$$

$$(E_i + \sum_{p=1}^{\infty} \lambda^p \varepsilon_{ip})(\Phi_i + \sum_{p=1}^{\infty} \lambda^p \chi_{ip}) \tag{12}$$

Now the coefficients of powers of λ in equation (12) can be equated to give

p=0: $$\mathfrak{H}_0 \Phi_i = E_i \Phi_i \tag{13a}$$

p=1: $$\mathfrak{H}_0 \chi_{i1} + \mathfrak{H}_1 \Phi_i = E_i \chi_{i1} + \varepsilon_{i1} \Phi_i \tag{13b}$$

p=2: $$\mathfrak{H}_0 \chi_{i2} + \mathfrak{H}_1 \chi_{i1} = E_i \chi_{i2} + \varepsilon_{i1} \chi_{i1} + \varepsilon_{i2} \Phi_i \tag{13c}$$

p=3: $$\mathfrak{H}_0 \chi_{i3} + \mathfrak{H}_1 \chi_{i2} = E_i \chi_{i3} + \varepsilon_{i1} \chi_{i2} + \varepsilon_{i2} \chi_{i1} + \varepsilon_{i3} \Phi_i \tag{13d}$$

p=4: $$\mathfrak{H}_0 \chi_{i4} + \mathfrak{H}_1 \chi_{i3} = E_i \chi_{i4} + \varepsilon_{i1} \chi_{i3} + \varepsilon_{i2} \chi_{i2} + \varepsilon_{i3} \chi_{i1} + \varepsilon_{i4} \Phi_i \tag{13e}$$

p=5: $$\mathfrak{H}_0 \chi_{i5} + \mathfrak{H}_1 \chi_{i4} = E_i \chi_{i5} + \varepsilon_{i1} \chi_{i4} + \varepsilon_{i2} \chi_{i3} + \varepsilon_{i3} \chi_{i2} + \varepsilon_{i4} \chi_{i1} + \varepsilon_{i5} \Phi_i \tag{13f}$$

$$\cdots$$

Generally, for p>1, these equations have the form

$$\mathfrak{H}_0 \chi_{ip} + \mathfrak{H}_1 \chi_{ip-1} = E_i \chi_{ip} + \sum_{q=1}^{p-1} \varepsilon_{iq} \chi_{ip-q} + \varepsilon_{ip} \Phi_i \tag{14}$$

Equations (13) and (14) constitute the basic equations of Rayleigh-Schrödinger perturbation theory.

2.3 Perturbed wave functions

From equations (13) and (14) given above, expressions for the perturbed wave functions, χ_{ip}, can be obtained. Without loss of generality, the perturbed wave functions can be taken to be orthogonal to the reference function

$$<\Phi_i | \chi_{ip}> = 0, \; p=1,2,... \tag{15}$$

If $\mathfrak{H}_0$ is self-adjoint then the eigenfunctions of the zero-order problem are orthonormal

$$<\Phi_i | \Phi_j> = \delta_{ij} \tag{16}$$

and so the reference function and the exact wave function satisfy the intermediate normalization condition

$$<\Phi_i | \Psi_i> = 1 \tag{17}$$

The first order perturbation equation, (13b), may be written

$$(E_i - \mathfrak{H}_0) \chi_{i1} = (\mathfrak{H}_1 - \varepsilon_{i1}) \Phi_i \tag{18}$$

Now, defining the projection operator onto the model function

$$\hat{P}_i = |\Phi_i> <\Phi_i| \tag{19}$$

and its orthogonal complement

$$\hat{Q}_i = 1 - \hat{P}_i \tag{20}$$

and noting that

$$\hat{P}_i \; |\chi_{ip}> = 0, \qquad p=1,2,... \tag{21}$$

$\hat{Q}_i$ may be applied to equation (18) from the left to give

$$(E_i - \mathfrak{H}_0) \hat{Q}_i \chi_{i1} = \hat{Q}_i (\mathfrak{H}_1 - \varepsilon_{i1}) \Phi_i = \hat{Q}_i \mathfrak{H}_1 \Phi_i \tag{22}$$

Furthermore, defining the resolvant, $\mathfrak{R}_i$, according to

$$\mathfrak{R}_i (E_i - \mathfrak{H}_0) \hat{Q}_i = \hat{Q}_i \tag{23}$$

multiplying equation (22) from the left by $\mathfrak{R}_i$ yields an expression for the first order wave function

$$\chi_{i1} = \mathfrak{R}_i \mathfrak{H}_1 \Phi_i \tag{24}$$

The second-order perturbation equation, (13c), may be written as

$$(E_i - \mathfrak{H}_0) \chi_{i2} = (\mathfrak{H}_1 - \varepsilon_{i1}) \chi_{i1} - \varepsilon_{i2} \Phi_i \tag{25}$$

or, multiplying from the left by $\hat{Q}_i$

$$(E_i - \mathfrak{H}_0) \hat{Q}_i \chi_{i2} = \hat{Q}_i (\mathfrak{H}_1 - \varepsilon_{i1}) \chi_{i1} \tag{26}$$

which, when multiplied from the left by $\mathfrak{R}_i$, yields

$$\chi_{i2} = \mathfrak{R}_i (\mathfrak{H}_1 - \varepsilon_{i1}) \chi_{i1} \tag{27}$$

or, using equation (24),

$$\chi_{i2} = \mathfrak{R}_i (\mathfrak{H}_1 - \varepsilon_{i1}) \mathfrak{R}_i \mathfrak{H}_1 \Phi_i \tag{28}$$

From the third-order perturbation equation, (13d), the third-order wave function may be obtained

$$\chi_{i3} = \mathfrak{R}_i (\mathfrak{H}_1 - \varepsilon_{i1}) \chi_{i2} - \varepsilon_{i2} \mathfrak{R}_i \chi_{i1} \tag{29}$$

which upon substituting for χ_{i1} and χ_{i2} from equations (24) and (28), respectively, becomes

$$\chi_{i3} = \mathfrak{R}_i (\mathfrak{H}_1 - \varepsilon_{i1}) \mathfrak{R}_i (\mathfrak{H}_1 - \varepsilon_{i1}) \mathfrak{R}_i \mathfrak{H}_1 \Phi_i - \varepsilon_{i2} \mathfrak{R}_i \mathfrak{H}_1 \Phi_i \tag{30}$$

The fourth-order and fifth-order wave functions can be shown, in a similar fashion, to have the forms

$$\chi_{i4} = \mathfrak{R}_i (\mathfrak{H}_1 - \varepsilon_{i1}) \chi_{i3} - \varepsilon_{i2} \mathfrak{R}_i \chi_{i2} - \varepsilon_{i3} \mathfrak{R}_i \chi_{i1}$$

and (31)

$$\chi_{i5} = \mathfrak{R}_i (\mathfrak{H}_1 - \varepsilon_{i1}) \chi_{i4} - \varepsilon_{i2} \mathfrak{R}_i \chi_{i3} - \varepsilon_{i3} \mathfrak{R}_i \chi_{i2} - \varepsilon_{i4} \mathfrak{R}_i \chi_{i1} \tag{32}$$

In general, the p th. order wave function can be written in terms of lower order wave functions

$$\chi_{ip} = \mathfrak{R}_i (\mathfrak{H}_1 - \varepsilon_{i1}) \chi_{ip-1} - \sum_{q=2}^{p-1} \varepsilon_{iq} \mathfrak{R}_i \chi_{ip-q} \tag{33}$$

From equation (33), it appears that the p th. order wave function depends on the perturbed energy coefficients up to and including order p-1.

2.4 Energy coefficients

The energy coefficients in the perturbation expansion (8) are obtained from the perturbation equations (13) and (14) by multiplying from the left by Φ_i and integrating over all the electronic coordinates. Recalling the orthogonality condition (15) and the orthonormality condition (16) leads immediately to the expressions

$$\varepsilon_{i1} = \langle \Phi_i | \mathfrak{H}_1 | \Phi_i \rangle \tag{34a}$$

$$\varepsilon_{i2} = \langle \Phi_i | \mathfrak{H}_1 | \chi_{i1} \rangle \tag{34b}$$

$$\varepsilon_{i3} = \langle \Phi_i | \mathfrak{H}_1 | \chi_{i2} \rangle \tag{34c}$$

$$\varepsilon_{i4} = \langle \Phi_i | \mathfrak{H}_1 | \chi_{i3} \rangle \tag{34d}$$

$$\varepsilon_{i5} = \langle \Phi_i | \mathfrak{H}_1 | \chi_{i4} \rangle \tag{34e}$$

or, generally,

$$\varepsilon_{ip} = \langle\Phi_i | \mathcal{H}_1 | \chi_{ip-1}\rangle, \quad p>1 \tag{35}$$

The p th. order energy coefficient thus depends on the p-1 th. order wave function.

Substituting the expressions for the perturbed wave functions obtained in section 2.3 into the above expressions for the energy coefficients gives, for the second-order energy

$$\varepsilon_{i2} = \langle\Phi_i | \mathcal{H}_1 \mathcal{R}_i \mathcal{H}_1 | \Phi_i\rangle \tag{36}$$

for the third-order energy

$$\varepsilon_{i3} = \langle\Phi_i | \mathcal{H}_1 \mathcal{R}_i (\mathcal{H}_1 - \varepsilon_{i1}) \mathcal{R}_i \mathcal{H}_1 | \Phi_i\rangle \tag{37}$$

for the fourth-order energy

$$\begin{aligned} \varepsilon_{i4} = &\langle\Phi_i | \mathcal{H}_1 \mathcal{R}_i (\mathcal{H}_1 - \varepsilon_{i1}) \mathcal{R}_i (\mathcal{H}_1 - \varepsilon_{i1}) \mathcal{R}_i \mathcal{H}_1 | \Phi_i\rangle \\ &- \varepsilon_{i2} \langle\Phi_i | \mathcal{H}_1 \mathcal{R}_i^2 \mathcal{H}_1 | \Phi_i\rangle \end{aligned} \tag{38}$$

and for the fifth- order energy

$$\begin{aligned} \varepsilon_{i5} = &\langle\Phi_i | \mathcal{H}_1 \mathcal{R}_i (\mathcal{H}_1 - \varepsilon_{i1}) \mathcal{R}_i (\mathcal{H}_1 - \varepsilon_{i1}) \mathcal{R}_i \\ &\qquad (\mathcal{H}_1 - \varepsilon_{i1}) \mathcal{R}_i \mathcal{H}_1 | \Phi_i\rangle \\ &- \varepsilon_{i2} \{ \langle\Phi_i | \mathcal{H}_1 \mathcal{R}_i (\mathcal{H}_1 - \varepsilon_{i1}) \mathcal{R}_i^2 \mathcal{H}_1 | \Phi_i\rangle + \\ &\qquad \langle\Phi_i | \mathcal{H}_1 \mathcal{R}_i^2 (\mathcal{H}_1 - \varepsilon_{i1}) \mathcal{R}_i \mathcal{H}_1 | \Phi_i\rangle \} \\ &- \varepsilon_{i3} \langle\Phi_i | \mathcal{H}_1 \mathcal{R}_i^2 \mathcal{H}_1 | \Phi_i\rangle \end{aligned} \tag{39}$$

Generally, the terms in the perturbation expansion for the energy are given by

$$\begin{aligned} \varepsilon_{ip} &= \langle\Phi_i | \mathcal{H}_1 | \chi_{ip-1}\rangle \\ &= \langle\Phi_i | \mathcal{H}_1 \mathcal{R}_i (\mathcal{H}_1 - \varepsilon_{i1}) | \chi_{ip-2}\rangle - \sum_{q=2}^{p-1} \varepsilon_{iq} \langle\Phi_i | \mathcal{H}_1 \mathcal{R}_i | \chi_{ip-q}\rangle \end{aligned} \tag{40}$$

The expressions obtained above may be expanded further in each order. For example, the third-order energy coefficient may be rewritten as a sum of 2 terms

$$\varepsilon_{i3} = \langle\Phi_i | \mathcal{H}_1 \mathcal{R}_i \mathcal{H}_1 \mathcal{R}_i \mathcal{H}_1 | \Phi_i\rangle - \varepsilon_{i1} \langle\Phi_i | \mathcal{H}_1 \mathcal{R}_i^2 \mathcal{H}_1 | \Phi_i\rangle \tag{41}$$

the fourth-order energy coefficient may be rewritten as a sum of 5 terms

$$\varepsilon_{i4} = \langle\Phi_i|\mathfrak{H}_1 \mathfrak{R}_i \mathfrak{H}_1 \mathfrak{R}_i \mathfrak{H}_1 \mathfrak{R}_i \mathfrak{H}_1|\Phi_i\rangle$$
$$- \varepsilon_{i1} \langle\Phi_i|\mathfrak{H}_1 \mathfrak{R}_i^2 \mathfrak{H}_1 \mathfrak{R}_i \mathfrak{H}_1|\Phi_i\rangle$$
$$- \varepsilon_{i1} \langle\Phi_i|\mathfrak{H}_1 \mathfrak{R}_i \mathfrak{H}_1 \mathfrak{R}_i^2 \mathfrak{H}_1|\Phi_i\rangle$$
$$- \varepsilon_{i1}^2 \langle\Phi_i|\mathfrak{H}_1 \mathfrak{R}_i^3 \mathfrak{H}_1|\Phi_i\rangle$$
$$- \varepsilon_{i2} \langle\Phi_i|\mathfrak{H}_1 \mathfrak{R}_i^2 \mathfrak{H}_1|\Phi_i\rangle \qquad (42)$$

and the fifth-order energy coefficient may be rewritten as a sum of 14 terms

$$\varepsilon_{i5} = \langle\Phi_i|\mathfrak{H}_1 (\mathfrak{R}_i \mathfrak{H}_1)^4|\Phi_i\rangle$$
$$- \langle\Phi_i|\mathfrak{H}_1|\Phi_i\rangle \{ \langle\Phi_i|\mathfrak{H}_1 \mathfrak{R}_i^2 \mathfrak{H}_1 \mathfrak{R}_i \mathfrak{H}_1 \mathfrak{R}_i \mathfrak{H}_1|\Phi_i\rangle$$
$$+ \langle\Phi_i|\mathfrak{H}_1 \mathfrak{R}_i \mathfrak{H}_1 \mathfrak{R}_i^2 \mathfrak{H}_1 \mathfrak{R}_i \mathfrak{H}_1|\Phi_i\rangle$$
$$+ \langle\Phi_i|\mathfrak{H}_1 \mathfrak{R}_i \mathfrak{H}_1 \mathfrak{R}_i \mathfrak{H}_1 \mathfrak{R}_i^2 \mathfrak{H}_1|\Phi_i\rangle \}$$
$$+ \langle\Phi_i|\mathfrak{H}_1|\Phi_i\rangle^2 \{ \langle\Phi_i|\mathfrak{H}_1 \mathfrak{R}_i^3 \mathfrak{H}_1 \mathfrak{R}_i \mathfrak{H}_1|\Phi_i\rangle$$
$$+ \langle\Phi_i|\mathfrak{H}_1 \mathfrak{R}_i^2 \mathfrak{H}_1 \mathfrak{R}_i^2 \mathfrak{H}_1|\Phi_i\rangle$$
$$+ \langle\Phi_i|\mathfrak{H}_1 \mathfrak{R}_i \mathfrak{H}_1 \mathfrak{R}_i^3 \mathfrak{H}_1|\Phi_i\rangle \}$$
$$- \langle\Phi_i|\mathfrak{H}_1|\Phi_i\rangle^3 \langle\Phi_i|\mathfrak{H}_1 \mathfrak{R}_i^4 \mathfrak{H}_1|\Phi_i\rangle$$
$$- \langle\Phi_i|\mathfrak{H}_1 \mathfrak{R}_i \mathfrak{H}_1|\Phi_i\rangle \{ \langle\Phi_i|\mathfrak{H}_1 \mathfrak{R}_i^2 \mathfrak{H}_1 \mathfrak{R}_i \mathfrak{H}_1|\Phi_i\rangle$$
$$+ \langle\Phi_i|\mathfrak{H}_1 \mathfrak{R}_i \mathfrak{H}_1 \mathfrak{R}_i^2 \mathfrak{H}_1|\Phi_i\rangle \}$$
$$+ 2 \langle\Phi_i|\mathfrak{H}_1|\Phi_i\rangle \langle\Phi_i|\mathfrak{H}_1 \mathfrak{R}_i \mathfrak{H}_1|\Phi_i\rangle \langle\Phi_i|\mathfrak{H}_1 \mathfrak{R}_i^3 \mathfrak{H}_1|\Phi_i\rangle$$
$$- \langle\Phi_i|\mathfrak{H}_1 \mathfrak{R}_i \mathfrak{H}_1 \mathfrak{R}_i \mathfrak{H}_1|\Phi_i\rangle \langle\Phi_i|\mathfrak{H}_1 \mathfrak{R}_i^2 \mathfrak{H}_1|\Phi_i\rangle$$
$$+ \langle\Phi_i|\mathfrak{H}_1|\Phi_i\rangle \langle\Phi_i|\mathfrak{H}_1 \mathfrak{R}_i^2 \mathfrak{H}_1|\Phi_i\rangle^2 \qquad (43)$$

In each order the energy coefficient consists of a principal term of the form

$$\langle\Phi_i|\mathfrak{H}_1 (\mathfrak{R}_i \mathfrak{H}_1)^{p-1}|\Phi_i\rangle \qquad (44)$$

and renormalization terms. The p th. order term is given by the sum of the principal term and the renormalization terms, which are obtained by inserting any number of bra-ket pairs into the principal term in all possible ways but with no adjacent bra-ket pairs. The sign $(-1)^m$ is assigned to each term where m is the number of bra-ket pairs. The total number of terms in the p th. order energy, including the principal term is

$$\mathfrak{T}(p) = (2p-2)!/(p!\,(p-1)!) \qquad (45)$$

Some typical values of $\mathfrak{T}(p)$ are displayed in Table 1. There is a rapid increase in the number of terms $\mathfrak{T}(p)$ with order p.

Table 1

Number of principal terms and total number terms in the Rayleigh-Schrödinger perturbation theory *p th.* order energy expression.

p	$\mathfrak{T}(p)$
1	1
2	1
3	2
4	5
5	14
6	46
7	132
8	429
9	1430
10	4862

2.5 Wigner's 2n+1 rule

It has been shown in section 2.4 that the n th. order energy coefficient can be written in terms of the n-1 th. order wave function (see equations (36) and (37)). Wigner's rule states that

> *the energy up to and including order* 2n+1 *can be written in terms of the wave function through* n *th. order.*

First, consider the third-order energy coefficient

$$\varepsilon_{i3} = \langle\Phi_i|\mathcal{H}_1|\chi_{i2}\rangle \tag{46}$$

From the first-order perturbation equation (13b)

$$\overset{*}{\Phi}_i \mathcal{H}_1 = \overset{*}{\chi}_{i1} (E_i - \mathcal{H}_0) + \varepsilon_{i1} \overset{*}{\Phi}_i \tag{47}$$

and, remembering (15), (46) becomes

$$\varepsilon_{i3} = \langle\chi_{i1}|E_i - \mathcal{H}_0|\chi_{i2}\rangle \tag{48}$$

Now, using the second-order perturbation equation (13c) in the form given in (25), equation (48) can be rewritten

$$\varepsilon_{i3} = \langle\chi_{i1}|\mathcal{H}_1|\chi_{i1}\rangle - \varepsilon_{i1} \langle\chi_{i1}|\chi_{i1}\rangle \tag{49}$$

showing that the third-order energy can be expressed in terms of the wave function up to first order.

The fourth-order energy coefficient

$$\varepsilon_{i4} = \langle\Phi_i|\mathcal{H}_1|\chi_{i3}\rangle \tag{50}$$

may be rewritten using equation (47) as

$$\varepsilon_{i4} = \langle\chi_{i1}|E_i - \mathcal{H}_0|\chi_{i3}\rangle \tag{51}$$

which, upon using equation (13d), may be cast in the form

$$\varepsilon_{i4} = <\chi_{i1}|\mathcal{H}_1 - \varepsilon_{i1}|\chi_{i2}> - \varepsilon_{i2} <\chi_{i1}|\chi_{i1}> \tag{52}$$

showing that the fourth order energy can be expressed in terms of the wave function up to second order.

Similarly, using equation (47) the fifth-order energy coefficient

$$\varepsilon_{i5} = <\Phi_i|\mathcal{H}_1|\chi_{i4}> \tag{53}$$

may be written

$$\varepsilon_{i5} = <\chi_{i1}|E_i - \mathcal{H}_0|\chi_{i4}> \tag{54}$$

which gives upon using the third-order perturbation equation

$$\varepsilon_{i5} = <\chi_{i1}|\mathcal{H}_1 - \varepsilon_{i1}|\chi_{i3}> - \varepsilon_{i2} <\chi_{i1}|\chi_{i2}> - \varepsilon_{i3} <\chi_{i1}|\chi_{i1}> \tag{55}$$

From the second-order perturbation equation (13c)

$$\chi_{i1}^{*} (\mathcal{H}_1 - \varepsilon_{i1}) = \chi_{i2}^{*} (E_i - \mathcal{H}_0) + \varepsilon_{i2} \Phi_i^{*} \tag{56}$$

which allows (55) to be written

$$\varepsilon_{i5} = <\chi_{i2}|E_i - \mathcal{H}_0|\chi_{i3}> - \varepsilon_{i2} <\chi_{i1}|\chi_{i2}> - \varepsilon_{i3} <\chi_{i1}|\chi_{i1}> \tag{57}$$

Using equation (13d), this gives

$$\varepsilon_{i5} = <\chi_{i2}|\mathcal{H}_1 - \varepsilon_{i1}|\chi_{i2}> - \varepsilon_{i2} \{<\chi_{i1}|\chi_{i2}> + <\chi_{i1}|\chi_{i2}>\} - \varepsilon_{i3} <\chi_{i1}|\chi_{i1}> \tag{58}$$

and thus the fifth-order energy can be expressed in terms of the wave function up to second order.

Generally, Wigner's rule allows the energy up to order 2p+1 to be expressed in terms of the wave function up to order p. Thus for ε_{i2p}

$$\varepsilon_{i2p} = <\chi_{i(p-1)}|\mathcal{H}_1 - \varepsilon_{i1}|\chi_{ip}> - \sum_{q=2}^{p} \varepsilon_{iq} \sum_{\substack{r,s=1 \\ r+s=2p-q}} <\chi_{ir}|\chi_{is}> \tag{59}$$

and for ε_{i2p+1}

$$\varepsilon_{i2p+1} = <\chi_{ip}|\mathcal{H}_1 - \varepsilon_{i1}|\chi_{ip}> - \sum_{q=2}^{p+1} \varepsilon_{iq} \sum_{\substack{r,s=1 \\ r+s=2p-q+1}} <\chi_{ir}|\chi_{is}> \tag{60}$$

If explicit expressions for the wave functions are substituted in equations (59) and (60) then the expressions for the energy coefficients (38) - (42) are recovered. For example, substituting the perturbed wave functions give in (24) and (28) into equation (58) for the fifth-order energy, yields equation (43) for the energy coefficient.

3. Double perturbation theory

3.1 Double Rayleigh-Schrödinger perturbation theory

The developments given in section 2 will be generalized in this section to the situation in which the total hamiltonian, $\mathfrak{H}$, is recovered when two perturbing operators, $\mathfrak{H}_1$ and $\mathfrak{H}_2$, are added to the zero-order operator, $\mathfrak{H}_0$. The total hamiltonian is

$$\mathfrak{H} = \mathfrak{H}_0 + \lambda_1 \mathfrak{H}_1 + \lambda_2 \mathfrak{H}_2 \qquad (61)$$

It is assumed that the total energy and the wave function for the particular state of interest are analytic functions of λ_1 and λ_2 and that the perturbations $\mathfrak{H}_1$ and $\mathfrak{H}_2$ neither introduce new singularities nor change the nature of an existing singularity.
Expanding the energy as

$$E_i(\lambda_1,\lambda_2) = E_i + \sum_{\substack{p=0 \\ (p+q)>0}}^{\infty} \sum_{q=0}^{\infty} \lambda_1{}^p \lambda_2{}^q \varepsilon_{ipq} \qquad (62)$$

where the coefficients are given by

$$\varepsilon_{ipq} = (p!q!)^{-1} \left[\partial^{p+q} E_i(\lambda_1,\lambda_2)/\partial \lambda_1{}^p \partial \lambda_2{}^q \right]_{\lambda_1=0,\lambda_2=0} \qquad (63)$$

and the wave function

$$\Psi_i(\lambda_1,\lambda_2) = \Phi_i + \sum_{\substack{p=0 \\ (p+q)>0}}^{\infty} \sum_{q=0}^{\infty} \lambda_1{}^p \lambda_2{}^q \chi_{ipq} \qquad (64)$$

where

$$\chi_{ipq} = (p!q!)^{-1} \left[\partial^{p+q} \Psi_i(\lambda_1, {}_2)/\partial \lambda_1{}^p \partial \lambda_2{}^q \right]_{\lambda_1=0,\lambda_2=0} \qquad (65)$$

the perturbed Schrödinger equation may be written in the form

$$(\mathcal{H}_0 + \lambda_1 \mathcal{H}_1 + \lambda_2 \mathcal{H}_2)\ (\Phi_i + \sum_{\substack{p=0 \\ p+q>0}}^{\infty} \sum_{q=0}^{\infty} \lambda_1{}^p \lambda_2{}^q \chi_{ipq}) =$$

$$(E_i + \sum_{\substack{p=0 \\ p+q>0}}^{\infty} \sum_{q=0}^{\infty} \lambda_1{}^p \lambda_2{}^q \varepsilon_{ip})\ (\Phi_i +$$

$$\sum_{\substack{p=0 \\ p+q>0}}^{\infty} \sum_{q=0}^{\infty} \lambda_1{}^p \lambda_2{}^q \chi_{ipq}) \qquad (66)$$

Equating powers of $\lambda_1{}^p \lambda_2{}^q$ in equation (66) leads to the basic equations of double Rayleigh-Schrödinger perturbation theory

p=0,q=0: $$\mathcal{H}_0 \Phi_i = E_i \Phi_i \qquad (67a)$$

p=1,q=0: $$\mathcal{H}_0 \chi_{i10} + \mathcal{H}_1 \Phi_i = E_i \chi_{i10} + \varepsilon_{i10} \Phi_i \qquad (67b)$$

p=0,q=1: $$\mathcal{H}_0 \chi_{i01} + \mathcal{H}_2 \Phi_i = E_i \chi_{i01} + \varepsilon_{i01} \Phi_i \qquad (67c)$$

p=1,q=1: $$\mathcal{H}_0 \chi_{i11} + \mathcal{H}_1 \chi_{i01} + \mathcal{H}_2 \chi_{i10} = E_i \chi_{i11} + \varepsilon_{i10} \chi_{i01} + \varepsilon_{i01} \chi_{i10} + \varepsilon_{i11} \Phi_i \qquad (67d)$$

p=2,q=0: $$\mathcal{H}_0 \chi_{i20} + \mathcal{H}_1 \chi_{i10} = E_i \chi_{i20} + \varepsilon_{i10} \chi_{i10} + \varepsilon_{i20} \Phi_i \qquad (67e)$$

p=0,q=2: $$\mathcal{H}_0 \chi_{i02} + \mathcal{H}_2 \chi_{i01} = E_i \chi_{i02} + \varepsilon_{i01} \chi_{i01} + \varepsilon_{i02} \Phi_i \qquad (69f)$$

p=2,q=1: $$\mathcal{H}_0 \chi_{i21} + \mathcal{H}_1 \chi_{i11} + \mathcal{H}_2 \chi_{i20} = E_i \chi_{i21} + \varepsilon_{i10} \chi_{i11} + \varepsilon_{i01} \chi_{i20} + \varepsilon_{i20} \chi_{i01} + \varepsilon_{i21} \Phi_i \qquad (67g)$$

p=1,q=2: $$\mathcal{H}_0 \chi_{i12} + \mathcal{H}_1 \chi_{i02} + \mathcal{H}_2 \chi_{i11} = E_i \chi_{i12} + \varepsilon_{i10} \chi_{i02} + \varepsilon_{i01} \chi_{i11} + \varepsilon_{i11} \chi_{i01} + \varepsilon_{i02} \chi_{i01} + \varepsilon_{i12} \Phi_i \qquad (67h)$$

p=3,q=0: $$\mathcal{H}_0 \chi_{i30} + \mathcal{H}_1 \chi_{i20} = E_i \chi_{i30} + \varepsilon_{i10} \chi_{i20} + \varepsilon_{i20} \chi_{i10} + \varepsilon_{i30} \Phi_i \qquad (67i)$$

p=0,q=3: $\mathcal{H}_0\, \chi_{i03} + \mathcal{H}_2\, \chi_{i02} = E_i\, \chi_{i03} + \varepsilon_{i01}\, \chi_{i02} +$

$$\varepsilon_{i02}\, \chi_{i01} + \varepsilon_{i03}\, \Phi_i \qquad (67j)$$

Generally, equating terms in $\lambda_1{}^p\, \lambda_2{}^q$ yields

$$\mathcal{H}_0\, \chi_{ipq} + \mathcal{H}_1\, \chi_{ip-1q} + \mathcal{H}_2\, \chi_{ipq-1} = E_i\, \chi_{ipq} + \sum_{r}\sum_{s} \varepsilon_{irs}\, \chi_{ip-rq-s} + \varepsilon_{ipq}\, \Phi_i$$

$$(p+q)>(r+s)>0 \qquad p,q \neq 0 \qquad (68)$$

$$\mathcal{H}_0\, \chi_{ip0} + \mathcal{H}_1\, \chi_{ip-10} = E_i\, \chi_{ip0} + \sum_{\substack{r \\ p>r>0}} \varepsilon_{ir0}\, \chi_{ip-r0} + \varepsilon_{ip0}\, \Phi_i \qquad p>0 \qquad (69)$$

$$\mathcal{H}_0\, \chi_{i0q} + \mathcal{H}_2\, \chi_{i0q-1} = E_i\, \chi_{i0q} + \sum_{\substack{s \\ q>s>0}} \varepsilon_{i0s}\, \chi_{i0q-s} + \varepsilon_{i0q}\, \Phi_i \qquad q>0 \qquad (70)$$

3.2 Perturbed wave functions

From equations (67) expressions for the perturbed wave functions in double Rayleigh-Schrödinger perturbation theory can be obtained

$$\chi_{i10} = \mathfrak{R}_i\, \mathcal{H}_1\, \Phi_i \qquad (71a)$$

$$\chi_{i01} = \mathfrak{R}_i\, \mathcal{H}_2\, \Phi_i \qquad (71b)$$

$$\chi_{i11} = \mathfrak{R}_i\, (\mathcal{H}_1 - \varepsilon_{i10})\, \chi_{i01} + \mathfrak{R}_i\, (\mathcal{H}_2 - \varepsilon_{i01})\, \chi_{i10} \qquad (71c)$$

$$\chi_{i20} = \mathfrak{R}_i\, (\mathcal{H}_1 - \varepsilon_{i10})\, \chi_{i10} \qquad (71d)$$

$$\chi_{i02} = \mathfrak{R}_i\, (\mathcal{H}_2 - \varepsilon_{i01})\, \chi_{i01} \qquad (71e)$$

$$\chi_{i21} = \mathfrak{R}_i\, (\mathcal{H}_1 - \varepsilon_{i10})\, \chi_{i11} + \mathfrak{R}_i\, (\mathcal{H}_2 - \varepsilon_{i01})\, \chi_{i20} - \varepsilon_{i20}\, \mathfrak{R}_i\, \chi_{i01} - \varepsilon_{i11}\, \mathfrak{R}_i\, \chi_{i10} \qquad (71f)$$

$$\chi_{i12} = \mathfrak{R}_i\, (\mathcal{H}_1 - \varepsilon_{i10})\, \chi_{i02} + \mathfrak{R}_i\, (\mathcal{H}_2 - \varepsilon_{i01})\, \chi_{i11} - \varepsilon_{i02}\, \mathfrak{R}_i\, \chi_{i10} - \varepsilon_{i11}\, \mathfrak{R}_i\, \chi_{i01} \qquad (71g)$$

$$\chi_{i30} = \mathfrak{R}_i\, (\mathcal{H}_1 - \varepsilon_{i10})\, \chi_{i20} - \varepsilon_{i20}\, \mathfrak{R}_i\, \chi_{i10} \qquad (71h)$$

$$\chi_{i03} = \mathfrak{R}_i\, (\mathcal{H}_2 - \varepsilon_{i01})\, \chi_{i02} - \varepsilon_{i02}\, \mathfrak{R}_i\, \chi_{i01} \qquad (71i)$$

where the resolvant $\mathfrak{R}_i$ is defined in equation (23). Generally,

$$\chi_{ipq} = \mathfrak{R}_i\,(\mathfrak{H}_1 - \varepsilon_{i10})\,\chi_{ip-1q} + \mathfrak{R}_i\,(\mathfrak{H}_2 - \varepsilon_{i01})\,\chi_{ipq-1} - \sum_{r}\sum_{s}{}_{(r+s)>1}\,\varepsilon_{irs}\,\mathfrak{R}_i\,\chi_{ip-rq-s} \qquad p,q>0 \qquad (72)$$

$$\chi_{ip0} = \mathfrak{R}_i\,(\mathfrak{H}_1 - \varepsilon_{i10})\,\chi_{ip-10} - \sum_{r=2}\varepsilon_{ir0}\,\mathfrak{R}_i\,\chi_{ip-r0} \qquad p>0 \qquad (73)$$

$$\chi_{i0q} = \mathfrak{R}_i\,(\mathfrak{H}_2 - \varepsilon_{i01})\,\chi_{i0q-1} - \sum_{s=2}\varepsilon_{i0s}\,\mathfrak{R}_i\,\chi_{i0q-s} \qquad q>0 \qquad (74)$$

3.3 Energy coefficients

Multiplying the basic double perturbation theory equations, equations (67), by Φ_i^* and integrating over the coordinates of all electrons leads to expressions for the energy coefficients. If it is required that the perturbed wave function be orthogonal to the reference function (a requirement that leads to no loss of generality), that is

$$\langle\Phi_i|\chi_{ipq}\rangle = 0, \quad p,q=1,2,..., \quad p+q>0 \qquad (75)$$

then the following expressions are obtained

$$\varepsilon_{i10} = \langle\Phi_i|\mathfrak{H}_1|\Phi_i\rangle \qquad (76a)$$

$$\varepsilon_{i01} = \langle\Phi_i|\mathfrak{H}_2|\Phi_i\rangle \qquad (76b)$$

$$\varepsilon_{i11} = \langle\Phi_i|\mathfrak{H}_1|\chi_{i01}\rangle + \langle\Phi_i|\mathfrak{H}_2|\chi_{i10}\rangle \qquad (76c)$$

$$\varepsilon_{i20} = \langle\Phi_i|\mathfrak{H}_1|\chi_{i10}\rangle \qquad (76d)$$

$$\varepsilon_{i02} = \langle\Phi_i|\mathfrak{H}_2|\chi_{i01}\rangle \qquad (76e)$$

$$\varepsilon_{i21} = \langle\Phi_i|\mathfrak{H}_1|\chi_{i11}\rangle + \langle\Phi_i|\mathfrak{H}_2|\chi_{i20}\rangle \qquad (76f)$$

$$\varepsilon_{i12} = \langle\Phi_i|\mathfrak{H}_1|\chi_{i02}\rangle + \langle\Phi_i|\mathfrak{H}_2|\chi_{i11}\rangle \qquad (76g)$$

$$\varepsilon_{i30} = \langle\Phi_i|\mathfrak{H}_1|\chi_{i20}\rangle \qquad (76h)$$

$$\varepsilon_{i03} = \langle\Phi_i|\mathfrak{H}_2|\chi_{i02}\rangle \qquad (76i)$$

Generally,

$$\varepsilon_{ipq} = \langle\Phi_i|\mathfrak{H}_1|\chi_{ip-1q}\rangle + \langle\Phi_i|\mathfrak{H}_2|\chi_{ipq-1}\rangle, \quad p,q>0 \qquad (77)$$

$$\varepsilon_{ip0} = \langle\Phi_i|\mathfrak{H}_1|\chi_{ip-10}\rangle, \quad p>0 \qquad (78)$$

$$\varepsilon_{i0q} = \langle\Phi_i|\mathfrak{H}_1|\chi_{ip-10}\rangle, \quad q>0 \qquad (79)$$

Substituting the perturbed wave functions, equations (71), into the expressions for the energy coefficients gives

$$\varepsilon_{i10} = <\Phi_i | \mathfrak{H}_1 | \Phi_i> \tag{80a}$$

$$\varepsilon_{i01} = <\Phi_i | \mathfrak{H}_2 | \Phi_i> \tag{80b}$$

$$\varepsilon_{i11} = <\Phi_i | \mathfrak{H}_1 \, \mathfrak{R}_i \, \mathfrak{H}_2 | \Phi_i> + <\Phi_i | \mathfrak{H}_2 \, \mathfrak{R}_i \, \mathfrak{H}_1 | \Phi_i>$$

$$= 2 <\Phi_i | \mathfrak{H}_1 \, \mathfrak{R}_i \, \mathfrak{H}_2 | \Phi_i> \tag{80c}$$

$$\varepsilon_{i20} = <\Phi_i | \mathfrak{H}_1 \, \mathfrak{R}_i \, \mathfrak{H}_1 | \Phi_i> \tag{80d}$$

$$\varepsilon_{i02} = <\Phi_i | \mathfrak{H}_1 \, \mathfrak{R}_i \, \mathfrak{H}_1 | \Phi_i> \tag{80e}$$

$$\varepsilon_{i21} = 2 <\Phi_i | \mathfrak{H}_1 \, \mathfrak{R}_i \, (\mathfrak{H}_1 - \varepsilon_{i10}) \, \mathfrak{R}_i \, \mathfrak{H}_2 | \Phi_i> +$$

$$<\Phi_i | \mathfrak{H}_1 \, \mathfrak{R}_i \, (\mathfrak{H}_2 - \varepsilon_{i01}) \, \mathfrak{R}_i \, \mathfrak{H}_2 | \Phi_i> \tag{80f}$$

$$\varepsilon_{i12} = 2 <\Phi_i | \mathfrak{H}_1 \, \mathfrak{R}_i \, (\mathfrak{H}_2 - \varepsilon_{i01}) \, \mathfrak{R}_i \, \mathfrak{H}_2 | \Phi_i> +$$

$$<\Phi_i | \mathfrak{H}_2 \, \mathfrak{R}_i \, (\mathfrak{H}_1 - \varepsilon_{i10}) \, \mathfrak{R}_i \, \mathfrak{H}_2 | \Phi_i> \tag{80g}$$

$$\varepsilon_{i30} = <\Phi_i | \mathfrak{H}_1 \, \mathfrak{R}_i \, (\mathfrak{H}_1 - \varepsilon_{i10}) \, \mathfrak{R}_i \, \mathfrak{H}_1 | \Phi_i> \tag{80h}$$

$$\varepsilon_{i03} = <\Phi_i | \mathfrak{H}_2 \, \mathfrak{R}_i \, (\mathfrak{H}_2 - \varepsilon_{i01}) \, \mathfrak{R}_i \, \mathfrak{H}_2 | \Phi_i> \tag{80i}$$

Generally, in order $\lambda_1{}^p \, \lambda_2{}^q$ there will be $\mathfrak{P}(p,q) = (p+q)!/(p!q!)$ principal terms of the form

$$<\Phi_i | \mathfrak{A}_1 \, \mathfrak{R}_i \, \mathfrak{A}_2 \, \mathfrak{R}_i \, \ldots \, \mathfrak{R}_i \, \mathfrak{A}_{p+q} | \Phi_i> \tag{81}$$

where p of the operators $\mathfrak{A}_k$ are $\mathfrak{H}_1$ and the remaining q operators are $\mathfrak{H}_2$. There will be a total of

$$\mathfrak{T}(p,q) = (2(p+q)-2)!/((p+q-1)! \; p! \; q!) \tag{82}$$

terms in order $\lambda_1{}^p \, \lambda_2{}^q$. This will be given by the sum of the principal terms and the renormalization terms, which are obtained by inserting and number of bra-ket pairs into the principal terms in all possible ways but with no adjacent bra-ket pairs. A factor of $(-1)^m$ is assigned to each term, m being the number of inserted bra-ket pairs. Both the number of terms in each order, $\mathfrak{T}(p,q)$, and the number of principal terms, $\mathfrak{P}(p,q)$, are shown in Table 2 for some typical values of p and q.

4. Sum-over-states formulae

4.1 Single Rayleigh-Schrödinger perturbation theory

The expressions given in the preceding section of this chapter for wave functions and energies depend upon the resolvant, $\mathfrak{R}_i$, which was defined in equation (23) as follows

$$\mathfrak{R}_i \, (E_i - \mathfrak{H}_0) \, \hat{Q}_i = \hat{Q}_i \tag{23}$$

Table 2

Number of principal terms and total number terms in the Rayleigh-Schrödinger double perturbation theory (p,q) th. order energy expression.

p	q	$\mathfrak{P}(p,q)$	$\mathfrak{T}(p,q)$
1	1	2	2
2	1	3	6
2	2	6	30
3	1	4	20
3	2	10	140
3	3	20	840
4	1	5	70
4	2	15	630
4	3	35	4620
4	4	70	30030

in which $\hat{Q}_i$ is the projector onto the orthogonal complement of the model function Φ_i. In terms of the eigenfunctions of the model eigenproblem, equation (2), this projector operator may be written

$$\hat{Q}_i = \sum_{k \neq i} |\Phi_k\rangle \langle\Phi_k| \tag{83}$$

Hence equation (23) becomes

$$\mathfrak{R}_i \, (E_i - \mathfrak{H}_0) \sum_{k \neq i} |\Phi_k\rangle \langle\Phi_k| = \sum_{k \neq i} |\Phi_k\rangle \langle\Phi_k| \tag{84}$$

$$\mathfrak{R}_i \sum_{k \neq i} (E_i - E_k) \, |\Phi_k\rangle \langle\Phi_k| = \sum_{k \neq i} |\Phi_k\rangle \langle\Phi_k| \tag{85}$$

or

$$\mathfrak{R}_i = \sum_{k \neq i} (E_i - E_k)^{-1} \, |\Phi_k\rangle \langle\Phi_k| \tag{86}$$

The resolvant can be expressed in terms of the eigenfunctions and eigenvalues of the zero-order problem. Substituting equation (86) into the expressions given in the preceding sections for the perturbed wave functions and energies gives the sum-over-states formulae most often used in practical applications of perturbation theory.

The first-order wave function in Rayleigh-Schrödinger perturbation theory for a single perturbation is given by equation (24)

$$\chi_{i1} = \mathfrak{R}_i \, \mathfrak{H}_1 \, \Phi_i \tag{24}$$

Using equation (86), this may be written as the sum-over-states expression

$$\chi_{i1} = \sum_{k \neq i} (E_i - E_k)^{-1} \, |\Phi_k\rangle \langle\Phi_k|\mathfrak{H}_1|\Phi_i\rangle \tag{87}$$

Similarly, the second-order wave function may be written as

$$\chi_{i2} = \sum_{k \neq i} \sum_{l \neq i} (E_i - E_k)^{-1} (E_i - E_l)^{-1} |\Phi_k\rangle \langle\Phi_k|\mathfrak{H}_1 - \delta_{kl}\, \varepsilon_{i1}|\Phi_l\rangle \langle\Phi_l|\mathfrak{H}_1|\Phi_i\rangle \tag{88}$$

For the second-order and third-order energy coefficients in Rayleigh-Schrödinger perturbation theory for a single perturbation, the sum-over-states formulae are

$$\varepsilon_{i2} = \sum_{k \neq i} (E_i - E_k)^{-1} \langle\Phi_i|\mathfrak{H}_1|\Phi_k\rangle \langle\Phi_k|\mathfrak{H}_1|\Phi_i\rangle \tag{89}$$

and

$$\varepsilon_{i3} = \sum_{k \neq i} \sum_{l \neq i} (E_i - E_k)^{-1} (E_i - E_l)^{-1} \langle\Phi_i|\mathfrak{H}_1|\Phi_k\rangle \langle\Phi_k|\mathfrak{H}_1|\Phi_l\rangle \langle\Phi_l|\mathfrak{H}_1|\Phi_i\rangle - \varepsilon_{i1} \sum_{k \neq i} (E_i - E_k)^{-2} \langle\Phi_i|\mathfrak{H}_1|\Phi_k\rangle \langle\Phi_k|\mathfrak{H}_1|\Phi_i\rangle \tag{90}$$

In fourth order we have

$$\begin{aligned}
\varepsilon_{i4} = & \sum_{k \neq i} \sum_{l \neq i} \sum_{m \neq i} (E_i - E_k)^{-1} (E_i - E_l)^{-1} (E_i - E_m)^{-1} \langle\Phi_i|\mathfrak{H}_1|\Phi_k\rangle \langle\Phi_k|\mathfrak{H}_1|\Phi_l\rangle \langle\Phi_k|\mathfrak{H}_1|\Phi_m\rangle \langle\Phi_m|\mathfrak{H}_1|\Phi_i\rangle \\
& -2\varepsilon_{i1} \sum_{k \neq i} \sum_{l \neq i} (E_i - E_k)^{-2} (E_i - E_l)^{-1} \langle\Phi_i|\mathfrak{H}_1|\Phi_k\rangle \langle\Phi_k|\mathfrak{H}_1|\Phi_l\rangle \langle\Phi_l|\mathfrak{H}_1|\Phi_i\rangle \\
& + \varepsilon_{i1}^2 \sum_{k \neq i} (E_i - E_k)^{-3} \langle\Phi_i|\mathfrak{H}_1|\Phi_k\rangle \langle\Phi_k|\mathfrak{H}_1|\Phi_i\rangle \\
& -\varepsilon_{i2} \sum_{k \neq i} (E_i - E_k)^{-2} \langle\Phi_i|\mathfrak{H}_1|\Phi_k\rangle \langle\Phi_k|\mathfrak{H}_1|\Phi_i\rangle
\end{aligned} \tag{91}$$

4.2 Double Rayleigh-Schrödinger perturbation theory

Similar sum-over-states expressions can be obtained from the perturbed wave functions and energies in double perturbation theory. For the low order wave functions we have the following

$$\chi_{i10} = \sum_{k\neq i} (E_i - E_k)^{-1} \,|\Phi_k\rangle \langle\Phi_k|\mathcal{H}_1|\Phi_i\rangle \tag{92}$$

$$\chi_{i01} = \sum_{k\neq i} (E_i - E_k)^{-1} \,|\Phi_k\rangle \langle\Phi_k|\mathcal{H}_2|\Phi_i\rangle \tag{93}$$

$$\chi_{i20} = \sum_{k\neq i}\sum_{l\neq i} (E_i - E_k)^{-1} (E_i - E_l)^{-1} |\Phi_k\rangle \langle\Phi_k|\mathcal{H}_1 - \delta_{kl}\,\varepsilon_{i10}|\Phi_l\rangle \langle\Phi_l|\mathcal{H}_1|\Phi_i\rangle \tag{94}$$

$$\chi_{i11} = \sum_{k\neq i}\sum_{l\neq i} (E_i - E_k)^{-1} (E_i - E_l)^{-1} |\Phi_k\rangle \langle\Phi_k|\mathcal{H}_1 - \delta_{kl}\,\varepsilon_{i10}|\Phi_l\rangle \langle\Phi_l|\mathcal{H}_2|\Phi_i\rangle + \sum_{k\neq i}\sum_{l\neq i} (E_i - E_k)^{-1} (E_i - E_l)^{-1} |\Phi_k\rangle \langle\Phi_k|\mathcal{H}_2 - \delta_{kl}\,\varepsilon_{i01}|\Phi_l\rangle \langle\Phi_l|\mathcal{H}_1|\Phi_i\rangle \tag{95}$$

$$\chi_{i02} = \sum_{k\neq i}\sum_{l\neq i} (E_i - E_k)^{-1} (E_i - E_l)^{-1} |\Phi_k\rangle \langle\Phi_k|\mathcal{H}_2 - \delta_{kl}\,\varepsilon_{i01}|\Phi_l\rangle \langle\Phi_l|\mathcal{H}_2|\Phi_i\rangle \tag{96}$$

The low order energy coefficients have the following sum-over-states expressions

$$\varepsilon_{i20} = \sum_{k\neq i} (E_i - E_k)^{-1} \langle\Phi_i|\mathcal{H}_1|\Phi_k\rangle \langle\Phi_k|\mathcal{H}_1|\Phi_i\rangle \tag{97}$$

$$\varepsilon_{i11} = \sum_{k\neq i} (E_i - E_k)^{-1} (\langle\Phi_i|\mathcal{H}_1|\Phi_k\rangle \langle\Phi_k|\mathcal{H}_2|\Phi_i\rangle + \langle\Phi_i|\mathcal{H}_1|\Phi_k\rangle \langle\Phi_k|\mathcal{H}_1|\Phi_i\rangle) \tag{98}$$

$$\varepsilon_{i02} = \sum_{k\neq i} (E_i - E_k)^{-1} \langle\Phi_i|\mathcal{H}_2|\Phi_k\rangle \langle\Phi_k|\mathcal{H}_2|\Phi_i\rangle \tag{99}$$

5. Degenerate Rayleigh-Schrödinger perturbation theory

5.1 Basic formalism

Consider now a system which is described in zero-order by the quantum mechanical eigenvalue problem

$$\mathcal{H}_0\,\Phi_{im} = E_i\,\Phi_{im},\ m=1,2,\ldots,g_i \tag{100}$$

in which the i th. state is g_i-fold degenerate. Any linear combination of the g_i functions Φ_{i1}, Φ_{i2}, ..., Φ_{ig_i} is also an eigenfunction of (134). Let the perturbed eigenproblem be

$$(\mathfrak{H}_0 + \lambda\, \mathfrak{H}_1)\, \Psi_k(\lambda) = E_k(\lambda)\, \Psi_k(\lambda) \tag{101}$$

for which the k th. state may or may not be degenerate.

When the perturbation is treated to first order, the degeneracy of the zero-order energies, E_i, is split into subsets, which are characterized by the first order energies. These subsets are themselves split once again in second order into smaller subsets, which are characterized by the second-order energies. This further splitting of the degeneracy in each order of the perturbation expansion continues until the perturbation is resolved, that is, either the subset if reduced so that it contains only one state, or else, no further splitting occurs in higher orders.

Let the g_i functions Φ_{im} span a subspace which is denoted by $\mathfrak{S}_{i0}$. In first order this is split into the subspaces

$$\mathfrak{S}_{i1(1)},\ \mathfrak{S}_{i1(2)},\ \ldots,\ \mathfrak{S}_{i1(n1)} \tag{102}$$

of degeneracy

$$g_{i1(1)},\ g_{i1(2)},\ \ldots,\ g_{i1(n1)} \tag{103}$$

respectively, such that

$$\sum_{p=1}^{n_1} g_{i1(p)} = g_i \tag{104}$$

In second order these subspaces are further split into the smaller subspaces

$$\begin{array}{llll} \mathfrak{S}_{i2(11)} & \mathfrak{S}_{i2(12)} & \cdots & \mathfrak{S}_{i2(1n21)} \\ \mathfrak{S}_{i2(21)} & \mathfrak{S}_{i2(22)} & \cdots & \mathfrak{S}_{i2(2n22)} \\ \quad\cdots & & & \\ \mathfrak{S}_{i2(n11)} & \mathfrak{S}_{i2(n12)} & \cdots & \mathfrak{S}_{i2(n_1 n2 n_1)} \end{array} \tag{105}$$

of dimension $g_{i2(pq)}$ where

$$\sum_{q} g_{i2(pq)} = g_{i1(p)} \tag{106}$$

The successive splitting of the subspaces in higher orders of the perturbation theory is shown graphically in Figure 1. After taking the perturbation expansion to n th. order the subspace $\mathfrak{S}_{i0}$ is split into subspaces which can be characterized by the genealogy of their construction $\mathfrak{S}_{in(p1p2\cdots)}$. The energies and perturbed wave functions can, therefore, be labeled according to the genealogy of the corresponding subspaces. Each energy coefficient is labeled as follows

$$\varepsilon_i^{(k)}{}_{p(\quad)m} \tag{107}$$

where i numbers the eigenvalues of the zero-order eigenproblem and k numbers the eigenvalues of the perturbed eigenproblem. p denotes the order of perturbation theory and m numbers the members of degenerate sets of energies in each order of perturbation theory. The parenthesis (   ) denotes the p indices which record the genealogy of a particular eigenvalue. This compound index labels the subspace associated with the corresponding eigenfunction(s). Generally, it takes the form

$$(j_1\, j_2 \cdots j_p) \tag{108}$$

It is convenient to introduce the following notation

(i) $(j_1\, j_2 \cdots j_{p-1}\, \circ)$ (109)

labels the subspaces having the same energy as $(j_1\, j_2 \cdots j_p)$ to order p-1 but not necessarily the same energy in order p.

(ii) $(j_1\, j_2 \cdots j_{p-1}\, \bullet)$ (110)

labels the subspaces having the same energy as $(j_1\, j_2 \cdots j_p)$ to order p-1 but a different energy in order p.

The resolution of the degeneracies in each order of the perturbation series can then be discussed in terms of the notation defined in equations (109) and (110). In nth. order the space labelled by

$$(j_1\, j_2 \cdots j_{n-1}\, \circ) \tag{111}$$

is split into

$$(j_1\, j_2 \cdots j_{n-1}\, j_n) \tag{112}$$

and

$$(j_1\, j_2 \cdots j_{n-1}\, \bullet) \tag{113}$$

The functions spanning the space labeled by $(j_1\, j_2 \cdots j_{n-1}\, \circ)$ are orthonormal linear combinations of the functions which span the space labeled by $(j_1\, j_2 \cdots j_{n-1})$.

Using the notation described above, the energy may now be expanded in the form

$$E_k(\lambda) = E_i + \sum_{p=1}^{\infty} \lambda^p\, \varepsilon_i^{(k)}{}_{p(\quad)m} \tag{114}$$

The wave function is expanded as

$$\Psi_k(\lambda) = \chi_{i0m} + \sum \lambda^p\, \chi_i^{(k)}{}_{p(\quad)m} \tag{115}$$

Substituting the expansions (114) and (115) into the perturbed eigenvalue problem leads to a set of perturbation equations which are identical to those given in the non-degenerate case, that is equations (13), except for the additional labels which are necessary.

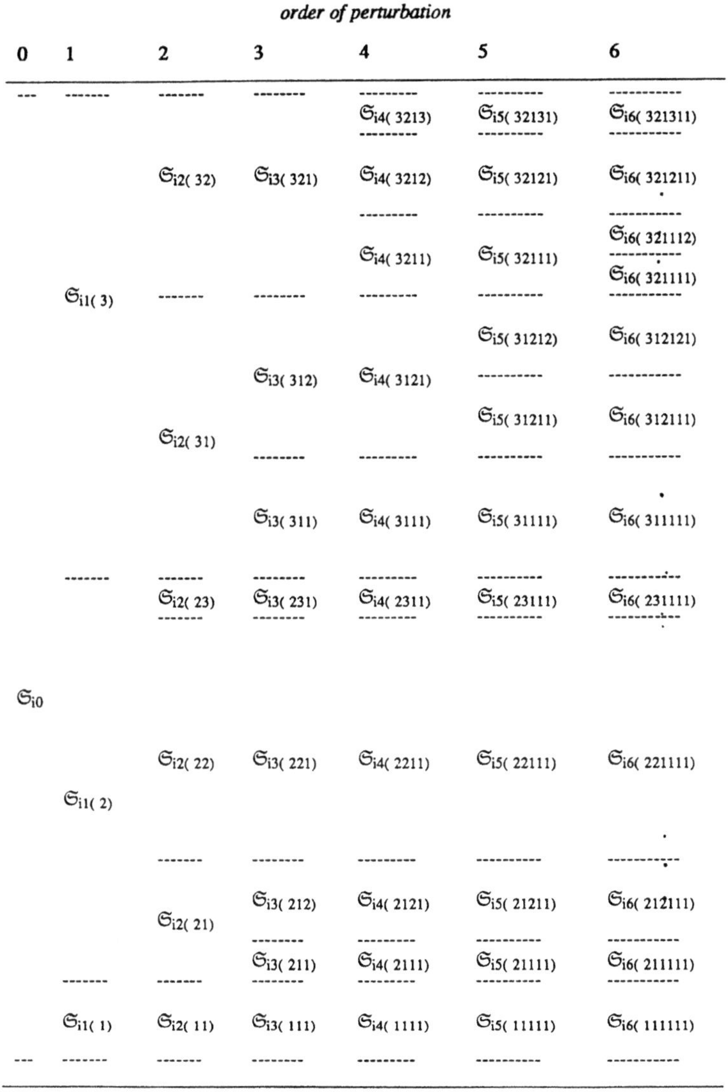

Figure 1. *Graphical representation of the successive splitting of subspaces in degenerate Rayleigh-Schrödinger perturbation theory.*

Having discussed the general features of degenerate Rayleigh-Schrödinger perturbation theory, attention will be given in the next three sections to low order degenerate perturbation expansions. The zero-order problem will be addressed in section 5.2, the first-order problem in section 5.3, and the second-order and higher-order problem in section 5.4.

5.2 Zero-order degenerate Rayleigh-Schrödinger perturbation theory.

The zero-order eigenvalue problem, equation (100), may be written

$$\mathfrak{H}_0\, \chi_i{}^{(k)}{}_{0m} = E_i\, \chi_i{}^{(k)}{}_{0m},\quad m=1,2,...,g_i \tag{116}$$

where the $\chi_i{}^{(k)}{}_{0m}$'s are linear combinations of the Φ_{im}'s

$$\chi_i{}^{(k)}{}_{0m} = \sum_{n=1}^{g_i} \Phi_{in}\, C_{iom\ in} \tag{117}$$

Both the Φ_{in} and the $\chi_i{}^{(k)}{}_{0m}$ can be taken to orthonormal. The expansion coefficients, $C_{iom\ in}$, cannot be defined at this stage.

5.3 First-order degenerate Rayleigh-Schrödinger perturbation theory.

By analogy with equation (13b) given in the discussion of non-degenerate Rayleigh-Schrödinger perturbation theory, the first-order perturbation equation in the presence of degeneracy assumes the form

$$(\mathfrak{H}_0 - E_i)\, \chi_i{}^{(k)}{}_{1(0)n} = (\varepsilon_i{}^{(k)}{}_{1(0)n} - \mathfrak{H}_1)\, \chi_i{}^{(k)}{}_{0n} \tag{118}$$

The first order function $\chi_i{}^{(k)}{}_{1(0)n}$ may be written as a linear combination of the zero-order functions

$$\chi_i{}^{(k)}{}_{1(0)n} = \sum_{m=1}^{g_i} \Phi_{im}\, C_{i1n\ im} + \sum_{j\neq i} \sum_{m=1}^{g_j} \Phi_{jm}\, C_{i1n\ jm} \tag{119}$$

where the degenerate set of zero-order functions Φ_{im} have been separated from other eigenfunctions of the zero-order hamiltonian Φ_{jm}, which may or may not be degenerate. Substituting the expansions (117) and (119) into the first order equation (118) gives

$$\sum_{j\neq i} \sum_{m=1}^{g_j} C_{i1n\ jm}\, (E_j - E_i)\, \Phi_{jm} =$$

$$\sum_{m=1}^{g_i} (\varepsilon_i{}^{(k)}{}_{1(0)n} - \mathfrak{H}_1)\, \Phi_{im}\, C_{ion\ im} \tag{120}$$

Multiplying from the left by Φ_{im}^* and integrating gives

$$\sum_{m=1}^{g_i} C_{i0n\ im} \left(\varepsilon_i^{(k)}{}_{1(0)n} - \langle \Phi_{im'} | \mathfrak{H}_1 | \Phi_{im} \rangle \right) = 0 \tag{121}$$

which may be written in matrix form as

$$(\mathfrak{H}_{i1} - \varepsilon_i^{(k)}{}_{1(0)n}) C_{i0n\ i} = \mathbf{0} \tag{122}$$

where $\mathfrak{H}_{i1}$ is the matrix

$$\mathfrak{H}_{i1m'm} = \langle \Phi_{im'} | \mathfrak{H}_1 | \Phi_{im} \rangle \tag{123}$$

and

$$C_{i0n\ i}^{\dagger} = (C_{ion\ i1}, C_{ion\ i2}, .. , C_{ion\ ig_i}) \tag{124}$$

The first order energies are obtained by solution of the secular equation

$$| \mathfrak{H}_{i1} - \varepsilon_i^{(k)}{}_{1(0)n} | = 0 \tag{125}$$

and are labeled $\varepsilon_i^{(k)}{}_{1(1)1}$, $\varepsilon_i^{(k)}{}_{1(1)2}$, ..., $\varepsilon_i^{(k)}{}_{1(2)1}$, $\varepsilon_i^{(k)}{}_{1(2)2}$, ... according to the scheme illustrated in Figure 1.

If the first order energy which is of interest is non-degenerate then solution of the eigenvalue equation (122) completely determines the corresponding perturbed wave function. If, on the other hand, the first order energy of interest is degenerate then equation (122) only determines a set of functions from which we are free to form arbitrary linear combinations. This degeneracy may or may not be removed when the perturbation series is taken to higher orders.

5.4 Second-order and higher-order degenerate Rayleigh-Schrödinger perturbation theory.

A practical method for the systematic determination of the second and higher order Rayleigh-Schrödinger energies and wave functions of perturbed degenerate states to arbitrary high orders has been presented by Hirschfelder and Certain [1]. Their method is applicable even when the degeneracies of the various states are resolved in arbitrary orders. The reader is referred to the cited publication for further details.

5.5 Quasi-degenerate Rayleigh-Schrödinger perturbation theory.

In section 2, we developed the Rayleigh-Schrödinger perturbation series with respect to a single determinantal reference function. If there are other determinants in the reference spectrum with energies close to the chosen zero-order determinant, that is there is a degree of quasi-degeneracy, the convergence properties of the non-degenerate perturbation expansion may deteriorate. One method for avoiding this problem is to employ a reference function which consists of a linear combination of determinants

$$\Psi = \sum_{i \in \mathfrak{S}} C_i \Phi_i \tag{126}$$

where $\mathfrak{S}$ denotes the space spanned by the set of determinants Φ_i. The zero-order space is then multidimensional and quasi-degenerate perturbation theory must be employed. The hamiltonian matrix is first constructed within a space of low dimensionality, referred to as the "model" space, interactions with the complementary space are calculated perturbatively, and then the effective hamiltonian matrix is diagonalized. In the present section, we describe the quasi-degenerate Rayleigh-Schrödinger perturbation theory in a form which affords a suitable starting point for the development of the quasi-degenerate many-body perturbation theory.

The total molecular hamiltonian, with eigenproblem $\mathfrak{H}\, \Psi_i = E_i\, \Psi_i$, is written in the form

$$\mathfrak{H} = \mathfrak{H}_0 + \mathfrak{H}_1 \tag{127}$$

with

$$\mathfrak{H}_0\, \Phi_i = \mathrm{E}_i\, \Phi_i \tag{128}$$

The projector onto the chosen "model" space $\mathfrak{S}$ is

$$P = \sum_{i \in \mathfrak{S}} |\Phi_i\rangle\langle\Phi_i| \tag{129}$$

and its complement is

$$\begin{aligned} Q &= I - P \\ &= \sum_{i \notin \mathfrak{S}} |\Phi_i\rangle\langle\Phi_i| \end{aligned} \tag{130}$$

Now, for an exactly degenerate "model" space with energy E_m, the energy correction is defined by

$$\Delta \mathrm{E} = E - \mathrm{E}_m \tag{131}$$

where that E is the total non-relativistic energy of the system. For a quasi-degenerate model space, the zero-order operator, $\mathfrak{H}_0$, can be shifted by a diagonal operator of the form

$$\mathfrak{V} = \sum_{i \in \mathfrak{S}} |\Phi_i\rangle\, (\mathrm{E}_m - \mathrm{E}_i)\, \langle\Phi_i| \tag{132}$$

giving

$$\tilde{\mathfrak{H}}_0 = \mathfrak{H}_0 + \mathfrak{V} \tag{133}$$

The perturbation operator is modified as follows

$$\tilde{\mathfrak{H}}_1 = \mathfrak{H}_1 - \mathfrak{V} \tag{134}$$

so that the hamiltonian $\mathfrak{H}$ is recovered when the modified zero-order hamiltonian is added to the modified perturbation. The modified zero-order problem

$$\tilde{\mathfrak{H}}_0\, \Phi_i = \mathrm{E}_m\, \Phi_i \tag{135}$$

is exactly degenerate. The formalism presented in sections 5.1 - 5.4 can therefore be employed to treat the Rayleigh-Schrödinger perturbation series based on the modified operators.

6. Rayleigh-Schrödinger perturbation theory for non-orthogonal functions.

The total hamiltonian in the quantum mechanical eigenvalue problem whose solutions we seek to approximate by using perturbation theory expansions is always hermitian. In the previous sections of this chapter it has been assumed that the zero-order hamiltonian is also hermitian. In this section the formalism of Rayleigh-Schrodinger perturbation theory is generalized to the case in which the total hamiltonian is hermitian but is split into components which are not hermitian, that is

$$\mathfrak{H} = \mathfrak{H}_0 + \lambda\, \mathfrak{H}_1, \tag{136}$$

$$\mathfrak{H}_0 \neq \mathfrak{H}_0^{\dagger} \tag{137}$$

The eigenbras and eigenkets of $\mathfrak{H}_0$ must now be clearly distinguished

$$\mathfrak{H}_0\, |\Phi_i\rangle = E_m\, |\Phi_i\rangle, \quad i = 0,1,2,... \tag{138}$$

$$\langle\Phi^i|\, \mathfrak{H}_0 = E_m \langle\Phi^i|, \quad i = 0,1,2,... \tag{139}$$

where the eigenbras and eigenkets satisfy the biorthogonality condition

$$\langle\Phi^i|\Phi_j\rangle = \delta_{ij} \tag{140}$$

in which δ_{ij}, the Kronecker delta, is unity if the indices i and j coincide and zero otherwise.

Following the development given in section 2, the perturbed eigenproblem is now written

$$(\mathfrak{H}_0 + \lambda\, \mathfrak{H}_1)|\Psi_i(\lambda)\rangle = \varepsilon_i(\lambda)\ |\Psi_i(\lambda)\rangle,\ i=0,1,2... \tag{141}$$

and it is assumed that the wave function and the energy are continuous functions of the parameter λ. Expanding the wave function in powers of λ

$$\Psi_i(\lambda) = \Phi_i + \sum_{p=1}^{\infty} \lambda^p\, \chi_{ip} \tag{142}$$

and the energy also

$$E_i(\lambda) = E_i + \sum_{p=1}^{\infty} \lambda^p\, \varepsilon_{ip} \tag{143}$$

equating powers of λ leads to the basic equations of perturbation theory

$$\mathfrak{H}_0\, \chi_{ip} + \mathfrak{H}_1\, \chi_{ip-1} = E_i\, \chi_{ip} + \sum_{q=1}^{p-1} \varepsilon_{iq}\, \chi_{ip-q} + \varepsilon_{ip}\, \Phi_i \tag{144}$$

The intermediate normalization condition

$$\langle\Phi^i|\Psi_i\rangle = 1 \tag{145}$$

together with the biorthogonality condition (140) implies that

$$\langle\Phi^i|\chi_{ip}\rangle = 0,\ p=1,2,... \tag{146}$$

Thus multiplying equation (144) from the left by $<\Phi^i|$ and integrating over all the electronic coordinates gives immediately

$$\epsilon_{ip} = <\Phi^i | \mathfrak{H}_1 | \chi_{ip-1}> \tag{147}$$

which should be compared with equation (35).

The perturbed wave functions for the Rayleigh-Schrödinger perturbation expansion using non-hermitian zero-order hamiltonian can be obtained in exactly the same way as in the case of hermitian $\mathfrak{H}_0$ with the exception that the projection operator $\mathfrak{P}_i$ defined in equation (19) is now defined as

$$\mathfrak{P}_i = |\Phi_i> <\Phi^i| \tag{148}$$

This results in a change in the definition of the orthogonal projector $\mathfrak{Q}_i$ and the resolvant $\mathfrak{R}_i$ which, in the sum-over-states form becomes

$$\mathfrak{R}_i = \sum_{k \neq i} |\Phi_k> <\Phi^k| \,/\, (E_i - E_k) \tag{149}$$

By analogy with equations (87) and (88), the first- and second-order wave functions become

$$\chi_{i1} = \sum_{k \neq i} (E_i - E_k)^{-1} \, |\Phi_k> <\Phi^k | \mathfrak{H}_1 | \Phi_i> \tag{150}$$

and

$$\chi_{i2} = \sum_{k \neq i} \sum_{l \neq i} (E_i - E_k)^{-1} (E_i - E_l)^{-1} |\Phi_k> <\Phi^k | \mathfrak{H}_1 - \delta_{kl} \, \varepsilon_{i1} | \Phi_l> <\Phi^l | \mathfrak{H}_1 | \Phi_i> \tag{151}$$

respectively. For the second- and third- order energy coefficients, the corresponding sum-over-states formulae are

$$\varepsilon_{i2} = \sum_{k \neq i} (E_i - E_k)^{-1} <\Phi^i | \mathfrak{H}_1 | \Phi_k> <\Phi^k | \mathfrak{H}_1 | \Phi_i> \tag{152}$$

and

$$\varepsilon_{i3} = \sum_{k \neq i} \sum_{l \neq i} (E_i - E_k)^{-1} (E_i - E_l)^{-1} <\Phi^i | \mathfrak{H}_1 | \Phi_k> <\Phi^k | \mathfrak{H}_1 | \Phi_l> <\Phi^l | \mathfrak{H}_1 | \Phi_i> - \varepsilon_{i1} \sum_{k \neq i} (E_i - E_k)^{-2} <\Phi^i | \mathfrak{H}_1 | \Phi_k> <\Phi^k | \mathfrak{H}_1 | \Phi_i> \tag{153}$$

7. Lennard-Jones Brillouin Wigner perturbation theory.

7.1 Lennard-Jones Brillouin Wigner perturbation expansion.

The perturbation theory first proposed by Lennard-Jones, Brillouin and Wigner is formally much simpler than the Rayleigh-Schrödinger theory which has been described in this article. It does not form the basis of a useful "many-body" theory but it does provide an alternative derivation of the Rayleigh-Schrödinger expansion.

Taking the zero-order eigenproblem to be

$$\mathfrak{H}_0 \, \Phi_i = E_i \, \Phi_i, \quad i=0,1,2,... \tag{154}$$

and the perturbed problem to be

$$\mathfrak{H} \, \Psi_i = \mathcal{E}_i \, \Psi_i, \; i=0,1,2,... \tag{155}$$

with $\mathfrak{H} = \mathfrak{H}_0 + \lambda \, \mathfrak{H}_1$, the Lennard-Jones- Brillouin - Wigner expansion is obtained by expanding the wave function Ψ_i in powers of λ

$$|\Psi_i(\lambda)\rangle = |\Phi_i\rangle + \sum_{p=1}^{\infty} \lambda \, |\chi_{ip}\rangle \tag{156}$$

to obtain

$$(\mathcal{E}_i - \mathfrak{H}_0) \left(|\Phi_i\rangle + \sum_{p=1}^{\infty} \lambda \, |\chi_{ip}\rangle\right) = \lambda \, \mathfrak{H}_1 \left(|\Phi_i\rangle + \sum_{p=1}^{\infty} \lambda \, |\chi_{ip}\rangle\right) \tag{157}$$

Equating powers of λ leads to the following set of equations

$$(\mathcal{E}_i - \mathfrak{H}_0) \, |\Phi_i\rangle = 0 \tag{158a}$$

$$(\mathcal{E}_i - \mathfrak{H}_0) \, |\chi_{i1}\rangle = \mathfrak{H}_1 \, |\Phi_i\rangle \tag{158b}$$

$$(\mathcal{E}_i - \mathfrak{H}_0) \, |\chi_{i2}\rangle = \mathfrak{H}_1 \, |\chi_{i1}\rangle \tag{158c}$$

...

$$(\mathcal{E}_i - \mathfrak{H}_0) \, |\chi_{ip}\rangle = \mathfrak{H}_1 \, |\chi_{ip-1}\rangle \tag{158d}$$

and introducing the resolvant

$$R_i = \sum_{k \neq i} (\mathcal{E}_i - E_k)^{-1} \, |\Phi_k\rangle \langle\Phi_k| \tag{159}$$

we can write the perturbed wave functions as

$$|\chi_{i1}\rangle = R_i \, \mathfrak{H}_1 \, |\Phi_i\rangle \tag{160a}$$

$$|\chi_{i2}\rangle = R_i \, \mathfrak{H}_1 \, |\chi_{i1}\rangle = (R_i \, \mathfrak{H}_1)^2 \, |\Phi_i\rangle \tag{160b}$$

...

$$|\chi_{ip}\rangle = (R_i \, \mathfrak{H}_1)^p \, |\Phi_i\rangle \tag{160c}$$

Multiplying the perturbed eigenvalue equation from the left by $\langle\Phi_i|$, integrating over all space and using intermediate normalization, that is putting, $\langle\Phi_i|\Psi_i\rangle = 1$, yields

$$\mathcal{E}_i = \langle\Phi_i|\mathfrak{H}|\Psi_i\rangle \tag{161}$$

which on substitution of the perturbation series for the wave function gives the Lennard-Jones - Brillouin - Wigner expansion for the energy

$$E_i = <\Phi_i|\mathcal{H}_0|\Phi_i> + <\Phi_i|\mathcal{H}_1|\Phi_i> + <\Phi_i|\mathcal{H}_1\, R_i\, \mathcal{H}_1|\Phi_i> + <\Phi_i|\mathcal{H}_1(R_i\, \mathcal{H}_1)^2|\Phi_i> + \quad ... \quad + <\Phi_i|\mathcal{H}_1(R_i\, \mathcal{H}_1)^p|\Phi_i> + ... \qquad (162)$$

The energy has the form of a geometric series in powers of $(\mathfrak{R}_i\, \mathcal{H}_1)$. However, it is not a simple power series in powers of $\mathcal{H}_1$, in that $\mathfrak{R}_i$ depends on E_i. Thus, in a sum-over-states form the low order terms in the Lennard-Jones - Brillouin - Wigner expansion are

$$E_i = <\Phi_i|\mathcal{H}_0|\Phi_i> + <\Phi_i|\mathcal{H}_1|\Phi_i> + \sum_{k\neq i} <\Phi_i|\mathcal{H}_1|\Phi_k> <\Phi_k|\mathcal{H}_1|\Phi_i> (E_i - \mathrm{E}_k)^{-1} + \sum_{k\neq i}\sum_{l\neq i} <\Phi_i|\mathcal{H}_1|\Phi_k> <\Phi_k|\mathcal{H}_1|\Phi_l> <\Phi_l|\mathcal{H}_1|\Phi_i> (E_i - \mathrm{E}_k)^{-1} (E_i - \mathrm{E}_l)^{-1} + \quad ... \qquad (163)$$

which clearly depends on the exact energy, E_i, and can only be determined by an iterative procedure.

Because of the energy dependent denominators which occur in the perturbation series of Lennard-Jones, Brillouin and Wigner, it does not form the basis of a satisfactory many-body theory. However, it does afford an alternative approach to the Rayleigh-Schrödinger perturbation expansion as will be shown in the following section.

7.2 Derivation of the Rayleigh-Schrödinger perturbation theory

The Rayleigh-Schrödinger perturbation theory can be derived from the Lennard-Jones - Brillouin - Wigner perturbation theory by expanding the energy-dependent denominators which occur in the latter. If the following resolvant operators are defined

$$\mathfrak{R}_i = \sum_{k\neq i} (\mathrm{E}_i - \mathrm{E}_k)^{-1}\; |\Phi_k> <\Phi_k| \qquad (164)$$

and

$$R_i = \sum_{k\neq i} (E_i - \mathrm{E}_k)^{-1}\; |\Phi_k> <\Phi_k| \qquad (165)$$

then it is clear that by writing

$$E_i = \mathrm{E}_i + \Delta\mathrm{E}_i, \qquad (166)$$

$$\Delta\mathrm{E}_i = \sum_{p=1}^{\infty} \varepsilon_{ip} \qquad (167)$$

one can obtain the relation

$$R_i = \mathfrak{R}_i + \mathfrak{R}_i\ (-\Delta E_i)\ R_i \qquad (168)$$

or

$$R_i = \mathfrak{R}_i + \mathfrak{R}_i\ (-\Delta E_i)\ \mathfrak{R}_i + \mathfrak{R}_i\ (-\Delta E_i)\ \mathfrak{R}_i\ (-\Delta E_i)\ \mathfrak{R}_i + \ \ldots \qquad (169)$$

Since $\mathfrak{R}_i$ is the resolvant which occurs in Rayleigh-Schrödinger perturbation theory and R_i that which occurs in the Lennard-Jones - Brillouin - Wigner theory, this expansion can be used to derive the former from the latter.

Reference

[1] J.O. Hirschfelder and P.R. Certain, J. Chem. Phys. **60**, 1118 (1974)

THE PERTURBATION THEORY OF ELECTRON CORRELATION: II. MANY-BODY PERTURBATION THEORY AND THE ALGEBRAIC APPROXIMATION

Stephen Wilson

Rutherford Appleton Laboratory
Chilton, Oxfordshire, OX11 0QX, England

1. Introduction

In this article we consider the many-body perturbation theory and its formulation within the algebraic approximation, that is using finite sets of analytic basis functions.

The linked diagram theorem of the many-body perturbation theory was developed by Brueckner [1], Goldstone [2], Hugenholtz [3] and others [4] from the quantum field theory of Feynman [5], Dyson [6] and Wick [7]. It facilitates the subdivision of a many-electron system into a number of comparatively small subsystems which can be considered independent of each other. We shall not provide a proof of the central pillar of many-body perturbation theory, the linked diagram theorem, since this is given in many text books (see, for example, reference [4]). Instead we concentrate first on the physical content of the theory and then on the actual mechanics of its computational realization.

For the idealized system of, say, a large number of non-interacting helium atoms the subsystems, i.e. individual helium atoms, are *completely independent* and the level shift, ΔE, (that is, the correction to the zero-order energy) is just the sum of the subsystem energy shifts, ΔE_i,

$$\Delta E = \sum_i \Delta E_i \qquad (1)$$

Other expectation values can also be calculated as the sum of contributions from each of the subsystems. The energy and other expectation values are thus strictly proportional to the number of electrons in the system, N; $\Delta E \propto N$.

For an actual many-electron system, an array of interacting helium atoms say, the subsystems are, of course, not completely independent. However, the correlation effects are described by an expansion in terms, each of which may be associated with a particular linked diagram and each of which behaves as if it were an independent subsystem. As Brandow [8] has emphasized "The essential point is that the interactions between different clusters are *strictly negligible*, by definition." Any interaction between different diagrammatic terms is associated with higher order terms involving all particles in the original diagrams. These interactions can be either dynamical, that is involving the

Methods in Computational Molecular Physics, Edited by Stephen Wilson and Geerd H.F. Diercksen, Plenum Press, New York, 1992

fluctuation potential, or statistical, that is through the common set of intermediate states. Thus for an actual many-electron system, the energy and other expectation values are thus also strictly proportional to the number of electrons in the system.

The theoretical importance of the linked diagram theorem and the consequential linear scaling of the energy and other expectation values with the number of electrons in the system is widely recognized in theoretical chemistry [9-16] and the majority of contemporary treatments of the correlation problem in molecules are based on a perturbation expansion [17-18]. The sum-over-states expressions arising in each of the terms in the perturbation series can be written as matrix multiplications and, therefore, implemented with near optimal efficiency on vector processing computers [19-20]. The linked diagram theorem also has important computational consequences which are particularly relevant when parallel processing machines are employed [13, 14, 21]. The many-electron problem is decoupled into completely independent subproblems each of which can be handled on a separate processor. We shall return to this aspect in the fourth article.

In this article, we concentrate on low order many-body perturbation theory and its formulation in the algebraic approximation. In section 2, we discuss the many-body perturbation theory and it relation with the Rayleigh-Schrödinger perturbation expansion discussed in the first article. The diagrammatic formulation of the many-body perturbation expansion is surveyed in section 3 with diagrammatic conventions being defined in section 3.1 and then the second-, third- and fourth- order energy expressions being given in sections 3.2, 3.3 and 3.4, respectively. Spin integration is briefly discussed in section 3.5. In section 4, we turn our attention to the algebraic approximation providing a brief overview of: *(1)* basis set truncation errors in molecular calculations in section 4.1, *(2)* universal systematic sequences of even-tempered basis sets in section 4.2, *(3)* a discussion of linear dependence in the two-electron integral matrix in section 4.3, and *(4)* some notes on the Cholesky decomposition of the two-electron integral matrix in section 4.4. In section 5, we turn our attention to the formulation of the many-body perturbation theory within the algebraic approximation discussing: *(1)* a model problem in section 5.1, *(2)* some diatomic molecule calculations in section 5.2, and *(3)* the preliminary results of a study of the contribution of fourth order triple excitation diagrams to the beryllium ground state in section 5.3.

Before describing the practical details of calculations based on many-body perturbation theory, it is useful to make three points about the method which are not, perhaps, widely appreciated:

(1) Many-body perturbation theory does not afford a useful approximation for the wave function. "To describe an extended system accurately we must know the wave function, not only locally, but over the entire region with equal accuracy and this effectively requires us to go to infinite order. But to determine the expectation value of a local operator (and calculation of the energy is that of finding the expectation value of a finite range two-body potential) we need to know accurately only the wave function in the immediate neighbourhood ($\sim$(range)^3) of the potential)" [4].

(2) The convergence of the many-body perturbation expansion is not always critical. "Even when the series does not converge, the first approximation obtained by means of it is usually fairly accurate." [22]. The sum of higher order terms may be small even though the terms themselves may be individually quite large. Divergent perturbation series, such as the anharmonic oscillator, having a radius of convergence of zero can be accurately summed using, for example, Padé approximants [23, 24].

(3) The equivalence of many-body perturbation theory formulated within the algebraic approximation [25, 26] and the Møller-Plesset perturbation theory [17, 27] is sometimes noted. Whilst this equivalence holds up to a point - the actual energy expressions in each order of perturbation theory are identical - in one very important respect the methods are radically different. The Møller-Plesset perturbation theory is a *first* quantized form of Rayleigh-Schrödinger perturbation theory with respect to a

Hartree-Fock reference. Many-body perturbation theory is *second* quantized. The cancellation of unphysical, non-linear terms is a feature of the many-body formalism whereas in the Møller-Plesset approach the "unlinked diagrams" must be canceled "by hand". In molecular electronic structure studies the use of the second quantized many-body perturbation theory is clearly preferable in studies of extended systems and mandatory in relativistic studies where one is dealing with an "infinitely many-bodied" problem.

2. Many-body perturbation theory

The many-body perturbation expansion is obtained by explicitly cancelling terms which would correspond to unlinked diagrams in each order of the Rayleigh-Schrödinger perturbation series.

For the second-order and third-order energy coefficients in Rayleigh-Schrödinger perturbation theory for a single perturbation, the sum-over-states formulae are

$$\varepsilon_{i2} = \sum_{k\neq i} (E_i - E_k)^{-1} <\Phi_i|\mathfrak{H}_1|\Phi_k> <\Phi_k|\mathfrak{H}_1|\Phi_i> \qquad (2)$$

and

$$\varepsilon_{i3} = \sum_{k\neq i}\sum_{l\neq i} (E_i - E_k)^{-1} (E_i - E_l)^{-1} <\Phi_i|\mathfrak{H}_1|\Phi_k> <\Phi_k|\mathfrak{H}_1|\Phi_l> <\Phi_l|\mathfrak{H}_1|\Phi_i> - \varepsilon_{i1} \sum_{k\neq i} (E_i - E_k)^{-2} <\Phi_i|\mathfrak{H}_1|\Phi_k> <\Phi_k|\mathfrak{H}_1|\Phi_i> \qquad (3)$$

When the single determinant many-electron functions Φ_p are constructed from canonical Hartree-Fock orbitals, the excited functions, Φ_k and Φ_l, are doubly excited with respect to the reference function, Φ_i, and it can be shown that the second term in the third order energy expression cancels diagonal components of the first term. The surviving terms may be associated with linked diagrams.

For the fourth-order energy coefficients in Rayleigh-Schrödinger perturbation theory for a single perturbation, the sum-over-states formula is

$$\varepsilon_{i4} = \sum_{k\neq i}\sum_{l\neq i}\sum_{m\neq i} (E_i - E_k)^{-1} (E_i - E_l)^{-1} (E_i - E_m)^{-1} <\Phi_i|\mathfrak{H}_1|\Phi_k> <\Phi_k|\mathfrak{H}_1|\Phi_l> <\Phi_k|\mathfrak{H}_1|\Phi_m> <\Phi_m|\mathfrak{H}_1|\Phi_i>$$

$$- 2\varepsilon_{i1} \sum_{k\neq i}\sum_{l\neq i} (E_i - E_k)^{-2} (E_i - E_l)^{-1} <\Phi_i|\mathfrak{H}_1|\Phi_k> <\Phi_k|\mathfrak{H}_1|\Phi_l> <\Phi_l|\mathfrak{H}_1|\Phi_i>$$

$$+ \varepsilon_{i1}^2 \sum_{k\neq i} (E_i - E_k)^{-3} <\Phi_i|\mathfrak{H}_1|\Phi_k> <\Phi_k|\mathfrak{H}_1|\Phi_i>$$

$$- \varepsilon_{i2} \sum_{k\neq i} (E_i - E_k)^{-2} <\Phi_i|\mathfrak{H}_1|\Phi_k> <\Phi_k|\mathfrak{H}_1|\Phi_i> \qquad (4)$$

Again certain diagonal components of the first term are canceled by the second, third and fourth terms. The surviving terms are again associated with linked diagrams.

3. Diagrammatic perturbation theory

3.1 Diagrammatic conventions

We use the indices i,j,k,... to label single particle state functions lying below the Fermi level (i.e. occupied spin orbitals or hole states) and the indices a,b,c,... to label single particle state functions lying above the Fermi level (i.e. unoccupied spin orbitals or particle states). p,q,r,... are used to label arbitrary single particle functions. The corresponding upper case indices are used to label the spatial orbitals.

The rules for translating the diagrams into algebraic expressions are as follows:-

(a) Each downward directed line is labelled with a unique "hole" line index: i,j,k,...

(b) Each upward directed line is labelled with a unique "particle" line index: a,b,c,...

(c) There is a summation over each unique hole and each unique particle index, covering all possible values of these indices.

(d) The numerator of the summand consists of a product of n one- and/or two-electron integrals in n order. The possible types of one-electron integrals which can arise are summarized in Figure 1. The two-electron integrals occur in antisymmetrized form

$$\langle pq|\hat{O}|rs\rangle = \iint dr_1\, dr_2\, \phi_p(r_1)\, \phi_q(r_2)\, r_{12}^{-1}\, \{\phi_r(r_1)\, \phi_s(r_2) - \phi_s(r_1)\, \phi_r(r_2)\} \quad (5)$$

There is an integral of this type for every interaction line, i.e. dashed horizontal line. The spin orbital indices p, q, r, s can be read from the diagram and correspond to the hole and particle lines entering or leaving the interaction in the order: left-out, right-out, left-in, right-in, respectively. The types of two-electron interactions which can arise are summarized in Figure 2.

(e) The denominator of the summand consists of a product of factors of the form

$$D_{ij...ab...} = (\varepsilon_i + \varepsilon_j + ...) - (\varepsilon_a + \varepsilon_b + ...) \quad (6)$$

where the first summation is over all hole lines that extend between adjacent interactions and the second summation is over all particle lines that also extend between the two adjacent interactions. ε_p denotes the one-electron orbital energy. There are n-1 factors of this type corresponding to an *nth* order diagram such that there is a denominator factor arising between adjacent interactions in the diagram.

(f) For each pair of equivalent lines there is a multiplicative factor of $\frac{1}{2}$. An equivalent pair of lines is defined as two lines, beginning at one interaction and ending at another and both going in the same direction.

*(g)*There is a multiplicative factor of $(-1)^p$ where

$$p = h + l \quad (7)$$

where h is the number of unique hole lines and l is the number of fermion loops. A fermion loop is determined by following the hole and particle lines in the direction of the arrows to form a continuous loop.

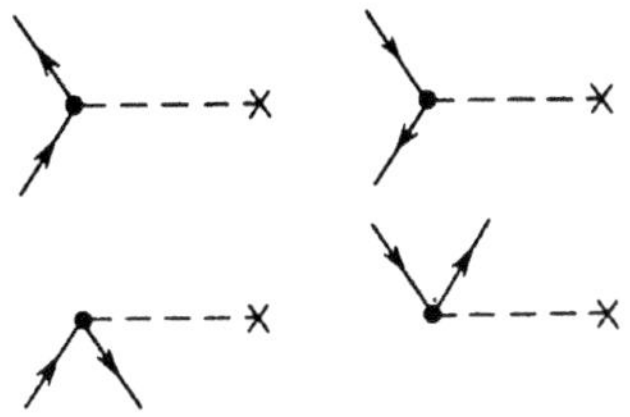

Figure 1. *Types of one-electron terms which can arise in diagrams*

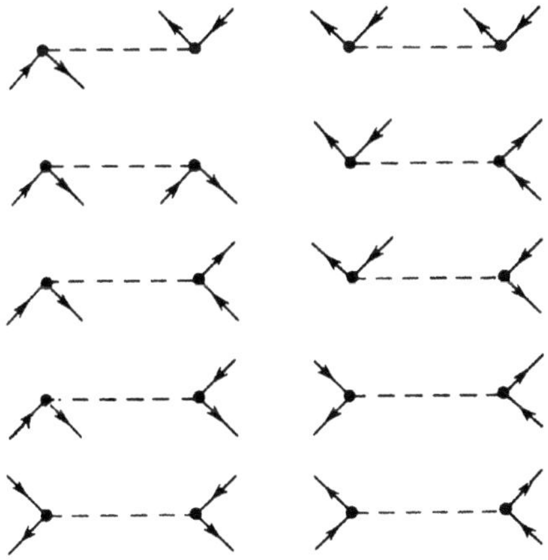

Figure 2. *Types of two-electron terms which can arise in diagrams*

3.2 Second-order energy

The only second order diagram which arises when a canonical Hartree-Fock reference function is used to describe a closed-shell system is shown in Figure 3. The diagram contains two interaction lines. The first describes the excitation of a pair of electrons creating hole states i and j below the Fermi level and particle states a and b above it. The second interaction describes the return to the vacuum state. The diagram contains two pairs of equivalent lines and two fermion loops. The corresponding algebraic expression can easily be shown to be

$$E_2(A_d) = \frac{1}{4} \sum_{ij} \sum_{ab} \frac{\langle ij|\hat{O}|ab\rangle \langle ab|\hat{O}|ij\rangle}{D_{ijab}} \qquad (8)$$

3.3 Third-order energy

The three third order diagrams which arise when a canonical Hartree-Fock reference function is used to describe a closed-shell system are shown in Figure 4. Each diagram describes a different double excitation process. All three diagrams describe the excitation of a pair of electrons creating hole states i and j below the Fermi level and particle states a and b above it. The third order diagrams differ in the nature of the second interaction which may involve a hole-hole interaction, a hole-particle interaction or a particle-particle interaction before the final interaction returns the system to the vacuum state.

3.3.1 Third-order hole-hole energy

The third-order energy component corresponding to the diagram with a hole-hole central interaction line takes the form:

$$E_3(A_d) = \frac{1}{8} \sum_{\substack{ijkl \\ ab}} \frac{\langle ij|\hat{O}|ab\rangle \langle kl|\hat{O}|ij\rangle \langle ab|\hat{O}|kl\rangle}{D_{ijab}\, D_{klab}} \qquad (9)$$

There are three pairs of equivalent lines and two fermion loops.

3.3.2 Third-order hole-particle energy

The third-order energy component corresponding to the diagram with a hole-particle central interaction line takes the form:

$$E_3(B_d) = \sum_{\substack{ijk \\ abc}} \frac{\langle ij|\hat{O}|ab\rangle \langle ak|\hat{O}|ic\rangle \langle bc|\hat{O}|jk\rangle}{D_{ijab}\, D_{jkbc}} \qquad (10)$$

There are no equivalent pairs of lines and two fermion loops.

3.3.3 Third-order particle-particle energy

The third-order energy component corresponding to the diagram with a particle-particle central interaction line takes the form:

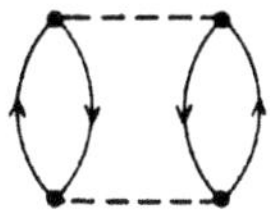

Figure 3. *Second order energy diagram.*

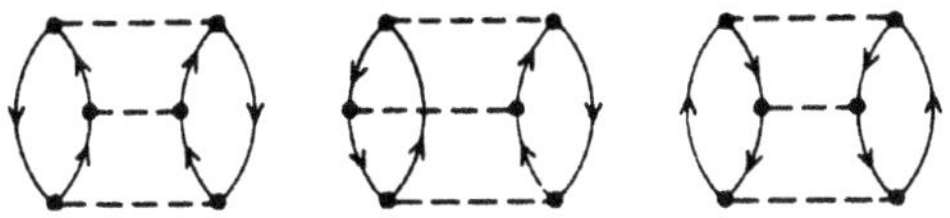

Figure 4. *Third order energy diagrams*

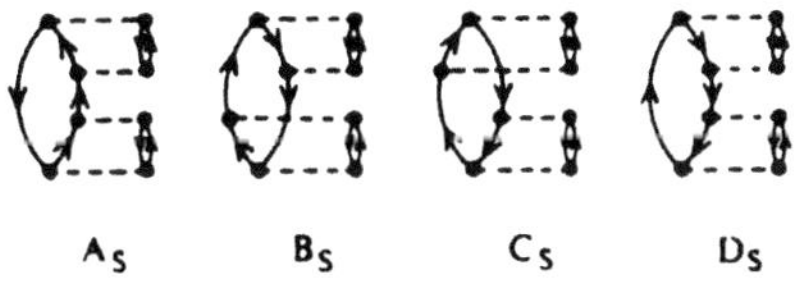

Figure 5. *Fourth order energy diagrams involving singly excited intermediate states*

$$E_3(C_d) = \frac{1}{8} \sum_{\substack{ij \\ abcd}} \frac{<ij|\hat{O}|ab> <ab|\hat{O}|cd> <cd|\hat{O}|ij>}{D_{ijab} D_{ijcd}} \tag{11}$$

There are three pairs of equivalent lines and two fermion loops.

3.4 Fourth-order energy

Fourth order diagrams contain four interaction lines. In all diagrams, the first interaction describes the excitation of a pair of electrons creating hole states i and j below the Fermi level and particle states a and b above it. The second interaction leads to a state which may be either singly, doubly, triply or quadruply excited with respect to the vacuum state. The third interaction sees the return to a doubly excited state before the fourth and final interaction returns the system to the vacuum state.

It is convenient to classify the fourth order energy components according the level of excitation achieved after the second interaction.

3.4.1 Fourth-order single excitation energy components

There are four fourth order diagrams which involve a singly excited intermediate state after the second interaction. They are shown in Figure 5. The corresponding energy expression can be shown to be as follows :-

$$E_4(A_s) = \frac{1}{4} \sum_{\substack{ijk \\ abcde}} \frac{<ij|\hat{O}|ab> <ab|\hat{O}|cj> <ck|\hat{O}|de> <de|\hat{O}|ik>}{D_{ijab} D_{ic} D_{ikde}}$$

$$E_4(B_s) = -\frac{1}{4} \sum_{\substack{ijkl \\ abcd}} \frac{<ij|\hat{O}|ab> <ab|\hat{O}|cj> <kl|\hat{O}|id> <cd|\hat{O}|kl>}{D_{ijab} D_{ic} D_{klcd}}$$

$$E_4(C_s) = -\frac{1}{4} \sum_{\substack{ijkl \\ abcd}} \frac{<ij|\hat{O}|ab> <kb|\hat{O}|ij> <al|\hat{O}|cd> <cd|\hat{O}|kl>}{D_{ijab} D_{ka} D_{klcd}}$$

$$E_4(D_s) = \frac{1}{4} \sum_{\substack{ijkl \\ abc}} \frac{<i|\hat{O}|ab> <kb|\hat{O}|ij> <lm|\hat{O}|kc> <ac|\hat{O}|lm>}{D_{ijab} D_{ka} D_{lmac}} \tag{12}$$

3.4.2 Fourth-order double excitation energy components

There are twelve fourth order diagrams which involve a doubly excited intermediate state after the second interaction. They are shown in Figure 6. The corresponding energy expression can be shown to be as follows :-

$$E_4(A_d) = \frac{1}{16} \sum_{\substack{ij \\ abcdef}} \frac{\langle ij|\hat{O}|ab\rangle \langle ab|\hat{O}|cd\rangle \langle cd|\hat{O}|ef\rangle \langle ef|\hat{O}|ij\rangle}{D_{ijab}\, D_{ijcd}\, D_{ijef}}$$

$$E_4(B_d) = \frac{1}{16} \sum_{\substack{ijkl \\ abcd}} \frac{\langle ij|\hat{O}|ab\rangle \langle ab|\hat{O}|cd\rangle \langle kl|\hat{O}|ij\rangle \langle cd|\hat{O}|kl\rangle}{D_{ijab}\, D_{ijcd}\, D_{klcd}}$$

$$E_4(C_d) = \frac{1}{16} \sum_{\substack{ijkl \\ abcd}} \frac{\langle ij|\hat{O}|ab\rangle \langle kl|\hat{O}|ij\rangle \langle ab|\hat{O}|cd\rangle \langle cd|\hat{O}|kl\rangle}{D_{ijab}\, D_{klab}\, D_{klcd}}$$

$$E_4(D_d) = \frac{1}{16} \sum_{\substack{ijklmn \\ ab}} \frac{\langle ij|\hat{O}|ab\rangle \langle kl|\hat{O}|ij\rangle \langle mn|\hat{O}|kl\rangle \langle ab|\hat{O}|mn\rangle}{D_{ijab}\, D_{klab}\, D_{mnab}}$$

$$E_4(E_d) = -\frac{1}{2} \sum_{\substack{ijk \\ abcde}} \frac{\langle ij|\hat{O}|ab\rangle \langle ab|\hat{O}|cd\rangle \langle kd|\hat{O}|ie\rangle \langle ce|\hat{O}|kj\rangle}{D_{ijab}\, D_{ijcd}\, D_{jkce}}$$

$$E_4(F_d) = -\frac{1}{2} \sum_{\substack{ijk \\ abcde}} \frac{\langle ij|\hat{O}|ab\rangle \langle kb|\hat{O}|ic\rangle \langle ac|\hat{O}|de\rangle \langle de|\hat{O}|kj\rangle}{D_{ijab}\, D_{jkac}\, D_{jkde}}$$

$$E_4(G_d) = -\frac{1}{2} \sum_{\substack{ijk \\ abcde}} \frac{\langle ij|\hat{O}|ab\rangle \langle kl|\hat{O}|ij\rangle \langle am|\hat{O}|cl\rangle \langle cb|\hat{O}|km\rangle}{D_{ijab}\, D_{klab}\, D_{kmbc}}$$

$$E_4(H_d) = -\frac{1}{2} \sum_{\substack{ijk \\ abcdc}} \frac{\langle ij|\hat{O}|ab\rangle \langle ak|\hat{O}|cj\rangle \langle lm|\hat{O}|ik\rangle \langle cb|\hat{O}|lm\rangle}{D_{ijab}\, D_{ikbc}\, D_{lmbc}}$$

$$E_4(I_d) = \sum_{\substack{ijkl \\ abcd}} \frac{\langle ij|\hat{O}|ab\rangle \langle ak|\hat{O}|cj\rangle \langle cl|\hat{O}|dk\rangle \langle db|\hat{O}|il\rangle}{D_{ijab}\, D_{ikbc}\, D_{ilbd}}$$

$$E_4(J_d) = -\sum_{\substack{ijkl \\ abcd}} \frac{\langle ij|\hat{O}|ab\rangle \langle kb|\hat{O}|cj\rangle \langle cl|\hat{O}|id\rangle \langle ad|\hat{O}|kl\rangle}{D_{ijab}\, D_{ikac}\, D_{klad}}$$

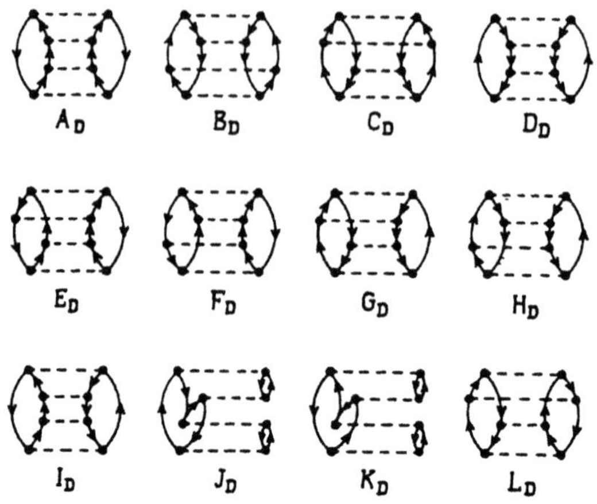

Figure 6. *Fourth order energy diagrams involving doubly excited intermediate states*

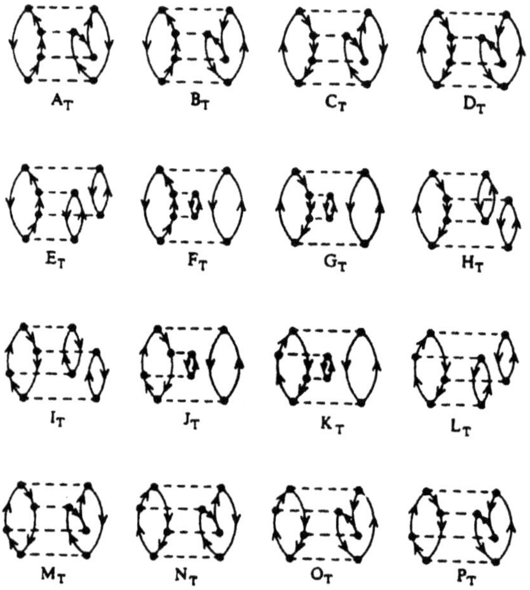

Figure 7. *Fourth order energy diagrams involving triply excited intermediate states*

$$E_4(K_d) = - \sum_{\substack{ijkl\\abcd}} \frac{\langle ij|\hat{O}|ab\rangle \langle kb|\hat{O}|cj\rangle \langle al|\hat{O}|kd\rangle \langle cd|\hat{O}|il\rangle}{D_{ijab}\, D_{ikac}\, D_{ilcd}}$$

$$E_4(L_d) = - \sum_{\substack{ijkl\\abcd}} \frac{\langle ij|\hat{O}|ab\rangle \langle kb|\hat{O}|ic\rangle \langle al|\hat{O}|dj\rangle \langle dc|\hat{O}|kl\rangle}{D_{ijab}\, D_{jkac}\, D_{klcd}} \tag{13}$$

3.4.3 Fourth-order triple excitation energy components

There are sixteen fourth order diagrams which involve a triply excited intermediate state after the second interaction. They are shown in Figure 7. The corresponding energy expression can be shown to be as follows :-

$$E_4(A_t) = -\frac{1}{2} \sum_{ijk} \sum_{abcde} \frac{\langle ij|\hat{O}|ab\rangle \langle ak|\hat{O}|cd\rangle \langle cb|\hat{O}|ek\rangle \langle ed|\hat{O}|ij\rangle}{D_{ijab}\, D_{ijkbcd}\, D_{ijde}}$$

$$E_4(B_t) = -\frac{1}{2} \sum_{ijk} \sum_{abcde} \frac{\langle ij|\hat{O}|ab\rangle \langle ak|\hat{O}|cd\rangle \langle cd|\hat{O}|ej\rangle \langle eb|\hat{O}|ik\rangle}{D_{ijab}\, D_{ijkbcd}\, D_{ijbe}}$$

$$E_4(C_t) = -\frac{1}{2} \sum_{ijklm} \sum_{abc} \frac{\langle ij|\hat{O}|ab\rangle \langle kl|\hat{O}|ic\rangle \langle mb|\hat{O}|kl\rangle \langle ac|\hat{O}|mj\rangle}{D_{ijab}\, D_{jklabc}\, D_{jmac}}$$

$$E_4(D_t) = -\frac{1}{2} \sum_{ijklm} \sum_{abc} \frac{\langle ij|\hat{O}|ab\rangle \langle kl|\hat{O}|ic\rangle \langle mc|\hat{O}|kj\rangle \langle ab|\hat{O}|ml\rangle}{D_{ijab}\, D_{jklabc}\, D_{jmab}}$$

$$E_4(E_t) = \sum_{ijk} \sum_{abcde} \frac{\langle ij|\hat{O}|ab\rangle \langle ak|\hat{O}|cd\rangle \langle cb|\hat{O}|ej\rangle \langle ed|\hat{O}|ik\rangle}{D_{ijab}\, D_{ijkbcd}\, D_{ikde}}$$

$$E_4(F_t) = \frac{1}{4} \sum_{ijk} \sum_{abcde} \frac{\langle ij|\hat{O}|ab\rangle \langle ak|\hat{O}|cd\rangle \langle cd|\hat{O}|ek\rangle \langle eb|\hat{O}|ij\rangle}{D_{ijab}\, D_{ijkbcd}\, D_{ijbe}}$$

$$E_4(G_t) = \frac{1}{4} \sum_{ijklm} \sum_{abc} \frac{\langle ij|\hat{O}|ab\rangle \langle kl|\hat{O}|ic\rangle \langle mc|\hat{O}|kl\rangle \langle ab|\hat{O}|mj\rangle}{D_{ijab}\, D_{jklabc}\, D_{jmab}}$$

$$E_4(H_t) = \sum_{ijklm} \sum_{abc} \frac{\langle ij|\hat{O}|ab\rangle \langle kl|\hat{O}|ic\rangle \langle mb|\hat{O}|kj\rangle \langle ac|\hat{O}|ml\rangle}{D_{ijab}\, D_{ijkbcd}\, D_{lmac}}$$

$$E_4(I_t) = -\frac{1}{4}\sum_{ijkl}\sum_{abcd} \frac{\langle ij|\hat{O}|ab\rangle \langle kl|\hat{O}|ic\rangle \langle ab|\hat{O}|dj\rangle \langle dc|\hat{O}|kl\rangle}{D_{ijab}\, D_{jklabc}\, D_{klcd}}$$

$$E_4(J_t) = -\sum_{ijkl}\sum_{abcd} \frac{\langle ij|\hat{O}|ab\rangle \langle kl|\hat{O}|ic\rangle \langle ac|\hat{O}|dl\rangle \langle db|\hat{O}|kj\rangle}{D_{ijab}\, D_{jklabc}\, D_{jkbd}}$$

$$E_4(K_t) = -\sum_{ijkl}\sum_{abcd} \frac{\langle ij|\hat{O}|ab\rangle \langle kl|\hat{O}|ic\rangle \langle ac|\hat{O}|dl\rangle \langle db|\hat{O}|kj\rangle}{D_{ijab}\, D_{jklabc}\, D_{jkbd}}$$

$$E_4(L_t) = -\frac{1}{4}\sum_{ijkl}\sum_{abcd} \frac{\langle ij|\hat{O}|ab\rangle \langle kl|\hat{O}|ic\rangle \langle ab|\hat{O}|dj\rangle \langle dc|\hat{O}|kl\rangle}{D_{ijab}\, D_{jklabc}\, D_{klcd}}$$

$$E_4(M_t) = \frac{1}{2}\sum_{ijkl}\sum_{abcd} \frac{\langle ij|\hat{O}|ab\rangle \langle kl|\hat{O}|ic\rangle \langle ab|\hat{O}|dl\rangle \langle dc|\hat{O}|kj\rangle}{D_{ijab}\, D_{jklabc}\, D_{jkcd}}$$

$$E_4(N_t) = \frac{1}{2}\sum_{ijkl}\sum_{abcd} \frac{\langle ij|\hat{O}|ab\rangle \langle kl|\hat{O}|ic\rangle \langle ab|\hat{O}|dl\rangle \langle dc|\hat{O}|kj\rangle}{D_{ijab}\, D_{jklabc}\, D_{jkcd}}$$

$$E_4(O_t) = \frac{1}{2}\sum_{ijkl}\sum_{abcd} \frac{\langle ij|\hat{O}|ab\rangle \langle kl|\hat{O}|ic\rangle \langle ac|\hat{O}|dj\rangle \langle db|\hat{O}|kl\rangle}{D_{ijab}\, D_{jklabc}\, D_{klbd}}$$

$$E_4(P_t) = \frac{1}{2}\sum_{ijkl}\sum_{abcd} \frac{\langle ij|\hat{O}|ab\rangle \langle ak|\hat{O}|cd\rangle \langle ld|\hat{O}|cj\rangle \langle cb|\hat{O}|lk\rangle}{D_{ijab}\, D_{ijkbcd}\, D_{klbc}} \qquad (14)$$

3.4.4 Fourth-order quadruple excitation energy components

There are seven fourth order diagrams which involve a quadruply excited intermediate state after the second interaction. They are shown in Figure 8. The corresponding energy expression can be shown to be as follows :-

$$E_4(A_q) = \sum_{\substack{ijkl\\abcd}} \frac{\langle ij|\hat{O}|ab\rangle \langle kl|\hat{O}|cd\rangle \langle cb|\hat{O}|il\rangle \langle ad|\hat{O}|kj\rangle}{D_{ijab}\, D_{ijklabcd}\, D_{jkad}}$$

$$E_4(B_q) = \frac{1}{16}\sum_{\substack{ijkl\\abcd}} \frac{\langle ij|\hat{O}|ab\rangle \langle kl|\hat{O}|cd\rangle \langle cd|\hat{O}|ij\rangle \langle ab|\hat{O}|kl\rangle}{D_{ijab}\, D_{ijklabcd}\, D_{klab}}$$

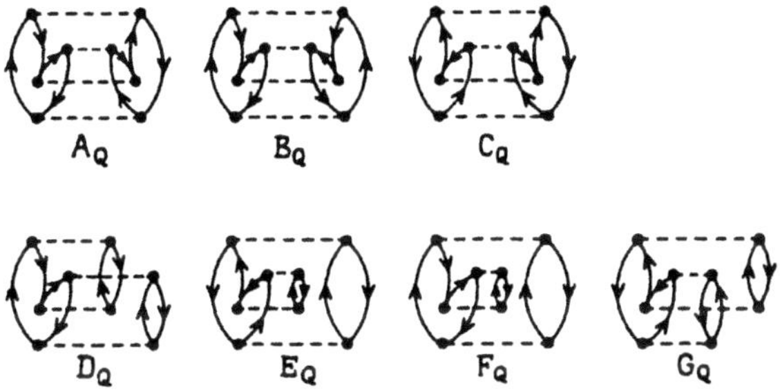

Figure 8. *Fourth order energy diagrams involving quadruply excited intermediate states*

$$E_4(C_q) = \frac{1}{16}\sum_{\substack{ijkl\\abcd}} \frac{\langle ij|\hat{O}|ab\rangle \langle kl|\hat{O}|cd\rangle \langle ab|\hat{O}|kl\rangle \langle cd|\hat{O}|ij\rangle}{D_{ijab}\; D_{ijklabcd}\; D_{ijcd}}$$

$$E_4(D_q) = -\frac{1}{4}\sum_{\substack{ijkl\\abcd}} \frac{\langle ij|\hat{O}|ab\rangle \langle kl|\hat{O}|ij\rangle \langle cb|\hat{O}|ij\rangle \langle ad|\hat{O}|kl\rangle}{D_{ijab}\; D_{ijklabcd}\; D_{klad}}$$

$$E_4(E_q) = -\frac{1}{4}\sum_{\substack{ijkl\\abcd}} \frac{\langle ij|\hat{O}|ab\rangle \langle kl|\hat{O}|cd\rangle \langle ad|\hat{O}|kl\rangle \langle cb|\hat{O}|ij\rangle}{D_{ijab}\; D_{ijklabcd}\; D_{ijbc}}$$

$$E_4(F_q) = -\frac{1}{4}\sum_{\substack{ijkl\\abcd}} \frac{\langle ij|\hat{O}|ab\rangle \langle kl|\hat{O}|cd\rangle \langle cd|\hat{O}|il\rangle \langle ab|\hat{O}|kj\rangle}{D_{ijab}\; D_{ijklabcd}\; D_{jkab}}$$

$$E_4(G_q) = -\frac{1}{4}\sum_{\substack{ijkl\\abcd}} \frac{\langle ij|\hat{O}|ab\rangle \langle kl|\hat{O}|cd\rangle \langle ab|\hat{O}|kj\rangle \langle cd|\hat{O}|il\rangle}{D_{ijab}\; D_{ijklabcd}\; D_{ilcd}}$$

(15)

Examination of the numerators in the expressions arising from diagrams B_q and C_q reveals that they contain the same two-electron integrals but in a different order. The denominators in the expressions arising from diagrams B_q and C_q can be added and manipulated as follows:

$$\frac{1}{D_{ijab}\; D_{ijklabcd}\; D_{klab}} + \frac{1}{D_{ijab}\; D_{ijklabcd}\; D_{ijcd}}$$

$$= \frac{1}{D_{ijab}\; D_{ijklabcd}}\left[\frac{1}{D_{klab}} + \frac{1}{D_{ijcd}}\right]$$

$$= \frac{1}{D_{ijab}\; D_{ijklabcd}}\left[\frac{D_{ijcd} + D_{klab}}{D_{klab}\; D_{ijcd}}\right]$$

$$= \frac{1}{D_{ijab}\; D_{klab}\; D_{ijcd}} \qquad (16)$$

Analogous results can be obtained by adding the expression corresponding to diagram A_q to itself, by adding the expressions corresponding to diagrams D_q and E_q and the expressions corresponding to F_q and G_q. The resulting energy expressions may then be cast in the form

$$E_4(A_q) = \frac{1}{2}\sum_{\substack{ijkl\\abcd}} \frac{<ij|\hat{O}|ab> <kl|\hat{O}|cd> <cb|\hat{O}|il> <ad|\hat{O}|kj>}{D_{ijab}\ D_{ilbc}\ D_{jkad}}$$

$$E_4(B_q+C_q) = \frac{1}{16}\sum_{\substack{ijkl\\abcd}} \frac{<ij|\hat{O}|ab> <kl|\hat{O}|cd> <cd|\hat{O}|ij> <ab|\hat{O}|kl>}{D_{ijab}\ D_{ijcd}\ D_{klab}}$$

$$E_4(D_q+E_q) = -\frac{1}{4}\sum_{\substack{ijkl\\abcd}} \frac{<ij|\hat{O}|ab> <kl|\hat{O}|cd> <cb|\hat{O}|ij> <ad|\hat{O}|kl>}{D_{ijab}\ D_{ijbc}\ D_{klad}}$$

$$E_4(F_q+G_q) = -\frac{1}{4}\sum_{\substack{ijkl\\abcd}} \frac{<ij|\hat{O}|ab> <kl|\hat{O}|cd> <cd|\hat{O}|il> <ab|\hat{O}|kj>}{D_{ijab}\ D_{jkab}\ D_{ilcd}} \qquad (17)$$

3.5 Spin integration

Computationally efficient schemes for evaluating the energy coefficients in any non-relativistic many-body perturbation theory calculation depend on explicitly carrying out the spin integrations. For the two-electron integral we have

$$<pq|\hat{O}|rs> = <PQ|r_{12}^{-1}|RS>\ \delta(\sigma_p,\sigma_r)\ \delta(\sigma_q,\sigma_s)$$

$$- <PQ|r_{12}^{-1}|SR>\ \delta(\sigma_p,\sigma_s)\ \delta(\sigma_q,\sigma_r) \qquad (18)$$

where δ is the Kronecker delta. The amount of computation required to evaluate the energy expressions given above can be significantly reduced by carrying out the spin integrations. We shall return to this point in the fourth article.

4. The algebraic approximation

Fundamental to the application of many-body perturbation theory to molecular systems is the algebraic approximation. In section 4.1, we discuss the nature of basis set truncation errors in atoms and molecular electronic structure calculations. In section 4.2, we discuss one approach to the reduction of this error, the universal systematic sequence of even-tempered basis sets. Sections 4.3 and 4.4 describe how the use of such sequences of basis set leads to a degree of linear dependence in the corresponding matrix of two-electron integrals and the exploitation of this fact by means of a Cholesky decomposition.

4.1 Basis set truncation errors

Huzinaga [28] has emphasized the *"quasi-empirical"* nature of the vast majority of contemporary *ab initio* electronic structure calculations associated with the error arising from basis set truncation. A similar point of view has been expressed in a review by Davidson and Feller [29] who comment that *"the smaller the basis the more ab initio calculations assume an empirical flavor"*. Clementi and his coworkers [30] observe that, because of the failure to address the problem of basis set construction in the 1970s, *"the chemical literature ... was flooded with computations which essentially compared experiments with basis set superposition error data"*. Even today, basis set truncation errors are a major and frequently dominant source of error in molecular electronic structure calculations. However, as long ago as 1963, Schwartz [31] emphasized the need to devise systematic schemes for extending basis sets employed in atomic and molecular electronic structure calculations and developing *"an orderly plan of attack"* which *"is ... essential for any mathematical analysis of the convergence rate. The old habit of picking the "best" (chosen by art) choice of a fixed number of terms is to be discarded if one wants to see how the problem converges"*. In more recent years, this point of view has been underlined by Ruedenberg and his coworkers and by Wilson and his coworkers. Following early work by Reeves [32], Ruedenberg and his colleagues [33-39] introduced the even-tempered basis sets which virtually eliminated at a stoke the tedious exponent optimization that had been the most time-consuming ingredient of many of the pioneering electronic structure calculations carried out in the 1960s. With the introduction of the algebraic approximation into powerful techniques based on many-body perturbation theory [25, 26] for handling the electron correlation problem in atoms and molecules, the basis set truncation error was seen to be the major source of error [40]. (A similar conclusion has been reached from more recently reported full configuration interaction calculations [41].) The need to use large flexible basis set in order to obtain high accuracy became apparent and the concept of the universal basis set was introduced [42-46]. These large and flexible basis sets are not only capable of yielding high accuracy but are also transferable from system to system without regard to the nuclear charges or molecular environment involved. The investigation of the convergence of calculated properties of atoms and molecules with increasing size of basis set was realized in practice in the work of Feller and Ruedenberg [47] and Schmidt and Ruedenberg [48], and ultimately led to the concept of a universal systematic sequence of even-tempered basis sets [49, 50]. Clementi and his coworkers [51, 52] have emphasized the simplifications which arise in evaluation of the two-electron integrals with universal even-tempered (or geometric) basis sets are employed. Wilson [53] has presented a scheme based on the Cholesky decomposition of the two-electron integral matrix which allows the effective generation, storage and processing of the integrals associated with a universal even-tempered basis set. The scheme become increasingly efficient as the size of the basis set is increased. Huzinaga [28] has stated that *"these remarkable universal geometric basis sets ... will certainly usher us to a new plateau of computational chemistry"*. More detailed reviews have been provided by the present author [54-56].

There appears to be a widely held belief (see, for example, Schaefer [57]) that finite difference approximations to atomic and molecular wave functions are superior to their finite basis set counterparts. Finite difference methods have, until recently, been the method of choice in atomic structure calculations [58, 59]. Furthermore, the finite difference approach has been applied to a range of diatomic molecules containing light atoms [60-64]. Numerical Hartree-Fock calculations for atoms and for diatomic molecules have been shown to afford sub-μhartree level of accuracy for total electronic energies. Over the past decade, it has also been demonstrated that finite basis set expansions can provide equivalent or superior approximations for Hartree-Fock wave functions for atoms and diatomic molecules. By using systematic sequences of even-tempered basis sets of either Gaussian-type functions [47, 48] or exponential-type functions [65-67], atomic self-consistent field energies which are, in fact, often more accurate than those obtained by finite difference techniques can be obtained. For diatomic molecules, the accuracy achieved in fully numerical self-consistent field calculations can be matched by calculations performed within the algebraic approximation when elliptical basis functions are employed [68-70]. Basis set expansions provide a compact representation of self-consistent field wave functions. Quiney *et al* [71] point out that *"The underlying reason for the success of the finite difference methods lies in the choice of coordinate system for the special cases in which they have found favour"*.

Systematically constructed sequences of basis sets can also be employed with advantage in studies of the effects of electron correlation in atomic and molecular systems. Use of basis sets yields a square-integrable representation of the complete spectrum and affords an effective quadrature over the continuum. Studies of a simple model problem (a hydrogenic atom perturbed by increasing the nuclear charge) using the techniques of many-body perturbation theory have established that integration over the continuum states can be carried out more accurately using systematically extended basis sets than by the numerical approach traditionally employed in atomic structure calculations. In relativistic electronic structure studies the use of the algebraic approximation has not only opened up the possibility of applications to molecules but has also provided representation of the negative energy branch of the spectrum which is require if one is to go beyond the so-called "no virtual pair" approximation which has been employed even in atomic structure calculations carried out by employing finite differences. Finally, the computational advantages of the algebraic approximation should be emphasized since it leads algorithms with basic linear algebra kernels which can be executed with great efficiency on vector processing machines and parallel processing computers.

For molecular systems without special symmetry properties, the use of atomic basis functions transfer a local coordinate system to the region where the wave function is most rapidly changing whereas the finite difference approach must, in practice, employ a Cartesian grid, which is a manifestly unsuitable coordinate system especially in the region of the nuclei.

4.2 Universal systematic sequences of even-tempered basis sets.

The orbital exponents, $\zeta_k(N)$ in an even-tempered basis set are taken to form a geometric progression, thus

$$\ln \zeta_k(N) = \ln \alpha(N) + k \ln \beta(N), \quad k = 1,2,\ldots,N \qquad (19)$$

where both $\alpha(N)$ and $\beta(N)$ depend on N, the number of functions in the basis set, as, therefore, do the orbital exponents. For the basis set to approach completeness as $N \to \infty$, α and β must be functions of N such that

$$\alpha(N) \to 0, \quad \beta(N) \to 1, \quad \beta(N)^N \to \infty \qquad (20)$$

Ruedenberg *et al.* [47, 48] have suggested empirical functional forms which ensure that these limits are satisfied. In particular, they suggested the forms

$$\ln \alpha(N) = a_0 \ln (\beta(N) - 1) + a_1 \qquad (21)$$

and

$$\ln \ln \beta(N) = b_0 \ln N + b_1 \qquad (22)$$

where a_0, a_1, b_0 and b_1 are constant for a given sequence. Basis sets generated according to these relation are termed systematic sequences of even-tempered basis sets. In the work of Ruedenberg *et al.* [47, 48], these constants are taken to be different for different atomic species. If they are taken to be independent of the particular system then this is termed a universal systematic sequence of even-tempered basis sets [49, 50].

The set of Gaussian-type basis functions $\exp(-\zeta_k(N)r^2)$ generated in this fashion can be shown to be asymptotically total in $L^2(R^+)$ in the limit of large N since the generalized Muntz-Szasz theorem [72-74].

$$\lim_{N\to\infty} \sum \zeta_k(N) \;/\; (1 + (\zeta_k(N))^2) = \infty \qquad (23)$$

holds. Note that for fixed α and β, the series of partial sums

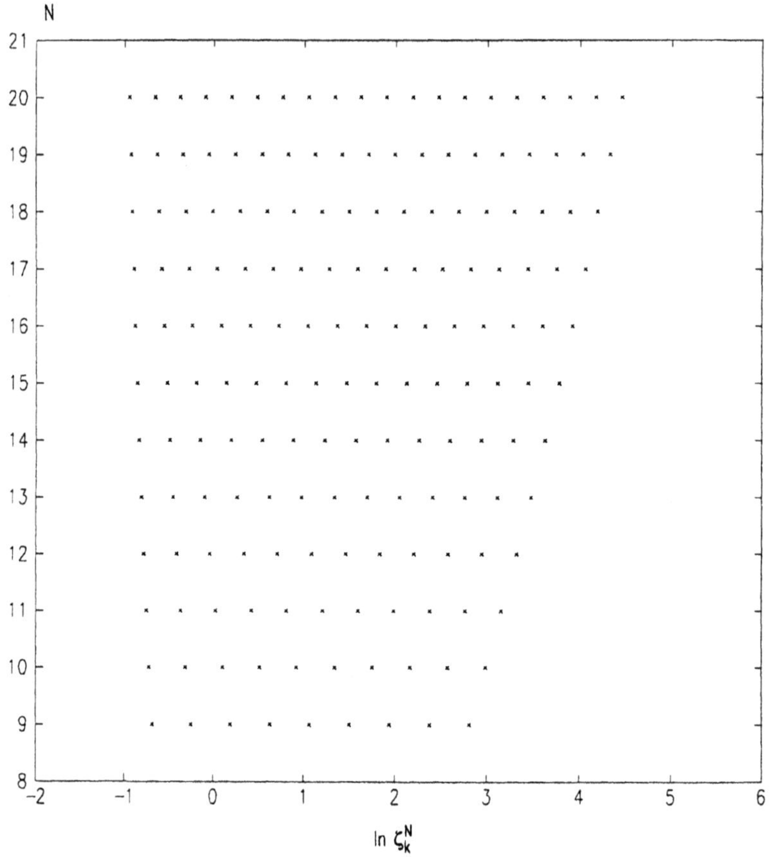

Figure 9. *Logarithm of the orbital exponents generated by using a universal systematic sequence of even-tempered basis functions. Each row of crosses in this figure corresponds to a given basis set in the sequence and is labelled by N, the number of functions it contains. Each of the crosses in a given row corresponds to an orbital exponent. (Based on the work of Wilson* [76]*).*

$$\mathfrak{S}(N) = \sum \zeta_k(N) \,/\, (1 + (\zeta_k(N))^2) \tag{24}$$

converge to a finite limit.

It is easily demonstrated that for fixed α and β this limit is finite. In Figure 9, the logarithms of the orbital exponents obtained with fixed α and β are compared with those generated by means of the recursions given above. Each of the rows in this Figure corresponds to a basis set in the sequence and is labelled by n, the number of functions it contains. It should be noted that for the exponents generated by the above recursions that the range of the exponents increases with n, the lower limit decreasing to zero and the upper limit becoming infinite. Furthermore, the distribution of the exponents between these two limits becomes increasingly dense as n is increased.

4.3 Linear dependence in the two-electron integral matrix

Now the two-electron integrals take the form

$$\int\int dr_1\, dr_2\, \varphi_i(r_1)\, \varphi_k(r_2)\, 1/r_{12}\, \varphi_j(r_1)\, \varphi_l(r_2) \tag{25}$$

where r_{12} is the interelectron distance and φ_k is a basis function. These integrals involve two charge distributions of the form

$$\rho_{jk}(r) = \varphi_j(r)\, \varphi_k(r) \tag{26}$$

In the case of an even-tempered basis set of exponential-type functions centred on the same point, these charge distributions assume the form

$$\exp(\zeta_j\, r) \exp(\zeta_k\, r) = \exp(-\{\zeta_j + \zeta_k\}\, r) \tag{27}$$

In Figure 8, the logarithm of the exponent arising in each of the charge distributions, $ln\ \{\zeta_j + \zeta_k\}$, is plotted for the same universal systematic sequence of even-tempered basis sets that was employed in the generation of Figure 9. Each row in Figure 10 corresponds to a particular basis set in the sequence and is labelled by the number of functions, N, it contains. Figure 10 suggests a great deal of linear dependence in the two-electron integral matrix. It should noted that whilst the exponents for the basis functions range from $\alpha\beta$ to $\alpha\beta^n$, the exponents for the charge distributions range from $2\alpha\beta$ to $2\alpha\beta^n$. However, there are n(n+1)/2 charge distribution exponents compared with only n orbital exponents, which implies that n(n-1)/2 of the former are redundant.

If the charge distribution exponents are defined as follows

$$\xi_{jk}(n) = \zeta_j(n) + \zeta_k(n) \tag{28}$$

then the set of charge distributions can be shown to be asymptotically complete in the limit $n \longrightarrow \infty$ since the generalized Muntz-Szasz theorem again holds

$$\lim_{n\to\infty} \sum \xi_{jk}(n)/(1 + \{\xi_{jk}(n)\}^2) = \infty \tag{29}$$

(n(n-1))/2 of the charge distribution exponents can be deleted and asymptotic completeness is maintained. One choice is to retain those distributions for which j=k giving

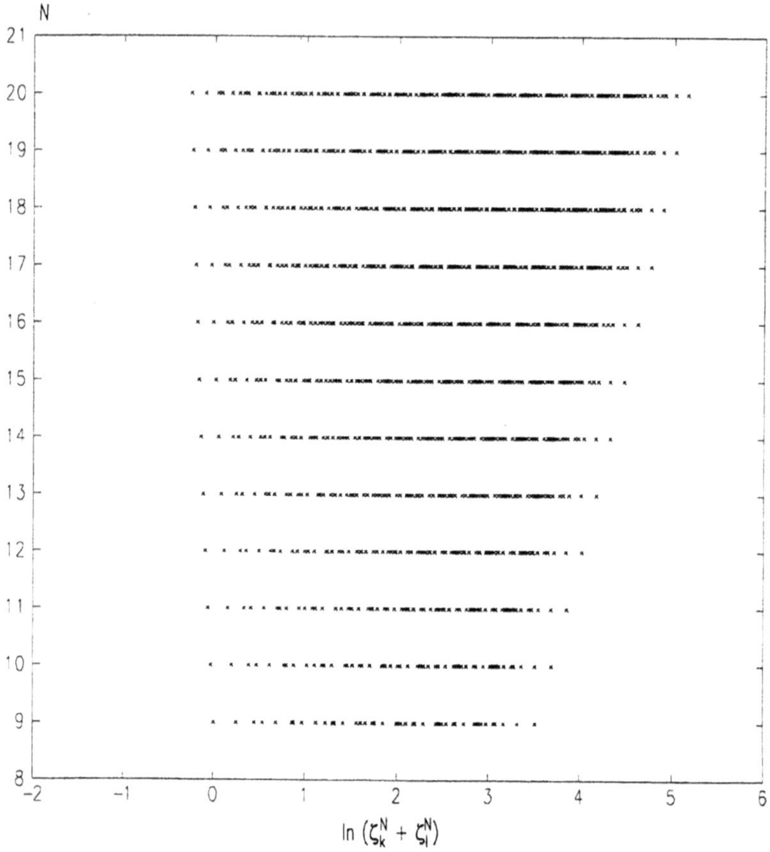

Figure 10. *Logarithm of the charge distribution exponents generated by using a universal systematic sequence of even-tempered basis functions. Each of the rows of crosses in this figure corresponds to a given basis set in the systematic sequence and is labelled by N, the number of functions is contains. Each of the crosses in a given row corresponds to a charge distribution exponent generated from the orbital exponents used in the construction of Figure 9. (Based on the work of Wilson* [76]*).*

$$\lim_{n\to\infty} \sum \xi_{jj}(n)/(1+\{\xi_{jj}(n)\}^2) =$$

$$\lim_{n\to\infty} \sum 2\zeta_j(n)/(1+4\{\zeta_j(n)\}^2) = \infty \quad (30)$$

4.4 Cholesky decomposition of the two-electron integral matrix

The two-electron integrals may be arranged in a symmetric, positive definite matrix of dimension (n(n+1))/2

$$V_{ij,kl} \qquad i\geq j \quad k\geq l \quad (ij)\geq(kl) \quad (31)$$

The computational linear dependence amongst the charge distributions involved in the two-electron integral matrix can be exploited by a Cholesky decomposition [75 - 77] in which the matrix $V_{ij,kl}$ is written in the form

$$V_{ij,kl} = \sum_{mn} L_{ij,mn} \; L^{\dagger}_{mn,kl} \quad (32)$$

where **L** is a lower triangular matrix and $\mathbf{L}^{\dagger}$ is its transpose. If linear dependence is present in **V** then the summation index mn will run for considerably less than its full range 1 to (n(n+1))/2. An approximation to **V** may be written

$$V'_{ij,kl} = \sum_{mn}^{\nu} L_{ij,mn} \; L^{\dagger}_{mn,kl} \quad (33)$$

where ν is the effective numerical rank of the two-electron integral matrix.

In Figure 11, the effective numerical rank of the two-electron integral matrix is shown for a universal systematic sequence of one-centre even-tempered basis sets of Gaussian-type functions of s symmetry. The upper curve shows the full rank of the two-electron integral matrix, that is

$$m = (n(n+1))/2 \quad (34)$$

as a function of the size of the basis set on a logarithmic scale. The lower curves show the effective numerical rank, ν, when a tolerance of 10^{-t}, t= 4,5,6,7,8,9,10 is demanded.

The total number of two-electron integrals arising in a calculation employing a basis set of n functions is

$$M = (m(m+1))/2 \quad (35)$$

The number of elements of **L** which have to be stored for a two-electron integral matrix of effective numerical rank ν is

$$N = M - (m - \nu)(m - \nu + 1)/2 \quad (36)$$

and the percentage storage requirements are thus

$$P = [N/M] \times 100\% \quad (37)$$

For a tolerance of 10^{-8}, P is shown as a function of the number of basis functions, n, in Table 1. It can be seen that as the basis set is systematically extended the percentage of the integral matrix that is actually required decreases monotonically. For a basis set of twenty functions the integral storage requirements are reduced to less than 25% of the length of the original integral list. The larger the basis set the more accurate is the calculation and the greater then percentage reduction in the computer storage demands.

Linear dependence in the two-electron integral matrix can be exploited in both non-relativistic and relativistic electronic structure calculations. It can reduce the number of integrals that have to be evaluated and stored in the integral evaluation phase of the

Table 1

Comparison of the total number of two-electron integrals, M, which arise in a calculation with n basis functions, with the number of elements, N, of the lower triangular matrix L required for a tolerance of 10^{-8}. *P* is the percentage storage requirements given by [N/M] x 100%

n	N	M	*P*
9	657	1035	63.5%
10	910	1540	59.1%
11	1176	2211	53.2%
12	1485	3081	48.2%
13	1840	4186	44.0%
14	2162	5565	38.8%
15	2604	7260	35.9%
16	3100	9316	33.3%
17	3653	11781	31.0%
18	4121	14706	28.0%
19	4779	18145	26.3%
20	5319	22155	24.0%

calculation. It can reduce the number of integrals that have to be processed in subsequent self-consistent field and many-body perturbation theory calculations.

The most demanding phase of a low order many-body perturbation theory calculation is the four-index transformation of the two-electron integrals

$$V_{pq,rs} = \sum_{ijkl} c_{pi}\, c_{qj}\, c_{rk}\, c_{sl}\, V_{ij,kl} \tag{38}$$

which is most efficiently performed as a series of partial transformations

$$V1_{ij,ks} = \sum_{l} c_{sl}\, V_{ij,kl} \tag{39}$$

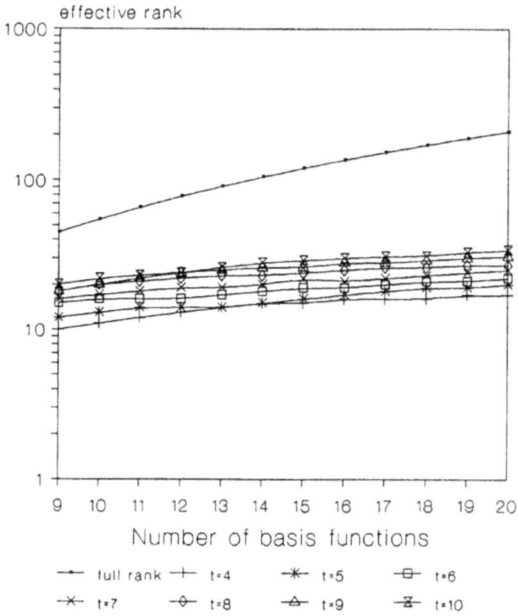

Figure 11. *Effective numerical rank of the two-electron integral matrix for a one-centre even-tempered basis set of Gaussian-type functions of s symmetry. (Based on the work of Wilson* [77]*).*

$$V2_{ij,rs} = \sum_{k} c_{rk} \, V1_{ij,ks} \tag{40}$$

$$V3_{iq,rs} = \sum_{j} c_{qj} \, V2_{ij,rs} \tag{41}$$

$$V_{pq,rs} = \sum_{i} c_{pi} \, V3_{iq,rs} \tag{42}$$

Introducing a Cholesky decomposition of the two-electron integral matrix, we can write the four-index transformation in the form

$$V_{pq,rs} = \sum_{mn} \sum_{ijkl} c_{pi} \, c_{qj} \, c_{rk} \, c_{sl} \, L_{ij,mn} \, L_{kl,mn} \tag{43}$$

Exploiting the computational linear dependence of the two-electron integral matrix restricts the summation over mn

$$V'_{pq,rs} = \sum_{mn}^{v} \sum_{ijkl} c_{pi} \, c_{qj} \, c_{rk} \, c_{sl} \, L_{ij,mn} \, L_{kl,mn} \tag{44}$$

which may be written in the form

$$V'_{pq,rs} = \sum_{mn}^{v} L_{pq,mn} \, L_{rs,mn} \tag{45}$$

$L_{pq,mn}$ is given by the two-index transformation

$$L_{pq,mn} = \sum_{ij} c_{pi} \, c_{qj} \, L_{ij,mn} \tag{46}$$

the computational demands of which scale as n^3.

5. Many-body perturbation theory and the algebraic approximation

5.1 A model problem

A study [78] of a simple model problem consisting of a hydrogenic atom with nuclear charge Z perturbed by increasing the nuclear charge to (Z+z) demonstrates quite clearly the accuracy that may be achieved within the algebraic approximation. For this model it is possible to evaluate the energy component corresponding to each diagram analytically using a technique due to Dalgarno and Lewis [79]. Although the total energy coefficient in each order beyond second order is identically zero, the components corresponding to each of the diagrams are non-zero. For example, in third order it can be shown that

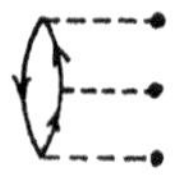

$$\Rightarrow \quad -0.75 \, z^3/Z \tag{47a}$$

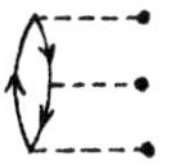

$$\Rightarrow \quad +0.75\, z^3/Z \qquad (47b)$$

Hence, for this model it is possible to investigate the dependence of each diagrammatic component of the perturbation expansion on the basis set employed. The results of such an investigation using a universal systematic sequence of even-tempered basis sets of exponential-type functions is presented in Table 2, where it can be seen that each component does indeed converge to the corresponding exact value. It is also of interest to compare the accuracy obtained in the calculation using basis sets with that achieved with numerical methods. It can be seen in Table 2 that the former approach is, in fact, more accurate. The reader is referred to references [78] and [80] for a discussion of the problems associated with the numerical approach in this context.

Table 2

Comparison of diagrammatic perturbation theory energy components obtained by using numerical methods with those obtained by employing basis set expansions of exponential type functions [1].

Energy components		Exact	Numerical method	Basis set expansion
(i)	Z=1, z=1			
	E_2	-0.5	-0.4996	-0.5000
	E_3(p)	-0.75	-0.7492	-0.7500
	E_3(pp)	-1.125	-1.1236	-1.1250
(ii)	Z=4, z=1			
	E_2	-0.5	-0.4987	-0.5000
	E_3(p)	-0.1875	-0.1866	-0.1875
	E_3(pp)	-0.0703125	-0.0698	-0.0703125

[1] based on results presented in references [78] and [80].

5.2 Diatomic molecule calculations

A number of calculations on diatomic molecules have demonstrated that quite high precision can be achieved by exploiting universal basis sets of exponential-type and Gaussian-type function in many-body perturbation theory calculations for diatomic molecules. Recent work using universal basis sets of elliptical functions also show promise. The interested reader is referred elsewhere for a review of this work [81].

5.3 Triple excitations in the beryllium ground state

The accuracy of the most complete theoretical calculations for helium-like systems often surpasses that of experiment. However, for few electron systems, such as the beryllium ground state, attaining an accuracy of μhartree or even 10 μhartree is not a trivial problem since many effects which are often neglected give contributions which are actually many times larger than the desired accuracy. A detailed comparison of theoretical data for the beryllium ground state with data derived from experiment has recently been presented by Mårtensson-Pendrill *et al.* [82]. They observe that the availability of a reliable calculation of the correlation energy component associated with triply excited intermediate states would result in a "very accurate estimate of the total correlation energy."

We have seen that the correlation energy component associated with triply excited intermediate states first arise in the many-body perturbation theory expansion at fourth order. They are particularly demanding from a computational standpoint leading to an algorithm which scales as the seventh power of the number of basis functions employed. Recent work, which will be described in more detail in the fourth article, has shown that the additive decomposition of correlation energy afforded by the linked diagram expansion of many-body perturbation theory leads to a very efficient algorithm for the determination of these terms, which can be effectively implemented on parallel processing machines such as the CRAY Y-MP/8.

Table 3

Convergence of the energy components associated with diagrams C_t and H_t with increasing size of basis set. Energies are given in μhartree.

Basis set	C_t	H_t
6s	+22.9762	-20.9638
8s	+29.1943	-27.2687
10s	+31.6814	-29.7751
12s	+32.0093	-30.0958
14s	+32.0077	-30.1017
16s	+32.0165	-30.1067
18s	+32.0190	-30.1102
20s	+32.0197	-30.1109
22s	+32.0194	-30.1105
24s	+32.0195	-30.1107
26s	+32.0196	-30.1108
28s	+32.0197	-30.1109
30s	+32.0198	-30.1110

Furthermore, for the beryllium ground state it is possible to perform a detailed angular momentum analysis of the fourth order triple excitation energy component [83]. For the beryllium ground state the occupied single particle state functions have s symmetry and this imposes restrictions on the symmetries of the unoccupied single particle state functions. For example, the energy components corresponding to diagrams C_t and H_t depend exclusively on basis functions of s symmetry.

We have performed calculations of the fourth order triple excitation energy components using systematic sequences of even-tempered basis sets of Gaussian-type functions. The energy components corresponding to diagrams C_t and H_t are displayed in Table 3 as a function of the size of the basis set employed. For both of these energy components sub-μhartree can be achieved with basis containing 12 or more functions. It appears that an accuracy of 10^{-10} can be approached without evidence of numerical instability of the results.

References

[1] K.A. Brueckner, Phys. Rev. **100**, 36 (1955)
[2] J. Goldstone, Proc. Roy. Soc. (London) A 239, 267 (1957)
[3] N.M. Hugenholtz, Physica 23, 481 (1957)
[4] N.H. March, W.H. Young and S. Sampanthar, *The many body problem in quantum mechanics*, Cambridge University Press, 1967
[5] R. Feynman, Phys. Rev. **76**, 769 (1949)
[6] F.J. Dyson, Phys. Rev. **75**, 1736 (1949)
[7] G.C. Wick, Phys. Rev. **80**, 268 (1950)
[8] B.H. Brandow, Rev. Mod. Phys. **39**, 771 (1967)
[9] P. Čársky and M. Urban, *Ab initio Calculations. Methods and Applications in Chemistry* (Springer, Berlin, 1980)
[10] S. Wilson, Specialist Periodical Reports: Theoretical Chemistry 4 1 (1981)
[11] R.J. Bartlett, Ann. Rev. Phys. Chem. **32**, 359 (1981)
[12] I. Lindgren and J. Morrison, *Atomic Many-Body Theory* (Springer, Berlin, 1982)
[13] S. Wilson, in: *Methods in Computational Molecular Physics*, eds. G.H.F. Diercksen and S. Wilson, (Reidel, Dordrecht, 1983)
[14] S. Wilson, Comput. Phys. Reports **2**, 389 (1985)
[15] K. Jankowski, Meth. Comput. Chem. 1, 1 (1987)
[16] M. Urban, I. Cernusak, V. Kello and J. Noga, Meth. Comput. Chem 1, 117 (1987)
[17] J.A. Pople, R. Krishnan, H.B. Schlegel and J.S. Binkley, Intern. J. Quantum Chem. 14 545 (1978)
[18] P.J. Knowles, K. Somasundram, N.C. Handy and K. Hirao, Chem. Phys. Lett. **113** 8 (1985)
[19] S. Wilson and V.R. Saunders, Comput. Phys. Commun. **19**, 293 (1980)
[20] M.F. Guest and S. Wilson, in: *Supercomputers in Chemistry*, ed. P. Lykos and I. Shavitt (American Chemical Society, Washington D.C., 1981) p.1
[21] S. Wilson, in *Proceedings of Fifth Seminar on Computational Problems in Quantum Chemistry*, Groningen, eds. P.Th. van Duijknen and W.C. Nieuwpoort, 1981
[22] P.A.M. Dirac, *The Principles of Quantum Mechanics*, Clarendon Press, Oxford, 4th edition, 1958
[23] B. Simon, Ann. Phys. 58, 76 (1970)
[24] C.M. Bender and T.T. Wu, Phys. Rev. **184**, 1231 (1969)
[25] U. Kaldor, Phys. Rev. A7, 427 (1973)
[26] S. Wilson and D.M. Silver, Phys. Rev. A14, 1949 (1976)
[27] C. Møller and M.S. Plesset, Phys. Rev. **46**, 618 (1934)
[28] S. Huzinaga, Comput. Phys. Repts. **2**, 279 (1985)
[29] E.R. Davidson and D. Feller, Chem. Rev. **86**, 681 (1986)
[30] S.J. Chakravorty, G. Corongiu, J.R. Flores, V. Sonnad, E. Clementi, V. Carravetta and I. Cacelli, *in* Modern Techniques in Computational Chemistry, *p.* 81,, ESCOM, Leiden, (1989)
[31] C. Schwartz, Meth. Comput. Phys. 2, 241 (1963)
[32] C.M. Reeves, J. Chem. Phys. 39, 11 (1963)
[33] K. Ruedenberg, R.C. Raffenetti and R. Bardo, *in* Energy, Structure and Reactivity, (Proc. 2nd Boulder Res. Conf. Theor. Chem.), Wiley, New York (1973)
[34] R.C. Raffenetti, J. Chem. Phys. 58, 4452, (1973)
[35] R.C. Raffenetti, J. Chem. Phys. 59, 5936, (1973)
[36] R.C. Raffenetti and K. Ruedenberg, J. Chem. Phys. 59, 5978, (1973)
[37] R. Bardo and K. Ruedenberg, J. Chem. Phys. 59, 5956 (1973)
[38] R. Bardo and K. Ruedenberg, J. Chem. Phys. 59, 5966 (1973)
[39] R. Bardo and K. Ruedenberg,J. Chem. Phys. 60, 918 (1974)
[40] S. Wilson and D.M. Silver, J. Chem. Phys. **66**, 5400 (1977)
[41] C.W. Bauschlicher and P.R. Taylor, J. Chem. Phys. 85, 2779 (1986)

[42] D.M. Silver, S. Wilson and W.C. Nieuwpoort, Intern. J. Quantum Chem. **14**, 635 (1978)
[43] D.M. Silver and S. Wilson, J. Chem. Phys. **69**, 3787 (1978)
[44] D.M. Silver and W.C. Nieuwpoort, Chem. Phys. Lett. **57**, 421 (1978)
[45] S. Wilson and D.M. Silver, J. Chem. Phys. **72**, 2159 (1980)
[46] S. Wilson and D.M. Silver, J. Chem. Phys. **77**, 3674 (1982)
[47] D.F. Feller and K. Ruedenberg, Theoret. chim. Acta. **52**, 231 (1979)
[48] M.W. Schmidt and K. Ruedenberg, J. Chem. Phys. **71**, 3951 (1979)
[49] S. Wilson, Theoret. chim. Acta. **57**, 53 (1980)
[50] S. Wilson, Theoret. chim. Acta. **58**, 31 (1980)
[51] E. Clementi, G. Corongiu, M. Gratarola, P. Habitz, C. Lupo, P. Otto and D. Vercauteren, Intern. J. Quantum Chem. Symp. **16**, 409 (1982)
[52] E. Clementi, G. Corongiu and S. Chakravorty, *in* Modern Techniques in Computational Chemistry, *p.* 343, ESCOM, Leiden, (1990)
[53] S. Wilson, Comput. Phys. Commun. **58**, 71 (1990)
[54] S. Wilson, Electron correlation in molecules, Clarendon Press, Oxford, (1984)
[55] S. Wilson, *in* Methods in Computational Molecular Physics, edited by G.H.F. Diercksen and S. Wilson, Reidel, Dordrecht, (1983)
[56] S. Wilson, Adv. Chem. Phys. **67**, 439 (1987)
[57] H.F. Schaefer III, Quantum Chemistry - The development of *ab initio* methods in molecular electronic structure theory, Clarendon Press, Oxford, (1984)
[58] C.F. Fischer, The Hartree-Fock Method for Atoms, Wiley, New York (1977)
[59] I.P. Grant, Meth. Comput. Chem. **2**, 1 (1988)
[60] L. Laaksonen, P. Pyykkö and D. Sundholm, Chem. Phys. Lett. **96**, 1 (1983)
[61] L. Laaksonen, P. Pyykkö and D. Sundholm, Intern. J. Quantum Chem. **23**, 309 (1983)
[62] L. Laaksonen, P. Pyykkö and D. Sundholm, Intern. J. Quantum Chem. **23**, 319 (1983)
[63] L. Laaksonen, P. Pyykkö and D. Sundholm, Comput. Phys. Repts. **4**, 313 (1986)
[64] P. Pyykkö, *in* Numerical Determination of the Electronic Structure of Atoms, Diatomic and Polyatomic Molecules, edited by M. Defranceschi and J. Delhalle, Kluwer, Dordrecht, (1989)
[65] D.L. Cooper and S. Wilson, J. Chem. Phys. **77**, 5053 (1982)
[66] D.L. Cooper and S. Wilson, J. Chem. Phys. **77**, 4551 (1982)
[67] D.L. Cooper and S. Wilson, J. Chem. Phys. **78**, 2456 (1983)
[68] B.H. Wells and S. Wilson, J. Phys. B: At. Mol. Phys. **18**, L731 (1985)
[69] B.H. Wells and S. Wilson, J. Phys. B: At. Mol. Phys. **19**, 17 (1986)
[70] J.W. Thompson and S. Wilson, J. Phys. B: At. Mol. Phys. **23**, 2205 (1990)
[71] H.M. Quiney, I.P. Grant and S. Wilson, *in* Numerical Determination of the Electronic Structure of Atoms, Diatomic and Polyatomic Molecules, edited by M. Defranceschi and J. Delhalle, Kluwer, Dordrecht, (1989)
[72] H. Muntz, Festschrift H.A. Schwartz, 303 (1914)
[73] O. Szasz, Math. Ann. 77 (1926)
[74] B.J. Klahn, J. Chem. Phys. **83**, 5749 (1985)
[75] N.H.F. Beebe and J. Linderberg, Intern. J. Quantum Chem. **12**, 683
[76] S. Wilson, Comput. Phys. Comm. **58**, 71 (1990)
[77] S. Wilson, *in* The Effects of Relativity in Atoms, Molecules and the Solid State, edited by S. Wilson, I.P. Grant and B.L. Gyorffy, Plenum Press (1991)
[78] H.M. Quiney, I.P. Grant and S. Wilson, J. Phys. B: At. Mol. Phys. **18**, 577 (1985)
[79] A. Dalgarno and J.T. Lewis, Proc. Roy. Soc. A **233**, 70 (1955)
[80] P.J. Rossky and M. Karplus, J. Chem. Phys. **67**, 5419 (1977)
[81] S. Wilson, Adv. Chem. Phys. **67**, 439 (1987)
[82] A.-M. Mårtensson-Pendrill, S.A. Alexander, L. Adamowicz, N. Oliphant, J. Olsen, P. Oster, H.M. Quiney, S. Salomonson and D. Sundholm, Phys. Rev. A**43**, 3355 (1991)
[83] A.-M. Mårtensson-Pendrill and S. Wilson *(work in progress)*

THE PERTURBATION THEORY OF ELECTRON CORRELATION. III. RELATIVISTIC MANY-BODY PERTURBATION THEORY

Stephen Wilson

Rutherford Appleton Laboratory
Chilton, Oxfordshire, OX11 0QX, England

1. Introduction

For systems which contain heavy atoms, the non-relativistic quantum mechanics on which almost all contemporary treatments of electron correlation are based, is inadequate because the mean speed of core electrons is a substantial fraction of the speed of light, so that a fully relativistic electronic structure theory is required. The proper treatment of relativity and quantum electrodynamic effects will demand increasing attention in the years ahead, since these may be more important than electron correlation in heavy elements and are thus an essential ingredient of an *ab initio* electronic structure studies for the lower rows of the Periodic Table. A great deal of research over the past decade has been directed towards the development of a relativistic many-body perturbation theory formulated within the algebraic approximation. Problems of principle have been largely resolved and the remaining problems are primarily of a computational nature.

In this article, we address some of the considerable computational demands of the relativistic electron correlation problem. Relativistic molecular structure calculations are sometimes regarded as two orders of magnitude more demanding than the corresponding atomic structure calculations [1]. Even non-relativistic calculations of the electronic structure of molecular systems containing heavy atoms demand considerable computational resources.

The outline of this paper is as follows:- In section 2, we describe the background to the relativistic electron correlation problem. An overview of the relativistic many-body perturbation theory is given in section 3 and in section 4 some recent results described. Section 5 contains some comments on future work.

2. The relativistic electron correlation problem

Following the development, in the mid 1950's, of the many-body perturbation theory, by Brueckner [2], Goldstone [3], Hugenholtz [4] and others (see, for example, reference [5]), from the quantum field theory of Feynman [6], Dyson [7] and Wick [8], the first applications to atoms were reported, in the early 1960's by Kelly [9]. He generated numerical solutions to the radial Hartree-Fock equations using finite

Methods in Computational Molecular Physics, Edited by Stephen Wilson and Geerd H.F. Diercksen, Plenum Press, New York, 1992

differences, which he then used as a reference for a perturbative treatment of correlation effects. The integration over the continuum states formed the most demanding part of these calculations.

In the late 1960's, the first applications of the many-body perturbation theory to the molecular electronic structure problem were also reported by Kelly [10]. He performed molecular calculations on systems containing only one non-hydrogenic atom, which was used as an expansion centre, and treated the hydrogen nuclei as additional perturbations.

Applications of the method to arbitrary molecular systems awaited the introduction of basis set expansion techniques, that is the algebraic approximation, of the type which are traditional in molecular electronic structure theory. This was achieved in the mid-1970's by Kaldor [11] and by Silver and his coworkers [12,13 and references therein].

Until the late 1970's, the "standard" approach to the correlation problem in molecules was the method of configuration mixing or configuration interaction. However, the rapid increase of the dimension of the hamiltonian matrix with increasing number of electrons in the system and number of basis functions, limited the applicability of the method to small systems and small basis sets unless the configuration expansion was severely truncated. A perturbative analysis indicates that only double excitations contribute to the energy in second and third order and thus limited "double excitation" configurations was a widely used approach. Today many-body perturbation theory has emerged as the standard approach. "Nowadays, many thousands of quantum chemistry calculations are performed on small and medium-sized molecules within the framework of perturbation theory. Such programs have become standard tools for many research workers ..." [14].

However, it soon became apparent that the major source of error in these calculations was attributable to basis set truncation effects. Research was first directed towards even-tempered and universal basis sets [15] and then systematic sequences of even-tempered basis sets and their completeness properties. For atoms and small molecules the errors associated with truncation of the basis set were reduced to the level where relativistic corrections became important for all but the lightest atoms and systems containing them.

Until recently relativistic effects have almost always been introduced into molecular electronic structure studies by means of perturbation theory. This approach is frequently employed in atomic studies also. This was the approach which we began to explore in the early 1980s [16, 17], but then became convinced that it is essential to use methods which build in relativity at the orbital level from the outset if reliable results are required. At that time, some Dirac-Hartree-Fock calculations were performed for atoms but they employed finite difference methods which preclude their application to arbitrary molecular systems [18]. Furthermore, because finite difference calculations do not afford a representation of the complete spectrum, a description of electron correlation effects was obtained by using a multiconfiguration wave function.

By invoking the algebraic approximation, that is by using a finite basis set, the relativistic electronic structure methods can not only be applied to molecules but also a representation of the complete Dirac spectrum is obtained allowing many-body perturbation theory treatment of correlation effects with all its theoretical and computational advantages. However, it has been known since the pioneering work of Kim [19] that the use of basis set expansion techniques in relativistic calculations requires particular care. In a series of papers with the general title "The Dirac equation in the algebraic approximation" [20-27], we have investigated this problem in detail.

The most frequently employed realization of the algebraic approximation in non-relativistic molecular calculations is the L.C.A.O. approximation in which molecular orbitals are approximated by some Linear Combination of Atomic Orbitals. In the relativistic case considerable care is required in the implementation of the algebraic approximation if the so-called "finite basis set disease" is to be avoided and a clean separation of the two branches of the spectrum is to be maintained. It emerges that this separation can be achieved if a spectral method is employed. According to the recent monograph by Mercier [28], the term "Spectral methods is the name given to a numerical approach to the solution of partial differential equations. In this approach the solution to the equation is approximated by a truncated series of special functions which are eigenfunctions of some differential operator.". The key to the successful implementation of the algebraic approximation in relativistic calculations is to use basis functions which are themselves solutions of a one-electron Dirac equation. Now the separation into the two branches of the Dirac spectrum is determined both by the finite value of the speed of light and the strength of the external field. (Problems occur for superheavy atomic nuclei with very high Z [29] but are not considered here.) The admissible forms for the radial basis functions are constrained by these requirements. The basis sets which seems to have the most general applicability to relativistic atomic structure theory consist of the exponential-type functions

$$f_i^T(r) = r^{\gamma}\exp(-\zeta_i r)\left\{ -(1-\delta_{0n})L_{n-1}^{2\gamma}(2\zeta_i r) \pm \frac{(N_n - \kappa)}{(n+2\gamma)} L_n^{2\gamma}(2\zeta_i r)\right\}$$

$$n = 0 \ \text{for} \ \kappa < 0 \qquad i = 1,2,\ldots,M_\kappa$$

$$n = 1 \ \text{for} \ \kappa > 0 \tag{1}$$

where the upper sign is taken for the large component (T=L) and the lower sign for the small component (T=S).

$$\gamma^2=(\kappa^2-(Z/c)^2) \tag{2}$$

and

$$N_n{}^2=n^2+2n\gamma+\kappa^2 \tag{3}$$

$L_n{}^{\alpha}(z)$ is the generalized Laguerre polynomial. The exponential parameters $\{\zeta_i\}$ are chosen so as to provide a sequence of basis sets which becomes complete in some well defined limit [30-34]. The non-relativistic limit of the relativistic exponential-type function basis set is a "kinetically balanced" set of exponential-type functions [35-37]. The dimension of the radial basis for symmetry-type κ is M_κ.

3. Relativistic many-body perturbation theory

3.1 A model problem

In our first relativistic diagrammatic perturbation theory calculation employing the algebraic approximation [23], it was demonstrated in a study of a simple hydrogenic model problem (the relativistic formulation of the model discussed in the previous article) that the discrete representation of the complete spectrum, comprising the bound states and the positive- and negative-energy continua, provided by the analytic basis set method greatly facilitates the evaluation of the algebraic sum-over-states expressions corresponding to each of the diagrams [38, 39].

In our preliminary study [23] of this model problem, we employed kinetically balanced basis sets of exponential-type functions to study unperturbed states of s symmetry. The more difficult class of symmetry types κ>0 requires the use of relativistic basis sets because of the additional singularities which are introduced into the radial functions [38, 39].

3.2 Many-electron systems

3.2.1 Choice of reference hamiltonian

In further work, we presented calculations within the independent electron model using both the Dirac-Coulomb and the Dirac-Breit hamiltonians [24, 25, 39, 40]. It has been known for many years that the frequency-independent Breit interaction may be incorporated into the self-consistent field equations for atoms, but this approach has until recently been regarded as too demanding computationally in finite-difference calculations; first-order perturbation theory has been the method of choice. In the algebraic approximation, we have demonstrated that the self-consistent treatment of the Breit interaction is preferred to the usual method which treats the Breit interaction as a first-order perturbation to the Dirac-Coulomb operator.

3.2.2 Relativistic correlation energy expansions

Relativistic many-body calculations based on the Dirac-Coulomb hamiltonian and including energy shifts which arise through the creation of virtual electron-positron pairs have been reported in [25], [35], [36] and [38]. Very recent calculations [41], which are discussed further below, reveal that the terms in the perturbation expansion corresponding to the self-consistent modification of the one-electron orbitals due to the Breit interaction result in the most significant relativistic many-body effect.

The principal advantage of the algebraic approximation technique described here is that the two-electron integrals over the basis functions which are required in many-body theory may be evaluated both efficiently and accurately. The transformation of the integrals over basis functions into relativistic matrix elements may be performed using linear algebra techniques, so that the scheme may be programmed for high performance on vector and parallel processing computers [42, 43]. Many of the developments embodied in non-relativistic electronic structure computer codes are transferred to the relativistic method without substantial modification. We suggest that this method has advantages over the alternatives proposed by Johnson and Sapirstein [44] and by Salomonsson and Öster [45] on the grounds of simplicity and computational efficiency.

The negative-energy states are taken to be filled in the true vacuum state. The events which occur in a relativistic many-body system are conveniently described within the particle-hole formalism in which the negative energy continuum and occupied positive energy states are taken to be below the Fermi level. Unlike the non-relativistic particle-hole formalism, the relativistic extension involves no restriction on particle number because virtual excitations from the negative energy sea must be considered. If we take the electron-electron interaction to be instantaneous, that is either just the Coulomb interaction or the sum of the Coulomb interaction and the frequency-independent Breit interaction, then the diagrammatic representation of the second-order energy for a closed-shell system is given in Figures 1 and 2. We find it convenient to use "open diagrams". In Figure 1 the second order diagrams involving singly excited intermediate states are shown whilst in Figure 2 the second order diagrams which involve doubly excited intermediate states are displayed. In order to keep the time ordering of events obvious when working beyond the no virtual pair approximation, it is most convenient to use "open" Goldstone diagrams. In these diagrams, upwards directed lines are used to represent single particle states in the positive-energy branch of the spectrum and downward directed lines represent states in the negative-energy branch. All of the diagrams given in Figures 1 and 2 arise in second order when an arbitrary single determinantal function is employed as a reference for the perturbation expansion.

Diagrams (a)-(d), which arise in the no virtual pair approximation, describe single excitations creating a hole in the positive-energy branch below the Fermi level and a

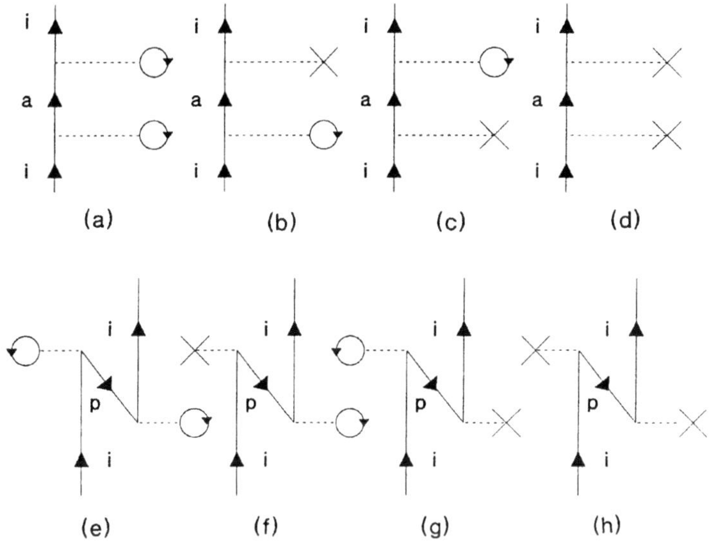

Figure 1. *Diagrammatic representation of some second-order energy contributions involving single excitations. Upwards directed lines represent single particle states in the positive energy branch of the spectrum whilst downward directed lines represent single particle states in the negative energy branch. The horizontal dashed line terminated by a cross represent a one-electron time-independent potential, such as the Dirac-Hartree-Fock (-Coulomb/-Breit) potential. The horizontal dashed line is taken to represent either the instantaneous Coulomb interaction or the sum of the Coulomb interaction and frequency-independent Breit interaction. The indices i,j,k,... label occupied positive energy states and the indices a,b,c,... label unoccupied states whilst the indices p,q,r,... label negative energy states.*

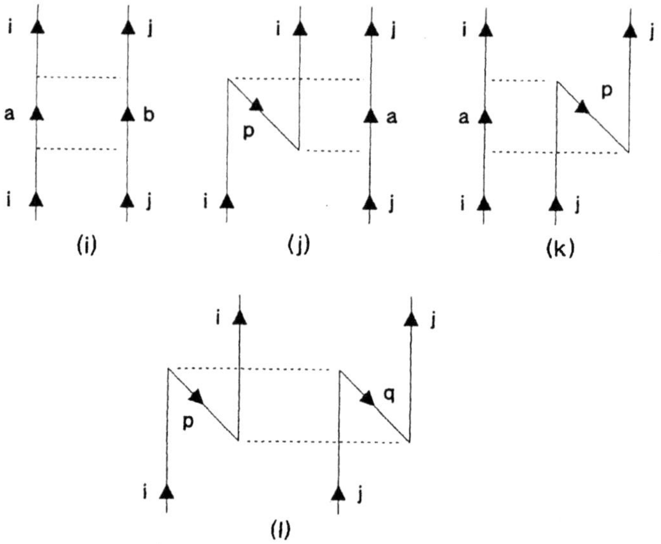

Figure 2. *Diagrammatic representation of some second-order energy contributions involving double excitations. Upwards directed lines represent single particle states in the positive energy branch of the spectrum whilst downward directed lines represent single particle states in the negative energy branch. The horizontal dashed line is taken to represent either the instantaneous Coulomb interaction or the sum of the Coulomb interaction and frequency-independent Breit interaction. The indices i,j,k,... label occupied positive energy states and the indices a,b,c,... label unoccupied states whilst the indices p,q,r,... label negative energy states.*

particle above the Fermi level. Diagrams (e)-(h), which represent effects that can be said to go beyond the no virtual pair approximation, describe the formation of a virtual electron-positron pair by creation of a hole in the negative-energy branch of the spectrum. Diagrams (i)-(l) involve double excitations. Diagram (i) arises in the no virtual pair approximation, diagrams (j) and (k) describes the creation of one virtual pair, and diagram (l) the creation of two virtual pairs.

For calculations using a bare nucleus reference function, the diagrams containing the single-particle potential, represented by a cross, do not arise. For calculations based on a Dirac-Hartree-Fock-Coulomb or Dirac-Hartree-Fock- Breit reference function there is not an exact cancellation between diagrams containing "bubbles" and "crosses" because the polarization of the negative-energy sea is not accommodated within the self-consistent field potential. Labzovskii [46] pointed out that although the familiar cancellation of single-particle insertions occurs within the positive-energy branch of the Dirac spectrum when the Dirac-Hartree-Fock (-Coulomb/-Breit) method is employed, this cancellation does not occur in the negative-energy branch.

The algebraic expressions corresponding to each of the diagrams shown in Figures 1 and 2 can be written in terms of the one-electron integrals

$$<\mu|V|\nu>, \qquad \mu,\nu = i,j,k,...,a,b,c,...,p,q,r,... \tag{4}$$

in which V is some potential (for example, the Dirac-Hartree-Fock(-Breit) potential), the two-electron integrals

$$<\mu\nu|\hat{O}|\sigma\tau>, \qquad \mu,\nu,\sigma,\tau = i,j,k,...a,b,c,...,p,q,r,... \tag{5}$$

where $\hat{O}$ is the electron-electron interaction operator including exchange, that is,

$$\hat{O} = (I-P_{12})\, g_{12} \tag{6}$$

where I and the identity and P_{12} permutes the coordinates of electrons 1 and 2. g_{12} is the electron-electron interaction, which in the Dirac-Coulomb hamiltonian is simply

$$g_{12} = 1/r_{12}, \tag{7}$$

and in the Dirac-Breit hamiltonian has the form

$$g_{12} = 1/r_{12} + B_{12} \tag{8}$$

where B_{12} is the Breit interaction

$$B_{12} = -(2R)^{-1}\,[\alpha_1.\alpha_2 + R^{-2}(\alpha_1.R)\,(\alpha_2.R)] \tag{9}$$

$\varepsilon_i,...,\ \varepsilon_a,...,\ \varepsilon_p,....,$ are the orbitals energies. We then have for diagrams (a)-(d), which arise in the "no pair" approximation, the following expressions

$$E(a) = \sum_{ijka} \frac{<ij|\hat{O}|aj>\ <ak|\hat{O}|ik>}{\varepsilon_i - \varepsilon_a} \tag{10a}$$

$$E(b) = \sum_{ija} \frac{<ij|\hat{O}|aj>\ \ <a|\hat{V}|i>}{\varepsilon_i - \varepsilon_a} \tag{10b}$$

$$E(c) = \sum_{ija} \frac{<i|\hat{V}|a> \ <aj|\hat{O}|ij>}{\varepsilon_i - \varepsilon_a} \tag{10c}$$

$$E(d) = \sum_{ia} \frac{<i|\hat{V}|a> \ <a|\hat{V}|i>}{\varepsilon_i - \varepsilon_a} \tag{10d}$$

For diagrams (e)-(h), which are single excitation diagrams involving the creation of one virtual electron-positron pair, we have

$$E(e) = -\sum_{ijkp} \frac{<pj|\hat{O}|ij> \ <ik|\hat{O}|pk>}{\varepsilon_i - \varepsilon_p} \tag{10e}$$

$$E(f) = -\sum_{ijp} \frac{<pj|\hat{O}|ij> \ <i|\hat{V}|p>}{\varepsilon_i - \varepsilon_p} \tag{10f}$$

$$E(g) = -\sum_{ijp} \frac{<p|\hat{V}|i> \ <ij|\hat{O}|pj>}{\varepsilon_i - \varepsilon_p} \tag{10g}$$

$$E(h) = -\sum_{ip} \frac{<p|\hat{V}|i> \ <i|\hat{V}|p>}{\varepsilon_i - \varepsilon_p} \tag{10h}$$

For the double excitation diagrams (i)-(l), we have

$$E(i) = \frac{1}{4}\sum_{ijab} \frac{<ij|\hat{O}|ab> \ <ab|\hat{O}|ij>}{\varepsilon_i + \varepsilon_j - \varepsilon_a - \varepsilon_b} \tag{10i}$$

$$E(j) = -\sum_{ijap} \frac{<pj|\hat{O}|ia> \ <ia|\hat{O}|pj>}{\varepsilon_i + \varepsilon_j - \varepsilon_a - \varepsilon_p} \tag{10j}$$

$$E(k) = -\sum_{ijap} \frac{<ip|\hat{O}|aj> \ <aj|\hat{O}|ip>}{\varepsilon_i + \varepsilon_j - \varepsilon_a - \varepsilon_p} \tag{10k}$$

$$E(l) = \frac{1}{4}\sum_{ijpq} \frac{<pq|\hat{O}|ij> \ <ij|\hat{O}|pq>}{\varepsilon_i + \varepsilon_j - \varepsilon_p - \varepsilon_q} \tag{10l}$$

Figures 3 and 4 show some typical third order diagrams. The diagrams in Figure 3 involve single excitations whilst those in Figure 4 are associated with doubly excited intermediate states. Diagram (m) arises in the no virtual pair approximation. Diagrams (n)-(o) represent effects beyond the no virtual pair approximation. Diagrams (p) arises in the no virtual pair approximation and diagram (q) represents effects beyond the no virtual pair approximation.

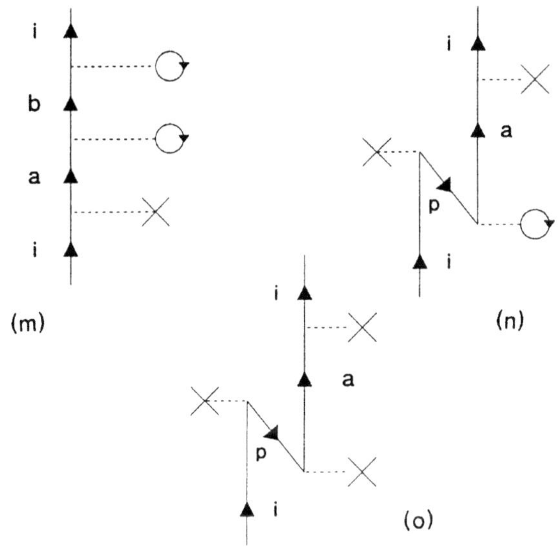

Figure 3. Diagrammatic representation of some third-order energy contributions involving single excitations. Upwards directed lines represent single particle states in the positive energy branch of the spectrum whilst downward directed lines represent single particle states in the negative energy branch. The horizontal dashed line terminated by a cross represent a one-electron time-independent potential, such as the Dirac-Hartree-Fock (-Coulomb/-Breit) potential. The horizontal dashed line is taken to represent either the instantaneous Coulomb interaction or the sum of the Coulomb interaction and frequency-independent Breit interaction. The indices i,j,k,... label occupied positive energy states and the indices a,b,c,... label unoccupied states whilst the indices p,q,r,... label negative energy states.

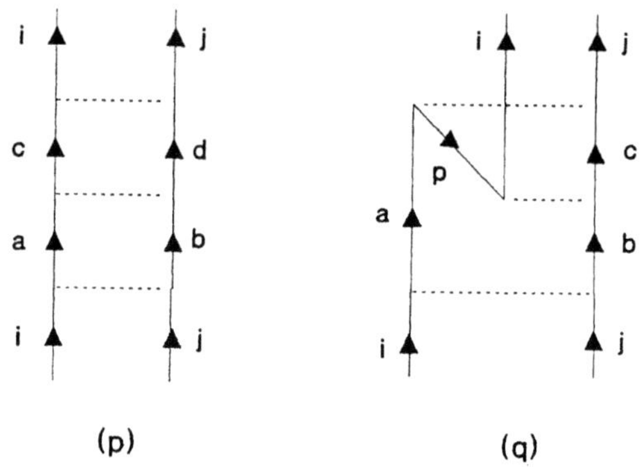

Figure 4. *Diagrammatic representation of some third-order energy contributions involving double excitations. Upwards directed lines represent single particle states in the positive energy branch of the spectrum whilst downward directed lines represent single particle states in the negative energy branch. The horizontal dashed line is taken to represent either the instantaneous Coulomb interaction or the sum of the Coulomb interaction and frequency-independent Breit interaction. The indices i,j,k,... label occupied positive energy states and the indices a,b,c,... label unoccupied states whilst the indices p,q,r,... label negative energy states.*

4. An application to the argon ground state

Recently, we have reported relativistic and non-relativistic many-body perturbation theory calculations using systematically constructed basis sets of exponential-type functions for the ground state of the argon atom [41]. The relativistic calculations employed both the Dirac-Coulomb and the Dirac-Breit hamiltonian.

At the independent electron model level, the difference between the Dirac-Hartree-Fock-Coulomb energy and the Dirac-Hartree-Fock-Breit energy was found to be to 0.132365 hartree. The difference between the sum of Dirac-Hartree-Fock-Coulomb energy and the first-order Breit energy and the Dirac-Hartree-Fock-Breit energy is only 40 μhartree. The effect of the Breit interaction is most pronounced in the spinor energies and in the shapes of the spinors as reflected in the expectation values of r^k. For the inner shell spinors, changes of about 0.1% are observed in the expectation values of r^k when the Breit interaction is included. A similar conclusion was reached by Lindroth et al [47] in studies of hyperfine interactions.

In Figure 5, a partial wave analysis of the second-order pair energies obtained from non-relativistic many-body perturbation theory calculations for the argon atom ground state using basis sets of s p and d symmetry are presented. Note that the ordinate has a logarithmic scale in microhartree. In Figures 6 and 7, a partial wave analysis of the second-order pair energies obtained from relativistic many-body perturbation theory calculations using the Dirac-Coulomb hamiltonian and the Dirac-Breit hamiltonian, respectively. Both relativistic calculations used the no virtual pair approximation.

Second-order pair correlation energies obtained from a non-relativistic calculation and from relativistic calculations employing basis sets of functions with s, p, d and f symmetry (the largest basis sets considered by Quiney, Grant and Wilson [41]) are displayed in Figures 8, 9 and 10, respectively. In these Figures, the ordinate has a linear scale in millihartree.

The differences between the second-order pair correlation energies for the ground state of the argon atom from calculations using the Schrödinger hamiltonian and those using the Dirac-Coulomb hamiltonian are shown in Figure 11. The basis sets employed in these calculations contained exponential-type functions of s, p, d and f symmetry type. In Figure 12, the differences between the second-order pair correlation energies from calculations using the Schrödinger hamiltonian and those using the Dirac-Breit hamiltonian are shown. The differences between the second-order pair correlation energies from calculations using the Dirac-Coulomb hamiltonian and those using the Dirac-Breit hamiltonian are shown in Figure 13.

Comparing the results of the relativistic calculation using the Dirac-Coulomb hamiltonian with the non-relativistic results, the effects of relativity on the correlation energy can be explained in terms of competing effects due to shielding and the relative phases of the large and small components of the relativistic wave functions. However, the Dirac-Coulomb hamiltonian accounts for only a small fraction of the total relativistic correction to the correlation energy, though it is, of course, this hamiltonian that has been used as a reference in all previous studies of relativistic correlation effects. The bulk of the relativistic correction to the correlation energy comes from single-particle insertions (one Coulomb and one Breit photon diagrams) which are not in the Dirac-Hartree-Fock- Coulomb basis. These diagrammatic terms are summed through all orders in the Dirac-Hartree-Fock-Breit approximation, which has the additional advantage that evaluation of terms in the relativistic many-body perturbation series is simplified, particularly in higher orders.

The terms in the partial wave expansion for for the relativistic shift associated with s^2 pairs vary approximately as $1/l^2$, where l is the orbital angular momentum of the intermediate states. This should be compared with the $1/l^4$ behaviour of the partial

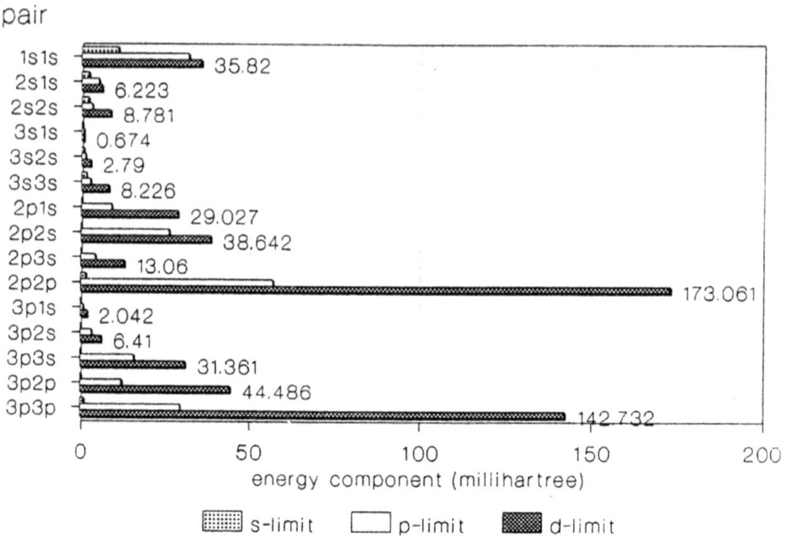

Figure 5. *Partial wave analysis of the second-order pair correlation energies for the ground state of the argon atom from calculations employing the Schrödinger hamiltonian and a basis set containing functions of s, p and d symmetry. (Based on the work of Quiney, Grant and Wilson (1990))*

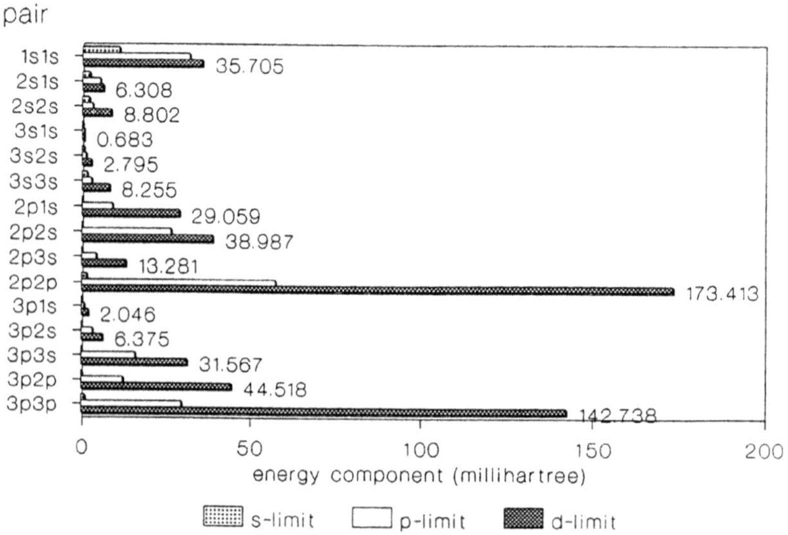

Figure 6. *Partial wave analysis of the second-order pair correlation energies for the ground state of the argon atom from calculations employing the Dirac-Coulomb hamiltonian and a basis set containing functions of s, p and d symmetry. (Based on the work of Quiney, Grant and Wilson (1990))*

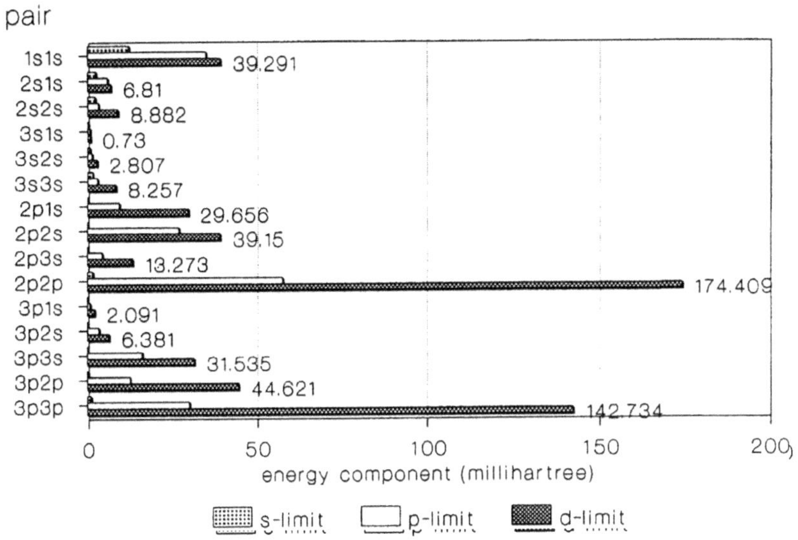

Figure 7. *Partial wave analysis of the second-order pair correlation energies for the ground state of the argon atom from calculations employing the Dirac-Breit hamiltonian and a basis set containing functions of s, p and d symmetry. (Based on the work of Quiney, Grant and Wilson (1990))*

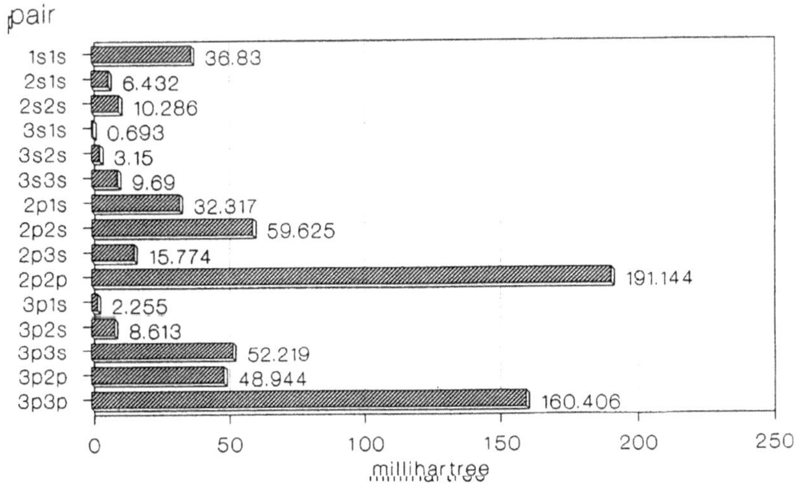

Figure 8. *Second-order pair correlation energies for the ground state of the argon atom from calculations employing the Schrödinger hamiltonian and a basis set containing functions of s, p, d and f symmetry. (Based on the work of Quiney, Grant and Wilson (1990))*

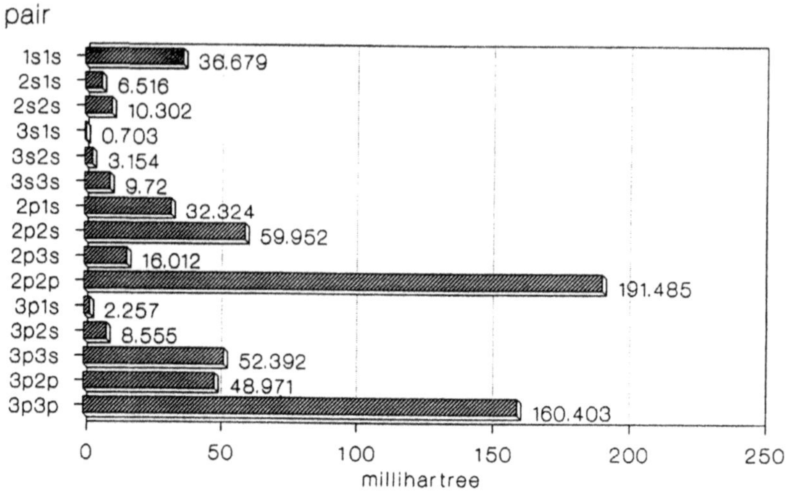

Figure 9. *Second-order pair correlation energies for the ground state of the argon atom from calculations employing the Dirac-Coulomb hamiltonian and a basis set containing functions of s, p, d and f symmetry. (Based on the work of Quiney, Grant and Wilson (1990))*

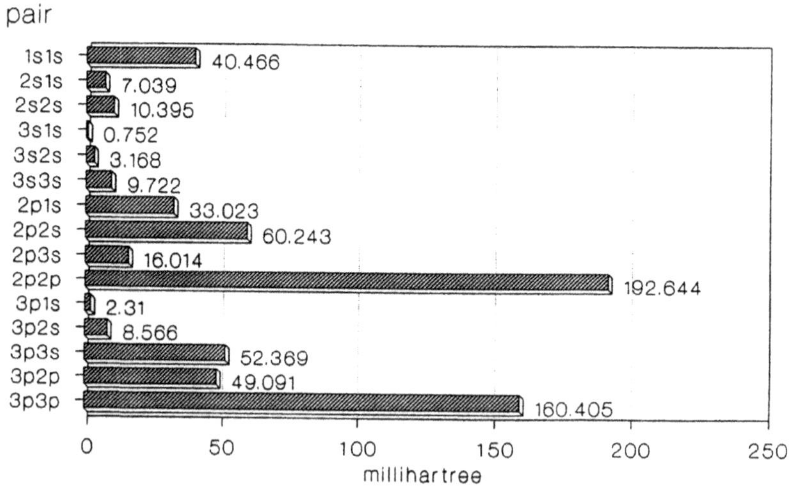

Figure 10. *Second-order pair correlation energies for the ground state of the argon atom from calculations employing the Dirac-Breit hamiltonian and a basis set containing functions of s, p, d and f symmetry. (Based on the work of Quiney, Grant and Wilson (1990))*

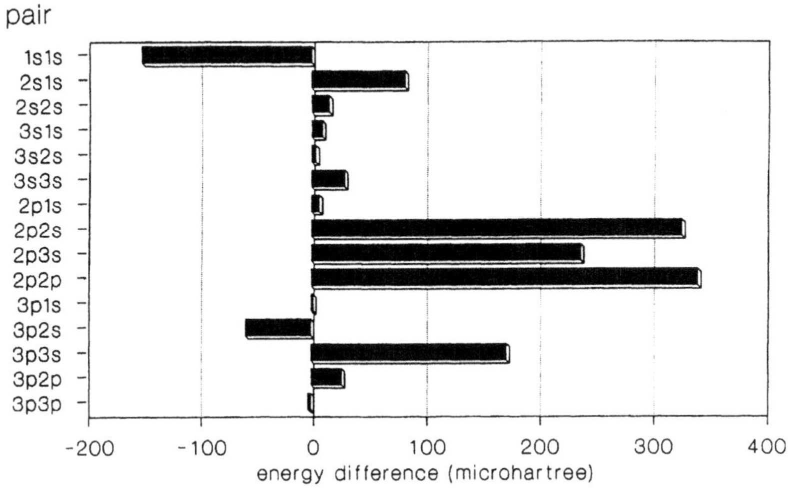

Figure 11. *Differences between second-order pair correlation energies for the ground state of the argon atom from calculations using the Schrödinger hamiltonian and those using the Dirac-Coulomb hamiltonian. The basis sets employed in these calculations contained exponential-type functions of s, p, d and f symmetry type. (Based on the work of Quiney, Grant and Wilson (1990))*

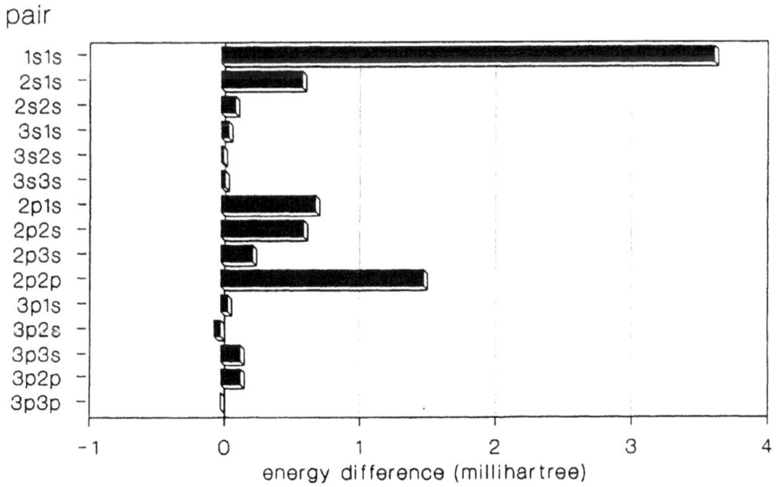

Figure 12. *Differences between second-order pair correlation energies for the ground state of the argon atom from calculations using the Schrödinger hamiltonian and those using the Dirac-Breit hamiltonian. The basis sets employed in these calculations contained exponential-type functions of s, p, d and f symmetry type. (Based on the work of Quiney, Grant and Wilson (1990))*

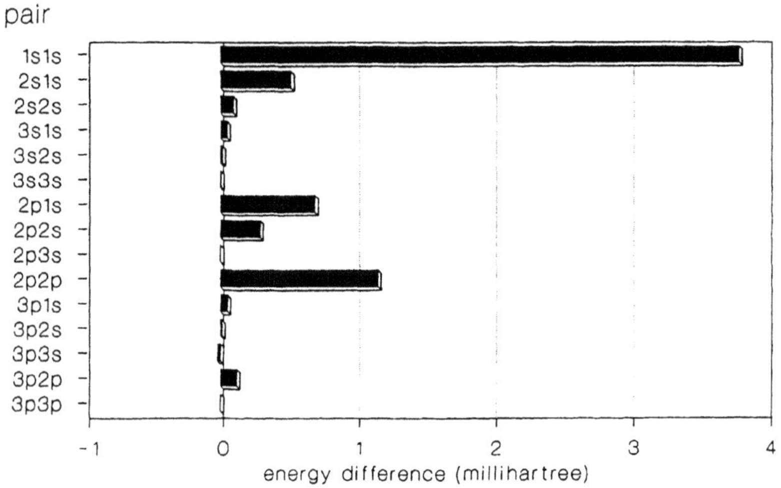

Figure 13. *Differences between second-order pair correlation energies for the ground state of the argon atom from calculations using the Dirac-Coulomb hamiltonian and those using the Dirac-Breit hamiltonian. The basis sets employed in these calculations contained exponential-type functions of s, p, d and f symmetry type. (Based on the work of Quiney, Grant and Wilson (1990))*

wave expansion for s^2 pairs in the non-relativistic case. The effects of relativity on the valence spinors are almost entirely accounted for by the Dirac-Coulomb hamiltonian. For the core spinors and for the interaction between the core and the valence spinors, it is necessary to use the Dirac-Breit hamiltonian.

The total non-relativistic second-order many-body perturbation theory energy at the f-limit is -638680 μhartree. Within the no virtual pair approximation, the Dirac-Coulomb second-order correlation energy is -639424 μhartree, so that the relativistic correction is -744 μhartree. This difference arises from the use of relativistic single-particle states, but the same two-body interaction. The corresponding relativistic calculation using the Dirac-Breit operator gives a total second-order energy of -646206 μhartree, which represents a relativistic shift of -7526 μhartree, which is an order of magnitude larger than the Dirac-Coulomb many-body correction. Although roughly half of the Dirac-Breit many-body correction is associated with the $1s^2$ pair, the remainder arises through subtle interactions between the core and the valence orbitals.

Quiney, Grant and Wilson [41] conclude that the terms in the perturbation expansion corresponding to the self-consistent modification of the one-electron orbitals due to the Breit interaction result in the most significant relativistic many-body effect. Furthermore, this studies underlines that fact that the relativistic many-body perturbation theory using the algebraic approximation is a tractable scheme giving a systematic approach to the calculation of relativistic correlation effects in atoms and opens up the possibility of molecular applications.

5. Future directions

The past decade has seen a growing interest in the effects of relativity in atoms, molecules and solids. On the one hand this can be seen as a result of the growing awareness of the importance of relativity in describing the properties of heavy atoms and systems containing them. This has been fueled by the inadequacy of physical models with either neglect relativity or which treat it as a small perturbation. On the other hand, it is dependent upon the technological developments which have resulted in computers powerful enough to make calculations on heavy atoms and on systems containing heavy atoms meaningful. Vector processing and, more recently, parallel processing techniques are playing an increasingly vital role in rendering the algorithms which arise in relativistic studies tractable. The recently published volume entitled *The Effects of Relativity in Atoms, Molecules and the Solid State* [48] provides a useful insight into the latest research in this area.

The relativistic many-body perturbation theory using the algebraic approximation affords a tractable scheme yielding a systematic approach to the calculation of relativistic correlation effects in atoms and molecules. Relativistic quantum chemistry has a vast range of potential applications in both the academic and the commercial sectors in fields as diverse as catalysis and pharmacology, molecular electronics and molecular biology. However, relativistic electronic structure calculations for molecules will remain demanding for some time to come both in the requirement for more efficient and reliable numerical methods and algorithms, and for the computer resources which these methods require.

References

[1] Y.-K. Kim, remark made at Relativistic Atomic Structure Meeting, Grenoble, October, 1989
[2] K.A. Brueckner, Phys. Rev. **100**, 36 (1955)
[3] J. Goldstone, J., Proc. Roy. Soc. A 239, 267 (1957)
[4] N.M. Hugenholtz, Physica 23, 481 (1957)
[5] N.H. March, W.H. Young, and S. Sampanthar, *The many-body problem in quantum mechanics*, Cambridge University Press (1967)

[6] R. Feynman, Phys. Rev. **76**, 769 (1949)
[7] F.J. Dyson, Phys. Rev. **75**, 1736 (1949)
[8] G.C. Wick, Phys. Rev. **80**, 268.(1950)
[9] H.P. Kelly, Phys. Rev. **131**, 684 (1963)
[10] H.P. Kelly, Phys. Rev. Lett. **23**, 455 (1969)
[11] U. Kaldor, J. Chem. Phys. **62**, 4634 (1975)
[12] S. Wilson, and D.M. Silver, Phys. Rev. **A14**, 1949 (1976)
[13] S. Wilson, and D.M. Silver, J. Chem. Phys. **66**, 5400 (1977)
[14] P.J. Knowles, K. Somasundram, N.C. Handy, and K. Hirao, Chem. Phys. Lett. **113** 8 (1985)
[15] *see previous article*
[16] D.L. Cooper, D.L., and S. Wilson, J. Phys. B: At. Mol. Phys. **15**, 493 (1982)
[17] D.L. Cooper, and S. Wilson, J. Chem. Phys. **76**, 6088 (1982)
[18] I.P. Grant, B. Mackenzie, D.F. Mayers and N.C. Pyper, Comput. Phys. Commun. **21**, 207 (1980)
[19] Y.-K. Kim, Phys. Rev. Phys. Rev. **154**, 17 (1968)
[20] K.G. Dyall, I.P. Grant, and S. Wilson, J. Phys. B: At. Molec. Phys. **17**, 493
[21] K.G. Dyall, I.P. Grant, and S. Wilson, J. Phys. B: At. Molec. Phys. **17**, L45 (1984)
[22] K.G. Dyall, K.G., I.P. Grant, and S. Wilson, J. Phys. B: At. Molec. Phys. **17**, L45 (1984)
[23] H.M. Quiney, I.P. Grant and S. Wilson, J. Phys. B: At. Mol. Phys. **18** 2805 (1985)
[24] H.M. Quiney, I.P. Grant and S. Wilson, J. Phys. B: At. Mol. Phys. **20** 1413 (1987)
[25] H.M. Quiney, I.P. Grant and S. Wilson, Physica Scripta 36 460 (1987)
[26] J. Wood, I.P. Grant, and S. Wilson, J. Phys. B: At. Mol. Phys. **18**, 3027 (1985)
[27] L. Laaksonen, I.P. Grant, and S. Wilson, J. Phys. B: At. Mol. Opt. Phys. 21, 1969 (1988)
[28] B. Mercier, *An introduction to the numerical analysis of spectral methods*, Lecture Notes in Physics 318, Springer, Berlin (1989)
[29] W. Greiner, B. Muller and J. Rafelski, *Quantum Electrodynamics of Strong Fields*, Springer, Berlin (1985)
[30] B. Klahn, and W.A. Bingel, Theoret. chim. Acta. **44** 9 (1977)
[31] B. Klahn, and W.A. Bingel, Theoret. chim. Acta. **44** 27 (1977)
[32] M.W. Schmidt and K. Ruedenberg, J. Chem. Phys. **71** 3951 (1979)
[35] R.E. Stanton, and S. Haviliak, J. Chem. Phys. 81 1910 (1984)
[33] S. Wilson, Theoret. chim. Acta. 57 53 (1980)
[34] S. Wilson, Theoret. chim. Acta. 58 31 (1980)
[36] H.M. Quiney, in *Relativistic effects in atoms and molecules*, ed. S.Wilson, Meth. Comput. Chem. 2 223 Plenum, New York (1988)
[37] S. Wilson, in *Relativistic effects in atoms and molecules*, ed. S.Wilson, Meth. Comput. Chem. 2 73 , Plenum, New York (1988)
[38] H.M. Quiney, I.P. Grant and S. Wilson, in *Numerical Determination of the Electronic Structure of Atoms, Diatomic and Polyatomic Molecules*, ed. M.Defranchesi and J.Delhalle,NATO Advanced Research Workshop, Versailles, April 1988, Reidel, Dordrecht (1989)
[39] H.M. Quiney, I.P. Grant and S. Wilson, in Many-Body Methods in Quantum Chemistry, Lecture Notes in Chemistry 52 331, ed. U. Kaldor, Springer, Berlin (1989)
[40] H.M. Quiney, I.P. Grant and S. Wilson, J. Phys. B: At. Mol. Opt. Phys. (Letters) 22 L15 (1989)
[41] H.M. Quiney, I.P. Grant and S. Wilson, J. Phys. B: At. Mol. Opt. Phys. (Letters) 23 L271 (1990)
[42] H.M. Quiney, in *Supercomputational Science*, edited by R.G. Evans and S.Wilson, Plenum, New York (1990)
[43] S. Wilson, in *Supercomputational Science*, edited by R.G. Evans and S.Wilson, Plenum, New York (1990)

[44] W.R. Johnson and J. Sapirstein, Phys. Rev. Lett. 57 1126 (1986)
[45] S. Salomonson S and P. Öster, Phys. Rev. A40 5559 (1989)
[46] L.N. Labzovskii, Sov. Phys. JETP 32 94 (1971)
[47] E. Lindroth, A-M Mårtensson-Pendrill, A. Ynnerman and P. Öster, J. Phys B: At. Mol. Opt. Phys. 22 2447 (1989)
[48] S. Wilson, I.P. Grant and B.L. Gyorffy, *The Effects of Relativity in Atoms, Molecules and the Solid State*, Plenum Press, New York (1991)

THE PERTURBATION THEORY OF ELECTRON CORRELATION IV

concurrent computation Many-Body Perturbation Theory

Stephen Wilson

Rutherford Appleton Laboratory
Chilton, Oxfordshire, OX11 0QX, England

1. Introduction

Computers play a central role in the practical applications of the vast majority of the apparatus of theoretical chemistry. Indeed, the outlook for performing quantum mechanical studies of the electronic structure of molecules changed radically in the 1950s with the advent of the electronic digital computer. Steady increase in the performance of these machines during the following decades was reflected in the increasing sophistication of computational techniques leading to higher precision in *ab initio* calculations and the ability to handle more extensive molecular systems. With the availability of vector processing computers, such as the CRAY 1 and CDC Cyber 205 in the late 70s, came a significant increase in performance; an improvement by a factor of 100 was observed in certain stages of electronic structure calculations. In more recent years, the trend in the development of computer architectures has been towards multiprocessor machines as exemplified by the CRAY X-MP and Y-MP computers. The successful exploitation of these new architectures requires that the effective algorithms be devised allowing parts of the computation to proceed concurrently [1-2].

The most demanding stage of an *ab initio* molecular electronic structure calculation is that associated with the description of corrections to independent electron models, that is electron correlation effects. Diagrammatic many-body perturbation theory is now one of the most effective and widely used approaches to the correlation problem in molecules [3] which, in addition to its well known theoretical properties [3], has the advantage that it can readily adapted for concurrent computation.

The first published many-body perturbation theory codes [4-6] for atomic and molecular electronic structure calculations using finite basis set expansions, which were employed in many of the early applications [7-12] of the method, were written for a scalar processing computer (the IBM 360/91). Furthermore, because of the limited memory available, the algorithms employed in these codes were constructed in such a way that only a small fraction of the two-electron integrals were required in the computer's central memory at any one time.

In this article, we described the *concurrent computation* Many-Body Perturbation Theory (*cc*MBPT). This is an adaption of the many-body perturbative formalism especially suited to implementation on vector and parallel processing machines with a

Methods in Computational Molecular Physics, Edited by Stephen Wilson and
Geerd H.F. Diercksen, Plenum Press, New York, 1992

large central memory shared by all of the processors. *cc*MBPT exploits the additive separability of the correlation energy components resulting from the linked diagram expansion to devise algorithms suitable for a parallel processing environment. A dynamic load balancing technique is employed to exploit the capabilities of parallel processing shared memory architecture such as the CRAY Y-MP. The performance of the new technique is demonstrated in calculations of the most computationally demanding of the fourth order energy components, those involving triply excited intermediate states.

The outline of the article is as follows: in section 2, we describe MBPT and its suitability for concurrent computation. In section 3, 4 and 5, we describe the calculation of the second-, third- and fourth-order terms, respectively. Section 6 describes the *cc*MBPT program and the results of benchmark tests on the CRAY Y-MP computer in both a multiuser and a dedicated environment. The performance of the *cc*MBPT algorithm is monitored as a function of the size of the basis set employed. Furthermore, since this machine has up to eight processors, we are able to monitor the *cc*MBPT algorithm as a function of the number of processors used. Finally, section 7 contains our concluding remarks.

2. MBPT and concurrent computation

The theoretical importance of the linked diagram theorem and the consequential linear scaling of the energy and other expectation values with the number of electrons in the system is widely recognized in theoretical chemistry [13-20] and the majority of contemporary treatments of the correlation problem in molecules are based on a perturbation expansion [21,22]. The sum-over-states expressions arising in each of the terms in the perturbation series can be written as matrix multiplications and, therefore, implemented with near optimal efficiency on vector processing computers [23,24]. The linked diagram theorem also has important computational consequences which are particularly relevant when parallel processing machines are employed [17,18,25] The many-electron problem is decoupled into completely independent subproblems each of which can be handled on a separate processor.

Computationally efficient schemes for evaluating the energy coefficients in any non-relativistic many-body perturbation theory calculation depend on explicitly carrying out the spin integrations. For the two-electron integral defined in (2) we have

$$\begin{aligned} \langle pq|\hat{O}|rs\rangle = {} & \langle PQ|r_{12}^{-1}|RS\rangle\, \delta(\sigma_p,\sigma_r)\, \delta(\sigma_q,\sigma_s) \\ & - \langle PQ|r_{12}^{-1}|SR\rangle\, \delta(\sigma_p,\sigma_s)\, \delta(\sigma_q,\sigma_r) \end{aligned} \qquad (1)$$

where δ is the Kronecker delta.

3. Second order diagrammatic term

The algebraic expressions corresponding to the second order diagram may be written as

$$E_2(A_d) = \frac{1}{4} \sum_{ij} \sum_{ab} \frac{\langle ij|\hat{O}|ab\rangle\, \langle ab|\hat{O}|ij\rangle}{D_{ijab}} \qquad (2)$$

A spin-free formulation may be obtained by explicitly carrying out the spin integrations. We use upper case indices to designate spatial orbitals as opposed to spin orbitals which are labeled by lower case indices. Consider the numerator in the summation involved in the evaluation of $E_2(A_d)$:

$$\mathfrak{T} = \langle i\,j\,|\hat{O}|\,ab\rangle \langle ab\,|\hat{O}|\,i\,j\rangle \tag{3}$$

This gives rise to three distinct spin types:

$$\mathfrak{T}_1 = \langle I\alpha\ J\alpha|\hat{O}|A\alpha\ B\alpha\rangle \langle A\alpha\ B\alpha|\hat{O}|\,I\alpha\ J\alpha\rangle \tag{4a}$$

$$\mathfrak{T}_2 = \langle I\alpha\ J\beta|\hat{O}|A\alpha\ B\beta\rangle \langle A\alpha\ B\beta|\hat{O}|\,I\alpha\ J\beta\rangle \tag{4b}$$

$$\mathfrak{T}_3 = \langle I\alpha\ J\beta|\hat{O}|A\beta\ B\alpha\rangle \langle A\beta\ B\alpha|\hat{O}|\,I\alpha\ J\beta\rangle \tag{4c}$$

together with three further terms in which the rôle of α and β are interchanged.

Explicitly carrying out the spin integrations we obtain

$$\mathfrak{T}_1 = (\langle IJ|\hat{r}_{12}^{-1}|AB\rangle - \langle IJ|\hat{r}_{12}^{-1}|BA\rangle)\ (\langle AB|\hat{r}_{12}^{-1}|\,IJ\rangle - \langle AB|\hat{r}_{12}^{-1}|\,JI\rangle) \tag{5a}$$

$$\mathfrak{T}_2 = \langle IJ|\hat{r}_{12}^{-1}|AB\rangle \langle AB|\hat{r}_{12}^{-1}|\,IJ\rangle \tag{5b}$$

$$\mathfrak{T}_3 = \langle IJ|\hat{r}_{12}^{-1}|BA\rangle \langle AB|\hat{r}_{12}^{-1}|\,JI\rangle \tag{5c}$$

Adding these three components, we get

$$\begin{aligned}\mathfrak{T}_1 + \mathfrak{T}_2 + \mathfrak{T}_3 = {} & 2\,\langle IJ|\hat{r}_{12}^{-1}|AB\rangle \langle AB|\hat{r}_{12}^{-1}|\,IJ\rangle \\ & - \langle IJ|\hat{r}_{12}^{-1}|AB\rangle \langle AB|\hat{r}_{12}^{-1}|\,JI\rangle \\ & - \langle IJ|\hat{r}_{12}^{-1}|AB\rangle \langle AB|\hat{r}_{12}^{-1}|\,IJ\rangle \\ & + 2\,\langle IJ|\hat{r}_{12}^{-1}|BA\rangle \langle AB|\hat{r}_{12}^{-1}|\,JI\rangle\end{aligned} \tag{6}$$

Defining following secondary intermediate quantities

$$T_1 = + \sum_{IJAB} \langle IJ|r_{ij}^{-1}|AB\rangle \langle AB|r_{12}^{-1}|\,IJ\rangle \ /\ D_{IJAB} \tag{7a}$$

$$T_2 = - \sum_{IJAB} \langle IJ|r_{ij}^{-1}|AB\rangle \langle AB|r_{12}^{-1}|\,JI\rangle \ /\ D_{IJAB} \tag{7b}$$

$$T_3 = - \sum_{IJAB} \langle IJ|r_{ij}^{-1}|BA\rangle \langle AB|r_{12}^{-1}|\,IJ\rangle \ /\ D_{IJAB} \tag{7c}$$

$$T_4 = + \sum_{IJAB} \langle IJ|r_{ij}^{-1}|BA\rangle \langle AB|r_{12}^{-1}|JI\rangle \,/\, D_{IJAB} \quad (7d)$$

we can write the total second order energy as

$$E_2(A_d) = 2\,\frac{1}{4}\,(\,2\,T_1 + T_2 + T_3 + 2\,T_4\,) \quad (8)$$

where the first factor of 2 arises from the interchanged of the α and β spin functions. Interchanging the particle indices A and B in the above summations we find that

$$T_1 = T_4 \;\text{ and }\; T_2 = T_3 \quad (9)$$

and so

$$E_2(A_d) = 2\,\frac{1}{4}\,(\,4\,T_1 + 2\,T_2\,) = 2\,T_1 - T_2 \quad (10)$$

which may be written

$$E_2(A_d) = + \sum_{IJAB} \langle IJ|r_{ij}^{-1}|AB\rangle \,(2\,\langle AB|r_{12}^{-1}|IJ\rangle - \langle AB|r_{12}^{-1}|JI\rangle)/D_{IJAB} \quad (11)$$

4. Third-order diagrammatic terms

The third order *hole-hole* diagram corresponds to the expression

$$E_3(A_d) = \frac{1}{8} \sum_{\substack{ijkl\\ab}} \frac{\langle ij|\hat{O}|ab\rangle \;\langle kl|\hat{O}|ij\rangle \;\langle ab|\hat{O}|kl\rangle}{D_{ijab}\; D_{klab}} \quad (12)$$

Again a spin-free formulation may be obtained by explicitly carrying out the spin integrations. Consider the numerator in the summation involved in the evaluation of $E_3(A_d)$:

$$\mathfrak{T} = \langle ij|\hat{O}|ab\rangle \langle kl|\hat{O}|ij\rangle \langle ab|\hat{O}|ij\rangle \quad (13)$$

This gives rise to five distinct spin types:

$$\mathfrak{T}_1 = \langle I\alpha\, J\alpha|\hat{O}|A\alpha\, B\alpha\rangle \langle K\alpha\, L\alpha|\hat{O}|\, I\alpha\, J\alpha\rangle \langle A\alpha\, B\alpha|\hat{O}|K\alpha\, L\alpha\rangle \quad (14a)$$

$$\mathfrak{T}_2 = \langle I\alpha\, J\beta|\hat{O}|A\alpha\, B\beta\rangle \langle K\alpha\, L\beta|\hat{O}|\, I\alpha\, J\beta\rangle \langle A\alpha\, B\beta|\hat{O}|K\alpha\, L\beta\rangle \quad (14b)$$

$$\mathfrak{T}_3 = \langle I\alpha\, J\beta|\hat{O}|A\alpha\, B\beta\rangle \langle K\beta\, L\alpha|\hat{O}|\, I\alpha\, J\beta\rangle \langle A\alpha\, B\beta|\hat{O}|K\beta\, L\alpha\rangle \quad (14c)$$

$$\mathfrak{T}_4 = \langle I\alpha\, J\beta|\hat{O}|A\beta\, B\alpha\rangle \langle K\beta\, L\alpha|\hat{O}|\, I\alpha\, J\beta\rangle \langle A\beta\, B\alpha|\hat{O}|K\beta\, L\alpha\rangle \quad (14d)$$

$$\mathfrak{T}_5 = \langle I\alpha\, J\beta|\hat{O}|A\beta\, B\alpha\rangle \langle K\alpha\, L\beta|\hat{O}|\, I\alpha\, J\beta\rangle \langle A\beta\, B\alpha|\hat{O}|K\alpha\, L\beta\rangle \tag{14e}$$

together with five further terms in which the rôle of α and β are interchanged. Explicitly carrying out the spin integrations we obtain

$$\mathfrak{T}_1 = (\langle IJ|\hat{r}_{12}^{-1}|AB\rangle - \langle IJ|\hat{r}_{12}^{-1}|BA\rangle)\,(\langle KL|\hat{r}_{12}^{-1}|IJ\rangle - \langle KL|\hat{r}_{12}^{-1}|JI\rangle)$$
$$(\langle AB|\hat{r}_{12}^{-1}|KL\rangle - \langle AB|\hat{r}_{12}^{-1}|LK\rangle) \tag{15a}$$

$$\mathfrak{T}_2 = (\langle IJ|\hat{r}_{12}^{-1}|AB\rangle)\,(\langle KL|\hat{r}_{12}^{-1}|IJ\rangle)\,(\langle AB|\hat{r}_{12}^{-1}|KL\rangle) \tag{15b}$$

$$\mathfrak{T}_3 = (\langle IJ|\hat{r}_{12}^{-1}|AB\rangle)\,(\langle KL|\hat{r}_{12}^{-1}|JI\rangle)\,(\langle AB|\hat{r}_{12}^{-1}|LK\rangle) \tag{15c}$$

$$\mathfrak{T}_4 = (\langle IJ|\hat{r}_{12}^{-1}|BA\rangle)\,(\langle KL|\hat{r}_{12}^{-1}|JI\rangle)\,(\langle AB|\hat{r}_{12}^{-1}|KL\rangle) \tag{15d}$$

$$\mathfrak{T}_5 = (\langle IJ|\hat{r}_{12}^{-1}|BA\rangle)\,(\langle KL|\hat{r}_{12}^{-1}|IJ\rangle)\,(\langle AB|\hat{r}_{12}^{-1}|LK\rangle) \tag{15e}$$

Adding these five components, we get

$$\begin{aligned}
\mathfrak{T}_1+\mathfrak{T}_2+\mathfrak{T}_3+\mathfrak{T}_4+\mathfrak{T}_5 &= 2(\langle IJ|\hat{r}_{12}^{-1}|AB\rangle \langle KL|\hat{r}_{12}^{-1}|IJ\rangle \langle AB|\hat{r}_{12}^{-1}|KL\rangle)\\
&- (\langle IJ|\hat{r}_{12}^{-1}|AB\rangle \langle KL|\hat{r}_{12}^{-1}|IJ\rangle \langle AB|\hat{r}_{12}^{-1}|LK\rangle)\\
&- (\langle IJ|\hat{r}_{12}^{-1}|AB\rangle \langle KL|\hat{r}_{12}^{-1}|JI\rangle \langle AB|\hat{r}_{12}^{-1}|KL\rangle)\\
&- (\langle IJ|\hat{r}_{12}^{-1}|BA\rangle \langle KL|\hat{r}_{12}^{-1}|IJ\rangle \langle AB|\hat{r}_{12}^{-1}|KL\rangle)\\
&+ 2(\langle IJ|\hat{r}_{12}^{-1}|AB\rangle \langle KL|\hat{r}_{12}^{-1}|JI\rangle \langle AB|\hat{r}_{12}^{-1}|LK\rangle)\\
&+ 2(\langle IJ|\hat{r}_{12}^{-1}|BA\rangle \langle KL|\hat{r}_{12}^{-1}|IJ\rangle \langle AB|\hat{r}_{12}^{-1}|LK\rangle)\\
&+ 2(\langle IJ|\hat{r}_{12}^{-1}|BA\rangle \langle KL|\hat{r}_{12}^{-1}|JI\rangle \langle AB|\hat{r}_{12}^{-1}|KL\rangle)\\
&- (\langle IJ|\hat{r}_{12}^{-1}|BA\rangle \langle KL|\hat{r}_{12}^{-1}|JI\rangle \langle AB|\hat{r}_{12}^{-1}|LK\rangle)
\end{aligned} \tag{16}$$

Defining the following secondary intermediate quantities

$$T_1 = +\sum_{\substack{IJKL\\AB}} \langle IJ|\hat{r}_{12}^{-1}|AB\rangle \langle KL|\hat{r}_{12}^{-1}|IJ\rangle \langle AB|\hat{r}_{12}^{-1}|KL\rangle / D_{IJAB}\, D_{KLAB}$$

$$T_2 = - \sum_{\substack{IJKL\\AB}} \langle IJ|\hat{r}_{12}^{-1}|AB\rangle \langle KL|\hat{r}_{12}^{-1}|IJ\rangle \langle AB|\hat{r}_{12}^{-1}|LK\rangle / D_{IJAB} D_{KLAB}$$

$$T_3 = - \sum_{\substack{IJKL\\AB}} \langle IJ|\hat{r}_{12}^{-1}|AB\rangle \langle KL|\hat{r}_{12}^{-1}|JI\rangle \langle AB|\hat{r}_{12}^{-1}|KL\rangle / D_{IJAB} D_{KLAB}$$

$$T_4 = - \sum_{\substack{IJKL\\AB}} \langle IJ|\hat{r}_{12}^{-1}|BA\rangle \langle KL|\hat{r}_{12}^{-1}|IJ\rangle \langle AB|\hat{r}_{12}^{-1}|KL\rangle / D_{IJAB} D_{KLAB}$$

$$T_5 = + \sum_{\substack{IJKL\\AB}} \langle IJ|\hat{r}_{12}^{-1}|AB\rangle \langle KL|\hat{r}_{12}^{-1}|JI\rangle \langle AB|\hat{r}_{12}^{-1}|LK\rangle / D_{IJAB} D_{KLAB}$$

$$T_6 = + \sum_{\substack{IJKL\\AB}} \langle IJ|\hat{r}_{12}^{-1}|BA\rangle \langle KL|\hat{r}_{12}^{-1}|IJ\rangle \langle AB|\hat{r}_{12}^{-1}|LK\rangle / D_{IJAB} D_{KLAB}$$

$$T_7 = + \sum_{\substack{IJKL\\AB}} \langle IJ|\hat{r}_{12}^{-1}|BA\rangle \langle KL|\hat{r}_{12}^{-1}|JI\rangle \langle AB|\hat{r}_{12}^{-1}|KL\rangle / D_{IJAB} D_{KLAB}$$

$$T_8 = - \sum_{\substack{IJKL\\AB}} \langle IJ|\hat{r}_{12}^{-1}|BA\rangle \langle KL|\hat{r}_{12}^{-1}|JI\rangle \langle AB|\hat{r}_{12}^{-1}|LK\rangle / D_{IJAB} D_{KLAB}$$

(17)

we can write the energy component as

$$E_3(A_d) = 2\,\frac{1}{8}\,(2T_1 + T_2 + T_3 + T_4 + 2T_5 + 2T_6 + 2T_7 + T_8\,)$$

(18)

where the first factor of 2 arises from the interchanged of the α and β spin functions. Interchanging K and L and/or A and B in the above summations we find that

$$T_1 = T_5 = T_6 = T_7 \text{ and } T_2 = T_3 = T_4 = T_8 \quad (19)$$

and so

$$E_3(A_d) = 2\,\frac{1}{8}\,(\,8\,T_1 + 4\,T_2\,) = 2\,T_1 - T_2 \quad (20)$$

which may be rewritten in the form

$$E_3(A_d) = + \sum_{\substack{IJKL \\ AB}} <IJ|r_{ij}^{-1}|AB> <KL|r_{ij}^{-1}|IJ>$$

$$(2 <AB|r_{12}^{-1}|KL> - <AB|r_{12}^{-1}|LK>)/D_{IJAB}\, D_{KLAB} \tag{21}$$

In a similar fashion, the spin integrations can be explicitly carried out for the third order hole-particle and particle-particle diagrammatic terms leading to the expressions

$$E_3(B_d) = + \sum_{\substack{IJKL \\ AB}} \Big[(2<IJ|r_{ij}^{-1}|AB>-<IJ|r_{ij}^{-1}|AB>)$$

$$(2<AK|r_{ij}^{-1}|IC>-<AK|r_{ij}^{-1}|CI>)(2<BC|r_{12}^{-1}|JK>-<BC|r_{12}^{-1}|KJ>)$$

$$-3 <IJ|r_{ij}^{-1}|BA><AK|r_{ij}^{-1}|CI><BC|r_{12}^{-1}|KJ>\Big]/D_{IJAB}\, D_{JKBC} \tag{22}$$

$$E_3(C_d) = + \sum_{\substack{IJKL \\ AB}} <IJ|r_{ij}^{-1}|AB> <AB|r_{ij}^{-1}|CD>$$

$$(2 <CD|r_{12}^{-1}|IJ> - <CD|r_{12}^{-1}|JI>)/D_{IJAB}\, D_{IJCD} \tag{23}$$

5. Fourth order diagrammatic terms

5.1 Fourth order diagrammatic terms involving single excitations

For diagram A_s the expression may be rearranged as follows

$$E_4(A_s) = \frac{1}{4} \sum_{ic} \left(\left\{ \sum_{jab} \frac{<ij|\hat{O}|ab> <ab|\hat{O}|cj>}{D_{ijab}} \right\} \left\{ \sum_{kcd} \frac{<ck|\hat{O}|de> <de|\hat{O}|ik>}{D_{ikde}} \right\} /D_{ic} \right) \tag{24}$$

which may be written

$$E_4(A_s) = \frac{1}{4} \sum_{ia} f_{ia} f_{ia}/D_{ia} \tag{25}$$

$$f_{ia} = \sum_{jbc} \frac{<ij|\hat{O}|bc> <bc|\hat{O}|aj>}{D_{ijbc}} \tag{26}$$

$$g_{ia} = \sum_{jkb} \frac{<jk|\hat{O}|ab> <ib|\hat{O}|jk>}{D_{jkab}} \tag{27}$$

Spin-free intermediates corresponding to f_{ia} and g_{ia} may be obtained by explicitly carrying out the spin integrations. They may be defined as follows

$$F_{IA}^{\alpha\alpha} = 2\, T_1^f + T_2^f + T_3^f + 2\, T_4^f \tag{28a}$$

$$G_{IA}^{\alpha\alpha} = 2\, T_1^g + T_2^g + T_3^g + 2\, T_4^g \tag{28b}$$

where the following secondary intermediate quantities have been used

$$T_1^f = + \sum_{JBC} <IJ|r_{ij}^{-1}|BC> <BC|r_{12}^{-1}|AJ> \, / \, D_{IJBC} \tag{29a}$$

$$T_2^f = - \sum_{JBC} <IJ|r_{ij}^{-1}|BC> <BC|r_{12}^{-1}|JA> \, / \, D_{IJBC} \tag{29b}$$

$$T_3^f = - \sum_{JBC} <IJ|r_{ij}^{-1}|CB> <BC|r_{12}^{-1}|AJ> \, / \, D_{IJBC} \tag{29c}$$

$$T_4^f = + \sum_{JBC} <IJ|r_{ij}^{-1}|CB> <BC|r_{12}^{-1}|JA> \, / \, D_{IJBC} \tag{29d}$$

$$T_1^g = + \sum_{JKB} <JK|r_{ij}^{-1}|AB> <IB|r_{12}^{-1}|JK> \, / \, D_{JKAB} \tag{29e}$$

$$T_2^g = - \sum_{JKB} <JK|r_{ij}^{-1}|AB> <IB|r_{12}^{-1}|KJ> \, / \, D_{JKAB} \tag{29f}$$

$$T_3^g = - \sum_{JKB} <JK|r_{ij}^{-1}|BA> <IB|r_{12}^{-1}|JK> \, / \, D_{IJAB} \tag{29g}$$

$$T_4^g = + \sum_{JKB} <JK|r_{ij}^{-1}|BA> <IB|r_{12}^{-1}|KJ> \, / \, D_{IJAB} \tag{29h}$$

However, by interchanging B and C we see that $T_1^f = T_4^f$ and $T_2^f = T_3^f$ and by interchanging J and K we have $T_1^g = T_4^g$ and $T_2^g = T_3^g$ so that

$$F_{IA}^{\alpha\alpha} = 4\, T_1^f + 2\, T_2^f \tag{30a}$$

$$G_{IA}^{\alpha\alpha} = 4\, T_1^g + 2\, T_2^g \tag{30b}$$

$$F^{\alpha\alpha}_{IA} = 2 \sum_{JBC} D^{-1}_{IJBC} \langle IJ|r^{-1}_{ij}|BC\rangle \, (2 \langle BC|r^{-1}_{12}|AJ\rangle - \langle BC|r^{-1}_{12}|JA\rangle)$$

(31a)

$$G^{\alpha\alpha}_{IA} = 2 \sum_{JKB} D^{-1}_{JKAB} \langle JK|r^{-1}_{ij}|AB\rangle \, (2 \langle IB|r^{-1}_{12}|JK\rangle - \langle IB|r^{-1}_{12}|KJ\rangle)$$

(31b)

The energy components corresponding to the diagrams A_s, B_s, C_s and D_s may then be written as

$$E_4(A_s) = \frac{1}{4} \sum_{IA} F^{\alpha\alpha}_{IA} . F^{\alpha\alpha}_{IA} \, / \, D_{IA} \tag{32a}$$

$$E_4(B_s) = -\frac{1}{4} \sum_{IA} F^{\alpha\alpha}_{IA} . G^{\alpha\alpha}_{IA} \, / \, D_{IA} \tag{32b}$$

$$E_4(C_s) = -\frac{1}{4} \sum_{IA} G^{\alpha\alpha}_{IA} . F^{\alpha\alpha}_{IA} \, / \, D_{IA} \tag{32c}$$

$$E_4(D_s) = \frac{1}{4} \sum_{IA} G^{\alpha\alpha}_{IA} . G^{\alpha\alpha}_{IA} \, / \, D_{IA} \tag{32d}$$

The following equality exist among the fourth order terms involving single excitations

$$E_4(B_s) = E_4(C_s) \tag{33}$$

5.2 Fourth order diagrammatic terms involving double excitations

For diagram A_d the energy expression may be rearranged as follows

$$E_4(A_d) = \frac{1}{16} \sum_{\substack{ij\\cd}} \left(\left\{ \sum_{ab} \frac{\langle ij|\hat{O}|ab\rangle \; \langle ab|\hat{O}|cd\rangle}{D_{ijab}} \right\} \left\{ \sum_{ef} \frac{\langle cd|\hat{O}|ef\rangle \; \langle ef|\hat{O}|ij\rangle}{D_{ijef}} \right\} / D_{ijcd} \right) \tag{34}$$

which may be written

$$E_4(A_d) = \frac{1}{16} \sum_{\substack{ij\\ab}} f_{ijab} f_{ijab} / D_{ijab} \tag{35}$$

The energy expressions corresponding to diagrams A_d - L_d may be written in terms of the following intermediates:

$$f_{ijab} = \sum_{cd} \frac{<ij|\hat{O}|cd> <cd|\hat{O}|ab>}{D_{ijcd}} \tag{36a}$$

$$g_{ijab} = \sum_{kl} \frac{<kl|\hat{O}|ab> <ij|\hat{O}|kl>}{D_{klab}} \tag{36b}$$

$$h_{ijab} = \sum_{kc} \frac{<ik|\hat{O}|ac> <jc|\hat{O}|kb>}{D_{ikac}} \tag{36c}$$

Spin-free intermediates corresponding to f_{ijab}, g_{ijab} and h_{ijab} may then be defined as follows

$$F^{\alpha\alpha\alpha\alpha}_{IJAB} = T^f_1 + T^f_2 + T^f_3 + T^f_4 \tag{37a}$$

$$F^{\alpha\beta\alpha\beta}_{IJAB} = T^f_1 + T^f_4 \tag{37b}$$

$$F^{\alpha\beta\beta\alpha}_{IJAB} = T^f_2 + T^f_3 \tag{37c}$$

$$G^{\alpha\alpha\alpha\alpha}_{IJAB} = T^g_1 + T^g_2 + T^g_3 + T^g_4 \tag{37d}$$

$$G^{\alpha\beta\alpha\beta}_{IJAB} = T^g_1 + T^g_4 \tag{37e}$$

$$G^{\alpha\beta\beta\alpha}_{IJAB} = T^g_2 + T^g_3 \tag{37f}$$

$$H^{\alpha\alpha\alpha\alpha}_{IJAB} = T^h_1 + T^h_2 + T^h_3 + T^h_4 \tag{37g}$$

$$H^{\alpha\beta\alpha\beta}_{IJAB} = T^h_1 + T^h_4 \tag{37h}$$

$$H^{\alpha\beta\beta\alpha}_{IJAB} = T^h_2 + T^h_3 \tag{37i}$$

where the following secondary intermediate quantities have been used

$$T^f_1 = + \sum_{CD} <IJ|r^{-1}_{ij}|CD> <CD|r^{-1}_{12}|AB> / D_{IJCD}$$

$$T^f_2 = - \sum_{CD} <IJ|r^{-1}_{ij}|CD> <CD|r^{-1}_{12}|BA> / D_{IJCD}$$

$$T^f_3 = - \sum_{CD} <IJ|r^{-1}_{ij}|DC> <CD|r^{-1}_{12}|AB> / D_{IJCD}$$

$$T_4^f = + \sum_{CD} \langle IJ | r_{ij}^{-1} | DC \rangle \langle CD | r_{12}^{-1} | BA \rangle \, / \, D_{IJCD}$$

$$T_1^g = + \sum_{KL} \langle KL | r_{ij}^{-1} | AB \rangle \langle IJ | r_{12}^{-1} | KL \rangle \, / \, D_{KLAB}$$

$$T_2^g = - \sum_{KL} \langle KL | r_{ij}^{-1} | BA \rangle \langle IJ | r_{12}^{-1} | KL \rangle \, / \, D_{KLAB}$$

$$T_3^g = - \sum_{KL} \langle KL | r_{ij}^{-1} | AB \rangle \langle IJ | r_{12}^{-1} | LK \rangle \, / \, D_{KLAB}$$

$$T_4^g = + \sum_{KL} \langle KL | r_{ij}^{-1} | BA \rangle \langle IJ | r_{12}^{-1} | LK \rangle \, / \, D_{KLAB}$$

$$T_1^h = + \sum_{KC} \langle IK | r_{ij}^{-1} | AC \rangle \langle JC | r_{12}^{-1} | KB \rangle \, / \, D_{IKAC}$$

$$T_2^h = - \sum_{KC} \langle IK | r_{ij}^{-1} | AC \rangle \langle JC | r_{12}^{-1} | BK \rangle \, / \, D_{IKAC}$$

$$T_3^h = - \sum_{KC} \langle IK | r_{ij}^{-1} | CA \rangle \langle JC | r_{12}^{-1} | KB \rangle \, / \, D_{IKAC}$$

$$T_4^h = + \sum_{KC} \langle IK | r_{ij}^{-1} | CA \rangle \langle JC | r_{12}^{-1} | BK \rangle \, / \, D_{IKAC} \qquad (38)$$

Interchanging C and D in the summations for T_i^f we see that $T_1^f = T_4^f$ and $T_2^f = T_3^f$ whilst interchanging K and L in the summations for T_i^g we see that $T_1^g = T_4^g$ and $T_2^g = T_3^g$ and so

$$F_{IJAB}^{\alpha\alpha\alpha\alpha} = 2\,(T_1^f + T_2^f) \qquad (39a)$$

$$F_{IJAB}^{\alpha\beta\alpha\beta} = 2\,T_1^f \qquad (39b)$$

$$F_{IJAB}^{\alpha\beta\beta\alpha} = 2\,T_2^f \qquad (39c)$$

$$G_{IJAB}^{\alpha\alpha\alpha\alpha} = 2\,(T_1^g + T_2^g) \qquad (39d)$$

$$G_{IJAB}^{\alpha\beta\alpha\beta} = 2\,T_1^g \qquad (39e)$$

$$G_{IJAB}^{\alpha\beta\beta\alpha} = 2\,T_2^g \qquad (39f)$$

$$F^{\alpha\alpha\alpha\alpha}_{IJAB} = F^{\alpha\beta\alpha\beta}_{IJAB} + F^{\alpha\beta\beta\alpha}_{IJAB} \tag{39g}$$

$$G^{\alpha\alpha\alpha\alpha}_{IJAB} = G^{\alpha\beta\alpha\beta}_{IJAB} + G^{\alpha\beta\beta\alpha}_{IJAB} \tag{39h}$$

The energy components corresponding to the diagram A_d, for example, may then be written as

$$E_4(A_d) = +\frac{1}{16}\sum_{IJAB}\left[F^{\alpha\alpha\alpha\alpha}_{IJAB}.F^{\alpha\alpha\alpha\alpha}_{IJAB} + F^{\alpha\beta\alpha\beta}_{IJAB}.F^{\alpha\beta\alpha\beta}_{IJAB} + F^{\alpha\beta\beta\alpha}_{IJAB}.F^{\alpha\beta\beta\alpha}_{IJAB}\right] / D_{IJAB} \tag{40}$$

Similar expressions can be written for diagrams B_d - L_d

5.3 Fourth order diagrammatic terms involving triple excitations

For diagram A_t the expression may be rearranged as follows

$$E_4(A_t) = -\frac{1}{2}\sum_{ijk}\sum_{bcd}\left(\left\{\sum_{a}\frac{<ij|\hat{O}|ab>\ <ak|\hat{O}|cd>}{D_{ijab}}\right\}\left\{\sum_{e}\frac{<cb|\hat{O}|ek>\ <ed|\hat{O}|ij>}{D_{ijde}}\right\}/D_{ijkbcd}\right) \tag{41}$$

which may be written

$$E_4(A_t) = -\frac{1}{2}\sum_{ijk}\sum_{abc} f_{ijk;abc}\, f_{ijk;acb} / D_{ijkabc} \tag{42}$$

with

$$f_{ijk;abc} = \sum_{d} <ij|\hat{O}|db>\ <dk|\hat{O}|ac>\ /\ D_{ijbd} \tag{43}$$

$$g_{ijk;abc} = \sum_{l} <jl|\hat{O}|ba>\ <ik|\hat{O}|lc>\ /\ D_{jlab} \tag{44}$$

There are nine spin-free intermediates for corresponding to $f_{ijk;abc}$:

$$F^{\alpha\alpha\alpha;\alpha\alpha\alpha}_{IJK;ABC} = T^f_1 + T^f_2 + T^f_3 + T^f_4$$

$$F^{\alpha\alpha\beta;\alpha\alpha\beta}_{IJK;ABC} = T^f_1 + T^f_3$$

$$F^{\alpha\alpha\beta;\beta\alpha\alpha}_{IJK;ABC} = T^f_2 + T^f_4$$

$$F^{\alpha\beta\alpha;\alpha\beta\alpha}_{IJK;ABC} = T^f_1 + T^f_2$$

$$F^{\alpha\beta\beta;\alpha\beta\beta}_{IJK;ABC} = T^f_1$$

$$F^{\alpha\beta\beta;\beta\beta\alpha}_{IJK;ABC} = T^f_2$$

$$F^{\beta\alpha\alpha;\alpha\beta\alpha}_{IJK;ABC} = T^f_3 + T^f_4$$

$$F^{\beta\alpha\beta;\alpha\beta\beta}_{IJK;ABC} = T^f_3$$

$$F^{\beta\alpha\beta;\beta\beta\alpha}_{IJK;ABC} = T^f_4 \tag{45}$$

where the following secondary intermediate quantities have been used

$$T^f_1 = + \sum_D <IJ|r^{-1}_{ij}|DB> <DK|r^{-1}_{12}|AC> / D_{IJBD}$$

$$T^f_2 = - \sum_D <IJ|r^{-1}_{ij}|DB> <DK|r^{-1}_{12}|CA> / D_{IJBD}$$

$$T^f_3 = - \sum_D <IJ|r^{-1}_{ij}|BD> <DK|r^{-1}_{12}|AC> / D_{IJBD}$$

$$T^f_4 = + \sum_D <IJ|r^{-1}_{ij}|BD> <DK|r^{-1}_{12}|CA> / D_{IJBD} \tag{46}$$

Nine spin-free intermediates corresponding to $g_{ijk;abc}$ can be defined in a similar fashion.

The energy component corresponding to diagrams F_t, for example, may then be cast in the form

$$\begin{aligned} E_4(F_t) = \frac{1}{2} \sum_{IJKABC} \{ & F^{\alpha\alpha\alpha;\alpha\alpha\alpha}_{IJK;ABC} \cdot F^{\alpha\alpha\alpha;\alpha\alpha\alpha}_{IJK;ABC} + F^{\alpha\alpha\beta;\alpha\alpha\beta}_{IJK;ABC} \cdot F^{\alpha\alpha\beta;\alpha\alpha\beta}_{IJK;ABC} + \\ & F^{\alpha\alpha\beta;\beta\alpha\alpha}_{IJK;ABC} \cdot F^{\alpha\alpha\beta;\beta\alpha\alpha}_{IJK;ABC} + F^{\alpha\beta\alpha;\alpha\beta\alpha}_{IJK;ABC} \cdot F^{\alpha\beta\alpha;\alpha\beta\alpha}_{IJK;ABC} + \\ & F^{\alpha\beta\beta;\alpha\beta\beta}_{IJK;ABC} \cdot F^{\alpha\beta\beta;\alpha\beta\beta}_{IJK;ABC} + F^{\alpha\beta\beta;\beta\beta\alpha}_{IJK;ABC} \cdot F^{\alpha\beta\beta;\beta\beta\alpha}_{IJK;ABC} + \\ & F^{\beta\alpha\alpha;\alpha\beta\alpha}_{IJK;ABC} \cdot F^{\beta\alpha\alpha;\alpha\beta\alpha}_{IJK;ABC} + F^{\beta\alpha\beta;\alpha\beta\beta}_{IJK;ABC} \cdot F^{\beta\alpha\beta;\alpha\beta\beta}_{IJK;ABC} + \\ & F^{\beta\alpha\beta;\beta\beta\alpha}_{IJK;ABC} \cdot F^{\beta\alpha\beta;\beta\beta\alpha}_{IJK;ABC} \} \end{aligned} \tag{47}$$

5.4 Fourth order diagrammatic terms involving quadruple excitations

The energy expression corresponding to diagram A_q may be cast in the form

$$E_4(A_q) = \frac{1}{2} \sum_{\substack{ik \\ bd}} \left(\sum_{lc} \frac{\langle kl|\hat{O}|cd\rangle \langle cb|\hat{O}|il\rangle}{D_{ilbc}} \right) \left(\sum_{ja} \frac{\langle ij|\hat{O}|ab\rangle \langle ad|\hat{O}|kj\rangle}{D_{ijab}\, D_{jkad}} \right) \tag{48}$$

or

$$E_4(A_q) = \frac{1}{2} \sum_{\substack{ij \\ ab}} f_{ijab}\, g_{ijab} \tag{49}$$

where

$$f_{ijab} = \sum_{kc} \frac{\langle ik|\hat{O}|ca\rangle \langle cb|\hat{O}|jk\rangle}{D_{ikac}} \tag{50}$$

and

$$g_{ijab} = \sum_{kc} \frac{\langle ik|\hat{O}|ca\rangle \langle cb|\hat{O}|jk\rangle}{D_{ikac}\, D_{jkbc}} \tag{51}$$

Spin-free intermediates corresponding to f_{ijab} and g_{ijab} may then be defined as follows

$$F^{\alpha\alpha\alpha\alpha}_{IJAB} = T^f_1 + T^f_2 + T^f_3 + 2\, T^f_4$$

$$F^{\alpha\alpha\beta\beta}_{IJAB} = T^f_1$$

$$F^{\alpha\beta\alpha\beta}_{IJAB} = T^f_2 + T^f_3 + 2\, T^f_4$$

$$G^{\alpha\alpha\alpha\alpha}_{IJAB} = T^g_1 + T^g_2 + T^g_3 + 2\, T^g_4$$

$$G^{\alpha\alpha\beta\beta}_{IJAB} = T^g_1$$

$$G^{\alpha\beta\alpha\beta}_{IJAB} = T^g_2 + T^g_3 + 2\, T^g_4$$

$$F^{\alpha\alpha\alpha\alpha}_{IJAB} = F^{\alpha\alpha\beta\beta}_{IJAB} + F^{\alpha\beta\alpha\beta}_{IJAB}$$

$$G^{\alpha\alpha\alpha\alpha}_{IJAB} = G^{\alpha\alpha\beta\beta}_{IJAB} + G^{\alpha\beta\alpha\beta}_{IJAB} \tag{52}$$

where the following secondary intermediates have been defined

$$T^f_1 = + \sum_{KC} \langle IK|r^{-1}_{ij}|CA\rangle \langle CB|r^{-1}_{12}|JK\rangle \;/\; D_{IKAC}$$

$$T^f_2 = - \sum_{KC} \langle IK|r^{-1}_{ij}|CA\rangle \langle CB|r^{-1}_{12}|KJ\rangle \;/\; D_{IKAC}$$

$$T^f_3 = - \sum_{KC} \langle IK|r^{-1}_{ij}|AC\rangle \langle CB|r^{-1}_{12}|JK\rangle \;/\; D_{IKAC}$$

$$T^f_4 = + \sum_{KC} \langle IK|r^{-1}_{ij}|AC\rangle \langle CB|r^{-1}_{12}|KJ\rangle \;/\; D_{IKAC}$$

$$T^g_1 = + \sum_{KC} \langle IK|r^{-1}_{ij}|CA\rangle \langle CB|r^{-1}_{12}|JK\rangle \;/\; (D_{IKAC}\, D_{JKBC})$$

$$T^g_2 = - \sum_{KC} \langle IK|r^{-1}_{ij}|CA\rangle \langle CB|r^{-1}_{12}|KJ\rangle \;/\; (D_{IKAC}\, D_{JKBC})$$

$$T^g_3 = - \sum_{KC} \langle IK|r^{-1}_{ij}|AC\rangle \langle CB|r^{-1}_{12}|JK\rangle \;/\; (D_{IKAC}\, D_{JKBC})$$

$$T^g_4 = + \sum_{KC} \langle IK|r^{-1}_{ij}|AC\rangle \langle CB|r^{-1}_{12}|KJ\rangle \;/\; (D_{IKAC}\, D_{JKBC})$$

(53)

Note that the T^f_i's only differ from the T^g_i's by the presence of an additional factor in the denominator.

The energy component corresponding to diagram A_q may then be written

$$E_4(A_q) = \sum_{IJ} \sum_{AB} \Big[F^{\alpha\alpha\alpha\alpha}_{IJAB} G^{\alpha\alpha\alpha\alpha}_{IJAB} + F^{\alpha\alpha\beta\beta}_{IJAB} G^{\alpha\alpha\beta\beta}_{IJAB} + F^{\alpha\beta\alpha\beta}_{IJAB} G^{\alpha\beta\alpha\beta}_{IJAB} \Big] \tag{54}$$

The expression corresponding to the sum of diagrams B_q and C_q may be cast in the form

$$E_4(B_q + C_q) = \frac{1}{16} \sum_{\substack{ij\\kl}} \left(\sum_{ab} \frac{\langle ij|\hat{O}|ab\rangle \;\langle ab|\hat{O}|kl\rangle}{D_{ijab}} \right)$$

$$\left(\sum_{cd} \frac{<kl|\hat{O}|cd> <cd|\hat{O}|ij>}{D_{ijcd} D_{klcd}} \right) \tag{55}$$

$$E_4(B_q+C_q) = \frac{1}{16} \sum_{\substack{ij\\kl}} f_{ijkl}\, g_{ijkl} \tag{56}$$

where

$$f_{ijkl} = \sum_{ab} \frac{<ij|\hat{O}|ab> <ab|\hat{O}|kl>}{D_{ijab}} \tag{57}$$

and

$$g_{ijkl} = \sum_{ab} \frac{<kl|\hat{O}|ab> <ab|\hat{O}|ij>}{D_{ijab} D_{klab}} \tag{58}$$

Spin-free intermediates corresponding to f_{ijkl} and g_{ijkl} may then be defined; we leave the details as a exercise for the reader.

The expression corresponding to the sum of diagrams D_q and E_q may be cast in the form

$$E_4(D_q+E_q) = -\frac{1}{4} \sum_{ac} \left(\sum_{kld} \frac{<kl|\hat{O}|cd> <ad|\hat{O}|kl>}{D_{klad}} \right)$$

$$\left(\sum_{ijb} \frac{<ij|\hat{O}|ab> <cb|\hat{O}|ij>}{D_{ijab} D_{ijbc}} \right) \tag{59}$$

or

$$E_4(D_q+E_q) = -\frac{1}{4} \sum_{ab} f_{ab}\, g_{ab} \tag{60}$$

where

$$f_{ab} = \sum_{ijc} \frac{<ij|\hat{O}|ac> <bc|\hat{O}|ij>}{D_{ijac}} \tag{61}$$

and

$$g_{ab} = \sum_{ijc} \frac{<ij|\hat{O}|bc> <ac|\hat{O}|ij>}{D_{ijbc} D_{ijac}} \tag{62}$$

Spin-free intermediates corresponding to f_{ab} and g_{ab} may then be defined; again we leave the details to the reader.

The sum of the expressions corresponding to diagrams F_q and G_q may be written in the form

$$E_4(F_q+G_q) = -\frac{1}{2}\sum_{ik}\left(\sum_{lcd}\frac{\langle kl|\hat{O}|cd\rangle\langle cd|\hat{O}|il\rangle}{D_{ilcd}}\right)\left(\sum_{jab}\frac{\langle ij|\hat{O}|ab\rangle\langle ab|\hat{O}|kj\rangle}{D_{ijab}D_{jkab}}\right) \quad (63)$$

or

$$E_4(F_q+G_q) = -\frac{1}{2}\sum_{ij} f_{ij}\, g_{ij} \quad (64)$$

where

$$f_{ij} = \sum_{kab}\frac{\langle jk|\hat{O}|ab\rangle\langle ab|\hat{O}|ik\rangle}{D_{ilab}} \quad (65)$$

and

$$g_{ij} = \sum_{kab}\frac{\langle ik|\hat{O}|ab\rangle\langle ab|\hat{O}|jk\rangle}{D_{ikab}D_{jkab}} \quad (66)$$

Spin-free intermediates corresponding to f_{ij} and g_{ij} may then be defined and again we leave the details to the interested reader.

6. *concurrent computation* Many body perturbation theory

6.1 *cc*MBPT

We illustrate the use of concurrent computation in many-body perturbation calculations by means of example. Consider the fourth order energy components corresponding to diagrams containing triply excited intermediate states. These terms are the most demanding of the fourth order terms leading to an algorithm which scales as the seventh power of the number of basis functions. In particular, consider the diagrams labelled F_t, G_t and J_t in the previous article (see also [26, 27]).

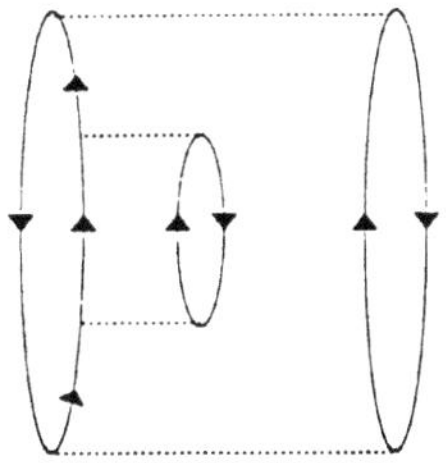

(67a)

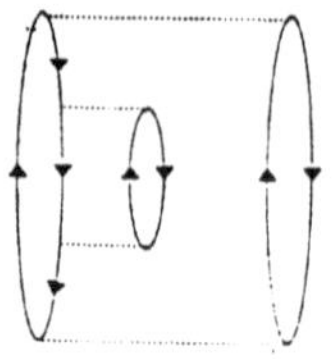

(67b)

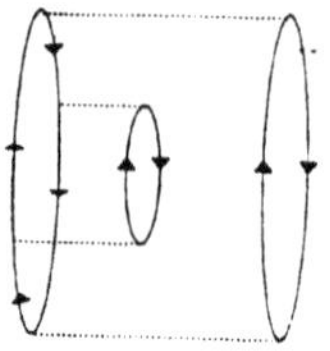

(67c)

Using the rules given in the previous article, the algebraic expressions for the energy components corresponding to diagrams F_t, G_t and J_t may be written as follows

$$E_4(F_t) = \frac{1}{4} \sum_{\substack{ijk \\ abcde}} \frac{\langle ij|\hat{O}|ab\rangle \langle ak|\hat{O}|cd\rangle \langle cd|\hat{O}|ek\rangle \langle eb|\hat{O}|ij\rangle}{D_{ijab}\, D_{ijkbcd}\, D_{ijbe}} \tag{68a}$$

$$E_4(G_t) = \frac{1}{4} \sum_{\substack{ijklm \\ abc}} \frac{\langle ij|\hat{O}|ab\rangle \langle kl|\hat{O}|ic\rangle \langle mc|\hat{O}|kl\rangle \langle ab|\hat{O}|mj\rangle}{D_{ijab}\, D_{jklabc}\, D_{jmab}} \tag{68b}$$

$$E_4(J_t) = - \sum_{\substack{ijkl \\ abcd}} \frac{\langle ij|\hat{O}|ab\rangle \langle ak|\hat{O}|cd\rangle \langle ld|\hat{O}|ik\rangle \langle cb|\hat{O}|lj\rangle}{D_{ijab}\, D_{ijkbcd}\, D_{jlbc}} \tag{68c}$$

Diagram F_t gives rise to summations over 3 occupied spin orbital indices and 5 virtual spin orbital indices, diagram G_t to summations over 5 occupied spin orbital indices and 3 virtual spin orbital indices, and diagram J_t leads to summations over 4 occupied and 4 virtual spin orbital indices.

In this work, we modify the rules presented previously in order to devise ccMBPT algorithms. We introduce an additional diagrammatic convention :-

Any particle or hole line which is labeled is not summed over and is not regarded as equivalent to any unlabeled line or any labeled line with a different label.

Thus, for example, the diagram

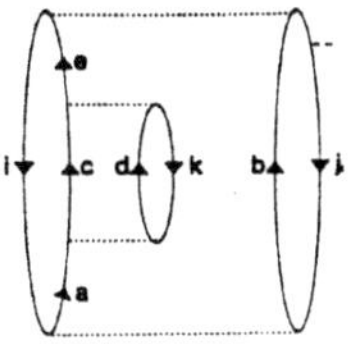

(69)

in which all lines are labelled corresponds to the algebraic expression.

$$\frac{\langle ij|\hat{O}|ab\rangle \; \langle ak|\hat{O}|cd\rangle \; \langle cd|\hat{O}|ek\rangle \; \langle eb|\hat{O}|ij\rangle}{D_{ijab}\; D_{ijkbcd}\; D_{ijbe}} \tag{70}$$

The partially labeled diagram

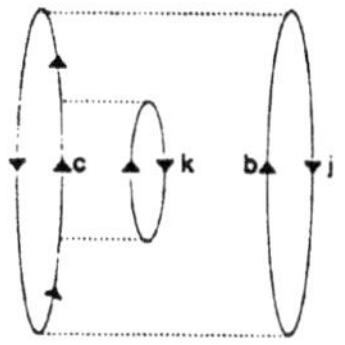

(71)

corresponds to the algebraic expression

$$\sum_{iad} \frac{\langle ij|\hat{O}|ab\rangle \; \langle ak|\hat{O}|cd\rangle \; \langle cd|\hat{O}|ek\rangle \; \langle eb|\hat{O}|ij\rangle}{D_{ijab}\; D_{ijkbcd}\; D_{ijbe}} \tag{72}$$

It is then clear that for diagram F_t we can write

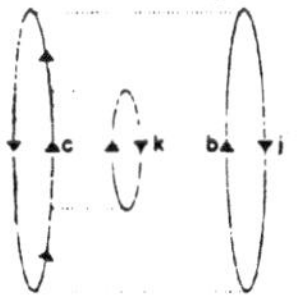 $= \sum_{iade}$ 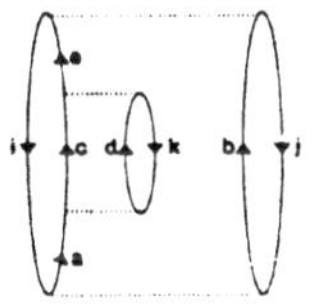 (73)

and

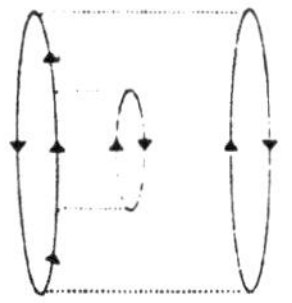 $= \frac{1}{4} \sum_{jkbc}$ 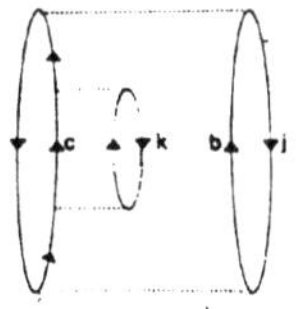 (74)

Similarly, for diagram G_t we can write

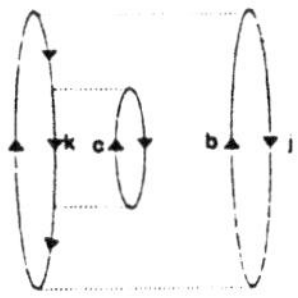 $= \sum_{ilma}$ 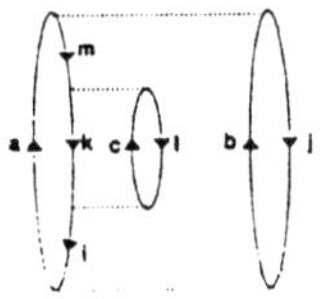 (75)

and

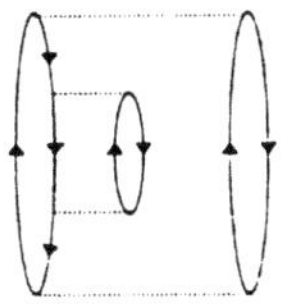 $= \frac{1}{4} \sum_{jkbc}$ 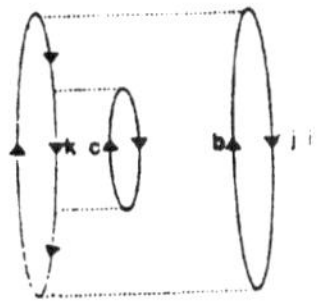 (76)

For diagram J_t we can put

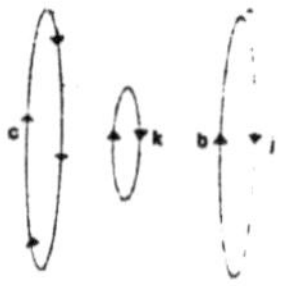 $= \sum_{ilad}$  (77)

and

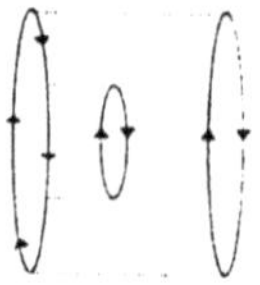 $= \frac{1}{4} \sum_{jkbc}$ 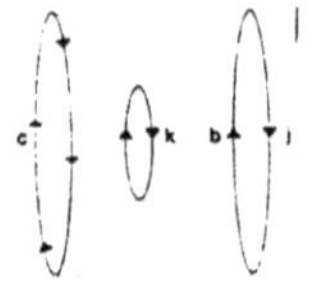 (78)

An algorithm for the evaluation of all fourth order triple excitation diagrams, which scales as the seventh power of the number of basis functions employed in the calculation, may be obtained by defining the intermediate f_{ijkabc} (see equation (43)), which involves a summation over a virtual spin orbital and, using the rule given above, may be represented diagrammatically as

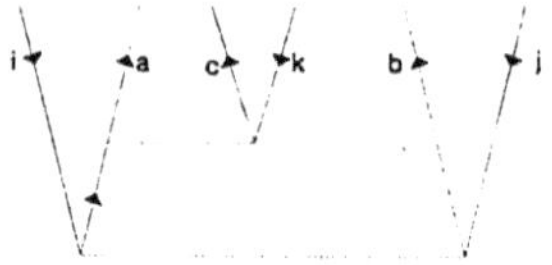

(79)

and the intermediate g_{ijkabc} (equation (44)), which involves a summation over an occupied spin orbital and may be represented by

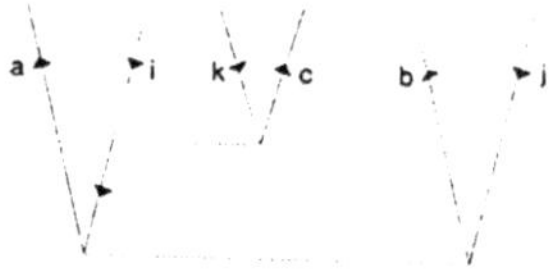

(80)

We emphasize that the kernel of the algorithm, the formation of the f_{ijkabc} and g_{ijkabc} intermediates, consists of matrix multiplications which can be executed very efficiently on vector processors.

6.2 Dynamic load balancing

In this article, we adopt the following notation: N denotes the number of parallel processes and T the number of tasks. I identifies a task and ranges from 1 to T. P identifies a process and, therefore, ranges from 1 to N.

Our algorithm for the evaluation of the contribution of the triply substituted configurations to the fourth order energy gives rise to a number of tasks of roughly equal size. Each task corresponds to the evaluation of the energy associated with the partially labeled diagram (71). Each task may proceed independently of the others. Orbital symmetries and the spatial disposition of the atoms upon which the basis functions are centred will influence the size of the tasks.

On shared memory architecture parallel computers (see Figure 1), static load balancing of the work assigned to each process is usually far from optimal. On a dedicated machine the tasks may not be of equal size and some idle time will inevitably accumulate. In a multiuser environment, each processor is not normally dedicated to a single user; instead its computing power is shared out amongst a number of users. The effective computing power delivered to a given user (or a parallel process of that user) by a given processor is largely unpredictable, because it depends upon what other users are doing at any given time. Thus instead of requiring that all processors carry out an identical workload, it is better to require that each processor be given a workload commensurate with its effective power; the latter is far from constant with time, so that it is necessary to feed work to the processors at a rate which is similarly time dependent.

This may be accomplished by the technique of "dynamic load balancing", using "global indices" under control of a "lock" [28, 29]. Standard FORTRAN code for the evaluation of the energy component corresponding to diagram F_t might take the following form:

```
      DO 1 I = 1,T
      Execute task I
  1   CONTINUE
```

where task I involves the evaluation of the partially labelled diagram (71). Using CRAY Y-MP multi-tasking software the dynamically balanced parallel version using N processes would be of the form shown in Figure 2. During the parallel phase of the job N copies of subroutine PROCESS are in execution. On the CRAY system PROCESS is made re-entrant, circumventing the need to actually make the copies. PROCESS then has the form shown in Figure 3. LOCKON is called to acquire unique access to the global index I which is being protected by LOCKI; if another parallel copy of PROCESS calls LOCKON whilst such unique access is granted, the second copy will be halted until the first copy releases access with a call to LOCKOFF. If I is greater than T then all tasks have been completed, and after releasing LOCKI, PROCESS returns control to the calling program. A local copy of the global index is taken in ITEMP, and the global index updated, after which LOCKI is released. The task indicated by ITEMP (note that the global index itself is not used in this phase since it may be updated by another parallel copy of PROCESS) is then carried out, after which the routine branches to the first executable statement to see if work remains. Obviously, if task execution is controlled by a nest of loops (in the present work there is a four-fold nest of loops over the tasks), then all loop indices associated with the nest should be treated as a shared locked data structure and accessed under control of a lock. The overheads associated with the above implementation are T O_r, where O_r is the time to acquire/release LOCKI. This is of the order of a few microseconds on the CRAY Y-MP system. The individual tasks are relatively large and, therefore, contention for access to the lock is negligible. Overheads can be largely ignored. Each process generates only as much work as its corresponding processor is able to deal with (it is dynamically balanced). Given a reasonably large number of tasks, dynamic balance produces an approximately even loading over dedicated processors even if the tasks are not of equal size. The major disadvantage of dynamic balance is that it requires the use of locks, hardware support for which is unavailable (and software emulations expensive - probably requiring that all processing of a lock be carried out by one processor which communicates its results to the other processors by message passing) on many current independent memory machines. Its use is at present restricted mainly to shared memory parallel architectures, which all provide hardware supported locks. We here observe that current shared memory systems such as are available from CRAY, IBM, Convex, Alliant, Sequent, Encore and FPS are capable of running each processor in multi-user mode (and thus require dynamic balance and the associated locks for optimal performance).

The final evaluation of the fourth order energy corresponding to diagram X_t can be written as:

$$\Delta E_{total}[X_t] = \sum_{I}^{T} \Delta E[X_t](I) \qquad (81)$$

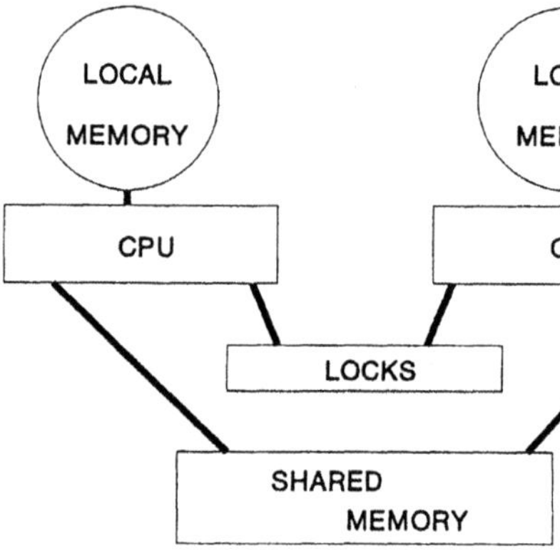

Figure 1. *Schematic diagram of a shared memory parallel system.*

where $\Delta E[X_t](I)$ is the contribution to $\Delta E_{total}[X_t]$ from the I *th* task. $\Delta E_{total}[X_t]$ is thus a shared locked data structure and must either be protected by a lock when it is updated, or else produced by the method of "multiple images" [38], and we have chosen the latter course. In this method partial contributions to $\Delta E_{total}[X_t]$ are accumulated in an array $\Delta E[X_t](P)$, P=1,N, by each process, the element of $\Delta E[X_t]$ to be used by a given process being indicated by the process identification index, P, which is passed via the argument list to the parallel routine in our implementation. $\Delta E[F_t](P)$ is a shared independent data structure; all other data structures manipulated by the central kernel of our algorithm are either shared independent (in fact read only) or local, and we therefore have a paradigm case of a sequence of completely independent tasks. When all parallel processing is complete, the fourth order energy is evaluated through:

$$\Delta E_{total}[X_t] = \sum_{P}^{N} \Delta E[X_t](P) \qquad (24)$$

We have written a computer program for the CRAY Y-MP machine which evaluates the energy components corresponding to each of the fourth order, triple excitation energy terms.

6.3 Results of benchmark tests

In the following benchmark tests the CRAY Y-MP/432 and CRAY Y-MP/8128 systems at the Florida State University (FSU) and Cray Research Incorporated (CRI), respectively, running under the UNICOS 5.1.10 operating system were used. These machines have four and eight central processing units, respectively, each with an 6 nanosecond cycle time. For the CRAY Y-MP/432 the central processors are symmetrically connected to 32 Mwords of shared memory organized into 128 banks. The CRAY Y-MP/8128 has 128 Mwords of shared memory organized into 256 banks.

The "theoretical peak" performance is given by (number of floating point operations/1 cycle) × (1 cycle/second) × (number of processors). Hence, for the CRAY Y-MP/432 the theoretical peak performance is 1333 MFLOPS (Millions of Floating Point Operations Per Second) whilst for the CRAY Y-MP/8128 the peak performance is 2667 MFLOPS. However, the vector registers on these machines have to be flushed after every 64 elements and this requires approximately 4.5 cycles so that an estimate of the "theoretical sustained" performance is 1245 and 2492 MFLOPS, respectively, for the four and eight processor machines. These rates of execution define the target performance for the present study.

The CFT77 Fortran compiler was chosen, and the CRAY-supplied routine MXMA was used to carry out the matrix multiply operations. The number of doubly occupied orbitals, N_{occ}, has consistently been set to 4. The number of unoccupied orbitals, N_{virt}, has been varied.

6.3.1 Benchmarks on dedicated machines [30]

In a dedicated machine benchmark, we have studied the dependence of the overall rate of computation on N_{virt}, using four parallel processes on the CRAY Y-MP/432 at FSU. We can confirmed that each case was extremely close to ideal balance. The results are displayed in Table 1. In this Table, ρ_{cpu} denotes the rate of computation in millions of floating point operations per central processing unit second, ρ_{wc} is the rate of computation in millions of floating point operations per wall clock second, and R is the ratio τ_{cpu}:τ_{wc}, where τ_{cpu} is the total central processing unit time and τ_{wc} is the total wall clock time. Ideally R should be equal to the number of processors.

Table 1

Rate of computation in millions of floating point operations per central processing unit second, ρ_{cpu}, and in millions of floating point operations per wall clock second, ρ_{wc}, as a function of the number of virtual orbitals , N_{virt}, using four occupied orbitals and four concurrent processes on a dedicated CRAY Y-MP/432 computer. R is the ratio $\rho_{wc}:\rho_{cpu}$.

N_{virt}	ρ_{cpu}	ρ_{wc}	R
16	150.8	209.4	1.39
32	221.3	672.2	3.04
48	254.3	901.4	3.55
64	272.3	1029.5	3.78
80	267.4	1027.5	3.84
96	280.5	1090.5	3.89
112	287.7	1123.7	3.91
128	293.2	1146.9	3.91

Table 2

Rate of computation in millions of floating point operations per central processing unit second, ρ_{cpu}, and in millions of floating point operations per wall clock second, ρ_{wc}, as a function of the number of processors, N_{cpu}, using four occupied orbitals and 112 virtual orbitals on a dedicated CRAY Y-MP/432 computer. R is the ratio $\rho_{wc}:\rho_{cpu}$.

N_{cpu}	ρ_{cpu}	ρ_{wc}	R
1	288.1	285.9	0.99
2	288.2	571.0	1.98
3	285.6	840.9	2.94
4	287.7	1123.7	3.91

Table 3

Rate of computation in millions of floating point operations per central processing unit second, ρ_{cpu}, and in millions of floating point operations per wall clock second, ρ_{wc}, as a function of the number of virtual orbitals , N_{virt}, using four occupied orbitals and eight concurrent processes on a dedicated CRAY Y-MP/8128 computer. R is the ratio $\rho_{wc}:\rho_{cpu}$.

N_{virt}	ρ_{cpu}	ρ_{wc}	R
16	151.5	136.9	0.90
32	223.2	825.9	3.70
48	255.7	1499.5	5.86
64	273.6	1879.2	6.87
80	268.3	1979.6	7.38
96	281.4	2133.7	7.58
112	288.4	2222.1	7.70
128	293.7	2284.1	7.78

```
      PARAMETER (NCPUMAX=8)
      COMMON /TASKID/ ITASK(3,NCPUMAX)
      COMMON /GLOBAL/ I, LOCKI, T
      EXTERNAL PROCESS
*
*     Initialize multitasking variables
*
      T=? : number of tasks
      N=? : number of processes
      CALL LOCKASGN(LOCKI)
      I=1
      DO 1 K=1,N
      ITASK(1,K)=3
      ITASK(3,K)=K
    1 CONTINUE
*
*     Create N-1 out-of-line processes and
*       one in-line process - FORK
*
      DO 2 K=2,N
      CALL TSKSTART(ITASK(1,K),PROCESS)
    2 CONTINUE
      CALL PROCESS
*
*     Wait until all processes have finished -
*       BARRIER
*
      DO 3 K=2,N
      CALL TSKWAIT(TASK(1,K))
    3 CONTINUE
```

Figure 2. *Calling sequence for dynamic balance. (Based on the work of Baker, Moncrieff, Saunders and Wilson* [29]*)*

```
      SUBROUTINE PROCESS
      COMMON /GLOBAL/ I, LOCKI, T
    1 CONTINUE
      CALL LOCKON(LOCKI)
      IF(I.GT.T) THEN
      CALL LOCKOFF(LOCKI)
      RETURN
      ELSE
      ITEMPI=I
      I=ITEMPI+1
      CALL LOCKOFF(LOCKI)
      ENDIF
      ...
      Execute Task ITEMP
      ...
      GOTO 1
      END
```

Figure 3. *Dynamically balanced parallel routine. (Based on the work of Baker, Moncrieff, Saunders and Wilson* [29]*)*

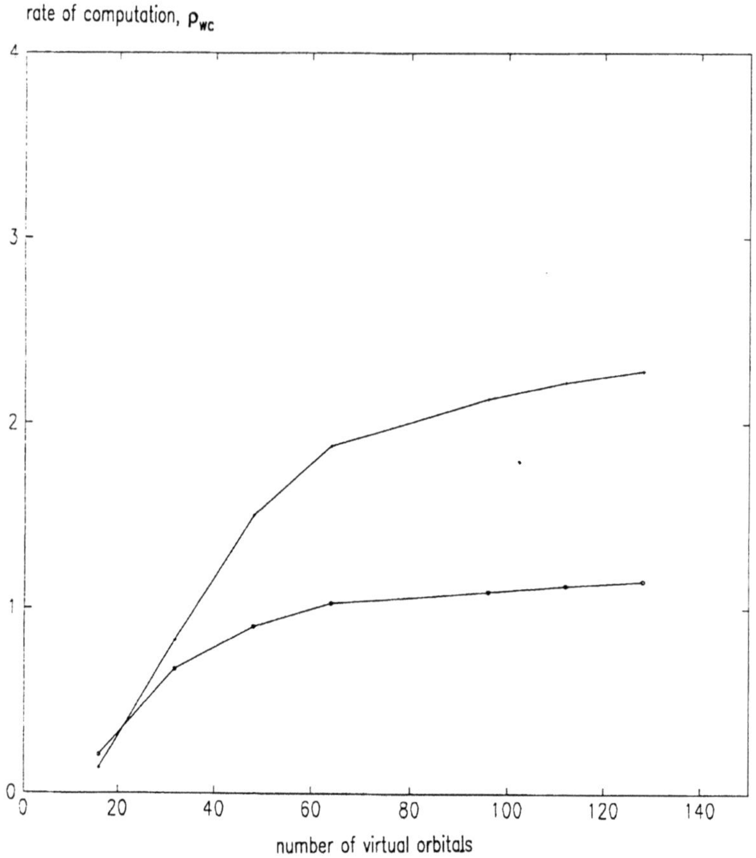

Figure 4. *Rate of computation as a function of the number of virtual orbitals using either four or eight parallel processes on a dedicated system. (Based on the work of Moncrieff, Saunders and Wilson* [33]*)*

The rates of computation ρ_{wc} shows a rapid increase as N_{virt} increases, reaching an initial peak of 1029.5 Mflops at N_{virt}=64, which corresponds to 77.2% of the theoretical peak performance of this machine and 82.7% of the theoretical sustained performance. The rate ρ_{wc} decreases to 1027.5 Mflops at N_{virt}=72 (this decrease is because the hardware vector length of the machine is 64), thereafter continuously rising to 1146.9 Mflops at N_{virt}=128 (the largest case studied). The represents 86.0% of the theoretical peak performance and 92.1% of the theoretical sustained performance. It is apparent that our algorithm is very well suited to both vector and parallel processing.

In Table 2, the rate of computation ρ_{cpu} and ρ_{wc} is given as a function of the number of processing units used, N_{cpu}, with fixed N_{virt} (=112) on a dedicated CRAY-Y-MP/432 together with the ratio R. The ratio R:N_{cpu} × 100% affords a measure of the efficiency of the parallel processing. This ratio takes the values 99%, 99%, 98% and 97.8% for N_{cpu}=1, 2, 3 and 4, respectively.

The results of a series of benchmark calculations for different values of N_{virt} on a CRAY Y-MP/8128 at CRI using eight parallel processes are displayed in Table 3. For N_{virt}=64 the rate of execution ρ_{wc} is 70.5% of the theoretical peak performance and 75.4% of the theoretical sustained performance of the eight processor machine. ρ_{wc} increases to 85.6% of the theoretical peak performance and 91.7% of the theoretical sustained performance for N_{virt}=128. The ratio R:N_{cpu} takes a value of 96.3% for the N_{virt}=112 case. For N_{cpu}=16 it is estimated that τ_{wc} would be of the order 28-30 seconds. For N_{cpu}=64 τ_{wc} may lie in the range 6-7 seconds whilst for N_{cpu}=256 a total wall clock time τ_{wc} of the order of 1-3 seconds is predicted.

In Figure 4, the rate of execution ρ_{wc} observed on the dedicated four processor machine is compared with that obtained on the eight processor computer.

6.3.2 Comments on a multi-user environment

The benefits of multi-processing on a CRAY Y-MP system in a multiuser environment have until now largely been seen in terms of the "1/N_{cpu} rule" [31, 32], where N_{cpu} denotes the number of processors in the system. In this model, the use of more than 1/N_{cpu} *th* of a scarce resource (for example, memory) requires that one uses more one processor, otherwise it is considered that other jobs will be unlikely to saturate the remaining processors because of shortage of the resource. We have presented previously an argument in favour of multi-processing even small memory jobs. We estimate that it is desirable to have about two executing processes per processor to ensure that no idle time accumulates, largely because of wait times for input/output. This is very difficult to ensure given the type of workload (many large memory heavily input/output bound jobs) often encountered. The presence of a single processor-bound small memory four-way multiprocessed job (in dynamic balance) running at low scheduling priority is capable of consuming all potentially idle cycles on all the processors of a CRAY Y-MP/4 system, with little disturbance to the flow of other work, and a greatly improved overall throughput [33].

7. Concluding remarks

We have presented a formulation of the many-body perturbation theory of electron correlation suitable for implementation in a computing environment in which parts of the calculation can be executed concurrently. We have shown that the linked diagram expansion leads to a formulation of the correlation problem which maps very effectively on to multiprocessor machines. We have considered in detail the evaluation of the most computationally demanding of the fourth order component of the correlation energy, the contribution of associated with triply substituted intermediate states.

We have shown that in each case the diagrammatic components can be decomposed into a set of partial contributions which can be computed by an algorithm structured around the matrix multiply operation, so that conventional vector processors can be driven nearly optimally. Thus on a CRAY Y-MP system running in one processor mode a rate of computation of ~293 Mflops is typical, compared with the theoretical sustainable maximum of ~311 Mflops.

We have also shown that the evaluation of the partial contributions may proceed entirely independently, so that the theory is well suited for implementation on a parallel processing system with either a distributed or shared memory. For typical shared memory systems, which normally operate in multi-user mode, dynamic balance allows one to partition the work amongst the processors according to their ability to deal with it.

The dynamic balance scheme has been implemented and tested on a CRAY Y-MP multi-processing system. In a dedicated environment close to linear speed-up with number of processors was observed, there being little degradation due to memory bank conflicts or other overheads of the parallel implementation, and rate of computations of greater than 2284 Mflops were observed. In a more typical multi-user environment it was found that there was little degradation in the rate of computation. However, the amount of work carried out by each processor was found to be not necessarily equal, and variable when the runs were repeated, a simple reflection of the fact that all processors are not equally busy at any given time.

The dynamic load balancing technique is essential to the efficient implementation of the relativistic many-body perturbation theory described in the previous article.

References

[1] S. Wilson, editor, *Concurrent computation in chemical calculations*, Methods in Computational Chemistry 3 Plenum Press, New York (1989)
[2] R.G. Evans and S. Wilson, editors, *Supercomputational Science* (Plenum Press, New York, 1990)
[3] S. Wilson, *Electron correlation in molecules* (Clarendon, Oxford, 1984), Russian translation (Mir, Moscow, 1987).
[4] D.M. Silver, 1978, Comput. Phys. Commun. 14, 71 (1978)
[5] D.M. Silver, 1978, Comput. Phys. Commun. 14 , 81 (1978)
[6] S. Wilson, 1978, Comput. Phys. Commun. 14, 91 (1978)
[7] S. Wilson and D.M. Silver, Phys. Rev. A14, 1949 (1976)
[8] S. Wilson and D.M. Silver, J. Chem. Phys. 66, 5400 (1977)
[8] R.J. Bartlett and D.M. Silver, J. Chem. Phys. 64, 4578 (1976)
[10] S. Wilson, D.M. Silver and R.J. Bartlett, Molec. Phys. 33, 1177 (1977)
[11] D.M. Silver, S. Wilson and R.J. Bartlett, Phys. Rev. A16, 477 (1977)
[12] S. Wilson, Intern. J. Quantum Chem. ; Molec. Phys. 35, 1 (1978)
[13] P. Čársky and M. Urban, *Ab initio Calculations. Methods and Applications in Chemistry* (Springer, Berlin, 1980)
[14] S. Wilson, Specialist Periodical Reports: Theoretical Chemistry 4 1 (1981)
[15] R.J. Bartlett, Ann. Rev. Phys. Chem. 32, 359 (1981)
[16] I. Lindgren and J. Morrison, *Atomic Many-Body Theory* (Springer, Berlin, 1982)
[17] S. Wilson, in: *Methods in Computational Molecular Physics*, eds. G.H.F. Diercksen and S. Wilson, (Reidel, Dordrecht, 1983)
[18] S. Wilson, Comput. Phys. Reports 2, 389 (1985)
[19] K. Jankowski, Meth. Comput. Chem. 1, 1 (1987)
[20] M. Urban, I. Cernusak, V. Kello and J. Noga, Meth. Comput. Chem 1, 117 (1987).
[21] J.A. Pople, R. Krishnan, H.B. Schlegel and J.S. Binkley, Intern. J. Quantum Chem. 14 545 (1978)

[22] P.J. Knowles, K. Somasundram, N.C. Handy and K. Hirao, Chem. Phys. Lett. **113** 8 (1985)
[23] S. Wilson and V.R. Saunders, Comput. Phys. Commun. **19**, 293 (1980)
[24] M.F. Guest and S. Wilson, in: *Supercomputers in Chemistry*, ed. P. Lykos and I. Shavitt (American Chemical Society, Washington D.C., 1981) p.1.
[25] S. Wilson, in *Proceedings of Fifth Seminar on Computational Problems in Quantum Chemistry*, Groningen, eds. P.Th. van Duijknen and W.C. Nieuwpoort, 1981.
[26] S. Wilson, in *Correlated Wavefunctions*, Proceeding of a Study Weekend, December 1977, edited by V.R. Saunders, Daresbury Laboratory, 1978.
[27] S. Wilson and D.M. Silver, Intern. J. Quantum Chem. **15**, 683 (1979)
[28 V.R. Saunders in *Supercomputational Science*, edited by R.G. Evans and S. Wilson, (Plenum, New York, 1990)
[29] D.J. Baker, D. Moncrieff, V.R. Saunders and S. Wilson, Comput. Phys. Commun. **62** 25 (1991)
[30] P.R. Taylor and C.W. Bauschlicher Theoret chim Acta **71**, 105 (1987)
[31] P.R. Taylor, C.W. Bauschlicher and D.W. Schwenke, Meth. Comput. Chem. **4** (1989)
[32] V.R. Saunders and S. Wilson, Parallel Computing *(in press)*
[33] D. Moncrieff, V.R. Saunders and S. Wilson, Intern. J. Supercomputer Applic. *(submitted)*

RESPONSE AND PROPAGATOR METHODS

Jens Oddershede

Department of Chemistry, Odense University
DK–5230 Odense M, Denmark

1. INTRODUCTION

The response, or reaction, of a system to external perturbations is described by response functions. The response is expanded in powers of the external perturbations and linear, quadratic, cubic etc. response functions are obtained. Many physical observables are either response functions or can be derived from response functions for particular choices of the external perturbations. In this series of lectures examples thereof will be given. However, the main emphasis will be on the theory of response functions, first of all defining the functions and secondly showing how to evaluate them at various levels of approximation. This lecture series will thus primarily serve as an introduction to the theory of response functions rather than attempting to present an exhaustive list of properties that may be calculated once we have a computational expression for the response function.

Response functions are often called polarization propagators or (particle-hole) Green's functions (Schirmer, 1982, Schirmer *et al.*, 1983). These names refer to different characteristics of the functions. One may view a response to an external perturbation as a process in which the system is disturbed at some initial time, the disturbance then propagates through the system until it is detected at a later time when a measurement is made. Since we only consider number conserving perturbations (polarization), the phrase *polarization propagators* was coined for these functions (Linderberg and Öhrn, 1973). This name, as well as *response functions*, emphasize the time-dependent aspects of the functions (Zubarev, 1960 and 1974). However, for practical applications in electronic structure theory it is often more convenient to

Methods in Computational Molecular Physics, Edited by Stephen Wilson and
Geerd H.F. Diercksen, Plenum Press, New York, 1992

work in the energy representation, that is, the Fourier transforms of the time-dependent quantities.

When the polarization is a particle-hole excitation, one refers to the response function as a particle-hole propagator or Green's function. Mathematically, the energy-dependent propagator is the Green's function [see *e.g.* Chapt. 2 of Linderberg and Öhrn (1973)]; hence, the name particle-hole Green's function. However, the polarization propagator and the particle-hole propagator are not exactly the same quantity. The two functions have the same poles but not the same residues. Therefore, they give the same excitation energies but not the same transition moments and response properties (*vide infra*). This point is discussed in more detail in an earlier publication (Oddershede, 1982) and applications showing the difference between the two propagators are given by Oddershede *et al.* (1978).

The basic characteristics, however, of the various propagators are the same, so the development to be outlined in this communication holds for all the propagators with very small modifications. The advantage of all of these functions is that they are closely related to the experimental situation. Often they *are* the observables and in other cases the residues and poles of the polarization propagators are directly related to experimental quantities. For instance, it is not necessary to subtract total energies to get excitation energies, or to perform sum over states to calculate second-order quantities, such as multipole polarizabilities and nuclear magnetic shieldings. Therefore, this approach to electronic structure calculations is often referred to as *direct methods.*

There are several reviews of this field that the reader may consult. The books by Jørgensen and Simons (1981) and McWeeny (1989) have chapters on propagators. Reviews by the present author, which also list work by other authors in the same field, include two papers summarizing the applications through 1977 (Oddershede, 1978) and from 1977 to 1986 (Oddershede, 1987) as well as a more methodological review by Oddershede *et al.* (1984) that also gives detailed computational expressions for several polarization propagators. Recently, Geertsen *et al.* (1991) have reviewed a method which utilized the coupled cluster ansatz for the wavefunction within the framework of the polarization propagator method. Oddershede and Sabin (1991) have written a review that discusses the relationship between properties observed in molecular spectroscopy and polarization propagators. The paper by Oddershede and Sabin (1991) also formulates 10 desirable features or criteria for general purpose methods to be applied to the calculation of a range of spectroscopic properties.

The lectures that I gave at the previous summer school in this series (Oddershede, 1983) will supplement the present lecture notes. The derivations and emphasizes are different but the final expressions and conclusions are the same.

2. EXACT POLARIZATION PROPAGATORS

In order to understand the underlying physical principles it is convenient first to consider exact propagators, that is, the response of an exact state to an external perturbation. We will describe this process using the Schrödinger picture: time-independent operators and time-dependent wavefunctions. In doing so we deviate from the 1983 lecture notes (Oddershede, 1983) but follow instead the derivation by Olsen and Jørgensen (1985).

2.A Definitions

Consider a system which is subjected to a time-dependent, but spatially uniform perturbation W(t), often referred to as the radiation field, about which we will assume that $W(-\infty) = 0$. The interaction operator between the system and W(t) will be denoted $V(\mathbf{r},t)$, and we expand $V(\mathbf{r},t)$ as

$$V(\mathbf{r},t) = \int_{-\infty}^{\infty} d\omega V^{\omega} \exp[(-i\omega+\epsilon)t] \tag{2.1}$$

where ϵ is positive infinitesimal ensuring that the interaction is zero at $t = -\infty$. Here, and in the following, it is helpful to have a particular application in mind when discussing the general case. The special case is that of an external electric field $\mathbf{F}(t)$, that is $W(t) = \mathbf{F}(t)$ and $V(\mathbf{r},t) = \mathbf{r}\cdot\mathbf{F}(t)$. We use atomic units throughout ($e = 1$) and the dipole operator in second quantization is thus

$$\boldsymbol{\mu} = -\mathbf{r} = -\sum_{ij} \mathbf{r}_{ij} a_i^{\dagger} a_j \tag{2.2}$$

where $a_i^{\dagger}$ and a_j are the creation and annihilation operators, respectively, and $\mathbf{r} = (x, y, z)$.

In the exact case we know the eigenvalues and eigenfunction for the unperturbed system ($t = -\infty$)

$$H_0|n\rangle = E_n|n\rangle \tag{2.3}$$

The total hamiltonian for the system at any time t is thus

$$H(t) = H_0 + V(r,t) \tag{2.4}$$

The problem at hand is now to determine the response of the system to the external perturbation W(t). We will thus calculate the change in the average value of any operator A due to the presence of W(t). Let us consider the average value in the state $|0(t)>$. Typically, $|0(t)>$ is the ground state, but the development is not restricted to ground state average values. Note also that only for $t = -\infty$ is $|0(t)> = |0>$. We now write

$$\begin{aligned} <0(t)|A|0(t)> &= <0|A|0> \\ &+ \int_{-\infty}^{\infty} d\omega <<A;V^{\omega}>>_{\omega} \exp[(-i\omega+\epsilon)t] \\ &+ \frac{1}{2}\int_{-\infty}^{\infty} d\omega_1 \int_{-\infty}^{\infty} d\omega_2 <<A;V^{\omega_1},V^{\omega_2}>>_{\omega_1,\omega_2} \exp[(-i\omega_1-i\omega_2+\epsilon+\delta)t] \\ &+ \ldots\ldots \end{aligned} \tag{2.5}$$

This equation, often referred as the Kubo (1959) relation, defines the *linear*, $<<A;V^{\omega}>>_{\omega}$, and the *quadratic*, $<<A;V^{\omega_1},V^{\omega_2}>>_{\omega_1,\omega_2}$, etc. response functions or polarization propagators. It is important to realize that the equation *defines* the response functions but it does not derive them. The derivation is outlined in sec. 2.B. The operators V^{ω_1}, V^{ω_2} etc. may be the same or different operators. We have allowed for different infinitesimals ϵ, δ .... In the case of an electric perturbation $V^{\omega} = \mathbf{r}\cdot\mathbf{F}^{\omega}$, $A = -\mathbf{r}$ and $<0(t)|-\mathbf{r}|0(t)> = \mu(t)$ is the induced dipole moment at time t. This induced dipole moment is normally expressed as

$$\begin{aligned} \mu(t) &= \mu_0 \\ &+ \int_{-\infty}^{\infty} d\omega\, \alpha(\omega) \mathbf{F}^{\omega} \exp[(-i\omega+\epsilon)]t \\ &+ \frac{1}{2}\int_{-\infty}^{\infty} d\omega_1 \int_{-\infty}^{\infty} d\omega_2\, \beta(\omega_1,\omega_2) \mathbf{F}^{\omega_1}\mathbf{F}^{\omega_2} \exp[(-i\omega_1-\omega_2+\epsilon+\delta)]t \\ &+ \ldots \end{aligned} \tag{2.6}$$

where $\alpha(\omega)$ and $\beta(\omega_1,\omega_2)$ are the frequency-dependent dipole polarizability and first hyperpolarizability, respectively. Using the form of V^{ω} given above, Eqs. (2.5) and (2.6) imply

$$\alpha(\omega) = - \langle\langle r;r\rangle\rangle_{\omega} \tag{2.7}$$

$$\beta(\omega_1,\omega_2) = - \langle\langle r;r,r\rangle\rangle_{\omega_1,\omega_2} \tag{2.8}$$

and we have here the two first examples of response functions. Other choices of A, V^{ω_1}, V^{ω_2} etc. will lead to different properties. A comprehensive list of properties is given in Table I of Oddershede and Sabin (1991).

2.B Spectral representations

It is possible to express the polarization propagators in terms of the eigenfunctions of the unperturbed hamiltonian, Eq. (2.3). These forms, often referred to as the spectral or the Lehman (1954) representations, are useful because they show how one may obtain excitation energies and transition moments from the propagators. The spectral forms are given in several places (Zubarev, 1974; Bloembergen, 1965; Olsen and Jørgensen, 1985; Dalgaard, 1982; Oddershede, 1983) and we shall not give a detailed derivation here. However, we will summarize the principles of the derivation.

A very convenient way (Olsen and Jørgensen, 1985) of deriving the spectral representation starts out with the observations that we may always write the perturbed wavefunction as

$$|0(t)\rangle = \exp[iP(t)]|0\rangle \tag{2.9}$$

where

$$P(t) = P^{+}(t) = \sum_{n} P_n(t)|n\rangle\langle 0| + P_n^{*}(t)|0\rangle\langle n| \tag{2.10}$$

The n = 0 term in Eq. (2.10) only has a non-vanishing real part which in turn is merely a phase factor on $|0(t)\rangle$ and thus does not affect the average value in Eq. (2.5) [and Eq. (2.21)]. In fact, this phase factor is the time-dependent phase factor that always must multiply the time-independent eigenfunction in the Schrödinger picture in order to obtain the time-dependent stationary solution (see Olsen and Jørgensen, 1985). Hence, without loss of generality, we may assume that n>0 in Eq. (2.10). The exponential expansion in Eq. (2.9) is a completely general, *i.e.* unrestricted, normalized expansion of $|0(t)\rangle$ in the unperturbed eigenstates. The orthonormality of $|0(t)\rangle$ follows from the hermiticity of P(t) while the unrestricted nature of the expansion can be seen as follows:

$$\exp[iP(t)]|0\rangle = (1+iP(t)-\frac{P^2(t)}{2}+\ldots)|0\rangle$$
$$= (1+i\sum_{n>0}\{P_n(t)|n\rangle\langle 0|+P_n^*(t)|0\rangle\langle n|\}$$
$$-\frac{1}{2}\sum_{n,m>0} P_n|n\rangle\langle m|P_m^* + \sum_n |P_n|^2\ |0\rangle\langle 0|+\ldots)|0\rangle$$
$$= \cos x|0\rangle+\sum_{n>0}(iP_n\frac{\sin x}{x})|n\rangle \tag{2.11}$$

where

$$x = \left[\sum_{n>0}|P_n|^2\right]^{\frac{1}{2}} \tag{2.12}$$

and where we have used the expansions

$$\cos x = 1-\frac{1}{2!}x^2+\ldots \tag{2.13}$$

$$\frac{\sin x}{x} = 1-\frac{1}{3!}x^2+\ldots \tag{2.14}$$

Equation (2.11) is consequently of the form

$$|0(t)\rangle = \sum_n a_n|n\rangle \tag{2.15}$$

that is, a general expansion of $|0(t)\rangle$ in terms of the unperturbed eigenfunctions. However, by requiring the relationship between a_0 and a_n $(n>0)$ implied by Eq. (2.11) we know that the expansion is normalized to all orders in $P(t)$.

The spectral representation is now obtained by expanding

$$P_n(t) = P_n^{(1)}(t) + P_n^{(2)}(t) + \ldots \tag{2.16}$$

in orders of the external perturbation [$P_n^{(0)} = 0$ since $|0\rangle$ is an eigenstate of H_0] which according to Eqs. (2.9)-(2.11) makes

$$|0(t)\rangle = |0\rangle + |0^{(1)}(t)\rangle + |0^{(2)}(t)\rangle + \ldots \tag{2.17}$$

Inserting Eq. (2.17) in the Ehrenfest (1927) equation of motion

$$\langle\dot{0}(t)|\Lambda|0(t)\rangle + \langle 0(t)|\Lambda|\dot{0}(t)\rangle = -i\langle 0(t)|[\Lambda,H]|0(t)\rangle \tag{2.18}$$

gives an equation of the required form, Eq. (2.5), as we shall argue below. In Eq. (2.18) Λ is any (Schrödinger) operator, H is the total hamiltonian in Eq. (2.9) and the dot indicates time differentiation. Choosing Λ as an arbitrary element of the $\{|n\rangle\langle 0|, |0\rangle\langle n|\}$ matrix, inserting Eq. (2.9) in Eq. (2.18) and collecting terms of the same order in the external perturbation give differential equations for $P_n^{(k)}(t)$, $k = 1,2,3 \ldots$, which may be solved. For example, one finds [Olsen and Jørgensen, 1985, Eq. (2.47)]

$$P_n^{(1)} = -i\int_{-\infty}^{\infty} d\omega \, \frac{\langle n|V^{\omega}|0\rangle}{\omega-E_n+E_0+i\epsilon} \exp[(-i\omega+\epsilon)t] \tag{2.19}$$

$$P_n^{(1)*} = -i\int_{-\infty}^{\infty} d\omega \, \frac{\langle 0|V^{\omega}|n\rangle}{\omega+E_n-E_0+i\epsilon} \exp[(-i\omega+\epsilon)t] \tag{2.20}$$

Note that $P_n^{(1)*}$ is the complex conjugate of $P_n^{(1)}$. That is easily verified using that $V(t) = V^{+}(t)$ or [see Eq. (2.1)] $V^{-\omega} = (V^{\omega})^{+}$. From Eq. (2.17) it follows that

$$\begin{aligned}\langle 0(t)|A|0(t)\rangle = {} & \langle 0|A|0\rangle \\ & + \langle 0|A|0^{(1)}\rangle + \langle 0^{(1)}|A|0\rangle \\ & + \langle 0|A|0^{(2)}\rangle + \langle 0^{(2)}|A|0\rangle + \langle 0^{(1)}|A|0^{(1)}\rangle \\ & + \ldots \end{aligned} \tag{2.21}$$

Comparing with Eq. (2.5) we see that the second line in the preceding equation determines the linear polarization propagator, the third line determines the quadratic polarization propagator, etc. Introducing in Eq. (2.21) $|0^{(k)}\rangle$ in terms of $P_n^{(m)}$ [see Eqs. (2.9) and (2.10)], such as for example the expressions for $P_n^{(1)}$ in Eqs. (2.19)-(2.20), we can identify the spectral representations

$$\langle\langle A;V^{\omega}\rangle\rangle_{\omega} = \sum_{n>0}\left\{\frac{\langle 0|A|n\rangle\langle n|V^{\omega}|0\rangle}{\omega-E_n+E_0+i\epsilon} - \frac{\langle 0|V^{\omega}|n\rangle\langle n|A|0\rangle}{\omega+E_n-E_0+i\epsilon}\right\} \tag{2.22}$$

and

$$<<A;V^{\omega_1},V^{\omega_2}>>_{\omega_1,\omega_2} = (1+P_{12}) \left\{ \sum_{k,n>0} \frac{<0|A|k><k|\overline{V^{\omega_1}}|n><n|V^{\omega_2}|0>}{(\omega_1+\omega_2-E_k+E_0+i\epsilon+i\delta)(\omega_2-E_n+E_0+i\delta)} \right.$$

$$+ \sum_{k,n>0} \frac{<0|V^{\omega_2}|n><n|\overline{V^{\omega_1}}|k><k|A|0>}{(\omega_1+\omega_2+E_k-E_0+i\epsilon+i\delta)(\omega_2+E_n-E_0+i\delta)}$$

$$\left. - \sum_{k,n>0} \frac{<0|V^{\omega_1}|k><k|\overline{A}|n><n|V^{\omega_2}|0>}{(\omega_1+E_k-E_0+i\epsilon)(\omega_2-E_n+E_0+i\delta)} \right\} \quad (2.23)$$

and similarly for higher order response functions (Olsen and Jørgensen, 1985, Sec. III.D). In Eq. (2.23) P_{12} is the permutation operator that interchanges ω_1 and ω_2, so the quadratic response function consists of 6 terms. We have also used the definition

$$<k|\overline{A}|n> = <k|A|n> - \delta_{kn}<0|A|0> \quad (2.24)$$

Comparing Eqs. (2.5), (2.19), (2.20) and (2.22) we see that the linear response function is simply the sums of the products $<0|A|n>P_n^{(1)}$ and $P_n^{(1)*}<n|A|0>$. One note about conventions: the operator, the response of which we are considering (A), is always placed as the first operator inside the double bracket (<<). This operator is separated by a semicolon to the operators ($V^{\omega_1},V^{\omega_2}$ ...) that cause the response and the perturbing operators are separated by commas. In this way we distinguish notationally between "the result and the cause".

Returning for a moment to the case of an electric perturbation, we see from Eq. (2.22)

$$\lim_{\epsilon\to 0} \mathrm{Re}<<r;r>>_\omega = -2 \sum_{n>0} \frac{<0|r|n><n|r|0>(E_n-E_0)}{(E_n-E_0)^2-\omega^2} \quad (2.25)$$

which is readily recognized as (the negative of) the usual sum over states expression for the frequency-dependent dipole polarizability tensor, in agreement with the identification in Eq. (2.7). A similar relation holds for $\beta(\omega_1,\omega_2)$.

2.C Time–dependent representations

The Fourier transformations of Eqs. (2.22) and (2.23) *define* the double-time polarization propagators

$$\langle\langle A(t);V(t')\rangle\rangle = \lim_{\epsilon\to 0} \frac{1}{2\pi}\int_{-\infty}^{\infty} d\omega \langle\langle A;V\rangle\rangle_{\omega} \exp[-i(\omega+i\epsilon)(t-t')] \tag{2.26}$$

and

$$\langle\langle A(t);V_1(t'),V_2(t'')\rangle\rangle =$$
$$\lim_{\epsilon,\delta\to 0} \frac{1}{4\pi^2}\int_{-\infty}^{\infty} d\omega_1 \int_{-\infty}^{\infty} d\omega_2 \langle\langle A;V_1,V_2\rangle\rangle_{\omega_1,\omega_2} \exp[-i(\omega_1+i\epsilon)(t-t')-i(\omega_2+i\delta)(t-t'')] \tag{2.27}$$

Inserting Eqs. (2.22) and (2.23) in Eqs. (2.26) and (2.27), respectively, and doing a bit of algebra lead to

$$\langle\langle A(t);V(t')\rangle\rangle = -i\langle 0|[A(t),V(t')]|0\rangle\Theta(t-t') \tag{2.28}$$

and

$$\begin{aligned}\langle\langle A(t);V_1(t'),V_2(t'')\rangle\rangle = \\ &- \langle 0|[[A(t),V_1(t')],V_2(t'')]|0\rangle\Theta(t-t')\Theta(t'-t'') \\ &- \langle 0|[[A(t),V_2(t'')],V_1(t')]|0\rangle\Theta(t-t'')\Theta(t''-t')\end{aligned} \tag{2.29}$$

where $\Theta(t)$ is the Heaviside step function

$$\Theta(t) = \begin{cases} 1 & t \geq 0 \\ 0 & t < 0 \end{cases} \tag{2.30}$$

The time-dependent operators in Eqs. (2.26)-(2.29) are in the interaction representation, *i.e.*

$$V(t) = \exp(iH_0t)V\exp(-iH_0t) \tag{2.31}$$

where H_0 is the hamiltonian of the unperturbed system and V and H_0 both are Schrödinger operators. The derivation of Eq. (2.28) from Eq. (2.22) can *e.g.* be found in Secs. 2 and 3 of Oddershede (1983).

It should be pointed out that ω-independent interaction operators are used in the definitions of the time-dependent polarization propagators in Eqs. (2.26) and (2.29). This is the simplest way of defining time-dependent propagators and the method used by Zubarev (1960, 1974). However, it is also possible to work with the time Fourier transforms with V replaced by V^{ω} (V^{ω_1},V^{ω_2}) in which case *e.g.* the Fourier transform in Eq. (2.26) will become an energy integration over a propagator of the form $\langle\langle A(t); V_s(t'-s)\rangle\rangle$ where

$$V_s(t) = \exp(iH_0t)V(s)\exp(-iH_0t) \tag{2.31a}$$

The response equation will then involve an integration over the extra time variable s. This line of derivation is followed by Oddershede *et al.* (1984).

In most texts on polarization propagators one uses Eqs. (2.28) and (2.29) as the definitions of the functions. That is *e.g.* the case in Zubarev's (1960) original formulation and that was also the approach used in my 1983 lecture notes (Oddershede, 1983). However, doing so it is not so straightforward to see the relationship to response theory and that is the reason I find the present mode of presentation more illustrative.

One advantage, though, of the time-dependent formulation is that it gives a good physical picture of the polarization propagators. Only for $t > t'$ is $\langle\langle A(t);V^{\omega}(t')\rangle\rangle$ non-vanishing. That is, one cannot observe the response (A) before the perturbation (V^{ω}) is switched on. Therefore, the propagators in Eqs. (2.28) and (2.29) are called the *retarded* polarization propagators and they are the time-dependent propagators that logically should be used to describe responses of a perturbed system. However, one often prefers to work with propagators that are non-vanishing for all $t-t'$ intervals. They are the so-called causal propagators. Their spectral representations are the same as Eqs. (2.22) and (2.23) except that $+i\epsilon$ becomes $-i\epsilon$ in the last term in Eq. (2.22) as well as similar changes in the imaginary parts of the non-linear response functions. As the interesting physical properties are obtained from the real parts of the polarization propagators [see *e.g.* Eq. (2.25)], there are only conceptual reasons for preferring one kind of propagators to another. In fact, in my 1983 lecture notes (Oddershede, 1983) I used the causal propagators. However, I have come to prefer the retarded propagators because of their more direct physical interpretation.

2.D Physical interpretation

Many observables can be obtained from the polarization propagators. We have already discussed, in some detail, that dynamic electric dipole polarizabilities and hyperpolarizabilities are response functions. However, several spectroscopic properties, Q, are evaluated from expressions like

$$Q(H',H'') = \sum_{n>0} \frac{<0|H'|n><n|H''|0>+<0|H''|n><n|H'|0>}{E_0-E_n} \tag{2.32}$$

Examples include the paramagnetic contribution to the nuclear magnetic shielding for which $H' = \ell_N/r_N^3$ (N is the magnetic nucleus and ℓ is the orbital angular momentum) and $H'' = \ell$ (Oddershede and Geertsen, 1990), and dipole moments for which $H' = \mathbf{p}$ and $H'' = \ell$ (Geertsen *et al.*, 1989). For a complete list, see Table I of Oddershede and Sabin (1991). Such static properties can be evaluated from the $\omega = 0$ limit of the linear response function in Eq. (2.22) since

$$\lim_{\epsilon \to 0} \mathrm{Re}<<H';H''>>_{\omega=0} = Q(H',H'') \tag{2.33}$$

The poles and residues of the spectral propagators also have simple physical interpretations. The real part of the (first order) poles are $\pm$ (E_n-E_0), *i.e.* excitation energies, and the residues of the linear response functions are transition moments between the reference state $|0>$ and the "excited" state $|n>$. From the double, first order residue of the quadratic response function, Eq. (2.23), we may also determine transition quantities between two "excited" states. As mentioned in Sec. 2.A, it is convenient, but not necessary, to choose the reference state (the state, the response of which we are considering) as the ground state. That is the reason for the quotation marks on "excited" in the previous sentence. For instance, when $A = V^{\omega_1} = V^{\omega_2} = \mathbf{r}$ we obtain electric dipole transition moments and since we also compute the excitation energies, it is straightforward to obtain the dipole length oscillator strengths

$$f^L_{n\to m} = \frac{2}{3}|<m|\mathbf{r}|n>|^2(E_m-E_n) \tag{2.34}$$

By using one or more of the operators A, V^{ω_1}, etc. equal to $\mathbf{p}$ rather than $\mathbf{r}$ we may compute some of the equivalent oscillator strengths (mixed, velocity etc.). That yields very useful information in regard to the completeness of the one-electron basis set as in the approximate methods we will discuss in the next section, the length, mixed and velocity forms of the oscillator strengths all must give identical results in a complete basis set.

In summary, the polarization propagators are response functions, that is, properties which in a state representation are often expressed as sum over states. The poles and residues of the propagator determine the individual, state to state, transition quantities.

3. APPROXIMATE POLARIZATION PROPAGATORS

In the preceding section we have described the properties of polarization propagators. In this section we will show how to evaluate the propagators at several levels of approximation and thus compute the molecular properties discussed in Sec. 2.D. The first problem at hand is to formulate the equations of motion obeyed by the response functions. Next, we will show how to obtain approximate solutions of these equations. An analogy with the state picture is the wave equation and its approximate solutions. For double-time propagators, Eqs. (2.28) and (2.29), one uses the Heisenberg equation of motion for the time-dependent operators in Eq. (2.31) to derive the equation of motion for the polarization propagators (Oddershede, 1983).

There are several routes to the determination of approximate polarization propagators. Perhaps the most straightforward way is to use various *approximate* forms for the exponential operator in Eqs. (2.9) and (2.10) and then use the Ehrenfest (1927) equation of motion, Eq. (2.18), just like we did in the exact case. That is the procedure used by Olsen and Jørgensen (1985) in their derivation of the multiconfigurational random-phase-approximation (MCRPA) for both linear and non-linear response functions. Instead of the Ehrenfest theorem, one may use the time-dependent Frenkel (1934) variation principle (Langhoff *et al.*, 1972) for response functions in the time-dependent formulation. This was the approach taken by Dalgaard (1982) who also showed (Dalgaard, 1981) that the Frenkel variation principle and the Ehrenfest theorem are equivalent if the exponential ansatz in Eq. (2.9) is used for the wavefunction.

In this lecture series, however, we shall choose an alternative way using the superoperator formulation, a method first pioneered by Goscinski and Lukman (1970) and Pickup and Goscinski (1973) within the field of propagators. We choose to do so because it gives a simple and transparent framework within which one may classify the various approximate polarization propagator methods.

3.A Equations of motion

Using the identity

$$\frac{\omega}{\omega \pm (E_n - E_0)} = 1 \mp \frac{E_n - E_0}{\omega \pm (E_n - E_0)} \tag{3.1}$$

it follows from Eq. (2.22) that

$$\omega\langle\langle A;V^{\omega}\rangle\rangle_{\omega} = \sum_{n>0}\left\{\langle 0|A|n\rangle\langle n|V^{\omega}|0\rangle - \langle 0|V^{\omega}|n\rangle\langle n|A|0\rangle\right\}$$
$$+ \sum_{n>0}\left\{\frac{\langle 0|A|n\rangle\langle n|V^{\omega}|0\rangle}{\omega-E_n+E_0} + \frac{\langle 0|V^{\omega}|n\rangle\langle n|A|0\rangle}{\omega+E_n-E_0}\right\}(E_n-E_0) \quad (3.2)$$

For notational simplicity we have absorbed $i\epsilon$ in ω, which then is complex. Since the $n = 0$ term in the first sum in Eq. (3.2) is zero, we find

$$\omega\langle\langle A;V^{\omega}\rangle\rangle_{\omega} = \langle 0|[A,V^{\omega}]|0\rangle + \langle\langle [A,H_0];V^{\omega}\rangle\rangle_{\omega} \quad (3.3)$$

The last term on the right hand side of Eq. (3.3) is obtained inserting the relation

$$\langle 0|A|n\rangle(E_n-E_0) = \langle 0|[A,H_0]|n\rangle \quad (3.4)$$

in the spectral form in Eq. (2.22). Using the identity in Eq. (3.1) and the same reformulations as above for Eq. (2.23) we find

$$(\omega_1+\omega_2)\langle\langle A;V^{\omega_1},V^{\omega_2}\rangle\rangle_{\omega_1,\omega_2} = (1+P_{12})\langle\langle [A,V^{\omega_1}];V^{\omega_2}\rangle\rangle_{\omega_2}$$
$$+ \langle\langle [A,H_0];V^{\omega_1},V^{\omega_2}\rangle\rangle_{\omega_1,\omega_2} \quad (3.5)$$

as one of several possible (Parkinson *et al.*, 1990) equations of motion for the quadratic polarization propagator.

Both equations of motion, Eqs. (3.3) and (3.5), have the same structure. They express the product of energy and the response function as the sum of a response function of one rank lower (rank: average value → linear response function → quadratic response function etc.) and a response function of the same rank, but in the latter propagator the "property operator" (A) commuted with the hamiltonian of the unperturbed system replaces the property operator of the original response function. Repeated use of Eqs. (3.3) and (3.5) gives that

$$\langle\langle A;V^{\omega}\rangle\rangle_{\omega} = \omega^{-1}\langle 0|[A,V^{\omega}]|0\rangle + \omega^{-2}\langle 0|[[A,H_0],V^{\omega}]|0\rangle + \ldots \quad (3.6)$$

and

$$\langle\langle A;V^{\omega_1},V^{\omega_2}\rangle\rangle_{\omega_1,\omega_2} = (1+P_{12})\{(\omega_1+\omega_2)^{-1}\langle\langle [A,V^{\omega_1}];V^{\omega_2}\rangle\rangle_{\omega_2}$$
$$+ (\omega_1+\omega_2)^{-2}\langle\langle [[A,H_0],V^{\omega_1}];V^{\omega_2}\rangle\rangle_{\omega_2}$$
$$+ \ldots.\} \quad (3.7)$$

We now introduce the superoperator formalism in which A, V^{ω_1}, etc. are construed as elements in an operator space with a binary product

$$(A|V^{\omega}) \equiv \langle 0|[A^{+},V^{\omega}]|0\rangle \tag{3.8}$$

In this operator space the identity operator is

$$\hat{I}A \equiv A \tag{3.9}$$

and the superoperator hamiltonian $\hat{H}_0$ is

$$\hat{H}_0 A \equiv [H_0, A] \tag{3.10}$$

With this notation Eq. (3.6) reads (Oddershede, 1983)

$$\langle\langle A;V^{\omega}\rangle\rangle_{\omega} = (A^{+}|(\omega\hat{I}-\hat{H}_0)^{-1}V^{\omega}) \tag{3.11}$$

where the superoperator resolvent is defined through its series expansion

$$(\omega\hat{I}-\hat{H}_0)^{-1} = \omega^{-1}\left\{\hat{I} + \sum_{n=1}^{\infty}\left[\frac{\hat{H}_0}{\omega}\right]^{n}\right\} \tag{3.12}$$

With the same condensed notation, Eq. (3.7) may be written

$$\langle\langle A;V^{\omega_1},V^{\omega_2}\rangle\rangle_{\omega_1,\omega_2} = (1+P_{12})\langle\langle[((\omega_1+\omega_2)\hat{I}+\hat{H}_0)^{-1}A,V^{\omega_1}];V^{\omega_2}\rangle\rangle_{\omega_2} \tag{3.13}$$

which in turn may be expressed as a reference state average value using eq. (3.11) and the definition in Eq. (3.8).

3.B Operator representations

It is virtually impossible to derive approximations to equations like Eqs. (3.11) and (3.13) because they involve inverse operators. Instead, it is useful to reformulate them in terms of inverse matrices which are much easier to handle. At present, such a reformulation exists for the linear response function (Oddershede and Jørgensen, 1977). The procedure consists of introducing a complete operator space

$$\mathbf{h} = \{\mathbf{h}_2,\ \mathbf{h}_4,\ \mathbf{h}_6,\ \ldots\} \tag{3.14}$$

where

$$h_2 = \{q^+, q\} = \{a_m^+ a_\alpha, \ a_\alpha^+ a_m\} \tag{3.15}$$

$$h_4 = \{q^+ q^+, \ qq\} = \{a_m^+ a_\alpha a_n^+ a_\beta, \ a_\beta^+ a_n a_\alpha^+ a_m\} \tag{3.16}$$

etc.

which are the particle-hole ($\mathbf{q}^+$) the hole-particle ($\mathbf{q}$), the two-particle-two-hole ($\mathbf{q}^+\mathbf{q}^+$) etc. excitations. Particle (Roman indices) and hole (Greek indices) refer to virtual and occupied Hartree-Fock orbitals, respectively. That this set of Hartree-Fock excitations in fact forms a complete set of operators is not self-evident. The proof of it was given by Dalgaard (1979). The only condition is that $\langle HF|0\rangle \neq 0$, where $|HF\rangle$ is the Hartree-Fock ground state and $|0\rangle$ is the reference state.

The resolvent $(\omega\hat{I}-\hat{H}_0)^{-1}$ is replaced by its inner projection (Löwdin, 1965) $|\mathbf{h})(\mathbf{h}|\omega\hat{I}-\hat{H}_0|\mathbf{h})^{-1}(\mathbf{h}|$ and Eq. (3.11) becomes

$$\langle\langle A;V^{\omega}\rangle\rangle_{\omega} = (A^+|\mathbf{h})(\mathbf{h}|\omega\hat{I}-\hat{H}_0|\mathbf{h})^{-1}(\mathbf{h}|V^{\omega}) \tag{3.17}$$

An alternative derivation of Eq. (3.17) is given in Eqs. (24)–(29) of Oddershede (1983).

Equation (3.17) is still an exact expression for the linear response function. A hierarchy of approximate solutions to it is obtained by choosing approximations to the *two*, in principle, exact quantities in Eq. (3.17), *i.e.* the operator space $\mathbf{h}$ and the reference state $|0\rangle$. The latter appears in Eq. (3.17) through the definition of the superoperator binary product in Eq. (3.8).

3.C Approximate linear response methods

We have discussed the specific form of the propagator in the various approximate methods at several places; see *e.g.* the 1983 lecture notes (Oddershede, 1983) and Oddershede *et al.* (1984). Therefore, we will take a more general discussion here and instead give the specific characteristics of the individual methods in Table 1.

Table 1. Description of Linear Response Methods.

Method	$\|0\rangle$	h	H	$t_A(\omega)$	Order in PT
Hartree-Fock	$\|HF\rangle$	$\mathbf{q}^+$	F	$(A^+\|\mathbf{q}^+)$	0
Mono-Excited CI (MECI), Tamm-Dancoff App. (TDA)	$\|HF\rangle$	$\mathbf{q}^+$	H_0	$(A^+\|\mathbf{q}^+)$	0–1
Random-Phase App. (RPA), Time-Dep. HF (TDHF), Coupled HF (CHF)	$\|HF\rangle$	$\mathbf{h}_2$	H_0	$(A^+\|\mathbf{h}_2)$	1
Higher RPA (HRPA)	$\|HF\rangle+\|0^{(1)}\rangle$[a]	$\mathbf{h}_2$	H_0	$(A^+\|\mathbf{h}_2)$	1–2
Second Order Pol. Prop. App. (SOPPA)	$\|HF\rangle+\|0^{(1+2')}\rangle$[b]	$\mathbf{h}_2,\mathbf{h}_4$	H_0	$(A^+\|\mathbf{h}_2)+$ $(A^+\|\mathbf{h}_4)(\omega-D)^{-1}C$[c]	2
Coupled Cluster Pol. Prop. App. (CCPPA)	$\|CCSD\rangle$	$\mathbf{h}_2,\mathbf{h}_4$	H_0	as above	2+[d]
Multiconfigurational Lin. Resp., Multiconfigurational TDHF	$\|MCSCF\rangle$	$\mathbf{h}_2,\boldsymbol{\Lambda}$[e]	H_0	$(A^+\|\mathbf{h}_2),(A^+\|\boldsymbol{\Lambda})$	N/A[d]

[a] $|0^{(1)}\rangle$ is the first order correction to the wavefunction.
[b] $|0^{(2')}\rangle$ is the second order correction to the wavefunction originating from single excitations only.
[c] The C,D matrices are defined, *e.g.*, in appendix C of Oddershede *et al.* (1984).
[d] See text for a discussion.
[e] $\boldsymbol{\Lambda} = \{|n\rangle\langle 0|, |0\rangle\langle n|\}$ and $\mathbf{h}_2$ includes in this case the non-redundant excitations from the inactive space to the active and secondary spaces as well as the excitations from the active to the secondary space.

Writing Eq. (3.17) in matrix form

$$\langle\langle A;V^{\omega}\rangle\rangle_{\omega} = \{(A^{+}|q^{+}),\ (A^{+}|q),\ \ldots\} \begin{bmatrix} (q^{+}|\omega\hat{I}-\hat{H}_0|q^{+}) & (q^{+}|\omega\hat{I}-\hat{H}_0|q) & \ldots \\ (q|\omega\hat{I}-\hat{H}_0|q^{+}) & (q|\omega\hat{I}-\hat{H}_0|q) & \ldots \\ \vdots & \vdots & \end{bmatrix}^{-1} \begin{bmatrix} (q^{+}|V^{\omega}) \\ (q|V^{\omega}) \\ \vdots \end{bmatrix} \tag{3.18}$$

we see that one way to approximate $\langle\langle A;V^{\omega}\rangle\rangle_{\omega}$ would be to truncate **h** after some specified level of excitation. That is in fact what is done in the random phase approximation (RPA) where only $\mathbf{h}_2$ [Eq. (3.15)] is kept. However, in higher order approximations it is often computationally advantageous to be guided by perturbation theory when truncating Eq. (3.18). We therefore perform a partitioning (Löwdin, 1963) of the inverse matrix in Eq. (3.18) whereby we fold the effect of the higher order excitations into the space of the particle-hole, hole-particle space. In order to determine the linear response function consistently through a given order n in perturbation theory it is not necessary to calculate all terms in the inverse matrix through order n. Only the particle-hole, hole-particle matrices will then be needed through order n. The structure of the partitioned propagator is [see *e.g.* Oddershede *et al.* (1984), Sec. 5]

$$\langle\langle A;V^{\omega}\rangle\rangle_{\omega} = \left\{t_A(\omega), \mp t_A(-\omega)\right\} P(\omega)^{-1} \begin{bmatrix} \tilde{t}_{V^{\omega}}(\omega) \\ \mp\ \tilde{t}_{V^{\omega}}(-\omega) \end{bmatrix} + W_4(\omega) \tag{3.19}$$

where the choice of the $\mp$ signs depends on the symmetry of the A and V^{ω} operators (Sauer *et al.*, 1991). The transition matrix elements, **t**, in various approximations are given in Table 1. Since neither the **t**'s nor $\mathbf{W}_4$ [Eq. (63) in Oddershede (1983)] have poles in ω, the poles of the linear response functions come solely from $\mathbf{P}(\omega)^{-1}$, sometimes called (Oddershede, 1982) the principal propagator. The poles of $\mathbf{P}(\omega)^{-1}$ *i.e.* the excitation energies [see Sec. 2.D] are in practice obtained by diagonalizing $\mathbf{P}(\omega)$ [see also Sec. 3.D]. In an n–th order polarization propagator approach **t**, **P** and $\langle\langle\ldots\rangle\rangle_{\omega}$ are all calculated through order n. The order in perturbation theory that we are referring to is defined such that Hartree-Fock is zeroth order and the perturbation operator is therefore the fluctuation potential, *i.e.* H_0 minus the sum of the one-electron Fock operators, F, for the reference state.

The most time-consuming step in a polarization propagator calculation is to set up the matrices that form $\mathbf{P}(\omega)$. However, from the structure of Eq. (3.19) we see that this step is independent of the property A and the perturbation V^{ω}. From one calculation of $\mathbf{P}(\omega)$ we may consequently evaluate a range of molecular properties at little extra cost by simply multiplying $\mathbf{P}(\omega)^{-1}$ by the appropriate transition matrices. We have used this property of the method on several occasions, most notably perhaps in the calculations of the indirect nuclear spin-spin coupling constants of NMR spectroscopy, where several terms contribute to each coupling constant (Geertsen and Oddershede, 1984).

Table 1 is organized such that the accuracy of the methods increase as one proceeds downwards in the Table ending up with the multiconfigurational linear response (MCLR) method (Yeager and Jørgensen, 1979) which is exact in the limit of a complete CI. The order concept is not applicable for MCLR, thus the N/A in the table. The three consistent approximations in terms of perturbation theory are HF (0), RPA (1) and SOPPA (2). Geertsen and Oddershede (1986) have proposed an extension of SOPPA using a coupled cluster wavefunction. This method, CCPPA, is formulated such that if the first iteration of the coupled cluster amplitudes is used, one obtains the SOPPA, but if converged amplitudes are used, several diagram series are included to infinite order. Still the method does not include all third order contributions; therefore the designation 2+ in Table 1. In terms of perturbation theory the relationship between SOPPA and CCPPA is the same as between MBPT2 and lineralized CCD calculations of correlation energies. At present we have not formulated a consistent third order polarization method which includes all third order terms in both $\mathbf{t}_A(\omega)$ and $\mathbf{P}(\omega)$. All third order terms in $\mathbf{P}(\omega)$ were identified some years ago (Oddershede and Jørgensen, 1977) and we have recently (Geertsen *et al.*, 1991) shown that the most important third order term in $\mathbf{P}(\omega)$ is in fact the third order term included in the CCPPA approach.

3.D Approximate quadratic response methods

Quadratic response functions have not yet been so widely applied as their linear counterpart and the only general method for deriving approximations to $<<A;V^{\omega_1},V^{\omega_2}>>_{\omega_1,\omega_2}$ that go beyond RPA is that by Olsen and Jørgensen (1985) as discussed in the introduction to this section. There still does not exist a general superoperator reformulation of Eq. (3.13) that is equivalent to the one used to derive Eq. (3.17) from Eq. (3.11). Actual applications have so far been restricted to the RPA, that is, the quadratic polarization propagator that is consistent through first order in the fluctuation potential. The RPA is hence defined equivalently (see Table 1) for both the quadratic and linear response functions (Dalgaard, 1982).

We have performed RPA calculations of the double residue of the quadratic response function, *i.e.* excited state-excited state transition probabilities, for a few simple diatomic molecules (Sengeløv and Oddershede, 1988; Oddershede and Sengeløv, 1989). The single residue of the quadratic response function is the two-photon transition moment and Parkinson *et al.* (1990) have reported the first consistent RPA calculations of this quantity. The full quadratic response function is the first hyperpolarizability, Eq. (2.8), and Parkinson and Oddershede (1991) have used this procedure to calculate $\beta(\omega,\omega)$ for H_2O and HF.

3.E Large Matrix problems

Very often the basis sets that one needs to use get so large that the particle-hole space, *i.e.* the dimension of $P(\omega)$ in Eq. (3.19), becomes too big for storing and/or inverting $P(\omega)$. In such cases we have to restore to large matrix techniques similar to those used in CI methods (Shavitt, 1977; Davidson, 1975). Instead of solving Eq. (3.19) we define

$$\chi_A(\omega) = P(\omega)^{-1} \begin{bmatrix} \tilde{t}_A(\omega) \\ \mp \tilde{t}_A(-\omega) \end{bmatrix} \tag{3.20}$$

or

$$P(\omega)\chi_A(\omega) = \begin{bmatrix} t_A(\omega) \\ \mp t_A(-\omega) \end{bmatrix} \tag{3.21}$$

such that Eq. (3.19) becomes

$$\langle\langle A;V^{\omega}\rangle\rangle_{\omega} = \left\{t_A(\omega), \mp t_A(-\omega)\right\}\chi_{V^{\omega}}(\omega) \tag{3.22}$$

Using the reduced linear equations techniques (Pople *et al.*, 1979; Pulay, 1980) we find the $\chi_A(\omega)$ that fulfills Eq. (3.21) without ever constructing the $P(\omega)$ matrix. We only compute the product of $P(\omega)$ and $\chi_A(\omega)$.

Rettrup (1982) gave the first description of such a method within the framework of RPA-like approximations and it was later reformulated into a computationally more efficient procedure by Olsen *et al.* (1988). Recently, Feyreisen *et al.* (1992) undertook a project using this direct technique in the atomic orbital basis. The integral driven construction of the product on the left hand side of Eq. (3.21) is written as a product of the atomic two-electron integrals and a modified matrix $\chi_A^{A0}(\omega)$. In doing so, the integral transformation step as well as the need for storing the atomic integrals on discs are eliminated. In our preliminary calculations we were able to use basis set as large as 600 atomic orbitals for the para-nitroanilin molecule and its dimer.

In closing, let us mention that the direct techniques can also be used for the quadratic response function, at least at the RPA and MCRPA levels (Olsen and Jørgensen, 1985; Parkinson and Zerner, 1989; Parkinson and Oddershede, 1991). Feyreisen *et al.* (1972) have also shown that at the RPA level it is even possible to have atomic orbital driven direct calculations of quadratic response properties.

4. SUMMARY

An introduction to the theory of response/polarization propagator methods is given. The main emphasize has been on understanding the relationship between the response of a system to an external perturbation and propagators or response functions. This point of view introduces the propagators in the energy or spectral representation and gives the most direct way of relating propagators to physical observables. In order to relate the present formulation to the original Zubarev (1960) exposition we have also discussed the time-dependent definition of Green's functions and how to go from one picture to the other.

The various approximate linear and quadratic response methods are discussed, mainly in tabular form [Table 1]. We end the lecture notes by a discussion of some of the more recent advances in propagator theory, including an atomic orbital driven technique for calculating RPA linear and quadratic response functions.

ACKNOWLEDGMENT

This work was supported by a grant from the Danish Natural Science Research Council (Grant no. 11–9004). The author is grateful to Direktør Ib Henriksen's Fond for the stay at their estate at Castellaras where this manuscript was written. Thank you also to several participants at the NATO Advanced Study Institute who helped improve the presentation at several places.

REFERENCES

Bloembergen, N., 1965, "Nonlinear Optics", Benjamin, Reading.

Dalgaard, E., 1979, *Intern. J. Quantum Chem.*, **15**:169.

Dalgaard, E., 1981, *in*: "Recent Developments and Applications of Multiconfiguration Hartree-Fock Methods," Natl. Res. Compt., Berkeley, California:136.

Dalgaard, E., 1982, *Phys. Rev. A*, **26**:42.

Davidson, E., 1975, *J. Comput. Phys.*, **17**:87.

Ehrenfest, P., 1927, *Z. Phys.*, **45**:455.

Feyreisen, M., Nichols, J., Oddershede, J., and Simons, J., 1992, work in progress.

Frenkel, J., 1934, "Wave Mechanics – Advanced General Theory", Oxford University Press, Oxford.

Geertsen, J., Eriksen, S., and Oddershede, J., 1991, *Adv. Quantum Chem.*, in press.

Geertsen, J., Jensen, F., and Scuseria, G. E., 1989, *J. Chem. Phys.*, **91**:364.

Geertsen, J., and Oddershede, J., 1984, *Chem. Phys.*, **90**:301.

Geertsen, J., and Oddershede, J., 1986, *J. Chem. Phys.*, **85**:2112.

Goscinski, O., and Lukman, B., 1970, *Chem. Phys. Lett.*, **7**:573.

Jørgensen, P., and Simons, J., 1981, "Second Quantization-Based Methods in Quantum Chemistry", Academic Press, New York.

Kubo, R., 1959, *J. Phys. Soc. Japan*, **12**:570.

Langhoff, P. W., Epstein, S. T., and Karplus, M., 1972, *Rev. Mod. Phys.*, **44**:602.

Lehman, H., 1954, *Nuovo Cimento* [9], **11**:342.

Linderberg, J., and Öhrn, Y., 1973, "Propagators in Quantum Chemistry", Academic Press, London.

Löwdin, P.–O., 1963, *J. Mol. Spectrosc.*, **10**:12.

Löwdin, P.–O., 1965, *Phys. Rev. A*, **139**:357.

McWeeny, R., 1989, "Methods of Molecular Quantum Mechanics", Academic Press, London.

Oddershede, J., 1978, *Adv. Quantum Chem.*, **11**:275.

Oddershede, J., 1982, *Intern. J. Quantum Chem., Quantum Chem. Symp.*, **16**:583.

Oddershede, J., 1983, *in*: "Methods in Computational Molecular Physics", G. H. F. Diercksen and S. Wilson, eds., D. Reidel, Dordrecht:249.

Oddershede, J., 1987, *Adv. Chem. Phys.*, **69**:201.

Oddershede, J., and Geertsen, J., 1990, *J. Chem. Phys.*, **92**:6036.

Oddershede, J., and Jørgensen, P., 1977, *J. Chem. Phys.*, **66**:1541.

Oddershede, J., Jørgensen, P., and Beebe, N. H. F., 1978, *J. Phys. B*, **11**:1.

Oddershede, J., Jørgensen, P., and Yeager, D. L., 1984, *Compt. Phys. Report*, **2**:33.

Oddershede, J., and Sabin, J. R., 1991, *Intern. J. Quantum Chem.*, **39**:371.

Oddershede, J., and Sengeløv, P. W., 1989, *in*: "Aspects of Many-Body Effects in

Molecules and Extended Systems", D. Mukherjee, ed., Springer-Verlag, Heidelberg:207.

Olsen, J., Jensen, H. J. Aa., and Jørgensen, P., 1988, *J. Comput. Phys.*, **74**:265.

Olsen, J., and Jørgensen, P., 1985, *J. Chem. Phys.*, **82**: 3235.

Parkinson, W. A., and Oddershede, J., 1991, *J. Chem. Phys.*, **94**:7251.

Parkinson, W. A., Sengeløv, P. W., and Oddershede, J., 1990, *Intern. J. Quantum Chem., Quantum Chem. Symp.*, **24**:487.

Pickup, B. T., and Goscinski, O., 1973, *Mol. Phys.*, **26**:1013.

Pople, J. A., Krishnan, R., Schlegel, H. B., and Binkley, J. S., 1979, *Intern. J. Quantum Chem., Quantum Chem. Symp.*, **13**:225.

Pulay, P., 1980, *Chem. Phys. Lett.*, **73**:393.

Rettrup, S., 1982, *J. Comput. Phys.*, **45**:100.

Sauer, S. P. A., Diercksen, G. H. F., and Oddershede, J., 1991, *Intern. J. Quantum Chem.*, **34**:667.

Schirmer, J., 1982, *Phys. Rev. A*, **26**:2395.

Schirmer, J., Cederbaum, L. S., and Walter, O., 1983, *Phys. Rev. A*, **28**:1237.

Sengeløv, P. W., and Oddershede, J., 1988, *Chem. Phys.*, **124**:371.

Shavitt, I., 1977, *in*: "Methods of Electronic Structure Theory", H. F. Schafer III and W. H. Miller, eds., Plenum Press, New York.

Yeager, D. L., and Jørgensen, P., 1979, *Chem. Phys. Lett.*, **65**:77.

Zubarev, D. N., 1960, *Usp. Fiz. Nauk.*, **71**:71 (Engl. Transl. *Sov. Phys. Usp.*, **3**:320).

Zubarev, D. N., 1974, "Nonequilibrium Statistical Thermodynamics", Consultants Bureau, New York.

VALENCE BOND THEORY

R.McWeeny

Dipartimento di Chimica e Chimica Industriale
Universitá di Pisa, Via Risorgimento 35, 56100 Pisa, Italy

1) Introduction

It was once claimed (Pauling and Wilson, 1935), and is still widely accepted, that the classic paper by Heitler and London (1927) "must be considered the greatest single contribution to the chemist's conception of valence that has been made since G. N. Lewis's suggestion in 1916 that the chemical bond between two atoms consists of *a pair of electrons held jointly by the two atoms*". In the years immediately following that paper, the idea of the 'electron-pair bond' was rapidly generalized to many-electron molecules and to more complicated bonding situations: the resultant *valence bond theory* won immediate acceptance and was applied with remarkable success (albeit in a semi-empirical form) not only to molecules of considerable complexity but even to the study of chemical reactions.

With the development of ab initio methods of computation, however, the situation has changed dramatically. Nowadays quantum chemistry is dominated by ideas drawn from the independent-particle model (IPM) and, in particular, from the concept of an *electron configuration* in which each electron occupies 'its own' orbital (a *molecular* orbital). When degeneracies occur, each configuration gives rise to a number of component states: and normally this 'configurational' description of molecular states is qualitatively rather accurate (one need only recall the immense value of the MO model in spectroscopy), the *mixing* of configurations (CI) being admitted essentially as a refinement – a means of recognizing electron correlation and obtaining wavefunctions of high numerical precision. Indeed, a newcomer to 'computational chemistry' may well be almost totally ignorant of a rich period of quantum chemistry (1927-1935) in which valence bond theory played the leading role.

The purpose of this chapter is to sketch the main ideas of VB theory, indicating the advantages and disadvantages of the approach; to study the mathematical framework of the theory and its relationship to topics that by now will be familiar (e.g. the symmetric and unitary groups and their representations); and to present several forms of modern *ab initio* VB theory that are computationally viable and may, in many areas, even be competitive with the established 'MO+CI' approach.

2) The Classical VB Approach

In the hydrogen molecule calculation of Heitler and London (1927) the two electrons are *not* regarded as quasi-independent particles; each occupies an *atomic* orbital (a or b, say, on the two centres) and at large distances a correct wavefunction, decribing independent *atoms*, would be

* Product functions will be written according to the natural order of the *variables* (e.g ba will mean $b(\mathbf{r}_1)a(\mathbf{r}_2)$); and normalizing factors will usually be omitted.

Methods in Computational Molecular Physics, Edited by Stephen Wilson and Geerd H.F. Diercksen, Plenum Press, New York, 1992

$a(\mathbf{r}_1)b(\mathbf{r}_2)$ or $a(\mathbf{r}_2)b(\mathbf{r}_1)$. Symmetry is then recognized by setting up*

$$\Phi_1 = ab + ba, \qquad \Phi_2 = ab - ba. \tag{1}$$

Spin is admitted by attaching a spin factor $\Theta_{0,0} = \alpha\beta - \beta\alpha$ (the subscripts indicating the quantum numbers S, M) to Φ_1, to obtain a singlet state; or one of the three factors $\Theta_{1,1} = \alpha\alpha$, $\Theta_{1,0} = \alpha\beta + \beta\alpha$, $\Theta_{1,-1} = \beta\beta$ to Φ_2, to obtain the three components of a triplet.

All the wavefunctions constructed in this way satisfy the Pauli principle and the following results are easily established:

(i) The energy expectation values in the singlet and triplet states are, respectively,

$$E_1 = \frac{Q+K}{1+S^2}, \qquad E_2 = \frac{Q-K}{1-S^2}, \tag{2}$$

where

$$Q = < ab|\hat{H}|ab >, \quad K = < ab|\hat{H}|ba >, \quad S = < a|b > . \tag{3}$$

The 'exchange integral' K is large and negative around the equilibrium geometry: the ground state is thus the singlet while the triplet shows repulsion at all internuclear distances.

(ii) Both the singlet and triplet energy curves behave correctly at large distance, going to the limit $E_\infty = 2E_{\rm H}$ for two isolated hydrogen atoms.

(iii) In calculating the energy, the spin factor is *irrelevant* (integration over spin variables giving only a normaization factor of unity); it serves only as an indicator of the *symmetry* of the spatial factor with which, through the Pauli principle, it is linked.

(iv) If overlap is neglected in the denominators, the energies E_1, E_2 are correctly reproduced using a spin-only *model* of the system, with a 'spin Hamiltonian'

$$\hat{H}_{\rm S} = Q' - 2K\hat{\mathbf{S}}(1)\cdot\hat{\mathbf{S}}(2) \qquad Q' = Q + \tfrac{1}{2}K \tag{4}$$

– for the scalar product operator, acting on a singlet or triplet spin function, multiplies it by $-3/4$ or $1/2$ and $\hat{H}_{\rm S}$ thus has eigenvalues $Q+K$ or $Q-K$, respectively.

The above features of the Heitler-London calculation can all be generalized. For a polyatomic molecule with N singly occupied valence AOs, $\phi_1, \phi_2, \ldots \phi_N$, the orbital product function

$$\Omega(\mathbf{r}_1, \mathbf{r}_2, \ldots \mathbf{r}_N) = \phi_1(\mathbf{r}_1)\phi_2(\mathbf{r}_2)\ldots\phi_N(\mathbf{r}_N) \tag{5}$$

(one of $N!$ possible permuted products) may be combined with 2^N possible spin products. There is thus a very high spin and permutation degeneracy. This may be partly resolved by introducing spin *eigenfunctions*, as for H_2, with a singlet factor $\theta = \alpha\beta - \beta\alpha$ for any pair (ij) with an assumed bonding interaction, but a factor $\alpha(s_k)$ for any electron whose spin belongs to a 'parallel-coupled' set. The permutation degeneracy can then be resolved (and the Pauli principle simultaneously satisfied) by applying the antisymmetrizer $\hat{A} = \sum_{\hat{P}} \epsilon_P \hat{P}$. Again, with neglect of overlap, one may define

$$Q = < \phi_1\phi_2\ldots\phi_N|\hat{H}|\phi_1\phi_2\ldots\phi_N >, \qquad K_{ij} = < \phi_1\phi_2..\phi_i..\phi_j..\phi_N|\hat{H}|\phi_1\phi_2..\phi_j..\phi_i..\phi_N > \tag{6}$$

and obtain for the energy expectation value

$$E = Q + \sum_{(ij)\ \text{paired}} K_{ij} - \frac{1}{2}\sum_{(ij)\ \text{uncoupled}} K_{ij} - \sum_{(ij)\ \text{parallel}} K_{ij} \tag{7}$$

where '(ij)' indicates summation over distinct pairs. As before, the same expression may be obtained as the expectation value of a *spin* Hamiltonian

$$\hat{H}_{\rm S} = Q' - 2\sum_{(ij)} \hat{\mathbf{S}}(i)\cdot\hat{\mathbf{S}}(j) \qquad (Q' = Q - \tfrac{1}{2}\textstyle\sum_{(ij)} K_{ij}), \tag{8}$$

– the expectation value of the spin scalar product giving the correct coefficients in (7) for any assumed spin coupling. It must be stressed that this 'Dirac-Van Vleck model' does not imply that there are in reality any spin-spin interactions between the electrons (for no such terms are present in the assumed Hamiltonian $\hat{H}$); it is simply a convenient way of representing the effects of antisymmetry, through a *fictional* spin-spin term, in a manner that can be easily visualized.

The formula (7) gives the 'perfect pairing' approximation to the energy: it corresponds to the use of one electron configuration (the orbitals appearing in Ω) and one VB 'structure' – corresponding to the spin-coupling scheme associated with a hypothetical (but plausible) allocation of chemical bonds. In many cases, with well chosen orbitals, such a structure can give a remarkably good approximation to the energy (often better than a comparable MO function, as in the case of the H_2 molecule; and the *dissociation limit is essentially correct*, with the valence orbitals of the separated atoms remaining singly occupied. This situation is to be compared with that in the standard molecular orbital approach, where a one-configuration approximation fails badly in bond-breaking geometries (expansion in AOs showing the presence of 'ionic' terms in which electrons are transferred from one atom to another). Often very extensive CI is necessary to correct this failure: indeed, in a study of symmetric dissociation of the water molecule Handy et al (1985) used many thousands of configurations and showed that many-body perturbation methods of including the CI did not always show satisfactory convergence, even in 50th order!

When a single structure does *not* give an adequate picture of the bonding (e.g. in the classic case of the benzene π electrons, where classical chemistry suggests two equally eligible structures) VB theory has a ready solution: all the plausible structures are admitted, with various 'pairing schemes', and the wavefunction is represented in the usual way as a linear combination – referred to as a 'resonance mixture'. The optimum mixing coefficients are obtained, as usual, by solving a matrix secular equation (1.12) and this entails the calculation of the matrix elements (1.13) – in which Φ_K, Φ_L are now typical structure functions.

Here again, VB theory in its classical form provides a simple rule for evaluating the matrix elements. Let us use $\Phi_\kappa, \Phi_\lambda$ to denote the structures associated with spin eigenfunctions $\Theta_\kappa, \Theta_\lambda$ in which some spins are paired and the others (if any) are parallel coupled. The numbers $1, 2, \ldots N$ are written in a ring and any 'pairs' (i, j) are represented by a link $i \rightarrow j$, unpaired numbers being left as dots. When two such 'Rumer diagrams' are superimposed, to give a 'superposition pattern', the corresponding matrix element of the spin operator in (8) is determined by the number of *islands* and the nature of any open *chains* formed by the links in the pattern. The most commonly used rules, due to Pauling (1933), apply to singlet states ($S = 0$ and no open chains: for two structures, Φ_κ and Φ_λ, the overlap and Hamiltonian matrix elements that result are given by

$$S_{\kappa\lambda} = 2^{n_{\kappa\lambda}-g}, \qquad H_{rs} = 2^{n_{\kappa\lambda}-g}(Q - x_{ij}^{\kappa\lambda} K_{ij}), \tag{9}$$

where $n_{\kappa\lambda}, g$ are the numbers of islands and links ($g = N/2$), while $x_{ij}^{\kappa\lambda}$ takes the value $+1$ for i, j in the same island, an odd number of links apart; -2 if they are an even number of links apart; and $-1/2$ if they are in different islands. These rules are easy to apply (see, for example, Coulson (1951)) and contain all that is needed for the application of VB theory in its primitive form to a considerable range of problems in both chemistry and physics. It will be recognised, for example, that the spin Hamiltonian (8), with its few-parameter form and simple pictorial appeal, is the basis of the Heisenberg model, widely used (even to this day) in the discussion of ferromagnetism, anti-ferromagnetism and other magnetic properties of materials, and even high-T_C superconductivity (see, for example, Anderson (1987), Mattis (1990)).

At this point it is perhaps not clear why an approach with such striking simplicity and generality should ever have been displaced from the position it occupied in the 'thirties. The explanation is contained in one word – *overlap*. The matrix element expressions given above are all *approximate* because orbital overlap has been neglected. This may seem to be a small blemish; but in fact it is profound and extremely difficult to eliminate. When quantum chemists started making *ab initio* calculations, these imperfections soon came to light and quickly lead to the almost universal adoption of IPM-based techniques in which the orbitals are solutions of a 1-electron eigenvalue equation and are thus automatically orthogonal: such orbitals lend themselves perfectly to the computer implementation of Slater's rules (1.26), even on a grand scale and with no neglect of 1- and 2-electron integrals.

The dilemma of VB theory is clear-cut: the negative values assumed for exchange terms K_{ij} (allowing their consequent association with chemical bonding) are a consequence solely of *orbital overlap*; but in deriving results such as (7) it was necessary to assume *zero* overlap to avoid the proliferation of terms corresponding to *multiple* exchange. To remove this fundamental inconsistency the alternatives are either (i) to orthogonalize the orbitals at the outset, so that the simple formulas will apply, or (ii) to accept non-orthogonality and the mathematical problems to which it leads. The difficulty with (i) is that, with only one orbital configuration, it is impossible to account for chemical bonding (the K_{ij} being essentially positive) as was first pointed out by Slater (1951); and to retrieve the bonding it is necessary to admit extensive CI with all kinds of 'ionic' structures (McWeeny 1954). The difficulty with (ii) is the famous 'N! problem' – to be considered in the next Section.

3) Modern Ab Initio Valence Bond Theory

Let us now examine more closely the evaluation of matrix elements such as (1.13), first for the case in which both structures belong to the single orbital configuration defined by the product function (5).

The antisymmetric function associated with Θ_κ will then be

$$\Phi_\kappa = \hat{A}[\Omega\Theta_\kappa], \tag{10}$$

where $\hat{A}$ is most conveniently chosen as the *idempotent* operator ($\hat{A}^2 = \hat{A}$)

$$\hat{A} = (N!)^{-1}\sum \epsilon_P \hat{P}, \tag{11}$$

in which every permutation $\hat{P}$ works on the indices of both space and spin variables. The matrix elements then reduce easily, giving (unnormalized)

$$H_{\kappa\lambda} = \sum_{\hat{P}} \epsilon_P H_P < \Theta_\kappa|\hat{P}|\Theta_\lambda >, \qquad H_P =< \phi_1\phi_2 \ldots \phi_N|\hat{H}\hat{P}|\phi_1\phi_2 \ldots \phi_N >, \tag{12}$$

while $M_{\kappa\lambda}$ follows on replacing the factor H_P by $M_P =< \phi_1\phi_2 \ldots \phi_N|\hat{P}\phi_1\phi_2 \ldots \phi_N >$ – which is simply a product of overlap integrals. Evidently $H_{\lambda\kappa}$ and $M_{\lambda\kappa}$ are both sums of $N!$ terms, a number which (even with present-day computing power) surely seems to preclude the calculation of **H** and **M** matrices for systems with more than, say, 20 electrons ($20! \approx 10^{12}$). This is the so-called 'N! problem'.

The first step towards meeting this difficulty is the realization that the number of electrons to be considered explicitly is not necessarily the total number of electrons in the molecule: N may, for example, be the number of *valence* electrons – or even of a *subset* of the valence electrons (e.g. the pi electrons of a conjugated system). To use MO-CI terminology, attention may be confined to an 'active space' which may be chosen orthogonal to a set of doubly-occpupied 'core' orbitals: the core may contain atomic inner-shells and, more generally, any orbitals of a closed-shell 'reference configuration' from which electron 'excitations' are not contemplated. With this reduction in the value of N, VB and MO-CI calculations are at once put on a par – the only reasonable goal in either case will usually be a 'full-CI' treatment of the electrons in the *active space*.

The presence of a set of core electrons, occupying orbitals $\phi_1^c, \phi_2^c, \ldots$, is recognized implicitly through the use of an *effective* 1-electron Hamiltonian (McWeeny 1955; McWeeny and Ohno 1960) for the remaining electrons of the active space: this operator has matrix elements*

$$< \phi_r|\hat{h}_{\text{eff}}|\phi_s > =< \phi_r|\hat{h}|\phi_s > + \sum_{i,j} P_{ij}^c[< \phi_r\phi_j^c|g|\phi_s\phi_i > -\frac{1}{2} < \phi_r\phi_j^c|g|\phi_i\phi_s >], \tag{13}$$

and simply replaces the usual 1-electron Hamiltonian $\hat{h}$, which contains only bare nuclear attraction terms. The 1-electron density matrix, with elements P^c_{ij}, is referred to the core orbitals alone; and in this form (11) remains valid (see, for example, MMQM Ch.14) for any type of (singlet) core function as long as the core and valence orbital sets are mutually orthogonal.

The calculation is now performed for the (valence) electrons of the active space alone and the total electronic energy, with no approximations other than those implicit in the form of the wavefunction, becomes

$$E_{\text{tot}} = E_{\text{core}} + E_{\text{val}}, \tag{14}$$

where E_{core} refers to the core electrons and is calculated as if the valence electrons were absent: all the core-valence interactions are included in E_{val}.

Let us now look more carefully into the evaluation of the matrix elements (12). The identity permutation obviously gives a leading term in $H_{\kappa\lambda}$ which coincides with the quantity Q given in (6); while for a transposition $\hat{P}_{ij}$ the corresponding $H_{\hat{P}}$ becomes the 'exchange integral' K_{ij}. For a single configuration of orthogonal orbitals, no other terms appear. The coefficient of K_{ij} will be

$$< \Theta_\kappa|\hat{P}_{ij}|\Theta_\lambda >= \tfrac{1}{2} < \Theta_\kappa|1 + 4\hat{\mathrm{S}}(i)\cdot\hat{\mathrm{S}}(j)|\Theta_\lambda >, \tag{15}$$

where we use the Dirac spin-exchange identity, and this leads to the vector-model result (8). The difficulties arise when the orbitals are *non*-orthogonal, because then there are no zero-overlap factors in the evaluation of H_P and consequently *all* permutations must be considered.

There are essentially two families of methods of implementing the calculation. The first is dependent on the theory of the symmetric group S_N of $N!$ permutations: for the coefficient of $H_{\hat{P}}$ may be written

$$< \Theta_\kappa|\hat{P}|\Theta_\lambda >= D_S(\hat{P})_{\kappa\lambda}, \tag{16}$$

which, for a complete orthonormal set of spin eigenfunctions of given S, M, is the $\kappa\lambda$-element of the matrix $\mathrm{D}_S(\hat{P})$ associated with permutation $\hat{P}$ in the *irreducible representation* D_S carried by the set*. Evidently, these 'irreps' may in principle be assumed 'known'; and various algorithms for obtaining the matrices are available. Thus, Kotani et al (1955), following the earlier work of Yamanouchi (1937), were able to tabulate the matrices of the *generators* of the group, from which all others may be obtained by simple matrix multiplication, for systems with up to 8 electrons. However, the number of matrices needed rises rapidly with N and there are obvious problems, even for $N = 10$, in storing and repeatedly processing $3\frac{1}{2}$ million matrices!

The second family of methods involves the explicit construction of the spin eigenfunctions Θ_κ in terms of simple spin products; the expansion of the antisymmetrized products $\hat{A}[\Omega\Theta_\kappa]$, which becomes a sum of Slater determinants; and the evaluation of the matrix elements $< \Phi_\kappa|\hat{H}|\Phi_\lambda >$ by summing the contributions from all pairs of determinants. This somewhat inelegant procedure was successfully used in early attempts (e.g. by Matsen and coworkers (Miller et al 1957)) to make ab initio VB calculations but is evidently hindered by the difficulty of using Löwdin's generalization of Slater's rules (Löwdin 1955) to determinants of non-orthogonal orbitals (which requires, in the general case, the evaluation of vast numbers of cofactors).

Before considering the implementation of any of these procedures, it must be noted that a VB calculation based on a single configuration of orbitals (which in the classical form of VB theory were assumed to be *atomic* orbitals, or possibly hybrids) will not necessarily lead to a good wavefunction. Ideally, the further generalizations needed are (i) the use of a *multi*-configuration reference function

---

* The core is assumed to be in a singlet state.

and (ii) the *optimization* of the orbitals to which it refers. At this level the approach will have points in common with multi-configuration (MC) SCF theory, differing mainly in the relaxation of orthogonality restrictions and in the choice of orbital forms (e.g. localized or delocalized) and spin-coupled structures to be employed.

4) Methods of implementation of the theory

In this Section we present some of the main features of several computational schemes at present in use, starting with those that involve expansion in determinants.

The direct approach: Slater determinants

As indicated above, the main problem in this type of method is the evaluation of matrix elements of the Hamiltonian between pairs of determinants (U_k, U_l, say), arising in the expansion of $< \Phi_\kappa|\hat{H}|\Phi_\lambda >$, when the orbitals are non-orthogonal. As remarked above, Slater's rules (Slater 1929,1931) must then be replaced by those due to Löwdin (1955). If we use, momentarily, $k_1, k_2, ... k_N$ and $l_1, l_2, ... l_N$ to stand for the spin-orbitals in two determinants, U_k, U_l, then the required matrix elements are given in terms of the basic determinant

$$D_{kl} = det| < k_1|l_1 >< k_2|l_2 > ... < k_N|l_N > | \tag{17}$$

(where, as usual, only the diagonal elements are shown) and its cofactors or 'signed' minors. Thus, for example,

$$\begin{aligned} < U_k|U_l > &= D_{kl}(D_{kk}D_{ll})^{-\frac{1}{2}} \\ < U_k|\sum_i \hat{h}(i)|U_l > &= (D_{kk}D_{ll})^{-\frac{1}{2}} \sum_{r,s=1}^{N} < k_r|\hat{h}|k_s > D_{kl}(k_r, l_s) \end{aligned} \tag{18}$$

where $D_{kl}(k_r, l_s)$ is the first-order *cofactor* of the element $< k_r|l_s >$ in D_{kl}, obtained by striking out the row and column containing $< k_r|l_s >$ and adding the sign $(-1)^{r+s}$. The corresponding matrix element for the 2-electron operator $\sum g(i,j)$ involves *second*-order cofactors, one for each 2-electron integral, and the number required is therefore large.

When D_{kl} is non-singular the explicit calculation of cofactors can be avoided by using Jacobi's ratio theorem (see MMQM Sect.3.3) and making a single matrix inversion. But in many cases, especially in multi-configuation calculations, the overlap matrix is singular and the evaluation of very large numbers of cofactors, which is computationally costly, is unavoidable.

In spite of such difficulties, direct methods have been successfully used, notably by Balint-Kurti and collaborators (e.g. Balint-Kurti and Karplus 1968; van Lenthe and Balint-Kurti 1983); see also Raimondi et al (1985). Inevitably, the evaluation and handling of the cofactors (King et al 1967; Prosser et al 1968; Hayes and Stone 1984; Figari and Magnasco 1985) involves very heavy data management problems and this has given added impetus to methods that do not involve the use of Slater determinants.

Group theoretical methods (I)

Many variants of the group-theoretical approach, in which matrix elements are evaluated using the irreps of the symmetric group (as indicated in Section 3), are currently in use. The first of these,

---

This follows because $\hat{P}\Theta_\lambda$ must then be expressible as a linear combination $\sum_\kappa \Theta_\kappa c_{\kappa\lambda}$ and taking a scalar product with Θ_κ then identifies the coefficients as elements of the representation matrix as in (16).

and incidentally the first in which *optimization* of the orbitals was included, is due to Goddard (1967, 1968). In its original form, it deals with one configuration of singly occupied orbitals and one spin coupling scheme i.e. a single VB structure; the most important advance was the recognition that the energy expression, which takes the form

$$E = \sum_{i,j} P_{ji} < \phi_i|\hat{h}|\phi_j > +(1/2) \sum_{i,j,k,l} < \phi_i\phi_j|g|\phi_k\phi_l > \tag{19}$$

in terms of the 1- and 2-electron (spinless) density matrices, could provide a basis for the variational optimization of the orbitals – even though the dependence on the orbitals is not confined to the integrals but is shared (via the overlap) by their coefficients. Goddard was able to show that the stationary-value condition could be cast in the form of a set of coupled pseudo eigenvalue equations, one for each occupied orbital, involving also the 3- and 4-electron density matrices. All the matrices depend on the irreps of the symmetric group and in its complete form the complexity of the problem is obvious. The form most commonly used applies when non-orthogonality is limited to *pairs* of orbitals (e.g. a system of mutually orthogonal electron-pair bonds), in which case the equations are greatly simplified.

Gerratt and co-workers (see, in particular, Gerratt 1971; Gerratt and Raimondi 1980; and, for a review, Cooper et al 1987) generalized Goddard's approach by admitting *all* spin coupling schemes and simultaneously optimizing both the orbitals and the coefficients of the various structures. Their procedures are feasible even with full non-orthogonality and have been applied quite widely. One of the most striking results of the calculations completed so far is that *when a multistructure function is fully optimized* it can give a highly accurate description of molecular electronic structure, in a very compact form, over a wide range of molecular geometries. Moreover, many of the principles adopted in the early semi-empirical applications of VB theory appear to be substantiated. In particular, the *optimized* orbitals usually *increase* their overlap in the bond regions – as would be expected on the basis of the 'principle of maximum overlap' – just as in the Coulson-Fischer calculation. A good example is the benzene pi-electron system (Cooper et al 1986) where the optimized orbitals are 'distorted AOs', each exhibiting a 'polarization' towards its nearest neighbours.

The details of Gerratt's approach, which depends on calculating the density matrices for 1,2,3,.. electrons by a recursive procedure will not be reproduced here (a simplified derivation is given in MMQM Ch.7); it involves the storage and processing of the matrices of the generating elements $\hat{P}_{n,n-1}$ and consequently requires quite powerful computing facilities. The optimization of the orbitals, through the pseudo eigenvalue equations, is performed by a 'stabilized Newton-Raphson' procedure which is efficient but also quite costly.

Group Theoretical Methods (II)

A radically different approach, employing group theoretical techniques within a spin-free formalism, was proposed and implemented by Gallup and coworkers (for a review see Gallup et al, 1982). In this approach the irreps of the permutation group are not explicitly required: instead, attention is focused on the *spatial factors* used in constructing the wavefunction in the form discussed in Section 4 of Ch.1. Only *one* such factor is needed, out of a degenerate set of f_S^N such functions – which, in principle, may be exact eigenfunctions of the spin-free Hamiltonian (1.2) – but that factor must be of pure symmetry species with respect to permutations of the spatial variables i.e. it must behave like a given basis vector (the κth, say) of the irrep D_{orb} with matrices $\mathbf{D}_{orb}(\hat{P}) = \epsilon \mathbf{D}_S(\hat{P})$. And any function of this species (S,κ) can be expanded in terms of 'symmetry-adapted' *orbital* functions of the same species. Thus an exact eigenfunction of the Hamiltonian, of given species, can be expanded in symmetry-adapted orbital products according to (1.19), namely

$$F^{S,\kappa} = \sum C_K F_K^{S,\kappa}, \tag{20}$$

where $F_K^{S,\kappa}$ is constructed from orbital products

$$\Omega_K(\mathbf{r}_1, \mathbf{r}_2, \dots \mathbf{r}_N) = \phi_{k_1}(\mathbf{r}_1)\phi_{k_2}(\mathbf{r}_2)\dots \phi_{k_N}(\mathbf{r}_N), \tag{21}$$

K indicating any choice of orbital indices from a complete set $\{\phi_k\}$. The symmetry-adapted product may be constructed from Ω_K by applying an appropriate projection operator, which makes all possible permutations of the variables (electron indices) and puts the results together with suitable coefficients. The fact that the 'exact' spatial eigenfunction, expanded in the form (20), makes no reference at all to spin (except for the naming of the irreps) means that Gallup's approach is essentially an application of the 'spin-free quantum chemistry' pioneered by Matsen (1964). Spin eigenfunctions play an important role, however, in the definition and construction of the irreps.

It will be recalled that, for a standard irrep, the species symbol κ corresponds to a particular path (sequence of 'up' or 'down' couplings) in the *branching diagram* (Fig.1) and that a choice of species therefore amounts to a choice of which path to use in constructing the symmetry-adapted function. Gallup chooses the *top* path, with spins parallel-coupled up to the apex and then parallel-coupled down to the end point (total spin S). This choice retains some contact with Slater determinants, because a system with all spins *parallel*-coupled can be represented by a determinant of *orbital* factors, multiplied by a single α-spin product; and the method of actually evaluating matrix elements such as $< F_K|\hat{H}|F_L >$ depends heavily on the theory of determinants. As for the method of projection, Gallup uses a 'Young idempotent' rather than a group-theoretical projection operator ('Wigner operator') of the usual form: this is a simple recipe, based on the idea of a 'Young tableau' in which the electron indices are assigned to a 'Young shape', and involves only *symmetrizing* with respect to the indices in each *row* and then *antisymmetrizing* with respect to those in each column. It can be shown (see Rutherford 1948; or Pauncz 1979) that the resultant function is of the desired symmetry species.

Some examples of Young tableaux for a five-electron system are shown in Fig.2: these 'standard tableaux' are obtained by inserting the electron indices into a two-row 'shape', with the numbers increasing along the rows (left to right) and down the columns. Evidently, starting from a spin product with 3 alpha factors for the first-row electrons and 2 beta factors for the second row, the operations of symmetrizing with respect to row indices and antisymmetrizing for the column indices will produce a spin eigenfunction of Weyl-Rumer form: that for the last of the tableaux in Fig.2 would give the prototype Weyl-Rumer function

$$(\alpha\beta - \beta\alpha)(\alpha\beta - \beta\alpha)\dots\alpha\alpha\dots, \tag{22}$$

in which the first few pairs are antiparallel-coupled and the rest parallel. In general, however, the branching diagram functions, which provide a standard orthogonal basis for the irrep are *not coincident* with the spin-paired functions: the relationship between a tableau and its corresponding branching-diagram function is indicated in Fig.3.

The significance of Gallup's choice of symmetry species is now clear: use of the 'top path' in the branching diagram is equivalent to using the Young idempotent for the *first* of the tableaux in Fig.2. And since the symmetry of the *spatial* wavefunction must differ by a parity factor, being antisymmetric where the spin factor is symmetric and vice versa, this means that the species of his spatial functions corresponds to a *two-column* standard tableau obtained by transposing the rows and columns of $T_1^{1/2}$ in Fig.2. Thus, if $K = k_1k_2...k_5$ is an orbital product, his spatial function is constructed by anti symmetrizing with respect to indices 1,2,3 (or orbitals k_1, k_2, k_3), similarly with respect to 4,5 (or k_4, k_5) and so on.

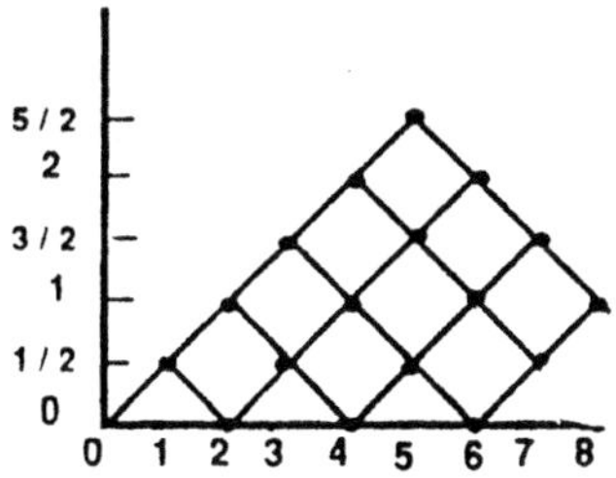

Fig.1 The branching diagram for an 8-electron system with $S = 1$.

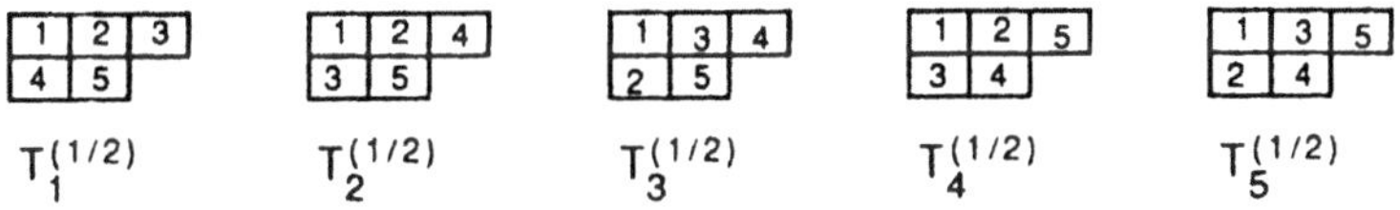

Fig.2 Standard Young tableaux for a 5-electron system.

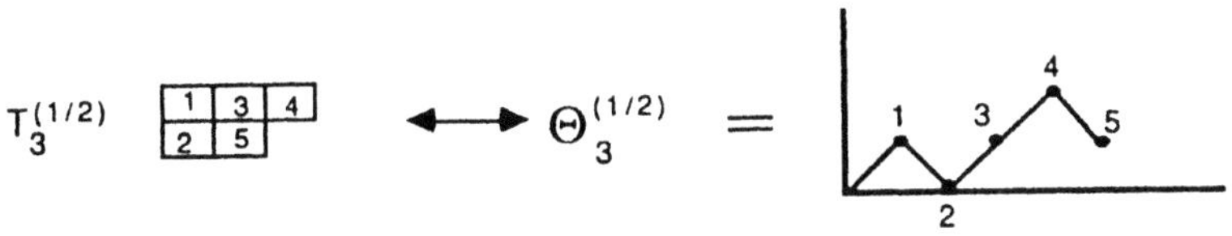

Fig.3 Relationship between a Young tableaux and a branching diagram spin-function.

When the *orbital* indices are inserted in the boxes of a tableau, in place of the electron indices, we obtain a 'Weyl tableau'; and since two orbital indices may coincide (double occupation) the number of independent Weyl tableaux may be much larger than the number of Young tableaux. This is the key to Gallup's formulation of CI in VB theory: for a given total spin (S) and a given orbital basis, of m functions, the number of *standard Weyl tableaux* (in which the orbital indices increase down the columns and are non-decreasing along the rows (double occupation permitted), is given by

$$D(m,N,S) = \frac{2S+1}{m+1}\binom{m+1}{\frac{1}{2}N+S+1}\binom{m+1}{\frac{1}{2}N-S}, \tag{23}$$

which is 'Weyl's dimension formula'. This number, which rapidly becomes enormous even for small basis sets, gives the dimension of a 'full-CI' calculation and is encountered again in other chapters dealing with the theory of the unitary group.

The approach outlined here does not lend itself readily to orbital optimization but is quite viable for extended CI calculations with non-orthogonal orbitals, as Gallup and collaborators have shown. Again, however, it does not lead immediately to wavefunctions of classical VB form; the wavefunction can only be expressed in terms of the Weyl-Rumer structures at the expense of an elaborate a posteriori transformation.

Group Theoretical Methods (III)

A second spin-free method (McWeeny 1988) appears to offer considerable advantages over the others: in particular, it retains contact with the classical approach by using spin-paired structures from the start; it is of such extreme simplicity that it may be implemented even on a small computer; and it lends itself quite readily to optimization of the orbitals, however many orbital configurations there may be in the reference function.

The essential idea is, as in Gallup's method, to start from an orbital product such as (21) and to produce symmetry adapted functions by applying a projection operator. In the approach used here, however, the Weyl-Rumer coupling scheme is fundamental; and since, as indicated above, this coupling scheme corresponds to the *last* of the tableaux (in the standard 'last letter' sequence used in Fig.2), or equivalently the *bottom* path in the branching diagram, the choice of projection operator is entirely different. In fact we write the standard functions in reverse order, so that the species index (S,κ) for the 'bottom path' will have $\kappa = 1$ (the 'first' basis function of the irrep). And instead of using a Young idempotent as the projection operator we use the appropriate Wigner operator, which takes the form (for $\kappa = 1$)

$$\rho = \frac{d_S}{N!}\sum_{\hat{P}} \epsilon_P D_S(\hat{P})_{11}\hat{P}. \tag{24}$$

The corresponding symmetry-adapted functions in a spin-free wavefunction of the form (18) will thus be

$$F_K = \rho\Omega_K, \tag{25}$$

which is simply a linear combination of permuted orbital products, transforming like the first basis vector (reverse order) of the standard irrep D_S. permutation group. Although spin-free, such a function is in one-to-one correspondence with a conventional VB structure in which the first two orbitals (with indices k_1, k_2) represent a 'bonded pair' in the usual sense, k_3 and k_4 a second pair, and so on.

At first sight, the explicit introduction of an irrep of the symmetric group may seem a backward step: but it will be noted that only *one element* of each representation matrix is required; and that this element may be obtained from a conventional superposition pattern for two VB structures, using a trivial algorithm. Thus $D_S(\hat{P})_{11}(=<\Theta_1|\hat{P}\Theta_1>)$ involves only two spin-paired functions, Θ_1 (whose Rumer diagram shows arrows connecting the first g pairs of points (1,2),(3,4)..., the remainder, $2g+1,...N$ being left free), and $\hat{P}\Theta_1$ (whose diagram follows on permuting the points, without detatching the arrows). An example is shown in Fig.4 for a triplet function of an 8-electron system.

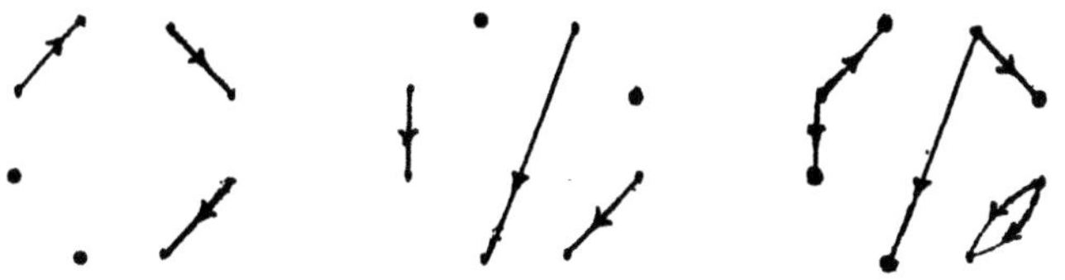

Fig.4 Two 8-electron triplet structures and their superposition pattern.

The scalar product follows from the superposition pattern, using the generalized Rumer-Pauling algorithm (see MMQM p.218) which applies for any value of S, namely

$$< \Theta_1|\hat{P}\Theta_1 >= (-1)^{\nu}\delta_E 2^{(n-g)}, \tag{26}$$

where ν is the number of arrow reversals needed to achieve head-to-head, tail-to tail matching; $\delta_E = 0$ if there are any open chains containing an *even* number of points (= 1 otherwise); and n is the number of islands.

When the spin-free wavefunction is represented as a 'resonance mixture' of the functions (25) – the spin-free equivalents of the classical VB structures – according to

$$F = \sum_K c_K F_K. \tag{27}$$

the variational energy eigenvalues, and corresponding expansion coefficients, follow from the secular equations in the usual way. Thus,

$$\mathbf{Hc} = E\mathbf{Mc}, \tag{28}$$

in which the matrix elements are

$$H_{KL} =< F_K|\hat{H}|F_L >, \qquad M_{KL} =< F_K|F_L > . \tag{29}$$

These elements easily reduce (using the idempotency of the Wigner operators and omitting normalization) giving, for example,

$$H_{KL} = \frac{d_S}{N!}\sum_{\hat{P}} \epsilon_P D_S(\hat{P})_{11} < \Omega_K|\hat{H}|\hat{P}\Omega_L > . \tag{30}$$

Such quantities may be evaluated easily by direct summation over all permutations. The factor $< \Omega_K|\hat{H}|\hat{P}\Omega_L >$ is trivial: it contains only orbital *products* and therefore reduces to sums of 1- and 2-electron integrals multiplied by chains of overlap integrals. The factor $D_S(\hat{P})_{11}$ has been dealt with already and it is now clear that the calculation of all matrix elements, of any type whatever, requires only simple algorithms for the efficient generation of permutations and for the processing of superposition patterns. All matrix elements can be generated simultaneously within a 'nest' of permutation loops, contributions being accumulated as the permutations are processed, and no storage of intermediate quantities is required.

The only other remarks necessary concern the choice of a suitable set of linearly independent structures (possibly for a full-CI calculation); and the possibility of optimizing the orbitals for a given, more limited, set of structures. With the usual definition of an 'orbital configuration' as a *selection* of N orbital indices from the basis, it is clear from the definition of the structures that different *orderings* of the given orbitals will in general lead to distinct projected functions. Obviously, all $N!$ choices will not give linearly independent results and the number to choose must be $f_S^{N'}$ where N' is the number of singly occupied orbitals. To obtain the total number of linearly independent structures, for a full-CI calculation, such numbers must be summed for all possible selections of the N orbitals from the basis: the result (Mulder 1966) must agree with Weyl's dimension formula (23). How the ordered sets may be chosen is discussed elsewhere (MMQM Sects. 7.6, 10.3); it is sufficient to write down all the *distinct* standard Weyl tableaux (as defined above) by putting the ith orbital index k_i into the box that contains i in a standard Young tableau; and then to read the orbital products, row by row, from all the resultant Weyl tableaux. This method produces the standard set of basis functions used in the theory of the unitary group, as discussed in other chapters: but it does *not* produce the VB structures, for the ordering of the orbital indices in the tableaux does not necessarily put 'chemically' paired orbitals in the same row – and the projected functions consequently lose all contact with chemistry. For present purposes it is therefore preferable to re-order the indices within the columns, to produce *non-standard* Weyl tableaux and a new set of orbital products, such that the paired indices correspond to Weyl-Rumer functions. It is sufficient to note that projection from these non-standard orbital products will lead to the usual VB structures, set up by drawing links between paired orbitals $(k_1, k_2)(k_3, k_4)$... and leaving any remaining (parallel-spin) orbitals as free points. Consequently, no deep familiarity with the symmetric and unitary groups is required: it is enough to draw the VB structures of all kinds ('covalent', 'singly ionic', etc. – corresponding to 0,1,.. doubly occupied orbitals), depending on chemical intuition to make a reasonable choice when only a limited subset is employed.

The optimization of the orbitals may be performed by a simple and robust method, applying equally well to single or multiconfiguration reference functions. This depends on the fact that variation of each orbital is equivalent, to first order, to admixture of the VB reference function with 'substituted functions' in which each structure is replaced by a new structure with a single orbital replacement (the new orbital belonging either to the reference set or the remaining functions of the basis); by performing a limited CI calculation it is thus possible to estimate, from the coefficients of the new structures, how the *orbital* should be changed to obtain the same improvement. This method, which is reminiscent of the 'super-CI' method (e. g. Grein and Chang 1971; Ruedenberg et al 1979) used in MC SCF calculations, has been found to give satisfactory convergence in a wide range of applications (see, for example, McWeeny 1989, 1990, 1991, 1992).

5) Conclusion

It remains only to summarize and restate the main advantages of the VB approach in its any of its modern ab initio forms and to indicate future prospects.

First of all, it must be emphasized that although the 'N! problem' remains (in one form or another) N refers essentially to the limited number of electrons in an 'active space': normally there is a passive 'core' (represented in MO-based methods by a set of doubly-occupied orbitals) whose effects can rigorously be absorbed into an effective 1-electron Hamiltonian for the active-space electrons. In practice, most extended CI calculations, by any method, apply to this active space; and it is usually assumed, with considerable experimental justification, that many processes of interest (such as electronic excitation or the breaking of localized bonds) can be adequately discussed without explicitly involving the electrons of a suitably defined core. At this level of description, both MO and VB approaches ultimately meet the same 'N! problem' – as a glance at equation (23) confirms. What really matters is *how fast the CI expansion converges* for that is essentially the factor that determines the feasibility of a calculation, the full-CI limit usually being attainable only for few-electron systems. In this context a VB approach can offer the following advantages:

(i) Wavefunctions of excellent quality can be obtained in a very compact form, involving a small number of classical VB structures, provided the orbitals (which are usually localized) are optimized.

(ii) The structures employed then retain close contact with chemistry, differing from those used in the semi-empirical VB theory only in the detailed form of the orbitals (which are 'overlap enhanced' when optimization is allowed but still essentially 'atomic' in nature); and this means that all the intuitions built up during the last 100 years, about alternative structures, the role of hybrid orbitals, and so on, can be fully exploited.

(iii) By using localized orbitals, optimized as a function of molecular geometry, the wavefunction passes smoothly into one describing the separate fragments as dissociation proceeds; bond breaking presents no problems and the theory is well-adapted to the calculation of complete reaction surfaces, both for the ground state and for excited states.

(iv) At the technical level, there seem to be no insuperable difficulties in dealing with systems of up to about 12-16 electrons in the active space even with full non-orthogonality; consequently a vast area of chemistry comes within the reach of VB methodology.

Against these many advantages, there is really only one disadvantage: in one way or another, all $N!$ permutations of the N electrons of the active space must be processed during the calculation. As noted above, this difficulty is common to all methods that attempt to approach a 'full-CI' limit; but in view of the compact form of the wave function, its close connection with chemistry, and the extreme simplicity and generality of the method (particularly in the form presented in Section 4), the balance appears to lie in favour of the VB approach. It will certainly provide a powerful tool for the calculations on reaction surfaces, transition state complexes, and all systems in general non-equilibrium conformations.

References

Anderson, P.W., 1987, Science 235, 1196

Balint-Kurti, G. G., and Karplus, M., 1960, J. Chem. Phys. 50, 478

Cooper, D. L., Gerratt, J., and Raimondi, M., 1987, Adv. Chem. Phys. 69, 319

Coulson, C. A., 1952, Valence, 1st edn. (Clarendon Press, Oxford)

Coulson C. A., and Fischer, I., 1949, Phil. Mag. 40, 306

Figari, G., and Magnasco, V., 1985, Mol. Phys. 55, 319

Gallup, G. A., Vance, R. L., Collins, J. R., and Norbeck, J. M., 1982, Adv. Quantum Chem. 69, 229

Gerratt, J., 1971, Adv. Atom Mol. Phys. 7, 141

Gerratt, J., and Raimondi, M., 1980, Proc. Roy. Soc. Lond. A371, 525

Goddard, W. A., 1967, Phys. Rev. 157, 73; ibid. 157, 81

Goddard, W. A., 1967, J. Chem. Phys. 48, 450; ibid. 48, 5337

Grein, F., and Chang, T. C., 1971, Chem. Phys. Lett. 12, 44

Handy, N. C., Knowles, P. J. and Somasundran, K. S., 1985, Theor. Chim. Acta 68, 87

Hayes, I. C., and Stone, A. J., 1981, Mol. Phys. 53, 69

Heitler, W., and London, F., 1927, Z. Phys., 44, 455

King, H. F., Stanton, Kim, H., Wyatt, R. E., and Parr, R. G., 1967, J. Chem. Phys., 47, 1936

Kotani, M., Ameniya, A., Ishiguro, E., and Kimura, T., 1955, Tables of Molecular Integrals (Maruzen, Tokyo)

Löwdin, P.-O., 1955, Phys. Rev. 97, 1474

Miller, J., Friedman, R., Hurst, R., and Matsen, F. A., 1957, J. Chem. Phys. 27, 1385

McWeeny, R., 1954, Proc. Roy. Soc. Lond. A233, 306

McWeeny, R., 1959, Proc. Roy. Soc. Lond. A253, 242

McWeeny, R., 1988, Int. J. Quantum Chem. 34, 25

McWeeny, R., 1989, Pure and Appl. Chem. 61, 2087

McWeeny, R., 1990, Int. J. Quantum Chem. Symp. 24, 733

McWeeny, R., 1991, J. Mol. Struct. (Theochem) 229, 29

McWeeny, R., 1992, Israeli J. Chem. (in press)

McWeeny, R., and Ohno, K., 1960, Proc. Roy. Soc. Lond. A255, 367

Matsen, F. A., 1964, Adv. Quantum Chem. 1, 59

Mattis, D. C., 1990, in Valence Bond Theory and Chemical Structure, eds. Klein, D. J., and Trinajstic, N. (Elsevier, Amsterdam)

Mulder, J. J. C., 1966, Mol. Phys. 10, 479

Pauling, L., and Wilson, E. B., 1935, Introduction to Quantum Mechanics (McGraw-Hill, New York)

Pauling, L., 1933, J. Chem. Phys. 1, 280

Prosser, F., and Hagstrom, S., 1968, Int. J. Quantum Cem., 2, 89

Pauncz, R., 1979, Spin Eigenfunctions (Plenum, New York)

Ruedenberg, K., Cheung, L.M., and Elbert, S. T., 1979, Int. J. Quantum Chem. 16, 1069

Rutherford, D. E., 1948, Substitutional Analysis (Edinburgh University Press)

Slater, J. C., 1951, Phys. Rev. 81, 385

Slater, J. C., 1929, Phys. Rev. 34, 1293

Slater, J. C., 1931, Phys. Rev. 38, 1109

van Lenthe, J. H., and Balint-Kurti, G. G., 1983, J. Chem. Phys. 78, 5699

Yamanouchi, T., 1937, Proc. Phys. Math. Soc. Jpn. 19, 436

MOLECULAR PROPERTIES

R.McWeeny

Dipartimento di Chimica e Chimica Industriale
Universitá di Pisa, Via Risorgimento 35, 56100 Pisa, Italy

1) Introduction

So far, the emphasis has been placed heavily on the computation of the wavefunctions and energies of molecular stationary states – with or without the inclusion of nuclear-motion effects. But what do we *do* with such information? Obviously we should ask how it may be used in the discussion of a very wide range of molecular *properties*, ranging from the fine and hyperfine structure of electronic energy levels (as revealed in ESR and NMR) to the properties that determine chemical reactivity and intermolecular forces – including of course the possible presence of an electromagnetic field and the phenomena of emission and absorption of radiation. In principle, once we have the free-molecule wavefunctions, it is possible to embark on a discussion of all such phenomena.

In the limited space available it will be possible to sketch only one or two unifying techniques, indicating a typical application here and there and hinting at others. Many examples are worked out in detail elsewhere (MMQM Chs. 11, 12). The effects of nuclear motion, usually secondary, will also be ignored though they can be admitted in broadly similar ways. To economize on space, it is also convenient to introduce static and dynamic properties together, the static properties arising as a 'zero-frequency limit' of the dynamic properties associated with an oscillating perturbation of the system.

First, then, the property *concept* must be established. Every property describes the *response* of the system to some kind of measurement – which in turn involves a *perturbation*. The dipole moment and the electric polarizability, for example, describe the response of the system to an applied electric field ($\mathbf{F}$); but whereas the former is associated with an energy change *linear* (i.e. first order) in $\mathbf{F}$, which can be calculated from an unperturbed wavefunction Ψ, the polarizability describes an energy response *quadratic* (second order) in $\mathbf{F}$ and thus involves the first-order change in Ψ. Moreover, the interpretation of such properties *at the molecular level* draws upon the concepts of classical physics; and one therefore seeks some 'picture' – or classical 'model' – of what is happening to the electron distribution.

The three aspects of the problem to be considered are thus (i) the nature of the perturbation associated with any given property; (ii) the development of a perturbation theory to determine its effect; and (iii) the interpretation of the origin of the property in terms of some electronic 'mechanism'.

2) Small Terms in the Hamiltonian

There are two types of 'small terms' in the Hamiltonian, so far ignored: those connected with

Methods in Computational Molecular Physics, Edited by Stephen Wilson and
Geerd H.F. Diercksen, Plenum Press, New York, 1992

externally imposed perturbations (e.g. applying fields or moving the nuclei); and those which are 'intrinsic' (e.g. terms of relativistic origin) but which have been discarded to obtain the simplified Hamiltonian used so far.

External fields will here be treated in a semi-classical way (i.e. without field quantization). This approximation is almost perfectly satisfactory and provides a basis for most of spectroscopy; its most notable blemish is that of not predicting *spontaneous emission* in which a system falls from an excited state into one of lower energy, even in the absence of any stimulation. Subject to this limitation, the presence of fields may be recognized by setting up the classical Hamiltonian and passing to the corresponding operator by the usual rules.

When an electric field $\mathbf{F}$ is applied, with associated electrostatic potential ϕ, the potential energy term $V(\mathbf{r}_i)$ for electron i changes by an amount $\delta V(\mathbf{r}_i) = -e\phi(\mathbf{r}_i)$. The perturbation of the Hamiltonian is thus

$$\hat{H}_{\text{elec}} = \sum_i \delta V(\mathbf{r}_i). \tag{1}$$

We note that this simple form also describes the effect of displacing a nucleus, or of taking into account a nuclear electric quadrupole, or of any change that simply modifies the electric field in which the electrons move.

A magnetic field, on the other hand, couples with the electronic *motion* and thus introduces a *velocity*-dependent perturbation. The kinetic energy term in $\hat{H}$ becomes in fact

$$\hat{T} = \sum_i \frac{1}{2m}\pi^2(i), \quad \pi(i) = \hat{\mathbf{p}}(i) + e\mathbf{A}(\mathbf{r}_i), \tag{2}$$

where π, which contains the magnetic vector potential $\mathbf{A}$, replaces the usual momentum operator $\hat{\mathbf{p}}$ and is called the 'gauge-invariant'momentum operator. On expanding the square, the perturbation of he Hamiltonian is seen to be

$$\hat{H}_{\text{mag}} = \hat{H}'_{\text{mag}} + \hat{H}''_{\text{mag}}, \tag{3}$$

where

$$\hat{H}'_{\text{mag}} = \sum_i \frac{1}{2m}[\hat{\mathbf{p}}(i)\cdot\mathbf{A}(\mathbf{r}_i) + \mathbf{A}(\mathbf{r}_i)\cdot\hat{\mathbf{p}}(i)], \qquad \hat{H}''_{\text{mag}} = \sum_i \frac{e^2A^2(\mathbf{r}_i)}{2m}. \tag{4}$$

For the frequently occurring case of a uniform field, of flux density $\mathbf{B}$, $\mathbf{A}(\mathbf{r}_i) = \frac{1}{2}\mathbf{B}\times\mathbf{r}_i$ and the two terms are of first and second order, respectively, in $\mathbf{B}$. For a non-uniform field (e.g. arising from spin magnetic dipoles), $\mathbf{A}$ may be a sum of several more complicated terms. The fact that the vector potential is arbitrary to within a term ∇f (f being a scalar function of position) is usually exploited by choosing f (if necessary) to make the two operator terms in $\hat{H}'_{\text{mag}}$ coincide. In this case one is said to be "working in coulomb gauge". We note that in the case of a time-dependent radiation field the electrostatic potential is negligible except very near the radiation source; and that transitions may then be associated with an oscillating vector potential term in $\hat{H}_{\text{mag}}$.

The 'intrinsic' small terms in the Hamiltonian are usually derived from the relativistic equations of Dirac (one-particle) and Breit (many-particles) in a somewhat *ad hoc* way (see, for example, Bethe and Salpeter (1957); Moss (1973); or MMQM App.4). They cannot be presented here but will form the substance of other chapters. We note only that in the so-called 'Pauli approximation', where the electrons are described using 2-component wavefunctions (α- and β-type spin-orbitals), there is a spin Zeeman term

$$\hat{H}_Z = \sum_i g\beta\mathbf{B}\cdot\hat{\mathbf{S}}(i), \tag{5}$$

where $\beta = e\hbar/2m$ is the Bohr magneton and $g \approx 2$. This term describes the coupling between the electron spin magnetic dipole and the magnetic field $\mathbf{B}$. There is a similar *nuclear* Zeeman term for every (magnetic) nucleus, with g_n (a nuclear g value) and β_p (a 'proton magneton') in place of g, β, and with a sign change. There are also many interaction terms, among which are $\hat{H}_{SL}$ (a spin-orbit interaction), $\hat{H}_{SS}$ (an electron spin-spin interaction), and an electron-nuclear 'contact' term

$$\hat{H}^{(n)}_{\rm cont} = \frac{g\beta g_n\beta_p}{(4\pi\epsilon_0)c^2}\frac{8\pi}{3}\sum_i \delta(\mathbf{r}_{ni})\mathbf{I}(n)\cdot\mathbf{S}(i) \qquad (6)$$

for each nucleus n. The manifold effects of terms of this kind are directly observable in ESR and NMR experiments: we must now set up the machinery for predicting such effects.

3) Formal Expressions for Response

In principle, the response of a molecule to any kind of perturbation can be determined by expanding the wavefunction in terms of the infinite set of exact stationary-state functions i.e solutions of the time-independent Schrödinger equation. In practice, such functions are not available – and a truncated set of approximate solutions must be used – but the possibility of such an expansion provides a good formal framework for discussing properties. We discuss the ground state though generalizations are possible.

For simplicity we consider a Hamiltonian

$$\hat{H} = \hat{H}_0 + \hat{H}'(t), \qquad \hat{H}'(t) = F(t)\hat{A}, \qquad (7)$$

where $\hat{H}'(t)$ typically contains a Hermitian operator $\hat{A}$ (any of the 'property terms' of Section 2), applied with a real time-dependent 'strength factor' $F(t)$. Let us suppose that $\Psi_0, \Psi_1, \Psi_2, ...$ are exact time-independent eigenfunctions of $\hat{H}_0$ and that $\hat{H}'$ is 'turned on' gradually, starting at $t = -\infty$. It is convenient to attach the usual oscillatory phase factors $\exp(-iE_nt/\hbar)$ to all functions, writing the *perturbed* ground state also with a factor $\exp(-iE_0t/\hbar)$, so that on dividing through by this factor the expansion becomes

$$\Psi_0' = c_0\Psi_0 + \sum_{n>0} c_n(t)\exp(-i\omega_{0n}t)\Psi_n, \qquad \omega_{0n} = (E_n - E_0)/\hbar, \qquad (8)$$

where $c_n(t)$ is small and varies relatively slowly with time. Substitution in (1.1)* then yields a set of coupled differential equations for the coefficients which, when integrated (with obvious boundary conditions at $t = -\infty$) give the first-order wavefunction at time t.

Normally the 'response' of the system at time t is 'detected' by a second operator $\hat{B}$, acting as a 'probe'. A simple calculation of the perturbed expectation value then shows a variation

$$\delta < \hat{B} > = < \hat{B} > - < \hat{B} >_0 = \int_{-\infty}^{t} K(\hat{B}\hat{A}|(t-t'))F(t')dt', \qquad (9)$$

where, with $\tau = t - t'$,

$$K(\hat{B}\hat{A}|\tau) = (i\hbar)^{-1}\sum_{n>0}[< 0|\hat{B}|n >< n|\hat{A}|0 > e^{-i\omega_{0n}\tau} - < 0|\hat{A}|n >< n|\hat{B}|0 > e^{i\omega_{0n}\tau}], \qquad (10)$$

and is a 'time correlation function of $\hat{B}$ with respect to $\hat{A}$': through the integral (9) it determines the accumulated *linear response* of $< \hat{B} >$, resulting from $\hat{A}$ acting at every instant t' $(= t - \tau)$ between $-\infty$ and t, with strength factor $F(t')$.

It is usually convenient to use the Fourier components of $F(t)$, namely $f(\omega) = \int_{-\infty}^{+\infty} F(t)e^{i\omega t}dt$, so that

$$F(t) = \frac{1}{2\pi}\int_{-\infty}^{+\infty} f(\omega)e^{-i\omega t}d\omega, \qquad (11)$$

* Equation 1 of Chapter 1

with $f(-\omega) = f(\omega)$ to keep $F(t)$ real. The perturbation in (7) is then equivalent to an integral of elemental terms

$$\hat{H}'(\omega) = \hat{A}\cos(\omega t) = \tfrac{1}{2}\hat{A}(e^{-i\omega t} + e^{i\omega t}) \tag{12}$$

applied with weight factor $f(\omega)$; and the linear response would apparently follow on obtaining the effect of (12), multiplying by $(2\pi)^{-1}f(\omega)$, and integrating over ω. Care must be taken, however, because the individual terms (12) do not vanish for $t \to -\infty$; and a convergence factor $\exp(\eta t)$ should therefore be added, the limit $\eta \to 0$ being taken after the integration. When this is done, the response due to (12) (which may be handled as indicated above to get the effect of an arbitrary $\hat{H}'(t)$) is found to be

$$\delta < \hat{B} >_\omega = \tfrac{1}{2}[\Pi(\hat{B}\hat{A}|\omega)e^{-i\omega t} + \Pi(\hat{B}\hat{A}|-\omega)e^{i\omega t}], \tag{13}$$

where the Π factor is (cf.(10))

$$\Pi(\hat{B}\hat{A}|\omega) = \text{Lim.}\ (\eta \to 0)\frac{1}{\hbar}\sum_{n>0}\left[\frac{<0|\hat{B}|n><n|\hat{A}|0>}{\omega + i\eta - \omega_{0n}} - \frac{<0|\hat{A}|n><n|\hat{B}|0>}{\omega + i\eta + \omega_{0n}}\right] \tag{14}$$

and is a 'frequency-dependent polarizability' (FDP) of $\hat{B}$ with respect to $\hat{A}$.

The effect of an arbitrary $\hat{H}'(t)$ in (7) now follows from (13) on multiplying by $(2\pi)^{-1}f(\omega)$ and integrating over all ω. It follows (using Cauchy's residue theorem to evaluate the ω-integral) that the FDP and the time correlation function are related by Fourier transformation:

$$\Pi(\hat{B}\hat{A}|\omega) = \int_{-\infty}^{+\infty}\theta(\tau)K(\hat{B}\hat{A}|\tau)e^{i\omega\tau}d\tau, \qquad \theta(\tau)K(\hat{B}\hat{A}|\tau) = \frac{1}{2\pi}\int_{-\infty}^{+\infty}\Pi(\hat{B}\hat{A}|\omega)e^{-i\omega\tau}d\omega, \tag{15}$$

where $\theta(\tau)$ is the step function (=0 for $\tau < 0$, =1 for $\tau > 0$) whose presence is linked to the switching factor $e^{\eta t}$, and consequently to the poles in (14). The step function ensures that the perturbation at time t' has an effect at time t *only when t' precedes t* (in keeping with the principle of causality).

It should be clear by now that the FDP is a particular type of *propagator*, describing how a disturbance $\hat{A}\cos\omega t$ 'propagates' and is 'detected' by the operator $\hat{B}$ which represents a 'probe'. Here $\hat{A}$ and $\hat{B}$ are number-conserving operators, associated with physical observables, and there is a simple connection with elementary perturbation theory: for $\hbar\omega_{0n} = E_n - E_0$ is an excitation energy for the unperturbed system and the zero-frequency limit of (14), used in (13), gives

$$\delta < \hat{B} > = \sum_{n>0}\left[\frac{<0|\hat{B}|n><n|\hat{A}|0> + <0|\hat{A}|n><n|\hat{B}|0>}{E_0 - E_n}\right]. \tag{16}$$

This result is well known from second-order perturbation theory for a non-degenerate ground state, in the case where $\hat{H}'$ is the sum of two terms $\hat{A}, \hat{B}$. It gives the second-order change in $<\hat{B}>$ due to the presence of the other perturbation $\hat{A}$ (or equivalently of $<\hat{A}>$ due to the presence of $\hat{B}$) i.e. the 'cross-term' in the second-order energy, linear in both perturbations.

4) Methods of Calculation

Let us first develop more fully the connection with time-independent perturbation theory, seeking ways of handling in practice the many-term perturbations referred to in Section 1. A typical perturbation will be

$$\hat{H}' = \lambda\hat{H}^\lambda + \nu\hat{H}^\nu + \dots, \tag{17}$$

in which the terms are switched on with parameters $\lambda, \nu, \dots$, convenient for separating the orders. Besides shifting the energies of non-degenerate states, such perturbations frequently *re-*

solve degeneracies and some generalization is therefore necessary. But, above all, we cannot assume an infinite set of exact unperturbed functions: we must be content with a finite set of variational approximations $\{\Phi_n\}$, which may be required only to satisfy $< \Phi_m|\Phi_n >= \delta_{mn}$ and $< \Phi_m|\hat{H}_0|\Phi_n >= 0 \quad (m \neq n)$.

It is convenient to use a partitioning technique (Löwdin 1951, 1962; MMQM Section 2.5) in which the expansion functions are divided into two subsets (A and B) so that the wavefunction after applying the perturbation $\hat{H}'$ may be approximated as

$$\Psi = \sum_n c_n \Phi_n = \sum_a c_a^A \Phi_a + \sum_b c_b^B \Phi_b, \tag{18}$$

Subset A comprises the 'states of interest' $\Phi_a, \Phi_{a'}, \dots$, usually an approximate non-degenerate or degenerate ground state, while subset B contains all the remaining functions. The coefficients then follow from the usual matrix equation $\mathbf{Hc} = E\mathbf{c}$, which may be written in the partitioned form

$$\begin{pmatrix} \mathbf{H}^{AA} & \mathbf{H}^{AB} \\ \mathbf{H}^{BA} & \mathbf{H}^{BB} \end{pmatrix} \begin{pmatrix} \mathbf{c}^A \\ \mathbf{c}^B \end{pmatrix} = E \begin{pmatrix} \mathbf{c}^A \\ \mathbf{c}^B \end{pmatrix}. \tag{19}$$

The functions $\{\Phi_a\}$ define a 'model space' and it is well known that (19), which may have very large dimensions, is entirely equivalent to a *small* secular equation *in the model space*, namely

$$\mathbf{H}_{\text{eff}}\mathbf{c}^A = E\mathbf{c}^A, \qquad \mathbf{H}_{\text{eff}} = \mathbf{H}^{AB}(E\mathbf{1} - \mathbf{H}^{BB})^{-1}\mathbf{H}^{BA}. \tag{20}$$

Solution of the secular equation in (20), which contains the *effective* Hamiltonian matrix $\mathbf{H}_{\text{eff}}$ may be trivial (e.g. it is 3×3 for a triplet state, 1×1 for a non-degenerate state!); and yet it gives the exact solutions of (19). When $\hat{H}$ includes the perturbation, the eigenvalues and eigenvectors give the perturbed states, the 'implicit' coefficients of the functions $\{\Phi_b\}$ following from the eigenvectors $\mathbf{c}^A$:

$$\mathbf{c}^B = (E\mathbf{1} - \mathbf{H}^{BB})^{-1}\mathbf{H}^{BA}\mathbf{c}^A. \tag{21}$$

The only trouble is that E is the energy we wish to calculate – and is consequently unknown! Fortunately, this is no real obstacle: the unperturbed energy E_a^0 may be used as the first approximation in an iterative procedure which usually converges nicely.

To develop a perturbation series we expand the inverse matrix in (20), remembering that off-diagonal elements of $\hat{H}_0$ between variationally determined expansion functions are zero. On taking just two terms in (17), the elements of the effective Hamiltonian become

$$H^{(1)}_{\text{eff},aa'} = \lambda < a|\hat{H}^\lambda|a' > + \nu < a|\hat{H}^\nu|a' >, \tag{22}$$

up to first order in the parameters, with further terms

$$\begin{aligned} H^{(2)}_{\text{eff},aa'} &= \lambda^2 \sum_b \frac{< a|\hat{H}^\lambda|b >< b|\hat{H}^\lambda|a' >}{E_a^0 - E_b^0} + (\nu^2 \text{ term}) \\ &+ \lambda\nu \sum_b \frac{< a|\hat{H}^\lambda|b >< b|\hat{H}^\nu|a' > + < a|\hat{H}^\nu >< b|\hat{H}^\lambda|a' >}{E_a^0 - E_b^0} \end{aligned} \tag{23}$$

giving the second-order contributions.

Equation (22) shows that the different perturbations may be treated separately to first order, each making its own effect: but (23) shows that in addition to additive second-order effects there are interesting 'cross-terms' depending jointly on both perturbations.

For a single non-degenerate state ($a = a' = 0$, say), the matrix elements above give directly the first- and second-order perturbation energies. And with $\lambda\hat{H}^\lambda = \hat{A}, \nu\hat{H}^\nu = \hat{B}$, the $\lambda\nu$-term reduces to (16) – the energy associated with A in the presence of B or (no distinction) B in the presence of A. The joint response is linear in each of the two perturbations; and this is why linear response theory gives correctly part of the *second*-order energy. The pure λ^2 and ν^2 terms do *not* result from linear response theory by simply putting $\hat{B} = \hat{A}$, for reasons fully discussed elsewhere (MMQM Sect. 11.4), but evidently require only a correction factor of $\frac{1}{2}$.

For a degenerate state the first- and second-order parts of $\mathbf{H}_{\text{eff}}$ are calculated from (22) and (23). The model-space secular equation in (20) is then solved to obtain the effect of the multiple perturbation (typically a sum of the small terms in Section 1) on the different states of the degenerate set. In the common situation where the degeneracy results from the presence of electron and nuclear spins it is often possible to go a step further, showing that the elements of $\mathbf{H}_{\text{eff}}$ coincide with those of a *spin Hamiltonian* (contining only spin operators and numerical 'coupling constants'), evaluated within a manifold of electron-nuclear spin states. This approach provides a satisfactory interface with the model Hamiltonians used in the interpretation of ESR and NMR experiments.

Besides being well adapted for numerical computation, the partitioning approach also provides for an interesting two-step analysis of the effect of a multiple perturbation, which is of considerable interpretative value. Thus, using a ket notation, the first-order effect of $\lambda \hat{H}^{\lambda}$ on the *wavefunction* may be indicated by

$$|a> \rightarrow \ |\tilde{a}_{\lambda}>=|a> +\lambda \sum_{b} \frac{|b><b|\hat{H}^{\lambda}|a>}{E_a - E_b} = |a> +\lambda|a_{\lambda}> \tag{24}$$

and the second-order terms in (23) may be expressed alternatively:

$$H^{(\lambda\lambda)}_{\text{eff},aa'} = \lambda^2 < a|\hat{H}^{\lambda}|a'_{\lambda}> = \lambda^2 < a_{\lambda}|\hat{H}^{\lambda}|a'> = \tfrac{1}{2} < \tilde{a}_{\lambda}|\lambda\hat{H}^{\lambda}|\tilde{a}'_{\lambda}>^{(\lambda\lambda)}, \tag{25}$$

where the superscript $(\lambda\lambda)$ means "taking terms in λ^2", and in a similar way it appears that

$$H^{(\lambda\nu)}_{\text{eff},aa'} =< \tilde{a}_{\lambda}|\nu\hat{H}^{\nu}|\tilde{a}'_{\lambda}>^{\lambda\nu}=< \tilde{a}_{\nu}|\lambda\hat{H}^{\lambda}|\tilde{a}'_{\nu}>^{\lambda\nu}. \tag{26}$$

The second result embodies an 'interchange theorem': one may calculate the cross-term as a matrix element of *either* perturbation, between wavefunctions perturbed to first order by the *other*. The λ^2 and ν^2 terms may also be obtained from the first-order perturbed functions, according to (25), provided the factor $\frac{1}{2}$ is included (MMQM Sect. 11.4).

The possibility of calculating second-order properties in terms of first-order perturbed functions is useful (apart from any computational considerations) in revealing the physical mechanism involved – as will become clear presently. As in response theory, the two-step analysis corresponds physically to switching on one perturbation operator and detecting its effect by means of another. It is also clear from (26) that the precise mode of determining the linear response to the first perturbation is immaterial. It is therefore possible to eliminate the large-basis expansion (18) from the discussion and to use a more compact variational wavefunction, the response then being described by parameter changes. This is the approach used, for example, in Hartree-Fock perturbation theory (Peng 1941; McWeeny 1961; Stevens et al 1963; Diercksen and McWeeny, 1966; see also MMQM Sect. 11.9-10), where the wavefunction remains a single determinant but the individual orbitals depend linearly on the perturbation.

To develop the variational approach in its most general form, suitable for dealing also with time-dependent (or frequency-dependent) properties, it is convenient to use the machinery set up in Ch.1 Sect. 5. We start from the variational principle due to Frenkel (1934),

$$< \delta\Psi| \left(\hat{H} - i\hbar\frac{\partial}{\partial t}\right) |\Psi>= 0, \tag{27}$$

which for completely arbitrary variations implies that Ψ satisfies (1.1). Otherwise, when Ψ is not exact but contains a number of time-dependent parameters, (27) provides a well-defined approximation. As in Sect. 2, it is convenient to extract a rapidly varying phase factor by using, instead of Ψ, a wavefunction of the form $g(t)\Psi(\mathbf{x};t)/\sqrt{<\Psi|\Psi>}$ in which the phase factor $g(t)$ is parameter independent and the remaining factor ensures normalization. In this way (Moccia,1974) $g(t)$ is easily determined and (27) reduces to a condition on the 'slowly varying' factor Ψ. Derivatives with respect to parameter variations may be defined as in Sect. 1.5 but now the stationary value condition will contain both the 'parameter displacements' $\mathbf{d}$ and their time derivatives $\dot{\mathbf{d}}$. The resultant equation (Moccia, 1973, 1974; see also MMQM Sect. 12.2-3) takes the form

$$\nabla H + \mathbf{M}\mathbf{d} + \mathbf{Q}\mathbf{d}^* = i\hbar\mathbf{V}\dot{\mathbf{d}}, \tag{28}$$

where $\nabla H, \mathbf{M}$ and $\mathbf{Q}$ are defined as in Ch. 1 Sect. 5 while $\mathbf{V}$ depends on the first derivatives of the wavefunction:

$$V_{ij} = <\Psi^i|\hat{Q}|\Psi^j>, \tag{29}$$

$\hat{Q}$ being the projection operator

$$\hat{Q} = <\Psi|\Psi>^{-1}[1 - |\Psi><\Psi|\Psi>^{-1}<\Psi|], \tag{30}$$

whose effect is to keep the variation $\delta\Psi$ orthogonal to Ψ.

On using $\hat{H} = \hat{H}_0 + \hat{H}'$, with $\hat{H}'$ of the oscillating form (12), the parameters will (in the steady state) also oscillate with the frequency ω and $\mathbf{d}$ may consequently be written

$$\mathbf{d} = \tfrac{1}{2}[\mathbf{X}e^{-i\omega t} + \mathbf{Y}^*e^{i\omega t}], \tag{31}$$

the star on the $\mathbf{Y}$ simply giving subsequent equations a nice symmetry.

At this point we assume that $\mathbf{d} = 0$ for parameter values $\mathbf{p}_0$, at which the energy functional $H_0 = <\Psi|\hat{H}_0|\Psi>$ has its stationary minimum value, and we first *remove* the oscillating perturbation that 'drives' the system: consequently $\nabla H = \nabla H_0 = 0$. Equation (29) then yields, substituting (31) and separating the terms in $e^{-i\omega t}$ and $e^{i\omega t}$,

$$\begin{pmatrix} \mathbf{M} & \mathbf{Q} \\ \mathbf{Q}^* & \mathbf{M}^* \end{pmatrix} \begin{pmatrix} \mathbf{X} \\ \mathbf{Y} \end{pmatrix} = \hbar\omega \begin{pmatrix} \mathbf{V} & 0 \\ 0 & \mathbf{V} \end{pmatrix} \begin{pmatrix} \mathbf{X} \\ \mathbf{Y} \end{pmatrix}, \tag{32}$$

where all derivatives are evaluated at the stationary point. If we use Ω to denote the square matrix on the left, Δ for that on the right, and $\mathbf{C}$ for the column $(\mathbf{X}/\mathbf{Y})$, this stationary value condition assumes the simple appearance

$$\Omega\mathbf{C} = \hbar\omega\Delta\mathbf{C}, \tag{33}$$

which is a type of eigenvalue equation to determine the 'natural frequencies' of possible parameter oscillations around their stationary-point values.

Equation (33) has immense importance. It possesses a set of 'paired' eigensolutions $(\mathbf{C}_\omega, \mathbf{C}_{-\omega})$, with eigenvalues $\omega, -\omega$ $(\omega > 0)$, which are orthogonal with metric Δ and may be normalized so that $\mathbf{C}_\omega\Delta\mathbf{C}_\omega = 1$ for the positive eigenvalues, changing sign for negative eigenvalues. *In terms of these eigensolutions all linear response properties are completely determined.*

Briefly, on adding a driving perturbation $\hat{H}' = \frac{1}{2}(\hat{A}e^{-i\omega t} + \hat{A}^{i\omega t})$, as in (12), equation (33) is replaced by

$$\nabla_A + \Omega\mathbf{C}' = \hbar\omega\Delta\mathbf{C}', \tag{34}$$

in which ∇_A contains gradient components of $<\hat{A}>$ (namely $\nabla A_i, A\nabla_i$) just like those of $<\hat{H}>$ used in (1.29), while $\mathbf{C}'$ is now a *forced*-oscillation vector. The solution of (34) is clearly

$$\mathbf{C}' = -(\Omega - \hbar\omega\mathbf{1})^{-1}\nabla_A = \Pi(\omega)\nabla_A, \tag{35}$$

where $\Pi(\omega)$ denotes a *response matrix* which is an *intrinsic property of the system* and depends in no way on the nature of the perturbation applied.

Once the response matrix is known, all linear response properties, for any pair of operators $\hat{A}, \hat{B}$, can be determined with no further calculation. Thus, the FDP of $\hat{B}$ with respect to $\hat{A}$, at frequency ω, is simply

$$\Pi(\hat{B}\hat{A}|\omega) = \nabla_B^\dagger \Pi(\omega) \nabla_A \tag{36}$$

and this is a best variational approximation to the quantity defined more formally in (14). The story is almost complete: to investigate the joint effect of any pair of perturbation terms in (17) it is sufficient to know a single reponse matrix for the system. It now remains only to discuss the physical interpretation of such effects and to present a few examples.

5) The Interpretation of Molecular Properties

When the terms in the Hamiltonian all have a 'semi-classical' interpretation, being obtained from a classical Hamiltonian by the usual recipe (possibly with the addition of spin magnetic dipoles etc.), it is generally possible to visualize the response in a corresponding way. Thus, an applied electric field results in the addition of a term (1) to the Hamiltonian, the combined potential energy change of all the electrons; and the first order response of the total electronic energy will be (for a non-degenerate state)

$$\delta E = < \Psi|\hat{H}_{\text{elec}}|\Psi > = \int \delta V(\mathbf{r}) P(\mathbf{r}) d\mathbf{r}, \tag{37}$$

where we use (2.11) to reduce the matrix element. This is exactly the same formula as in classical physics – provided the distribution of electrons is regarded as a 'smeared out' charge of density $P(\mathbf{r})$ (in electrons/unit volume). For $P(\mathbf{r})d\mathbf{r}$ will then be the number of electrons associated with $d\mathbf{r}$ at point $\mathbf{r}$, while $\delta V(\mathbf{r})$ is the potential energy in the applied field of *one* electron at that point; the integral thus gives the total potential energy of the whole 'charge cloud'. A vast range of molecular properties can be discussed in similar terms (see, for example, McWeeny, 1989; and MMQM Ch. 11). Here we simply give one example before indicating further generalizations.

Example 1. Application of a uniform electric field

One of the simplest possible perturbations is the application of a uniform electric field $\mathbf{F}$ to a system in a non-degenerate ground state. In this case $\hat{H}_{\text{elec}}$ in (1) takes the form

$$\hat{H}_{\text{elec}} = -\sum_i \mathbf{F} \cdot \mathbf{d}(i), \tag{38}$$

$\mathbf{d}(i)$ being the electronic part of the dipole moment operator, with components $d_x(i) = -ex_i$, etc.

The field components (F_x, F_y, F_z) may be adopted as the parameters $\lambda, \nu, \ldots$ in (17) and the first-order energy (22) becomes

$$\delta E^{(1)} = -\mathbf{F} \cdot \mu^0 = -(F_x \mu_x^0 + F_y \mu_y^0 + F_z \mu_z^0), \tag{39}$$

where, for example,

$$\mu_x^0 = < \Psi|\sum_i d_x(i)|\Psi > = \int x[-eP(\mathbf{r})] d\mathbf{r}, \tag{40}$$

as in (37). This formula of course gives the x-moment of a continuous distribution of negative charge of density $-eP(\mathbf{r})$. Once $P(\mathbf{r})$ has been computed from quantum mechanics, the effect of the field can be visualized (and calculated) in a wholly classical manner. This is true, as anticipated in (37), for *any* kind of perturbation that can be represented as a potential energy change δV: if δV arises from displacement of δx_n of nucleus n the limit $\delta E/\delta x_n$ will give the force component F_{nx} exerted by the electrons on the nucleus; if δV corresponds to the presence of a 'test charge' at a point outside the molecule δE will give the electrostatic potential at that point; if δV derives from a nuclear electric quadrupole δE will allow us to infer the electric field gradient at the nucleus; and so on. It will be recognized, of course, that the validity of this first-order approximation expresses the Hellmann-Feynman theorem (see, for example, Epstein et al 1967) and rests upon the assumption that the unperturbed wavefunction has been variationally optimized.

In second order, direct application of (23) yields a term that may be written in the form (cf. classical physics)

$$\delta E^{(2)} = -\tfrac{1}{2}\mathbf{F}\cdot\alpha\cdot\mathbf{F} = -\tfrac{1}{2}\sum_{\lambda,\nu} F_\lambda \alpha_{\lambda\nu} F_\nu, \tag{41}$$

where the components of the *electric polarizability tensor* are given by, for example,

$$\alpha_{xy} = \sum_b \frac{< a|d_x|b >< b|d_y|a > + < a|d_y|b >< b|d_x|a >}{E_a \; - \; E_b} = \alpha_{yx}. \tag{42}$$

The presence of the infinite summation over intermediate states (b), which makes second-order effects so difficult to calculate, is associated with perturbation of the *wavefunction*. The physical interpretation of the result is also obscured because the off-diagonal matrix elements in the numerator refer to 'transitions' and can only be expressed in terms of *transition densities*, $\rho(ab|\mathbf{x};\mathbf{x}')$ and their spinless counterparts $P(ab|\mathbf{r};\mathbf{r}')$ (cf. (2.9), with Ψ, Ψ^* replaced by Ψ_a, Ψ_b^*, respectively).

To obtain a classical interpretation of the tensor components $\alpha_{\lambda\nu}$ they must be related to the electron density *in the presence of the field* – which *polarizes* the density. Thus, in classical physics, α_{xy} may be defined as the x-component of an *induced electric moment* due to unit applied field in the y direction. The same result may be established in quantum physics by making use of (25) and (26), instead of (23), for this leads to the alternative result

$$\delta E^{(2)} = -\frac{1}{2}\sum_\lambda F_\lambda \mu_\lambda^{\mathrm{ind}}, \tag{43}$$

where, for example,

$$\mu_x^{\mathrm{ind}} = \int x[-eP^{\mathrm{ind}}(\mathbf{r})]d\mathbf{r}, \tag{44}$$

and $P^{\mathrm{ind}}(\mathbf{r})$ is the polarization (i.e. density change) induced by the applied field. This polarization density is obtained, up to terms linear in field components, by using the perturbed wavefunction $|\tilde{a}>=|a>+F_x|a_x>+F_y|a_y>+F_z|a_z>$ to calculate the perturbed density in the form

$$\tilde{P}(\mathbf{r}) = P(\mathbf{r}) + P^{\mathrm{ind}}(\mathbf{r}) = P(\mathbf{r}) + F_x P_x(\mathbf{r}) + F_y P_y(\mathbf{r}) + F_z P_z(\mathbf{r}). \tag{45}$$

Consequently, from (44) and (45), $\mu_x^{\mathrm{ind}} = \sum_\nu \alpha_{x\nu} F_\nu$ where, for example,

$$\alpha_{xy} = \int x[-eP_y(\mathbf{r})]d\mathbf{r} \tag{46}$$

is the x-moment of the density change produced by unit field in the y direction.

Generalizations

Many of the small terms in the Hamiltonian give rise to *magnetic* effects, and these are due not only to the application of an external magnetic field but also to the presence of electron and nuclear spins and their associated magnetic dipoles. All such effects can be discussed along the lines already indicated, using perturbation methods to obtain numbers and electron distribution functions to understand pictorially 'what is happening'. Among the new functions required for this purpose are the *spin density* and the *current density*, for dealing with small terms of 1-electron type, and various 'coupling' densities for the terms that involve 2-electron operators (e.g. spin-spin dipole couplings). The coupling densities were first defined by McWeeny and Mizuno (1961) and were used (McWeeny, 1965) in connection with the calculation of spin Hamiltonian parameters (see also MMQM Sect. 5.9, 11.7-8); they are essentially linear combinations of the spinless components of the 2-electron density matrix (one particular combination giving the pair function $\Pi(\mathbf{r}_1, \mathbf{r}_2)$ already mentioned) and have interesting general properties (McWeeny and Kutzelnigg, 1968). Here, however, the discussion will be confined to the effect of *one*-electron terms, for which the coupling densities are not required.

In general, a 'property density' (sometimes referred to (Hirschfelder 1977) as a 'sub-observable') may be defined by writing an expectation value as an integral over all space of an appropriate density function: thus, the electron (number) density $P(\mathbf{r})$ integrates to give the total number of electrons N, while $P(\mathbf{r})d\mathbf{r}$ is the number associated with volume element $\mathbf{r}$. When the expectation value of the total spin z component, $\hat{S}_z = \sum_i \hat{S}_z(i)$ is written in the form (cf. (2.11))

$$< \hat{S}_z >= \int_{\mathbf{x}_1' = \mathbf{x}_1} \hat{S}_z(1)\rho(\mathbf{x}_1; \mathbf{x}_1')d\mathbf{x}_1 \tag{47}$$

a simple reduction gives at once, for a spin eigenstate with $< \hat{S}_z >= M$,

$$< \hat{S}_z >= \int Q_z(\mathbf{r}_1)d\mathbf{r}_1 = M, \tag{48}$$

where $Q_z(\mathbf{r}_1)$ is the 'diagonal element' ($\mathbf{r}_1' = \mathbf{r}_1$) of the *spin density matrix*

$$Q_z(\mathbf{r}_1; \mathbf{r}_1') = \tfrac{1}{2}[P_\alpha(\mathbf{r}_1; \mathbf{r}_1') - P_\beta(\mathbf{r}_1; \mathbf{r}_1')]. \tag{49}$$

Equation (48) shows that the electron distribution carries not only *charge* but also *spin angular momentum*, of which an amount $Q_z(\mathbf{r})d\mathbf{r}$ is associated with the spatial volume element $d\mathbf{r}$; and integration yields the *total* spin z component M. Since the electron spin carries a magnetic dipole of moment $g\beta$, the system may be visualized as a distribution of charge, of density $-eP(\mathbf{r})$, and of *magnetization*, of density $g\beta Q_z(\mathbf{r})$; and the two densities derive, respectively, from the sum and the difference of 'up-spin' and 'down-spin' densities. For a pure spin eigenstate, this picture has an appealing simplicity. In general, however, the spin density is a three-component quantity describing a *pseudovector field*: for spin quantization along the z axis, only one component (Q_z) is non-zero and the direction of the spin is thus uniform at all points in space; but more generally (e.g. in the presence of spin-orbit coupling) the spin distribution may be 'warped' in an interesting way (McWeeny, 1974). It should be noted that (for a spin eigenstate) the actual spin angular momentum density is proportional to the quantum number M and may be written $Q_z(\mathbf{r}) = MD_S(\mathbf{r})$ where $D_S(\mathbf{r})$ is a *normalized* spin density.

The second density we consider is the *current density*. This is also characterized by three components at every point in space and may be defined most generally (McWeeny 1986) as

$$J_\lambda(\mathbf{r}) = m^{-1}\text{Re.}[\hat{\pi}_\lambda P(\mathbf{r}; \mathbf{r}')]_{\mathbf{r}'=\mathbf{r}}, \tag{50}$$

where $\hat{\pi}_\lambda$ is a component of the gauge-invariant momentum operator (on dividing by m, a *velocity* operator) and the variables are identified *after* the operation. For a system with no resultant momentum in the λ direction, the integral of (50) over all space is of course zero; but when the wavefunction is complex there is in general a non-zero *local* current density at every point in space. In the absence of a magnetic field, the wavefunction may always be written in real form and the current vanishes; but, on applying a magnetic field, circulating currents are *induced* in the electron distribution, just as an electric field leads to a polarization, and these currents result in an induced *magnetic* moment at every point in space. The magnetization produced generates, of course, a secondary field; and this secondary field may be 'seen' in NMR experiments, where a magnetic nucleus exhibits a Zeemann coupling with a field $\mathbf{B}_{\text{eff}} = \mathbf{B} - \mathbf{B}^{\text{ind}}$ instead of the applied field $\mathbf{B}$. The remarkable fact is that, once the current density has been defined as in (50), all first and second-order properties involving the applied field and other perturbations can be calculated using classical equations.

To illustrate the use of spin and current densities let us consider a molecule with one or more magnetic nuclei, in an electronic singlet state, in the presence of a uniform magnetic field. First we consider the application of the field; the perturbation due to the nuclear dipoles may be added later to give second-order terms linear in both spins and field components – and consequently a nuclear Zeemann term containing a 'screened' magnetic field.

Example 2. Application of a uniform magnetic field

With a uniform magnetic field, vector potential $\mathbf{A} = \frac{1}{2}\mathbf{B} \times \mathbf{r}$, the term $\hat{H}'_{\text{mag}}$ in (4) assumes the form

$$\hat{H}'_{\text{mag}} = -\sum_i \mathbf{B} \cdot \mathbf{m}(i), \tag{51}$$

where $\mathbf{m}$ is a *magnetic* moment operator, analogous to $\mathbf{d}$ in (38), with components

$$m_\lambda(i) = -\beta \hat{L}_\lambda(i). \tag{52}$$

The first-order energy change then becomes

$$\delta E^{(1)} = -\mathbf{B} \cdot \mu_m^0, \tag{53}$$

where μ_m^0 is the magnetic analogue of the electric moment μ^0, computed in the absence of the field, with typical component

$$\mu_{m,x}^0 = -\beta \int_{\mathbf{r}'=\mathbf{r}} [\hat{L}_x P(\mathbf{r};\mathbf{r}')] d\mathbf{r}.$$

Since $\hbar \hat{L}_x = (\mathbf{r} \times \mathbf{p})_x$ and $\beta = e\hbar/2m$, this may be expressed in terms of the current density (if any) before applying the field:

$$\mu_{m,x}^0 = \frac{1}{2} \int [\mathbf{r} \times (-e\mathbf{J}^0(\mathbf{r})]_x d\mathbf{r}, \tag{54}$$

where $\mathbf{J}^0$ is obtained from (50) with the field-free $\hat{p}$ in place of $\hat{\pi}$. Such expressions are well known in classical physics: they give the components of the magnetic moment due to a distribution of electric currents $(-e\mathbf{J}^0)$ in a continuous conducting medium. For a molecule, in the absence of an applied field, such currents usually vanish; but for a system with a resultant orbital angular momentum there is a circulating current around the quantization axis, giving rise to 'orbital paramagnetism'.

The second-order energy change may be handled as in the case of the electric field, but more care is needed as the perturbation itself now contains the field-quadratic term $\hat{H}''_{\text{mag}}$ in (4). We write the first-order perturbed density matrix in the form (cf. (45))

$$\tilde{P}(\mathbf{r};\mathbf{r}') = P(\mathbf{r};\mathbf{r}') + \sum_\nu B_\nu P_\nu(\mathbf{r};\mathbf{r}'), \tag{55}$$

where $P_\nu(\mathbf{r};\mathbf{r}'))$ gives the density change produced by unit field in the ν direction, and collect all field-quadratic terms in the expressions deriving from (25) and (26). The results are exactly parallel to those for the electric field:

$$\delta E^{(2)} = -\tfrac{1}{2}\mathbf{B} \cdot \mu_m^{\text{ind}} \tag{56}$$

where the induced magnetic moment is related to the applied field through a *magnetic polarizability tensor*, according to

$$\mu_m^{\text{ind}} = \sum_\nu \alpha_{\lambda\nu}^m B_\nu. \tag{57}$$

The typical component is

$$\alpha_{\lambda\nu}^{\text{ind}} = \frac{1}{2}\int [\mathbf{r}\times(-e\mathbf{J}_{\nu}^{\text{ind}}]_{\lambda} d\mathbf{r}, \tag{58}$$

where $\mathbf{J}_{\nu}^{\text{ind}}(\mathbf{r})$ is the current density *induced* by unit field in the ν direction and is obtained from the ν term in the perturbed density (55). The induced dipole components are thus obtained from the usual classical formula (54) on replacing 'permanent' currents by induced currents.

Example 3. Inclusion of a nuclear spin (magnetic shielding)

Consider next the addition of a nuclear magnetic dipole term

$$\hat{H}_{Z}(n) = -g_{n}\beta_{p}\mathbf{B}\cdot\hat{\mathbf{I}}(n) \tag{59}$$

(see (5) et seq.). In this case, even for an electronically non-degenerate ground state, there will be a *nuclear-spin* degeneracy and the kets $|a>, |b>$ in the general treatment of Section 4 will be replaced by $|a\sigma>, |b\sigma>$, where σ labels different nuclear spin states (in general nuclear spin product functions with one factor for each nucleus); but the derivation of an effective Hamiltonian proceeds in essentially the same way. To discuss magnetic shielding effects we look for a second-order term similar to (59), linear in the field and nuclear spin components. This will arise from $\hat{H}'_{\text{mag}}$, producing the field-proportional currents already considered, and a term linear in the nuclear spin operators, representing (through a further term in the vector potential) the energy of the electrons in the field of the nuclear magnetic dipole. The full derivation is given elsewhere (MMQM Sect. 11.8) but the result is very simple: the matrix elements of the effective Hamiltonian $\mathbf{H}_{\text{eff}}$ contain a spin-field contribution of the form

$$H_{\text{eff},\sigma\sigma'}^{(2)} = -g_{n}\beta_{p} < \sigma|\hat{\mathbf{I}}(n)\cdot\mathbf{B}_{n}^{\text{ind}}|\sigma'>, \tag{60}$$

connecting the different nuclear-spin states of the ground-state manifold and not depending explicitly on the electronic wavefunction. The electronic wavefunction enters only in determining the *secondary field* $\mathbf{B}_{n}^{\text{ind}}$ at the position of nucleus n; this field has components (μ_0 being the magnetic permeability of free space)

$$B_{n,\lambda}^{\text{ind}} = \frac{\mu_0}{4\pi}\int\frac{[\mathbf{r}_n\times(-e\mathbf{J}^{\text{ind}}(\mathbf{r})]_{\lambda}}{r_n^3}d\mathbf{r}, \tag{61}$$

where $\mathbf{r}_n$ is the vector distance from nucleus n to point $\mathbf{r}$. The induced current has been discussed already: what (61) shows is that the secondary field produced at any point in space is correctly given by the classical Biot-Savart formula.

The magnetic shielding tensor is obtained on writing $\mathbf{J}^{\text{ind}} = \sum_{\nu} F_{\nu}\mathbf{J}_{\nu}^{\text{ind}}$, where $\mathbf{J}_{\nu}^{\text{ind}}$ is the current induced by unit field in the ν direction. On using the result in (60) and adding the *first*-order term associated with (59) we obtain an effective Hamiltonian whose elements coincide with those of a pure nuclear-spin operator

$$\hat{H}_{s}(\text{nuc. Zeeman}) = -g_{n}\beta_{p}\hat{\mathbf{I}}(n)\cdot(\mathbf{1}-\sigma_{\mathbf{n}})\cdot\mathbf{B}, \tag{62}$$

which is the experimentally observed form of the nuclear Zeeman interaction. The shielding tensor has components

$$\sigma_{\lambda\nu}^{(n)} = -\frac{\mu_0}{4\pi}\int\frac{[\mathbf{r}_n\times(-e\mathbf{J}^{\nu}(\mathbf{r})]_{\lambda}}{r_n^3}d\mathbf{r}, \tag{63}$$

and this expression provides a sound basis both for computation and for pictorial interpretation of the effects of the induced currents.

Example 4. Inclusion of a nuclear spin (hyperfine coupling)

The importance of the spin density concept is revealed in ESR experiments (Weissman 1956; McConnell 1958), where there is also an *electronic* degeneracy of $2S+1$ for a multiplet of spin S. In this case the individual states of the A-set may be represented as $|a, M_a, \sigma>$ where $M_a = S_a, S_a - 1, \ldots -S_a$ runs over the $2S_a + 1$ electronic states and σ runs over the nuclear spin states, as in the previous Example. The observed coupling between the nuclear spin and the total electronic spin can be represented by a spin-Hamiltonian term usually written in the form

$$\hat{H}_n(\text{cont.} = hA_n^{\text{cont}}\hat{\mathbf{I}}(n)\cdot\hat{\mathbf{S}} \tag{64}$$

(in general there is one such term for each magnetic nucleus). This is not a direct magnetic dipole interaction because it is present even for molecules in the fluid phase, where dipolar interactions average to zero: it arises in fact from the 'contact' term (6) and serves to resolve, even in first order, the levels which are already shifted by the electron and nuclear Zeeman terms, (5) et seq. The matrix elements of (6), within the given manifold of states, are easily reduced (MMQM Sect.11.7) and are indeed reproduced by the simple spin operator (64) provided the coupling constant is determined according to

$$A_n^{\text{cont}} = \frac{8\pi}{3}\frac{gg_n\beta\beta_p}{(4\pi\epsilon_0)c^2h}D_S(\mathbf{R}_n), \tag{65}$$

where $D_S(\mathbf{R}_n)$ is the normalized spin density evaluated at the position $\mathbf{R}_n$ of nucleus n. The nuclei thus act as 'probes', giving information on the quality of the wavefunction at corresponding points in space.

6) Conclusion

The calculation of electronic properties, often from wavefunctions of very high precision, is nowadays a vast field of activity. In this Chapter the aim has been simply to reveal the underlying coherence of the bewildering variety of techniques to be found in the literature; and to show that the mathematics and computation, heavy as it may be, is only the vehicle for carrying relatively primitive physical ideas.

Further generalizations, and other properties (including the 'dynamic' properties whose importance has only been hinted at), are discussed in the references cited and are taken up in other Chapters.

References

Bethe, H. and Salpeter, E., 1957, "Quantum Mechanics of One- and Two-Electron Systems" (Springer-Verlag, Berlin)

Diercksen, G. H. F. and McWeeny, R., 1966, J. Chem. Phys. **44**, 3554

Epstein, S. T., Hurley, A. C. and Parr, R. G., 1967, J. Chem. Phys. **47**, 1275

Frenkel, J., 1934, "Wave Mechanics: Advanced General Principles" (Clarendon Press, Oxford)

Hirschfelder, J. O., 1977, J. Chem. Phys. **67**,5477

Löwdin, P.-O., 1951, J. Chem. Phys. **19**, 1396

Löwdin, P.-O., 1962, J. Math. Phys. **3**, 969; *ibid.* **3**, 1171

McWeeny, R., 1961, Phys. Rev. **126**, 1028

McWeeny, R., 1965, J. Chem. Phys. **42**, 1717

McWeeny, R., 1973, "Quantum Mechanics: Methods and Basic Applications" (Pergamon Press, Oxford)

McWeeny, R., 1986, Proc. Ind. Acad. Sci. **96**, 263

McWeeny, R., 1989, in " Molecules in Physics, Chemistry and Biology Vol.3" (ed. J. Maruani) (Kluwer, Amsterdam)

McWeeny, R. and Kutzelnigg, W.,1968, Int. J. Quantum Chem. **2**, 187

McWeeny, R. and Mizuno, Y., 1961, Proc. Roy. Soc. Lond. bf A259, 554

Moccia, R., 1973, Int. J. Quantum Chem. **7**, 779

Moccia, R., 1974, Int. J. Quantum Chem. **8**, 293

Moss, R. E., 1973, "Advanced Molecular Quantum Mechanics" (Chapman-Hall, London)

Peng, M., 1941, Proc. Roy. Soc. Lond. **A178**, 449

Stevens, R. M., Pitzer, R. M. and Lipscomb, W. N., 1963, J. Chem. Phys. **38**, 550

CALCULATION OF GEOMETRICAL DERIVATIVES IN MOLECULAR ELECTRONIC STRUCTURE THEORY

Trygve Helgaker
Department of Chemistry
University of Oslo
P.O.B. 1033 Blindern
N-0315 Oslo 3
Norway

Poul Jørgensen
Department of Chemistry
Aarhus University
DK-8000 Aarhus C
Denmark

The analytical calculation of geometrical derivatives in molecular electronic structure theory is reviewed. The simplifications arising from the use of variational wave functions (Wigner's 2n+1 rule) are described and it is shown how energies for non-variational wave functions may be written in a variational form to take advantage of these simplifications. After a discussion on Hamiltonian derivatives, general expressions are given for the molecular gradients and Hessians for the following *ab initio* electronic structure techniques: self-consistent field (SCF) and multiconfigurational (MC) SCF wave functions, configuration interaction (CI) and coupled cluster (CC) wave functions, and Møller-Plesset (MP) perturbation theory. Finally, use of translational and rotational symmetries is discussed.

I. INTRODUCTION

The development of and computational implementation of analytic expressions for the derivatives of ab initio electronic energy surfaces has undergone rapid growth in recent years (Pulay, 1987, Gaw and Handy 1984 and Helgaker and Jørgensen 1988). This growth reflects the central role these derivatives play in the understanding of chemical reactions and many spectroscopic experiments (Fogarasi and Pulay 1984, Hess *et al.* 1986, Amos 1987 and Schlegel 1987).

The energy derivatives may be calculated by numerical techniques such as finite differences or polynomial fitting. The numerical approach is simple in the sense that usually no special quantum chemical computational techniques are needed except the ability to calculate the wave function. However, it suffers from difficulties related to numerical accuracy and computational efficiency. For non totally symmetric energy de-

Methods in Computational Molecular Physics, Edited by Stephen Wilson and Geerd H.F. Diercksen, Plenum Press, New York, 1992

rivatives energy calculations must also be carried out at a lower point group symmetry. Alternatively we may derive analytical expressions for the energy derivatives and calculate the derivatives directly from these expressions. This requires considerable programming efforts but the initial investments are compensated for since the analytical methods yield the properties with greater speed, precision and convenience than the numerical approach.

In the following we consider energy geometrical derivatives only but the methods presented are quite general and may be used for most other properties with small modifications. For magnetic properties we need to be careful to ensure that the calculated properties are gauge invariant (London 1937). Our emphasis will be on theoretical and technical aspects. We set up a mathematical framework within which the expressions for the various energy derivatives can be conveniently derived and discuss efficient computer implementations of these analytical expressions.

In Section II we first discuss in general terms the calculation of derivatives for variational wave functions. We demonstrate that the variational property of the energy significantly simplifies the calculation of its derivatives and show that the response of the wave function to n'th order is sufficient to determine the energy derivatives to order 2n+1. This is known as the 2n+1 rule. To extend this rule to non-variational wave functions we use Lagrange's method of undetermined multipliers to construct an energy function (Lagrangian) which is variational in all parameters, even though the wave function itself is non-variational. We show explicitly that the wave function parameters in the Lagrangian satisfy the 2n+1 rule and that the Lagrangian parameters satisfy a 2n+2 rule (Helgaker and Jørgensen 1989 and Jørgensen and Helgaker 1988). We derive expressions for the molecular gradient and Hessian in terms of this Lagrangian.

In Section III we discuss the geometrical derivatives of the Hamiltonian operator in second quantization (Jørgensen and Simons 1983 and Helgaker and Almlöf 1984). We show how the geometry dependence which occurs as a result of using finite nuclear fixed basis sets can be isolated in the Hamiltonian and thus separated from the physical relaxation of the wave function in the presence of a geometrical perturbation. The basis set dependence on the geometry is treated at a level common to all wave-functions and there is therefore no need to give this geometry dependence special considerations when the specific wave functions are considered in Section IV. This makes the exposition and derivation of the derivatives much more transparent.

In Section IV we give expressions for the energy derivatives calculated using Hartree-Fock (HF), multiconfigurational self-consistent field (MCSCF), configuration interaction (CI), coupled cluster (CC) and second-order Møller-Plesset (MP2) theories. For each wave function we construct a variational Lagrangian and use the general expressions of Section III to arrive at the appropriate response equations and derivatives. The calculation of these derivatives is discussed in Section V where we also compare the derivatives of the different wave functions. Energy derivatives are related by translational and rotational symmetries and this aspect is discussed in Section VI. Section VII contains some concluding remarks.

II. DERIVATIVES OF VARIATIONAL ENERGIES

In this section we discuss some general properties of energy derivatives. We start in Sec. A by making the distinction between variational and non-variational wave functions. In Sec. B we discuss the calculation of derivatives for variational wave functions and show how the variational property simplifies the calculation. Non-variational wave functions are treated in Sec. C where we demonstrate how energies of such wave functions may be recast in a variational form by using Lagrange's method of undetermined multipliers. Derivatives for non-variational wave functions may then be calculated in the same way as for variational wave functions. Section D gives a more general treatment of derivatives. We prove the 2n+1 and 2n+2 rules, derive the general form of the response equations and discuss numerical stabilities.

A. Variational wave functions

Wave functions may be classified as variational and non-variational. A wave function is referred to as *variational* if the electronic energy function $E(x,\lambda)$ fulfils the condition

$$\frac{\partial E(x,\lambda)}{\partial \lambda} = 0 \text{ for all } x \tag{2.1}$$

where x is the molecular geometry or any other perturbational parameter and λ the electronic (wave function) parameters. Variational wave functions are therefore characterized by the fact that the gradient with respect to the electronic parameters (the electronic gradient) vanishes at all geometries. *Non-variational wave functions* are those wave functions for which the variational condition Eq. (2.1) is not satisfied by some or all of the electronic parameters λ.

The variational condition Eq. (2.1) determines λ as a function of x and we obtain the molecular electronic energy $\varepsilon(x)$ by inserting this value of λ in the energy function $E(x,\lambda)$. For the lowest state of a given symmetry we may use the equivalent condition

$$\varepsilon(x) = \min_{\lambda} E(x,\lambda) \quad \text{for all } x \tag{2.2}$$

and determine the energy by minimization. The variational condition Eq. (2.1) is more general and is used in the following.

Let us clarify by some examples. The Hartree-Fock wave function is variational since the orbital rotation parameters κ are variational:

$$\frac{\partial E_{SCF}(x,\kappa)}{\partial \kappa} = 0 \tag{2.3}$$

MCSCF wave functions are also variational since the variational condition is fulfilled for both the orbital rotation parameters κ and the state transfer parameters p[1]

$$\frac{\partial E_{MC}(x,\kappa,p)}{\partial \kappa} = 0 \tag{2.4}$$

$$\frac{\partial E_{MC}(x,\kappa,p)}{\partial p} = 0 \tag{2.5}$$

For CI wave functions the variational condition is fulfilled by the state transfer parameters P but not by the orbital parameters κ:

$$\frac{\partial E_{CI}(x,\kappa,P)}{\partial \kappa} \neq 0 \tag{2.6}$$

$$\frac{\partial E_{CI}(x,\kappa,P)}{\partial P} = 0 \tag{2.7}$$

---

[1]The state transfer parameters p are a set of independent parameters used to describe the CI configuration state. The usual CI coefficients may also be used to describe the CI state but they are not independent.

The orbitals are determined in an (MC)SCF calculation prior to the calculation of the CI wave function. The orbitals satisfy condition Eq. (2.3) or Eq. (2.4) and are therefore obtained by applying the variational principle to the (MC)SCF energy rather than to the CI energy. The CI wave function is therefore not variational. CI wave functions are often referred to as variational since the configuration coefficients are calculated from the variational principle. However, CI wave functions may only be considered variational if we disregard orbital variations.

Having classified wave functions as variational and non-variational we go on to discuss the derivatives of the energy with respect to the geometrical parameters x. The importance of the distinction between variational and non-variational wave functions then becomes clear.

B. Derivatives for variational wave functions

Let us consider a variational wave function and calculate the first derivative with respect to the geometry or any other time-independent perturbation. We obtain

$$\frac{d\varepsilon}{dx} = \frac{\partial E}{\partial x} + \frac{\partial E}{\partial \lambda}\frac{\partial \lambda}{\partial x} \tag{2.8}$$

since the energy depends both explicitly and implicitly on the geometry. Since we assume that the energy is variational

$$\frac{\partial E}{\partial \lambda} = 0 \tag{2.9}$$

the last term in Eq. (2.8) vanishes and the total derivative may be calculated as the partial derivative with respect to the geometry

$$\frac{d\varepsilon}{dx} = \frac{\partial E}{\partial x} \tag{2.10}$$

In other words, *to calculate first derivatives of variational energies we do not need the response of the wave function* $\partial\lambda/\partial x$. For example, when calculating SCF or MCSCF molecular gradients we do not need the variation of the wave function with geometry. This is an important simplification since the calculation of such responses may be time-consuming.

Equation (2.10) is closely related to the Hellmann-Feynman theorem. If we assume that the energy function may be written as an expectation value

$$E(x,\lambda) = \langle \lambda | H(x) | \lambda \rangle \tag{2.11}$$

where $|\lambda\rangle$ is a variational state we obtain from Eq. (2.10)

$$\frac{d\varepsilon(x)}{dx} = \left\langle \lambda \left| \frac{\partial H}{\partial x} \right| \lambda \right\rangle \tag{2.12}$$

This is the *Hellmann-Feynman theorem.* Although originally stated for geometrical distortions, it holds for any perturbation.

We have seen how the variational condition simplifies the calculation of gradients. We now consider how this simplification carries over to higher derivatives. Differentiating the energy twice with respect to the geometry and using the variational condition Eq. (2.9) to eliminate the term containing $\partial^2\lambda/\partial x^2$ we obtain

$$\frac{d^2\varepsilon}{dx^2} = \left(\frac{\partial^2 E}{\partial x^2}\right) + 2\left(\frac{\partial^2 E}{\partial x \partial \lambda}\right)\left(\frac{\partial \lambda}{\partial x}\right) + \left(\frac{\partial^2 E}{\partial \lambda^2}\right)\left(\frac{\partial \lambda}{\partial x}\right)^2 \tag{2.13}$$

Hence we only need the *first-order* response of the wave function to calculate the energy to *second-order*. To determine this response we note that the variational condition Eq. (2.9) holds at all geometries. We therefore differentiate this condition with respect to the geometry and obtain a set of linear equations

$$\left(\frac{\partial^2 E}{\partial \lambda^2}\right)\left(\frac{\partial \lambda}{\partial x}\right) = -\left(\frac{\partial^2 E}{\partial x \partial \lambda}\right) \tag{2.14}$$

from which the first-order response may be determined. We refer to these equations as the *response equations*. Higher-order responses are obtained by further differentiation.

It is clear that the variational property greatly simplifies the calculation of derivatives. We shall discuss variational derivatives in a more general way in Sec. D. In particular we shall prove *the 2n+1 rule*: The derivatives of the wave function to order n determine the derivatives of the energy to order 2n+1. But first we discuss derivatives for non-variational wave functions.

C. Derivatives for non-variational wave functions

As an example we consider the gradient of the CI energy. The CI wave function depends on two sets of parameters: the state transfer operators P and the orbital rotations κ. The CI energy is variational with respect to the configuration coefficients

$$\frac{\partial E_{CI}}{\partial P} = 0 \tag{2.15}$$

but non-variational with respect to the orbital rotations

$$\frac{\partial E_{CI}}{\partial \kappa} \neq 0 \tag{2.16}$$

Therefore, if we differentiate the energy function with respect to the geometry we do not obtain the simplifications of the 2n+1 rule:

$$\frac{d\varepsilon_{CI}}{dx} = \frac{dE_{CI}}{dx} = \frac{\partial E_{CI}}{\partial x} + \frac{\partial E_{CI}}{\partial \kappa}\frac{\partial \kappa}{\partial x} \tag{2.17}$$

and it appears that we also need the first-order response of the orbitals Eq. (2.14):

$$\frac{\partial^2 E_{SCF}}{\partial \kappa^2}\frac{\partial \kappa}{\partial x} = -\frac{\partial^2 E_{SCF}}{\partial \kappa \partial x} \tag{2.18}$$

There is one set of equations for each perturbation, i.e. for each independent geometrical distortion.

The calculation of the first-order orbital responses may be avoided by using *Lagrange's method of undetermined multipliers*. The orbitals fulfil the condition

$$\frac{\partial E_{SCF}}{\partial \kappa} = 0 \tag{2.19}$$

since they are determined in a separate SCF (or MCSCF) calculation. The variational condition Eq. (2.19) may be considered as a set of *constraints* in the optimization of the CI energy function with respect to configuration coefficients *and* orbital rotations. Following a standard method for constrained optimization we introduce the Lagrangian function

$$L_{CI} = E_{CI} + \bar{\kappa}\frac{\partial E_{SCF}}{\partial \kappa} \tag{2.20}$$

where $\bar{\kappa}$ are the Lagrange multipliers. (In the following we use overbars $\bar{\lambda}$ for the multipliers corresponding to λ.) The functional form of L_{CI} is different from that of E_{CI} but it gives the same energy when condition Eq. (2.19) is fulfilled. We adjust the multipliers so that this Lagrangian is variational in *all* variables:

$$\frac{\partial L_{CI}}{\partial P} = \frac{\partial E_{CI}}{\partial P} = 0 \tag{2.21}$$

$$\frac{\partial L_{CI}}{\partial \kappa} = \frac{\partial E_{CI}}{\partial \kappa} + \bar{\kappa}\,\frac{\partial^2 E_{SCF}}{\partial \kappa^2} = 0 \tag{2.22}$$

$$\frac{\partial L_{CI}}{\partial \bar{\kappa}} = \frac{\partial E_{SCF}}{\partial \kappa} = 0 \tag{2.23}$$

Equation (2.21) is the usual variational condition for the state transfer parameters. The second expression Eq. (2.22) determines the Lagrange multipliers in such a way that the Lagrangian is variational with respect to the orbital rotations, and Eq. (2.23) is the usual SCF variational condition. When all variational conditions are fulfilled the Lagrangian gives the CI energy. The idea behind the Lagrangian formulation is to introduce a new set of variables (the multipliers) which are variational [Eq. (2.23)] and to adjust these multipliers in such a way that the original variables also become variational [Eq. (2.22)].

Therefore, if we use the CI Lagrangian rather than the original CI expectation value we have a *completely variational formulation of the CI energy*. We may then take advantage of the 2n+1 rule for variational energies. The price we pay for this is a larger number of variables: one Lagrange multiplier is introduced for each constraint.

The total derivative of the Lagrangian Eq. (2.20) is given by

$$\frac{dL_{CI}}{dx} = \frac{\partial E_{CI}}{\partial x} + \frac{\partial E_{CI}}{\partial P}\frac{\partial P}{\partial x} + \left(\frac{\partial E_{CI}}{\partial \kappa} + \bar{\kappa}\frac{\partial^2 E_{SCF}}{\partial \kappa^2}\right)\frac{\partial \kappa}{\partial x} + \frac{\partial \bar{\kappa}}{\partial x}\frac{\partial E_{SCF}}{\partial \kappa} + \bar{\kappa}\,\frac{\partial^2 E_{SCF}}{\partial x \partial \kappa} \tag{2.24}$$

Since Eqs. (2.21) - (2.23) are satisfied, all terms on the right-hand side of Eq. (2.24) except the first and last vanish and we obtain the following expression for the CI gradient

$$\frac{d\varepsilon_{CI}}{dx} = \frac{dL_{CI}}{dx} = \frac{\partial E_{CI}}{\partial x} + \bar{\kappa}\frac{\partial^2 E_{SCF}}{\partial x \partial \kappa} \tag{2.25}$$

where the multipliers are calculated from the equations

$$\frac{\partial^2 E_{SCF}}{\partial \kappa^2}\bar{\kappa} = -\frac{\partial E_{CI}}{\partial \kappa} \tag{2.26}$$

The gradient Eq. (2.25) is simply the partial derivative of the Lagrangian with respect to x as expected for a variational function. Equations (2.25) and (2.26) may be compared with those derived by differentiating the CI energy Eqs. (2.17) and (2.18). We have replaced the first-order equations for the orbitals by a set of linear equations for the orbital multipliers. There is now only one set of linear equations to solve for all perturbations.

The method of Lagrange multipliers is quite general. It includes the technique introduced by Handy and Schaefer (1984) to simplify the calculation of derivatives for non-variational energies and gives the Hylleraas functional (Epstein 1974) when applied to perturbational energy expressions.

D. Variational perturbation theory

We begin by reviewing the properties of unperturbed variational energies. We assume that the energy is described by a set of parameters λ (the variational parameters describing the electronic state) and expand the energy $E(\lambda)$ in a Taylor series

$$E(\lambda) = E^{(0)} + E^{(1)}\lambda + \frac{1}{2}E^{(2)}\lambda^2 + \cdots \tag{2.27}$$

where

$$E^{(n)} = \left.\frac{\partial^n E(\lambda)}{\partial \lambda^n}\right|_{\lambda=0} \tag{2.28}$$

We then require the first-order variation with respect to λ to vanish

$$\frac{\partial E}{\partial \lambda} = E^{(1)} + E^{(2)}\lambda + \frac{1}{2}E^{(3)}\lambda^2 + \cdots = 0 \tag{2.29}$$

This is a set of non-linear equations in λ which we solve to obtain the stable value λ_* and then calculate the energy as[2]

$$\varepsilon = E(\lambda_*) = E^{(0)} + E^{(1)}\lambda_* + \frac{1}{2}E^{(2)}\lambda_*^2 + \cdots \tag{2.30}$$

The non-linear equations Eq. (2.29) are solved iteratively. Usually the energy Eq. (2.27) is reparametrized such that

$$\lambda_* = 0 \tag{2.31}$$

This is not essential to the method but often simplifies the calculation of derivatives. In the following we do not assume that Eq. (2.31) holds unless we explicitly state so.

We now investigate what happens to the energy when the system is perturbed in some way, for example by a geometrical distortion. The electronic energy is expanded in orders of the perturbation

$$\varepsilon(x) = \varepsilon^{(0)} + \varepsilon^{(1)}x + \frac{1}{2}\varepsilon^{(2)}x^2 + \cdots \tag{2.32}$$

Here $\varepsilon^{(0)}$ is the total energy in the absence of the perturbation and our purpose is to derive expressions for the perturbed energies

$$\varepsilon^{(n)} = \left.\frac{\partial^n \varepsilon(x)}{\partial x^n}\right|_{x=0} \tag{2.33}$$

We proceed in the following manner. In Sec. 1 we consider the dependence of the energy function Eq. (2.27) on the perturbation. We consider the generalization of the variational condition Eq. (2.29) to perturbed systems in Sec. 2. This leads to a hierarchy of perturbed variational conditions (response equations) from which the perturbed wave functions (wave function responses) are obtained. In Sec. 3 we discuss the calculation of the perturbed energies Eq. (2.32) and the 2n+1 rule. Numerical errors are treated in Sec. 4 and finally in Sec. 5 we discuss the calculation of derivatives from variational Lagrangians.

---

[2] In this section we need to distinguish carefully between a variable and its stable value. We therefore add an asterisk to the stable value of any variable.

1. Perturbed energy functions

Starting with the expansion of the energy function Eq. (2.27) we note that each term now depends on the perturbation. In fact, we have one expansion for each value of the perturbational parameter x:

$$E(x,\lambda) = E^{(0)}(x) + E^{(1)}(x)\,\lambda(x) + \tfrac{1}{2}E^{(2)}(x)\,\lambda^2(x) + \cdots \tag{2.34}$$

where

$$E^{(n)}(x) = \left.\frac{\partial^n E(x,\lambda)}{\partial \lambda^n}\right|_{\lambda=0} \tag{2.35}$$

The Taylor coefficients in Eq. (2.34) can be expanded in the perturbation

$$E^{(n)}(x) = E^{(0n)} + E^{(1n)}x + \tfrac{1}{2}E^{(2n)}x^2 + \cdots \tag{2.36}$$

where double indices indicate differentiation with respect to x and λ

$$E^{(mn)} = \left.\frac{\partial^{m+n}E}{\partial x^m \partial \lambda^n}\right|_{\substack{x=0\\ \lambda=0}} \tag{2.37}$$

The variational parameters in Eq. (2.34) may also be expanded in the perturbation

$$\lambda(x) = \lambda^{(0)} + \lambda^{(1)}x + \tfrac{1}{2}\lambda^{(2)}x^2 + \cdots \tag{2.38}$$

and from now on we treat $\lambda^{(n)}$ as the variational parameters [Sellers (1986)]. *There is thus one set of variational parameters $\lambda^{(n)}$ for each order n in the perturbation rather than one set $\lambda(x)$ for each value x of the perturbation.* Therefore, instead of determining the stable values of λ(x) for each x we determine the stable values of $\lambda^{(n)}$ for each order n. If we insert the expansions Eqs. (2.36) and (2.38) in Eq. (2.34) and collect terms of the same order in x we obtain

$$E(x,\lambda) = E^{[0]} + E^{[1]}x + \tfrac{1}{2}E^{[2]}x^2 + \cdots \tag{2.39}$$

using square brackets [n] to indicate order in perturbation

$$E^{[n]}(\lambda^{(0)}, \lambda^{(1)}, \lambda^{(2)}, \ldots) = \left. \frac{\partial^n E(x,\lambda)}{\partial x^n} \right|_{x=0} \tag{2.40}$$

to avoid confusion with order in λ as in Eq. (2.29). It is clear that each term $E^{[n]}$ [Eq. (2.40)] is a polynomial in $\lambda^{(k)}$ in the same way that the unperturbed energy $E(\lambda)$ [Eq. (2.27)] is a polynomial in λ. Before we consider the generalization of the variational condition Eq. (2.29) to perturbed energies $E^{[n]}$ we investigate in more detail the explicit form of these polynomials.

To determine the functional form of the polynomial $E^{[n]}$ we differentiate with respect to $\lambda^{(k)}$ and obtain

$$\frac{\partial E^{[n]}}{\partial \lambda^{(k)}} = \left. \frac{\partial^n}{\partial x^n} \frac{\partial}{\partial \lambda^{(k)}} E(x,\lambda) \right|_{x=0} = \left. \frac{\partial^n}{\partial x^n} \frac{\partial \lambda}{\partial \lambda^{(k)}} \frac{\partial E}{\partial \lambda} \right|_{x=0} \tag{2.41}$$

using the chain rule. Inserting the expansion Eq. (2.38) we find

$$\frac{\partial E^{[n]}}{\partial \lambda^{(k)}} = \left. \frac{1}{k!} \frac{\partial^n}{\partial x^n} x^k \frac{\partial E}{\partial \lambda} \right|_{x=0} \tag{2.42}$$

Finally, using the binomial expansion and the fact that we differentiate at x=0 we obtain

$$\frac{\partial E^{[n]}}{\partial \lambda^{(k)}} = \begin{cases} 0 & k > n \\ \left. \binom{n}{k} \frac{\partial^{n-k}}{\partial x^{n-k}} \frac{\partial E}{\partial \lambda} \right|_{x=0} & k \leq n \end{cases} \tag{2.43}$$

We conclude that $E^{[n]}$ is independent of $\lambda^{(k)}$ if $k>n$ and we may write

$$E^{[n]}(\lambda^{(0)}, \lambda^{(1)}, \ldots \lambda^{(n)}) \tag{2.44}$$

Next we establish whether $E^{[n]}$ is linear or non-linear in $\lambda^{(k)}$ for $k \leq n$. Differentiating Eq. (2.43) with respect to $\lambda^{(m)}$ we find

$$\frac{\partial^2 E^{[n]}}{\partial \lambda^{(k)} \partial \lambda^{(m)}} = \begin{cases} 0 & k + m > n \\ \left. \binom{n}{k} \binom{n-k}{m} \frac{\partial^{n-k-m}}{\partial x^{n-k-m}} \frac{\partial^2 E}{\partial \lambda^2} \right|_{x=0} & k + m \leq n \end{cases} \tag{2.45}$$

We see that $E^{[n]}$ is non-linear in $\lambda^{(k)}$ if $k \le n/2$ and linear if $k > n/2$. Also, there are no quadratic or higher terms containing $\lambda^{(k)}$ and $\lambda^{(m)}$ if $k > n/2$ and $m > n/2$. Differentiating further we find that

$$\frac{\partial^p E^{[n]}}{\partial \lambda^{(k)\,p}} \equiv 0\,, \qquad pk > n \tag{2.46}$$

which means that $\lambda^{(k)}$ does not enter $E^{[n]}$ to orders higher than n/k. The only parameter which enters $E^{[n]}$ to infinite order is $\lambda^{(0)}$. We may therefore write

$$E^{[n]}(\underbrace{\lambda^{(0)}}_{\substack{\text{infinite}\\\text{order}}},\ \underbrace{\lambda^{(1)},\ldots\lambda^{(\mathrm{int}(n/2))}}_{\substack{\text{non-linear}\\\text{parameters}}},\ \underbrace{\lambda^{(\mathrm{int}(n/2)+1)},\ldots\lambda^{(n)}}_{\substack{\text{linear}\\\text{parameters}}}) \tag{2.47}$$

where int(m) is the largest integer smaller than m. The importance of the linear dependence of the higher order perturbations $\lambda^{(k)}$ will soon become apparent.[3]

Equation (2.47) is easily verified for the lower-order energies $E^{[n]}$. Inserting expansions Eqs. (2.36) and (2.38) in Eq. (2.34) and collecting terms to same order in x we find

$$E^{[0]} = E^{(00)} + E^{(01)}\lambda^{(0)} + \cdots \tag{2.48}$$

$$E^{[1]} = E^{(10)} + E^{(11)}\lambda^{(0)} + E^{(01)}\lambda^{(1)} + E^{(02)}\lambda^{(0)}\lambda^{(1)} + \cdots \tag{2.49}$$

$$\begin{aligned} E^{[2]} = {} & E^{(20)} + E^{(21)}\lambda^{(0)} + 2E^{(11)}\lambda^{(1)} + 2E^{(12)}\lambda^{(0)}\lambda^{(1)} \\ & + E^{(01)}\lambda^{(2)} + E^{(02)}\lambda^{(0)}\lambda^{(2)} + E^{(02)}\lambda^{(1)}\lambda^{(1)} \\ & + E^{(03)}\lambda^{(0)}\lambda^{(1)}\lambda^{(1)} + \cdots \end{aligned} \tag{2.50}$$

neglecting all terms which are non-linear in $\lambda^{(0)}$. In accordance with Eq. (2.47) we see that $E^{[1]}$ and $E^{[2]}$ are linear in $\lambda^{(1)}$ and $\lambda^{(2)}$, respectively. The remaining dependencies are non-linear. In Eqs. (2.48) - (2.50) the terms are collected in orders of the perturbation and $\lambda^{(k)}$ are order-dependent variational parameters. Equations (2.48) - (2.50) are the perturbed versions of Eq. (2.27).

2. Perturbed variational conditions

In the unperturbed case we required the energy function $E(\lambda)$ to be stable against variations in λ, see Eq. (2.29). We now require *the perturbed functions* $E^{[n]}$ *to be stable against variations in all* $\lambda^{(k)}$

$$\frac{\partial E^{[n]}}{\partial \lambda^{(k)}} = 0, \quad k \leq n \tag{2.51}$$

These conditions are not all independent. Using Eq. (2.43) we may write Eq. (2.51)

$$\binom{n}{k}^{-1} \frac{\partial E^{[n]}}{\partial \lambda^{(k)}} = \frac{\partial^{n-k}}{\partial x^{n-k}} \frac{\partial E}{\partial \lambda}\bigg|_{x=0} = 0 \tag{2.52}$$

Hence, except for a multiplicative factor the variational conditions depend only on the difference between n and k, n–k. For example, from Eq. (2.52) the condition on $E^{[2]}$ with respect to $\lambda^{(1)}$ is equivalent to the condition on $E^{[1]}$ with respect to $\lambda^{(0)}$:

$$\frac{\partial E^{[2]}}{\partial \lambda^{(1)}} = 2\frac{\partial E^{[1]}}{\partial \lambda^{(0)}} \tag{2.53}$$

Therefore, Eq. (2.52) shows that the independent conditions to order n are

$$\frac{\partial^p}{\partial x^p} \frac{\partial E}{\partial \lambda}\bigg|_{x=0} = 0, \qquad p = 0, 1, \cdots n \tag{2.54}$$

These are the perturbed versions of Eq. (2.29).

If we differentiate the expression on the left hand side of Eq. (2.54) with respect to $\lambda^{(k)}$ we obtain in the same way as before [using the chain rule, expansion Eq. (2.38), the binomial expansion and x=0]

[3]Note that $E^{[n]}$ is linear in all $\lambda^{(k)}$ (k>n) collectively, i.e. there are no cross terms containing different $\lambda^{(k)}$ and $\lambda^{(m)}$ where both k>n and m>n.

$$\frac{\partial}{\partial \lambda^{(k)}} \frac{\partial^p}{\partial x^p} \frac{\partial E}{\partial \lambda}\bigg|_{x=0} \equiv \begin{cases} 0 & k > p \\ E^{(02)} & k = p \\ \binom{p}{k} \frac{\partial^{p-k}}{\partial x^{p-k}} \frac{\partial^2 E}{\partial \lambda^2}\bigg|_{x=0} & k < p \end{cases} \tag{2.55}$$

Further differentiation with respect to $\lambda^{(k)}$ must give zero for $k \geq p$. We therefore conclude that the variational conditions Eq. (2.54) may be written

$$\frac{\partial^p}{\partial x^p} \frac{\partial E}{\partial \lambda}\bigg|_{x=0} = E^{(02)} \lambda^{(p)} + R^{(p)} (\lambda^{(0)}, \cdots \lambda^{(p-1)}) \tag{2.56}$$

where $R^{(p)}$ is a polynomial in $\lambda^{(k)}$ $(k<p)$. Except for $\lambda^{(0)}$ all parameters occur to finite order in Eq. (2.56). Inserting Eq. (2.56) in Eq. (2.54) we obtain *the response equations*

$$E^{(02)} \lambda^{(p)} + R^{(p)} (\lambda^{(0)}, \cdots \lambda^{(p-1)}) = 0, \qquad p = 0, 1, \cdots n \tag{2.57}$$

from which the responses $\lambda_*^{(k)}$ to order n may be calculated. We obtain the following hierarchy of linear equations

$$E^{(02)} \lambda_*^{(1)} = - R^{(1)} (\lambda_*^{(0)}) \tag{2.58}$$

$$E^{(02)} \lambda_*^{(2)} = - R^{(2)} (\lambda_*^{(0)}, \lambda_*^{(1)}) \tag{2.59}$$

$$\vdots$$

$$E^{(02)} \lambda_*^{(n)} = - R^{(n)} (\lambda_*^{(0)}, \cdots \lambda_*^{(n-1)}) \tag{2.60}$$

which must be solved in the order written. Note that the coefficient matrix $E^{(02)}$ is independent of the order of the equations as well as the nature of the perturbation. *Only the right-hand side depends on the nature and order of perturbation.*

The response equations are all *linear*. The reason for this is that the highest order perturbation always occurs linearly in the perturbed energy functions Eq. (2.47). This is to be contrasted with the unperturbed variational condition Eq. (2.29) which in general is non-linear since the variational parameters occur non-linearly in the unperturbed energy function Eq. (2.27).

Explicit expressions for the lower order equations are obtained by differentiating $E^{[2]}$ with respect to $\lambda^{(2)}$, $\lambda^{(1)}$, and $\lambda^{(0)}$ and setting the results equal to zero. We obtain

$$E^{(01)} = 0 \tag{2.61}$$

$$E^{(02)} \lambda_*^{(1)} = - E^{(11)} \tag{2.62}$$

$$E^{(02)} \lambda_*^{(2)} = - (E^{(21)} + 2 E^{(12)} \lambda_*^{(1)} + E^{(03)} \lambda_*^{(1)} \lambda_*^{(1)}) \tag{2.63}$$

assuming that $\lambda_*^{(0)} = 0$. Note that Eq. (2.61) is simply the variational condition on the unperturbed energy [Eq (2.29)].

3. Perturbed energies

The functional dependence of $E^{[n]}$ on $\lambda^{(k)}$ is given in Eq. (2.47) and the optimal (stable) parameters $\lambda_*^{(k)}$ are obtained from Eq. (2.54). The perturbed energies therefore become

$$\varepsilon^{(n)} = E^{[n]} (\underbrace{\lambda_*^{(0)}, \lambda_*^{(1)}, \ldots \lambda_*^{(\mathrm{int}(n/2))}}_{\text{non-linear}}, \underbrace{\lambda_*^{(\mathrm{int}(n/2)+1)}, \ldots \lambda_*^{(n)}}_{\text{linear}}) \tag{2.64}$$

which are the perturbed versions of Eq. (2.30). Since $\varepsilon^{(n)}$ represents a point where $E^{[n]}$ is stationary with respect to $\lambda^{(k)}$, the functional form of $\varepsilon^{(n)}$ must be such that $\varepsilon^{(n)}$ has no terms linear in $\lambda^{(k)}$. This means that $\varepsilon^{(n)}$ in Eq. (2.64) must be completely independent of $\lambda_*^{(k)}$ for $k>n/2$ since there are no quadratic or higher terms in these responses. We may therefore remove all terms in $\varepsilon^{(n)}$ containing these linear parameters and calculate the perturbed energies according to

$$\varepsilon^{(n)} = E^{[n]} (\lambda_*^{(0)}, \lambda_*^{(1)}, \ldots \lambda_*^{(\mathrm{int}(n/2))}) \tag{2.65}$$

This important result is referred to as the 2n+1 rule since it implies that *the wave function responses to order n determine the energy to order 2n+1*. Returning to the low-order cases we may now write the energies [see Eqs. (2.49) and (2.50)] in the form

$$\varepsilon^{(1)} = E^{(10)} \tag{2.66}$$

$$\varepsilon^{(2)} = E^{(20)} + 2 E^{(11)} \lambda_*^{(1)} + E^{(02)} \lambda_*^{(1)} \lambda_*^{(1)} \tag{2.67}$$

assuming according to Eq. (2.31) that $\lambda_*^{(0)} = 0$.

We may use the response equations to rewrite the perturbed energies. For example, if we substitute the response Eq. (2.62)

$$\lambda_*^{(1)} = -\left[E^{(02)}\right]^{-1} E^{(11)} \tag{2.68}$$

into the Hessian Eq. (2.67) we obtain the following alternative expressions:

$$\varepsilon^{(2)} = E^{(20)} + E^{(11)} \lambda_*^{(1)} \tag{2.69}$$

$$\varepsilon^{(2)} = E^{(20)} - E^{(11)} \left[E^{(02)}\right]^{-1} E^{(11)} \tag{2.70}$$

These expressions look simpler but do not necessarily imply any computational advantages. For example, we shall see below that although Eq. (2.69) is more compact than the original expression Eq. (2.67), it is also more sensitive to numerical errors in $\lambda_*^{(1)}$ (Helgaker *et al.* 1986, Sellers 1986). In addition the term which has been eliminated is computationally inexpensive. In Eq. (2.70) we have eliminated the responses completely. This form of the Hessian suggests that we first invert $E^{(02)}$. The dimension of the electronic Hessian $E^{(02)}$ is often so large that inverting it becomes impossible. Instead we solve the response equations Eq. (2.62) iteratively and calculate the Hessian from Eq. (2.67)

4. Numerical errors

We now examine the error in the derivatives Eqs. (2.64) and (2.65) due to errors in the responses. Let $\lambda_*^{(p)}$ be the exact solution to the response equations Eq. (2.57) and $\lambda_a^{(p)}$ an approximate solution. The error in the responses are thus

$$\Delta\lambda^{(p)} = \lambda_a^{(p)} - \lambda_*^{(p)} \tag{2.71}$$

We expand $E^{[n]}$ *before* the 2n+1 rule has been applied [Eq. (2.64)] in a Taylor series about the exact solutions

$$\begin{aligned} E^{[n]}(\lambda_a^{(0)}, \lambda_a^{(1)}, \cdots \lambda_a^{(n)}) &= E^{[n]}(\lambda_*^{(0)}, \lambda_*^{(1)}, \cdots \lambda_*^{(n)}) \\ &+ \sum_{k=0}^{n} \left.\frac{\partial E^{[n]}}{\partial \lambda^{(k)}}\right|_* \Delta\lambda^{(k)} + \sum_{k,i}^{n} \left.\frac{\partial^2 E^{[n]}}{\partial \lambda^{(k)} \partial \lambda^{(i)}}\right|_* \Delta\lambda^{(k)} \Delta\lambda^{(i)} + \cdots \end{aligned} \tag{2.72}$$

The variational conditions Eq. (2.51) implies that the second term on the right-hand side of Eq. (2.72) vanishes and $E^{[n]}$ therefore has errors which are *quadratic* in the errors of the responses. To determine the errors in the derivatives *after* the 2n+1 rule has been applied we expand $E^{[n]}_{2n+1}$ [Eq. (2.65)] around the exact responses $\lambda_*^{(p)}$ and obtain

$$E^{[n]}_{2n+1}(\lambda_a^{(0)}, \lambda_a^{(1)}, \cdots \lambda_a^{(int(n/2))}) = E^{[n]}_{2n+1}(\lambda_*^{(0)}, \lambda_*^{(1)}, \cdots \lambda_*^{(int(n/2))})$$

$$+ \sum_{k=0}^{int(n/2)} \left.\frac{\partial E^{[n]}_{2n+1}}{\partial \lambda^{(k)}}\right|_* \Delta\lambda^{(k)} + \sum_{k,i}^{int(n/2)} \left.\frac{\partial^2 E^{[n]}_{2n+1}}{\partial \lambda^{(k)} \partial \lambda^{(i)}}\right|_* \Delta\lambda^{(k)}\Delta\lambda^{(i)} + \cdots \quad (2.73)$$

The linear coefficients in Eq. (2.73) are not necessarily zero since the functional form of $E^{[n]}_{2n+1}$ Eq. (2.65) is different from that of $E^{[n]}$ Eq. (2.64). The linear coefficients in $E^{[n]}_{2n+1}$ may be written

$$\left.\frac{\partial E^{[n]}_{2n+1}}{\partial \lambda^{(k)}}\right|_* = \frac{\partial}{\partial \lambda^{(k)}} \left[E^{[n]} - \sum_{i>n/2}^{n} \frac{\partial E^{[n]}}{\partial \lambda^{(i)}} \lambda_*^{(i)} \right]\Bigg|_*$$

$$= \left.\frac{\partial E^{[n]}}{\partial \lambda^{(k)}}\right|_* - \sum_{i>n/2}^{n} \left.\frac{\partial^2 E^{[n]}}{\partial \lambda^{(i)} \partial \lambda^{(k)}}\right|_* \lambda_*^{(i)} \quad (2.74)$$

$$= - \sum_{i>n/2}^{n} \left.\frac{\partial^2 E^{[n]}}{\partial \lambda^{(i)} \partial \lambda^{(k)}}\right|_* \lambda_*^{(i)}$$

To arrive at the first identity we have added and subtracted the higher-order linear terms in $E^{[n]}$. The last identity follows since $E^{[n]}$ is variational with respect to all responses. Combining Eqs. (2.73) and (2.74) we see that

$$E^{[n]}_{2n+1}(\lambda_a^{(j)}) = E^{[n]}(\lambda_*^{(j)}) - \sum_{\substack{k\leq n/2 \\ i>n/2}} \left.\frac{\partial^2 E^{[n]}}{\partial \lambda^{(i)} \partial \lambda^{(k)}}\right|_* \lambda_*^{(i)} \Delta\lambda^{(k)} + \cdots \quad (2.75)$$

For k=n/2 all terms in the summation vanish according to Eq. (2.45) since k+i>n. Therefore the linear dependence on $\Delta\lambda^{(n/2)}$ vanishes. However, if k<n/2 there is at least one non-vanishing term in the summation and the term linear in $\Delta\lambda^{(k)}$ does not vanish. We therefore conclude that $E^{[n]}_{2n+1}$ depends linearly on the errors in all responses except $\lambda^{(n/2)}$. Note that $\lambda^{(n/2)}$ occurs in derivatives of even orders only.

Let us illustrate this by considering the molecular gradient and Hessian

$$\varepsilon^{(1)} = E^{(10)} + \left[E^{(01)} + E^{(02)}\lambda_a^{(0)}\right]\lambda_a^{(1)} + E^{(11)}\lambda_a^{(0)} \tag{2.76}$$

$$\begin{aligned}\varepsilon^{(2)} = E^{(20)} &+ 2E^{(11)}\lambda_a^{(1)} + E^{(02)}\lambda_a^{(1)}\lambda_a^{(1)} \\ &+ \left[E^{(01)} + E^{(02)}\lambda_a^{(0)}\right]\lambda_a^{(2)} \\ &+ \left[E^{(21)} + 2E^{(12)}\lambda_a^{(1)} + E^{(03)}\lambda_a^{(1)}\lambda_a^{(1)}\right]\lambda_a^{(0)}\end{aligned} \tag{2.77}$$

where we have omitted higher-order terms in $\lambda_a^{(0)}$. These expressions have errors which are quadratic in the errors $\lambda_*^{(p)}$ since the 2n+1 rule has not yet been applied. The zero-order equation (2.61) may be used to eliminate $\lambda^{(1)}$ in the gradient and $\lambda^{(2)}$ in the Hessian. We then obtain

$$\varepsilon^{(1)} = E^{(10)} + E^{(11)}\lambda_a^{(0)} \tag{2.78}$$

$$\begin{aligned}\varepsilon^{(2)} = E^{(20)} &+ 2E^{(11)}\lambda_a^{(1)} + E^{(02)}\lambda_a^{(1)}\lambda_a^{(1)} \\ &+ \left[E^{(21)} + 2E^{(12)}\lambda_a^{(1)} + E^{(03)}\lambda_a^{(1)}\lambda_a^{(1)}\right]\lambda_a^{(0)}\end{aligned} \tag{2.79}$$

If Eq. (2.78) is used to calculate the gradient it will have errors proportional to the errors in $\lambda_a^{(0)}$. If the Hessian in calculated from Eq. (2.79) it will have errors linear in the errors in $\lambda_a^{(0)}$ but quadratic in $\lambda_a^{(1)}$. If we assume that $\lambda_*^{(0)}$ is zero we obtain

$$\varepsilon^{(1)} = E^{(10)} \tag{2.80}$$

$$\varepsilon^{(2)} = E^{(20)} + 2E^{(11)}\lambda_a^{(1)} + E^{(02)}\lambda_a^{(1)}\lambda_a^{(1)} \tag{2.81}$$

where we have errors *linear* in the error of the zero-order responses (the wave function) since the equations for these have been used to eliminate $\lambda_*^{(1)}$ and $\lambda_*^{(2)}$. Nevertheless, this is a small price to pay for the computational savings obtained by the 2n+1 rule.

To summarize, when $\lambda_*^{(k)}$ is eliminated from $\varepsilon^{(n)}$ using the (n−m)'th order response equation, the resulting expression depends linearly on the errors in $\lambda_*^{(n-k)}$. When the 2n+1 rule is used to eliminate all higher-order responses, we find that odd-order derivatives $\varepsilon^{(2n+1)}$ are linear in *all* the remaining responses and that even-order derivatives $\varepsilon^{(2n)}$ are linear in all responses *except* $\lambda_*^{(n)}$.

	Responses needed	Dependency on errors
$\varepsilon^{(0)}$	$\lambda^{(0)}$	quadratic
$\varepsilon^{(1)}$	$\lambda^{(0)}$	linear
$\varepsilon^{(2)}$	$\lambda^{(0)}$	linear
	$\lambda^{(1)}$	quadratic
$\varepsilon^{(3)}$	$\lambda^{(0)}$	linear
	$\lambda^{(1)}$	linear
$\varepsilon^{(4)}$	$\lambda^{(0)}$	linear
	$\lambda^{(1)}$	linear
	$\lambda^{(2)}$	quadratic

5. Non-variational wave functions and Lagrangians

We assume that we have an energy function $E(x,\lambda)$ where the parameters λ are determined from the equation

$$e(x,\lambda) = 0 \tag{2.82}$$

The Lagrangian becomes

$$L(x,\lambda,\bar{\lambda}) = E(x,\lambda) + \bar{\lambda}\, e(x,\lambda) \tag{2.83}$$

where $\bar{\lambda}$ are the Lagrange multipliers. The variational conditions are

$$\frac{\partial L(x,\lambda,\bar{\lambda})}{\partial \bar{\lambda}} = e(x,\lambda) = 0 \tag{2.84}$$

$$\frac{\partial L(x,\lambda,\bar{\lambda})}{\partial \lambda} = \frac{\partial E(x,\lambda)}{\partial \lambda} + \bar{\lambda}\frac{\partial e(x,\lambda)}{\partial \lambda} = 0 \tag{2.85}$$

From these expressions we may obtain the response equations and the derivatives in exactly the same way as before. The only difference is that the zero-order Lagrange multipliers $\bar{\lambda}^{(0)}$ are not zero.

Since we may use the same procedure as above we simply give the response equations to zero order

$$e^{(00)} = 0 \tag{2.86}$$

$$E^{(01)} + \bar{\lambda}_*^{(0)} e^{(01)} = 0 \tag{2.87}$$

and first order

$$e^{(10)} + e^{(01)} \lambda_*^{(1)} = 0 \tag{2.88}$$

$$E^{(11)} + E^{(02)} \lambda_*^{(1)} + \bar{\lambda}_*^{(0)} e^{(11)} + \bar{\lambda}_*^{(0)} e^{(02)} \lambda_*^{(1)} + \bar{\lambda}_*^{(1)} e^{(01)} = 0 \tag{2.89}$$

These equations reduce to the response equations for variational wave functions Eqs. (2.62) and (2.63) if we omit all terms containing the constraints. The derivatives may also be obtained as before. Using the 2n+1 rule we obtain

$$\varepsilon^{(0)} = E^{(00)} + \bar{\lambda}_*^{(0)} e^{(00)} \tag{2.90}$$

$$\varepsilon^{(1)} = E^{(10)} + \bar{\lambda}_*^{(0)} e^{(10)} \tag{2.91}$$

$$\begin{aligned} \varepsilon^{(2)} = {} & E^{(20)} + 2E^{(11)} \lambda_*^{(1)} + E^{(02)} \lambda_*^{(1)} \lambda_*^{(1)} \\ & + \bar{\lambda}_*^{(0)} \left[e^{(20)} + 2e^{(11)} \lambda_*^{(1)} + e^{(02)} \lambda_*^{(1)} \lambda_*^{(1)} \right] \\ & + 2\bar{\lambda}_*^{(1)} \left[e^{(10)} + e^{(01)} \lambda_*^{(1)} \right] \end{aligned} \tag{2.92}$$

We note that $\bar{\lambda}_*^{(1)}$ in the expression for the Hessian is multiplied by the first-order response equation of the constraints. It therefore vanishes. This is an example of the 2n+2 rule: *The derivatives of the Lagrange multipliers to order n determine the derivatives of the energy to order 2n+2* (Helgaker and Jørgensen 1989 and Jørgensen and Helgaker 1988). This rule differs from the 2n+1 rule only for derivatives of even orders: For derivatives of order 2n+1 (odd order) we need the response of the variables and the multipliers to order n. For derivatives of order 2n (even order) we need the response of the variables to order n and of the multipliers to order n−1.

To see that we may always eliminate $\bar{\lambda}_*^{(n)}$ from $L^{[2n]}$, we first note from Eq. (2.83) that $L^{[2n]}$ is linear in $\bar{\lambda}_*^{(n)}$. Therefore, *before* the 2n+1 rule is applied the variational condition on $\bar{\lambda}_*^{(n)}$ ensures that $L^{[2n]}$ is independent of $\bar{\lambda}_*^{(n)}$. Next from Eq. (2.45)

$$\frac{\partial^2 L^{[2n]}}{\partial\bar{\lambda}^{(n)}\,\partial\lambda^{(k)}} \equiv 0, \qquad k > n \tag{2.93}$$

we note that there are no terms in $L^{[2n]}$ containing both $\bar{\lambda}_*^{(n)}$ and $\lambda_*^{(k)}$ with k>n. The application of the 2n+1 rule therefore does not affect any terms containing $\bar{\lambda}_*^{(n)}$ and the variational condition on $\bar{\lambda}_*^{(n)}$ is still satisfied *after* the 2n+1 rule has been applied. Therefore, $L^{[2n]}$ is still be independent of $\bar{\lambda}_*^{(n)}$ and $\bar{\lambda}_*^{(n)}$ may be removed. Hence the Lagrange multipliers follow the 2n+2 rule. The reason we cannot remove $\lambda_*^{(n)}$ from $L^{[2n]}$ is that these responses occur quadratically in $L^{[2n]}$.

Applying the 2n+2 rule we obtain the final expressions for differentiated Lagrangians

$$\varepsilon^{(0)} = E^{(00)} \tag{2.94}$$

$$\varepsilon^{(1)} = E^{(10)} + \bar{\lambda}_*^{(0)}\, e^{(10)} \tag{2.95}$$

$$\begin{aligned}\varepsilon^{(2)} &= E^{(20)} + 2E^{(11)}\lambda_*^{(1)} + E^{(02)}\lambda_*^{(1)}\lambda_*^{(1)} \\ &+ \bar{\lambda}_*^{(0)}\left[e^{(20)} + 2e^{(11)}\lambda_*^{(1)} + e^{(02)}\lambda_*^{(1)}\lambda_*^{(1)}\right]\end{aligned} \tag{2.96}$$

Each derivative contains a contribution (capital letters) which is identical to the one that would be obtained if $E(x,\lambda)$ was variational in the parameters λ. The non-variational character of $E(x,\lambda)$ is corrected for by adding the terms containing the multipliers. These terms are similar to the variational expressions for $E(x,\lambda)$ (capital letters replaced by small).

In many cases the wave function is constructed both from variational and non-variational parameters:

$$\lambda^{(1)} = \begin{bmatrix} \lambda_v^{(1)} \\ \lambda_n^{(1)} \end{bmatrix} \tag{2.97}$$

Since the non-variational parameters enter the constraints they are completely determined by Eq. (2.88)

$$e^{(01)} \lambda_{n\,*}^{(1)} = -e^{(10)} \tag{2.98}$$

The variational parameters do not enter the constraints and are instead determined from Eq. (2.89). This equation now simplifies since the last three terms vanish for the variational parameters. Using the notation

$$E^{(11)} = \begin{bmatrix} E_v^{(11)} \\ E_n^{(11)} \end{bmatrix} \tag{2.99}$$

$$E^{(02)} = \begin{bmatrix} E_{vv}^{(02)} & E_{vn}^{(02)} \\ E_{nv}^{(02)} & E_{nn}^{(02)} \end{bmatrix} \tag{2.100}$$

we find that the variational parameters are obtained from the equations

$$E_{vv}^{(02)} \lambda_{v\,*}^{(1)} = -E_v^{(11)} - E_{vn}^{(02)} \lambda_{n\,*}^{(1)} \tag{2.101}$$

Therefore, the response of the non-variational parameters modifies the equations for the variational parameters. The variational parameters depend on the non-variational and are determined from Eq. (2.101) but the non-variational are independent and are determined from Eq. (2.98).

We have previously found that derivatives of variational energies are linear in the errors of the responses $\Delta\lambda^{(k)}$ except that even-order derivatives $\varepsilon^{(2n)}$ depend quadratically on $\Delta\lambda^{(n)}$. In contrast, Lagrangian derivatives depend linearly on all responses including $\Delta\lambda^{(n)}$ in $\varepsilon^{(2n)}$. This follows from the fact that we have eliminated the term containing $\bar{\lambda}_*^{(n)}$ using the n'th order constraint equation which contains $\lambda_*^{(n)}$. If the wave function contains both variational and non-variational parameters, then $\varepsilon^{(2n)}$ depends quadratically on $\Delta\lambda_v^{(n)}$ and linearly on all other parameters.

To summarize, we have found that the 2n+1 rule greatly simplifies the calculation of derivatives for variational wave functions since it reduces the number of response equations needed. We have also seen that the 2n+1 rule may be extended to non-variational wave functions by using Lagrange's method of undetermined multipliers and that the Lagrange multipliers obey the stronger 2n+2 rule. The derivatives depend linearly on the errors in the responses except for derivatives of even order $\varepsilon^{(2n)}$ which depend quadratically on the errors in variational parameters of order n, $\Delta\lambda_v^{(n)}$.

III. DERIVATIVES OF THE HAMILTONIAN

Before considering the derivatives for the various wave function approximations we must determine the derivatives of the Hamiltonian operator. This may appear to be a trivial matter but *complications arise whenever the basis set depends explicitly on the perturbation.* This is true for geometrical perturbations when the atomic orbitals are fixed on the nuclei. Another example of basis sets that depend explicitly on the perturbation are the so-called gauge invariant atomic orbitals (GIAOs) or London orbitals (London 1937) which depend explicitly on the vector potential representing an externally applied magnetic field. Such orbitals are used in the calculation of for example nuclear magnetic shieldings and magnetic susceptibilities to ensure that the calculated results are gauge independent. In this section we consider geometrical perturbations only, but the method may easily be extended to any perturbation that changes the orbital basis.

A. The Hamiltonian at the reference geometry

The basic elements of the second-quantization formalism are the *annihilation and creation operators.* The annihilation operator a_m annihilates an electron in orbital ϕ_m , while the creation operator a_m^+ (the conjugate of a_m) creates an electron in orbital ϕ_m. These operators satisfy the anticommutation relations

$$[a_m, a_n]_+ = 0 \tag{3.1}$$

$$[a_m^+, a_n^+]_+ = 0 \tag{3.2}$$

$$[a_m^+, a_n]_+ = \delta_{mn} \tag{3.3}$$

provided the orbitals ϕ_m are orthonormal. Omitting the nuclear-nuclear repulsion part, the Hamiltonian is

$$H = \sum_{mn} h_{mn} E_{mn} + \frac{1}{2} \sum_{mnpq} g_{mnpq}\, e_{mnpq} \tag{3.4}$$

where

$$E_{mn} = \sum_{\sigma} a_{m\sigma}^+ a_{n\sigma} \tag{3.5}$$

are the generators of the unitary group and

$$e_{mnpq} = E_{mn}E_{pq} - \delta_{pn} E_{mq} \tag{3.6}$$

In Eq. (3.5) the summation is over spin. In atomic units the one- and two-electron integrals are obtained by integrating over electronic coordinates

$$h_{mn} = \left\langle \phi_m(r_1) \middle| h(r_1) \middle| \phi_n(r_1) \right\rangle \tag{3.7}$$

$$g_{mnpq} = \left\langle \phi_m(r_1)\, \phi_p(r_2) \middle| \frac{1}{r_{12}} \middle| \phi_n(r_1)\, \phi_q(r_2) \right\rangle \tag{3.8}$$

where r_1 denotes the coordinates of electron 1 and r_{12} is the distance between electrons 1 and 2. The one-electron operator h(r) is given by

$$h(r) = -\frac{1}{2}\nabla_r^2 - \sum_A \frac{Z_A}{|r - x_A|} \tag{3.9}$$

where Z_A is the charge and x_A the position of nucleus A. In the linear combination of atomic orbitals (LCAO) approach the orbitals ϕ_m are expanded in a set of non-orthogonal atomic orbitals χ_μ fixed on the atomic nuclei:

$$\phi_m = \sum_\mu C_{m\mu}\, \chi_\mu \tag{3.10}$$

The integrals are calculated in terms of the atomic orbitals (AO) and are subsequently transformed to the orthonormal basis. In some cases it may be more efficient to evaluate the expressions in the non-orthogonal AO basis. We return to this problem when we consider the calculation of the individual derivatives. For the time being we assume that the Hamiltonian is expressed in the orthonormal molecular orbital (MO) basis. The second-quantized Hamiltonian Eq. (3.4) is a projection of the full Hamiltonian onto the space spanned by the molecular orbitals ϕ_m , i.e., the space in which the calculations are carried out.

B. Orbital connections

So far we have considered the Hamiltonian at one geometry, as appropriate for single-point calculations of the energy. However, if we wish to calculate the derivatives of the energy with respect to variations in the geometry we must consider the geometry dependence of the Hamiltonian. This introduces certain complications which are treated in the remainder of this section.

The Hamiltonian integrals depend on the molecular geometry in two ways. The first is trivial and arises because the Coulomb interactions between the electrons and the nuclei depend on the geometry, see Eq. (3.9). The second is more complicated and arises because the orbitals are themselves functions of the geometry. The reason for this is that the MOs are expanded in a finite set of AOs fixed on the nuclear centers. A consequence of using a finite set of AOs is that we are faced with a different basis set at each geometry.

To study the geometry dependence of the Hamiltonian in any detail, *we must first find a way to relate or "connect" orbitals at neighboring geometries*. In other words, we must establish a one-to-one correspondence between orbitals at neighboring points. Rules that accomplish this are called *orbital connections* (Helgaker 1986). In the LCAO approach the situation may be pictured as follows. At each geometry we have a set of AOs from which an infinite set of orthogonally equivalent orbital bases can be constructed. As the geometry changes we must pick out exactly one of these orbital bases at each geometry x. In this way an orthogonal orbital connection is established. (A connection is called orthogonal if it preserves orthonormality between the orbitals.)

One may wish to impose an additional requirement on the connection, namely that it is translationally and rotationally invariant. This may seem to be a trivial requirement. However, a connection is conveniently defined in terms of atomic Cartesian displacements rather than in terms of a set of nonredundant internal coordinates. This implies that each molecular geometry may be described in an infinite number of translationally and rotationally equivalent ways. The corresponding connections may be different and therefore not translationally and rotationally invariant. In other words, the orbital basis is not necessarily uniquely determined by the internal coordinates when the connections are defined in terms of Cartesian coordinates. Conversely, a rotationally invariant connection picks up the same basis set regardless of how the rotation is carried out and so the basis is uniquely defined by the internal coordinates. (For a discussion of translationally and rotationally invariant connections see Carlaca and McIver 1986).

In many cases it does not matter what particular orbital basis is chosen at a given geometry as long as the connection is orthogonal. For example, at any geometry an MCSCF wave function can be constructed from an arbitrary orthonormal basis by parametrizing the MCSCF state in terms of exponential orthogonal configuration and orbital transformations of a non-optimized reference state. The parameters of the orthogonal transformations are determined from the variational principle. Usually we change the orbital basis in each iteration of MCSCF wave function optimizations, and the final MCSCF state is expressed in a basis for which the exponential orthogonal transformation corresponds to the identity transformation. In principle, however, we can carry out an MCSCF calculation at any geometry using an arbitrary set of orthonormal orbitals. From this it follows that we are free to choose whatever orthogonal connection we find most convenient.

Again we note that the same remarks apply to any perturbation that affects the basis set. For example, when considering variations in an external magnetic field one must connect orbitals at different values of the field. The only difference is that in this case we must allow for complex orbitals.

C. Geometry dependence of the Hamiltonian

First $x = x_0$ is chosen as the *reference geometry* and a set of orthonormal MOs are determined as linear combinations of AOs:

$$\phi_m(x_0) = \sum_{\mu} C_{m\mu}^{(0)} \chi_\mu(x_0) \tag{3.11}$$

The orthonormality of the MOs implies

$$S_{mn}(x_0) = \left\langle \phi_m(r;\, x_0) \,\middle|\, \phi_n(r;\, x_0) \right\rangle = \sum_{\mu\nu} C_{m\mu}^{(0)*} C_{n\nu}^{(0)} S_{\mu\nu}^{AO}(x_0) = \delta_{mn} \tag{3.12}$$

where integration is over the electronic coordinates r. When the geometry changes the *unmodified molecular orbitals* $\phi_m(x)$ (UMOs) are no longer orthonormal:

$$S_{mn}(x) = \sum_{\mu\nu} C_{m\mu}^{(0)*} C_{n\nu}^{(0)} S_{\mu\nu}^{AO}(x) \neq \delta_{mn} \tag{3.13}$$

Provided the overlap matrix S(x) is nonsingular we can define a set of *orthonormalized molecular orbitals* (OMOs)

$$\psi(x) = \phi(x)\, S^{-1/2}(x)\, U(x) \tag{3.14}$$

where $S^{-1/2}(x)$ ensures that the connection is orthogonal (Helgaker and Almlöf 1984) and the orthogonal matrix U(x) makes the connection general (Helgaker 1986) so that we can pick out any set of orthonormal orbitals. For simplicity we set U(x) equal to the identity matrix since any orthogonal connection can be used, and the OMOs $\psi(x)$ are then defined completely in terms of the unmodified molecular orbitals through the *connection matrix* $S^{-1/2}$.

Using the OMOs we may construct a Hamiltonian matrix which is defined at all geometries (Jørgensen and Simons 1983, Helgaker and Almlöf 1984, Simons *et al.* 1984)

$$H(x) = \sum_{mn} \tilde{h}_{mn}(x)\, E_{mn}(x) + \frac{1}{2} \sum_{mnpq} \tilde{g}_{mnpq}(x)\, e_{mnpq}(x) \tag{3.15}$$

where the geometry dependence of $E_{mn}(x)$ and $e_{mnpq}(x)$ is discussed later. The integrals which appear in this expression are given in terms of the UMOs as

$$\tilde{h}_{mn}(x) = \sum_{m'n'} h_{m'n'}(x) \left[S^{-1/2}(x)\right]_{mm'} \left[S^{-1/2}(x)\right]^*_{nn'} \tag{3.16}$$

$$\tilde{g}_{mnpq}(x) =$$

$$\sum_{m'n'p'q'} g_{m'n'p'q'}(x) \left[S^{-1/2}(x)\right]_{mm'} \left[S^{-1/2}(x)\right]^*_{nn'} \left[S^{-1/2}(x)\right]_{pp'} \left[S^{-1/2}(x)\right]^*_{qq'} \tag{3.17}$$

The UMO integrals in Eqs. (3.16) and (3.17) are given as [compare the expression for the overlap matrix Eq. (3.12)]

$$h_{mn}(x) = \sum_{\mu\nu} C^{(0)*}_{m\mu}\, C^{(0)}_{n\nu}\, h^{AO}_{\mu\nu}(x) \tag{3.18}$$

$$g_{mnpq}(x) = \sum_{\mu\nu\rho\sigma} C^{(0)*}_{m\mu}\, C^{(0)}_{n\nu}\, C^{(0)*}_{p\rho}\, C^{(0)}_{q\sigma}\, g^{AO}_{\mu\nu\rho\sigma}(x) \tag{3.19}$$

Note that in these expressions the geometry dependence is isolated in the atomic integrals.

We are now in a position to discuss the expansion of the Hamiltonian Eq. (3.15) in geometrical distortions. Both the integrals and the operators in this Hamiltonian are geometry dependent. In particular the geometry dependence of the operators $E_{mn}(x)$ and $e_{mnpq}(x)$ arises due to the geometry dependence of the orbitals. However, for the purposes of calculating geometrical derivatives of Born-Oppenheimer adiabatic energies we may treat these operators as geometry independent (Jørgensen and Simons 1983, Helgaker and Almlöf 1984). This can be seen in the following way.

The creation and annihilation operators always appear in expectation and transition densities such as $\langle \mu | a_m^\dagger a_n^\dagger a_q a_p | \nu \rangle$ where $|\mu\rangle$ and $|\nu\rangle$ are configuration state functions (CSFs). The elements of $\langle \mu | a_m^\dagger a_n^\dagger a_q a_p | \nu \rangle$ may therefore be written as a vacuum expectation value of a string of operators such as $\langle \text{vac} | a_{\mu_1} a_{\mu_2} \cdots a_{\mu_N} a_m^\dagger a_n^\dagger a_q a_p a_{\nu_N}^\dagger \cdots a_{\nu_2}^\dagger a_{\nu_1}^\dagger | \text{vac} \rangle$. According to Wick's theorem such vacuum expectation values can be expressed as a sum over all totally contracted terms, each of which depends only on the overlap between the orbitals. Since these are the Kronecker delta for orthonormal orbitals, we see that the vacuum expectation values are geometry independent, provided the orbitals are orthonormal at all geometries. Since we use orthogonal connections this condition is fulfilled, and *we may ignore the geometry dependence of the creation and annihilation operators* in our derivations. This is not the case when we consider non-adiabatic matrix elements.

We may now expand the Hamiltonian Eq. (3.15) in geometrical distortions about the reference geometry

$$H(x_0 + \Delta x) = H^{(0)} + H^{(1)} \Delta x + \frac{1}{2} H^{(2)} \Delta x \, \Delta x + \cdots \tag{3.20}$$

where, following the discussion above, each term has the form

$$H^{(n)} = \sum_{mn} \tilde{h}_{mn}^{(n)} E_{mn} + \frac{1}{2} \sum_{mnpq} \tilde{g}_{mnpq}^{(n)} e_{mnpq} \tag{3.21}$$

The geometry dependence of the Hamiltonian is isolated in the integrals. It only remains to determine the explicit form of these integrals.

D. Derivatives of the integrals

The expressions Eqs. (3.16) and (3.17) for the OMO integrals are not convenient of differentiation. To write these integrals in a more convenient form we first introduce the class of integrals (denoted by caret)

$$\hat{h}_{mn}(t) = \sum_{pq} h_{pq} [\exp(t \ln S)]_{mp} [\exp(t \ln S)]_{nq}^{*} \tag{3.22}$$

which includes as special cases the UMO and OMO integrals

$$\hat{h}_{mn}(0) = \sum_{pq} h_{pq} [S^0]_{mp} [S^0]_{nq}^{*} = h_{mn} \tag{3.23}$$

$$\hat{h}_{mn}(-1/2) = \sum_{pq} h_{pq} [S^{-1/2}]_{mp} [S^{-1/2}]^*_{nq} = \tilde{h}_{mn} \tag{3.24}$$

Although lnS appears in Eq. (3.22) the explicit calculation of the logarithm is avoided as shown below. We now expand the caret integrals Eq. (3.22) in t around 0 and set t = –1/2. Differentiating the caret integrals with respect to t at the expansion point we obtain

$$\left.\frac{d\hat{h}_{mn}}{dt}\right|_{t=0} = \{\ln S, h\}_{mn} \tag{3.25}$$

$$\left.\frac{d^2\hat{h}_{mn}}{dt^2}\right|_{t=0} = \{\ln S, \ln S, h\}_{mn} \tag{3.26}$$

and so on. In Eq. (3.25) we have introduced a brace notation for *one-index transformations*

$$\{T, h\}_{mn} = \sum_{o} \left(T_{mo}\, h_{on} + T^*_{no}\, h_{mo} \right) \tag{3.27}$$

and in Eq. (3.26) we use the following notation for symmetrized multiple one-index transformations

$$\{T_1, T_2, \cdots T_n, h\} = \frac{1}{n!} P(1,2,\cdot\cdot n) \{T_1, \{T_2, \cdots \{T_n, h\}\}\} \tag{3.28}$$

where P(1,2,...n) is the permutation operator. We may now write the caret integrals in terms of UMO one-index transformations

$$\hat{h}(t) = h + \{\ln S, h\}\, t + \tfrac{1}{2}\{\ln S, \ln S, h\}\, t^2 + \cdots \tag{3.29}$$

and setting t = –1/2 we obtain the desired expression for the OMO integrals:

$$\tilde{h} = h - \tfrac{1}{2}\{\ln S, h\} + \tfrac{1}{8}\{\ln S, \ln S, h\} - \cdots \tag{3.30}$$

This expression is equivalent to Eq. (3.16) but more convenient since there are *no references to individual indices*. Before taking derivatives, we note that

$$\ln S(x_0) = [\ln S]^{(0)} = 0 \tag{3.31}$$

at reference geometry. Therefore n'th order derivatives with respect to **x** contain no higher than n'th order one-index transformations:

$$\tilde{h}^{(0)} = h^{(0)} \tag{3.32}$$

$$\tilde{h}^{(1)} = h^{(1)} - \frac{1}{2}\{[\ln S]^{(1)}, h^{(0)}\} \tag{3.33}$$

$$\tilde{h}^{(2)} = h^{(2)} - \tfrac{1}{2}\{[\ln S]^{(2)}, h^{(0)}\} - \{[\ln S]^{(1)}, h^{(1)}\} + \tfrac{1}{4}\{[\ln S]^{(1)}, [\ln S]^{(1)}, h^{(0)}\} \tag{3.34}$$

Note that higher-order terms are constructed from those of lower orders. To arrive at the final expressions we must determine the derivatives of lnS. Expanding lnS in a power series we obtain

$$\ln S = (S-1) - \frac{(S-1)^2}{2} + \frac{(S-1)^3}{3} - \cdots \tag{3.35}$$

which gives us

$$[\ln S]^{(1)} = S^{(1)} \tag{3.36}$$

$$[\ln S]^{(2)} = S^{(2)} - S^{(1)} S^{(1)} \tag{3.37}$$

Inserting these in Eqs. (3.33) and (3.34) above we obtain the final expressions

$$\tilde{h}^{(0)} = h^{(0)} \tag{3.38}$$

$$\tilde{h}^{(1)} = h^{(1)} - \frac{1}{2}\{S^{(1)}, h^{(0)}\} \tag{3.39}$$

$$\tilde{h}^{(2)} = h^{(2)} - \tfrac{1}{2}\{S^{(2)} - S^{(1)}S^{(1)}, h^{(0)}\} - \{S^{(1)}, h^{(1)}\} + \tfrac{1}{4}\{S^{(1)}, S^{(1)}, h^{(0)}\} \tag{3.40}$$

for the derivatives of the OMO integrals in terms of the derivatives of the UMO integrals. In the two-electron case the one-index transformations are given by

$$\{T, g\}_{mnpq} = \sum_{o} \left(T_{mo}\, g_{onpq} + T^{*}_{no}\, g_{mopq} + T_{po}\, g_{mnoq} + T^{*}_{qo}\, g_{mnpo} \right) \tag{3.41}$$

but otherwise the expressions are the same as for the one-electron integrals. We finally note that the UMO integral derivatives are given by

$$S_{mn}^{(n)} = \sum_{\mu\nu} C_{m\mu}^{(0)*} C_{n\nu}^{(0)} \left[S_{\mu\nu}^{AO}\right]^{(n)} \tag{3.42}$$

$$h_{mn}^{(n)} = \sum_{\mu\nu} C_{m\mu}^{(0)*} C_{n\nu}^{(0)} \left[h_{\mu\nu}^{AO}\right]^{(n)} \tag{3.43}$$

$$g_{mnpq}^{(n)} = \sum_{\mu\nu\rho\sigma} C_{m\mu}^{(0)*} C_{n\nu}^{(0)} C_{p\rho}^{(0)*} C_{q\sigma}^{(0)} \left[g_{\mu\nu\rho\sigma}^{AO}\right]^{(n)} \tag{3.44}$$

The calculation of the AO integral derivatives is discussed elsewhere.

There is an alternative way to arrive at expressions for the differentiated integrals. We introduce the operator

$$\nabla_{S^{-1/2}}\{A,\ I\} = \frac{d\{A,\ I\}}{dx} - \frac{1}{2}\left\{\frac{d\ln S}{dx},\ A,\ I\right\} \tag{3.45}$$

where I stands for any UMO integral. By successive application of this operator on the UMO integrals it is possible to generate the Hamiltonian derivatives to any order, for example

$$\tilde{I}^{(1)} = \nabla_{S^{-1/2}} I \Big|_{x=x_0} \tag{3.46}$$

$$\tilde{I}^{(2)} = \nabla_{S^{-1/2}}^{2} I \Big|_{x=x_0} \tag{3.47}$$

as is easily verified. These derivatives are reminiscent of the *covariant derivatives* which appear in tensor theory and more generally in differential geometry. This is no coincidence. The purpose of correction terms in covariant derivatives is always to preserve some property of the space they work on: On Riemannian manifolds the correction terms preserve parallelism between vectors, in gauge theories the corrections preserve the gradient of the wave function under a gauge transformation. As seen from the relation

$$\nabla_{S^{-1/2}} S \Big|_{x=x_0} = 0 \tag{3.48}$$

our differentiation operator preserves orthonormality between orbitals.

The underlying reason for this analogy is that our orbital basis has the structure of a *fiber bundle*. Roughly speaking, a fiber bundle consists of a base space (here the set of nuclear configurations) and a fiber (here the MO vector space) attached to each point on the base space. In fact, Newtonian spacetime has the same structure, time being the base space and space the fiber. Note that there is no natural relation between points on different fibers: In spacetime observers moving with respect to each other will disagree on what constitutes a fixed point. In the same way there is no natural relation between orbitals at different geometries. We have to define such a relation by setting up a connection, and this connection later appears as a correction to the derivatives, ensuring that we pick out the right orbitals as we take the derivatives.

E. Hamiltonian commutators

We will frequently use the effective Hamiltonians

$$J^{(1)} = H^{(1)} + \left[\kappa^{(1)}, H^{(0)} \right] \tag{3.49}$$

$$J^{(2)} = H^{(2)} + 2\left[\kappa^{(1)}, H^{(1)} \right] + \left[\kappa^{(1)}, \kappa^{(1)}, H^{(0)} \right] \tag{3.50}$$

where $\kappa^{(1)}$ is the first-order orbital rotation operator

$$\kappa^{(1)} = \sum_{m>n} \kappa^{(1)}_{mn} [E_{mn} - E_{nm}] \tag{3.51}$$

The $J^{(1)}$ represents the first derivative of the Hamiltonian operator including orbital rotations. The $J^{(2)}$ operator is the second derivative of the Hamiltonian with $\kappa^{(2)}$ omitted since according to the 2n+1 rule $\kappa^{(2)}$ does not contribute to molecular Hessians.

The commutator of the Hamiltonian with the orbital rotation operator may be expressed in terms of one-index transformations of the integrals

$$\left[\kappa^{(1)}, H^{(i)} \right] = \sum_{mn} \left\{ \kappa^{(1)}, \tilde{h}^{(i)} \right\}_{mn} E_{mn} + \frac{1}{2} \sum_{mnpq} \left\{ \kappa^{(1)}, \tilde{g}^{(i)} \right\}_{mnpq} e_{mnpq} \tag{3.52}$$

We may therefore write the effective Hamiltonians

$$J^{(i)} = \sum_{mn} J^{(i)}_{mn} E_{mn} + \frac{1}{2} \sum_{mnpq} j^{(i)}_{mnpq} e_{mnpq} \tag{3.53}$$

where

$$J^{(1)} = \tilde{h}^{(1)} + \{ \kappa^{(1)}, \tilde{h}^{(0)} \} \quad (3.54)$$

$$J^{(2)} = \tilde{h}^{(2)} + 2\{ \kappa^{(1)}, \tilde{h}^{(1)} \} + \{ \kappa^{(1)}, \kappa^{(1)}, \tilde{h}^{(0)} \} \quad (3.55)$$

and similarly for the two-electron integrals. Combining these expressions with the formulas for differentiated OMO integrals we obtain

$$J^{(1)} = h^{(1)} + \left\{ \kappa^{(1)} - \frac{1}{2} S^{(1)}, h^{(0)} \right\} \quad (3.56)$$

$$J^{(2)} = h^{(2)} + \{ 2\kappa^{(1)} - S^{(1)}, h^{(1)} \} - \frac{1}{2}\{ S^{(2)} - S^{(1)}S^{(1)}, h^{(0)} \}$$
$$+ \{ \kappa^{(1)}, \kappa^{(1)}, h^{(0)} \} + \frac{1}{4}\{ S^{(1)}, S^{(1)}, h^{(0)} \} - \{ \kappa^{(1)}, \{ S^{(1)}, h^{(0)} \}\} \quad (3.57)$$

and similarly for the two-electron integrals. Hence we may calculate $J^{(1)}$ and $J^{(2)}$ by a series of one-index transformations using $\kappa^{(1)}$, $S^{(1)}$ and $S^{(2)}$ as transformation matrices.

IV. DERIVATIVES FOR SPECIFIC WAVE FUNCTIONS

We now consider the derivatives for the various wave functions in more detail. In each case we first set up the energy function and if necessary construct the Lagrangian. Next we give the formulas for molecular gradients and Hessians along with the necessary response equations. We start by treating the variational MCSCF wave functions and then go on to consider the non-variational CI, CC and MP2 energies. Computational details and comparisons between the various wave functions are given in the following section (Sec. V).

A. MCSCF derivatives

We write the unperturbed MCSCF state as

$$|MC\rangle = \sum_i C_i^{MC} |\phi_i\rangle \quad (4.1)$$

where C^{MC} are the expansion coefficients of configuration state functions (CSFs) constructed from the MCSCF orbitals. The orthogonal complement states to $|MC\rangle$ may be written

$$|k\rangle = \sum_i C_i^k |\phi_i\rangle \tag{4.2}$$

where the coefficients $C = \{C^{MC}, C^k\}$ form an orthogonal matrix. The MCSCF state at a displaced geometry x may be obtained by an orthogonal transformation of the state at the reference geometry (Dalgaard 1979)

$$|MC(x)\rangle = \exp(-\kappa)\exp(-p)|MC\rangle \tag{4.3}$$

where the state transfer operator

$$p = \sum_{k \neq MC} p_k(|k\rangle\langle MC| - |MC\rangle\langle k|) = \sum_{k \neq MC} p_k(r_k^+ - r_k) \tag{4.4}$$

describes transformations in the configuration space and

$$\kappa = \sum_{m>n} \kappa_{mn}(E_{mn} - E_{nm}) = \sum_{m>n} \kappa_{mn} E_{mn}^- \tag{4.5}$$

describes transformations in the orbital space (Jørgensen and Linderberg 1970, Levy 1970). Only nonredundant orbital rotations are included in κ (Dalgaard and Jørgensen 1978, Hoffmann *et al.* 1984). With the ansatz of Eq. (4.3) we may express an arbitrary state spanned by the orbital and configuration space in terms of a particular set of nonredundant orbital κ_{mn} and state p_k parameters. One of these sets represents the optimized state at the displaced geometry (Jørgensen and Simons 1983, Helgaker and Almlöf 1984).

To obtain a more compact expression we arrange the nonredundant parameters and operators as column vectors

$$\lambda = \begin{bmatrix} \kappa \\ p \end{bmatrix} \tag{4.6}$$

$$\omega = \begin{bmatrix} E^+ - E \\ r^+ - r \end{bmatrix} = \begin{bmatrix} E^- \\ r^- \end{bmatrix} \tag{4.7}$$

where E^+ is the adjoint of E. The wave function Eq. (4.3) at an arbitrary geometry may now be written

$$|MC(x)\rangle = \exp(-\lambda)|MC\rangle \tag{4.8}$$

in terms of the operator

$$\lambda = \sum_i \lambda_i \omega_i \tag{4.9}$$

and the unperturbed state. Note that from the discussion in Section III.C the operators ω_i are geometry independent. All geometry dependence has been isolated in the parameters λ_i.

Combining the wave function Eq. (4.8) and the Hamiltonian Eq. (3.15) the MCSCF energy at an arbitrary geometry may be written

$$\begin{aligned} E(x,\lambda) &= \langle MC(x) | H(x) | MC(x) \rangle \\ &= \langle MC | \exp(\lambda) H(x) \exp(-\lambda) \, | MC \rangle \end{aligned} \tag{4.10}$$

where the parameters λ_i depend implicitly and the Hamiltonian H(x) explicitly on the geometry. Since the MCSCF energy is variational we may simply substitute Eq. (4.10) in the expressions for variational derivatives and response functions of Section II.D. Using Eqs. (2.66) and (2.67) we obtain for the molecular gradient and Hessian

$$\varepsilon_{MC}^{(1)} = \langle MC | H^{(1)} | \, MC \rangle \tag{4.11}$$

$$\begin{aligned} \varepsilon_{MC}^{(2)} = \langle MC | H^{(2)} | \, MC \rangle + 2 \langle MC | [\lambda^{(1)}, H^{(1)}] | \, MC \rangle \\ + \langle MC | [\lambda^{(1)}, \lambda^{(1)}, H^{(0)}] | \, MC \rangle \end{aligned} \tag{4.12}$$

The molecular gradient Eq. (4.11) consists of a single term: the expectation value of the first derivative of the Hamiltonian in accord with the Hellmann-Feynman theorem. The molecular Hessian Eq. (4.12) contains three terms. The first is similar to the molecular gradient and gives the contribution to the Hessian from the unperturbed wave function. The remaining two terms describe the relaxation of the wave function in the perturbed field and has no counterpart in the molecular gradient.

We obtain the first-order response equations by substituting Eq. (4.10) into the general expression Eq. (2.62):

$$\sum_j \langle MC | [\omega_i, \omega_j, H^{(0)}] | \, MC \rangle \, \lambda_j^{(1)} = -\langle MC | [\omega_i, H^{(1)}] | \, MC \rangle \tag{4.13}$$

These equations are linear in the response and except for the differentiated Hamiltonian on the right-hand side are identical to the Newton equations for optimizing the MCSCF wave function.

Hartree-Fock derivatives are obtained from the MCSCF expressions Eqs. (4.11) - (4.13) by omitting the state-transfer operators. We obtain

$$\varepsilon_{HF}^{(1)} = \langle HF | H^{(1)} | HF \rangle \tag{4.14}$$

$$\varepsilon_{HF}^{(2)} = \langle HF | H^{(2)} | HF \rangle + 2\langle HF | [\kappa^{(1)}, H^{(1)}] | HF \rangle + \langle HF | [\kappa^{(1)}, \kappa^{(1)}, H^{(0)}] | HF \rangle \tag{4.15}$$

$$\sum_{tj} \langle HF | [E_{si}^{-}, E_{tj}^{-}, H^{(0)}] | HF \rangle \kappa_{tj}^{(1)} = -\langle HF | [E_{si}^{-}, H^{(1)}] | HF \rangle \tag{4.16}$$

for the SCF molecular gradient, Hessian, and first-order responses. Note that Eq. (4.16) determines only the non-redundant orbital rotations. It differs from the coupled-perturbed Hartree-Fock (CPHF) equations (Bratoz 1958, Gerratt and Mills 1968) which determine the geometry dependence of the canonical orbitals and orbital energies.

Except for the differentiated Hamiltonians all terms in the MCSCF molecular gradient and Hessian are used in second-order optimization of the wave function. It is therefore a relatively simple task to program MCSCF derivatives once the differentiated Hamiltonians have been constructed as described in Section III. We return to the calculation of MCSCF derivatives later.

The expression for HF gradients was first derived by Bratoz (1958) and the first practical implementation was reported by Pulay (1969, 1970). HF Hessians were first implemented by Thomsen and Swanstrøm (1973) based on the work by Bratoz (1958) and Gerratt and Mills (1968) The first practical implementation was reported by Pople *et al.* (1979).

The MCSCF gradient was first derived by Pulay (1977). The MCSCF Hessian and first anharmonicity expressions were derived by Pulay (1983), Jørgensen and Simons (1983) and Simons and Jørgensen (1983). The MCSCF first order response equations were first derived by Dalgaard and Jørgensen (1978). For geometrical perturbations these equations were derived by Osamura *et al.* (1984a). The MCSCF gradient was first implemented by Goddard *et al.* (1979) and Kato and Morokuma (1979). The first implementation of general MCSCF hessians were reported by Hoffmann *et al.* (1984) and shortly after by Page *et al.* (1984a). A program capable of calculating MCSCF Hessians for large configuration spaces has been reported by Helgaker *et al.* (1986).

B. CI derivatives

The calculation of CI wave functions proceeds in two steps. First an MCSCF calculation is carried out to determine the molecular orbitals. Next the CI energy is optimized with respect to the configuration coefficients, keeping the orbitals fixed. We therefore parametrize the CI wave function as

$$|\,\mathrm{CI(x)}\,\rangle = \exp(-\kappa)\exp(-P)\,|\,\mathrm{CI}\,\rangle \tag{4.17}$$

where the orbital rotation parameters fulfil the Brillouin condition[4]

$$\langle \mathrm{MC}\,|\,[\omega_i, \exp(\lambda)H(x)\exp(-\lambda)]\,|\,\mathrm{MC}\,\rangle = 0 \tag{4.18}$$

or the equivalent SCF condition. The state transfer operator is

$$P = \sum_{K \neq CI} P_K(|K\rangle\langle CI| - |CI\rangle\langle K|) = \sum_{K \neq CI} P_K(R_K^\dagger - R_K) \tag{4.19}$$

using capital letters to distinguish it from the MCSCF counterpart Eq. (4.4). The parameters in Eq. (4.19) fulfil the CI variational condition

$$\frac{\partial E_{CI}}{\partial P} = 0 \tag{4.20}$$

Again we use a more compact notation

$$\Lambda = \sum_i \Lambda_i\, \Omega_i \tag{4.21}$$

where

$$\Lambda = \begin{bmatrix} \kappa \\ P \end{bmatrix} \tag{4.22}$$

$$\Omega = \begin{bmatrix} E^+ - E \\ R^+ - R \end{bmatrix} = \begin{bmatrix} E^- \\ R^- \end{bmatrix} \tag{4.23}$$

[4]This condition is equivalent to the requirement $\partial E_{MC}/\partial\lambda = 0$.

In this notation the CI wave function Eq. (4.17) at an arbitrary geometry may be written

$$|\,CI(x)\,\rangle = \exp(-\Lambda)|\,CI\,\rangle \tag{4.24}$$

where the geometry dependence is isolated in the parameters κ and P which fulfil Eqs. (4.18) and (4.20).

The energy of the MRCI state at the displaced geometry is therefore

$$E_{CI}(x,\Lambda) = \langle\,CI\,|\,\exp(\Lambda)H(x)\,\exp(-\Lambda)\,|\,CI\,\rangle \tag{4.25}$$

Since this energy is not variational with respect to orbital rotations we construct the variational Lagrangian

$$\begin{aligned} L_{CI}(x,P,\lambda,\bar{\lambda}) &= \langle\,CI\,|\,\exp(\Lambda)H(x)\exp(-\Lambda)\,|\,CI\,\rangle \\ &+ \sum_i \bar{\lambda}_i \langle\,MC\,|\,[\omega_i,\,\exp(\lambda)H(x)\exp(-\lambda)]\,|\,MC\,\rangle \end{aligned} \tag{4.26}$$

There is one constraint and therefore one multiplier for each MCSCF parameter. Since only the orbital rotations of the MCSCF parameters enter the non-variational CI energy the Lagrangian Eq. (4.26) has in fact more multipliers than there are constrained variables in the original CI energy Eq. (4.25).

It is now straightforward to derive the expressions for the CI molecular gradient and Hessian. Substituting Eq. (4.26) into the general expressions for variational Lagrangians Eqs. (2.95)-(2.96) we obtain

$$\varepsilon_{CI}^{(1)} = \langle\,CI\,|\,H^{(1)}\,|\,CI\,\rangle + \langle\,MC\,|\,[\bar{\lambda}^{(0)},\,H^{(1)}]\,|\,MC\,\rangle \tag{4.27}$$

$$\begin{aligned} \varepsilon_{CI}^{(2)} &= \langle\,CI\,|\,H^{(2)}\,|\,CI\,\rangle + \langle\,MC\,|\,[\bar{\lambda}^{(0)},\,H^{(2)}]\,|\,MC\,\rangle \\ &+ 2\langle\,CI\,|\,[\Lambda^{(1)},\,H^{(1)}]\,|\,CI\,\rangle + 2\langle\,MC\,|\,[\bar{\lambda}^{(0)},\,\lambda^{(1)},\,H^{(1)}]\,|\,MC\,\rangle \\ &+ \langle\,CI\,|\,[\Lambda^{(1)},\,\Lambda^{(1)},\,H^{(0)}]\,|\,CI\,\rangle + \langle\,MC\,|\,[\bar{\lambda}^{(0)},\,\lambda^{(1)},\,\lambda^{(1)},\,H^{(0)}]\,|\,MC\,\rangle \end{aligned} \tag{4.28}$$

The CI molecular gradient Eq. (4.27) consists of two terms. The first is similar to the MCSCF gradient, the second is a correction term which arises from the constraints. We shall later see how the CI gradient may be calculated in much the same way as the MCSCF gradient by combining the two terms of Eq. (4.27) in a special way. The CI molecular Hessian Eq.

(4.28) contains six terms. The first three are similar to the MCSCF molecular Hessian and the last three arise from the constraints.

If Hartree-Fock orbitals are used the molecular Hessian Eq. (4.28) may be written in the form

$$\varepsilon^{(2)}_{HFCI} = \langle CI | J^{(2)} | CI \rangle + \langle HF | [\bar{\kappa}^{(0)}, J^{(2)}] | HF \rangle + 2\langle CI | [P^{(1)}, J^{(1)}] | CI \rangle + \langle CI | [P^{(1)}, P^{(1)}, H^{(0)}] | CI \rangle \tag{4.29}$$

using the effective Hamiltonians

$$J^{(1)} = H^{(1)} + [\kappa^{(1)}, H^{(0)}] \tag{4.30}$$

$$J^{(2)} = H^{(2)} + 2[\kappa^{(1)}, H^{(1)}] + [\kappa^{(1)}, \kappa^{(1)}, H^{(0)}] \tag{4.31}$$

discussed in Section III.E. Inserting Eq. (4.19) into Eq. (4.29) we obtain

$$\varepsilon^{(2)}_{HFCI} = \langle CI | J^{(2)} | CI \rangle + \langle HF | [\bar{\kappa}^{(0)}, J^{(2)}] | HF \rangle - 4\langle P^{(1)} | J^{(1)} | CI \rangle + 2\langle P^{(1)} | H^{(0)} | P^{(1)} \rangle - 2\varepsilon^{(0)}_{HFCI} | P^{(1)} |^2 \tag{4.32}$$

using the notation

$$| P^{(1)} \rangle = \sum_{K \neq CI} P^{(1)}_{K} | K \rangle \tag{4.33}$$

The corresponding expression for the MRCI Hessians is slightly more involved since it has more terms containing Lagrange multipliers.

The zero-order Lagrange multipliers are obtained by substituting the Lagrangian Eq. (4.26) into the general expression Eq. (2.87):

$$\langle MC | [\omega, \omega, H^{(0)}] | MC \rangle \bar{\lambda}^{(0)} = -\begin{bmatrix} \langle CI | [E^{-}, H^{(0)}] | CI \rangle \\ 0 \end{bmatrix} \tag{4.34}$$

The zero elements on the right-hand side of Eq. (4.34) occur since the CI energy is independent of the MCSCF state transfer operators. The first-order response equations may be obtained from Eqs. (2.88) and (2.89) but since the CI energy contains variational as well as non-variational parameters we may use the simpler expressions Eqs. (2.98) and (2.101). We obtain the following set of equations

$$\langle MC|[\omega, \omega, H^{(0)}]|MC\rangle \lambda^{(1)} = -\langle MC|[\omega, H^{(1)}]|MC\rangle \quad (4.35)$$

$$\langle CI|[R^-, R^-, H^{(0)}]|CI\rangle P^{(1)} = -\langle CI|[R^-, H^{(1)}]|CI\rangle - \langle CI|[R^-, \kappa^{(1)}, H^{(0)}]|CI\rangle \quad (4.36)$$

which determine the orbital and configuration responses. Equation (4.35) is simply the MCSCF response equations Eq. (4.13). After some algebra the equations for the state transfer parameters Eq. (4.36) reduce to the expression

$$\sum_L [\langle K|H^{(0)}|L\rangle - \delta_{KL}\langle CI|H^{(0)}|CI\rangle] P_L^{(1)} = \langle K|J^{(1)}|CI\rangle \quad (4.37)$$

The responses Eqs. (4.35) - (4.37) are needed for the molecular Hessian only.

The CI gradient expressions were derived by Krishnan *et al.* (1980) and Brooks *et al.* (1980) The generalization to MRCI is due to Osamura *et al.* (1981, 1982a, b) The Hessian expression was derived by Jørgensen and Simons (1983) and implemented by Fox *et al.* (1983). A more efficient implementation has been reported by Lee *et al.* (1986). The first order CI response equations were derived by Jørgensen and Simons (1983) and reformulated by Fox *et al.* (1983).

C. CC derivatives

The coupled cluster energy at an arbitrary geometry is given by

$$E_{CC}(x,t,\kappa) = \langle HF|\exp(-T)\exp(\kappa)H(x)\exp(-\kappa)\exp(T)|HF\rangle = \langle HF|\exp(\kappa)H(x)\exp(-\kappa)|CC\rangle \quad (4.38)$$

where κ is the usual orbital rotation operator Eq. (4.5) and T the cluster operator

$$T = T_1 + T_2 + \cdots \quad (4.39)$$

$$T_1 = \sum_{si} t_i^s a_s^+ a_i \quad (4.40)$$

$$T_2 = \sum_{\substack{st \\ ij}} t_{ij}^{st} a_s^+ a_i a_t^+ a_j \tag{4.41}$$

Using a shorthand notation we write

$$T = \sum_{\mu} t_\mu \tau_\mu \tag{4.42}$$

where t_μ are the amplitudes and τ_μ the operators. The amplitudes Eq. (4.39) are obtained from the coupled cluster equations

$$\langle \mu | \exp(-T) \exp(\kappa) H(x) \exp(-\kappa) \exp(T) | HF \rangle = 0 \tag{4.43}$$

where

$$\langle \mu | = \langle HF | \tau_\mu^\dagger \tag{4.44}$$

and the orbital rotations fulfil the Brillouin condition

$$\left\langle HF \middle| \left[E_{si}^-, \exp(\kappa) H(x) \exp(-\kappa)\right] \middle| HF \right\rangle = 0 \tag{4.45}$$

The CC Lagrangian is therefore

$$\begin{aligned} L_{CC}(x,t,\kappa) = {} & \langle HF | \exp(-T) \exp(\kappa) H(x) \exp(-\kappa) \exp(T) | HF \rangle \\ & + \sum_{\mu} \bar{t}_\mu \langle \mu | \exp(-T) \exp(\kappa) H(x) \exp(-\kappa) \exp(T) | HF \rangle \\ & + \sum_{si} \bar{\kappa}_{si} \left\langle HF \middle| \left[E_{si}^-, \exp(\kappa) H(x) \exp(-\kappa)\right] \middle| HF \right\rangle \end{aligned} \tag{4.46}$$

and we obtain the molecular derivatives from the general expressions for variational Lagrangians in Section II.D in the usual way. We find

$$\varepsilon_{CC}^{(1)} = \langle HF\bar{t} | H^{(1)} | CC \rangle + \langle HF | [\bar{\kappa}^{(0)}, H^{(1)}] | HF \rangle \tag{4.47}$$

$$\begin{aligned} \varepsilon_{CC}^{(2)} = {} & \langle HF\bar{t} | J^{(2)} | CC \rangle + \langle HF | [\bar{\kappa}^{(0)}, J^{(2)}] | HF \rangle \\ & - 2\langle HF\bar{t} | [T^{(1)}, J^{(1)}] | CC \rangle + \langle HF\bar{t} | [T^{(1)}, T^{(1)}, H^{(0)}] | CC \rangle \end{aligned} \tag{4.48}$$

using the notation

$$\langle \mathrm{HF\bar{t}}| = \langle \mathrm{HF}| + \sum_{\mu} \bar{t}_{\mu}^{(0)} \langle \mu | \exp[-T^{(0)}] \tag{4.49}$$

to simplify the final expressions.

Again we note that the molecular gradient Eq. (4.47) may be calculated from the unperturbed wave function after the zero-order Lagrange multipliers have been obtained. The amplitude multipliers have been absorbed in the state $\langle \mathrm{HF\bar{t}}|$ Eq. (4.49) and the orbital multipliers enter the molecular gradient Eq. (4.47) in a commutator with the differentiated Hamiltonian. The molecular Hessian Eq. (4.49) contains as always static contributions and relaxation terms. The orbital relaxation has been absorbed in the effective Hamiltonian operators $J^{(1)}$ and $J^{(2)}$ Eqs. (3.49) and (3.50) for convenience. Note the similarity between Eq. (4.49) and the corresponding MCSCF and CI expressions in Sections IV.A and IV.B.

The zero-order response equations are obtained from Eq. (2.87):

$$\sum_{\nu} \bar{t}_{\nu}^{(0)} \langle \nu | \exp[-T^{(0)}] \left[\tau_{\mu}, H^{(0)} \right] | \mathrm{CC} \rangle = -\langle \mathrm{HF} | [\tau_{\mu}, H^{(0)}] | \mathrm{CC} \rangle \tag{4.50}$$

$$\sum_{si} \left\langle \mathrm{HF} \left| \left[E_{si}^{-}, \left[E_{tj}^{-}, H^{(0)} \right] \right] \right| \mathrm{HF} \right\rangle \bar{\kappa}_{tj}^{(0)} = -\left\langle \mathrm{HF\bar{t}} \left| \left[E_{si}^{-}, H^{(0)} \right] \right| \mathrm{CC} \right\rangle \tag{4.51}$$

In the equations for the amplitude multipliers Eq. (4.50) the coefficients are the CC Jacobian matrix used for Newton optimization of the wave function. In terms of the state $\langle \mathrm{HF\bar{t}}|$ we may write Eq. (4.50) as

$$\left\langle \mathrm{HF\bar{t}} \left| \left[\tau_{\mu}, H^{(0)} \right] \right| \mathrm{CC} \right\rangle = 0 \tag{4.52}$$

which has a structure similar the Brillouin theorem. The linear equations for the orbital multipliers Eq. (4.51) are identical to the Hartree-Fock response equations Eq. (4.16) except for the right-hand side.

Inserting the Lagrangian Eq. (4.46) into Eq. (2.57) we obtain the first-order SCF response equations Eq. (4.16) and the first-order amplitudes:

$$\sum_{\nu} \langle \mu | \exp[-T^{(0)}] \left[\tau_{\nu}, H^{(0)} \right] | \mathrm{CC} \rangle \, t_{\nu}^{(1)} = -\langle \mu | \exp[-T^{(0)}] J^{(1)} | \mathrm{CC} \rangle \tag{4.53}$$

As in Eq. (4.50) the coefficient matrix is the Jacobian but this time it is multiplied from the the right. These equations are identical to the CC

Newton equations except that $J^{(1)}$ replaces $H^{(0)}$ on the right-hand side of Eq. (4.53).

The CC molecular gradient and Hessian expressions were derived by Jørgensen and Simons (1983). Simplified expressions for the gradient were obtained by Adamowicz *et al.* (1984) and Bartlett (1986). CC molecular gradients have been reported by Fitzgerald *et al.* (1985) and Lee *et al.* (1987). CC Hessian implementations have been reported by Koch *et al.* (1990).

D. Møller-Plesset derivatives

We write the second-order Møller-Plesset energy (strictly speaking the second-order correction to the Hartree-Fock energy) as

$$E_{MP2}(x) = \langle HF(x) | H(x) | t(x) \rangle \tag{4.54}$$

where

$$| t(x) \rangle = \sum_{\mu} t_{\mu} | \mu(x) \rangle \tag{4.55}$$

is a linear combination of doubly excited states whose amplitudes satisfy the equation

$$\begin{aligned} \sum_{\nu} \langle \mu(x) | G(x) | \nu(x) \rangle t_{\nu} &= -\langle \mu(x) | H(x) - {}^{i}F(x) | HF(x) \rangle \\ &= -\langle \mu(x) | H(x) | HF(x) \rangle \end{aligned} \tag{4.56}$$

$$\langle \mu(x) | G(x) | \nu(x) \rangle = \langle \mu(x) | {}^{i}F(x) | \nu(x) \rangle - \delta_{\mu\nu} \langle HF(x) | {}^{i}F(x) | HF(x) \rangle \tag{4.57}$$

We have here introduced the inactive Fock operator

$${}^{i}F(x) = \sum_{mn} {}^{i}F_{mn}(x)\, E_{mn}(x) \tag{4.58}$$

$${}^{i}F_{mn}(x) = \frac{1}{2} \sum_{\sigma} \langle HF(x) | [a^{+}_{m\sigma}(x), [a_{n\sigma}(x), H(x)]]_{+} | HF(x) \rangle \tag{4.59}$$

which is the zero-order Hamiltonian in Møller-Plesset theory. The fluctuation operator $H(x) - {}^{i}F(x)$ in Eq. (4.56) is the perturbation operator. The annihilation and creation operators in Eqs. (4.58) and (4.59) are geometry dependent

$$a^{+}_{m\sigma}(x) = \exp(-\kappa)\, a^{+}_{m\sigma} \exp(\kappa) \tag{4.60}$$

since the Fock matrix is calculated from the orbitals at geometry x.

The above expression for the Møller-Plesset energy Eq. (4.54) does not require the Fock matrix Eq. (4.59) to be diagonal. However, at the unperturbed geometry the Fock matrix is usually assumed diagonal (canonical orbitals)

$$^{i}F_{mn}(x_o) = h_{mn} + \sum_{i} (2g_{mnii} - g_{miin}) = \delta_{mn}\, \varepsilon_m \tag{4.61}$$

If we now use the biorthonormal representation of the excited states

$$\left| {}^{st}_{ij} \right\rangle = \frac{1}{3}\left(1 + \delta_{st}\delta_{ij}\right)^{-1} \left(2\, E_{si}\, E_{tj} + E_{sj}E_{ti}\right)\left| HF \right\rangle \quad (si \geq tj) \tag{4.62}$$

$$\left\langle {}^{st}_{ij} \right| = \frac{1}{2}\left\langle HF \right| E_{is}E_{jt} \quad (si \geq tj) \tag{4.63}$$

we find that the matrix in Eq. (4.56) is diagonal

$$\left\langle {}^{st}_{ij} \right| G \left| {}^{st}_{ij} \right\rangle = \varepsilon_s - \varepsilon_i + \varepsilon_t - \varepsilon_j \tag{4.64}$$

and that the amplitudes are given by

$$t_{isjt} = \frac{k_{isjt}}{\varepsilon_i - \varepsilon_s + \varepsilon_j - \varepsilon_t} \tag{4.65}$$

$$k_{mnpq} = 2\, g_{mnpq} - g_{mqpn} \tag{4.66}$$

The Møller-Plesset energy correction now reduces to the well-known expression

$$\varepsilon^{(0)}_{MP2} = \sum_{isjt} \frac{g_{isjt}\, k_{isjt}}{\varepsilon_i - \varepsilon_s + \varepsilon_j - \varepsilon_t} \tag{4.67}$$

in terms of integrals and orbital energies.

At displaced geometries we do not insist on the canonical representation since degeneracies among the occupied or virtual orbitals may lead to instabilities in the derivatives. Instead we require the orbitals to fulfil the Brillouin condition at displaced geometries:

$$\left\langle \mathrm{HF}(x) \middle| \left[E_{si}^{-}(x), H(x) \right] \middle| \mathrm{HF}(x) \right\rangle = 0 \tag{4.68}$$

We may now construct the Lagrangian, taking Eq. (4.54) as the energy and Eqs. (4.56) and (4.68) as the constraints on the amplitudes and the orbitals. We obtain

$$\begin{aligned} L_{MP2}(x) = {} & \langle \mathrm{HF} | \exp(\kappa) H(x) \exp(-\kappa) | t \rangle \\ & + \langle \bar{t} | \exp(\kappa)\, G(x) \exp(-\kappa) | t \rangle + \langle \bar{t} | \exp(\kappa) H(x) \exp(-\kappa) | \mathrm{HF} \rangle \\ & + \sum_{si} \bar{\kappa}_{si} \left\langle \mathrm{HF} \middle| \left[E_{si}^{-}, \exp(\kappa)\, H(x) \exp(-\kappa) \right] \middle| \mathrm{HF} \right\rangle \end{aligned} \tag{4.69}$$

where

$$\langle \bar{t} | = \sum_{\mu} \bar{t}_{\mu} \langle \mu | \tag{4.70}$$

from which we may obtain molecular derivatives and response equations in the usual way. Differentiating Eq. (4.69) with respect to t_μ we find that

$$\sum_{\nu} \bar{t}_{\nu} \langle \nu(x) | G(x) | \mu(x) \rangle = -\langle \mathrm{HF}(x) | H(x) | \mu(x) \rangle \tag{4.71}$$

and we may also calculate the energy according to the expression

$$E_{MP2}(x) = \langle \bar{t}(x) | H(x) | \mathrm{HF}(x) \rangle \tag{4.72}$$

Equation (4.71) shows that the Lagrange multipliers are related to the amplitudes in a simple fashion and in the canonical representation we find that

$$\bar{t}_{isjt} = \frac{2}{1 + \delta_{st}\delta_{ij}} \frac{g_{isjt}}{\varepsilon_i - \varepsilon_s + \varepsilon_j - \varepsilon_t} \tag{4.73}$$

which should be compared with Eq. (4.65). Also note the symmetric way in which the amplitudes and their multipliers enter the Lagrangian Eq.

(4.69). In fact, except for the orbital multiplier term this is the Hylleraas function.

Using the Lagrangian Eq. (4.69) in Eq. (2.87) we obtain the zero-order multipliers for amplitudes and orbitals

$$\sum_{\nu} \bar{t}_{\nu}^{(0)} \langle \nu | G^{(0)} | \mu \rangle = -\langle HF | H^{(0)} | \mu \rangle \tag{4.74}$$

$$\sum_{tj} \langle HF | [E_{si}^{-}, [E_{tj}^{-}, H^{(0)}]] | HF \rangle \bar{\kappa}_{tj}^{(0)} = -2 \langle HF | [E_{si}^{-}, H^{(0)}] | t^{(0)} \rangle - \langle \bar{t}^{(0)} | G[[E_{si}^{-}, H^{(0)}]] | t^{(0)} \rangle \tag{4.75}$$

where G[A] means that the matrix elements are calculated from the Hamiltonian A rather than H. The first-order response equations Eq. (2.57) give us

$$\sum_{tj} \langle HF | [E_{si}^{-}, [E_{tj}^{-}, H^{(0)}]] | HF \rangle \kappa_{tj}^{(1)} = -\langle HF | [E_{si}^{-}, H^{(1)}] | HF \rangle \tag{4.76}$$

$$\sum_{\nu} \langle \mu | G^{(0)} | \nu \rangle t_{\nu}^{(1)} = - \langle \mu | J^{(1)} | HF \rangle - \langle \mu | G[J^{(1)}] | t^{(0)} \rangle \tag{4.77}$$

$$\sum_{\nu} \bar{t}_{\nu}^{(1)} \langle \nu | G^{(0)} | \mu \rangle = - \langle HF | J^{(1)} | \mu \rangle - \langle \bar{t}^{(0)} | G[J^{(1)}] | \mu \rangle \tag{4.78}$$

We have here given the equations for both the amplitudes and their multipliers since they are closely related. The expressions for the molecular gradient and Hessian Eqs. (2.95) and (2.96) now become

$$\varepsilon_{MP2}^{(1)} = 2 \langle HF | H^{(1)} | t^{(0)} \rangle + \langle \bar{t}^{(0)} | G[H^{(1)}] | t^{(0)} \rangle + \langle HF | [\bar{\kappa}^{(0)}, H^{(1)}] | HF \rangle \tag{4.79}$$

$$\varepsilon_{MP2}^{(2)} = 2 \langle HF | J^{(2)} | t^{(0)} \rangle + \langle \bar{t}^{(0)} | G[J^{(2)}] | t^{(0)} \rangle + \langle HF | [\bar{\kappa}^{(0)}, J^{(2)}] | HF \rangle + 2 \langle \bar{t}^{(1)} | G^{(0)} | t^{(1)} \rangle \tag{4.80}$$

We have used the first-order response of the amplitude multipliers Eq. (4.78) as well as the amplitudes themselves Eq. (4.76) in the Hessian Eq. (4.80) since it gives a more symmetric expression. The multipliers can easily be eliminated according to the 2n+2 rule but this does not lead to computational simplifications since the amplitudes and their multipliers are trivially related.

The equations may be simplified slightly if we introduce the density matrix

$$D^{MP}_{mn} = \langle \bar{t}^{(0)} | E_{mn} | t^{(0)} \rangle - \langle \bar{t}^{(0)} | t^{(0)} \rangle \langle HF | E_{mn} | HF \rangle \tag{4.81}$$

whose only non-vanishing blocks are the occupied-occupied and virtual-virtual blocks and the effective densities

$$\rho_{mn} = D^{MP}_{mn} + \frac{1}{2}\begin{bmatrix} 0 & \bar{\kappa}_{si} \\ \bar{\kappa}_{si} & 0 \end{bmatrix} \tag{4.82}$$

which contain the orbital multipliers in the off-diagonal blocks. The orbital multiplier equations Eq. (4.75) now become

$$\sum_{tj} \left\langle HF \left| \left[E^-_{si}, \left[E^-_{tj}, H^{(0)} \right] \right] \right| HF \right\rangle \bar{\kappa}^{(0)}_{tj} = -2 \left\langle HF \left| \left[E^-_{si}, H^{(0)} \right] \right| t^{(0)} \right\rangle - \mathrm{Tr}\, D^{MP}\, {}^{i}F\left\{ \left[E^-_{si}, H^{(0)} \right] \right\} \tag{4.83}$$

and the molecular gradient Eq. (4.79) and Hessian Eq. (4.80) may be written

$$\varepsilon^{(1)}_{MP2} = 2 \langle HF | H^{(1)} | t^{(0)} \rangle + \mathrm{Tr}\rho\, {}^{i}F[H^{(1)}] \tag{4.84}$$

$$\varepsilon^{(2)}_{MP2} = 2 \langle HF | J^{(2)} | t^{(0)} \rangle + \mathrm{Tr}\rho\, {}^{i}F[J^{(2)}] + 2 \langle \bar{t}^{(1)} | G^{(0)} | t^{(1)} \rangle \tag{4.85}$$

where ${}^{i}F[A]$ is the inactive Fock matrix obtained by replacing H by A. Again we note the similarity between the expressions for the molecular gradient Eq. (4.84) and Hessian Eq. (4.85). In fact, the first two contributions to the Hessian are identical to the gradient except that $H^{(1)}$ is replaced by $J^{(2)}$. The last term contains the relaxation of the amplitudes. The orbital relaxation is incorporated into the first two terms in the effective operator $J^{(2)}$.

The second order Møller-Plesset gradient expressions were derived and implemented by Pople *et al.* (1979). Second derivative expressions were given by Jørgensen and Simons (1983). Handy *et al.* (1985, 1986) and

Harrison *et al.* (1986) simplified the gradient and Hessian expressions and reported implementations of these expressions.

V. CALCULATION OF DERIVATIVES

In Section IV we derived expressions for the molecular gradient and Hessian for a variety of wave functions. We have seen that the derivatives are calculated mostly from the same ingredients that we use for optimizing the wave function and calculating the energy. In many cases the only new ingredient is the differentiated Hamiltonian discussed in Section III. We may then obtain the derivatives simply by plugging these Hamiltonians into the derivative expressions of Section IV. Although this gives working programs with little effort, such programs are not efficient. For example, our expressions for the molecular gradient require the construction of the differentiated Hamiltonian in the MO basis. This is expensive but may be avoided as discussed in this section, the purpose of which is to consider the expressions in Section IV from a computational point of view.

A. Lagrange multipliers

The general equation for zero-order Lagrange multipliers Eq. (2.87) is

$$\bar{\lambda}^{(0)} e^{(01)} = - E^{(01)} \tag{5.1}$$

The coefficient matrix $e^{(01)}$ is the Jacobian of the constraint equations, i.e. the derivatives of the constraints with respect to the electronic parameters λ. If the constraints correspond to the electronic gradient, then the Jacobian becomes symmetric and reduces to the electronic Hessian. The right-hand side of Eq. (5.1) is simply the electronic gradient of the non-variational energy.

In the CI case [Eq. (4.34)] the coefficient matrix is the MCSCF electronic Hessian and the right hand side is the gradient of the CI energy with respect to the constrained orbitals:

$$\langle MC | [\omega, \omega, H^{(0)}] | MC \rangle \bar{\lambda}^{(0)} = - \begin{bmatrix} \langle CI | [E^{-}, H^{(0)}] | CI \rangle \\ 0 \end{bmatrix} \tag{5.2}$$

Note that the number of MCSCF constraints is larger than the number of constrained variables in the CI energy. Therefore, some of the elements on the right-hand side are zero. If Hartree-Fock orbitals are used no zero elements occur

$$\langle HF|[E^-, E^-, H^{(0)}]|\, HF\rangle\, \bar{\kappa}^{(0)} = -\langle CI|[E^-, H^{(0)}]|\, CI\rangle \tag{5.3}$$

since the number of constraints now equals the number of constrained variables in the CI wave function.

For coupled cluster functions the coefficient matrix for the amplitude and orbital multipliers are the CC Jacobian (non-symmetric) [Eq. (4.50)] and SCF electronic Hessian (symmetric) [Eq. (4.51)], respectively:

$$\sum_{\nu} \bar{t}_{\nu}^{(0)} \langle \nu|\, \exp[-T^{(0)}]\big[\, \tau_{\mu}, H^{(0)}\,\big]\big|\, CC\big\rangle = -\big\langle HF\big|\big[\tau_{\mu}, H^{(0)}\big]\big|\, CC\big\rangle \tag{5.4}$$

$$\sum_{tj} \Big\langle HF\, \Big|\Big[E_{si}^{-}, \big[\, E_{tj}^{-}, H^{(0)}\big]\Big]\Big|\, HF\Big\rangle\, \bar{\kappa}_{tj}^{(0)} = -\Big\langle HF\bar{t}\, \Big|\big[E_{si}^{-}, H^{(0)}\big]\Big|\, CC\Big\rangle \tag{5.5}$$

The right hand side of Eq. (5.4) is the gradient of the CC energy with respect to the amplitudes. Note, however, that the right-hand side of the orbital equations Eq. (5.5) is not simply the gradient of the CC energy with respect to orbital rotations as we would expect from the general equation Eq. (5.1). To understand this note that the general equations Eq. (2.87) are a set of equations for *all* multipliers, the amplitude multipliers as well as the orbital multipliers. When the CC Lagrangian Eq. (4.46) is used, these equations reduce to two smaller sets of equations, one for the amplitudes Eq. (5.4) and one for the rotations Eq. (5.5). The amplitudes are calculated first and are then used to construct the right-hand side of the orbital equations. In other words, we first make the energy variational with respect to the amplitudes and then with respect to the orbitals. This is possible since the orbitals are independent of the amplitudes so that when the orbital corrections are added they do not affect variations with respect to the amplitudes.

As expected the Møller-Plesset multiplier equations Eqs. (4.74) and (4.83) appear as simplified versions of the coupled cluster equations:

$$\sum_{\nu} \bar{t}_{\nu}^{(0)} \langle \nu|\, G^{(0)}\, |\mu\rangle = -\langle HF|\, H^{(0)}\, |\mu\rangle \tag{5.6}$$

$$\sum_{tj} \left\langle HF \left| \left[E^-_{si}, \left[E^-_{tj}, H^{(0)} \right] \right] \right| HF \right\rangle \bar{\kappa}^{(0)}_{tj} = -2 \left\langle HF \left| \left[E^-_{si}, H^{(0)} \right] \right| t^{(0)} \right\rangle$$

$$- \operatorname{Tr} D^{MP} F\left\{ \left[E^-_{si}, H^{(0)} \right] \right\} \quad (5.7)$$

The amplitude equations Eq. (5.6) are now trivial since the Jacobian is not only symmetric but also diagonal if we use canonical orbitals. The right-hand side of the orbital equations Eq. (5.7) is not simply the gradient of the MP2 energy with respect to the orbitals but contains corrections from the amplitudes as discussed above.

Since the coefficients of the multiplier equations are simply the Jacobian matrix of the constraints we may solve these equations in the same manner as when the constrained variables are obtained by Newton's method. For example, the orbital multipliers Eq. (5.5) may be solved using the algorithm we use for second-order optimization of the orbitals and the amplitude multipliers Eq. (5.4) may be obtained using the algorithm for second-order optimization of the amplitudes. We therefore do not discuss the solution of the multiplier equations in more detail.

B. Variational densities

From Section II.D we know that energies may be calculated according to the equation

$$\varepsilon^{(0)} = E^{(00)} + \bar{\lambda}^{(0)} e^{(00)} \quad (5.8)$$

which is stable with respect to errors in $\lambda^{(0)}$ and $\bar{\lambda}^{(0)}$. This follows since even-order derivatives $\varepsilon^{(2n)}$ are stable with respect to $\lambda^{(n)}$ and since we have not used the 2n+2 rule to eliminate $\bar{\lambda}^{(0)}$. The variational expressions derived in Section IV are

$$\varepsilon^{(0)}_{MC} = \left\langle MC \middle| H^{(0)} \middle| MC \right\rangle \quad (5.9)$$

$$\varepsilon^{(0)}_{CI} = \left\langle CI \middle| H^{(0)} \middle| CI \right\rangle + \left\langle MC \left| \left[\bar{\lambda}^{(0)}, H^{(0)} \right] \right| MC \right\rangle \quad (5.10)$$

$$\varepsilon^{(0)}_{CC} = \left\langle HF\bar{t} \middle| H^{(0)} \middle| CC \right\rangle + \left\langle HF \left| \left[\bar{\kappa}^{(0)}, H^{(0)} \right] \right| HF \right\rangle \quad (5.11)$$

$$\varepsilon^{(0)}_{MP2} = 2 \left\langle HF \middle| H^{(0)} \middle| t^{(0)} \right\rangle + \operatorname{Tr} \rho^{i} F\left[H^{(0)} \right] \quad (5.12)$$

These energies may all be calculated as

$$\varepsilon^{(0)} = \sum_{mn} \mathcal{D}_{mn}\, \tilde{h}_{mn}^{(0)} + \frac{1}{2} \sum_{mnpq} d_{mnpq}\, \tilde{g}_{mnpq}^{(0)} \tag{5.13}$$

where $\mathcal{D}_{mn}$ and d_{mnpq} are a set of effective densities which we refer to as *variational densities* (Rice and Amos 1985). As we shall see the variational densities are important for the efficient calculation of molecular derivatives. For the wave functions considered by us their explicit form is

$$\mathcal{D}_{mn}^{MC} = D_{mn}^{MC} \tag{5.14}$$

$$\mathcal{D}_{mn}^{CI} = D_{mn}^{CI} - 2 \sum_{k \neq MC} \bar{p}_k \langle k | E_{mn} | MC \rangle - \{ \bar{\kappa}, D^{MC} \}_{mn} \tag{5.15}$$

$$\mathcal{D}_{mn}^{CC} = D_{mn}^{CC} + \sum_{\mu} \bar{t}_{\mu}^{(0)} \langle \mu | \exp[-T^{(0)}]\, E_{mn} | CC \rangle - \{ \bar{\kappa}, D^{HF} \}_{mn} \tag{5.16}$$

$$\mathcal{D}_{mn}^{MP} = \rho_{mn} \tag{5.17}$$

where

$$D_{mn}^{HF} = \langle HF | E_{mn} | HF \rangle \tag{5.18}$$

$$D_{mn}^{MC} = \langle MC | E_{mn} | MC \rangle \tag{5.19}$$

$$D_{mn}^{CI} = \langle CI | E_{mn} | CI \rangle \tag{5.20}$$

$$D_{mn}^{CC} = \langle HF | E_{mn} | CC \rangle \tag{5.21}$$

The two-electron densities are all calculated in the same way except the Møller-Plesset densities which are obtained from the equation

$$d_{mnpq}^{MP} = 2 t_{mnpq}^{MP} + 2 \mathcal{D}_{mn}^{MP} D_{pq}^{HF} - \frac{1}{2} \mathcal{D}_{mp}^{MP} D_{nq}^{HF} - \frac{1}{2} \mathcal{D}_{mq}^{MP} D_{pn}^{HF} \tag{5.22}$$

where the only non-vanishing parts of t_{mnpq}^{MP} are given by Eq. (4.65). We do not consider the calculation of these densities in any detail but note that they may be calculated straightforwardly.

Except for numerical errors we get exactly the same results when we calculate the energy from the original densities and from the variational densities. If we write the corrections to the original densities

$$\Delta_{mn} = \mathcal{D}_{mn} - D_{mn} \tag{5.23}$$

$$\delta_{mnpq} = \mathcal{d}_{mnpq} - d_{mnpq} \tag{5.24}$$

the expectation value corresponding to these densities gives zero energy:

$$\sum_{mn} \Delta_{mn} h_{mn} + \frac{1}{2} \sum_{mnpq} \delta_{mnpq} g_{mnpq} = 0 \tag{5.25}$$

We have therefore added corrections to our original densities which do not change the energy but merely corrects for the non-variational character of the wave function. As discussed in Section II, these corrections make the error in the energy depend quadratically on the errors in the wave function and the Lagrange multipliers. For example, CI energies calculated from the usual CI densities are quadratic in the error in the CI coefficients but linear in the error in the orbitals. If we calculate the energy from the variational densities the error depends quadratically on all variables, including the multipliers.

Although it is nice to have energies with quadratic errors, this fact alone does not justify the calculation of variational densities. It is easier just to converge the wave function better. However, when it comes to the calculation of properties the variational densities are very useful as discussed in the following sections.

C. Molecular gradients

The general expression for molecular gradients Eq. (2.95) is

$$\varepsilon^{(1)} = E^{(10)} + \bar{\lambda}^{(0)} e^{(10)} \tag{5.26}$$

The expressions for the wave functions considered by us are

$$\varepsilon^{(1)}_{MC} = \langle MC | H^{(1)} | MC \rangle \tag{5.27}$$

$$\varepsilon^{(1)}_{CI} = \langle CI | H^{(1)} | CI \rangle + \langle MC | [\bar{\lambda}^{(0)}, H^{(1)}] | MC \rangle \tag{5.28}$$

$$\varepsilon^{(1)}_{CC} = \langle HF\bar{t} | H^{(1)} | CC \rangle + \langle HF | [\bar{\kappa}^{(0)}, H^{(1)}] | HF \rangle \tag{5.29}$$

$$\varepsilon^{(1)}_{MP2} = 2\langle HF|H^{(1)}|t^{(0)}\rangle + \mathrm{Tr}\rho\, {}^{i}F[H^{(1)}] \tag{5.30}$$

These are similar to the expressions for variational energies Eqs. (5.9-12), the only difference is the Hamiltonian. They may therefore be calculated from the same variational densities. Introducing a shorthand notation for the summations we obtain

$$\begin{aligned}\varepsilon^{(1)} &= \sum_{mn} \mathcal{D}_{mn}\, \tilde{h}^{(1)}_{mn} + \frac{1}{2}\sum_{mnpq} d_{mnpq}\, \tilde{g}^{(1)}_{mnpq} \\ &= \mathrm{Tr}\, \mathcal{D}\tilde{h}^{(1)} + \frac{1}{2}\mathrm{Tr}\, d\tilde{g}^{(1)}\end{aligned} \tag{5.31}$$

The variational densities are obviously very useful: Since they are independent of the perturbation we may use them to calculate any first-order property, for example molecular gradients and multipole moments. Note, however, that Eq. (5.31) is linear in the errors of the wave function and the multipliers in contrast to the energy Eq. (5.13) which is quadratic in the same errors.

We now discuss how to calculate expectation values of the differentiated OMO integrals as in Eq. (5.31). From Sec. III the integrals are given by

$$\tilde{h}^{(1)} = h^{(1)} - \frac{1}{2}\{S^{(1)}, h^{(0)}\} \tag{5.32}$$

$$\tilde{g}^{(1)} = g^{(1)} - \frac{1}{2}\{S^{(1)}, g^{(0)}\} \tag{5.33}$$

The gradients may therefore be written

$$\varepsilon^{(1)} = \mathrm{Tr}\, \mathcal{D}h^{(1)} + \frac{1}{2}\mathrm{Tr}\, d g^{(1)} - \frac{1}{2}\mathrm{Tr}\, \mathcal{D}\{S^{(1)}, h^{(0)}\} - \frac{1}{4}\mathrm{Tr}\, d\{S^{(1)}, g^{(0)}\} \tag{5.34}$$

where the first two terms are the expectation value of the differentiated UMO integrals, and the last two the expectation value of one-index transformed integrals. Such expectation values may be calculated by taking the trace of the transformation matrix multiplied by the generalized Fock matrix constructed from untransformed integrals:

$$\mathrm{Tr}\, \mathcal{D}\{A, h\} + \frac{1}{2}\mathrm{Tr}\, d\{A, g\} = 2\mathrm{Tr}A\mathcal{F} \tag{5.35}$$

$$\mathcal{F}_{mn} = \sum_{o} \mathcal{D}_{mo}\, h_{no} + \sum_{opq} d_{mopq}\, g_{nopq} \tag{5.36}$$

The molecular gradient Eq. (5.34) may therefore be written

$$\varepsilon^{(1)} = \mathrm{Tr}\, \mathcal{D} h^{(1)} + \frac{1}{2} \mathrm{Tr}\, d g^{(1)} - \mathrm{Tr} S^{(1)} \mathcal{F}^{(0)} \tag{5.37}$$

where $\mathcal{F}^{(0)}$ is the Fock matrix Eq. (5.36) constructed from variational densities and undifferentiated integrals.

The gradient Eq. (5.37) is written in terms of molecular orbitals. We may rewrite it in the AO basis and obtain

$$\varepsilon^{(1)} = \mathrm{Tr}\, \mathcal{D}_{ao}\, h^{(1)}_{AO} + \frac{1}{2} \mathrm{Tr}\, d_{ao}\, g^{(1)}_{AO} - \mathrm{Tr} S^{(1)}_{AO}\, \mathcal{F}^{(0)}_{ao} \tag{5.38}$$

The one-electron densities and integrals in the AO and MO basis are related in the following manner:

$$[\mathcal{D}_{ao}]_{\mu\nu} = \sum_{mn} C^{(0)}_{m\mu}\, C^{(0)}_{n\nu}\, \mathcal{D}_{mn} \tag{5.39}$$

$$h^{(1)}_{mn} = \sum_{\mu\nu} C^{(0)}_{m\mu}\, C^{(0)}_{n\nu} \left[h^{(1)}_{AO}\right]_{\mu\nu} \tag{5.40}$$

Note that the AO densities and integrals are related to the MO quantities in different ways. In tensor theory we say that in Eq. (5.38) the integrals are the covariant components of the AO integrals and the densities are the contravariant components of the AO densities. We distinguish these by using AO for covariant components and ao for contravariant components.

Equation (5.38) is convenient for calculating gradients since it avoids completely storage and transformation of the differentiated integrals. We first construct the variational densities in the AO basis from the Lagrange multipliers. Next we calculate the differentiated integrals and multiply them by densities on the fly. No integrals are written on disk. The calculation of gradients is therefore very efficient for all wave functions (Rice and Amos 1985).

D. Response equations

The expressions derived in Section II.D for first-order response equations are

$$e^{(01)} \lambda_n^{(1)} = -e^{(10)} \tag{5.41}$$

for non-variational parameters and

$$E_{vv}^{(02)} \lambda_v^{(1)} = -E_v^{(11)} - E_{vn}^{(02)} \lambda_n^{(1)} \tag{5.42}$$

for variational parameters. The coefficient matrix of the variational parameters Eq. (5.42) is the electronic Hessian with respect to these parameters. The Hessian is symmetric and the responses are therefore obtained by solving one set of linear symmetric equations for each perturbation.

The coefficient matrix of the non-variational parameters Eq. (5.41) is the Jacobian of the constraints. This matrix is symmetric only when the constraints correspond to a gradient and the non-variational responses are therefore in general obtained by solving one set of non-symmetric equations for each perturbation.

The non-variational responses Eq. (5.41) are independent of the variational parameters and must therefore be determined first. The variational parameters respond to changes both in the Hamiltonian and in the non-variational parameters [see Eq. (5.42)] and are therefore solved last.

The above discussion is illustrated by the response equations for the wave functions in Section IV. The SCF and MCSCF equations [Eqs. (4.16) and (4.13)]

$$\sum_{tj} \left\langle HF \middle| \left[E_{si}^{-}, \left[E_{tj}^{-}, H^{(0)} \right] \right] \middle| HF \right\rangle \kappa_{tj}^{(1)} = -\left\langle HF \middle| \left[E_{si}^{-}, H^{(1)} \right] \middle| HF \right\rangle \tag{5.43}$$

$$\sum_{j} \left\langle MC \middle| \left[\omega_i, \omega_j, H^{(0)} \right] \middle| MC \right\rangle \lambda_j^{(1)} = -\left\langle MC \middle| \left[\omega_i, H^{(1)} \right] \middle| MC \right\rangle \tag{5.44}$$

are symmetric since they are obtained by differentiating the SCF and MCSCF gradients. The right-hand sides are the partial derivatives of these gradients with respect to the perturbations. They may therefore be calculated in the same way as these once the Hamiltonian $H^{(1)}$ has been constructed as discussed in Section III.D. The CI response equations Eq. (4.37) are also symmetric

$$\sum_{L}\left[\langle K|\,H^{(0)}\,|L\rangle-\delta_{KL}\langle CI|\,H^{(0)}\,|CI\rangle\right]P_L^{(1)}=\langle K|\,J^{(1)}\,|CI\rangle \tag{5.45}$$

and may be solved using standard CI techniques for optimization. The coefficient matrix is simply the CI Hamiltonian with the energy subtracted from the diagonal elements. The right-hand side is the CI electronic gradient calculated from the effective Hamiltonian $J^{(1)}$. This Hamiltonian contains the first-order response of the orbitals since the orbitals are constrained to satisfy the conditions Eqs. (5.43) and (5.44). The CI coefficients Eq. (5.45) respond to the perturbation and therefore also to the change in the orbitals induced by the perturbation. The orbital equations must therefore be solved first.

The coupled cluster energy contains no variational parameters and the response equations Eq. (5.41) are used. These equations separate and we first calculate the orbital responses from Eq. (5.43) and then the first-order amplitudes:

$$\sum_{\nu}\langle\mu|\,\exp[-T^{(0)}][\,\tau_\nu,H^{(0)}\,]|CC\rangle\,t_\nu^{(1)}=-\langle\mu|\,\exp[-T^{(0)}]\,J^{(1)}|CC\rangle \tag{5.46}$$

The coefficient matrix is the non-symmetric CC Jacobian used for optimizing the wave functions. The right-hand side are the CC constraints calculated from the effective Hamiltonian $J^{(1)}$. As for CI wave functions Eq. (5.45) the orbital responses are needed to construct the right-hand side.

The MP2 response equations are similar to the CC equations but simpler. Again the full set of equations Eq. (5.41) separates into the usual SCF equations Eq. (5.43) and a set of equations for the amplitudes:

$$\sum_{\mu}\langle\mu|\,G^{(0)}|\nu\rangle\,t_\nu^{(1)}=-\langle\mu|\,J^{(1)}|HF\rangle-\langle\mu|\,G[J^{(1)}]|\,t^{(0)}\rangle \tag{5.47}$$

where the orbital responses are needed to construct the right-hand side. Since the coefficient matrix is diagonal in the canonical representation it is trivial to solve Eq. (5.47) after the right-hand side has been constructed. In the biorthonormal representation Eqs. (4.62) and (4.63) the right-hand side may be calculated from the expression

$$\left\langle {}^{st}_{ij} \middle| J^{(1)} \middle| HF \right\rangle + \left\langle {}^{st}_{ij} \middle| G[J^{(1)}] \middle| t^{(0)} \right\rangle = 2\left(R^{(1)}_{isjt} - R^{(1)}_{itjs} \right) \tag{5.48}$$

$$R^{(1)}_{isjt} = \tilde{j}^{(1)}_{isjt} - \sum_k \left({}^{i}F[J^{(1)}]_{ik} \, \bar{t}^{(0)}_{ksjt} + {}^{i}F[J^{(1)}]_{jk} \, \bar{t}^{(0)}_{iskt} \right)$$
$$+ \sum_u \left({}^{i}F[J^{(1)}]_{su} \, \bar{t}^{(0)}_{iujt} + {}^{i}F[J^{(1)}]_{tu} \, \bar{t}^{(0)}_{isju} \right) \tag{5.49}$$

from the inactive Fock matrix Eq. (4.59) constructed from $J^{(1)}$ and the zero-order amplitude multipliers.

The response equations have the same structure as the Newton equations for optimization of the wave function. In fact, we may regard the Newton equations as the zero-order response equations. We may therefore use the same methods for solving the response equations and the Newton equations. However, the response equations are slightly more complicated than the Newton equations.

First, there is one set of response equations for each perturbation. We may of course solve each of these separately in the same way as the Newton equations are solved. But the equations are nearly always solved iteratively by expanding the solution in a linear space of trial vectors. Therefore, if we solve the equations for all perturbations simultaneously the equations may converge faster since they share the same set of trial vectors.

Second, the perturbed Hamiltonian of the response equations need not be totally symmetric while the unperturbed Hamiltonian of the Newton equations is always totally symmetric. Therefore, the algorithm for the response equations must be more general than the algorithm for the Newton equations since it must also handle non-symmetric perturbations.

E. Molecular Hessians

Comparing the expressions for the molecular gradient and Hessian derived in Section II.D [Eqs. (2.95) and (2.96)]

$$\varepsilon^{(1)} = E^{(10)} + \bar{\lambda}^{(0)} e^{(10)} \tag{5.50}$$

$$\varepsilon^{(2)} = \left[E^{(20)} + \bar{\lambda}^{(0)} e^{(20)}\right] + 2\left[E^{(11)} + \bar{\lambda}^{(0)} e^{(11)}\right] \lambda^{(1)}$$
$$+ \left[E^{(02)} + \bar{\lambda}^{(0)} e^{(02)}\right] \lambda^{(1)} \lambda^{(1)} \tag{5.51}$$

we see that the first terms in the Hessian are the same as the gradient except that second derivatives are taken. We may therefore calculate this part of the Hessian in the same way as the gradient but from the second-order Hamiltonian. This is also seen from the expressions derived in Section IV:

$$\varepsilon_{MC}^{(2)} = \langle MC|H^{(2)}|MC\rangle + 2\langle MC|[\lambda^{(1)}, H^{(1)}]|MC\rangle + \langle MC|[\lambda^{(1)}, \lambda^{(1)}, H^{(0)}]|MC\rangle \tag{5.52}$$

$$\varepsilon_{HFCI}^{(2)} = \langle CI|J^{(2)}|CI\rangle + \langle HF|[\bar{\kappa}^{(0)}, J^{(2)}]|HF\rangle - 4\langle P^{(1)}|J^{(1)}|CI\rangle + 2\langle P^{(1)}|H^{(0)}|P^{(1)}\rangle - 2\,\varepsilon_{HFCI}^{(0)}|P^{(1)}|^2 \tag{5.53}$$

$$\varepsilon_{CC}^{(2)} = \langle HF\bar{t}|J^{(2)}|CC\rangle + \langle HF|[\bar{\kappa}^{(0)}, J^{(2)}]|HF\rangle - 2\langle HF\bar{t}|[T^{(1)}, J^{(1)}]|CC\rangle + \langle HF\bar{t}|[T^{(1)}, T^{(1)}, H^{(0)}]|CC\rangle \tag{5.54}$$

$$\varepsilon_{MP2}^{(2)} = 2\langle HF|J^{(2)}|t^{(0)}\rangle + \mathrm{Tr}\rho\,\bar{F}[J^{(2)}] + 2\langle \bar{t}^{(1)}|G^{(0)}|t^{(1)}\rangle \tag{5.55}$$

Comparing these with the molecular gradients Eqs. (5.27-30) we find that the first one or two terms are indeed identical to the gradients except for a different Hamiltonian. We may therefore use the variational densities of Section V.B and contract these with the second-derivative integrals.

In the Hessians of the non-variational energies Eqs. (5.53)-(5.55) the terms corresponding to the molecular gradient contain $J^{(2)}$ rather than $H^{(2)}$ since we have absorbed the orbital relaxation terms in this operator. This is not done in the (MC)SCF Hessian Eq. (5.52) since here the orbitals are variational and their response is taken care of by the last two terms. In principle, we could calculate SCF Hessians from the expression

$$\varepsilon_{HF}^{(2)} = \langle HF|J^{(2)}|HF\rangle \tag{5.56}$$

rather than

$$\varepsilon_{HF}^{(2)} = \langle HF|H^{(2)}|HF\rangle + 2\langle HF|[\kappa^{(1)}, H^{(1)}]|HF\rangle + \langle HF|[\kappa^{(1)}, \kappa^{(1)}, H^{(0)}]|HF\rangle \tag{5.57}$$

but Eq. (5.57) is faster since it only requires a few extra dot products to be taken after the response equations have been solved.

After the Hamiltonian integrals have been obtained we may calculate the contributions to the Hessian by contracting the integrals with the variational densities

$$\langle MC|H^{(2)}|MC\rangle = \mathrm{Tr}\,\mathcal{D}^{MC}\,\tilde{h}^{(2)} + \frac{1}{2}\mathrm{Tr}\,d^{MC}\,\tilde{g}^{(2)} \tag{5.58}$$

$$\langle CI|J^{(2)}|CI\rangle + \langle HF|[\bar{\kappa}^{(0)}, J^{(2)}]|HF\rangle = \mathrm{Tr}\,\mathcal{D}^{CI}\,\tilde{J}^{(2)} + \frac{1}{2}\mathrm{Tr}\,d^{CI}\,\tilde{j}^{(2)} \tag{5.59}$$

$$\langle HF\bar{t}|J^{(2)}|CC\rangle + \langle HF|[\bar{\kappa}^{(0)}, J^{(2)}]|HF\rangle = \mathrm{Tr}\,\mathcal{D}^{CC}\,\tilde{J}^{(2)} + \frac{1}{2}\mathrm{Tr}\,d^{CC}\,\tilde{j}^{(2)} \tag{5.60}$$

$$2\langle HF|J^{(2)}|t^{(0)}\rangle + \mathrm{Tr}\,\rho\,{}^{i}F[J^{(2)}] = \mathrm{Tr}\,\mathcal{D}^{MP}\,\tilde{J}^{(2)} + \frac{1}{2}\mathrm{Tr}\,d^{MP}\,\tilde{j}^{(2)} \tag{5.61}$$

but this scheme requires the transformation of all derivative integrals to the MO basis. We therefore transform the densities to the AO basis. At the same time we must also take into account all terms involving orbital responses and overlap derivatives in the integrals. We find that

$$\begin{aligned}\mathrm{Tr}\mathcal{D}\tilde{J}^{(2)} + \frac{1}{2}\mathrm{Tr}d\tilde{j}^{(2)} &= \mathrm{Tr}\mathcal{D}_{ao}\,h^{(2)}_{AO} + \frac{1}{2}\mathrm{Tr}\,d_{ao}\,g^{(2)}_{AO}\\ &- \mathrm{Tr}\,S^{(2)}_{AO}\,\mathcal{F}_{ao}[I^{(0)}] + \mathrm{Tr}\,S^{(1)}S^{(1)}\,\mathcal{F}[I^{(0)}]\\ &+ 2\,\mathrm{Tr}\,\kappa^{(1)}\,\mathcal{F}[2\,I^{(1)} - \{S^{(1)}, I^{(0)}\} + \{\kappa^{(1)}, I^{(0)}\}]\\ &- 2\,\mathrm{Tr}\,S^{(1)}\,\mathcal{F}[I^{(1)} - \frac{1}{4}\{S^{(1)}, I^{(0)}\}]\end{aligned} \tag{5.62}$$

where $\mathcal{F}[A]$ is the generalized Fock matrix Eq. (5.36) calculated from the variational densities and the Hamiltonian A. The above expression may be simplified further for Hartree-Fock and MP2 derivatives by taking into account the structure of their densities.

It now remains to discuss the relaxation terms in the Hessian. In the MCSCF case these terms are easily calculated after the response equations have been solved since all ingredients are available from the solution of the response equations and it only remains to take a few dot products. It might be thought that using the alternative expression

$$\varepsilon_{MC}^{(2)} = \langle MC | H^{(2)} | MC \rangle + \langle MC | [\lambda^{(1)}, H^{(1)}] | MC \rangle \tag{5.63}$$

is faster but this expression depends linearly on the response errors and saves very little time compared to the stable expression Eq. (5.52). The CI relaxation in Eq. (5.53) is also easily calculated from the responses. Again the more compact expression

$$\varepsilon_{HFCI}^{(2)} = \langle CI | J^{(2)} | CI \rangle + \langle HF | [\bar{\kappa}^{(0)}, J^{(2)}] | HF \rangle - 2 \langle P^{(1)} | J^{(1)} | CI \rangle \tag{5.64}$$

is not to be preferred since it is less stable with respect to numerical errors.

In contrast to the MCSCF and CI wave functions, the CC relaxation terms are rather difficult to calculate. For example, the last term in Eq. (5.54) contains the CC amplitude Hessian. This Hessian is not needed for optimizing the wave function or for solving the response equations. It must therefore be calculated exclusively for this contribution to the Hessian. Also, we cannot eliminate this term as we did for MCSCF and CI wave functions Eqs. (5.63) and (5.64) since the amplitudes are not variational. Both relaxation terms in Eq. (5.54) must be calculated.

The calculation of the relaxation term in the MP2 molecular Hessian Eq. (5.55) is simple. Since all responses must be available at the same time this term may require substantial disk storage. This problem is also shared by all wave functions containing a large number of parameters.

VI. TRANSLATIONAL AND ROTATIONAL SYMMETRIES

For technical reasons geometrical derivatives are calculated with respect to Cartesian displacements of the nuclei. Because of translational and rotational symmetries, such derivatives are not independent. For example, in an isolated molecule only 3N-m Cartesian forces are independent [N is the number of atoms and m is the number of translational and rotational degrees of freedom (5 or 6)]. The remaining forces can be determined from the translational and rotational invariance of the energy. This has two useful applications. First, the number of derivatives that must be calculated *ab initio* may be reduced. This is especially useful for small molecules. Alternatively, one may calculate all derivatives *ab initio* and simply use translational and rotational symmetries as an easy check on the calculation.

Over the years many discussions of translational and rotational symmetries have appeared in connection with calculation of energy derivatives (Page *et al.* 1984b, Helgaker *et al.* 1986b, Lee *et al.* 1986). However, all of these have been restricted to special cases. We present here a method which is both simple and general.

Let $F^{(n)}$ denote the $(3N)^n$ n'th-order geometrical derivatives of some molecular property, for example the energy or a component of the dipole moment. Let us assume that we know $(3N-m)^{n+1}$ appropriately chosen components of $F^{(n+1)}$ and all components of $F^{(n)}$. We wish to derive an expression which allows us to calculate the remaining $(3N)^{n+1}-(3N-m)^{n+1}$ components of $F^{(n+1)}$ from the translational and rotational symmetries of $F^{(n)}$. For the sake of simplicity we assume that the molecule is non-linear, so that m is equal to 6.

We denote the translational and rotational coordinates by $\{Q_x,Q_y,Q_z,R_x,R_y,R_z\}$ or collectively by $\{r_i;i=1,6\}$. The 3N Cartesian coordinates are denoted by $\{X_i,Y_i,Z_i;i=1,N\}$ or $\{\zeta_i;i=1,3N\}$. Using the chain rule for differentiation we obtain

$$\frac{\partial F^{(n)}_{i_1 i_2 \cdots i_n}}{\partial r_j} = \sum_{k=1}^{3N} \frac{\partial F^{(n)}_{i_1 i_2 \cdots i_n}}{\partial \zeta_k} \frac{\partial \zeta_k}{\partial r_j} \tag{6.1}$$

We may write this equation is matrix form

$$T(6\times 3N)\, A(3N) = B(6) \tag{6.2}$$

where the dimensions are indicated in parentheses and

$$T_{jk} = \frac{\partial \zeta_k}{\partial r_j} \tag{6.3}$$

$$A_k = \frac{\partial F^{(n)}_{i_1 i_2 \cdots i_n}}{\partial \zeta_k} = F^{(n+1)}_{i_1 i_2 \cdots i_n k} \tag{6.4}$$

$$B_j = \frac{\partial F^{(n)}_{i_1 i_2 \cdots i_n}}{\partial r_j} \tag{6.5}$$

Equation (6.2) relates $F^{(n+1)}$ to the translational and rotational derivatives of $F^{(n)}$. There is one set of equations for each component of $F^{(n)}$.

Selecting 6 independent Cartesian coordinates, we may reorder the columns of T so that the first 6 are linearly independent. We may then write T as

$$T(6\times 3N) = \left[\; T_D(6\times 6) \quad T_I\{6\times(3N-6)\} \; \right] \tag{6.6}$$

where T_D is non-singular. Reordering the rows of A in the same way

$$A(3N) = \begin{bmatrix} A_D(6) \\ A_I(3N-6) \end{bmatrix} \tag{6.7}$$

we find that Eq. (6.2) may be written as

$$T_D A_D + T_I A_I = B \tag{6.8}$$

or equivalently

$$A_D = T_D^{-1} (B - T_I A_I) \tag{6.9}$$

This equation relates 6 dependent derivatives A_D to 3N–6 independent derivatives A_I in a linear way. The transformation matrix $T_D^{-1} T_I$ is independent of the property and can be calculated once and for all. It only remains to describe the calculation of T and B.

The T matrix contains the partial derivatives of the Cartesian coordinates with respect to the translational and rotational coordinates. Its elements may be determined from the expressions

$$\frac{\partial}{\partial Q_x} = \sum_k \frac{\partial}{\partial X_k} \tag{6.10}$$

$$\frac{\partial}{\partial Q_y} = \sum_k \frac{\partial}{\partial Y_k} \tag{6.11}$$

$$\frac{\partial}{\partial Q_z} = \sum_k \frac{\partial}{\partial Z_k} \tag{6.12}$$

$$\frac{\partial}{\partial R_x} = \sum_k \left[Y_k \frac{\partial}{\partial Z_k} - Z_k \frac{\partial}{\partial Y_k} \right] \tag{6.13}$$

$$\frac{\partial}{\partial R_y} = \sum_k \left[Z_k \frac{\partial}{\partial X_k} - X_k \frac{\partial}{\partial Z_k} \right] \tag{6.14}$$

$$\frac{\partial}{\partial R_z} = \sum_k \left[X_k \frac{\partial}{\partial Y_k} - Y_k \frac{\partial}{\partial X_k} \right] \tag{6.15}$$

where the summations are over all atoms. The following columns of T correspond to the coordinates of atom k:

$$T_k = \begin{bmatrix} 1 & 0 & 0 \\ 0 & 1 & 0 \\ 0 & 0 & 1 \\ 0 & -Z_k & Y_k \\ Z_k & 0 & -X_k \\ -Y_k & X_k & 0 \end{bmatrix} \tag{6.16}$$

Note that T depends on the geometry only.

To obtain expressions for the elements of B we first consider the commutators between the partial derivatives with respect to r_j and Z_i. From Eqs. (6.10) and (6.15) we find that the only non-vanishing commutators are

$$\left[\frac{\partial}{\partial R_x}, \frac{\partial}{\partial Y_i} \right] = -\frac{\partial}{\partial Z_i} \tag{6.17}$$

$$\left[\frac{\partial}{\partial R_x}, \frac{\partial}{\partial Z_i} \right] = \frac{\partial}{\partial Y_i} \tag{6.18}$$

$$\left[\frac{\partial}{\partial R_y}, \frac{\partial}{\partial X_i} \right] = \frac{\partial}{\partial Z_i} \tag{6.19}$$

$$\left[\frac{\partial}{\partial R_y}, \frac{\partial}{\partial Z_i} \right] = -\frac{\partial}{\partial X_i} \tag{6.20}$$

$$\left[\frac{\partial}{\partial R_z}, \frac{\partial}{\partial X_i} \right] = -\frac{\partial}{\partial Y_i} \tag{6.21}$$

$$\left[\frac{\partial}{\partial R_z}, \frac{\partial}{\partial Y_i} \right] = \frac{\partial}{\partial X_i} \tag{6.22}$$

We use the notation

$$\left[\frac{\partial}{\partial r_j}, \frac{\partial}{\partial \zeta_i}\right] = a_{ji}\frac{\partial}{\partial \zeta_{[j,i]}} \tag{6.23}$$

for a general commutator, where a_{ji} is 0, −1, or +1. Repeated use of this equation gives

$$\frac{\partial F^{(n)}_{i_1 i_2 \cdots i_n}}{\partial r_j} = \sum_{k=1}^{n}\left(a_{ji_k}\, F^{(n)}_{i_1 i_2 \cdots i_{k-1}[j,i_k]\, i_{k+1} i_n}\right) + \left(\frac{\partial F}{\partial r_j}\right)^{(n)}_{i_1 i_2 \cdots i_n} \tag{6.24}$$

in an obvious notation.

From expression Eq. (6.24) we see that the derivatives of $F^{(n)}$ in general contain n+1 terms. For translations, the first n vanish since the commutators are zero. For rotations, they are obtained by rotating one index at a time according to the commutator relationships Eqs. (6.17)-(6.22) between the rotational and Cartesian coordinates. The last term is determined from the translational and rotational properties of F. For the energy this term vanishes. For other properties such as the dipole moment it is non-vanishing and must be obtained from the translational and rotational symmetries of that property.

As an illustration, we consider the derivative of ε_{xyz} with respect to rotation about the z axis. Assuming all coordinates belong to the same atom we obtain

$$\begin{aligned}\frac{\partial \varepsilon_{xyz}}{\partial R_z} &= \varepsilon_{[z,x]\,yz} + \varepsilon_{x\,[z,y]\,z} + \varepsilon_{xy\,[z,z]} + \left(\frac{\partial \varepsilon}{\partial R_z}\right)_{xyz} \\ &= -\,\varepsilon_{yyz} + \varepsilon_{xxz}\end{aligned} \tag{6.25}$$

where the last two terms vanish.

When A represents the energy gradient $\varepsilon^{(1)}$, Eq. (6.9) reduces to

$$\varepsilon^{(1)}_D = T_D^{-1}\, T_I\, \varepsilon^{(1)}_I \tag{6.26}$$

B vanishes since the energy is independent of the orientation of the molecule. When A corresponds to the part of the Hessian which is obtained by differentiating the Cartesian forces f_x, f_y, and f_z on a single nucleus (i.e., a matrix g consisting of three columns), Eq. (6.9) becomes

$$g_D = T_D^{-1}[B - T_I g_I] \tag{6.27}$$

where B is the matrix [use Eq. (6.24)]

$$B = \begin{bmatrix} 0 & 0 & 0 \\ 0 & 0 & 0 \\ 0 & 0 & 0 \\ 0 & -f_z & f_y \\ f_z & 0 & -f_x \\ -f_y & f_x & 0 \end{bmatrix} \tag{6.28}$$

B vanishes at equilibrium since the gradient is zero at this geometry. There is one set of linear equations Eq. (6.27) for each atom in the molecule.

The dependent elements of the dipole gradient may be calculated from a similar set of equations Eq. (6.27). The three-column matrix g now contains the gradients of the dipole components p_x, p_y, and p_z, and B becomes [use Eq. (6.24)]

$$B = \begin{bmatrix} q & 0 & 0 \\ 0 & q & 0 \\ 0 & 0 & q \\ 0 & -p_z & p_y \\ p_z & 0 & -p_x \\ -p_y & p_x & 0 \end{bmatrix} \tag{6.29}$$

where q is the total molecular charge. Equation (6.29) corresponds to the translational and rotational sum rules of the dipole moment.

From the above examples we see that whenever a new property is considered we only has to set up a new B matrix based on Eq. (6.24) and the translational and rotational symmetries of the property in question.

VII. CONCLUDING REMARKS

The main purpose of these notes has been to demonstrate that it is possible to derive compact expressions for the molecular derivatives of any electronic wave function. Important features of our approach are (i) the use of symmetrically orthonormalized orbitals to isolate the geometry dependence of the basis set in the Hamiltonian; (ii) the use of Lagrangians to extend the 2n+1 rule to non-variational wave functions; (iii) unconstrained exponential parametrization of the wave functions. These features allow us to derive the derivatives in a simple way by expanding the wave functions and Hamiltonian in geometrical distortions and collecting terms to same order in the displacements. The techniques presented here may also be applied to other time-independent properties when the correct Hamiltonian has been calculated.

REFERENCES

Adamowicz, L, Laidig, W.D., and Bartlett, R.J. (1984). *Int. J. Quantum Chem., Quantum Chem. Symp.* No. 18, 245.

Amos, R.D. (1987). *Adv. Chem. Phys.* **67**, 99.

Bartlett, R.J. (1986). In "Geometrical Derivatives of Energy Surfaces and Molecular Properties" (P. Jørgensen and J. Simons, eds), p. 35. Reidel, Dordrecht, Netherlands.

Bratož, S. (1958). *Collq. Int. C.N.R.S.* No. 82, 287.

Brooks, B.R., Laidig, W.D., Saxe, P., Goddard, J.D., Yamaguchi, Y., and Schaefer, H.F., III (1980). *J. Chem. Phys.* **72**, 4652.

Carlacci, L, and McIver, J.W., Jr. (1986). *J. Chem. Phys.* **85**, 634.

Dalgaard, E. (1979). *Chem. Phys. Lett.* **65**, 559.

Dalgaard, E., and Jørgensen, P. (1978). *J. Chem. Phys.* **69**, 3833.

Epstein, S.T. (1974). *The variation method in quantum chemistry*, Academic Press.

Fitzgerald, G., Harrison, R., Laidig, W.D., and Bartlett, R.J. (1985). *Chem. Phys. Lett.* **117**, 433.

Fogarasi, G., and Pulay, P. (1984). *Annu. Rev. Phys. Chem.* **35**, 191.

Fox, D.J., Osamura, Y., Hoffmann, M.R., Gaw, J.F., Fitzgerald, G., Yamaguchi, Y., and Schaefer, H.F., III (1983). *Chem. Phys. Lett.* **102**, 17.

Gaw, J.F., and Handy, N.C. (1984). *R. Soc. Chem. Annu. Rep., Sect. C* p. 291.

Gerratt, J., and Mills, I.M. (1968). *J. Chem. Phys.* **49**, 1719.

Goddard, J.D., Handy, N.C., and Schaefer, H.F.-, III (1979). *J. Chem. Phys.* **81**, 1525.

Handy, N.C., and Schaefer, H.F., III (1984). *J. Chem. Phys.* **81**, 5031.

Handy, N.C., Amos, R.D., Gaw, J.F., Rice, J.E., and Simandiras, E.D. (1985). *Chem. Phys. Lett.* **120**, 151.

Handy, N.C., Amos, R.D., Gaw, J.F., Rice, J.E., Simandiras, E.D., Lee, T.J., Harrison, R.J., Laidig, W.D., Fitzgerald, G.B., and Bartlett, R.J. (1986). *In* "Geometrical Derivatives of Energy Surfaces and Molecular Properties" (P. Jørgensen and J. Simons, eds.), p. 179. Reidel, Dordrecht, Netherlands.

Helgaker, T.U. (1986). *In* "Geometrical Derivatives of Energy Surfaces and Molecular Properties" (P. Jørgensen and J. Simons, eds), p.1. Reidel, Dordrecht, Netherlands.

Helgaker, T.U., and Almlöf, J. (1984). *Int. J. Quantum Chem.* **26**, 275.

Helgaker, T.U., Almlöf, J. Jensen, H.J. Aa., and Jørgensen, P. (1986). *J. Chem. Phys.* **84**, 6266.

Helgaker, T.U. and Jørgensen, P. (1988). *Adv. Quant. Chem.* **19**, 183.

Helgaker, T. and Jørgensen, P. (1989). *Theoret. Chim. Acta* **75**, 111.

Hess, B.A., Jr., Schaad, L.J., Čársky, P., and Zahradnik, R. (1986). *Chem. Rev..* **86**, 709.

Hoffmann, M.R., Fox, D.J., Gaw, J.F., Osamura, Y., Yamaguchi, Y., Grev, R.S., Fitzgerald, G., Schaefer, H.F., III, Knowles, P.J., and Handy, N.C. (1984). *J. Chem. Phys.* **80**, 2660.

Jørgensen, P., and Linderberg, J. (1970). *Int. J. Quantum Chem.* **4**, 587.

Jørgensen, P., and Simons, J. (1983). *J. Chem. Phys.* **79**, 334.

Jørgensen, P. and Helgaker, T.U. (1988). *J. Chem. Phys.* **89**, 3654.

Kato, S., and Morokuma, K. (1979). *Chem. Phys. Lett.* **65**, 19.

Koch, H., Jensen, H.J. Aa., Helgaker, T.U., Jørgensen, P., Schuseria, G.E., and Schaefer, H.F. III (1990). *J. Chem. Phys.* **92**, 4924.

Krishnan, R., Schlegel, H.B., and Pople, J.A. (1980). *J. Chem. Phys.* **72**, 4654.

Lee, T.J., Handy, N.C., Rice, J.E., Scheiner, A.C., and Schaefer, H.F., III (1986). *J. Chem. Phys.* **85**, 3930.

Lee, T.J., Schuseria, G.E., Rice, J.E., Scheiner, A.C., and Schaefer, H.F., III (1987). *Chem. Phys. Lett.* **139**, 134.

Levy, B. (1970). *Int. J. Quantum Chem.* **4**, 297.

London, F. (1937). *J. Phys. Radium* **8**, 397.

Osamura, Y., Yamaguchi, Y., and Schaefer, H.F., III (1981). *J. Chem. Phys.* **75**, 2919.

Osamura, Y., Yamaguchi, Y., and Schaefer, H.F., III (1982a) *J. Chem. Phys.* **77**, 383.

Osamura, Y., Yamaguchi, Y., Saxe, P., Vincent, M.A., Gaw, J.F., and Schaefer, H.F., III (1982b). *Chem. Phys.* **72**, 131.

Page, M., Saxe, P., Adams, G.F., and Lengsfield, B.H., III (1984a). *J. Chem. Phys.* **81**, 434.

Page, M., Saxe, P., Adams, G.F., and Lengsfield, B.H., III (1984b). *Chem. Phys. Lett.* **104**, 587.

Pople, J.A., Krishnan, R., Schelgel, H.B., and Binkley, J.S. (1979). *Int. J. Quantum Chem.* **S13**, 225.

Pulay, P. (1969). *Mol. Phys.* **17**, 197.

Pulay, P. (1970). *Mol. Phys.* **18**, 473.

Pulay, P. (1977). *In* "Applications of Electronic Structure Theory" (H.F. Schaefer, III,ed.), p. 153. Plenum, New York.

Pulay, P. (1983). *J. Chem. Phys.* **78**, 5043.

Pulay, P. (1987). *Adv. Chem. Phys.* **69**, 241.

Rice, J.E., and Amos, R.D. (1985). *Chem. Phys. Lett.* **122**, 585.

Sellers, H. (1986). Int. J. Quantum Chem. 30, 433

Schlegel, H.B. (1987). *Adv. Chem. Phys.* **67**, 249.

Simons, J., and Jørgensen, P. (1983). *J. Chem. Phys.* **79**, 3599.

Simons, J., Jørgensen, P., and Helgaker, T.U. (1984). *Chem. Phys.* **86** 413.

Thomsen, K., and Swanstrøm, P. (1973). *Mol. Phys.* **26**, 735.

CALCULATION OF MOLECULAR ROTATION-VIBRATION ENERGIES DIRECTLY FROM THE POTENTIAL ENERGY FUNCTION

Per Jensen

Physikalisch-Chemisches Institut, Justus-Liebig-Universität Giessen
Heinrich-Buff-Ring 58, W-6300 Giessen, Germany

1. INTRODUCTION

We consider a molecule consisting of N nuclei with masses m_α and charges $Z_\alpha e$, $\alpha = 1, 2, \ldots, N$, and n electrons, each with the mass m_e and charge $-e$. The molecule is not influenced by external fields. We initially describe each nucleus by means of its Cartesian coordinates $(X_\alpha, Y_\alpha, Z_\alpha)$ in an axis system rigidly attached to the laboratory frame [a *laboratory fixed* axis system], and each electron by its coordinates (x_i, y_i, z_i), $i = 1, 2, \ldots, n$, in the same axis system, and we denote the set of all Cartesian coordinates for the nuclei (X_1, Y_1, Z_1, X_2, Y_2, Z_2, ..., X_N, Y_N, Z_N) as $\mathbf{R}_n$ and the set of all Cartesian coordinates for the electrons, (x_1, y_1, z_1, x_2, y_2, z_2, ... x_n, y_n, z_n) as $\mathbf{r}_e$. If we consider only Coulomb interaction between the particles, the non-relativistic Schrödinger equation for the molecule can be written as

$$\left[\widehat{T}_n + \widehat{T}_e + V(\mathbf{R}_n, \mathbf{r}_e)\right] \psi_{ne}(\mathbf{R}_n, \mathbf{r}_e) = E_{ne}\psi_{ne}(\mathbf{R}_n, \mathbf{r}_e) \tag{1}$$

with the molecular wavefunction $\psi_{ne}(\mathbf{R}_n, \mathbf{r}_e)$ and its corresponding eigenvalue E_{ne}. In Eq. (1), the operator $\widehat{T}_n$ representing the kinetic energy of the nuclei is given by

$$\widehat{T}_n = -\frac{\hbar^2}{2}\sum_{\alpha=1}^{N}\frac{1}{m_\alpha}\left[\frac{\partial^2}{\partial X_\alpha^2} + \frac{\partial^2}{\partial Y_\alpha^2} + \frac{\partial^2}{\partial Z_\alpha^2}\right], \tag{2}$$

the operator $\widehat{T}_e$ representing the kinetic energy of the electrons is given by

$$\widehat{T}_e = -\frac{\hbar^2}{2m_e}\sum_{i=1}^{n}\left[\frac{\partial^2}{\partial x_i^2} + \frac{\partial^2}{\partial y_i^2} + \frac{\partial^2}{\partial z_i^2}\right] \tag{3}$$

and the Coulomb potential energy function $V(\mathbf{R}_n, \mathbf{r}_e)$ is given by

$$\begin{aligned} V(\mathbf{R}_n, \mathbf{r}_e) = &-\sum_{\alpha=1}^{N}\sum_{i=1}^{n}\frac{Z_\alpha e^2}{\sqrt{(X_\alpha - x_i)^2 + (Y_\alpha - y_i)^2 + (Z_\alpha - z_i)^2}} \\ &+ \sum_{i<i'}\frac{e^2}{\sqrt{(x_i - x_{i'})^2 + (y_i - y_{i'})^2 + (z_i - z_{i'})^2}} \\ &+ \sum_{\alpha<\alpha'}\frac{Z_\alpha Z_{\alpha'} e^2}{\sqrt{(X_\alpha - X_{\alpha'})^2 + (Y_\alpha - Y_{\alpha'})^2 + (Z_\alpha - Z_{\alpha'})^2}}. \end{aligned} \tag{4}$$

Methods in Computational Molecular Physics, Edited by Stephen Wilson and Geerd H.F. Diercksen, Plenum Press, New York, 1992

At the present time, all practical schemes for solving Eq. (1) are two-stage procedures: First an *ab initio* calculation is carried out, that is, the *electronic Schrödinger equation*

$$\left[\hat{T}_e + V(\mathbf{R}_n^{(0)}, \mathbf{r}_e)\right] \psi_e(\mathbf{R}_n^{(0)}, \mathbf{r}_e) = V_{\text{ai}}(\mathbf{R}_n^{(0)}) \, \psi_e(\mathbf{R}_n^{(0)}, \mathbf{r}_e) \tag{5}$$

is solved for fixed nuclear coordinates $\mathbf{R}_n^{(0)}$. By solving Eq. (5) for many values of $\mathbf{R}_n^{(0)}$, we obtain the effective potential energy function $V_{\text{ai}}(\mathbf{R}_n)$, and in the second step of the calculation we can solve the Schrödinger equation for the nuclear motion

$$[\hat{T}_n + V_{\text{ai}}(\mathbf{R}_n)]\psi_n(\mathbf{R}_n) = E_{ne}\psi_n(\mathbf{R}_n) \tag{6}$$

where $\psi_n(\mathbf{R}_n)$ is the nuclear wavefunction. In the approximation defined by Eqs. (5) and (6), the total molecular wavefunction is then taken to be

$$\psi_{ne}(\mathbf{R}_n, \mathbf{r}_e) = \psi_e(\mathbf{R}_n, \mathbf{r}_e)\psi_n(\mathbf{R}_n). \tag{7}$$

This approximation is usually credited to Born and Oppenheimer (1927).

Most other chapters of this book are concerned with the solution of Eq. (5), i.e., the *ab initio* calculation that delivers the potential energy function $V_{\text{ai}}(\mathbf{R}_n)$ and the electronic wavefunction $\psi_e(\mathbf{R}_n, \mathbf{r}_e)$. In the present chapter we assume that the *ab initio* calculation has been carried out and that consequently the potential energy function $V_{\text{ai}}(\mathbf{R}_n)$ is known. With the known $V_{\text{ai}}(\mathbf{R}_n)$ we then develop procedures for solving Eq. (6) in order to obtain the molecular rovibronic energies E_{ne} and the corresponding wavefunctions $\psi_n(\mathbf{R}_n)$.

2. THE CHOICE OF ROTATION-VIBRATION COORDINATES

The first problem that one encounters when developing a method for solving the nuclear Schrödinger equation (6), is to choose a suitable set of coordinates for describing the nuclear motion. This problem does not exist in *ab initio* theory, where all existing methods appear to describe the motion of the electrons through the Cartesian coordinates (x_i, y_i, z_i) [i = 1, 2, ..., n]. These coordinates are well suited to express the atomic-orbital-type basis functions used in *ab initio* calculations, and consequently they constitute a rather natural choice for the coordinates of the electronic problem. However, the analogous nuclear Cartesian coordinates (X_α, Y_α, Z_α) [α = 1, 2, ..., N] are entirely inconvenient for setting up Eq. (6) in a form that allows it to be solved in an efficient manner. Consequently, we must rewrite the total nuclear Hamiltonian

$$\hat{H}_{\text{total}} = \hat{T}_n + V_{\text{ai}}(\mathbf{R}_n) \tag{8}$$

by introducing new, more suitable coordinates. There is no natural, generally applicable choice for these coordinates, and the currently available methods for solving Eq. (6) employ different strategies for choosing them. We shall discuss some of these strategies in the present section.

2.1 Translation

In general, we obviously need $3N$ coordinates to describe completely a system consisting of N nuclei. All methods for solving Eq. (6) have in common that they use three of these coordinates to describe the *translational* motion of the molecule. During a translation of the molecule, all nuclei move uniformly along straight lines without changing their relative positions. Usually the three translational coordinates (X_0, Y_0, Z_0) are chosen as the laboratory-fixed coordinates of the nuclear center of mass

$$\begin{Bmatrix} X_0 \\ Y_0 \\ Z_0 \end{Bmatrix} = \frac{1}{\sum_{\alpha=1}^{N} m_\alpha} \sum_{\alpha=1}^{N} m_\alpha \begin{Bmatrix} X_\alpha \\ Y_\alpha \\ Z_\alpha \end{Bmatrix} \tag{9}$$

where m_α is the mass of atom α.[1] The remaining $3N - 3$ coordinates used to describe the nuclear motion are socalled *internal* coordinates; they are invariant under uniform translation of all nuclei. It is customary to define initially such coordinates $(t_{1,X}, t_{1,Y}, t_{1,Z}, t_{2,X}, t_{2,Y}, t_{2,Z}, \ldots t_{N-1,X}, t_{N-1,Y}, t_{N-1,Z})$ through $N - 1$ equations [see Sutcliffe and Tennyson (1991)]

$$\left\{ \begin{array}{c} t_{i,X} \\ t_{i,Y} \\ t_{i,Z} \end{array} \right\} = \sum_{\alpha=1}^{N} V_{\alpha i} \left\{ \begin{array}{c} X_\alpha \\ Y_\alpha \\ Z_\alpha \end{array} \right\}, \quad i = 1, 2, \ldots, N-1. \tag{10}$$

The condition that the coordinates $t_{i,A}$ $[A = X, Y, \text{or } Z]$ be invariant under uniform translation of the molecule can be expressed through the equation

$$\sum_{\alpha=1}^{n} V_{\alpha i} = 0 \text{ for all } i = 1, 2, \ldots, N-1. \tag{11}$$

Further, it necessary that the $V_{\alpha i}$ be such that the coordinate transformation defined through Eqs. (9) and (10),

$$\begin{aligned} &(X_1, Y_1, Z_1, X_1, Y_1, Z_1, \ldots, X_N, Y_N, Z_N) \\ &\rightarrow (X_0, Y_0, Z_0, t_{1,X}, t_{1,Y}, t_{1,Z}, t_{2,X}, t_{2,Y}, t_{2,Z}, \ldots t_{N-1,X}, t_{N-1,Y}, t_{N-1,Z}) \end{aligned} \tag{12}$$

can be *inverted* so that it is possible to determine the coordinates $(X_1, Y_1, \ldots, Z_N)$ from the coordinates $(X_0, Y_0, Z_0, t_{1,X}, \ldots, t_{N-1,Z})$. Traditionally, the $t_{i,A}$ coordinates have been chosen so that they define the positions of the nuclei in an axis system with axes parallel to the laboratory fixed XYZ system but with origin in the nuclear center of mass [see Chapter 6 of Bunker (1979) or Chapter 11 of Wilson *et al.* (1955)]. This leads to the following values for the $V_{\alpha i}$ matrix elements

$$V_{\alpha i} = \delta_{\alpha i} - \frac{m_\alpha}{\sum_{\alpha=1}^{N} m_\alpha}. \tag{13}$$

In this case, the $V_{\alpha i}$ matrix elements depend on the nuclear masses and hence the $t_{i,A}$ coordinates become *mass dependent* or *isotope dependent*. This means that if we consider two isotopic molecules whose nuclei are arranged such that the two molecules have the same values of the laboratory-fixed coordinates $(X_1, Y_1, \ldots, Z_N)$, these two molecules will be described by *different* values of the coordinates $(t_{1,X}, t_{1,Y}, \ldots, t_{N-1,Z})$. It is not necessary that the $V_{\alpha i}$ matrix elements depend on the nuclear masses, and in many cases it is advantageous to choose them to be mass-independent.

It can be shown that we can write the nuclear kinetic energy operator $\widehat{T}_n$ as

$$\begin{aligned} \widehat{T}_n = &-\frac{\hbar^2}{2\sum_{\alpha=1}^{N} m_\alpha} \left\{ \frac{\partial^2}{\partial X_0^2} + \frac{\partial^2}{\partial Y_0^2} + \frac{\partial^2}{\partial Z_0^2} \right\} \\ &+ \widehat{T}_{\text{rot-vib}} \left(-i\hbar\frac{\partial}{\partial t_{1,X}}, -i\hbar\frac{\partial}{\partial t_{1,Y}}, -i\hbar\frac{\partial}{\partial t_{1,Z}}, \ldots, -i\hbar\frac{\partial}{\partial t_{N-1,Z}} \right) \end{aligned} \tag{14}$$

where, as indicated, the kinetic energy operator $\widehat{T}_{\text{rot-vib}}$ [which describes the *rotation* and *vibration* of the molecule] only depends on the momenta $-i\hbar\partial/\partial t_{i,A}$ conjugate to the $t_{i,A}$ coordinates. Since in the absence of external fields, the effective potential energy function $V_{\text{ai}}(\mathbf{R}_n)$ cannot depend on the coordinates (X_0, Y_0, Z_0) [if we let the molecule undergo a uniform translation of all nuclei corresponding to a change in the center-of-mass coordinates

[1] In rotation-vibration calculations it is accepted practice to use atomic masses rather than nuclear masses. As explained by Bunker and Moss (1980) this compensates to some extent for the breakdown of the Born-Oppenheimer approximation [see also Gordy and Cook (1970)].

(X_0, Y_0, Z_0), $V_{ai}(\mathbf{R}_n)$ cannot change], this means that we can write $\hat{H}_{total}$ [Eq. (8)] as the sum of two commuting operators,

$$\hat{H}_{total} = \hat{T}_{translation} + \hat{H}_{rot-vib} \tag{15}$$

where

$$\hat{T}_{translation} = -\frac{\hbar^2}{2\sum_{\alpha=1}^{N} m_\alpha}\left\{\frac{\partial^2}{\partial X_0^2} + \frac{\partial^2}{\partial Y_0^2} + \frac{\partial^2}{\partial Z_0^2}\right\} \tag{16}$$

and

$$\hat{H}_{rot-vib} = \hat{T}_{rot-vib} + V_{ai}(\mathbf{R}_n). \tag{17}$$

Consequently the eigenvalues of $\hat{H}_{total}$ can be written as

$$E_{total} = E_{translation} + E_{rot-vib} \tag{18}$$

where $E_{translation}$ is an eigenvalue of $\hat{T}_{translation}$ and $E_{rot-vib}$ is an eigenvalue of $\hat{H}_{rot-vib}$. We are not interested in the energy contribution $E_{translation}$ from the translational motion. The reason is that this contribution is of no importance in the prediction and interpretation of spectroscopic experiments; such experiments do not involve transitions between translational states. We thus simply ignore $\hat{T}_{translation}$ and concentrate on the solution of the eigenvalue problem

$$\hat{H}_{rot-vib}\psi_{rot-vib} = E_{rot-vib}\psi_{rot-vib} \tag{19}$$

which involves the $t_{i,A}$ coordinates only.

2.2 Rotation and vibration

We have been led to introduce the $3N - 3$ coordinates $t_{i,A}$ to describe the *translation-free* nuclear motion. These coordinates, however, are not ideal for this purpose, and we must carry out further coordinate transformations before we can attempt to solve Eq. (19). We can think of the nuclei as carrying out a superposition of two kinds of motion:

- uniform *rotation* of the molecule as a whole. When the molecule rotates without the relative positions of the nuclei changing, the effective potential function $V_{ai}(\mathbf{R}_n)$ obviously does not change when there are no external fields. Consequently, the rotation is *free*, i.e., it is not influenced by $V_{ai}(\mathbf{R}_n)$.

- motion of the nuclei relative to each other, which we will call *vibration*. Such motion obviously is influenced by the effective potential function $V_{ai}(\mathbf{R}_n)$.

When we define the coordinates $t_{i,A}$ as linear combinations of laboratory-fixed coordinates [Eq. (10)], the displacement of any one of the $t_{i,A}$ coordinates will normally involve both rotation of the molecule as a whole *and* some relative motion of the nuclei. Consequently, $V_{ai}(\mathbf{R}_n)$ will in general depend on all $3N - 3$ $t_{i,A}$'s. We would like to define a new set of $3N - 3$ coordinates that reflect correctly the separation of the nuclear motion into rotation and vibration. The general strategy for doing this, which is common for all methods of rotation-vibration calculations, is as follows: A Cartesian axis system [the *molecule fixed* axis system] with origin at the nuclear center of mass is attached to the molecule so that it follows the rotational motion [in a sense that we shall discuss below]. The orientation of the molecule fixed axis system relative to the laboratory fixed XYZ axis system is defined by three Euler angles (θ, ϕ, χ) [see Appendix A of Papoušek and Aliev (1982)]. We use these angles as *rotational coordinates*. Further, we define $3N - 6$ new coordinates (q_1, q_2, q_3, $\ldots$, q_{3N-6}) which are invariant not only under uniform translation of the molecule [as were the $t_{i,A}$'s] but also under uniform rotation of the molecule.[2] Obviously we must select the scheme for attaching the molecule fixed coordinate system to the molecule and define the q_i coordinates such that the coordinate transformations

$$(t_{1,X}, t_{1,Y}, t_{1,Z}, t_{2,X}, t_{2,Y}, t_{2,Z}, \ldots t_{N-1,X}, t_{N-1,Y}, t_{N-1,Z}) \\ \leftrightarrow \quad (\theta, \phi, \chi, q_1, q_2, q_3, \ldots, q_{3N-6}) \tag{20}$$

can be carried out in both directions, so that if we know a set of values of the coordinates $t_{i,A}$, we can calculate the values of the coordinates $(\theta, \phi, \chi, q_1, \ldots, q_{3N-6})$ and vice versa.

We have now reached the crucial point in setting up a solution scheme for Eq. (19): the definition of the *vibrational* coordinates $(q_1, q_2, q_3, \ldots, q_{3N-6})$. There are two fundamentally different ways of defining these coordinates, and we will briefly discuss these two possibilities here since it will help us classify the practical methods for solving Eq. (19) which will be discussed in the following sections.

2.3 Hamiltonians in mass-dependent vibrational coordinates

The traditional, spectroscopic approach to solving Eq. (19) [Eckart, 1935; Wilson, 1939; Nielsen, 1951; Wilson *et al.*, 1955; Watson, 1968*a*; Hoy *et al.*, 1972; Papoušek and Aliev, 1982; Aliev and Watson, 1985] uses vibrational coordinates chosen in order to simplify the expression for the kinetic energy operator $\widehat{T}_{\text{rot-vib}}$. This operator can be written as

$$\begin{aligned} \widehat{T}_{\text{rot-vib}} &= \widehat{T}_{\text{rot}}\left(\theta, \phi, \chi, \widehat{P}_\theta, \widehat{P}_\phi, \widehat{P}_\chi\right) \\ &+ \widehat{T}_{\text{vib}}\left(q_1, q_2, \ldots, q_{3N-6}, \widehat{P}_{q_1}, \widehat{P}_{q_2}, \ldots, \widehat{P}_{q_{3N-6}}\right) \\ &+ \widehat{T}_{\text{rest}}\left(\theta, \phi, \chi, \widehat{P}_\theta, \widehat{P}_\phi, \widehat{P}_\chi, q_1, q_2, \ldots, q_{3N-6}, \widehat{P}_{q_1}, \widehat{P}_{q_2}, \ldots, \widehat{P}_{q_{3N-6}}\right) \end{aligned} \tag{21}$$

[where $\widehat{P}_a$ denotes the momentum conjugate to a coordinate a], and the rotational and vibrational coordinates are defined such that the *rotation-vibration interaction term* $\widehat{T}_{\text{rest}}$ is minimized. The background for this is the following: It follows from the discussion given above that the effective potential energy function V_{ai} can only depend on the $3N - 6$ coordinates q_i. When we write the rotation-vibration Hamiltonian $\widehat{H}_{\text{rot-vib}}$ as

$$\widehat{H}_{\text{rot-vib}} = \widehat{T}_{\text{rot}} + \left\{\widehat{T}_{\text{vib}} + V_{\text{ai}}\right\} + \widehat{T}_{\text{rest}}, \tag{22}$$

$\widehat{T}_{\text{rot}}$ will commute with $\widehat{T}_{\text{vib}} + V_{\text{ai}}$ since these two operators depend on different coordinates. In the event that we can neglect $\widehat{T}_{\text{rest}}$ we thus have the remaining Hamiltonian written as a sum of two commuting operators whose eigenvalue problems we can solve separately and then construct the rotation-vibration wavefunctions as the product

$$\psi_{\text{rot-vib}}(\theta, \phi, \chi, q_1, q_2, \ldots, q_{3N-6}) = \psi_{\text{rot}}(\theta, \phi, \chi)\, \psi_{\text{vib}}(q_1, q_2, \ldots, q_{3N-6}). \tag{23}$$

In practice, we cannot neglect $\widehat{T}_{\text{rest}}$, but if we can minimize it through our coordinate choice we might be able to treat it as a perturbation.

The coordinate choice outlined here aims at eliminating rotation-vibration interaction terms from the kinetic energy operator. Since these terms depend on the nuclear masses, the coordinates $(\theta, \phi, \chi, q_1, q_2, q_3, \ldots, q_{3N-6})$ which are chosen to eliminate them also become mass-dependent. However, the potential energy function is not naturally expressed in

[2] In the "standard theory" for a molecule whose equilibrium geometry is linear, 2 rotational coordinates, θ and ϕ, and $3N - 5$ vibrational coordinates are employed [see Papoušek and Aliev (1982)]. This is a special case which we will not consider here.

these mass-dependent coordinates. We normally use a parameterized function of the nuclear coordinates to express the effective potential function V_{ai}; we would typically obtain the values of the parameters by fitting this function through a set of *ab initio* energies with corresponding nuclear geometries. As noted by Hoy *et al.* (1972), the values of the parameters in V_{ai} are independent of isotopic substitution provided that we express V_{ai} as a function of $3N - 6$ *mass-independent, geometrically defined* coordinates which we might call ($\mathcal{R}_1$, $\mathcal{R}_2$, $\mathcal{R}_3$, ..., $\mathcal{R}_{3N-6}$). These coordinates are typically instantaneous values of internuclear distances ["bond lengths"] or angles $\angle$(ABC) defined by three nuclei A, B, and C ["bond angles"]. They are different from the mass-dependent coordinates (q_1, q_2, ..., q_{3N-6}) which we have introduced to simplify the kinetic energy operator, so before we can solve Eq. (19) we must express the effective potential function in terms of the coordinates chosen for the kinetic energy operator through *coordinate transformations* of the type

$$\mathcal{R}_i = \mathcal{R}_i(q_1, q_2, \ldots, q_{3N-6}). \tag{24}$$

These relations are inserted in the effective potential energy function V_{ai} and this function can now be seen as depending on the coordinates q_i. It is normally possible to derive exact relations of the type given in Eq. (24), i.e., the $\mathcal{R}_i$ coordinates can be expressed exactly in terms of the q_j coordinates. However, inserting these relations in the effective potential energy function leads to a mathematically intractable form of this function. Consequently, one normally resorts to an approximate coordinate transformation, in which the $\mathcal{R}_i$ coordinates are expressed as power series in the q_j coordinates [see Eq. (3) of Hoy *et al.* (1972)]:

$$\mathcal{R}_i \approx \mathcal{R}_{i0} + \sum_{j=1}^{3N-6} B_i^{(j)} q_j + \sum_{j=1}^{3N-6} \sum_{k=1}^{3N-6} B_i^{(jk)} q_j q_k + \ldots. \tag{25}$$

In the preceding paragraphs we have outlined the choice of what one might call "spectroscopic coordinates". The form of these coordinates is dictated by the kinetic energy operator, because they are defined so that they minimize the rotation-vibration interaction term $\hat{T}_{\text{rest}}$ in Eq. (21). Thus it is necessary to transform the *potential energy function* to make it depend on the spectroscopic coordinates. In the beginnings of rotation-vibration theory 50-60 years ago, one only had hopes of calculating approximative eigenvalues for a rotation-vibration Hamiltonian in which rotation and vibration were *separated* [i.e., in which $\hat{T}_{\text{rest}}$ could be neglected], and consequently this type of coordinates constituted the only feasible choice. We shall see in the next section, however, that the "standard" separation of rotation and vibration is possible only if the molecule carries out its vibration in one deep "well" of the potential energy surface. This well defines one equilibrium structure, and the amplitudes of the vibrational motion must be small compared to the linear dimensions of this structure. It was realized early [Sayvetz, 1939] that there existed molecules which carried out vibrations between multiple minima of the potential energy function or whose vibrational amplitudes were not small compared to the linear dimensions of their equilibrium structures, and that in such cases the "standard" treatment must be extended. In modern times, the ideas of Sayvetz were revived by Hougen, Bunker, and Johns (1970) and inspired the *nonrigid bender* approach [Hoy and Bunker, 1974; Hoy and Bunker, 1979; Jensen and Bunker, 1983; Jensen, 1983; Jensen and Bunker, 1986]. We shall discuss this theoretical model in some detail below. It allows the rotation-vibration energies of a triatomic molecule to be calculated directly from an *ab initio* potential energy function or the parameters of the potential energy function to be refined on the basis of experimental data. The nonrigid bender approach is an extension of the standard approach in the sense that its assumptions about the vibrational motion are less restrictive than those made in the standard treatment. It does, however, still use "spectroscopic" coordinates chosen to simplify the kinetic energy operator and facilitate the computation of the rotation-vibration eigenvalues.

2.4 Hamiltonians in geometrically defined vibrational coordinates

The advent of supercomputers has made it feasible to solve the Schrödinger equation for the molecular rotation and vibration using as vibrational coordinates the geometrically defined coordinates ($\mathcal{R}_1$, $\mathcal{R}_2$, $\mathcal{R}_3$, ..., $\mathcal{R}_{3N-6}$) discussed above in connection with the effective potential energy function. This approach is, in a sense, the exact opposite of the "spectroscopic" approach. While the form of the spectroscopic coordinates was dictated by the kinetic energy operator, and the effective potential energy function was transformed, in this new approach the coordinates are dictated by the potential energy function, and the *kinetic energy operator* is transformed to depend on the geometrically defined coordinates. The transformation is carried out in the expression for the rotation-vibration Hamiltonian depending on the momenta conjugate to the $t_{i,A}$ coordinates [Eq. (14)]

$$\hat{T}_{\mathrm{rot-vib}}\left(-i\hbar\frac{\partial}{\partial t_{1,X}},-i\hbar\frac{\partial}{\partial t_{1,Y}},-i\hbar\frac{\partial}{\partial t_{1,Z}},\ldots,-i\hbar\frac{\partial}{\partial t_{N-1,Z}}\right) \tag{26}$$

using the *chain rule* [see Sutcliffe (1982) and Carter and Handy (1986)]

$$-i\hbar\frac{\partial}{\partial t_{i,A}} = -i\hbar\left\{\frac{\partial\theta}{\partial t_{i,A}}\frac{\partial}{\partial\theta}+\frac{\partial\phi}{\partial t_{i,A}}\frac{\partial}{\partial\phi}+\frac{\partial\chi}{\partial t_{i,A}}\frac{\partial}{\partial\chi}+\sum_{k=1}^{3N-6}\frac{\partial\mathcal{R}_k}{\partial t_{i,A}}\frac{\partial}{\partial\mathcal{R}_k}\right\}. \tag{27}$$

Hamiltonians derived along these lines have been known in the literature [see, for example, Istomin *et al.* (1977), Zhilinskii *et al.* (1978), and Sørensen (1979)], but most work of this type carried out for tri- and four-atomic molecules at the present time [Carter and Handy, 1986; Tennyson, 1986; Miller and Tennyson, 1987; Miller and Tennyson, 1988*a*; Miller and Tennyson, 1989; Miller *et al.*, 1989; Carter *et al.*, 1990] has been triggered by the work of Sutcliffe (1982) who suggested a practical scheme for carrying out the transformation given by Eq. (27). In the Sutcliffe approach, no assumptions are made about the nature of the vibrational motion and thus it is not attempted to simplify the form of the kinetic energy operator through the choice of the coordinates. Since the transformation given by Eq. (27) can be carried out without introducing any approximation, an *exact* kinetic energy operator can be derived. This, of course, gives a very high degree of generality which must be paid for through the computational effort necessary to obtain the rotation-vibration eigenvalues [see Sutcliffe *et al.* (1987)]. The Hamiltonians for triatomic molecules developed using the methods of Sutcliffe have also been employed by Bačić and Light (1986, 1987, 1989) in the socalled Discrete Variable Representation - Distributed Gaussian Basis [DVR-DGB] approach.

The molecular ion H_3^+ is the prime example of a molecule for which the separation of rotation and vibration, as attempted by the standard approach, is not possible. H_3^+ has a rather pathological potential energy function [Meyer *et al.*, 1986] and its nuclei are very light, so the amplitudes of its vibrations are not small compared to the linear dimensions of the equilateral triangular equilibrium configuration. Thus the standard theory breaks down for this molecule. In fact, the only truly successful description of the rotation and vibration for H_3^+ is the very general approach by Sutcliffe as developed by Tennyson, Sutcliffe, and Miller [see Drossart *et al.*, 1989; Tennyson, 1986; Miller and Tennyson, 1987; Miller and Tennyson, 1988*a*; Miller and Tennyson, 1989; Miller *et al.*, 1989]. However as mentioned above, their methods are computationally very expensive and one might suspect that for a more usual molecule, H_2O say, a cheaper approach could yield accurate results. In the MORBID (Morse Oscillator Rigid Bender Internal Dynamics) model for triatomic molecules [Jensen, 1988*a*; 1988*b*; 1988*c*] this author has sought to develop such an approach by combining some of the advantages of the Hougen-Bunker-Johns theory [the separation of rotation and vibration] with the advantages of geometrically defined vibrational coordinates such as those used in the Sutcliffe scheme.

3. THE STANDARD APPROACH

The standard approach to rotation-vibration theory [Eckart, 1935; Wilson and Howard, 1936; Wilson, 1939; Darling and Dennison, 1940; Nielsen, 1951; Wilson *et al.*, 1955; Watson, 1968*a*; Hoy *et al.*, 1972; Papoušek and Aliev, 1982; Aliev and Watson, 1985] describes molecules whose potential energy surfaces have one minimum only, and the molecule fixed coordinate system [see Section 2.2] is defined so that the configuration of the molecule corresponding to the one minimum [the equilibrium configuration] is rigidly attached to the molecule fixed coordinate system. Hence the equilibrium configuration is described in the molecule fixed axis system through *constant* position vectors $\mathbf{a}_\alpha$ for the nuclei $\alpha = 1, 2, \ldots, N$:

$$\mathbf{a}_\alpha = \begin{Bmatrix} a_{\alpha x} \\ a_{\alpha y} \\ a_{\alpha z} \end{Bmatrix}. \tag{28}$$

Vibrational motion is then defined as the instantaneous diplacements of the nuclei from the positions given by the vectors $\mathbf{a}_\alpha$, measured along the axes of the molecule fixed coordinate system. As already mentioned, the standard approach assumes that these displacements are small compared to the linear dimensions of the equilibrium configuration.

3.1 The Eckart equations

We have still not given a precise definition of the molecule fixed coordinate system. If we think of a non-linear molecule [i.e., a molecule whose equilibrium structure is non-linear], we imagine for a moment that we could photograph an instantaneous configuration of the nuclei. On the photograph we would see the nuclei as points, and we could in principle determine their Cartesian, laboratory-fixed coordinates $(X_\alpha, Y_\alpha, Z_\alpha)$, but we would initially have no way of deciding to which extent the molecule on the picture was displaced from its equilibrium structure, and we would have no way of determining the Euler angles θ, ϕ, and χ. We obviously need a mathematical "recipe" allowing us to position the molecule fixed axis system [and the equilibrium geometry attached to it] relative to the instantaneous positions of the nuclei. In the standard approach, this recipe is given through the *Eckart equations* (Eckart 1935).

In order to formulate the Eckart equations, we define the following column vectors

$$\mathbf{r}_\alpha = \begin{Bmatrix} x_\alpha \\ y_\alpha \\ z_\alpha \end{Bmatrix}, \quad \mathbf{R}_\alpha = \begin{Bmatrix} X_\alpha \\ Y_\alpha \\ Z_\alpha \end{Bmatrix}, \quad \text{and } \mathbf{R}_0 = \begin{Bmatrix} X_0 \\ Y_0 \\ Z_0 \end{Bmatrix} \tag{29}$$

where $(x_\alpha, y_\alpha, z_\alpha)$ are the coordinates of nucleus α in the *molecule fixed* coordinate system. Since the molecule fixed coordinate system is rotated by the Euler angles relative to the laboratory fixed axis system, we have that

$$\mathbf{r}_\alpha = \mathbf{S}(\theta, \phi, \chi)\,\{\mathbf{R}_\alpha - \mathbf{R}_0\} \tag{30}$$

where $\mathbf{S}(\theta, \phi, \chi)$ is a 3×3 matrix effecting the transformation between the molecule fixed and laboratory fixed coordinate system; the elements of this matrix are simple trigonometric functions of the angles θ, ϕ, and χ [see Appendix A and Table A.1 of Papoušek and Aliev (1982)].

We now define the instantaneous displacements of the nuclei from the equilibrium configuration, measured along the axes of the molecule fixed coordinate system, as column vectors

$$\mathbf{d}_\alpha = \begin{Bmatrix} d_{\alpha x} \\ d_{\alpha y} \\ d_{\alpha z} \end{Bmatrix} = \mathbf{r}_\alpha - \mathbf{a}_\alpha, \quad \alpha = 1, 2, \ldots, N. \tag{31}$$

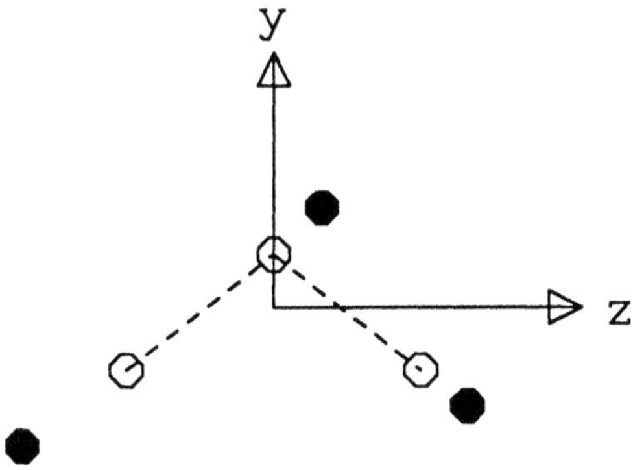

Figure 1. The definition of the molecule fixed coordinate system for a triatomic molecule [the x axis is chosen so that the x, y, and z axes form a right-handed coordinate system]. The open circles are the positions of the nuclei in the equilibrium configuration; the filled circles are the instantaneous positions of the nuclei.

The vector components $d_{\alpha x}$, $d_{\alpha y}$, and $d_{\alpha z}$, $\alpha = 1, 2, \ldots, N$, are the vibrational coordinates. However, we have seen in Section 2.2 that there exist only $3N - 6$ independent vibrational coordinates and consequently the $3N$ components of the $\mathbf{d}_\alpha$-vectors cannot all be independent. Six of the $3N$ components are redundant, and consequently there must exist six equations linking all $3N$ components. We already know three of these equations: The molecule fixed coordinate system is defined to have its origin at the nuclear center of mass [Section 2.2], both when the molecule is in its equilibrium configuration and when it is in an arbitrary, displaced configuration. One can easily show that this leads to the equations

$$\sum_{\alpha=1}^{N} m_\alpha \mathbf{a}_\alpha = \sum_{\alpha=1}^{N} m_\alpha \mathbf{d}_\alpha = \mathbf{O} \tag{32}$$

where $\mathbf{O}$ is a zero, three-component column vector. Equation (32) delivers three equations connecting the components of the vectors $\mathbf{d}_\alpha$. We need to choose three additional equations connecting these components, and in order to understand the choice that is customarily made we consider the *classical* kinetic energy T of the molecular rotation and vibration. This energy is given by Eq. (3) in Section 11-1 of Wilson *et al.* (1955):

$$2T = \sum_{\alpha=1}^{N} m_\alpha (\omega \times \mathbf{r}_\alpha) \cdot (\omega \times \mathbf{r}_\alpha) + \sum_{\alpha=1}^{N} m_\alpha \dot{\mathbf{r}}_\alpha \cdot \dot{\mathbf{r}}_\alpha + 2\omega \cdot \sum_{\alpha=1}^{N} m_\alpha \mathbf{r}_\alpha \times \dot{\mathbf{r}}_\alpha \tag{33}$$

where the vector ω contains the components of the instantaneous angular velocity of the molecule and, for an arbitrary quantity x, $\dot{x}$ is the time derivative dx/dt.

The last term on the right hand side of Eq. (33) is the Coriolis coupling term which represents the dominant part of the interaction between rotation and vibration. We would like to choose the coordinates so that this term disappears completely, i.e.,

$$\sum_{i=1}^{N} m_\alpha \mathbf{r}_\alpha \times \dot{\mathbf{r}}_\alpha = \mathbf{O} \tag{34}$$

where $\dot{\mathbf{r}}_\alpha$ is the velocity of nucleus α relative to the molecule fixed axes. This would mean that the molecule would have no angular momentum relative to the molecule fixed axes. It turns out that we cannot in general choose coordinates that cause the angular momentum in the molecule fixed axis system [and hence the Coriolis coupling term] to disappear. The best we can do is to minimize this angular momentum through the *Eckart equations* [Eckart, 1935] which we can express as

$$\sum_{\alpha=1}^{N} m_\alpha \mathbf{a}_\alpha \times \mathbf{d}_\alpha = \mathbf{O}. \tag{35}$$

Using Eq. (35) we can easily show that classically

$$\sum_{\alpha=1}^{N} m_\alpha \mathbf{r}_\alpha \times \dot{\mathbf{r}}_\alpha = \sum_{\alpha=1}^{N} m_\alpha \mathbf{d}_\alpha \times \dot{\mathbf{d}}_\alpha \tag{36}$$

so that the Coriolis coupling term vanishes in the equilibrium configuration when all $\mathbf{d}_\alpha = \mathbf{O}$. Since we assume that the molecule carries out small oscillations around the equilibrium configuration, the angular momentum in the molecule fixed coordinate system is then always small and a high degree of rotation-vibration separation is achieved.

When we insert Eqs. (30) and (31) in Eq. (35) we obtain

$$\sum_{\alpha=1}^{N} m_\alpha \mathbf{a}_\alpha \times [\mathbf{S}(\theta,\phi,\chi)\{\mathbf{R}_\alpha - \mathbf{R}_0\}] = \mathbf{O}. \tag{37}$$

In a given instantaneous configuration of the molecule, we can in principle measure the Cartesian laboratory fixed coordinates contained in the vectors $\mathbf{R}_\alpha$ and use Eq. (9) to obtain $\mathbf{R}_0$. If we assume that we know the constant vectors $\mathbf{a}_\alpha$, Eq. (37) then delivers three equations in the Euler angles θ, ϕ, and χ, and we can solve the equations to obtain these angles as functions of the laboratory-fixed Cartesian coordinates.

The first term on the right hand side of Eq. (33) is the rotational kinetic energy, and we can rewrite it as

$$\sum_{\alpha=1}^{N} m_\alpha (\omega \times \mathbf{r}_\alpha) \cdot (\omega \times \mathbf{r}_\alpha) = \omega^{\mathrm{T}} \mathbf{I} \omega \tag{38}$$

where a superscript T denotes transposition and the 3×3 matrix $\mathbf{I}$ contains the instantaneous moments of inertia and products of inertia [see, for example, Section 4.1 of Papoušek and Aliev (1982)]. When the molecule is in its equilibrium configuration, i.e. when all $\mathbf{r}_\alpha = \mathbf{a}_\alpha$, we have that $\mathbf{I} = \mathbf{I}_e$, and the elements of $\mathbf{I}_e$ depend on the components of the constant vectors $\mathbf{a}_\alpha$ only. These vectors must be chosen such that when all $\mathbf{r}_\alpha = \mathbf{a}_\alpha$ the relative positions of the nuclei correspond to the minimum of the potential energy function. However in defining the $\mathbf{a}_\alpha$ we are free to rotate them relative to the molecule fixed axis system since this does not change the relative positions of the nuclei. We can use this freedom of choice to make the molecule fixed axis system a *principal axis system* for the molecule in its equilibrium configuration, i.e., to make $\mathbf{I}_e$ diagonal. Hence for the molecule in its equilibrium configuration we have

$$\omega^{\mathrm{T}} \mathbf{I}_e \omega = I_{e,xx}\, \omega_x^2 + I_{e,yy}\, \omega_y^2 + I_{e,zz}\, \omega_z^2. \tag{39}$$

Figure 1 shows how the molecule fixed axis system can be defined for a triatomic molecule.

3.2 Normal coordinates

We have now defined the rotational Euler angle coordinates through the Eckart equations (35). We now need to define the $3N - 6$ vibrational coordinates [which we called (q_1, q_2,

$q_3, \ldots, q_{3N-6})$ in Section 2.2]. Through this definition we aim at simplifying the *vibrational* Hamiltonian. The classical vibrational energy is the sum of the vibrational kinetic energy, which is given through the second term on the right hand side of Eq. (33):

$$T_{\text{vib}} = \frac{1}{2}\sum_{\alpha=1}^{N} m_\alpha \dot{\mathbf{r}}_\alpha \cdot \dot{\mathbf{r}}_\alpha = \frac{1}{2}\sum_{\alpha=1}^{N} m_\alpha \dot{\mathbf{d}}_\alpha \cdot \dot{\mathbf{d}}_\alpha = \frac{1}{2}\sum_{\alpha=1}^{N} m_\alpha \left\{\dot{d}_{\alpha x}^2 + \dot{d}_{\alpha y}^2 + \dot{d}_{\alpha z}^2\right\} = \frac{1}{2}\dot{\mathbf{d}}^T \mathbf{M}\dot{\mathbf{d}} \quad (40)$$

and the potential energy $V_{\text{ai}}(\mathbf{R}_n)$. In Eq. (40), $\mathbf{d}$ is a column vector with $3N$ rows containing the components of the vectors $\mathbf{d}_\alpha$: d_{1x}, d_{1y}, d_{1z}, d_{2x}, ..., d_{Nz}, and $\mathbf{M}$ is a $3N \times 3N$ diagonal matrix with $M_{11} = M_{22} = M_{33} = m_1$, $M_{44} = M_{55} = M_{66} = m_2$ and so on.

In the standard approach the potential energy $V_{\text{ai}}(\mathbf{R}_n)$ is taken to be a Taylor expansion in the geometrically defined coordinates of Section (2.3):

$$\begin{aligned} V_{\text{ai}}(\mathcal{R}_1, \mathcal{R}_2, \ldots, \mathcal{R}_{3N-6}) &= \frac{1}{2}\sum_{j,k} f_{jk}\mathcal{R}_j\mathcal{R}_k \\ &+ \frac{1}{6}\sum_{j,k,\ell} f_{jk\ell}\mathcal{R}_j\mathcal{R}_k\mathcal{R}_\ell + \frac{1}{24}\sum_{j,k,\ell,m} f_{jk\ell}\mathcal{R}_j\mathcal{R}_k\mathcal{R}_\ell\mathcal{R}_m + \ldots \end{aligned} \quad (41)$$

where the coordinates $\mathcal{R}_i$ are defined as *displacements* from equilibrium values so that the configuration in which all $\mathcal{R}_i = 0$ corresponds to the equilibrium geometry [where all first derivatives are zero and where we choose $V_{\text{ai}} = 0$].

In order to define the vibrational motion of the molecule, we now define $3N - 6$ independent, *linearized internal coordinates* S_i through the transformation

$$S_i = \sum_{\alpha=1}^{N}\sum_{a=x,y,z} B_{i,\alpha a} d_{\alpha a}, \quad i = 1, 2, 3, \ldots, 3N-6, \quad (42)$$

with the elements of the $(3N - 6) \times 3N$ matrix $\mathbf{B}$ given as

$$B_{i,\alpha a} = \frac{\partial \mathcal{R}_i}{\partial d_{\alpha a}}; \quad (43)$$

the elements of this matrix can be derived using purely geometrical arguments. In these coordinates the kinetic energy of the vibrations can be expressed as [see Wilson *et al.* (1955)]

$$T_{\text{vib}} = \frac{1}{2}\dot{\mathbf{S}}^T \mathbf{G}^{-1}\dot{\mathbf{S}} \quad (44)$$

with $\mathbf{S}$ being a column vector with $3N - 6$ rows containing the S_i coordinates and

$$\mathbf{G} = \mathbf{B}\mathbf{M}^{-1}\mathbf{B}^T. \quad (45)$$

Through the definition of the $\mathbf{B}$ matrix we have for small values of the $d_{\alpha a}$ [i.e., configurations close to equilibrium] that $S_i \approx \mathcal{R}_i$, and we can then write

$$\frac{1}{2}\sum_{j,k} f_{jk}\mathcal{R}_j\mathcal{R}_k \approx \frac{1}{2}\sum_{j,k} f_{jk}S_jS_k = \frac{1}{2}\mathbf{S}^T\mathbf{F}\mathbf{S} \quad (46)$$

where the matrix $\mathbf{F}$ has the elements f_{jk}. Using a standard technique of classical mechanics, the **GF**-*calculation*, we can change to so-called normal coordinates Q_k given through the transformation

$$d_{\alpha a} = \frac{1}{\sqrt{m_\alpha}}\sum_{k=1}^{3N-6} \ell_{\alpha a,k}\, Q_k. \quad (47)$$

The normal coordinates are defined such that

$$T_{\text{vib}} + \frac{1}{2}\sum_{j,k} f_{jk}S_jS_k = \frac{1}{2}\dot{\mathbf{S}}^T\mathbf{G}^{-1}\dot{\mathbf{S}} + \frac{1}{2}\mathbf{S}^T\mathbf{F}\mathbf{S} = \frac{1}{2}\sum_{k=1}^{3N-6}\left\{\dot{Q}_k^2 + \lambda_k Q_k^2\right\}. \tag{48}$$

In Eqs. (47) and (48), the λ_k are the eigenvalues of the matrix **GF**, and the coefficients $\ell_{\alpha a,k}$ can be obtained from the eigenvectors of this matrix [see Eq. (3.3.10) of Papoušek and Aliev (1982)]. When we approximate the vibrational energy by the expression in Eq. (48) [that is, when we neglect expansion terms of third and higher order from Eq. (41)] we can write this energy as the sum of $3N-6$ independent contributions, each one only depending on one Q_k coordinate and we have thus simplified considerably the form of the vibrational energy.

We have now defined the coordinates used in the standard approach, and during this process we have made three choices:

- We have chosen the Eckart equations (35) to define the Euler angles θ, ϕ, and χ. The direct reason for making this choice was to minimize the Coriolis coupling term [the third term on the right hand side of Eq. (33)] which contains the major contribution to the rotation-vibration interaction.
- We have rotated the constant $\mathbf{a}_\alpha$-vectors relative to the molecule fixed axis system so that the molecule fixed axes are principal axes when the molecule is in its equilibrium configuration. Thus we obtain that the products of inertia vanish at equilibrium [Eq. (39)].
- We have chosen the vibrational coordinates to be the normal coordinates Q_k, thus eliminating terms coupling different vibrational modes of the molecule from the vibrational kinetic energy T_{vib} and from the *harmonic* potential energy [Eq. (46)].

Clearly each choice was made in order to eliminate certain terms in the classical energy. Since the eliminated terms in the kinetic energy are mass-dependent, the resulting coordinates are also all mass-dependent in the sense defined in Section 2.1.

When the classical Hamiltonian function is expressed in terms of the coordinates (θ, ϕ, χ, Q_1, Q_2, Q_3, $\ldots Q_{3N-6}$) and converted to quantum mechanical form using the *Podolsky trick* [see Chapter 7 of Bunker (1979)] the following quantum mechanical operator is obtained:

$$\begin{aligned}\widehat{H}_{\text{rot-vib}} &= \frac{1}{2}\sum_{\delta,\gamma=x,y,z}\left(\widehat{J}_\delta - \widehat{p}_\delta\right)\mu_{\delta,\gamma}(Q_1,Q_2,\ldots,Q_{3N-6})\left(\widehat{J}_\gamma - \widehat{p}_\gamma\right)\\ &+ U(Q_1,Q_2,\ldots,Q_{3N-6}) + \frac{1}{2}\sum_{k=1}^{3N-6}\left\{\widehat{P}_k^2 + \lambda_k Q_k^2\right\}\\ &+ \frac{1}{6}\sum_{j,k,\ell}\Phi_{jk\ell}Q_jQ_kQ_\ell + \frac{1}{24}\sum_{j,k,\ell,m}\Phi_{jk\ell m}Q_jQ_kQ_\ell Q_m + \ldots\end{aligned} \tag{49}$$

where the operators $\widehat{J}_\delta$, $\delta = x$, y, z, are the components of the total angular momentum of the nuclei relative to the molecule fixed axes, the functions $\mu_{\delta,\gamma}(Q_1,Q_2,\ldots,Q_{3N-6})$ are elements of the matrix $\mu = (\mathbf{I}')^{-1}$, where the elements of $\mathbf{I}'$ are given by Eq. (10) in Section 11-2 of Wilson *et al.* (1955). The matrix $\mathbf{I}'$ is almost, but not quite, equal to the matrix $\mathbf{I}$ containing the instantaneous values of the moments and products of inertia. Further, $\widehat{P}_k = -i\hbar\partial/\partial Q_k$ is the momentum conjugate to Q_k, the vibrational angular momentum operators $\widehat{p}_\delta$ are given by

$$\widehat{p}_\delta = \sum_{k,\ell=1}^{3N-6}\zeta_{k,\ell}^{\delta}Q_k\widehat{P}_\ell \tag{50}$$

where the *Coriolis coupling constants* $\zeta_{k,\ell}^{\delta}$ depend on the matrix elements $\ell_{\alpha\alpha,k}$ [Eq. (47)] through Eq. (4.1.7) of Papoušek and Aliev (1982). The function U is given by [Watson, 1968*a*]

$$U(Q_1, Q_2, \ldots, Q_{3N-6}) = -\frac{\hbar^2}{8} \sum_{\delta=x,y,z} \mu_{\delta\delta}(Q_1, Q_2, \ldots, Q_{3N-6}), \tag{51}$$

and the parameters $\Phi_{jk\ell\ldots}$ are the expansion coefficients of the potential energy V_{ai} expanded in terms of the Q_k coordinates. The derivation of these expansion coefficients is the subject of Hoy *et al.* (1972).

The standard procedure for obtaining the eigenvalues of $\hat{H}_{\text{rot-vib}}$ is to construct its matrix representation in a basis set of rigid rotor functions $|J, K, M\rangle$ [which depend on the Euler angles θ, ϕ, and χ; see Chapter 6 of Papoušek and Aliev (1982)] multiplied by products of harmonic oscillator functions depending on the normal coordinates (Q_1, Q_2, ..., Q_{3N-6}). We assume that there exist n_2 *degeneracies* between the λ_k-parameters, that is, there are n_2 pairs, λ_m and λ_n, for which $\lambda_m = \lambda_n$ [we will not consider here the special case, encountered for spherical top molecules, for which *three* λ_k-values are equal]. In this case we can write the harmonic oscillator products as

$$|v_1 v_2 \ldots v_{n_1}; v_{n_1+1}^{\ell_{n_1+1}} v_{n_1+2}^{\ell_{n_1+2}} \ldots v_{n_1+n_2}^{\ell_{n_1+n_2}}\rangle = \prod_{k=1}^{n_1} |v_k\rangle \times \prod_{i=n_1+1}^{n_1+n_2} |v_i^{\ell_i}\rangle \tag{52}$$

where $n_1 + 2n_2 = 3N - 6$ [see Chapter 5 of Papoušek and Aliev (1982)]. The function $|v_1 v_2 \ldots v_{n_1}; v_{n_1+1}^{\ell_{n_1+1}} v_{n_1+2}^{\ell_{n_1+2}} \ldots v_{n_1+n_2}^{\ell_{n_1+n_2}}\rangle$ diagonalizes the operator

$$\hat{H}_{\text{vib}}^{(0)} = \frac{1}{2} \sum_{k=1}^{3N-6} \left\{ \hat{P}_k^2 + \lambda_k Q_k^2 \right\} \tag{53}$$

with the eigenvalues

$$E_{\text{vib}}^{(0)}(v_1 v_2 \ldots v_{n_1}; v_{n_1+1}^{\ell_{n_1+1}} v_{n_1+2}^{\ell_{n_1+2}} \ldots v_{n_1+n_2}^{\ell_{n_1+n_2}}) = \sum_{k=1}^{n_1} \hbar\sqrt{\lambda_k}\left(v_k + \frac{1}{2}\right) + \sum_{i=n_1+1}^{n_1+n_2} \hbar\sqrt{\lambda_i}\,(v_i + 1) \tag{54}$$

As already mentioned, the standard approach to rotation-vibration theory attempts to diagonalize the matrix representation of $\hat{H}_{\text{rot-vib}}$ in the basis set

$$|\phi_{\text{basis}}\rangle = |J, K, M\rangle \times |v_1 v_2 \ldots v_{n_1}; v_{n_1+1}^{\ell_{n_1+1}} v_{n_1+2}^{\ell_{n_1+2}} \ldots v_{n_1+n_2}^{\ell_{n_1+n_2}}\rangle. \tag{55}$$

In order to calculate the matrix elements $\langle\phi_{\text{basis}}|\hat{H}_{\text{rot-vib}}|\phi'_{\text{basis}}\rangle$ it is customary to approximate the Q_k dependent functions $\mu_{\delta,\gamma}$ and U appearing in $\hat{H}_{\text{rot-vib}}$ as Taylor series in the Q_k, e.g.

$$\mu_{\delta,\gamma}(Q_1, Q_2, \ldots, Q_{3N-6}) = \mu_{\delta,\gamma}^{e} + \sum_{k=1}^{3N-6} \mu_{\delta,\gamma}^{(k)} Q_k + \frac{1}{2} \sum_{k,\ell=1}^{3N-6} \mu_{\delta,\gamma}^{(k\ell)} Q_k Q_\ell + \ldots. \tag{56}$$

Such expansions are normally truncated after the second order terms. The reason for making this approximation is that matrix elements of the type $\langle\phi_{\text{basis}}|Q_k^{n_k} Q_\ell^{n_\ell} \ldots|\phi'_{\text{basis}}\rangle$ can be calculated *analytically* [see Chapter 5 of Papoušek and Aliev (1982)]. Without the approximation it would be necessary to calculate the integrals over the Q_k coordinates using expensive multidimensional numerical integration. Whitehead and Handy (1975) have developed a method to obtain the eigenvalues of $\hat{H}_{\text{rot-vib}}$ *variationally*, that is through direct diagonalization in the basis set given by Eq. (55) using multidimensional numerical integration to calculate the matrix elements without making the approximation defined by Eq. (56). In the standard approach the eigenvalues are obtained using *perturbation theory*, and we shall here briefly discuss this approach.

3.3 The van Vleck transformation

If we let $\mathbf{H}$ be the matrix representing $\widehat{H}_{\mathrm{rot-vib}}$ in the basis set given by Eq. (55), we can in principle diagonalize this matrix. This means that we determine a similarity transformation effected by the unitary matrix $\mathbf{C}$ so that

$$\mathbf{D} = \mathbf{C}\mathbf{H}\mathbf{C}^{-1} \tag{57}$$

is diagonal, and the diagonal elements of $\mathbf{D}$ are the eigenvalues of $\mathbf{H}$. Now the idea of perturbation theory is to use an unitary matrix $\mathbf{U}$ so that

$$\widetilde{\mathbf{H}} = \mathbf{U}\mathbf{H}\mathbf{U}^{-1} \tag{58}$$

is *approximately* diagonal, i.e., in carrying out the similarity transformation given by Eq. (58) we obtain that the matrix $\widetilde{\mathbf{H}}$ has smaller off-diagonal elements than the matrix $\mathbf{H}$ so that we can neglect the off-diagonal elements of $\widetilde{\mathbf{H}}$ and use its diagonal elements as approximations for the eigenvalues of $\mathbf{H}$.

One version of perturbation theory is the socalled van Vleck transformation [van Vleck, 1929; Jordahl, 1934; see also Gordy and Cook (1970)] whose principle we briefly present here. For clarity we denote here by $\phi_j^{(0)}$ the basis functions given by Eq. (55), where j is an index numbering the functions. To each basis function corresponds a zero order energy

$$E_j^{(0)} = H_{jj} = \langle \phi_j^{(0)} | \widehat{H}_{\mathrm{rot-vib}} | \phi_j^{(0)} \rangle. \tag{59}$$

We consider a set of ℓ basis functions $\phi_i^{(0)}$, $i = i_0, i_0+1, \ldots, i_0+\ell-1$, that are energetically close, while all other basis function are energetically far away. That is, for $i_0 \leq i < i_0+\ell$, $i_0 \leq i' < i_0+\ell$, and $k < i_0$ or $k \geq i_0+\ell$ we have

$$|E_i^{(0)} - E_{i'}^{(0)}| \ll |E_k^{(0)} - E_i^{(0)}|. \tag{60}$$

In the van Vleck theory it is assumed that between all eigenvalues of $\widehat{H}_{\mathrm{rot-vib}}$ there exists a group of ℓ eigenvalues "near" the zero order energies $E_i^{(0)}$, $i_0 \leq i < i_0+\ell$, and that we can obtain these eigenvalues through diagonalization of an $\ell \times \ell$ matrix with elements

$$\begin{aligned} h_{ii'} &= \langle \phi_i^{(0)} | \widehat{H}_{\mathrm{rot-vib}} | \phi_{i'}^{(0)} \rangle \\ &+ \frac{1}{2} \sum_k \langle \phi_i^{(0)} | \widehat{H}_{\mathrm{rot-vib}} | \phi_k^{(0)} \rangle \langle \phi_k^{(0)} | \widehat{H}_{\mathrm{rot-vib}} | \phi_{i'}^{(0)} \rangle \left[\frac{1}{E_i^{(0)} - E_k^{(0)}} + \frac{1}{E_{i'}^{(0)} - E_k^{(0)}} \right], \end{aligned} \tag{61}$$

where $i_0 \leq i < i_0+\ell$ and $i_0 \leq i' < i_0+\ell$, and the sum runs over all $k < i_0$ or $k \geq i_0+\ell$. The entire interaction between the basis functions $\phi_i^{(0)}$, $i = i_0, i_0+1, \ldots, i_0+\ell-1$, and all other basis functions $|\phi_k^{(0)}\rangle$ is treated through the second term in Eq. (61). If $\ell = 1$ [i.e., $i = i' = i_0$] Eq. (61) yields the known expression from second order perturbation theory:

$$h_{i_0 i_0} = \langle \phi_{i_0}^{(0)} | \widehat{H}_{\mathrm{rot-vib}} | \phi_{i_0}^{(0)} \rangle + \sum_{k \neq i_0} \frac{\langle \phi_{i_0}^{(0)} | \widehat{H}_{\mathrm{rot-vib}} | \phi_k^{(0)} \rangle \langle \phi_k^{(0)} | \widehat{H}_{\mathrm{rot-vib}} | \phi_{i_0}^{(0)} \rangle}{E_{i_0}^{(0)} - E_k^{(0)}}. \tag{62}$$

In setting up the matrix $\mathbf{H}$ in terms of the basis functions given in Eq. (55) one can easily show that the matrix is block diagonal in the angular momentum quantum numbers J and M. Further, the matrix elements are independent of M. In the standard approach it is assumed that basis functions with different v_k quantum numbers are widely separated in energy so that the ideas of the van Vleck transformation can be employed. The group of basis functions $\phi_i^{(0)}$,

$i = i_0, i_0+1, \ldots, i_0+\ell-1$, is thus taken to have a common J and M values and common values for all v_k quantum numbers but different values for the quantum numbers K and ℓ_k. Normally the dependence of the zero order energy [used to construct the energy denominators] on the rotational quantum numbers is neglected, i.e. one takes

$$E^{(0)}(v_1 v_2 \ldots v_{n_1}; v_{n_1+1}^{\ell_{n_1+1}} v_{n_1+2}^{\ell_{n_1+2}} \ldots v_{n_1+n_2}^{\ell_{n_1+n_2}}) = \sum_{k=1}^{n_1} \hbar\sqrt{\lambda_k}\left(v_k + \frac{1}{2}\right) + \sum_{i=n_1+1}^{n_1+n_2} \hbar\sqrt{\lambda_i}\,(v_i+1). \quad (63)$$

Though the van Vleck transformation one thus obtains a matrix block characterized by J and the v_k [the value of M is immaterial], and this matrix block can be diagonalized. When this technique is employed for an asymmetric top molecule, the Watson Hamiltonian [or *Watsonian*] is obtained [Watson, 1967; 1968*b*; 1968*c*; 1977]. The formalism can be modified to account for the situation when to sets of basis functions characterized by the v_k quantum numbers (v_1, v_2, ..., $v_{n_1+n_2}$) and (v_1', v_2', ..., $v_{n_1+n_2}'$), respectively, are energetically close [*an accidental resonance*]. In this case one simply takes the group of basis functions $\phi_i^{(0)}$, $i = i_0, i_0+1, \ldots, i_0+\ell-1$, to contain all K and ℓ_k functions corresponding to both these sets of v_k-values.

Brodersen and Lolck (1987) [see also Brodersen (1988, 1991)] have developed a method for diagonalizing the "standard" rotation-vibration Hamiltonian for methane-type spherical top molecules. In this approach, the $J = 0$ energies and wavefunctions are initially obtained through diagonalization of the Hamiltonian in a truncated basis of harmonic oscillator functions. The matrix blocks for $J > 0$ are then constructed in a truncated basis consisting of products of the calculated $J = 0$ functions and rigid rotor functions. The size of these matrix blocks is now reduced in that selected off-diagonal matrix elements are removed through the van Vleck transformation given by Eq. (61), and the reduced matrix blocks are finally diagonalized. This method is highly analogous to the MRDCI *ab initio* method [Buenker and Peyerimhoff, 1974; Buenker *et al.*, 1978; Buenker, 1980; Buenker, 1981; Buenker and Phillips, 1985].

3.4 The contact transformation

There is an alternative, equivalent way to think about the van Vleck transformation. This is the socalled *contact transformation* which is often mentioned in spectroscopic work. In the van Vleck transformation described in Section 3.3 we construct the matrix representation $\mathbf{H}$ of $\widehat{H}_{\text{rot-vib}}$ and then we manipulate the *matrix* $\mathbf{H}$ through Eq. (61) [which is equivalent to carrying out the similarity transformation given in Eq. (58)] to obtain another *matrix* $\widetilde{\mathbf{H}}$ whose off-diagonal elements are smaller than those of $\mathbf{H}$. The "normal" way of thinking of this would be to say that $\widetilde{\mathbf{H}}$ is the matrix representation of $\widehat{H}_{\text{rot-vib}}$ in a new basis set φ_i obtained from the old basis set through the transformation

$$\varphi_i = \sum_j \left(\mathbf{U}^{-1}\right)_{ij} \phi_j \quad (64)$$

In the contact transformation, however, one considers $\widetilde{\mathbf{H}}$ to be the matrix representation of a *transformed Hamiltonian* $\widetilde{H}_{\text{rot-vib}}$ in *the old* basis set $|\phi_i\rangle$. The transformed operator is obtained by interpreting Eq. (58) as an operator equation

$$\widetilde{H}_{\text{rot-vib}} = \widehat{P}\widehat{H}_{\text{rot-vib}}\widehat{P}^{-1} \quad (65)$$

where $\widehat{P}$ is an operator whose matrix elements in the basis set $\phi_i^{(0)}$ are the elements of the $\mathbf{U}$ matrix

$$U_{ij} = \langle\phi_i^{(0)}|\widehat{P}|\phi_j^{(0)}\rangle. \quad (66)$$

The basic problem is the contact transformation is now to determine the operator $\widehat{P}$ to fulfill Eq. (66) and then to derive $\widetilde{H}_{\text{rot-vib}}$ from Eq. (65). The traditional way of doing this is to write $\widehat{P}$ as

$$\widehat{P} = \exp(i\widehat{S}) \tag{67}$$

where $\widehat{S}$ is a hermitian operator, and one takes the exponential as its Taylor series:

$$\exp(i\widehat{S}) = 1 + i\widehat{S} - \frac{1}{2}\widehat{S}^2 - \frac{i}{6}\widehat{S}^3 \dots . \tag{68}$$

In practice, one derives in the contact transformation $\widehat{S}$ and not $\widehat{P}$. One can then write Eq. (65) as

$$\widetilde{H}_{\text{rot-vib}} = \exp(i\widehat{S})\widehat{H}_{\text{rot-vib}}\exp(-i\widehat{S}) \tag{69}$$

and one can derive $\widetilde{H}_{\text{rot-vib}}$ from Eq. (68). The advantage of the contact transformation is that one can obtain the transformed Hamiltonian $\widetilde{H}_{\text{rot-vib}}$ once and for all, that is for *all* vibrational states of the molecule. In the van Vleck transformation one would have to derive the appropriate matrix block for each (v_1, v_2, ..., $v_{n_1+n_2}$) state separately. It should be emphasized that van Vleck- and contact transformations are only appropriate if the energy separations between the vibrational states are much larger than the matrix elements coupling these states. We shall see below that for most molecules, the perturbation theory approaches outlined here are poor approximations.

4. THE HOUGEN-BUNKER-JOHNS APPROACH

As mentioned above, the Hougen-Bunker-Johns approach [henceforth: the HBJ approach] is a generalization of the standard approach described in the previous sections. It describes the situation when not *all* vibrational amplitudes are small compared to the linear dimensions of the equilibrium structure of the molecule. The basic idea behind the HBJ approach was suggested by Sayvetz (1939), but the idea could not be exploited in practice until digital computers became available, and the modern research in this field originated with the pioneering work of Hougen, Bunker, and Johns (1970). In the case when one or more vibrational modes have amplitudes comparable with the linear dimensions of the molecule [i.e., they are *large amplitude motions*], it would clearly not be reasonable to expand the Hamiltonian as a power series in the normal coordinates with starting point in a rigid equilibrium configuration. Such a Taylor expansion would converge unacceptably slowly.

4.1 Coordinates for the large amplitude motion

In the HBJ approach the vibrational motions are divided into large amplitude motions and small amplitude motions. We will here consider the example of a triatomic molecule with a large amplitude bending motion and small amplitude stretching of the bond distances. In order to eliminate the translational motion from the Schrödinger problem [see Section 2.1] we introduce the center-of-mass coordinates (X_0, Y_0, Z_0) in exactly the same way as it is done in the standard approach [Eq. (9)]. We still aim at defining three Euler angles θ, ϕ und χ through a molecule fixed axis system which follows the rotation of the molecule. We do not, however, describe the vibrational motion as displacements from one rigid equilibrium structure. Instead we introduce a variable *reference structure* which follows the rotation *and* the large amplitude motion. For the triatomic molecule that we use as example here, we choose the reference structure as having fixed internuclear distances r_{12}^e [the equilibrium distance between the "outer" nucleus 1 and the center nucleus 2] and r_{32}^e [the equilibrium distance between the outer nucleus 3 and the center nucleus 2], but with a variable bond angle α. The coordinate $\rho = \pi - \alpha$ is chosen to describe the large amplitude motion, and the stretching motions are described as displacements from the reference geometry. Mathematically this means that we cannot introduce *constant* vectors $\mathbf{a}_\alpha$ that define the equilibrium structure in the molecule fixed axis system. Instead *variable* vectors $\mathbf{a}_\alpha(\rho)$ are defined; these vectors depend on the large amplitude coordinate ρ. We define displacement vectors $\mathbf{d}_\alpha$ as $\mathbf{d}_\alpha = \mathbf{r}_\alpha - \mathbf{a}_\alpha(\rho)$ in analogy with Eq. (31); $\mathbf{r}_\alpha$ is the position vector of nucleus α in the molecule fixed axis system.

The classical velocity of nucleus α relative to the laboratory fixed coordinate system is given by

$$\mathbf{V}_\alpha = \dot{\mathbf{R}}_0 + \omega \times [\mathbf{d}_\alpha + \mathbf{a}_\alpha(\rho)] + \dot{\mathbf{d}}_\alpha + \frac{d\mathbf{a}_\alpha}{d\rho}\dot{\rho} \tag{70}$$

where the vector $\mathbf{R}_0$ is the laboratory fixed position vector of the nuclear center of mass [see Eq. (29)], the vector ω contains the components of the instantaneous angular velocities exactly as in the standard approach, and $\dot{\rho} = d\rho/dt$.

In order to understand the definition of the HBJ coordinates we consider the expression for the classical kinetic energy in the HBJ approach [cf. Eq. (33)]. We can derive this expression on the basis of Eq. (70). If we use that the origin of the molecule fixed axis system is always at the nuclear center of mass so that

$$\sum_{\alpha=1}^{N} m_\alpha \mathbf{a}_\alpha(\rho) = \sum_{\alpha=1}^{N} m_\alpha \mathbf{d}_\alpha = \mathbf{O} \tag{71}$$

we obtain for the classical, translation-free energy

$$\begin{aligned} 2T = & \sum_{\alpha=1}^{N} m_\alpha(\omega \times \mathbf{r}_\alpha)\cdot(\omega \times \mathbf{r}_\alpha) + \dot{\rho}^2 \sum_{\alpha=1}^{N} m_\alpha \frac{d\mathbf{a}_\alpha}{d\rho}\cdot\frac{d\mathbf{a}_\alpha}{d\rho} \\ - & \; 2\dot{\rho}\omega\cdot\sum_{\alpha=1}^{N} m_\alpha \frac{d\mathbf{a}_\alpha}{d\rho} \times [\mathbf{a}_\alpha(\rho) + \mathbf{d}_\alpha] + 2\dot{\rho}\sum_{\alpha=1}^{N} m_\alpha \frac{d\mathbf{a}_\alpha}{d\rho}\cdot\dot{\mathbf{d}}_\alpha \\ + & \sum_{\alpha=1}^{N} m_\alpha \dot{\mathbf{d}}_\alpha\cdot\dot{\mathbf{d}}_\alpha + 2\omega\cdot\sum_{\alpha=1}^{N} m_\alpha [\mathbf{a}_\alpha(\rho) + \mathbf{d}_\alpha] \times \dot{\mathbf{d}}_\alpha. \end{aligned} \tag{72}$$

The last term on the right hand side of Eq. (72) is the Coriolis coupling term containing products of rotational velocities ω_a and velocities of the small amplitude vibrations $\dot{\mathbf{d}}_\alpha$. We aim at minimizing this term through the Eckart equations

$$\sum_{\alpha=1}^{N} m_\alpha \mathbf{a}_\alpha(\rho) \times \mathbf{d}_\alpha = \mathbf{O}. \tag{73}$$

The time derivatives of these equations yield

$$\sum_{\alpha=1}^{N} m_\alpha \mathbf{a}_\alpha(\rho) \times \dot{\mathbf{d}}_\alpha = -\dot{\rho}\sum_{\alpha=1}^{N} m_\alpha \frac{d\mathbf{a}_\alpha}{d\rho} \times \mathbf{d}_\alpha. \tag{74}$$

The second term in the second line of Eq. (72) describes the interaction between large amplitude bending motion [$\dot{\rho}$] and small amplitude stretching motion [$\dot{\mathbf{d}}_\alpha$]. This term is simplified by means of the *Sayvetz equation* [Sayvetz, 1939]

$$\sum_{\alpha=1}^{N} m_\alpha \frac{d\mathbf{a}_\alpha(\rho)}{d\rho}\cdot\mathbf{d}_\alpha = \mathbf{O} \tag{75}$$

whose time derivative yields

$$\sum_{\alpha=1}^{N} m_\alpha \frac{d\mathbf{a}_\alpha(\rho)}{d\rho}\cdot\dot{\mathbf{d}}_\alpha = -\dot{\rho}\sum_{\alpha=1}^{N} m_\alpha \frac{d^2\mathbf{a}_\alpha}{d\rho^2}\cdot\mathbf{d}_\alpha. \tag{76}$$

A final choice we have to make here concerns the orientation of the reference structure relative to the molecule fixed coordinate system. We used the analogous choice in the standard approach to make the inertial tensor **I** diagonal when the molecule is in its equilibrium configuration. We could make a similar choice here, but instead we use the definitions of the $\mathbf{a}_\alpha$ vectors to simplify the first term in the second line of Eq. (72), which describes the interaction between the large amplitude bending [$\dot{\rho}$] and the rotation, given through the angular velocity ω. We define the orientation of the vectors $\mathbf{a}_\alpha$ vectors so that they fulfill the equation

$$\sum_{\alpha=1}^{N} m_\alpha \frac{d\mathbf{a}_\alpha}{d\rho} \times \mathbf{a}_\alpha(\rho) = \mathbf{O}. \tag{77}$$

Figure 2 shows the definition of the molecule fixed axis system in the HBJ approach.

When we insert the equations (74–77) in Eq. (72) [and introduce the instantaneous inertial tensor **I**] we obtain the following simplified kinetic energy

$$\begin{aligned} 2T &= \omega^T \mathbf{I}\omega + \dot{\rho}^2 \sum_{\alpha=1}^{N} m_\alpha \left\{ \frac{d\mathbf{a}_\alpha}{d\rho} \cdot \frac{d\mathbf{a}_\alpha}{d\rho} - 2\frac{d^2\mathbf{a}_\alpha}{d\rho^2} \cdot \mathbf{d}_\alpha \right\} \\ &+ \sum_{\alpha=1}^{N} m_\alpha \dot{\mathbf{d}}_\alpha \cdot \dot{\mathbf{d}}_\alpha - 4\dot{\rho}\omega \cdot \sum_{\alpha=1}^{N} m_\alpha \frac{d\mathbf{a}_\alpha}{d\rho} \times \mathbf{d}_\alpha + 2\omega \cdot \sum_{\alpha=1}^{N} m_\alpha \mathbf{d}_\alpha \times \dot{\mathbf{d}}_\alpha, \end{aligned} \tag{78}$$

where all interaction terms between different types of motion vanish in the reference configuration when all $\mathbf{d}_\alpha = \mathbf{O}$.

Following the procedure already outlined for the standard approach we can insert Eq. (30) in the equations (73) and (75) to obtain

$$\sum_{\alpha=1}^{N} m_\alpha \mathbf{a}_\alpha(\rho) \times \mathbf{d}_\alpha = \sum_{\alpha=1}^{N} m_\alpha \mathbf{a}_\alpha(\rho) \times \mathbf{S}(\theta,\phi,\chi)\{\mathbf{R}_\alpha - \mathbf{R}_0\} = \mathbf{O} \tag{79}$$

and

$$\sum_{\alpha=1}^{N} m_\alpha \frac{d\mathbf{a}_\alpha(\rho)}{d\rho} \cdot \mathbf{d}_\alpha = \sum_{\alpha=1}^{N} m_\alpha \frac{d\mathbf{a}_\alpha(\rho)}{d\rho} \cdot [\mathbf{S}(\theta,\phi,\chi)\{\mathbf{R}_\alpha - \mathbf{R}_0\} - \mathbf{a}_\alpha(\rho)] = 0. \tag{80}$$

The functions $\mathbf{a}_\alpha(\rho)$ are known, and hence Eqs. (79) and (80) deliver four equations in θ, ϕ, χ und ρ. Using these four equations we can determine the instantaneous position of the molecule fixed axis system [given through θ, ϕ, and χ] and the instantaneous value of ρ in an arbitrary geometry.

It is important to appreciate here that the coordinate ρ is defined through the Eckart and Sayvetz conditions and is *mass dependent*. For the triatomic molecule with large amplitude bending motion and small amplitude stretching motions it is almost, but not quite, equal to $\bar{\rho} = \pi - \bar{\alpha}$, where $\bar{\alpha}$ is the instantaneous value of the bond angle. We use the geometrically defined coordinate $\bar{\rho}$ together with the bond length displacements $\Delta r_{j2} = r_{j2} - r_{j2}^e$ (where r_{j2} is the instantaneous value of the distance between the outer nucleus j, $j = 1$ or 3, and the central nucleus 2 of the triatomic molecule) to express the potential energy function. By analogy with the standard approach, it is assumed that the nuclear potential energy function can be written as a rapidly converging power series in Δr_{12} and Δr_{32}:

$$\begin{aligned} V_{\mathrm{ai}}(\Delta r_{12}, \Delta r_{32}, \bar{\rho}) &= V_0(\bar{\rho}) + \sum_j \bar{F}_j(\bar{\rho})\Delta r_{j2} + \frac{1}{2}\sum_{j,k} \bar{F}_{jk}(\bar{\rho})\Delta r_{j2}\Delta r_{k2} \\ &+ \frac{1}{6}\sum_{j,k,m} \bar{F}_{jkm}(\bar{\rho})\Delta r_{j2}\Delta r_{k2}\Delta r_{m2} + \frac{1}{24}\sum_{j,k,m,n} \bar{F}_{jkmn}(\bar{\rho})\Delta r_{j2}\Delta r_{k2}\Delta r_{m2}\Delta r_{n2}, \end{aligned} \tag{81}$$

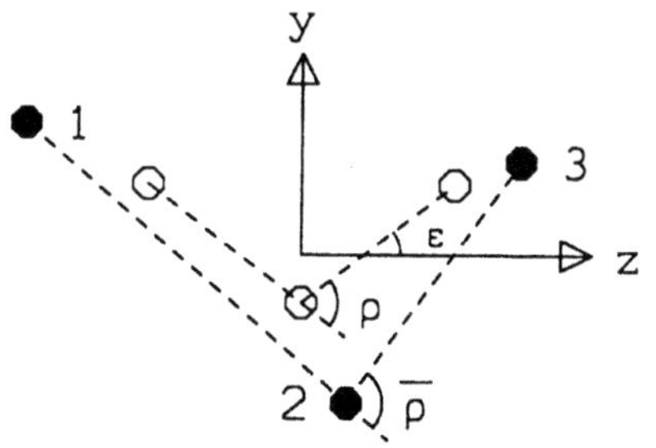

Figure 2. The definition of the molecule fixed coordinate system for a triatomic molecule in the HBJ approach [the x axis is chosen so that the x, y, and z axes form a right-handed coordinate system]. The open circles are the positions of the nuclei in the reference configuration; the filled circles are the instantaneous positions of the nuclei. The angle ϵ is defined so that Eq. (77) is fulfilled.

where the indices j, k, m, and n assume the values 1 or 3. The expansion coefficients $\bar{F}_{jk\ldots}$ are chosen as

$$\begin{aligned}\bar{F}_j(\bar{\rho}) &= f_{j2}(\rho_e - \bar{\rho}) + \frac{1}{2} f_{j22}(\rho_e - \bar{\rho})^2 + \frac{1}{6} f_{j222}(\rho_e - \bar{\rho})^3, \\ \bar{F}_{jk}(\bar{\rho}) &= f_{jk} + f_{jk2}(\rho_e - \bar{\rho}) + \frac{1}{2} f_{jk22}(\rho_e - \bar{\rho})^2, \\ \bar{F}_{jkm}(\bar{\rho}) &= f_{jkm} + f_{jkm2}(\rho_e - \bar{\rho}), \\ \bar{F}_{jkmn}(\bar{\rho}) &= f_{jkmn}\end{aligned} \tag{82}$$

where the $f_{jk\ldots}$ are usual force constants, and ρ_e is the equilibrium value of $\bar{\rho}$. The function $V_0(\bar{\rho})$ is the potential energy function for the molecule bending with its bond lengths fixed at their equilibrium values [i.e., bending in the reference configuration]. For molecules with a non-zero barrier to linearity, the most popular choice is the harmonic function with a Lorentzian barrier to linearity rewritten in the form given by Barrow *et al.* (1974):

$$V_0(\bar{\rho}) = \frac{H(1 + c\bar{\rho}^2) f_{\alpha\alpha}(\bar{\rho}^2 - \rho_e^2)^2}{\{f_{\alpha\alpha}\rho_e^4 + [8H(1 + c\bar{\rho}^2) - f_{\alpha\alpha}\rho_e^2]\bar{\rho}^2\}}. \tag{83}$$

In Eq. (83) H is the height of the barrier to linearity, $f_{\alpha\alpha}$ the harmonic bending force constant at $\bar{\rho} = \rho_e$, and c is a small constant which allows the anharmonicity around $\bar{\rho} = \rho_e$ to be adjusted.

4.2 Coordinates for the small amplitude motion

The description of the small amplitude stretching motions in the HBJ approach is almost exactly as in the standard approach. Normal coordinates are introduced through the equation

$$d_{\alpha a} = \frac{1}{\sqrt{m_\alpha}} \sum_{r=1}^{3N-7} \ell_{\alpha a,r}(\rho)\, Q_r; \quad a = x, y, \text{ or } z, \tag{84}$$

where the ρ-dependent transformation matrix elements $\ell_{\alpha a,r}(\rho)$ are obtained through a **GF** calculation designed to re-express the harmonic vibrational energy involving the vibrational kinetic energy from Eq. (72)

$$T_{\rm vib} = \frac{1}{2}\sum_{\alpha=1}^{N} m_\alpha \dot{\mathbf{d}}_\alpha \cdot \dot{\mathbf{d}}_\alpha \tag{85}$$

and the harmonic potential energy from Eq. (81)

$$V_{\rm harmonic} = \frac{1}{2}\sum_{j,k} \bar{F}_{jk}(\rho)\Delta r_{j2}\Delta r_{k2} \tag{86}$$

in the following simplified form

$$E_{\rm vib} = \frac{1}{2}\sum_{k=1}^{3N-7}\left\{\dot{Q}_k^2 + \lambda_k(\rho)Q_k^2\right\}. \tag{87}$$

The details of this coordinate transformation [which is carried out via linearized internal coordinates S_i as in the standard approach] are given by Hougen *et al.* (1970) [see also Hoy *et al.* (1972)]. Equation (87) is highly analogous to Eq. (48) but there are now only $3N - 7$ normal coordinates as we are using ρ as one of the vibrational coordinates.

4.3 The HBJ Hamiltonian

When we

- insert the transformation (84) in Eq. (78) and convert the resulting kinetic energy to quantum mechanical form using the *Podolsky trick* [see Chapter 7 of Bunker (1979)],
- convert the *potential energy function* to depend on ρ and the normal coordinates Q_k,

we obtain the following quantum mechanical Hamiltonian:

$$\begin{aligned}\hat{H}_{\rm HBJ} = {} & \frac{1}{2}\sum_{\delta,\gamma=x,y,z,\rho}\left(\hat{J}_\delta - \hat{p}_\delta\right)\mu_{\delta,\gamma}(\rho,Q_1,Q_2,\ldots,Q_{3N-7})\left(\hat{J}_\gamma - \hat{p}_\gamma\right) \\ & + V_0(\rho) + U_0(\rho,Q_1,Q_2,\ldots,Q_{3N-7}) + \frac{1}{2}\sum_{k=1}^{3N-7}\left\{\hat{P}_k^2 + \lambda_k(\rho)\,Q_k^2\right\} \\ & + \frac{1}{2}(\mu)^{1/4}\left[\hat{J}_\rho,\mu_{\rho\rho}\,(\mu)^{-1/2}\left[\hat{J}_\rho,(\mu)^{1/4}\right]\right] \\ & + \frac{1}{6}\sum_{j,k,\ell}\Phi_{jk\ell}(\rho)Q_jQ_kQ_\ell + \frac{1}{24}\sum_{j,k,\ell,m}\Phi_{jk\ell m}(\rho)Q_jQ_kQ_\ell Q_m + \ldots\end{aligned} \tag{88}$$

where

- the operators $\hat{J}_\delta$, $\delta = x$, y, z, are the components of the total angular momentum of the nuclei relative to the molecule fixed axes,
- $\hat{J}_\rho = -i\hbar\partial/\partial\rho$ is the momentum conjugate to ρ,
- the functions $\mu_{\delta,\gamma}(\rho,Q_1,Q_2,\ldots,Q_{3N-7})$ are elements of the 4×4 matrix $\mu = (\mathbf{I}')^{-1}$, where the elements of $\mathbf{I}'$ are given by Eqs. (4.8–4.20) of Jensen (1983). The quantity μ in the commutator term is the determinant of the $\mu_{\delta,\gamma}$ matrix.
- $\hat{P}_k = -i\hbar\partial/\partial Q_k$ is the momentum conjugate to Q_k,

- the vibrational angular momentum operators $\hat{p}_\delta$ are given by

$$\hat{p}_\delta = \sum_{k,\ell=1}^{3N-6} \zeta_{k,\ell}^{(\delta)}(\rho)\, Q_k \hat{P}_\ell \tag{89}$$

where the ρ-dependent functions $\zeta_{k,\ell}^{(\delta)}(\rho)$ are given by Eq. (4.5–4.6) of Jensen (1983),

- the pseudopotential function $U_0(\rho, Q_1, Q_2, \ldots, Q_{3N-7})$ is derived by Sarka and Bunker (1987), and
- the parameters $\Phi_{jk\ell\ldots}(\rho)$ are the ρ-dependent expansion coefficients of the potential energy V_{ai} expanded in terms of the Q_k coordinates. The details of the coordinate transformations involved are given by Hoy and Bunker (1974) and are reviewed by Jensen (1983).

4.4 The nonrigid bender Hamiltonian

In a first crude approximation we can write the Schrödinger equation for the small amplitude vibration as

$$\left[\sum_{k=1}^{3N-7} \left\{\hat{P}_k^2 + \lambda_k(\rho) Q_k^2\right\}\right] \left[\prod_{k=1}^{3N-7} |v_k\rangle\right] = \left[\sum_{k=1}^{3N-7} \hbar\sqrt{\lambda_k(\rho)}\left\{v_k + \frac{1}{2}\right\}\right] \left[\prod_{k=1}^{3N-7} |v_k\rangle\right] \tag{90}$$

where the $|v_k\rangle$ are usual harmonic oscillator eigenfunctions. We assume that all $\lambda_k(\rho)$ are different so that we do not have to consider two-dimensional harmonic oscillators involving an ℓ quantum number. Clearly we can solve Eq. (90) for any value of ρ [which enters the equation through the ρ-dependence of the λ_k parameters]. As the first step in obtaining the eigenvalues of $\hat{H}_{\mathrm{HBJ}}$, a van Vleck transformation [Section 3.3] is carried out using the harmonic oscillator functions $\prod_{k=1}^{3N-7} |v_k\rangle$ as basis functions. This transformation aims at removing all matrix elements coupling different small amplitude states (v_1, v_2, v_3, ..., v_{3N-7}). In order to carry out the transformation the Q_k-dependent functions in $\hat{H}_{\mathrm{HBJ}}$ are expanded as power series in the Q_k with ρ-dependent expansion coefficients, e.g.,

$$\mu_{\delta,\gamma}(\rho, Q_1, Q_2, \ldots, Q_{3N-7}) \approx \mu_{\delta,\gamma}^{(0)}(\rho) + \sum_{k=1}^{3N-7} \mu_{\delta,\gamma}^{(k)}(\rho) Q_k + \frac{1}{2}\sum_{k,\ell=1}^{3N-7} \mu_{\delta,\gamma}^{(k\ell)}(\rho) Q_k Q_\ell \tag{91}$$

by analogy with Eq. (56). The expansion is truncated after the second order terms. This approximation makes it possible to obtain analytical expressions for the matrix elements necessary for the van Vleck transformation.

When the van Vleck transformation is carried out for a triatomic molecule with a large amplitude bending vibration and two small amplitude stretching vibrations characterized by two harmonic oscillator quantum numbers v_1 and v_3 [v_2 is customarily reserved for the bending quantum number], the result is an *effective Hamiltonian* $\hat{H}_{\mathrm{rb}}^{(\mathrm{eff})}$ for one particular (v_1, v_3) state. The derivation and subsequent improvements of $\hat{H}_{\mathrm{rb}}^{(\mathrm{eff})}$ is described by Hoy and Bunker (1973, 1979) and by Jensen and Bunker (1983). A review is given by Jensen (1983), and Eqs. (5.24) and (5.29) of that paper contain the final expression for $\hat{H}_{\mathrm{rb}}^{(\mathrm{eff})}$. The Hamiltonian $\hat{H}_{\mathrm{rb}}^{(\mathrm{eff})}$ depends on the three Euler angles and the bending coordinate ρ only, since the dependence on Q_1 and Q_3 has been removed through the van Vleck transformation. We call $\hat{H}_{\mathrm{rb}}^{(\mathrm{eff})}$ the *nonrigid bender* Hamiltonian.

Table 1. Calculated vibrational energies for $^{12}CH_2$ (in cm^{-1}).

Ref.	Method[a]	ν_1	ν_2	ν_3
Jensen *et al.* (1982)	Fit, NRB0	3021	963	3153
Bunker and Jensen (1983)	Fit, NRB0	2948	963	3078
Bunker *et al.* (1986)	Fit, NRB1	3025	963	3466
Bunker *et al.* (1986)	Fit, NRB2	3357	963	3444
McLean *et al.* (1987)	*ab initio*, NRB2	2985	967	3205

[a]NRB0 is original nonrigid bender model by Hoy and Bunker (1974, 1979).

In the formulation of the nonrigid bender theory described in the previous paragraph, the zero order energies used to calculate the energy denominators in the van Vleck transformation [see Eq. (61)] are taken to be the small amplitude vibrational eigenvalues given in Eq. (90). Hence in this theory the energy denominators do not contain any contributions from the bending or the rotational motion. We call this formulation of the nonrigid bender theory NRB1. An improved formulation, NRB2, in which the energy denominators are allowed to depend on the bending quantum number v_2 and on K, the rotational quantum number describing rotation around the axis with smallest moment of inertia, has been derived by Jensen and Bunker (1986).

In order to obtain the eigenvalues of $\hat{H}_{\mathrm{rb}}^{(\mathrm{eff})}$, we write this Hamiltonian as the sum of two terms

$$\hat{H}_{\mathrm{rb}}^{(\mathrm{eff})} = \hat{H}_{\mathrm{rb}}^{(\mathrm{eff})}(1) + \hat{H}_{\mathrm{rb}}^{(\mathrm{eff})}(2). \tag{92}$$

The Hamiltonian $\hat{H}_{\mathrm{rb}}^{(\mathrm{eff})}(1)$ is given by Eq. (6.2) of Jensen (1983) and describes the bending motion and the rotation around the axis with smallest moment of inertia. The eigenvalues of this Hamiltonian are obtained using *Numerov-Cooley* integration [Numerov, 1933; Cooley, 1961; see also Jensen (1983)] along the coordinate ρ. Finally the eigenvalues of the total Hamiltonian $\hat{H}_{\mathrm{rb}}^{(\mathrm{eff})}$ are obtained *variationally* through direct diagonalization in a basis set consisting of products of eigenfunctions of $\hat{H}_{\mathrm{rb}}^{(\mathrm{eff})}(1)$ and usual rigid rotor functions. The details of this calculation scheme are given by Jensen (1983).

Špirko and co-workers [Papoušek *et al.*, 1973; Špirko, 1983] have used the HBJ approach to treat the large amplitude inversion motion in ammonia-type pyramidal molecules. This work has led to the nonrigid inverter Hamiltonian [Špirko, 1983] which is analogous to the nonrigid bender Hamiltonian.

4.5 Nonrigid bender calculations

The development of the nonrigid bender approach carried out during the period 1982-1986 was motivated by the need for theoretical models to describe new experimental results for the methylene radical CH_2 in its electronic ground state $\tilde{X}^3B_1$. The first high-resolution rotation-vibration spectra of $\tilde{X}^3B_1$ CH_2 were obtained by Sears *et al.* (1981) [for references

Table 2. Perturbation corrections for the lowest vibrational (v_1, v_2, v_3) states of $^{12}CH_2$ (in cm^{-1}) calculated with the *nonrigid bender*-program NRB2 on the basis of the *ab initio* potential energy surface from McLean *et al.* (1987).

	(v_1, v_3)		
v_2	(0,0)	(1,0)	(0,1)
0	−110.1	−378.2	−404.9
1	−123.3	−392.6	−419.8
2	−149.6	−439.6	−460.1
3	−216.0	−555.7	−576.7
4	−300.0	−703.0	−714.7
5	−401.0	−863.3	−868.2
6	−508.9	−1020.5	−1029.0

to later experimental and theoretical work see Jensen and Bunker (1988)]. The CH_2 radical is *quasilinear* in its electronic ground state in that it has a barrier to linearity of approximately 2000 cm^{-1}. The standard approach cannot describe this molecule satisfactorily since it employs different Hamiltonians for molecules with a linear equilibrium structure and molecules with a bent equilibrium structure [see Papoušek and Aliev (1982)], and $\tilde{X}^3B_1$ CH_2 does not fit into any of these two categories. The HBJ approach [and the nonrigid bender theory based on it] makes no assumptions about the nature of the bending motion and hence describes linear, bent, and quasilinear molecules equally well. The nonrigid bender approach described by Jensen (1983) and by Jensen and Bunker (1986) was used to interpret the observed spectra of CH_2 and to do predictions [for references to this work, see Jensen and Bunker (1988)]. It was found, however, that the calculations were not entirely satisfactory. The symptoms of the problems are presented in Table 1. This table shows the energies of the fundamental levels of $^{12}CH_2$ calculated in several different ways: Jensen *et al.* (1982) used a slightly modified version of the original nonrigid bender model by Hoy and Bunker (1974, 1979) to refine the force constants of Eqs. (81) and (82), using as input the 26 rotation-vibration transitions measured for $^{12}CH_2$ at that time. With the potential function obtained in this way they could then predict ν_1, ν_2, and ν_3. Bunker and Jensen (1983) essentially repeated this calculation but obtained the force constant values on the basis of 61 experimental data points [for $^{12}CH_2$, $^{12}CD_2$, and $^{13}CH_2$] that had become available. Bunker *et al.* (1986) had 152 data points [for $^{12}CH_2$, $^{12}CD_2$, and $^{13}CH_2$] at their disposal and used new versions of the nonrigid bender Hamiltonian and computer programs, NRB1 and NRB2 [see above] to refine the force constant values and do the predictions. Finally, a purely *ab initio* study, using NRB2 to obtain the rotation-vibration energies from the potential energy function, was carried out by McLean *et al.* (1987). The experimental value for the ν_2 (bending) fundamental energy is known as 963.1 cm^{-1} [Sears *et al.*, 1982; Marshall and McKellar, 1986] and was used as input for all the fittings presented in Table 1. Consequently all calculations based on fittings yield this value in perfect agreement with experiment. For ν_2 there is also good agreement between experiment and the *ab initio* value from McLean *et al.* (1987). However, the values for the ν_1 and ν_3

(stretching) fundamental energies have not been experimentally measured, and it is seen that the different calculations in Table 1 yield very different values for these energies. For ν_1, there is a significant discrepancy between the NRB1 and NRB2 values from Bunker *et al.* (1986), and for ν_3 there is a significant discrepancy between the "experimental", fitted values from Bunker *et al.* (1986) and the *ab initio* value from McLean *et al.* (1987). It was thus obvious that the nonrigid bender programs did not calculate the stretching energies very accurately.

The reason for the considable deviations between the different calculations for ν_1 and ν_3 is illustrated by Table 2. As mentioned above, in the nonrigid bender approach it is attempted to remove matrix elements coupling different stretching (v_1, v_3) states through van Vleck transformations [Section 3.3]. In the NRB2 formulation, the corrections to the matrix elements in a given (v_1, v_3) block are calculated numerically. We list in Table 2 such *perturbation corrections* to diagonal matrix elements calculated for $^{12}CH_2$. If the assumptions behind the perturbation theory treatment inherent in the van Vleck transformation are fulfilled, these corrections should be small compared to the energy spacing between two zero order [harmonic oscillator] stretching states in $^{12}CH_2$ which is approximately 3500 cm^{-1}. Clearly, especially for the fundamental stretching states this is not the case. Hence the inaccuracies encountered in the nonrigid bender approach are due to the fact that the perturbation theory employed has not converged. Similar effects were noted for H_2O by Jensen (1988*b*) and are in keeping with the observations by Whitehead and Handy (1975) that their vibrational energies calculated from the "standard" Hamiltonian of Eq. (49) through a *variational* procedure, i.e., without using perturbation theory, could differ 50–60 cm^{-1} from energies calculated in the standard approach from the same Hamiltonian which, however, was diagonalized using perturbation theory.

5. THE MORBID APPROACH

We have seen in the previous section that poor convergence of the perturbation theory used in the nonrigid bender approach caused the calculated values of the stretching energies to be inaccurate. A closer analysis of calculation producing the the perturbation corrections of Table 2 shows that their large values are predominantly due to the cubic anharmonic stretching terms in the potential energy function, i.e., the terms $1/6\ \sum_{j,k,\ell} \Phi_{jk\ell}(\rho) Q_j Q_k Q_\ell$ in Eq. (88). Consequently, the difficulties encountered with the nonrigid bender approach are almost entirely due to the fact that we are trying to account for these cubic anharmonic terms using perturbation theory. The other essential approximation made in the nonrigid bender theory, the expansion of the Q_k-dependent functions in the HBJ Hamiltonian as power series in the Q_k [see Eq. (91)], appears to be a reasonable approximation for molecules like CH_2 and H_2O and does not contribute significantly to the breakdown of the theory.

In developing the MORBID Hamiltonian and computer program [Jensen 1988*a*; 1988*b*; 1988*c*] to calculate the rotation-vibration energies of a triatomic molecule we have tried to learn from the difficulties encountered with the nonrigid bender approach. In the HBJ approach the zero order approximation for the small amplitude stretching motion is defined by Eq. (90). The zero order Hamiltonian contains only the *harmonic* (second order) terms in the stretching coordinates. Since we have seen that it is a poor approximation to treat the cubic anharmonic terms as perturbations, it would clearly be desirable to develop a theory where these were already treated in the zero order approximation. One way of achieving this would be to use as stretching basis functions *Morse oscillator functions*, that is eigenfunctions of the one-dimensional Schrödinger equation

$$\left[-\frac{\hbar^2}{2\mu_i}\frac{d^2}{dr_i^2} + V^{(i)}_{\mathrm{Morse}}(r_i)\right]\psi_{n_i}(r_i) = E_{n_i}\psi_{n_i}(r_i). \tag{93}$$

In Eq. (93), $\mu_i = m_A m_B/(m_A + m_B)$ where we assume that the internuclear distance r_i

connects two nuclei with masses m_A and m_B, respectively. Further,

$$V_{\text{Morse}}^{(i)}(r_i) = D_i\{1 - \exp[-a_i(r_i - r_i^e)]\}^2. \tag{94}$$

Equation (93) has, in a very good approximation, analytical solutions characterized by a principal quantum number n_i [see, for example, Wallace (1976), Efremov (1977), and Tipping and Ogilvie (1983)]. The eigenfunctions are given by

$$\psi_{n_i}(r_i) = |n_i\rangle = N_{n_i} e^{-\frac{x}{2}} x^{\frac{\alpha}{2}} L_{n_i}^{\alpha}(x), \tag{95}$$

where N_{n_i} is a normalization factor, $L_{n_i}^{\alpha}(x)$ is a generalized Laguerre polynomial,

$$k_i = 2\frac{\sqrt{\mu_i D_i}}{a_i\hbar}, \tag{96}$$

$$x = 2k_i \exp[-a_i(r_i - r_i^e)], \tag{97}$$

and

$$\alpha = 2k_i - 2n_i - 1. \tag{98}$$

The energy eigenvalue is

$$E_{n_i} = 4D_i\left[\frac{n_i + 1/2}{2k_i} - \left(\frac{n_i + 1/2}{2k_i}\right)^2\right]. \tag{99}$$

It is obvious that the Morse potential energy function $V_{\text{Morse}}^{(i)}(r_i)$ approaches the constant energy D_i for long internuclear distances. For short distances it assumes large positive values and thus it gives a reasonable description of the strong, short-range repulsion between the nuclei. Hence it is reasonable to use the Morse oscillator eigenfunctions as basis functions for the stretching motions.

The derivation of the MORBID Hamiltonian is given in detail by Jensen (1988*a*), and we will only give a brief outline here. The central idea is the introduction of a new analytical function to represent the potential energy:

$$\begin{aligned} V_{\text{ai}}(\Delta r_{12}, \Delta r_{32}, \bar\rho) &= V_0(\bar\rho) + \sum_j F_j(\bar\rho) y_j \\ &+ \sum_{j\le k} F_{jk}(\bar\rho) y_j y_k + \sum_{j\le k\le m} F_{jkm}(\bar\rho) y_j y_k y_m + \sum_{j\le k\le m\le n} F_{jkmn}(\bar\rho) y_j y_k y_m y_n, \end{aligned} \tag{100}$$

where all of the indices j, k, m, and n assume the values 1 or 3. In Eq. (100),

$$y_j = 1 - \exp(-a_j \Delta r_{j2}), \tag{101}$$

where the definition of Δr_{j2} is identical with that given for the HBJ approach (Section 4.1). The $F_{jkm...}$ expansion coefficients of Eq. (100) are functions of $\bar\rho$ [the instantaneous value of the supplement of the bond angle; see Section 4.1]. The function $V_0(\bar\rho)$ is the pure bending potential energy function, i.e. the potential energy for the molecule bending with its bond lengths fixed at their equilibrium values. The total potential energy function given by Eq. (100) has correct asymptotics at long bond lengths [it approaches a constant], and for short bond lengths its asymptotics are physically reasonable as it approaches a very large [but not infinite] value.

The $\bar\rho$-dependent functions entering into Eq. (100) are defined as general cosine expansions:

$$\begin{aligned} F_j(\bar\rho) &= \sum_{i=1}^{4} f_j^{(i)}(\cos\rho_e - \cos\bar\rho)^i, \\ F_{jk...}(\bar\rho) &= f_{jk...}^{(0)} + \sum_{i=1}^{N} f_{jk...}^{(i)}(\cos\rho_e - \cos\bar\rho)^i \end{aligned} \tag{102}$$

with ρ_e being the equilibrium value of $\bar{\rho}$. The function $F_{jk}(\bar{\rho})$ has $N = 3$, $F_{jkl}(\bar{\rho})$ has $N = 2$, and $F_{jklm}(\bar{\rho})$ has $N = 1$. Symmetry relations exist between the potential energy parameters for the symmetrical ABA molecule to ensure that V_{ai} is totally symmetric under the interchange of y_1 and y_3. For $V_0(\bar{\rho})$, we use two different parameterizations,

$$V_0(\bar{\rho}) = \sum_{i=2}^{8} f_0^{(i)}(\cos\rho_e - \cos\bar{\rho})^i, \tag{103}$$

or

$$V_0(\bar{\rho}) = \sum_{i=1}^{8} f_0^{(i)}(1 - \cos\bar{\rho})^i. \tag{104}$$

The parameterization given by Eq. (103) has ρ_e as an independent parameter and is appropriate for *bent* molecules, i.e. molecules with $\rho_e \neq 0$. Equation (104) has $f_0^{(1)}$ as an independent parameter, whereas ρ_e is determined as a function of the $f_0^{(i)}$ quantities. This parameterization is appropriate for both linear and bent molecules.

The parameterized function given by Eqs. (100–104) has now been used to describe the potential energy function in isolated electronic states of many triatomic molecules: NH_2^+ [Jensen *et al.*, 1987], HOC^+ [Jensen and Kraemer, 1988*a*], C_3 [Jensen and Kraemer, 1988*a*; Jensen, 1989*b*], CNC^+ and CCN^+ [Jensen and Kraemer, 1988*b*], H_2O [Jensen, 1988*d*; 1989*a*], LiOH [Bunker *et al.*, 1989], CH_2 [Jensen, 1988*a*; Jensen and Bunker, 1988; Comeau *et al.*, 1989], NH_2 [Jensen *et al.*, 1990] and H_2F^+ [Bunker *et al.*, 1990]. In all cases it was found sufficiently flexible to represent accurately the potential energy surface. The form of the chosen $\bar{\rho}$-dependent functions ensure that the potential energy function also has correct asymptotics as a function of $\bar{\rho}$: It is an even, periodic function of $\bar{\rho}$ with period 2π, and thus it has zero slope at $\bar{\rho} = 0$ and $\bar{\rho} = \pi$.

The kinetic energy operator used in the MORBID approach is based on the same ideas as were used in the nonrigid bender approach [Sections 4.1-4.2]. Again, we introduce a nonrigid reference configuration with fixed bond lengths r_{12}^e and r_{32}^e and variable bond angle supplement ρ that approximately follows the bending motion, and again ρ is chosen as one of the three vibrational coordinates. The angle ρ and the Euler angles θ, ϕ, and χ are defined by Eqs. (79)-(80) exactly as in the HBJ approach. In the MORBID approach, however, we do *not* use two mass-dependent normal coordinates Q_1 and Q_3 to describe the stretching motion; instead we express the stretching dependence of the kinetic energy operator directly by using the bond length displacements Δr_{12} and Δr_{32}. The final Hamiltonian is given by

$$\begin{aligned}\hat{H}_{\text{MORBID}} &= \frac{1}{2}\sum_{\delta,\gamma=x,y,z,\rho} (\hat{J}_\delta - \hat{p}_\delta)\mu_{\delta\gamma}(\Delta r_{12}, \Delta r_{32}, \rho)(\hat{J}_\gamma - \hat{p}_\gamma)\\ &+ \frac{1}{2}\sum_{j,k=1,3} \hat{P}_j G_{jk}^{(r)}(\Delta r_{12}, \Delta r_{32}, \rho)\hat{P}_k\\ &+ U_1(\Delta r_{12}, \Delta r_{32}, \rho) + U_0(\rho) + V_{ai}(\Delta r_{12}, \Delta r_{32}, \rho)\end{aligned} \tag{105}$$

where $\hat{J}_\delta$, $\delta = x, y, z$, are the components of the total angular momentum along the molecule fixed axes, $\hat{J}_\rho = -i\hbar\partial/\partial\rho$, $\mu_{\alpha\beta}$ (α, $\beta = x, y, z, \rho$) are the elements of the 4×4 "inverse inertial" matrix defined by Eqs. (45-50) of Jensen (1988*a*), $\hat{P}_j = -i\hbar\partial/\partial\Delta r_{j2}$ is the momentum conjugate to Δr_{j2}, the vibrational angular momentum $\hat{p}_\alpha$ is given by

$$\hat{p}_\alpha = \frac{1}{2}\sum_{j,k=1,3}\left\{\hat{P}_j G_{jk}^{(r)}\chi_k^{(\alpha)} + \chi_k^{(\alpha)} G_{jk}^{(r)}\hat{P}_j\right\}, \tag{106}$$

the $\chi_k^{(\alpha)}$ being Coriolis coupling coefficients, and the $G_{jk}^{(r)}$ matrix elements are defined by Eq. (40) of Jensen (1988*a*). The potential function V_{ai} is given by Eq. (100), transformed so that instead of depending on Δr_{12}, Δr_{32}, and the instantaneous value of the bond angle supplement $\bar{\rho}$ [see Section 4.1], it now depends on Δr_{12}, Δr_{32}, and the mass dependent coordinate ρ. The transformation is carried out using basic geometrical identities as shown in Eqs. (34-36) of Jensen (1988*a*). The U_1 and U_0 functions are pseudopotential terms defined in Eqs. (42-44) of Jensen (1988*a*). The dependence of U_0 on the stretching coordinates Δr_{12} and Δr_{32} has been neglected; this possibly explains why sometimes splittings due to Fermi resonances are calculated to be too small by the MORBID program [see Bunker *et al.* (1990)].

The functions $\mu_{\alpha\beta}$, $G_{jk}^{(r)}$, and $\chi_k^{(\alpha)}$ all depend on the coordinates Δr_{12}, Δr_{32}, and ρ, and we expand them as polynomials in y_1 and y_3 with ρ-dependent coefficients, e.g.

$$\mu = \mu_0(\rho) + \sum_j \mu_j(\rho) y_j + \sum_{j\le k} \mu_{jk}(\rho) y_j y_k + \sum_{j\le k\le m} \mu_{jkm}(\rho) y_j y_k y_m + \sum_{j\le k\le m\le n} \mu_{jkmn}(\rho) y_j y_k y_m y_n. \quad (107)$$

The expansion of the μ-tensor elements are truncated after the quartic terms, the expansions of the $\chi_k^{(\alpha)}$ functions are truncated after the cubic terms, and the expansions of the $G_{jk}^{(r)}$ matrix elements are truncated after the quadratic terms. In this way, we obtain an expansion to fourth order in the vibrational operators y_j and $\widehat{P}_k$.

We obtain the eigenvalues of $\widehat{H}$ *variationally*, i.e. by setting up its matrix in a truncated set of basis functions and diagonalizing this matrix block. The basis functions are symmetrized products of *stretching functions*, *bending functions*, and *rotational functions*. As stretching basis functions we choose the eigenfunctions $|N_{\mathrm{Vib}}\Gamma_{\mathrm{Sym}}\rangle$ [see Section V of Jensen (1988*a*)] of

$$\widehat{H}_{\mathrm{Stretch}} = \frac{1}{2}\left[\frac{1}{m_1} + \frac{1}{m_2}\right]\widehat{P}_1^2 + \frac{1}{2}\left[\frac{1}{m_3} + \frac{1}{m_2}\right]\widehat{P}_3^2 - \frac{1}{m_2}\cos\rho_e \widehat{P}_1 \widehat{P}_3 + \sum_{j\le k} f_{jk}^{(0)} y_j y_k + \sum_{j\le k\le m} f_{jkm}^{(0)} y_j y_k y_m + \sum_{j\le k\le m\le n} f_{jkmn}^{(0)} y_j y_k y_m y_n, \quad (108)$$

i.e. the Hamiltonian for a molecule stretching with its bond angle fixed at the value $\pi - \rho_e$. In Eq. (108), m_α is the mass of atom α. We obtain these eigenfunctions [and the corresponding eigenvalues, which are our zero order stretching energies] by setting up the matrix of $\widehat{H}_{\mathrm{Stretch}}$ in a basis of symmetrized Morse oscillator functions as described in Section V of Jensen (1988*a*).

As bending basis functions we use the eigenfunctions $|v_2, K\rangle$ [see Section V of Jensen (1988*a*)] of

$$\widehat{H}_{\mathrm{Bend}} = \frac{1}{2}\widehat{J}_\rho \mu_{\rho\rho}^0(\rho)\widehat{J}_\rho + \frac{1}{2}\mu_{zz}^0(\rho)\widehat{J}_z^2 + \frac{1}{2}\hbar^2\mu_{\rho\rho}^0(\rho)\, g(\rho) + \frac{1}{2}(\mu^0)^{-1/4}[\widehat{J}_\rho, \mu_{\rho\rho}^0][\widehat{J}_\rho, (\mu^0)^{1/4}] + V_0(\rho), \quad (109)$$

the Hamiltonian for the molecule bending and rotating around the z-axis with its bond lengths fixed at the equilibrium values r_{12}^e and r_{32}^e. In Eq. (109), the $\mu_{\alpha\alpha}^0$, $\alpha = \rho$ or z, are elements of the μ matrix (and μ^0 is its determinant) determined at $\Delta r_{12} = \Delta r_{32} = 0$, and $g(\rho)$ is given by Eq. (43) of Jensen (1988*a*). The bending functions are calculated using the Numerov-Cooley numerical integration method as explained in Section 6 of Jensen (1983). When the stretching and bending basis functions are calculated, they are combined with usual rigid rotor functions [see Eq. (7.1) of Jensen (1983)] to form the complete basis to diagonalize the $\widehat{H}$ Hamiltonian of Eq. (22). This is described in Section V of Jensen (1988*a*).

When we let $|J, K, M, \tau\rangle$ denote symmetrized rigid rotor functions [see Eq. (7.1) of Jensen (1983)] we can express the eigenfunctions of the MORBID Hamiltonian [Eq. (105)] as linear

Table 3. Vibrational energies of the radical $^{12}CH_2$ (in cm^{-1}) predicted with the MORBID program.

Ref.	Method	ν_1	ν_2	ν_3
Jensen (1988*a*)[a]	*ab initio*	2994	973	3212
Comeau *et al.* (1989)	*ab initio*	3015	974	3236
Jensen and Bunker (1988)	Fit	2992	963	3214

[a]Potential energy surface from McLean *et al.* (1987).

combinations of the product basis functions:

$$\psi_{\text{MORBID}} = \sum_{K,v_2,N_{\text{Vib}},\Gamma_{\text{Sym}}} c_{K,v_2,N_{\text{Vib}},\Gamma_{\text{Sym}}} |J,K,M,\tau\rangle |v_2,K\rangle |N_{\text{Vib}}\Gamma_{\text{Sym}}\rangle; \tag{110}$$

the expansion coefficients $c_{K,v_2,N_{\text{Vib}},\Gamma_{\text{Sym}}}$ and the corresponding eigenvalues are determined through standard matrix diagonalization. In order to construct the matrix to be diagonalized we need to calculate the matrix elements of $\widehat{H}_{\text{MORBID}}$ between the basis functions defined by Eq. (110). The vibrational factors in these matrix elements are in principle three-dimensional integrals over the coordinates ρ, Δr_{12} and Δr_{32}. Since we have made the approximation defined by Eq. (107) we can calculate all integrals over Δr_{12} and Δr_{32} *analytically* using Morse oscillator matrix elements given by Špirko *et al.* (1985, 1987). This leads to a considerable saving of computer time. The integrals over ρ are calculated numerically using Simpson's rule [see Jensen (1983)].

Table 3 shows three sets of fundamental energies for $^{12}CH_2$, calculated using the MORBID approach: One calculated by Jensen (1988*a*) on the basis of the *ab initio* potential surface of McLean *et al.* (1987), one calculated from the *ab initio* surface by Comeau *et al.* (1989), and one calculated from a potential energy surface obtained by Jensen and Bunker (1988) by fitting to 152 experimental data points for $^{12}CH_2$, $^{13}CH_2$, and $^{12}CD_2$ [these data points are identical to those used in the nonrigid bender fittings by Bunker *et al.* (1986)]. The table shows that the MORBID calculations have led to consistent values for ν_1 and ν_3 of CH_2. Thus we can now be confident that we can predict these energies reliably.

6. HAMILTONIANS USING GEOMETRICALLY DEFINED COORDINATES

The type of Hamiltonians treated in the present section is the subject of several review articles [Carter and Handy, 1986; Tennyson, 1986; Bačić and Light, 1989; Hutson, 1990] so only the highlights of these methods will be given here.

As already mentioned in Section 2.4, the basic idea is to choose geometrically defined coordinates to describe the vibrational motion. These coordinates are typically the bond lengths and bond angles ($\mathcal{R}_1$, $\mathcal{R}_2$, $\mathcal{R}_3$, ..., $\mathcal{R}_{3N-6}$) which we introduced in Section 2.3; they enter naturally into the theory when we define an analytical function representing the potential energy function.

We have seen in the Sections 3 and 4 how in the standard approach and the Hougen-Bunker-Johns approach elaborate coordinate definitions were introduced in order to simplify the form of the Hamiltonian. In the approaches considered in the present section little effort is made to achieve such simplifications. The vibrational coordinates are chosen in a simple way by considering the geometry of the molecule and the Hamiltonian consistent with these coordinates is derived. The only freedom in the coordinate choice lies in the definition of the molecule fixed axis system. This axis system is said to be a *body-fixed* axis system, i.e., it is rigidly attached to the *instantaneous* configuration of the molecule [so that, for example, axes can be chosen to be parallel to a bond or to bisect a bond angle]. The orientation of the molecule fixed axis system relative to the instantaneous configuration of the molecule can be chosen freely, and this choice can be used to reduce rotation-vibration interaction terms in the Hamiltonian. Sutcliffe, Tennyson and co-workers use the word "embedding" for the procedure of defining the orientation of the molecule coordinate system relative to the molecule [see, for example, Tennyson (1986)].

6.1 A Hamiltonian for a triatomic molecule

We have already mentioned in Section 2.4 that the Hamiltonians in geometrically defined vibrational coordinates are derived using the chain rule [Eq. (27)] as described by Sutcliffe (1982). A complete derivation of the kinetic energy operator for a triatomic molecule ABC is given by Carter and Handy (1986), and we give this operator here as an example. Carter and Handy use the vibrational coordinates r_{12}, r_{32}, and $\bar{\alpha}$ defined in Section 4.1 [where nucleus A is labeled 1, nucleus B is labeled 2, and nucleus C is labeled 3]. Their molecule fixed coordinate system [see Fig. 3] is defined such that the y axis is the bisector of the angle $\angle$(ABC). Its orientation is such that A and C have positive y coordinates. The x axis also lies in the molecular plane; it is perpendicular to the y axis and its orientation is such that nucleus A is in the positive xy quadrant. The z axis completes the triad and is perpendicular to the molecular plane. With these definitions, the following expression for the kinetic energy of the rotation and vibration is obtained [Eqs. (3.48)–(3.50) of Carter and Handy (1986)]:

$$\hat{T} = \hat{T}_{\mathrm{V}} + \hat{T}_{\mathrm{VR}} \tag{111}$$

with

$$\begin{aligned} \hat{T}_{\mathrm{V}} = & -\frac{1}{4}\left[\left(\frac{1}{\mu_1 r_{12}^2} + \frac{1}{\mu_2 r_{32}^2} - \frac{2\cos\bar{\alpha}}{m_2 r_{12} r_{32}}\right), \left(\frac{\partial^2}{\partial\bar{\alpha}^2} + \cot\bar{\alpha}\frac{\partial}{\partial\bar{\alpha}}\right)\right]_+ \\ & - \frac{1}{2\mu_1}\frac{\partial^2}{\partial r_{12}^2} - \frac{1}{2\mu_2}\frac{\partial^2}{\partial r_{32}^2} - \frac{\cos\bar{\alpha}}{m_2}\frac{\partial^2}{\partial r_{12}\partial r_{32}} \\ & + \frac{1}{m_2}\left(\frac{1}{r_{12}}\frac{\partial}{\partial r_{32}} + \frac{1}{r_{32}}\frac{\partial}{\partial r_{12}}\right)\left(\sin\bar{\alpha}\frac{\partial}{\partial\bar{\alpha}} + \cos\bar{\alpha}\right) \end{aligned} \tag{112}$$

and

$$\begin{aligned} \hat{T}_{\mathrm{VR}} = & \frac{1}{8\cos^2(\bar{\alpha}/2)}\left(\frac{1}{\mu_1 r_{12}^2} + \frac{1}{\mu_2 r_{32}^2} + \frac{2}{m_2 r_{12} r_{32}}\right)\hat{\Pi}_z^2 \\ & + \frac{1}{8\sin^2(\bar{\alpha}/2)}\left(\frac{1}{\mu_1 r_{12}^2} + \frac{1}{\mu_2 r_{32}^2} - \frac{2}{m_2 r_{12} r_{32}}\right)\hat{\Pi}_x^2 \\ & + \frac{1}{8}\left(\frac{1}{\mu_1 r_{12}^2} + \frac{1}{\mu_2 r_{32}^2} - \frac{2\cos\bar{\alpha}}{m_2 r_{12} r_{32}}\right)\hat{\Pi}_y^2 - \frac{1}{4\sin\bar{\alpha}}\left(\frac{1}{\mu_1 r_{12}^2} - \frac{1}{\mu_2 r_{32}^2}\right)\left(\hat{\Pi}_z\hat{\Pi}_x + \hat{\Pi}_x\hat{\Pi}_z\right) \\ & - \frac{i}{2}\left(\frac{1}{\mu_1 r_{12}^2} - \frac{1}{\mu_2 r_{32}^2}\right)\left(\frac{\partial}{\partial\bar{\alpha}} + \frac{\cot\bar{\alpha}}{2}\right)\hat{\Pi}_y - \frac{\sin\bar{\alpha}}{2m_2}\left(\frac{1}{r_{32}}\frac{\partial}{\partial r_{12}} - \frac{1}{r_{12}}\frac{\partial}{\partial r_{32}}\right)\hat{\Pi}_y, \end{aligned} \tag{113}$$

where for two operators $\hat{A}$ and $\hat{B}$, $[\hat{A}, \hat{B}]_+ = \hat{A}\hat{B} + \hat{B}\hat{A}$. In Eqs. (112-113), the masses $m_1 = m_A$, $m_2 = m_B$, and $m_3 = m_C$, $\mu_1 = m_1 m_2/(m_1 + m_2)$, $\mu_2 = m_3 m_2/(m_3 + m_2)$, and the

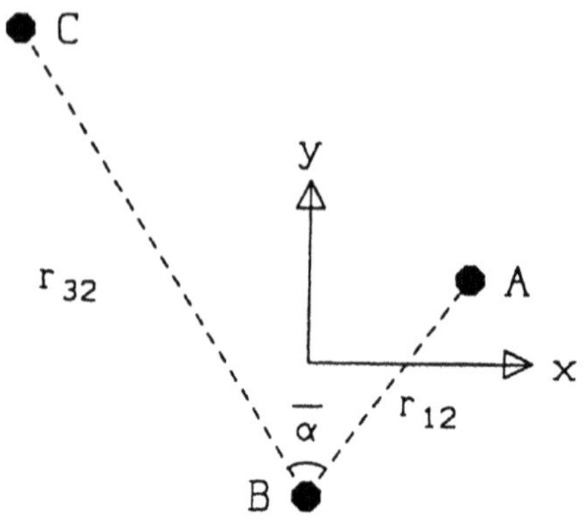

Figure 3. The definition of the vibrational coordinates (r_{12}, r_{32}, $\bar{\alpha}$) and the molecule fixed axis system employed by Carter and Handy (1986).

operators ($\hat{\Pi}_x$, $\hat{\Pi}_y$, $\hat{\Pi}_z$) are angular momentum operators with respect to the molecule fixed axes; they are chosen with a sign convention [see Eq. (3.31) of Carter and Handy (1986)] that makes them obey the standard commutation relations for angular momenta [see Section 6.1 of Papoušek and Aliev (1982)]. Additional examples of kinetic energy operators for triatomic molecules derived in terms of different types of geometrically defined coordinates are given by Sutcliffe and Tennyson (1991). It should be emphasized that these operators are *exact* in that no approximations have been made in deriving them.

6.2 Variational calculations

When the kinetic energy operator $\hat{T}$ has been obtained in terms of a chosen set of geometrically defined coordinates, it is obviously necessary to choose a suitable analytical function V_{ai} depending on these coordinates to represent the potential energy operator. The total Hamiltonian is then given as $\hat{H} = \hat{T} + V_{ai}$. It would clearly not be reasonable to attempt a calculation of the eigenvalues of $\hat{H}$ through a perturbation theory approach since no effort has been taken to separate the different types of motion carried out by the molecule. Consequently, in all work using geometrically defined vibrational coordinates, these eigenvalues and the corresponding eigenfunctions are obtained variationally, i.e., through direct numerical diagonalization of the matrix representation of the Hamiltonian in a suitable basis set. This is the procedure also used in the MORBID approach [Section 5].

Obviously it is now necessary to choose a set of basis functions. The normal procedure is to choose the vibrational basis functions as products of socalled single mode functions, i.e., functions depending on one vibrational coordinate. Hence for the triatomic molecule used as example above one would write a vibrational basis function as

$$\Phi_{n_1 n_2 n_3} = \phi^{(s)}_{n_1}(r_{12})\, \phi^{(s)}_{n_3}(r_{32})\, \phi^{(b)}_{n_2}(\bar{\alpha}) \tag{114}$$

where the single mode functions are divided into stretching functions $\phi^{(s)}_{n_s}(r)$ and bending [or angular] functions $\phi^{(b)}_{n_b}(\bar{\alpha})$. For the stretching functions the $\phi^{(s)}_{n_s}(r)$ the most popular choice is the Morse oscillator functions given by Eq. (95) [see Carter and Handy (1986) and Tennyson (1986)]. If the behaviour of the basis function near $r = 0$ is important, the Morse-like functions given by Eq. (3.25) of Tennyson (1986) can be used. It is also possible to use harmonic oscillator functions [Carter and Handy (1986)] but as explained in Section 5 these are clearly less physically realistic than the Morse oscillator functions. As angular basis functions $\phi_{n_b}^{(b)}(\bar{\alpha})$ Legendre polynomials can be used [Carter and Handy, 1986; Tennyson, 1986]

$$\phi_{n_b}^{(b)}(\bar{\alpha}) = N_{n_b} P_{n_b}(\cos\bar{\alpha}). \tag{115}$$

The vibrational basis functions are multiplied by rotational functions, normally chosen as rigid rotor functions $|J, K, M\rangle$, to obtain the rotation-vibration basis.

The calculation of the matrix elements of the Hamiltonian $\widehat{H}$ between the rotation-vibration basis functions can be carried out according to different schemes. In the case of the triatomic molecule used as example above, such a matrix element can generally be expressed as

$$\begin{aligned}
&\left\langle n_1 n_2 n_3 J K M | \widehat{H} | n_1' n_2' n_3' J K' M \right\rangle \\
&= \int dr_{12} \int dr_{32} \int \sin\bar{\alpha} d\bar{\alpha} \int \sin\theta d\theta \int d\phi \int d\chi \; \Phi_{n_1 n_2 n_3}(r_{12}, r_{32}, \bar{\alpha})^* \\
&\times \; \psi_{JKM}(\theta, \phi, \chi)^* \widehat{H} \psi_{JK'M}(\theta, \phi, \chi) \Phi_{n_1' n_2' n_3'}(r_{12}, r_{32}, \bar{\alpha})
\end{aligned} \tag{116}$$

with $\psi_{JKM}(\theta, \phi, \chi) = |J, K, M\rangle$. The integration over θ, ϕ, and χ which involves the angular momentum operators ($\widehat{\Pi}_x$, $\widehat{\Pi}_y$, $\widehat{\Pi}_z$) in $\widehat{H}$ can be carried out analytically [see, for example, Papoušek and Aliev (1982)], so we are left with the problem of integrating over the vibrational coordinates r_{12}, r_{32}, and $\bar{\alpha}$. Carter and Handy (1986) suggest that the entire three-dimensional integral over these three coordinates be calculated numerically using Gaussian quadrature [see Eq. (6.19) of Carter and Handy (1986)]. This procedure is obviously computationally cumbersome, so they have also implemented procedures in which the matrix elements of $\widehat{T}_V$ are expressed exactly as sums of terms, each term being the product of three one-dimensional integrals [one over r_{12}, one over r_{32}, and and one over $\bar{\alpha}$]. These matrix elements can then be calculated exactly using one-dimensional quadrature.

Tennyson (1986) describes integration procedures for triatomic molecules in which all angular integration over the kinetic *and* potential energy operators is carried out analytically. For the potential energy operator, the analytical integration is made possible by expressing this operator as

$$V_{ai}(r_{12}, r_{32}, \bar{\alpha}) = \sum_{\lambda} V_\lambda(r_{12}, r_{32}) P_\lambda(\cos\bar{\alpha}) \tag{117}$$

where $P_\lambda(\cos\bar{\alpha})$ is a Legendre polynomial and $V_\lambda(r_{12}, r_{32})$ is a suitable [in principle arbitrary] function of r_{12} and r_{32}. Part of the integration over r_{12} and r_{32} involving the kinetic energy operator $\widehat{T}$ can also be carried out analytically, but this socalled radial integration involving V_{ai} [and part of the radial integration involving $\widehat{T}$] must be carried out using two-dimensional numerical quadrature. When the matrix representation of the Hamiltonian has been calculated, it must be diagonalized. Tennyson (1986) has discussed possible diagonalization methods.

6.3 Close-coupling calculations

Hutson (1990, 1991*a*) employs another approach, the socalled close-coupling calculation, to the determination of the eigenvalues for a Hamiltonian expressed in terms of geometrically defined vibrational coordinates. For example, this method is developed for atom-diatom van der Waals complexes. The coordinates used for the complex A – BC are defined as follows [see Fig. 4]: The vector from the center of mass of the diatom BC to the nucleus A is denoted

R and has the length R. The vector between the nuclei B and C is **r** (with length r), it is usually taken to originate on the heavier of the nuclei B and C. The angle between **R** and **r** is τ. The orientations relative to the laboratory fixed axis system of the vectors **R** and **r** are given by the Euler angles (θ_R, ϕ_R) and (θ_r, ϕ_r), respectively. The quantities (r, R, θ_R, ϕ_R, θ_r, ϕ_r) are used as coordinates. The Hamiltonian in terms of these coordinates is given by Eqs. (11-12) of Hutson (1991*a*) [see also Tennyson and Sutcliffe (1982)]. The basic idea behind the approach is to represent the wavefunction in the form

$$\psi_\alpha(r, R, \theta_R, \phi_R, \theta_r, \phi_r) = r^{-1}\, R^{-1} \sum_{va} \phi_{vj}(r)\, \Phi_a(\theta_R, \phi_R, \theta_r, \phi_r)\, \chi^{\alpha}_{va}(R). \tag{118}$$

In this expression, the functions $\phi_{vj}(r)$ and $\Phi_a(\theta_R, \phi_R, \theta_r, \phi_r)$ are known basis functions. The function $\phi_{vj}(r)$ is a free-diatom stretching function and $\Phi_a(\theta_R, \phi_R, \theta_r, \phi_r)$ is an angle dependent function defined on the basis of spherical harmonic functions [see Hutson (1991*a*)]. The functions $\chi^{\alpha}_{va}(R)$, however, are initially unknown. Equation (118) is inserted in the Schrödinger equation for the system, and the resulting equation is multiplied from the left by $[r^{-1}\, R^{-1}\, \phi_{v'j'}(r)\, \Phi'_a(\theta_R, \phi_R, \theta_r, \phi_r)]^*$ and integrated over all coordinates except R. This yields a set of *coupled equations* for the functions $\chi^{\alpha}_{va}(R)$. These equations are now solved through a numerical integration procedure in which their values are obtained in a grid of discrete R-values: $R_k = R_0 + k\Delta R$, $k = 1, 2, \ldots, k_{max}$ [see Johnson (1978) and Manolopoulos (1986, 1988)]. This numerical approach is a generalization of the Numerov-Cooley integration employed in the nonrigid bender and MORBID approaches [see Section 4.4]. Hutson (1991*a*) has also developed the close-coupling approach for diatom-diatom complexes. For such molecules, the diatomic moieties are considered to be rigid, and the stretching between the two diatomic molecules plays the role of the R stretching for the atom-diatom complex. Very recently, Hutson's program BOUND [Hutson, 1991*b*] was used by Marshall *et al.* (1991) and by Jensen *et al.* (1991*a*) for calculating the lowest vibrational energies of the dimers $(HF)_2$ and $(HCl)_2$, respectively. One notable feature of these calculations is their extreme cost. The calculation of *one* $J = 0$ energy typically used around 90 minutes CPU time on a Cray Y-MP2/232 installation. Obviously it would be possible to obtain the rotation-vibration energies through a standard variational calculation in which the matrix representation of the Hamiltonian would be constructed in a set of known basis functions. In such a calculation one would choose, in addition to the basis functions $\phi_{vj}(r)$ and $\Phi_a(\theta_R, \phi_R, \theta_r, \phi_r)$ already used in the close-coupling calculations, known, R-dependent basis functions, $\phi_{N_R}(R)$ say, and the wavefunctions would be expressed as a linear combination of the product basis functions

$$\phi_{\text{basis}} = \phi_{vj}(r)\, \Phi_a(\theta_R, \phi_R, \theta_r, \phi_r)\, \phi_{N_R}(R). \tag{119}$$

This author feels that in 90 minutes of CPU time on a Cray Y-MP2/232 it would be possible to construct and diagonalize a matrix representation of the Hamiltonian involving an immense number of the basis functions given by Eq. (119) and thus determine several vibrational energies instead of the one energy that results from the close-coupling calculation. Hence he surmises that the standard variational calculation would be more efficient for this type of problem. However, it might be difficult to choose suitable basis functions $\phi_{N_R}(R)$ for the R motion.

6.4 The DVR-DGB approach

The methods for calculating rotation-vibration energies presented here all have in common that they seek to represent the rotation-vibration wavefunction $\psi_{\text{rot-vib}}$ as a linear combination of known basis functions φ_j:

$$\psi_{\text{rot-vib}} = \sum_j c_j \varphi_j. \tag{120}$$

The methods presented so far all employ functions φ_j which we might term *global* basis functions. Each such basis function could in principle be the wavefunction of the molecule in that its amplitude extends over a significant part of molecular configuration space. The DVR-DGB

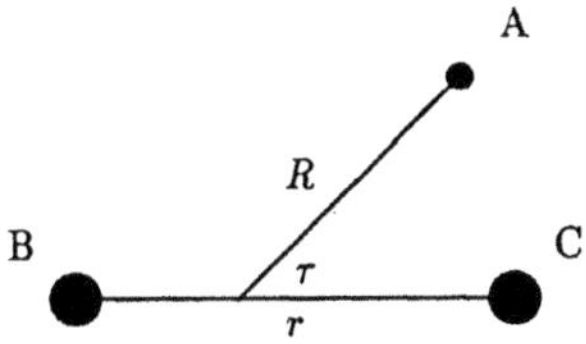

Figure 4. The definition of the vibrational coordinates (r, R, τ) employed in close-coupling calculations for atom-diatom van der Waals complexes.

[Discrete Variable Representation–Distributed Gaussian Basis] method for triatomic molecules suggested by Bačić and Light (1986, 1987, 1989) [see also Light and Bačić (1987)] uses a different philosophy for representing some of the basis functions. We will give only a brief description of the DVR-DGB method here. For details, the reader is referred to the original papers [in particular Bačić and Light (1986)].

Bačić and Light develop their approach on the basis of the Hamiltonian derived by Tennyson and Sutcliffe (1982) in terms of the three vibrational coordinates (r, R, τ) defined above for the atom-diatom complex. The $J = 0$ wavefunction [which does not depend on the Euler angles (θ, ϕ, χ)] is expressed as a linear combination of Legendre polynomials:

$$\psi_{\text{vib}}(R, r, \tau) = \sum_{\ell=0}^{\ell_{\max}} f_\ell(r, R) P_\ell(\cos\tau) \tag{121}$$

with initially unknown "expansion coefficients" $f_\ell(r, R)$ depending on r and R.

This expression is inserted in the Schrödinger equation for the system which is then multiplied from the left by $P_k(\cos\tau)^*$, and integration over τ is carried out. This resembles the technique used by Hutson in the close-coupled calculations. The result of the integration is a set of $\ell_{\max} + 1$ coupled equations in the functions $f_\ell(r, R)$; these equations are labeled by the index ℓ of the Legendre polynomials forming the angular basis. This set of coupled equations is subjected to an orthogonal transformation in which the transformation matrix is determined on the basis of the quadrature points and weights in the ($\ell_{\max}$+1)-point Gauss-Legendre quadrature over the angle τ [or rather over $\cos\tau$ as this is the argument of the Legendre polynomial in Eq. (121)]. This transformation yields a new set of $\ell_{\max} + 1$ coupled equations, each new equation "belonging" to one particular quadrature point defined through a value τ_α of the angle τ. The unknown quantities in the new coupled equations are ($\ell_{\max}$+1) (r,R)-dependent functions $g_\alpha(r, R)$ [where $g_\alpha(r, R)$ is associated with the Gauss quadrature point τ_α]; these functions are linear combinations of the original expansion coefficients $f_\ell(r, R)$. The fact that each new coupled equation corresponds to a discrete value τ_α of the angle τ is referred to as a *discretization* of this angle, and the new set of coupled equations are said to be in the Discrete Variable Representation (DVR) [see Lill *et al.* (1982), Light *et al.* (1985), and Lanczos (1956)].

The idea is now to expand the functions $g_\alpha(r, R)$ in a basis of Distributed Gaussian Basis (DGB) functions:

$$g_\alpha(r, R) = \sum_{i=1}^{m_\alpha} c_i^{(\alpha)} \phi_i^\alpha(r, R) \tag{122}$$

with

$$\phi_i^\alpha(R,r) = \sqrt{\frac{2A_i^\alpha}{\pi}} \exp\left\{-A_i^\alpha\left[(R-R_i^\alpha)^2 + (r-r_i^\alpha)^2\right]\right\}. \tag{123}$$

The parameters A_i^α and the number of basis functions m_α may vary from one angle τ_α to another. It is clear that the function $\phi_i^\alpha(R,r)$ only has significant amplitude in a very small region of (r, R) space around the point (r_i^α, R_i^α). We might call such a function a local basis function. In order to obtain a reasonable representation of the function $g_\alpha(r,R)$ through Eq. (122) it is obviously necessary to introduce a grid of points (r_i^α, R_i^α) in (r, R) space. It seems evident, however, that if we are able to employ a large number of grid points, the expansion of the wavefunction in terms of local basis functions offers more flexibility than the expansions in global basis functions used in the methods described above.

Through the equations (121) and (122) the DVR-DGB matrix representation of the vibrational Hamiltonian can be constructed. The eigenvalues can then be obtained through usual diagonalization; we will not give the details of this procedure here. Calculations based on the DVR principle have recently been presented by Tennyson and Henderson (1989).

7. CONCLUSIONS

The preceding sections give an exposé of some current methods for calculating molecular rotation-vibration energies on the basis of a nuclear potential energy function. We can classify the methods in two main groups:

1. the "spectroscopic" approaches in which we choose the coordinates on the basis of what we surmise about the molecular motion. Further, the coordinate definitions are heavily used to influence the form of the Hamiltonian and approximations are made to facilitate the calculation of its eigenvalues. These approaches comprise the standard approach, the nonrigid bender approach, and the MORBID approach.

2. the "exact-Hamiltonian" methods described in Section 6 which employ an exact kinetic energy operator derived in terms of geometrically defined vibrational coordinates and which obtain the eigenvalues of the Hamiltonian variationally.

Clearly the latter methods are the most accurate as they involve less approximation than the spectroscopic approaches. However, the "exact-Hamiltonian" methods are computationally heavy and have only been developed for triatomic and tetraatomic molecules. For larger molecules it is still necessary to resort to spectroscopic approaches. Another consideration, which we shall discuss further in Section 8, is the fact that even though it is possible to solve the Schrödinger equation for the rotation and vibration [Eq. (19)] with high precision, the agreement between the calculated results and the corresponding experimental values will still be limited by the accuracy of the potential energy surface used as input for the rotation-vibration calculation. If rotation-vibration energies are calculated from an *ab initio* potential energy function, the discrepancy between theory and experiment will normally be predominantly due to inaccuracies of the potential energy surface. There are of course exceptions to this rule, the most prominent one being the two-electron molecule H_3^+ which, as already mentioned, has been treated with great success by Tennyson, Sutcliffe, and Miller [see Drossart *et al.*, 1989; Tennyson, 1986; Miller and Tennyson, 1987; Miller and Tennyson, 1988*a*; Miller and Tennyson, 1989; Miller *et al.*, 1989] using the methods described in Section 6. For more usual molecules, however, it is desirable to be able to refine the parameters of an analytical representation of the potential energy function through fitting to experimental data and it would appear that "spectroscopic" approaches to the calculation of the rotation-vibration energies can play a role in this context. For example, Fernley *et al.* (1991) have very recently used "exact-Hamiltonian" calculations to compare four available potential energy surfaces for water, all of which were obtained through fitting to experimental data, and the best potential was found to be one fitted by this author using the "spectroscopic" MORBID approach [Jensen, 1989*a*].

8. FITTING TO EXPERIMENTAL DATA

As mentioned above it is often desirable to refine the parameters of an analytical representation of the potential function V_{ai} through least squares fitting to experimental data. In discussing the nonrigid bender and MORBID calculations for the methylene radical CH_2 we have already mentioned this type of calculations. Examples of analytical representations of the potential energy surface given in the present work are Eqs. (81–83) and Eqs. (100–104).

As starting point for the fitting to experimental data we have a computational method to obtain the molecular energies $E(Q; P_1, P_2, \ldots, P_K)$ as functions of the molecular state (here symbolized by Q) and as functions of the potential energy parameters (P_1, P_2, ..., P_K). The problem is now to refine the parameters P_k so that a set of energy differences

$$\Delta E(Q', Q; P_1, P_2, \ldots, P_K) = E(Q'; P_1, P_2, \ldots, P_K) - E(Q; P_1, P_2, \ldots, P_K) \tag{124}$$

reproduce the corresponding set of experimentally measured values in the best possible way.

If ΔE were a *linear* function of the P_k, this problem could be solved through standard least squares methods [see, for example, Albritton *et al.* (1976)]. This is not the case, however, so in practice one has to approximate the true expression for ΔE by a linear expression in the form of a first-order Taylor series:

$$\Delta E(Q', Q; P_1, P_2, \ldots, P_K) \approx \Delta E(Q', Q; P_1^{(0)}, P_2^{(0)}, \ldots, P_K^{(0)}) + \sum_{k=1}^{K} \frac{\partial \Delta E}{\partial P_k} \left(P_k - P_k^{(0)}\right) \tag{125}$$

where the ($P_1^{(0)}$, $P_2^{(0)}$, ..., $P_K^{(0)}$) are initial guesses for the parameter values. On the basis of Eq. (125) we can carry out a linear least squares fit and obtain the differences $P_k - P_k^{(0)}$. The new parameter values P_k are of course not the solution to our problem because Eq. (125) is not exactly valid. We can, however, generally obtain the solution through an iteration, i.e., we use the parameters P_k as new starting values to be refined, and we repeat this procedure until self-consistent P_k-values have been obtained.

To carry out the least squares fitting we need values of the derivatives $\partial \Delta E / \partial P_k$ which we can calculate as

$$\frac{\partial \Delta E(Q', Q)}{\partial P_k} = \frac{\partial E(Q')}{\partial P_k} - \frac{\partial E(Q)}{\partial P_k}. \tag{126}$$

The derivatives $\partial E(Q)/\partial P_k$ can be obtained from the Hellman-Feynman theorem:

$$\frac{\partial E}{\partial P_k}(Q; P_1^{(0)}, P_2^{(0)}, \ldots, P_k^{(0)}, \ldots, P_K^{(0)}) = \left\langle \psi_Q \left| \frac{\partial \hat{H}}{\partial P_k} \right| \psi_Q \right\rangle \tag{127}$$

[where ψ_Q is the rotation-vibration eigenfunction corresponding to the energy $E(Q; P_1^{(0)}$, $P_2^{(0)}$, ..., $P_K^{(0)})$ and $\hat{H}$ is the rotation-vibration Hamiltonian] or through primitive approximations, for instance

$$\begin{aligned} &\frac{\partial E}{\partial P_k}(Q; P_1^{(0)}, P_2^{(0)}, \ldots, P_k^{(0)}, \ldots, P_K^{(0)}) \\ &\approx \frac{1}{\Delta_k}\Big[E(Q; P_1^{(0)}, P_2^{(0)}, \ldots, P_k^{(0)} + \Delta_k, \ldots, P_K^{(0)}) \\ &- E(Q; P_1^{(0)}, P_2^{(0)}, \ldots, P_k^{(0)}, \ldots, P_K^{(0)})\Big] \end{aligned} \tag{128}$$

where the increments Δ_k must be selected large enough that the subtraction in the numerator of Eq. (128) can be carried out without introducing unreasonable round-off errors and small enough that the approximation for $\partial E(Q)/\partial P_k$ is reasonable.

It is obvious that a least squares fitting of the kind described here is computationally more expensive than a single, "forward" calculation from a set of parameters ($P_1^{(0)}$, $P_2^{(0)}$, ..., $P_K^{(0)}$) to the corresponding set of rotation-vibration energies. In addition to a more or less expensive calculation of the derivatives $\partial \Delta E/\partial P_k$ the least squares fitting involves a number of iterations, each of which contains one forward calculation using the current set of refined parameters. The computational cost is presumably the reason that in the few calculations of this type which have been carried out using the variational principle and exact kinetic energy operators, only energy spacings involving $J = 0$ have been fitted. Two such calculations are known to this author, one for H_2O by Carter and Handy (1987) and one for HCN by Carter *et al.* (1990).

As already mentioned, the approximations made in the kinetic energy operator of the MORBID approach makes it possible to calculate the matrix element integrals over the stretching coordinates analytically, leading to a considerable saving in computer time. This has made it feasible to carry out perturbation-theory-free least squares fittings of levels involving $N \leq 10$ for CH_2 [Jensen and Bunker, 1988], of levels involving $J \leq 2$ for H_2O [Jensen, 1989*a*], and of levels involving $J \leq 1$ for C_3 [Jensen, 1989*b*]. The fitting for H_2O was especially extensive as it involved 103 vibrational states [with energies up to 19000 cm^{-1} above the vibrational ground state] for six isotopic species of H_2O. As already mentioned in Section 7, Fernley *et al.* (1991) have used "exact-Hamiltonian" calculations of the rotation-vibration energies to compare the fitted potential for H_2O with three other potential energy functions obtained in an analogous manner: The one by Carter and Handy (1987) determined using an "exact-Hamiltonian" approach, one by Halonen and Carrington Jr. (1988) and one by Kauppi and Halonen (1990). The two latter potentials were determined by means of Hamiltonians obtained through drastic approximations. The "exact-Hamiltonian" rotation-vibration energies obtained from the four potential energy surfaces were compared with experimental values, and the potential obtained through the MORBID approach [Jensen, 1989*a*] was found to give the best agreement.

By using the method for methane-type spherical top molecules mentioned in Section 3.3 [which makes use of perturbation theory in the form of van Vleck transformations to reduce the size of the $J > 0$ matrix blocks], Brodersen (1991) has fitted levels with $J \leq 70$ for CF_4 in vibrational states up to about 1400 cm^{-1} above the vibrational ground state. The standard deviation of these fittings was only slightly larger than the experimental uncertainty. On the basis of predictions carried out with the fitted potential the complicated rotational structure in the ν_3 R branch of CF_4 could be assigned [Mortensen *et al.*, 1991].

It would seem that in order to carry out least squares fittings of rotation-vibration energy spacings involving $J > 0$, it is at the present time necessary to use approximate methods for calculating the rotation-vibration energy levels such as the MORBID approach. It is of course to be expected that because of the approximations, the fitted potential energy function will deviate from a hypothetical function obtained through fitting with an exact Hamiltonian. We can investigate the extent of this "contamination" by using the approximate fitted potential energy function as input for a "forward" calculation on the basis of an exact Hamiltonian. Table 4 shows an example of such a comparison. Using the fitted potential energy function for H_2O obtained by Jensen (1989*a*) with the MORBID approach, vibrational energies have been calculated with the MORBID approach and with an "exact-Hamiltonian" approach developed by Schwenke (1991). Schwenke uses an exact kinetic energy operator expressed in socalled Radau coordinates [see Johnson and Reinhardt (1986) and Sutcliffe and Tennyson (1991)]. He obtains the eigenvalues of this Hamiltonian by diagonalizing its matrix representation in a basis consisting of products of single mode vibrational functions and symmetric top basis functions.

The single mode vibrational functions are determined from an SCF procedure on the ground vibrational state for a particular J and K. The deviations between the MORBID results and the "exact-Hamiltonian" results in Table 4 are not zero, but they are sufficiently small that also when the "forward" calculation of the rotation-vibration energies is carried out with an "exact-Hamiltonian" method, the fitted potential allows non-observed vibrational energies to be predicted with much higher precision than it would generally have been possible from a pure *ab initio* potential energy function. This effect can be appreciated from the following example: In 1988, this author used the highly precise *ab initio* potential energy surface of Bartlett *et al.* (1987) and the MORBID program to calculate the ν_2 fundamental energy of the HTO molecule to be 1350 cm^{-1} [Table 3 of Jensen (1988*d*)]. The results of Table 4 indicate that the "exact-Hamiltonian" result would differ insignificantly from the MORBID value. There existed already an experimental estimate of this energy ν_2 = 1324 cm^{-1} [Staats *et al.*, 1956]. This value was obtained from low resolution spectra and consequently not very accurate. In the fitting of H_2O energies carried out by this author [Jensen, 1989*a*] no data for HTO were used as input. On the basis of the fitted potential function, however, the ν_2 energy for HTO could be predicted as ν_2 = 1332.3 cm^{-1} [Table IV of Jensen (1989*a*)]. Very recently, Ulenikov *et al.* (1991) have analysed high-resolution spectra of HTO to obtain the experimental value ν_2 = 1332.480 cm^{-1}. The *ab initio* prediction [Jensen, 1988*d*] deviates 18 cm^{-1} [or 1.3 %] from the experimental value, whereas the predicted value calculated on the basis of the fitted potential energy function deviates 0.2 cm^{-1} [or 0.014 %] from experiment.

Table 4. Vibrational energies of water $H_2{}^{16}O$ (in cm^{-1})

(v_1,v_2,v_3)	Exp.	MORBID	Schwenke (1991)
(0,0,0)	0.0	0.0	0.0
(0,1,0)	1594.7475[a]	1594.2503	1594.3184
(0,2,0)	3151.6307[b]	3151.8808	3152.0030
(1,0,0)	3657.0542[b]	3656.3370	3656.4813
(0,0,1)	3755.9242[b]	3755.7210	3755.9254
(0,3,0)	4666.8080[b]	4667.5078	4667.6771
(1,1,0)	5234.9851[b]	5233.8650	5234.2804
(0,1,1)	5331.2409[b]	5331.4955	5332.0574
(0,4,0)	6134.030[c]	6133.8710	6134.0832
(1,2,0)	6775.095[d]	6774.3910	6775.0098
(0,2,1)	6871.5214[a]	6872.6203	6873.4573
(2,0,0)	7201.540[d]	7202.1900	7202.6592

[a]from Camy-Peyret *et al.* (1977).

[b]from Pugh and Rao (1973).

[c]from Camy-Peyret *et al.* (1980).

[d]from Toth and Margolis (1975).

The example for HTO suggests that when experimental data for a molecule are available, a considerable improvement in the prediction of non-observed energy levels can be achieved if the *ab initio* potential energy function is refined to reproduce in the best possible way the existing experimental data. It should be noted, however, that even with the approximations made in the MORBID approach to facilitate the numerical computation, the refinement of the potential energy parameters is a formidable numerical task. The final least squares fitting for H_2O, for example, used 37.25 hrs CPU time on a CDC Cyber 860 computer. In order to carry out such fittings it is very important to have good starting values for the potential energy parameters from *ab initio* calculations.

From the experience of this author it would seem fair to say that for "usual" small molecules like CH_2 and H_2O, typical deviations between experimental vibrational energies and values calculated from *ab initio* potential energy surfaces using variational approaches lie around 1 %. We have already mentioned that the molecular ion H_3^+ and its isotopic species constitute an exception from this rule as they have two electrons only so that their potential energy surface can be calculated *ab initio* with high precision [Meyer *et al.*, 1986]. Hence it has been possible for Tennyson, Miller and Sutcliffe [see Drossart *et al.*, 1989; Tennyson, 1986; Miller and Tennyson, 1987; Miller and Tennyson, 1988*a*; Miller and Tennyson, 1989; Miller *et al.*, 1989] to predict the rotation-vibration energy levels of this this molecule directly from the *ab initio* potential energy surface with an accuracy of 0.02 % [Tennyson, 1990]. Even with highly advanced methods for calculating the rotation-vibration energy it would be problematic to obtain such accuracy in first principles calculations for molecules with many electrons as the *ab initio* calculation of the potential energy function is the limiting step. For such molecules, the only hope of attaining a comparable accuracy in the prediction of energies lies in extrapolation of already existing experimental values using the techniques outlined here. However, it should perhaps be noted here that Smith *et al.* (1991) have recently voiced certain doubts about the accuracy of *intensities* calculated on the basis on fitted potential energy functions. They compared intensities for HCN, calculated on the basis of

1. a potential energy function determined by them by fitting to experimental data. In the fitting, the vibrational energies were calculated using a perturbation-theory-free approach, but the rotational energy spacings were calculated using perturbation theory,
2. the fitted potential energy surface determined by Carter *et al.* (1990) using a perturbation-theory-free approach, and
3. an *ab initio* potential energy surface by Botschwina (1988),

and they found the *ab initio* potential energy surface to yield intensities in best agreement with experiment.

9. CALCULATION OF MOLECULAR PROPERTIES

In the previous sections we have discussed methods for the calculation of molecular rotation-vibration energies. Clearly in such calculations one also necessarily determines the rotation-vibration wavefunctions. These wavefunctions can be used together with electronic integrals from *ab initio* calculations to obtain observable properties of molecules such as rotation-vibration intensities, hyperfine structure parameters etc. If we consider an operator $\hat{O}$ representing an observable quantity, for example a component of the molecular dipole moment, this operator can be averaged over the electronic coordinates in the *ab initio* calculation:

$$\hat{O}_{\text{nuc}} = \left\langle \psi_e \middle| \hat{O} \middle| \psi_e \right\rangle \tag{129}$$

where integration has taken place over the electronic coordinates only so that $\hat{O}_{\text{nuc}}$ is a function of the nuclear coordinates. In the rotation-vibration calculation we are then interested in carrying out the integration over the nuclear coordinates:

$$O_{ij} = \left\langle \psi^{(i)}_{\text{rot-vib}} \middle| \hat{O}_{\text{nuc}} \middle| \psi^{(j)}_{\text{rot-vib}} \right\rangle \tag{130}$$

where $|\psi^{(i)}_{\text{rot-vib}}\rangle$ and $|\psi^{(j)}_{\text{rot-vib}}\rangle$ are rotation-vibration wavefunctions.

If the rotation-vibration calculation involved perturbation theory in the form of a van Vleck transformation or a contact transformation, it is very complicated to carry out the integration given by Eq. (130). The reason is that the rotation-vibration Hamiltonian [or its matrix representation] was subjected to a transformation to obtain an effective Hamiltonian [or its matrix representation] so that averaging over some part of the nuclear motion has taken place. It is now necessary to subject the operator $\hat{O}_{\text{nuc}}$ to *exactly* the same transformation to obtain an effective $\hat{O}_{\text{nuc}}$ which can then be used to calculate matrix elements between the eigenfunctions of the effective Hamiltonian. The development of such a theory for rotation-vibration intensities has been described in Section 21 of Papoušek and Aliev (1982) [see also Aliev and Mikhailov (1984)], and we will not discuss it further here.

When the rotation-vibration eigenvalues and eigenfunctions have been obtained through a variational calculation, i.e., without the use of perturbation theory, the situation is much simpler. In this case the rotation-vibration eigenstates can be written as linear combinations of basis functions $|\varphi^{(k)}_{\text{rot-vib}}\rangle$

$$|\psi^{(i)}_{\text{rot-vib}}\rangle = \sum_k c_{ik} |\varphi^{(k)}_{\text{rot-vib}}\rangle \tag{131}$$

where the expansion coefficients c_{ik} are determined through a matrix diagonalization. When we insert Eq. (131) in Eq. (130) we obtain

$$O_{ij} = \sum_k \sum_{k'} c^*_{ik} c_{jk'} \left\langle \varphi^{(k)}_{\text{rot-vib}} \middle| \hat{O}_{\text{nuc}} \middle| \varphi^{(k')}_{\text{rot-vib}} \right\rangle \tag{132}$$

where the superscript asterisk denotes complex conjugation.

If we assume that we can find a way of calculating the matrix elements between basis functions $\langle \varphi^{(k)}_{\text{rot-vib}} | \hat{O}_{\text{nuc}} | \varphi^{(k')}_{\text{rot-vib}} \rangle$ it is of course simple to calculate O_{ij} in Eq. (132). However, special considerations are required when $\hat{O}_{\text{nuc}}$ is a component of a vectorial or tensorial operator whose definition is linked to a particular axis system. As an example of this we discuss here the calculation of molecular rotation-vibration intensities.

9.1 Rotation-vibration intensities

The integrated absorption coefficient for a transition from an initial state $|\psi^{(i)}_{\text{rot-vib}}\rangle$ to a final state $|\psi^{(f)}_{\text{rot-vib}}\rangle$ is given by [Smith *et al.*, 1985]

$$I(f \leftarrow i) = \frac{8\pi^3 N_A g_{ns} \omega_{if} \exp(-E_i/kT)[1 - \exp(-hc\omega_{if}/kT)]}{3hcQ} S(f \leftarrow i) \tag{133}$$

where N_A is Avogadro's number, g_{ns} is the nuclear spin statistical weight, ω_{if} is the wavenumber of the transition (in cm^{-1}), E_i is the energy of the initial state, k is the Boltzmann constant, T is the absolute temperature, h is Planck's constant, c is the vacuum velocity of light and the partition function Q is given by

$$Q = \sum_w g_w \exp(-E_w/kT). \tag{134}$$

In Eq. (134), E_w is the energy and g_w the total degeneracy of the state w and the summation extends over all such states of the molecule. The line strength $S(f \leftarrow i)$ of an electric dipole transition is defined as

$$S(f \leftarrow i) = \sum_{A=X,Y,Z} \sum_{M_i,M_f} \left| \left\langle \psi^{(f)}_{\text{rot-vib}} \right| \mu_A \left| \psi^{(i)}_{\text{rot-vib}} \right\rangle \right|^2 \tag{135}$$

where μ_A is the component of the molecular dipole moment operator along the A axis (A = X, Y, or Z) of the laboratory fixed coordinate system and M_i and M_f are the rotational M quantum numbers (quantizing the Z component of the total angular momentum) of the initial and final states, respectively.

Obviously the line strength is determined by the components of the dipole moment along the *laboratory fixed* axes. These components depend on the instantaneous orientation of the molecule in space, that is on the molecular rotation. If we do an *ab initio* calculation, however, in order to obtain electronically averaged dipole moment components computed according to Eq. (129), the *ab initio* dipole moment values are clearly independent of molecular rotation. The *ab initio* calculation delivers dipole moment components along axes rigidly attached to the framework of clamped-down nuclei, and such axes obviously belong to a *molecule fixed* axis system. Hence we see that the *ab initio* calculation allows us to obtain the dipole moment components along the axes of the molecule fixed coordinate system, whereas we need the components along the laboratory fixed axes in order to calculate the line strengths.

The trick used for the necessary transformation from molecule fixed to laboratory fixed axes is called *irreducible tensor theory* [see Wigner (1959) and Tinkham (1964)]. In the present case we introduce two sets of socalled irreducible tensors: In the laboratory fixed axis system, where the dipole moment components are called (μ_X, μ_Y, μ_Z) we define

$$\begin{aligned} \mathcal{T}^{(1)}_{\pm 1} &= \frac{1}{\sqrt{2}}[\mp\mu_X + i\mu_Y] \\ \mathcal{T}^{(1)}_0 &= \mu_Z \end{aligned} \tag{136}$$

and in the molecule fixed coordinate system [where the dipole moment components are called (μ_x, μ_y, μ_z)] we make analogous definitions

$$\begin{aligned} T^{(1)}_{\pm 1} &= \frac{1}{\sqrt{2}}[\mp\mu_x + i\mu_y] \\ T^{(1)}_0 &= \mu_z . \end{aligned} \tag{137}$$

The quantities $\mathcal{T}^{(1)}_\sigma$ [or $T^{(1)}_\sigma$] span the irreducible representation $D^{(1)}$ of the full rotation group [see Wigner (1959) and Tinkham (1964)] and their transformation properties under rotations in space are known.

The point is now that we can obtain the laboratory fixed axis system [or at least an axis system with axes parallel to those of the laboratory fixed system and origin at the nuclear center of mass] by rotating the molecule fixed axis system through the Euler angles ($-\theta$, $-\phi$, $-\chi$). Hence we can use the known transformation properties of $T^{(1)}_\sigma$ to relate these operators to the operators $\mathcal{T}^{(1)}_\sigma$:

$$\mathcal{T}^{(1)}_\sigma = \hat{P}_R \, T^{(1)}_\sigma \, \hat{P}_R^{-1} = \sum_{\sigma'=-1}^{1} T^{(1)}_{\sigma'} D^{(1)}_{\sigma'\sigma} \tag{138}$$

where $\hat{P}_R$ is an operator effecting the rotation that makes the axes of the molecule fixed axis system parallel to those of the laboratory fixed axis system, and $D^{(1)}_{\sigma'\sigma}$ is an element of a representation matrix belonging to the irreducible representation $D^{(1)}$ of the full rotation group. This matrix element depends on the Euler angles (θ, ϕ, χ).

In all the methods for calculating rotation-vibration energies presented here the basis functions $|\varphi^{(k)}_{\text{rot-vib}}\rangle$ are taken to be products of vibrational basis functions and rotational basis functions. The common choice for the rotational basis functions are the symmetric top functions $|J, K, M\rangle$ [see Chapter 6 of Papoušek and Aliev (1982)] which depend on the Euler angles (θ, ϕ, χ). The choice of the vibrational basis functions depends on the model employed for the calculation of the rotation-vibration energies; we will here denote vibrational basis functions as $|V\rangle$. Hence a rotation-vibration wavefunction can be written as [cf. Eq. (131)]

$$|\psi^{(i)}_{\text{rot-vib}}\rangle = \sum_{VK} c_{i,VK}|V\rangle|J,K,M\rangle \tag{139}$$

It is easy to show from Eqs. (135) and (136) that

$$S(f \leftarrow i) = \sum_{\sigma=-1}^{1} \sum_{M_i,M_f} \left|\left\langle \psi^{(f)}_{\text{rot-vib}} \left| T^{(1)}_{\sigma} \right| \psi^{(i)}_{\text{rot-vib}} \right\rangle\right|^2 \tag{140}$$

and we can insert Eqs. (138) and (139) in Eq. (140) to obtain

$$S(f \leftarrow i) = \sum_{\sigma=-1}^{1} \sum_{M_i,M_f} \left| \sum_{\sigma'=-1}^{1} \sum_{V'K'} \sum_{VK} c^{*}_{f,V'K'} c_{i,VK} \right.$$
$$\times \left. \left\langle V'; J_f, K', M_f \left| T^{(1)}_{\sigma'} D^{(1)}_{\sigma'\sigma} \right| V; J_i, K, M_i \right\rangle \right|^2 \tag{141}$$

We can now use the fact that the functions $|V\rangle$, $|V'\rangle$, and $T^{(1)}_{\sigma'}$ all depend on the vibrational coordinates whereas $|J_f, K', M_f\rangle$, $|J_i, K', M_i\rangle$, and $D^{(1)}_{\sigma'\sigma}$ all depend on the Euler angles to simplify Eq. (141) further

$$S(f \leftarrow i) = \sum_{\sigma=-1}^{1} \sum_{M_i,M_f} \left| \sum_{\sigma'=-1}^{1} \sum_{V'K'} \sum_{VK} c^{*}_{f,V'K'} c_{i,VK} \right.$$
$$\times \left. \left\langle V' \left| T^{(1)}_{\sigma'} \right| V \right\rangle \left\langle J_f, K', M_f \left| D^{(1)}_{\sigma'\sigma} \right| J_i, K, M_i \right\rangle \right|^2 . \tag{142}$$

The next step is to introduce known analytical expressions for the rotational matrix elements $\left\langle J_f, K', M_f \left| D^{(1)}_{\sigma'\sigma} \right| J_i, K, M_i \right\rangle$. However, the derivation of these expressions leaves room for endless confusion for the following reasons:

1. the functions $|J, K, M\rangle$ are the eigenfunctions of the operators $\hat{\mathbf{J}}^2$, $\hat{J}_Z$, and $\hat{J}_z$ [see Chapter 6 of Papoušek and Aliev (1982)] and consequently they are only defined to within a phase factor.
2. the matrix elements $D^{(1)}_{\sigma'\sigma}$ are only defined to within a similarity transformation.

Before one can derive expressions for the matrix elements $\left\langle J_f, K', M_f \left| D^{(1)}_{\sigma'\sigma} \right| J_i, K, M_i \right\rangle$ one must introduce phase conventions for both the functions $|J, K, M\rangle$ and the matrix elements $\left\langle J_f, K', M_f \left| D^{(1)}_{\sigma'\sigma} \right| J_i, K, M_i \right\rangle$. One literature source where, to the knowledge of this author, such a derivation is carried out correctly, is the paper by Rasmussen and Brodersen (1968).

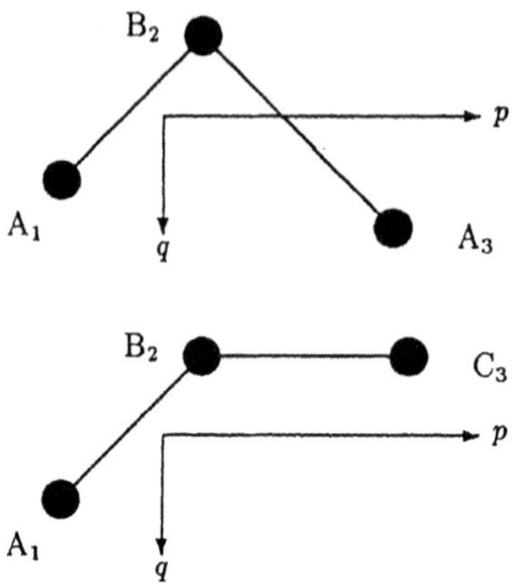

Figure 5. The definition of the *pq* axis system for symmetric ABA and unsymmetric ABC molecules.

They assume the functions $|J, K, M\rangle$ to have relative phases such that all matrix elements of the ladder operators

$$\hat{J}_{\pm} = \hat{J}_x \pm i\,\hat{J}_y \tag{143}$$

are real and positive, and the matrix elements $D^{(1)}_{\sigma'\sigma}$ to be as given by Tinkham (1964). With these definitions, the following expression is obtained

$$\begin{aligned}&\left\langle J_f, K', M_f \left| D^{(1)}_{\sigma'\sigma} \right| J_i, K, M_i \right\rangle \\ &= \delta_{K',K+\sigma'}\,\delta_{M_f,M_i+\sigma}\sqrt{\frac{J_f+1}{J_i+1}}\,C(J_i 1 J_f; K\sigma')\,C(J_i 1 J_f; M_i\sigma)\end{aligned} \tag{144}$$

where $C(J_i 1 J_f; ab)$ is a Clebsch-Gordan coefficient [see Tinkham (1964)]. In terms of $3j$-symbols [see Tinkham (1964)] Eq. (144) becomes

$$\begin{aligned}&\left\langle J_f, K', M_f \left| D^{(1)}_{\sigma'\sigma} \right| J_i, K, M_i \right\rangle \\ &= (-1)^{J_f - M_f}\sqrt{(J_f+1)(J_i+1)}\begin{pmatrix} J_f & 1 & J_i \\ -K' & \sigma' & K \end{pmatrix}\begin{pmatrix} J_f & 1 & J_i \\ -M_f & \sigma & M_i \end{pmatrix}.\end{aligned} \tag{145}$$

These expressions can now be inserted in Eq. (142) so that the final expression for $S(f \leftarrow i)$ can be obtained.

Examples of this type of calculations are given by Jensen and Špirko (1986), Jensen (1988*c*), and by Miller and Tennyson (1988*b*).

Jensen (1988*c*) used the wavefunctions obtained for triatomic molecules in the MORBID approach [see Section 5] to calculate the line strengths and integrated absorption coefficients of rotation-vibration transitions. In that work, the matrix elements $\langle V'|T^{(1)}_{\sigma}|V\rangle$ of Eq. (142) were calculated as follows: A *pq* axis system was defined for the triatomic molecule [see Fig. 5]. This axis system has origin at the nuclear center of mass and the *p* and *q* axes are in the plane defined by the three nuclei. For an unsymmetrical ABC molecule, the *p* axis is parallel to the bond between nucleus 2 and nucleus 3 and points so that the p coordinate of nucleus 1 is negative. The q axis is perpendicular to the p axis and points so that the q coordinate of nucleus 1 is positive. For a symmetical ABA molecule, the q axis is the bisector of the angle $\angle(\mathrm{ABA})$ and points such that the two A nuclei have positive q-coordinates. The p axis is perpendicular to the q axis and points so that the p coordinate of nucleus 3 is positive. The dipole moment component along the p axis, $\bar{\mu}^{(p)}$, is expanded as:

$$\begin{aligned}\bar{\mu}^{(p)}(\Delta r_{12}, \Delta r_{32}, \bar{\rho}) &= \mu_0^{(p)}(\bar{\rho}) + \sum_j \mu_j^{(p)}(\bar{\rho})\Delta r_{j2} \\ &+ \sum_{j\leq k} \mu_{jk}^{(p)}(\bar{\rho})\Delta r_{j2}\Delta r_{k2} + \sum_{j\leq k\leq m} \mu_{jkm}^{(p)}(\bar{\rho})\Delta r_{j2}\Delta r_{k2}\Delta r_{m2} \\ &+ \sum_{j\leq k\leq m\leq n} \mu_{jkmn}^{(p)}(\bar{\rho})\Delta r_{j2}\Delta r_{k2}\Delta r_{m2}\Delta r_{n2},\end{aligned} \quad (146)$$

where all of the indices j, k, m, and n assume the values 1 or 3 and

$$\mu_{jk\ldots}^{(p)}(\bar{\rho}) = \sum_{i=0}^{N} p_{jk\ldots}^{(i)}(\cos\rho_e - \cos\bar{\rho})^i. \quad (147)$$

The function $\mu_0^{(p)}(\bar{\rho})$ has $N = 8$, $\mu_j^{(p)}(\bar{\rho})$ has $N = 4$, $\mu_{jk}^{(p)}(\bar{\rho})$ has $N = 3$, $\mu_{jkm}^{(p)}(\bar{\rho})$ has $N = 2$, and $\mu_{jkmn}^{(p)}(\bar{\rho})$ has $N = 1$.

The dipole moment component along the q axis, $\bar{\mu}^{(q)}$, was represented by

$$\begin{aligned}\bar{\mu}^{(q)}(\Delta r_{12}, \Delta r_{32}, \bar{\rho}) &= \sin\bar{\rho}\,[\mu_0^{(q)}(\bar{\rho}) + \sum_j \mu_j^{(q)}(\bar{\rho})\Delta r_{j2} \\ &+ \sum_{j\leq k} \mu_{jk}^{(q)}(\bar{\rho})\Delta r_{j2}\Delta r_{k2} + \sum_{j\leq k\leq m} \mu_{jkm}^{(q)}(\bar{\rho})\Delta r_{j2}\Delta r_{k2}\Delta r_{m2} \\ &+ \sum_{j\leq k\leq m\leq n} \mu_{jkmn}^{(q)}(\bar{\rho})\Delta r_{j2}\Delta r_{k2}\Delta r_{m2}\Delta r_{n2}].\end{aligned} \quad (148)$$

Again, all of the indices j, k, m, and n assume the values 1 or 3. The equations for the $\mu_{jkm\ldots}^{(q)}(\bar{\rho})$ functions are obtained when p is replaced by q in Eq. (147).

The values of the parameters $p_{jk\ldots}^{(i)}$ and $q_{jk\ldots}^{(i)}$ in Eqs. (146-148) were obtained by fitting the analytical expressions through *ab initio* dipole moment values, i.e. values calculated as outlined in Eq. (129). As explained by Jensen (1988c) the dipole moment components can be transformed into polynomials in the quantities $y_k = 1 - \exp(-a_k\Delta r_{k2})$ [with expansion coefficients depending on the bending coordinate ρ] and the matrix elements of these expansions between vibrational basis functions [that is, the matrix elements $\langle V'|T_\sigma^{(1)}|V\rangle$ needed in Eq. (142)] can be straightforwardly calculated using Morse oscillator matrix elements from Špirko *et al.* (1985).

9.2 Other molecular properties

The problem of transforming a vector or tensor from the molecule fixed axis system to the laboratory fixed axis system is often encountered in calculations of molecular properties. It can generally be solved using irreducible tensor theory in combination with the Wigner-Eckart theorem [see Tinkham (1964)] following the ideas outlined for the intensity calculation above. Examples of such studies would be calculations of Raman intensities [where one needs to transform the polarisability tensor from the molecule fixed to the laboratory fixed axis system] and calculations of quadrupole hyperfine structure [where one needs to transform the electric field gradient tensor from the molecule fixed to the laboratory fixed axis system]. For an extensive treatment of the theory of quadrupole hyperfine structure, the reader is referred to Gordy and Cook (1970) and to the excellent review article by Cook and De Lucia (1971). In a somewhat exploratory study in this field, Jensen *et al.* (1991*b*) have used *ab initio* values for the electric field gradient to calculate the nuclear quadrupole hyperfine effects for the molecules H_2D^+, HD_2^+ and D_3^+.

10. ACKNOWLEDGMENTS

I am grateful to P.R. Bunker and B.P. Winnewisser for critically reading the manuscript, to a number of participants at the NATO Advanced Study Institute for pointing out misprints in the first version of the notes, and to D.W. Schwenke and J.T. Tennyson for communicating their results prior to publication. Since 1989, my work has been supported by the Deutsche Forschungsgemeinschaft (through grants no. Je 144/2-1, Je 144/2-2 and Je 144/2-3). Additional funding has been provided by the Fonds der Chemischen Industrie, the Dr. Otto Röhm Gedächtnisstiftung, and the Fritz Thyssen-Stiftung.

11. REFERENCES

Albritton, D.L., Schmeltekopf, A.L., and Zare, R.N., 1976, in: "Molecular Spectroscopy: Modern Research", Vol. II, K. Narahari Rao, ed., Academic Press, New York.

Aliev, M.R., and Mikhailov, V.M., 1984, Acta Physica Hungarica 55:293.

Aliev, M.R., and Watson, J.K.G., 1985, in: "Molecular Spectroscopy: Modern Research", Vol. III, K. Narahari Rao, ed., Academic Press, Orlando.

Bačić, Z., and Light, J.C., 1986, J. Chem. Phys. 85:4594.

Bačić, Z., and Light, J.C., 1987, J. Chem. Phys. 86:3065.

Bačić, Z., and Light, J.C., 1989, Ann. Rev. Phys. Chem. 40:469.

Barrow, T., Dixon, R.N., and Duxbury, G., 1974, Mol. Phys. 27:1217.

Bartlett, R.J., Cole, S.J., Purvis, G.D., Ermler, W.C., Hsieh, H.C., and Shavitt, I., 1987, J. Chem. Phys. 87:6579.

Born, M., and Oppenheimer, J.R., 1927, Ann. der Physik 84:457.

Botschwina, P., 1988, J. Chem. Soc., Faraday Trans. 2 84:1263.

Brodersen, S., 1988, J. Chem. Soc., Faraday Trans. 2 84:1365.

Brodersen, S., 1991, J. Mol. Spectrosc. 145:331.

Brodersen, S., and Lolck, J.-E., 1987, J. Mol. Spectrosc. 126:405.

Buenker, R.J., 1980, in: "Proceedings of Workshop on Quantum Chemistry and Molecular Physics in Wollongong, Australia", P. Burton, ed., University Press, Wollongong.

Buenker, R.J., 1981, in: "Studies in Physical and Theoretical Chemistry, Current Aspects of Quantum Chemistry", Vol. 21, R. Carbó, ed., Elsevier, Amsterdam.

Buenker, R.J., and Peyerimhoff, S.D., 1974, Theor. Chim. Acta (Berl.) 35:33.

Buenker, R.J., Peyerimhoff, S.D., and Butscher, W., 1978, Mol. Phys. 35:771.

Buenker, R.J., and Phillips, R.A., 1985, J. Mol. Struct. THEOCHEM 123:291.

Bunker, P.R., 1979, "Molecular Symmetry and Spectroscopy", Academic Press, New York.

Bunker, P.R., and Jensen, P., 1983, J. Chem. Phys. 79:1224.

Bunker, P.R., Jensen, P., Karpfen, A., and Lischka, H., 1989, J. Mol. Spectrosc. 135:89.

Bunker, P.R., Jensen, P., Kraemer, W.P., and Beardsworth, R., 1986, J. Chem. Phys. 85:3724.

Bunker, P.R., Jensen, P., Wright, J.S., and Hamilton, I.P., 1990, J. Mol. Spectrosc. 144:310.

Bunker, P.R., and Moss, R.E., 1980, J. Mol. Spectrosc. 80:217.

Camy-Peyret, C., Flaud, J.-M., Maillard, J.-P., and Guelachvili, G., 1977, Mol. Phys. 33:1641.
Camy-Peyret, C., Flaud, J.-M., and Maillard, J.-P., 1980, J. Phys. Lett. 41:L23.
Carter, S., and Handy, N.C., 1986, Comp. Phys. Rep. 5:115.
Carter, S., and Handy, N.C., 1987, J. Chem. Phys. 87:4294.
Carter, S., Handy, N.C., and Mills, I.M., 1990, Phil. Trans. R. Soc. Lond. A 332:309.
Comeau, D.C., Shavitt, I., Jensen, P., and Bunker, P.R., 1989, J. Chem. Phys. 90:6491.
Cook, R.L., and De Lucia, F.C., 1971, Am. J. Phys. 39:1433.
Cooley, J.W., 1961, Math. Comp. 15:363.
Darling, B.T., and Dennison, D.M., 1940, Phys. Rev. 57:128.
Drossart, P., Maillard, J.-P., Caldwell, J., Kim, S.J., Watson, J.K.G, Majewski, W.A., Tennyson, J., Miller, S., Atreya, S., Clarke, J., Waite Jr., J.H., and Wagener, R., 1989, Nature 340:539.
Eckart, C., 1935, Phys. Rev. 47:552.
Efremov, Yu. S., 1977, Opt. Spectrosc. (USSR) 43:693.
Fernley, J.A., Miller, S., and Tennyson, J.T., 1991, J. Mol. Spectrosc., submitted for publication.
Gordy, W., and Cook, R.L., 1970, "Microwave Molecular Spectra", John Wiley & Sons, New York.
Halonen, L., and Carrington Jr., T., 1988, J. Chem. Phys. 88:4171.
Hougen, J.T., Bunker, P.R., and Johns, J.W.C., 1970, J. Mol. Spectrosc. 34:136.
Hoy, A.R., and Bunker, P.R., 1974, J. Mol. Spectrosc. 52:439.
Hoy, A.R., and Bunker, P.R., 1979, J. Mol. Spectrosc. 74:1.
Hoy, A.R., Mills, I.M., and Strey, G., 1972, Mol. Phys. 24:1265.
Hutson, J.M., 1990, in: "Dynamics of Polyatomic van der Waals Molecules", N. Halberstadt and K.C. Janda, eds., Plenum, New York.
Hutson, J.M., 1991*a*, Advances in Molecular Vibrations and Collision Dynamics, in press.
Hutson, J.M., 1991*b*, BOUND computer code, distributed via Collaborative Computational Project No. 6, of the U.K. Science and Engineering Research Council, on Heavy Particle Dynamics.
Istomin, V.A., Stepanov, N.F., and Zhilinskii, B.I., 1977, J. Mol. Spectrosc. 67:265.
Jensen, P., 1983, Comp. Phys. Rep. 1:1.
Jensen, P., 1988*a*, J. Mol. Spectrosc. 128:478.
Jensen, P., 1988*b*, J. Chem. Soc., Faraday Trans. 2 84:1315.
Jensen, P., 1988*c*, J. Mol. Spectrosc. 132:429.
Jensen, P., 1988*d*, J. Mol. Structure 190:149.
Jensen, P., 1989*a*, J. Mol. Spectrosc. 133:438.
Jensen, P., 1989*b*, Collection of Czechoslovak Chemical Communications 54:1209.
Jensen, P., Buenker, R.J., Hirsch, G., and Rai, S.N., 1990, Mol. Phys. 70:443.
Jensen, P., and Bunker, P.R., 1983, J. Mol. Spectrosc. 99:348.
Jensen, P., and Bunker, P.R., 1986, J. Mol. Spectrosc. 118:18.
Jensen, P., and Bunker, P.R., 1988, J. Chem. Phys. 89:1327.
Jensen, P., Bunker, P.R., and Hoy, A.R., 1982, J. Chem. Phys. 77:5370.
Jensen, P., Bunker, P.R., and McLean, A.D., 1987, Chem. Phys. Lett. 141:53.
Jensen, P., and Kraemer, W.P., 1988*a*, J. Mol. Spectrosc. 129:172.
Jensen, P., and Kraemer, W.P., 1988*b*, J. Mol. Spectrosc. 129:216.
Jensen, P., Marshall, M.D., Bunker, P.R., and Karpfen, A., 1991*a*, Chem. Phys. Lett. 180:594.
Jensen, P., Paidarová, I., Vojtík, J., and Špirko, V., 1991*b*, J. Mol. Spectrosc., in press.
Jensen, P., and Špirko, V., 1986, J. Mol. Spectrosc. 118:208.
Johnson, B.R., 1978, J. Chem. Phys. 69:4678.

Johnson, B.R., and Reinhardt, W.P, 1986, J. Chem. Phys. 85:4538.
Jordahl, O.M., 1934, Phys. Rev. 45:87.
Kauppi, E., and Halonen, L., 1990, J. Phys. Chem. 94:5779.
Lanczos, C., 1956, "Applied Analysis", Prentice-Hall, New York.
Light, J.C., and Bačić, Z., 1987, J. Chem. Phys. 87:4008.
Light, J.C., Hamilton, I.P., and Lill, J.V., 1985, J. Chem. Phys. 82:1400.
Lill, J.V., Parker, G.A., and Light, J.C., 1982, Chem. Phys. Lett. 89:483.
Manolopoulos, D.E., 1986, J. Chem. Phys. 85:6425.
Manolopoulos, D.E., 1988, Ph.D. Thesis, Cambridge University.
Marshall, M.D., Jensen, P., and Bunker, P.R., 1991, Chem. Phys. Lett. 176:255.
Marshall, M.D., and McKellar, A.R.W., 1986, J. Chem. Phys. 85:3716.
McLean, A.D., Bunker, P.R., Escribano, R.M., and Jensen, P., 1987, J. Chem. Phys. 87:2166.
Meyer, W., Botschwina, P., and Burton, P., 1986, J. Chem. Phys. 84:891.
Miller, S., and Tennyson, J., 1987, J. Mol. Spectrosc. 126:183.
Miller, S., and Tennyson, J., 1988*a*, J. Mol. Spectrosc. 128:530.
Miller, S., and Tennyson, J., 1988*b*, Astrophys. J. 335:486.
Miller, S., and Tennyson, J., 1989, J. Mol. Spectrosc. 136:223.
Miller, S., Tennyson, J., and Sutcliffe, B.T., 1989, Mol. Phys. 66:429.
Mortensen, L., Brodersen, S., and Krivtsun, V.M., 1991, J. Mol. Spectrosc. 145:352.
Nielsen, H.H., 1951, Rev. Mod. Phys. 23:90.
Numerov, B., 1933, Publs. Observatoire Central. Astrophys. Russ. 2:188.
Papoušek, D., and Aliev, M.R., 1982, "Molecular Vibrational-Rotational Spectra", Elsevier, Amsterdam.
Papoušek, D., Stone, J.M.R., and Špirko, V., 1973, J. Mol. Spectrosc. 48:17.
Pugh, L.A., and Rao, K. Narahari, 1973, J. Mol. Spectrosc. 47:403.
Rasmussen, F., and Brodersen, S., 1968, J. Mol. Spectrosc. 25:166.
Sarka, K., and Bunker, P.R., 1987, J. Mol. Spectrosc. 122:259.
Sayvetz, A., 1939, J. Chem. Phys. 7:383.
Schwenke, D.W., 1991, private communication.
Sears, T.J., Bunker, P.R., McKellar, A.R.W., 1981, J. Chem. Phys. 75:4731.
Sears, T.J., Bunker, P.R., McKellar, A.R.W., 1982, J. Chem. Phys. 77:5363.
Smith, A.M., Klemperer, W., and Lehmann, K.K., 1991, J. Chem. Phys. 94:5040.
Smith, M.A.H., Rinsland, C.P., Fridovich, B., and Rao, K. Narahari, 1985, in: "Molecular Spectroscopy: Modern Research", Vol. III, K. Narahari Rao, ed., Academic Press, Orlando.
Sørensen, G.O., 1979, in: "Topics in Current Chemistry," M.J.S. Dewar *et al.*, eds., Vol. 82, Springer-Verlag, Heidelberg.
Špirko, V., 1983, J. Mol. Spectrosc. 101:30.
Špirko, V., Jensen, P., Bunker, P.R., and Čejchan, A., 1985, J. Mol. Spectrosc. 112:183.
Špirko, V., Čejchan, A., and Jensen, P., 1987, J. Mol. Spectrosc. 124:430.
Staats, P.A., Morgan, H.W., and Goldstein, J.H., 1956, J. Chem. Phys. 24:916.
Sutcliffe, B.T., 1982, in: "Current Aspects of Quantum Chemistry", Vol. 21, R. Carbó, ed., Elsevier, Amsterdam.
Sutcliffe, B.T., Tennyson, J., and Miller, S., 1987, Theor. Chim. Acta 72:265.
Sutcliffe, B.T., and Tennyson, J., 1991, Int. J. Quant. Chem. XXXIX:183.
Tennyson, J., 1986, Comp. Phys. Rep. 4:1.
Tennyson, J., 1990, Proceedings of the "XIth International Conference on High Resolution Infrared Spectroscopy", Dobříš near Prague, Czechoslovakia, September 1990, Lecture LM3.

Tennyson, J., and Henderson, J.R., 1989, J. Chem. Phys. 91:3815.
Tennyson, J., and Sutcliffe, B.T., 1982, J. Chem. Phys. 77:4061.
Tinkham, M., 1964, "Group Theory and Quantum Mechanics", McGraw-Hill, New York.
Tipping, R.H., and Ogilvie, J.F., 1983, J. Chem. Phys. 79:2537.
Toth, R.A., and Margolis, J.S., 1975, J. Mol. Spectrosc. 55:229.
Ulenikov, O.N., Cherepanov, V.N., and Malikova, A.B., 1991, J. Mol. Spectrosc. 146:97.
van Vleck, J.H., 1929, Phys. Rev. 33:467.
Wallace, R., 1976, Chem. Phys. Lett. 37:115.
Watson, J.K.G., 1967, J. Chem. Phys. 46:1935.
Watson, J.K.G., 1968*a*, Mol. Phys. 15:479.
Watson, J.K.G., 1968*b*, J. Chem. Phys. 48:181.
Watson, J.K.G., 1968*c*, J. Chem. Phys. 48:4517.
Watson, J.K.G., 1977, in: "Vibrational Spectra and Structure", J. Durig, ed., Vol. 6, Elsevier, Amsterdam.
Whitehead, R.J., and Handy, N.C., 1975, J. Mol. Spectrosc. 55:356.
Wigner, E.P., 1959, "Group Theory and it Application to the Quantum Mechanics of Atomic Spectra", Academic Press, New York.
Wilson, E.B., 1939, J. Chem. Phys. 7:1047.
Wilson, E.B., and Howard, J.B., 1936, J. Chem. Phys. 4:260.
Wilson, E.B., Decius, J.C., and Cross, P., 1955, "Molecular Vibrations", McGraw-Hill, New York.
Zhilinskii, B.I., Istomin, V.A., and Stepanov, N.F., 1978, Chem. Phys. 31:413.

MANY-BODY PERTURBATION THEORY FOR VIBRATIONAL ELECTRONIC MOLECULAR HAMILTONIAN

Ivan Hubač

Division of Chemical Physics, Faculty of Mathematics and Physics Komensky University, Mlynská dolina, 842 15 Bratislava, Czechoslovakia,

Michal Svrček

Institute of Chemistry, Division of Chemistry and Physics of Biomolecules, Komensky University, Kalinčiakova 8, 832 33 Bratislava, Czechoslovakia

1 Introduction

The analytical evaluation of energy derivatives is one of the faremost achievements of modern ab initio quantum chemistry [1-3]. Derivatives of the energy with respect to external field perturbation parameters are directly associated with the one-electron properties of the molecule such as moments, field gradients, spin densities etc., (for details see ref.) [1].

Consider the Schrödinger equation corresponding to a molecular system, where the electronic Hamiltonian $H(\chi)$, its eigenfunctions $\psi_i(\chi)$, and its eigenvalues $E_i(\chi)$ are dependent upon some parameter χ [2]:

$$H(\chi)\psi_i(\chi) = E_i(\chi)\psi_i(\chi) \tag{1.1}$$

$$H(\chi) = -1/2\sum_i \nabla_i^2 - \sum_{i,\alpha} Z_\alpha r_{\alpha i}^{-1} + \sum_i \chi w_i + \sum_{i>j} r_{ij}^{-1} \tag{1.2}$$

Methods in Computational Molecular Physics, Edited by Stephen Wilson and Geerd H.F. Diercksen, Plenum Press, New York, 1992

or

$$H(\chi) = h + v + \sum_i \chi w_i, \tag{1.3}$$

w'iere h is the one-particle operator

$$h = -1/2 \sum_i \nabla_i^2 - \sum_{i,\alpha} Z_\alpha r_{\alpha i}^{-1} \tag{1.4}$$

and v is the two-paricle operator

$$v = \sum_{i>j} r_{ij}^{-1}. \tag{1.5}$$

The parameter χ may control the displacement of an atomic nucleus in the molecular system, or it may be the strength of an external one-electron potential energy perturbation w_i , such as an uniform electric field applied in the z - direction $(\chi w_i = \chi Z_i)$.
Using the formalism of second quantization [4] we can write the Hamiltomian (1.3) in the form

$$H(\chi) = \sum_{PQ} < P|h|Q > a_P^+ a_Q + \frac{1}{2} \sum_{PQRS} < PQ|v|RS > a_P^+ a_Q^+ a_S a_R + \\ + \sum_{PQ} < P| \sum_i \chi w_i |Q > a_P^+ a_Q, \tag{1.6}$$

where $a_P^+(a_Q)$ are the creation (annihilation) operators defined on the one-particle basis set $\{|P>, |Q>; \ldots\}$. Using the concept of normal product (N - product) we can rewrite the Hamiltonian (1.6) into the N-product form [2,4] .

$$H_N(\chi) = < \Phi_0(\chi) \mid H(\chi) | \Phi_0(\chi) > + \sum_{PQ} < P(\chi) | f(\chi) \mid Q(\chi) > . \\ . N[a_P^+(\chi) a_Q(\chi)] + \frac{1}{2} \sum_{PQRS} < P(\chi) Q(\chi) | v(\chi) | R(\chi) S(\chi) > . \\ . N[a_P^+(\chi) a_Q^+(\chi) a_S(\chi) a_R(\chi)], \tag{1.7}$$

where N [ ] is the normal product of creation $(a_P^+(\chi))$ and annihilation $(a_Q(\chi))$ operators, $| \Phi_0(\chi) >$ is the χ dependent reference state vector and f is the χ dependent Hartree-Fock operator. In the last two decades, the many-body methods e.g. many-body perturbation theory (MBPT) as well as coupled cluster (CC) methods [5-13] became very successful in calculation of correlation energy of atomic or molecular system, i.e. they permited to perform the calculations beyond the Hartree-Fock method with very high efficiency. Therefore it became of primarily interest to introduce the analytic gradient techniques into MBPT and/or for CC methods (see e.g. ref.) [2] and references therein). The important step in this respect was the introduction of an efficient method for the solution of the coupled perturbed Hartree-Fock (CPHF) equations [14-16]. Pople et al. [16] deal with the calculation of MP2 (second order of MBPT) derivatives. We are not going to discuss here the algorithm how to calculate the MP2 gradients. The efficient way how to calculate these quantities is discussed in many papers (see. e.g.) [1,17]. Evaluation of gradients in CC method and in higher orders of MBPT is discussed in [2,18,19]. Recently the procedure for the calculation of Moller-Plesset energy derivatives based on the theory of Lagrangians was developed by Jorgensen, Helgaker and Handy [20,21]. In our contribution we present an alternative approach based on the technique of canonical transformations. To explain our approach more clearly let us imagine purely electronic Hamiltonian without the presence of the external perturbation $\sum_i \chi w_i$.

$$H^e = z + v \,. \tag{1.8}$$

The N - product form of this Hamiltonian is

$$H_N^e = < \Phi_0|H^e|\Phi_0 > + \sum_{PQ} < P|f|Q > N[a_P^+ a_Q] + \frac{1}{2} \sum_{PQRS} < PQ|v|RS > N[a_P^+ a_Q^+ a_S a_R], \tag{1.9}$$

where f is the Hartree-Fock operator.

If we ask our basis set $\{|P>\}$ to be the solution of the Hartree-Fock equations

$$f|P> = \varepsilon_P|P> \tag{1.10}$$

and split the Hamiltonian (1.9) into unperturbed part H_0 and the perturbation H'

$$H_N^e = \epsilon_0 + H_0 + H' \tag{1.11}$$

where

$$\epsilon_0 = < \Phi_0|H^e|\Phi_0 >, \tag{1.12}$$

$$H_0 = \sum_P \varepsilon_P N\left[a_P^+ a_P\right] \tag{1.13}$$

and

$$H' = \frac{1}{2} \sum_{PQRS} < PQ|v|RS > N\left[a_p^+ a_Q^+ a_S a_R\right] \tag{1.14}$$

then we have very well known Moller-Plesset splitting of the Hamiltonian (1.9) [22]. Last two decades this Hamiltonian was very successfully used for the calculation of correlation energy of atoms and molecules either using MBPT or CC methods [5-13]. In MBPT we proceede in this simple way.

Let us have the perturbed (exact) Schrödinger equation

$$H_N^e \psi_i = E_i \psi_i \tag{1.15}$$

and the unperturbed Schrödinger equation

$$H_0|\Phi_i> = \epsilon_i|\Phi_i> \tag{1.16}$$

The eigenvalue E_i can be found using the non-degenerate Rayleigh-Schrödinger perturbation expansion in the form

$$E_i = <\Phi_i|H_0|\Phi_i> + <\Phi_i|H'|\Phi_i> + <\Phi_i|H'Q_iH'|\Phi_i> + \cdots \tag{1.17}$$

The great advantage of MBPT formulation is that the individual orders of perturbation expansion (1.17) can be treated diagrammatically using the technique of the Feynman like diagrams which makes the theory rather transparent.

Let us compare now the Hamiltonians (1.9) and (1.7). We see that they are formally identical except for the fact that all quantities in the Hamiltonian (1.7) are χ - dependent. If we further ask

$$f(\chi)|P(\chi)> = \varepsilon_P(\chi)|P(\chi)> \tag{1.18}$$

we can split the Hamiltonian (1.7) into the Moller-Plesset form with

$$\epsilon_0(\chi) = <\Phi_0(\chi)|H(\chi)|\Phi_0(\chi)>, \tag{1.19}$$

$$H_0(\chi) = \sum_P \varepsilon_P(\chi) N\left[a_P^+(\chi)a_P(\chi)\right] \tag{1.20}$$

and

$$H'(\chi) = \frac{1}{2}\sum_{PQRS} <P(\chi)Q(\chi)|v(\chi)|R(\chi)S(\chi)> . \\ . N\left[a_P^+(\chi)a_Q^+(\chi)a_S(\chi)a_R(\chi)\right]. \tag{1.21}$$

In order to get the eigenvalue $E_i(\chi)$ from equation (1.1) we can use the standard non-degenerate Rayleigh-Schrödinger perturbation theory expansion (1.17) in the from

$$E_i(\chi) = <\Phi_i(\chi)|H_0(\chi)|\Phi_i(\chi)> + <\Phi_i(\chi)|H'|\Phi_i'(\chi)> + \\ + <\Phi_i(\chi)|H'(\chi)Q_i(\chi)H'(\chi)|\Phi_i(\chi)> + \cdots. \tag{1.22}$$

To satisfy the equations (1.18) can be accomplished by solving the CPHF equations [16] . In order to get the creation $(a_P^+(\chi))$ and annihilation $(a_Q(\chi))$ operators we shall use the technique of canonical transformations. In our derivation we shall study such case in which the external perturbation $\sum_i \chi w_i$ will control the displacement of an atomic nucleus in the molecular system. The theory can be easily generalized. Further we shall work with the total vibrational-electronic (V-E) Hamiltonian. If the reader wants to follow the derivation only with the electronic Hamiltonian it vill be sufficcient to work in what follows only with the Hamiltonian H_A from our total V-E Hamiltonian. The quantization of V-E Hamiltonian is done hierarchically in a way that we first work in crude representation then using the technique of canonical transformations we work in adiabatic representation. The theory is presented in details. Diagrammatic rules for the Feynman like diagrams which can be used in MBPT are also developed. The total V-E Hamiltonian is then used to derive the method for the calculation of the energy of the first vibrational transitions in molecules based on MBPT. The present theory permits to derive the explicit formulae for the energy of the first vibrational transition in molecules. The correction to harmomic frequences $\hbar\,\omega$ calculated on the SCF level are due to electron correlation, anharmonicity effects and the non-adiabatic effects [23,24] .

2 The molecular vibrational-electronic (V-E) Hamiltonian

In this chapter we shall pay attention to quantization of total molecular V-E Hamiltonian. Our aim will be to bring this Hamiltonian into the form suitable for many-body perturbation theory calculations.

Our molecular V-E Hamiltonian is

$$H = H_{NN}(R) + H_{EN}(r,R) + H_{EE}(r) \tag{2.1}$$

where $H_{NN}(R)$ is the nuclei term

$$H_{NN}(R) = T_N(R) + E_{NN}(R) \tag{2.2}$$

representing the kinetic energy of the nuclei (T_N) and interaction between nuclei (E_{NN}); R denotes the nuclear coordinates. The last two terms of the Hamiltonian (2.1) correspond to standard electronic Hamiltonian (1.8)

$$H_{EN}(r,R) + H_{EE}(r) = h + \nu^0, \tag{2.3}$$

where h is the one-electron part representing the kinetic energy of the electrons and electron-nuclear atraction term, and v^0 is the two electron part of the Hamiltonian corresponding to electron-electron repulsion term and r denotes the electron coordinates. For the purpose of diagrammatic many-body perturbation theory it will be efficient to work in second quantization formalism [4,7,9]. The electronic Hamiltonian (2.3) has the form

$$H_{EN} + H_{EE} = \sum_{PQ} < P|h|Q > a_P^+ a_Q + + \frac{1}{2} \sum_{PQRS} < PQ|v^0|RS > a_P^+ a_Q^+ a_S a_R, \tag{2.4}$$

where $a_P^+(a_Q)$ is the creation (annihilation) operator for electrons in the spinorbital basis $|P>, |Q>, \cdots$. If we apply the Wick theorem to (2.4) we can rewrite (2.4) as

$$\begin{aligned} H_{EN} + H_{EE} &= \sum_I h_{II} + \frac{1}{2} \sum_{IJ} \left(v^0_{IJIJ} - v^0_{IJJI} \right) + \\ &+ \sum_{PQ} h_{PQ} N\left[a_P^+ a_Q\right] + \sum_{PQI} \left(v^0_{PIQI} - v^0_{PIIQ} \right) N\left[a_P^+ a_Q\right] + \\ &+ \frac{1}{2} \sum_{PQRS} v^0_{PQRS} N\left[a_P^+ a_Q^+ a_S a_R\right], \end{aligned} \tag{2.5}$$

where $v^0_{ABAB}(v^0_{ABBA})$ denotes the coulomb (exchange) integral. Note that the operators (2.2) and (2.5) depend on the nuclear coordinate R. The operators (2.2) and (2.5) together represent the complete molecular V-E Hamiltonian.

In order to proceede further we have to specify the one-particle basis set $|P>, |Q>, \cdots$ in which we define the electron creation and annihilation operators. One possibility is to use the fixed basis set in crude representation. In this approximation the spinorbital basis set $|P>, |Q, \cdots$ is determined at some fixed (equilibrium) nuclear coordinate R_0.

Of course using the fixed basis set does not describe properly the physical situation, because we need to have the R dependent basis set in which the electrons will follow adiabatically the motion of nuclei. Using the formalism of second quantization we shall do this by the technique of canonical transformations. Therefore we first quantize the V-E Hamiltonian in crude representation and then through the canonical transformation we shall re-quantize it in adiabatic representation.

Let us fix the nuclear coordinate R at some point R_0. We shall divide the individual terms of the Hamiltonian (2.1) into two parts :

(1) Terms that are determined at point R_0

(2) Terms that represent the shift with respect to point R_0 (we shall use prime to denote these terms).

We can rewrite the electronic part of V-E Hamiltonian (2.5) in the following way

$$H_{EN} + H_{EE} = E^0_{SCF} + h'_{SCF} + \sum_P \varepsilon_P N\left[a^+_P a_P\right] + \\ + \sum_{PQ} h'_{PQ} N\left[a^+_P a_Q\right] + \frac{1}{2}\sum_{PQRS} v^0_{PQRS} N\left[a^+_P a^+_Q a_S a_R\right] , \tag{2.6}$$

where E^0_{SCF} is the Hartree-Fock energy calculated at the point R_0 , and h'_{SCF} is the shift in the Hartree-Fock energy with respect to the point other than R_0 . The same is true for one-particle operator of (2.6), where ε_P are the one-particle Hartree-Fock energies calculated at point R_0 . The correlation operator is not changed because it does not depend on nuclear coordinates R . Furthermore we shall use the following notation

$$h'_{SCF} = u'_{SCF} \tag{2.7}$$

$$h'_{PQ} = u'_{PQ} \tag{2.8}$$

where

$$u = \sum_{ij} \frac{-Z_j e^2}{|\mathbf{r_i} - \mathbf{R_j}|} \tag{2.9}$$

$$u_{PQ} = < P| \sum_j \frac{-Z_j e^2}{|\mathbf{r} - \mathbf{R_j}|} |Q > \tag{2.10}$$

and

$$u_{SCF} = \sum_I u_{II} \tag{2.11}$$

Let us perform the Taylor expansion for the energies E_{NN} and u_{SCF} about the point R_0.

$$E_{NN} = E^{(0)}_{NN} + E'_{NN} = \sum_{i=0}^{\infty} E^{(i)}_{NN} \tag{2.12}$$

and

$$u_{SCF} = u^{(0)}_{SCF} + u'_{SCF} = \sum_{i=0}^{\infty} u^{(i)}_{SCF} \tag{2.13}$$

Using (2.12) and (2.13) we can rewrite our Hamiltonian (2.1) in the following way

$$H = E^{(0)}_{NN} + E^{(0)}_{SCF} + T_N + E^{(2)}_{NN} + u^{(2)}_{SCF} + \sum_P \varepsilon_P N\left[a^+_P a_P\right] \\ + \frac{1}{2}\sum_{PQRS} v^0_{PQRS} N\left[a^+_P a^+_Q a_S a_R\right] + E'_{NN} - E^{(2)}_{NN} + \\ + u'_{SCF} - u^{(2)}_{SCF} + \sum_{PQ} u'_{PQ} N\left[a^+_P a_Q\right] \tag{2.14}$$

Nothing is changed in Hamiltonian (2.1) and therefore Hamiltonian (2.14) represents the complete V-E Hamiltonian (2.1).

Now we can define the harmonic oscilators through which we define the boson (phonon) creation (b^+) and boson (phonon) annihilation (b) operators.

$$T_N + E_{NN}^{(2)} + u_{SCF}^{(2)} = \sum_r \hbar\omega_r \left(b_r^+ b_r + \frac{1}{2}\right) \tag{2.15}$$

where ω_r is the frequency of the harmonic oscilator.

This is an important step and we note that this is not the only possibility to define the harmonic oscilators in the V-E Hamiltonian. Our aim will be to split the V-E Hamiltonian (2.14) onto the unperturbed part and the perturbation and the natural separation will be to take the harmonic oscilators as unperturbed part of the Hamiltonian. Defining the harmonic oscilators by different ways we have different perturbations.

Using (2.15) we can rewrite Hamiltonian (2.14) to the form

$$\begin{aligned} H &= E_{NN}^{(0)} + E_{SCF}^{(0)} + \sum_r \hbar\omega_r \left(b_r^+ b_r + \frac{1}{2}\right) + \sum_P \varepsilon_P N\left[a_P^+ a_P\right] \\ &+ \frac{1}{2} \sum_{PQRS} v_{PQRS}^0 N\left[a_P^+ a_Q^+ a_S a_R\right] + E_{NN}^{'} - E_{NN}^{(2)} + \\ &+ u_{SCF}^{'} - u_{SCF}^{(2)} + \sum_{PQ} u_{PQ}^{'} N\left[a_P^+ a_Q\right] . \end{aligned} \tag{2.16}$$

In order to use the perturbation theory we have to split the Hamiltonian (2.16) onto the unperturbed part (H_0) and the perturbation $(H^{'})$

$$H = H_0 + H^{'} \tag{2.17}$$

Due to the crude approximation, we can partition the Hamiltonian (2.16) in the following way

$$H_0 = E_{NN}^{(0)} + E_{SCF}^{(0)} + \sum_P \varepsilon_P N\left[a_P^+ a_P\right] + \sum_r \hbar\omega_r \left(b_r^+ b_r + \frac{1}{2}\right) \tag{2.18}$$

and

$$H^{'} = H_E^{'} + H_F^{'} + H_I^{'}. \tag{2.19}$$

We do not specify explicitly the terms in (2.19) this will be done later, but let

$H_E^{'}$ corresponds to electron correlation operator,

$H_F^{'}$ represents the vibrational operator and

$H_I^{'}$ represents the electron-vibrational operator.

In equation (2.16) all quantities were defined through the cartesian coordinates. For further purposes it will be natural to work in normal coordinates $\{Q_r\}$. The normal coordinate in second quantized formalism is given in the following way :

$$Q_r \sim (b_r + b_r^+). \tag{2.20}$$

Therefore in order to work with the Hamiltonian (2.16) and in order to use the Taylor expansion (2.12) and (2.13) in normal coordinate space, we shall have to work with products of $(b_r + b_r^+)$ operators.
Let us first analyze the product of two normal coordinates

$$Q_r.Q_s \approx (b_r + b_r^+).(b_s + b_s^+) \tag{2.21}$$

and then we generalize it to any number of Q_r operators in order to describe any order of Taylor expansion in normal coordinates space. According to the Wick theorem we have.

$$\begin{aligned}(b_r + b_r^+)(b_s + b_s^+) &= N\left[(b_r + b_r^+).(b_s + b_s^+)\right] + \\ + N\left[(\underline{b_r + b_r^+).(b_s} + b_s^+)\right] &+ N\left[(\underline{b_r + b_r^+).(b_s + b_s^+}))\right] + \\ + N\left[(b_r + \underline{b_r^+).(b_s} + b_s^+)\right] &+ N\left[(b_r + \underline{b_r^+).(b_s + b_s^+})\right]\end{aligned} \tag{2.22}$$

Since for boson creation and annihilation operators we have the following commutation relations

$$\left[b_r, b_s^+\right] = \delta_{rs} \tag{2.23}$$

$$\left[b_r, b_s\right] = 0 \tag{2.24}$$

$$\left[b_r^+, b_s^+\right] = 0 \tag{2.25}$$

and the contraction of two operators is defined as

$$\underline{M_k M_l} = M_k M_l - N\left[M_k M_l\right] \tag{2.26}$$

we can write

$$Q_r Q_s = N[Q_r Q_s] + N\left[\underline{Q_r Q_s}\right]. \tag{2.27}$$

To generalize (2.27) and therefore to develop the Wick theorem for products of $(b_r + b_r^+)$ operators we proceede in the following way.
Let us denote the vector boson operators as :

(1) Vector boson annihilation operator

$$\mathbf{b} = \{b_r\} \tag{2.28}$$

(2) Vector boson creation operator

$$\mathbf{b}^+ = \{b_r^+\} \tag{2.29}$$

Let us define the operator $\mathbf{B}$ as

$$\mathbf{B} = \mathbf{b} + \mathbf{b}^+ \tag{2.30}$$

Furthermore, we denote by $\mathbf{P}^{(n)}$ the tensor of the order n and by $\mathbf{Q}^{(m)}$ the tensor of the order m. The tensor product we denote as $\mathbf{P}^{(n)}\mathrm{x}\mathbf{Q}^{(m)}$ and the scalar product we denote as $\mathbf{P}^{(n)}.\mathbf{Q}^{(m)}$. We also define

$$P^{(n)k} = \underbrace{P^{(n)}\mathrm{x}P^{(n)}\mathrm{x}\cdots\mathrm{x}P^{(n)}}_{k} \tag{2.31}$$

If we denote by $S^{(n)}$ the totally symmetric tensor of the n-th order which stands for the partial derivatives with respect to the normal coordinate then we can express generally the n-th order of Taylor expansion in normal coordinates space as

$$\begin{aligned} \mathbf{S}^{(n)} \, . \, \mathbf{B}^n &= \mathbf{S}^{(n)} \, . \, N\left[\mathbf{B}^n\right] + \binom{n}{2} \mathbf{S}^{(n)} \, . \, N\left[\underline{\mathbf{B}\mathrm{x}\mathbf{B}}\mathrm{x}\mathbf{B}^{n-2}\right] \\ &+ \binom{n}{4} \frac{4!}{2!2^2}\mathbf{S}^{(n)} \, . \, N\left[(\underline{\mathbf{B}\mathrm{x}\mathbf{B}})^2\mathrm{x}\mathbf{B}^{n-4}\right] + \cdots \end{aligned} \tag{2.32}$$

$$\left\{ \begin{array}{l} + \frac{n!}{(n/2)!2^{n/2}}\mathbf{S}^{(n)} \, . \, N\left[(\underline{\mathbf{B}\mathrm{x}\mathbf{B}})^{n/2}\right] \text{ for } n - \text{even} \\ + \frac{n(n-1)!}{((n-1)/2)!2^{(n-1)/2}}\mathbf{S}^{(n)} \, . \, N\left[(\underline{\mathbf{B}\mathrm{x}\mathbf{B}})^{(n-1)/2}\mathrm{x}\mathbf{B}\right] \text{ for } n - \text{odd} \end{array} \right.$$

Equation (2.32) can be further simplified as

$$\mathbf{S}^{(n)} \, . \, \mathbf{B}^{(n)} = \sum_{k=0}^{[n/2]} \binom{n}{2k} \frac{(2k)!}{k!2^k}\mathbf{S}^{(n)} \, . \, N\left[(\underline{\mathbf{B} \mathrm{~x~} \mathbf{B}})^k \mathrm{~x~} \mathbf{B}^{n-2k}\right] \tag{2.33}$$

The quantity

$$\underline{\mathbf{B} \mathrm{~x~} \mathbf{B}} = \mathbf{I}^{(2)} \tag{2.34}$$

is the identity tensor of the second order, and from $[n/2]$ we take only the integer part. Let us further introduce

$$\mathbf{B}^{(n)} = \{B_{r_1} \cdots r_n\} = \frac{1}{n!}N[\mathbf{B}^n] \tag{2.35}$$

we can finally write

$$\begin{aligned} \mathbf{S}^{(n)} \, . \, \mathbf{B}^n &= \sum_{k=0}^{[n/2]} \frac{n!}{(2k)!(n-2k)!} \frac{(2k)!}{k!2^k}\mathbf{S}^{(n)} \, . \, (N[\mathbf{B}^{n-2k}] \mathrm{~x~} \mathbf{I}^{(2)k}) = \\ &= n! \sum_{k=0}^{[n/2]} \frac{1}{k!2^k}\mathbf{S}^{(n)} \, . \, (\mathbf{B}^{(n-2k)} \mathrm{~x~} \mathbf{I}^{(2)k}) \end{aligned} \tag{2.36}$$

We can further simplify the equation (2.36) if we take into account the Taylor expansion of function of classical normal coordinates.

Let Q be the set of classical normal coordinates $\{Q_r\}$, and let $A(Q)$ be the function of classical normal coordinates that we write as Taylor expansion in the following way

$$A(Q) = \sum_{n=0}^{\infty} A^{(n)}(Q) \tag{2.37}$$

If we put now $\mathbf{B} = \mathbf{Q}$ we can write in second quantization

$$A^{(n)}(Q) = \frac{1}{n!}\mathbf{A}^{(n)} \, . \, \mathbf{B}^n \tag{2.38}$$

where

$$\mathbf{A}^{(n)} = \{A^{r_1 \dots r_n}\} = \left\{ \frac{\partial^n A(Q)}{\partial Q_{r_1} \dots \partial Q_{r_n}} |_{Q_{r_1} = \dots = Q_{r_n} = 0} \right\} \tag{2.39}$$

Using (2.36), (2.37) and (2.38) we can write

$$\mathbf{A}^{(n)}(Q) = \mathbf{A}^{(n)} \, . \sum_{k=0}^{[n/2]} \frac{1}{k!2^k} (\mathbf{B}^{(n-2k)} \times \mathbf{I}^{(2)k}) \, . \tag{2.40}$$

If we define

$$\mathbf{A}^{(k,m)} = \{A^{\overbrace{00\dots0}^{2k} r_1 r_2 \dots r_m}\} =$$

$$= \left\{ \frac{1}{k!2^k} \sum_{g_1, g_2 \cdots g_k} \mathbf{A}^{g_1 g_1 g_2 g_2 \cdots g_k g_k r_1 r_2 \cdots r_m} \right\} \tag{2.41}$$

we finally arrive at the expression

$$\mathbf{A}(Q) = \sum_{n=0}^{\infty} \sum_{k=0}^{[n/2]} \mathbf{A}^{(k,n-2k)} \, . \, \mathbf{B}^{(n-2k)} \, . \tag{2.42}$$

Using (2.42) we can finally write the Hamiltonian (2.17) in explicit form i.e.

$$H = E_{NN}^{(0)} + E_{SCF}^{(0)} + \sum_P \varepsilon_P N\left[a_P^+ a_P\right] + \sum_r \hbar\omega_r \left(b_r^+ b_r + \frac{1}{2}\right) +$$

$$+ H_E' \left\{ \equiv \frac{1}{2} \sum_{PQRS} v_{PQRS}^0 N\left[a_P^+ a_Q^+ a_S a_R\right] \right\} + \tag{2.43}$$

$$+ H_F'\{ \equiv \sum_{\substack{n=1 \\ n \neq 2}}^{\infty} \sum_{k=0}^{[n/2]} \left(\mathbf{E}_{NN}^{(k,n-2k)} + \mathbf{u}_{SCF}^{(k,n-2k)}\right) . \mathbf{B}^{(n-2k)}\} +$$

$$+ H_I' \left\{ \equiv \sum_{n=1}^{\infty} \sum_{k=0}^{[n/2]} \sum_{PQ} \mathbf{u}_{PQ}^{(k,n-2k)} . \mathbf{B}^{(n-2k)} N[a_P^+ a_Q] \right\} .$$

The term H_E' is the electron correlation operator, the term H_F' corresponds to phonon-phonon interaction and H_I' corresponds to electron-phonon interaction.

Let us comment on this partitioning of V-E Hamiltonian. Since we have used the crude approximation the electrons are fixed at R_0 i.e. at the minimum of the electronic energy. Due to this fact this approximation does not describe properly the physical situation (the electrons do not feel the R dependence due to the nuclei term) the perturbation term H_I which corresponds to electron-phonon interaction will be too large and the perturbation theory based on this partitioning of V-E Hamiltonian will not converge. Therefore in the next chapter we shall study the partitioning of V-E Hamiltonian in which we couple together the electronic and vibrational motions and make thus the electrons feel the R dependence due to the presence of nuclei terms.

3 Canonical transformations defined for the mixed system of fermions and bosons

In this chapter we shall define the group of canonical transformations for creation - $a_P^+(\chi)$ and annihilation - $a_Q(\chi)$ operators. These transformations will couple together the electronic and vibrational motions. We shall study the properties of these transformations and we shall also compare our approach with that based on gradient techniques. Finally, we develop the Feynman like diagrammatic technique which will make the theory rather transparent. For detail understanding of techniques of canonical transformations we refer the reader to the book [25].

As we have said in the introduction, the term $\sum_i \chi w_i$ in our Hamiltonian (1.2) will control the displacement of atomic nuclei.

Therefore we want to have the creation (annihilation) operators which create (annihilate) particles on the R - dependent basis set. So we have to pass from the basis set fixed at point $R_0\{|P>\}$ to the generally R dependent basis set $\{|P(R)>\}$.

$$|P> \rightarrow |P(R)> = \sum_Q C_{QP}|Q> \tag{3.1}$$

We shall define the new set of fermion second quantized operators $\{\bar{a}_P\}$, $\{\bar{a}_Q^+\}$ as well as the new set of boson second quantized operators $\{\bar{b}_r\}$, $\{\bar{b}_r^+\}$. Since in our transformation we shall use the mixed set of old second quantized operators we shall look at this tranformation as on the quasiparticle transformation [25] and instead of electrons and phonons when speaking about the new set of second quantized operators we shall speak about fermions and bosons.

Furthermore we shall ask that the operators of the new fermions will commute with the new set of boson operators. This condition will make the formulation of MBPT for V-E Hamiltonian relatively simple.

We define the set of new fermion annihilation operators through the old set of electron annihilation operators in the follovning way. Annihilation operator $\bar{a}_P$:

$$\bar{a}_P = a_P + \sum_Q \sum_{k=1}^{\infty} \frac{1}{k!} \sum_{r_1 \ldots r_k} C_{PQ}^{r_1 \ldots r_k} B_{r_1} \ldots B_{r_k} a_Q , \tag{3.2}$$

where B_r are defined through (2.30) and therefore have the meaning of normal coordinate operators, $C_{PQ}^{r_1 \ldots r_k}$ is the set of unknown coefficients which will be determined. Since it holds that

$$[B_r , B_s] = 0 , \tag{3.3}$$

we can choose the coefficients $C_{PQ}^{r_1 \ldots r_k}$ as totally symmetric tensors in indices $r_1 \ldots r_k$. Therefore we can write

$$\frac{1}{k!} C_{PQ}^{r_1 \ldots r_k} B_{r_1} \ldots B_{r_k} = \frac{1}{k!} \mathbf{C}_{PQ}^{(k)} . \mathbf{B}^k = C_{PQ}^{(k)} , \tag{3.4}$$

where in the tensor $\mathbf{C}_{PQ}^{(k)}$ the symbol (k) means the set of k indices $r_1, \ldots, r_k$ in which the tensor is totally symmetric, and in scalar $C_{PQ}^{(k)}$ the symbol (k) means only the order of the Taylor expansion.

Transformation (3.2) can be therefore simplified as

$$\begin{aligned} \bar{a}_P = \sum_Q \sum_{k=0}^{\infty} \frac{1}{k!} \mathbf{C}_{PQ}^{(k)} \mathbf{B}^k a_Q &= \sum_Q \sum_{k=0}^{\infty} C_{PQ}^{(k)} a_Q = \\ &= \sum_Q C_{PQ} a_Q , \end{aligned} \tag{3.5}$$

where

$$C_{PQ} = \sum_{k=0}^{\infty} C_{PQ}^{(k)} = \delta_{PQ} + \sum_{k=1}^{\infty} C_{PQ}^{(k)} . \tag{3.6}$$

The transformation for the new set of fermion creation operators $\bar{a}_P^+$ is given as a hermitean conjugate to (3.2)

$$\bar{a}_P^+ = a_P^+ + \sum_Q \sum_{k=1}^{\infty} \frac{1}{k!} \sum_{r_1 \ldots r_k} C_{PQ}^{r_1 \ldots r_k *} B_{r_1} \ldots B_{r_k} a_Q^+ \tag{3.7}$$

or using (3.5)

$$\bar{a}_P^+ = \sum_Q \sum_{k=0}^{\infty} \frac{1}{k!} C_{PQ}^{(k)*} \cdot \mathbf{B}^k a_Q^+ = \sum_Q \sum_{k=0}^{\infty} C_{PQ}^{(k)+} a_Q^+ = \tag{3.8}$$

$$= \sum_Q C_{PQ}^+ a_Q^+ \ .$$

Note that we used

$$\mathbf{B} = \mathbf{B}^+ \ . \tag{3.9}$$

Further we have to guarantee that our new set of second quantized operators are fermions i.e. they satisfy the folloving anticommutation relations

$$\{\bar{a}_P, \bar{a}_Q\} = 0 \ ; \ \{\bar{a}_P^+, \bar{a}_Q^+\} = 0 \tag{3.10}$$

$$\{\bar{a}_P^+, \bar{a}_Q\} = \delta_{PQ} \ .$$

Substituting (3.2) and (3.7) into (3.10) will give us some restrictions on the coefficients C_{PQ} .
If we denote the commutator by [A, B] and use the commutator algebra

$$[A, BC] = [A, B]C + B[A, C] = \tag{3.11}$$

$$= \{A, B\}C - B\{A, C\}$$

$$\{A, BC\} = \{A, B\}C - B[A, C] = [A, B]C + B\{A, C\} \tag{3.12}$$

we find the following relations for the coefficients C_{PQ} .

$$[C_{PQ}, C_{RS}] = 0 \ ; \ [C_{PQ}, C_{RS}^+] = 0 \ ; \tag{3.13}$$

$$[C_{PQ}, a_R] = 0 \ ; \ [C_{PQ}, a_R^+] = 0$$

And the unitary condition

$$\sum_R C_{PR} C_{QR}^+ = \sum_R C_{PR}^+ C_{QR} = \sum_R C_{RP} C_{RQ}^+ = \tag{3.14}$$

$$= \sum_R C_{RP}^+ C_{RQ} = \delta_{PQ} \ .$$

Since our new set of fermion second quantized operators satisfies the anticommutation rules and the transformation is unitary (3.14) our transformation (3.2) and (3.7) is the canonical transformation.

Because our aim is to quantize the whole V-E Hamiltonian (2.16) we introduce the canonical transformation for the new set of boson second quantized operators.

$$\bar{b}_r = b_r + \sum_{PQ}\sum_{k=0}^{\infty}\frac{1}{k!}\sum_{s_1\ldots s_k} d_{rPQ}^{s_1\ldots s_k} B_{s_1}\ldots B_{s_k} a_P^+ a_Q \ , \tag{3.15}$$

where b_r is the old set of phonon annihilation operators, and $d_{rPQ}^{s_1\ldots s_k}$ are new coefficients which have to be determined. We can choose them as a totally symmetric tensor in indices $s_1 \ldots s_k$. Then we can write

$$\frac{1}{k!}\sum_{s_1\ldots s_k} d_{rPQ}^{s_1\ldots s_k} B_{s_1}\ldots B_{s_k} = \frac{1}{k!}\mathbf{d}_{rPQ}^{(k)} \cdot \mathbf{B}^k = d_{rPQ}^{(k)} \ . \tag{3.16}$$

Similarly as in the case of (3.5) the transformation (3.15) can be simplified as

$$\begin{aligned}\bar{b}_r &= b_r + \sum_{PQ}\sum_{k=0}^{\infty}\frac{1}{k!}\mathbf{d}_{rPQ}^{(k)} \cdot \mathbf{B}^k a_P^+ a_Q = \\ &= b_r + \sum_{PQ}\sum_{k=0}^{\infty} d_{rPQ}^{(k)} a_P^+ a_Q = \\ &= b_r + \sum_{PQ} d_{rPQ} a_P^+ a_Q \ .\end{aligned} \tag{3.17}$$

By making the hermiteam conjugate we get the new set of boson creation operators

$$\bar{b}_r^+ = b_r^+ + \sum_{PQ}\sum_{k=0}^{\infty}\frac{1}{k!}\sum_{s_1\ldots s_k} d_{rPQ}^{s_1\ldots s_k *} B_{s_1}\ldots B_{s_k} a_Q^+ a_P \tag{3.18}$$

or

$$\bar{b}_r^+ = b_r^+ + \sum_{PQ} d_{rQP}^+ a_P^+ a_Q \ . \tag{3.19}$$

For the new set of boson second quantized operators we ask that they satisfy the commutation relations

$$\begin{aligned}[\bar{b}_r, \bar{b}_s] &= 0 \ , \quad [\bar{b}_r^+, \bar{b}_s^+] = 0 \\ [\bar{b}_r, \bar{b}_s^+] &= \delta_{rs} \ .\end{aligned} \tag{3.20}$$

These give us some restrictions on the coefficients d_{rPQ}. Further we shall ask that our new set of fermion and boson second quantized operators will commute, i.e.

$$[\bar{a}_P, \bar{b}_r] = 0 \tag{3.21}$$

$$[\bar{a}_P, \bar{b}_r^+] = 0 \tag{3.22}$$

The conditions (3.21, 3.22) will make the formulation of the many-body perturbation theory simple because we shall be able to split the V-E Hamiltonian in a way that the unperturbed wave function will be the product of the fermion and boson wave function. Furthermore the conditions (3.21) and (3.22) couple together the C_{PQ} coefficients with d_{rPQ} coefficients in the folloving way

$$[C_{PQ}, b_r] + \sum_R C_{PR} d_{rRQ} = 0 \tag{3.23}$$

Using the unitary condition (3.14) we get

$$d_{rPQ} = \sum_R C^+_{RP} [b_r, C_{RQ}] , \tag{3.24}$$

from which we get important relation

$$d_{rPQ} = - d^+_{rQP} \tag{3.25}$$

Similarly using (3.22) we get

$$[C_{PQ}, b^+_r] + \sum_R C_{PR} d^+_{rQR} = 0 \tag{3.26}$$

Finally the relations (3.20) lead to

$$[d_{rPQ}, b_s] + [b_r, d_{sPQ}] + \sum_R (d_{rPR} d_{sRQ} - d_{rRQ} d_{sPR}) = 0 \tag{3.27}$$

and

$$[d_{rPQ}, b^+_s] + [b_r, d^+_{sQP}] + \sum_R (d_{rPR} d^+_{sQR} - d_{rRQ} d^+_{sRP}) = 0 \tag{3.28}$$

If we introduce in analogy with (2.30)

$$\bar{B}_r = \bar{b}_r + \bar{b}^+_r \tag{3.29}$$

and the total number operator

$$\bar{N} = \sum_P \bar{a}^+_P \bar{a}_P \tag{3.30}$$

then substituting the canonical transformations (3.5), (3.8), (3.17) and (3.19) into (3.29) and (3.30) we find that we have two invariants of the transformation

$$\bar{\mathbf{B}} = \mathbf{B} \tag{3.31}$$

and

$$\bar{N} = N . \tag{3.32}$$

This means that according to (3.31) by the canonical transformation we do not change the normal coordinate and according to (3.32) we do not change the number of fermions.
The relation (3.32) means that we can redefine the Fermi vacuum. To say simply by our canonical transformations we do the following. Instead of using the former electrons and phonons as a building blocks of the molecule, we use the new building blocks, namely fermions and bosons where the former electrons and phonons are coupled together by the canonical transformation.

4 Adiabatic representation of V-E Hamiltonian. Diagrammatic technique

The canonical transformation introduced in the previous chapter permit us to requantize the total V-E Hamiltonian (2.43). Instead of using the electrons and phonons as a building blocks we shall use our new quasiparticles introduced by canonical transformations which are fermions and bosons.
Crucial step in requantization of our V-E Hamiltonian in adiabatic representation lies in choosing the $V_N^{(2)}$ potential.
Let us introduce the new effective potential $V_N^{(2)}(B)$ which will be specified later through which we introduce the harmonic oscilators

$$H_{NN}^{(2)}(B) + V_N^{(2)}(B) = \sum_r \hbar\omega_r \left(b_r^+ b_r + \frac{1}{2}\right) . \tag{4.1}$$

By the dependence on (B) we specify that we work with normal coordinates.
Let us divide our V-E Hamiltonian (2.16) into two parts

$$H = H_A + B_B \tag{4.2}$$

where

$$H_A = E_{NN}(B) - E_{NN}^{(2)}(B) - V_N^{(2)}(B) + \sum_{PQ} h_{PQ}(B) a_P^+ a_Q + \frac{1}{2} \sum_{PQRS} v_{PQRS}^0 a_P^+ a_Q^+ a_S a_R \tag{4.3}$$

and

$$H_B = \sum_r \hbar\omega_r \left(b_r^+ b_r + \frac{1}{2}\right) . \tag{4.4}$$

The Hamiltonian (4.2) with (4.3) and (4.4) is our total V-E Hamiltonian in crude representation. To requantize it we have to substitute for the electron creation (a_P^+) and annihilation (a_Q) operators as well as for the phonon creation (b_r^+) and annihilation (b_r) operators from our canonical transformations. Since our transformations form the group there exist the inverse transformations.

Let us denote the inverse transformations in the following way

$$a_P = \sum_Q \bar{C}_{PQ}(B)\, \bar{a}_Q \tag{4.5}$$

$$a_P^+ = \sum_Q \bar{C}_{PQ}(B)\, \bar{a}_Q^+ \tag{4.6}$$

$$b_r = \bar{b}_r + \sum_{PQ} \bar{d}_{rPQ}(B)\, \bar{a}_P^+ \bar{a}_Q \tag{4.7}$$

$$b_r^+ = \bar{b}_r^+ + \sum_{PQ} \bar{d}_{rQP}(B)\, \bar{a}_P^+ \bar{a}_Q\ , \tag{4.8}$$

where a_P is the annihilation operator for the old set of electrons and $\bar{a}_P$ is the annihilation operator for the new set of fermions, etc.

From the group properties of our transformations it follows that [23].

$$\bar{C}_{PQ} = C_{QP}^+\ . \tag{4.9}$$

First let us requantize the operator H_A (4.3). Substituting (4.5) and (4.6) into (4.3) we have

$$H_A = E_{NN}(B) - E_{NN}^{(2)}(B) - V_N^{(2)}(B) + \sum_{PQRS} h_{RS}(B)\, \bar{C}_{RP}(B) \cdot \bar{C}_{SQ}(B)\, \bar{a}_P^+ \bar{a}_Q + \frac{1}{2} \sum_{\substack{PQRS \\ TUVW}} v^0_{TUVW}\, \bar{C}_{TP}(B)\, \bar{C}_{UQ}(B)\, \bar{C}_{VR}(B) \cdot \bar{C}_{WS}(B)\, \bar{a}_P^+ \bar{a}_Q^+ \bar{a}_S \bar{a}_R\ . \tag{4.10}$$

Since we found that the total number operator N (3.32) is the invariant of the transformation we can redefine the vacuum state to be the Fermi vacuum of our new fermion quasiparticle operators. We shall use the indices $I,\ J,\ K,\ L$ for occupied states; $A,\ B,\ C,\ D$ for virtual states, and $P,\ Q,\ R,\ S,\ T,\ U,\ V,\ W$ for either of them. To simplify the notation we further omit the writing of the dependence on the operator B, and we also omit the bar sign on top of the coefficients C and operators $a(a^+)$. Using the Wick theorem [4] we can write the operator (4.10) in normal product form.

$$
\begin{aligned}
H_A &= E_{NN} - E_{NN}^{(2)} - V_N^{(2)} + \sum_{RSI} h_{RS} C_{RI} C_{SI} + \\
&+ \frac{1}{2} \sum_{RSTUIJ} \left(v_{RTSU}^0 - v_{RSTU}^0 \right) C_{RI} C_{SI} C_{TJ} C_{UJ} + \\
&+ \sum_{PQRS} h_{RS} C_{RP} C_{SQ} N \left[a_P^+ a_Q \right] + \sum_{PQRSTUI} \left(v_{RTSU}^0 - v_{RSTU}^0 \right) \\
&. C_{RP} C_{SQ} C_{TI} C_{UI} N \left[a_P^+ a_Q \right] + \sum_{PQRSTUVW} v_{TUVW}^0 C_{TP} C_{UQ} C_{VR} C_{WS} \, . \\
&. N \left[a_P^+ a_Q^+ a_S a_R \right]
\end{aligned} \tag{4.11}
$$

Using the concept of quasiparticle Fermi vacuum we can define the new quasiparticle Hartree-Fock energy

$$
\begin{aligned}
E_{SCF} &= \sum_{RSI} h_{RS} C_{RI} C_{SI} + \\
&+ \frac{1}{2} \sum_{RSTUIJ} \left(v_{RTSU}^0 - v_{RSTU}^0 \right) C_{RI} C_{SI} C_{TJ} C_{UJ} \, ,
\end{aligned} \tag{4.12}
$$

the new Hartree-Fock operator f with the matrix elements

$$
\begin{aligned}
f_{PQ} &= \sum_{RS} h_{RS} C_{RP} C_{SQ} + \\
&+ \sum_{RTSUI} \left(v_{RTSU}^0 - v_{RSTU}^0 \right) C_{RP} C_{SQ} C_{TI} C_{UI}
\end{aligned} \tag{4.13}
$$

and the new two-particle integral

$$
v_{PQRS} = \sum_{TUVW} v_{TUVW}^0 C_{TP} C_{UQ} C_{VR} C_{WS} \, . \tag{4.14}
$$

This permits us to write the operator (4.10) in the form

$$
\begin{aligned}
H_A &= E_{NN} - E_{NN}^{(2)} - V_N^{(2)} + E_{SCF} + \sum_{PQ} f_{PQ} N \left[a_P^+ a_Q \right] + \\
&+ \frac{1}{2} \sum_{PQRS} v_{PQRS} N \left[a_P^+ a_Q^+ a_S a_R \right]
\end{aligned} \tag{4.15}
$$

Except for the first three terms which are due to the nuclei interaction the rest of the operator H_A (4.15) is exactly the Moller - Plesset splitting [22] of the new quasiparticle fermion Hamiltonian. Note that Hamiltonian (4.15) has exactly the form (1.7) we discussed in the introduction. The quantities (4.12), (4.13) and (4.14) are analogous to (1.19), (1.18) and (1.21).

In order to develop the many-body perturbation theory for the Hamiltonian H_A (4.15) we need to split it into the unperturbed part and the perturbation. Let us split the quantities E_{NN}, E_{SCF} and f_{PQ} in the following way :

$$E_{NN} = E^0_{NN} + E'_{NN} \tag{4.16}$$

$$E_{SCF} = E^0_{SCF} + E'_{SCF} \tag{4.17}$$

$$f_{PQ} = f^0_{PQ} + f'_{PQ} ; \tag{4.18}$$

where the superscript 0 corresponds to the equilibrium (or reference) internuclear distance.
We have

$$E^0_{SCF} = \sum_I h^0_{II} + \frac{1}{2}\sum_{IJ}\left(v^0_{IJIJ} - v^0_{IJJI}\right) \tag{4.19}$$

$$f^0_{PQ} = h^0_{PQ} + \sum_I \left(v^0_{PIQI} - v^0_{PQII}\right) = \epsilon_P \delta_{PQ} . \tag{4.20}$$

Let us denote

$$h_{PQ} = h^0_{PQ} + u'_{PQ} \tag{4.21}$$

$$w'_{PQ} = u'_{PQ} + \sum_{RI}\left(2v^0_{PRQI} - v^0_{PRIQ} - v^0_{PIRQ}\right) C'_{RI} \tag{4.22}$$

$$z'_{PQ} = \sum_{RS}\left(v^0_{PRQS} - v^0_{PRSQ}\right) C'_{RI} C'_{SI} \tag{4.23}$$

Using (4.21) - (4.23) we get for E'_{SCF}

$$\begin{aligned} E'_{SCF} = & \sum_I \left(u'_{II} + 2\epsilon_I C'_{II}\right) + \sum_{RI}\left(u'_{IR}C'_{RI} + w'_{IR}C'_{RI} + \right. \\ & \left. + \epsilon_R C'_{RI}C'_{RI}\right) + \sum_{RSI}\left(w'_{RS}C'_{RI}C'_{SI} + \frac{1}{2}z'_{RS}C'_{RI}C'_{SI}\right) . \end{aligned} \tag{4.24}$$

Using (3.33) we have

$$C'_{II} = -\frac{1}{2}\sum_R C'_{RI}C'_{RI} \tag{4.25}$$

and therefore

$$\begin{aligned} E'_{SCF} = & \sum_I u'_{II} + \sum_{RI}\left[(\epsilon_R - \epsilon_I)C'_{RI} + u'_{RI} + w'_{RI}\right] C'_{RI} + \\ & + \sum_{RSI}\left(w'_{RS} + \frac{1}{2}z'_{RS}\right) C'_{RI}C'_{SI} . \end{aligned} \tag{4.26}$$

Similarly using (3.33) we get

$$f'_{PQ} = (\epsilon_P - \epsilon_Q)C'_{PQ} + \sum_R (\epsilon_R - \epsilon_Q)C'_{RP}C'_{RQ} +$$

$$+ w'_{PQ} + z'_{PQ} + \sum_R \left[\left(w'_{RQ} + z'_{RQ}\right) C'_{RP} + h\,.\,c\,.\right] + \tag{4.27}$$

$$+ \sum_{RS} \left(w'_{RS} + z'_{RS}\right) C'_{RP}C'_{SQ}$$

where $h\,.\,c\,.$ stands for hermitean conjugate. Now we are able to split the Hamiltonian H_A into the unperturbed part H_A^0 and the perturbation H'_A

$$H_A = H_A^0 + H'_A\,, \tag{4.28}$$

$$H_A^0 = E_{NN}^0 + E_{SCF}^0 + \sum_P \epsilon_P N\left[a_P^+ a_P\right]\,. \tag{4.29}$$

Let us denote for simplicity

$$E' = E'_{NN} - E_{NN}^{(2)} - V_N^{(2)} + E'_{SCF} \tag{4.30}$$

than using (4.26), (4.27), (4.30) we can find for the perturbed part of the Hamiltonian (4.28) the following three terms

$$H'_{A^0} = E' \tag{4.31}$$

$$H'_{A'} = \sum_{PQ} f'_{PQ} N\left[a_P^+ a_Q\right] \tag{4.32}$$

$$H'_{A''} = \frac{1}{2}\sum_{PQRS} v_{PQRS} N\left[a_P^+ a_Q^+ a_S a_R\right]\,. \tag{4.33}$$

Since all these three terms are given as the expansion through the normal coordinates we can use the expression (2.42) to obtain the final result

$$H'_{A^0} = \sum_{n=1}^{\infty} H'_{A_n^0} = \sum_{n=1}^{\infty}\sum_{k=0}^{[n/2]} \mathbf{E}^{(k,n-2k)}\,.\,\mathbf{B}^{(n-2k)} \tag{4.34}$$

$$H'_{A'} = \sum_{n=1}^{\infty} H'_{A'_n} = \sum_{n=1}^{\infty}\sum_{k=0}^{[n/2]}\sum_{PQ} \mathbf{f}_{PQ}^{(k,n-2k)}\,\mathbf{B}^{(n-2k)}\,N\left[a_P^+ a_Q\right] \tag{4.35}$$

$$H'_{A''} = \sum_{n=0}^{\infty} H'_{A''_n} =$$

$$= \frac{1}{2}\sum_{n=0}^{\infty}\sum_{k=0}^{[n/2]}\sum_{PQRS} \mathbf{v}_{PQRS}^{(k,n-2k)}\,.\,\mathbf{B}^{(n-2k)}\,N\left[a_P^+ a_Q^+ a_S a_R\right]\,. \tag{4.36}$$

Since the term (4.34) contains only boson N – product $B^{(n-2k)}$ this term corresponds to boson-boson interaction. The term (4.35) contains the one-particle (fermion) N – product and boson N – product $B^{(n-2k)}$ and it representes the one-particle (fermion) - boson interaction. The last term (4.36) is due to two - particle - boson interaction. These three terms (4.34) - (4.36) we shall represent diagrammaticaly using the Feynman like diagrams as follows :

To the n-th order of the term H'_{A^0} (4.34) we ascribe the following diagram

$$H_{A^0_n} = \sum_{k=0}^{[n/2]} \mathbf{E}^{(k,n-2k)} \,.\, \mathbf{B}^{(n-2k)} \rightarrow \tag{4.37}$$

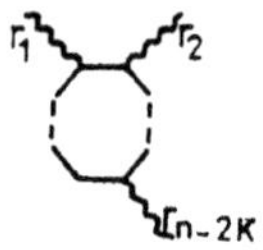

To the n-th order of the term $\mathbf{H}'_{A'}$ (4.35) we ascribe the diagram

$$\mathbf{H}'_{A'_n} = \sum_{k=0}^{[n/2]} \sum_{PQ} \mathbf{f}_{PQ}^{(k,n-2k)} \,.\, \mathbf{B}^{(n-2k)} \, N\left[a_P^+ a_Q\right] \rightarrow \tag{4.38}$$

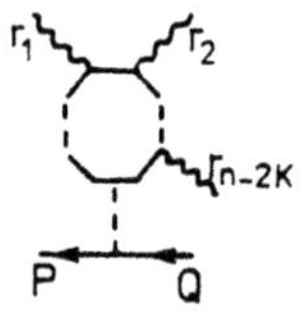

and to the n-th order of $H'_{A''}$ (4.36) the diagram

$$\mathbf{H}'_{A''_n} = \frac{1}{2} \sum_{n=0}^{[n/2]} \sum_{PQRS} \mathbf{v}_{PQRS}^{(k,n-2k)} \,.\, \mathbf{B}^{(n-2k)} \, N\left[a_P^+ a_Q^+ a_S a_R\right] \rightarrow \tag{4.39}$$

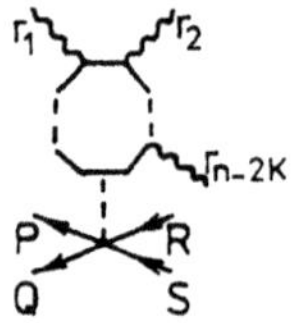

Since the n-th order of individual perturbations is given through the polygon we shall use such a convention that the first order will be denoted by small circle "o" , the second order by small abscissa ' - ' , the third order by small triangle "Δ" etc. Thus for example the diagram

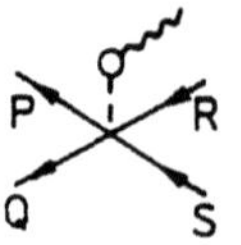

(4.40)

corresponds to the first derivative of the two-particle integral with respect to the normal coordinates. In order to make the perturbations as small as possible we can ask the first order of the term $H'_{A_1^0}$ to be equal to zero, i.e.,

$$H'_{A_1^0} = E^{(1)} = E^{(1)}_{NN} + E^{(1)}_{SCF} = 0 \tag{4.41}$$

This is nothing else than the condition that we optimize the geometry of the molecule on the SCF level. Of course we can choose also the other condition. This will be discucced later.
We further can ask the second order of $H'_{A_2^0}$ to be equal to zero, i.e.,

$$H'_{A_2^0} = E^{(2)} = E^{(2)}_{SCF} - V^{(2)}_N = 0 \tag{4.42}$$

This means that we choose the force constant $V^{(2)}_N$ to be equal to the second derivative of the SCF energy with respect to the normal coordinates. Furthermore we can try to simplify the perturbation $H'_{A'}$. This can be achieved by making the terms $f^{(1)}_{PQ}, f^{(2)}_{PQ}, \ldots$ diagonal. For our practical applications it will be sufficient to diagonalize the term f'_{PQ} to the first or to the second order of the expansion. This can be done by solving the $CPHF$ equations [14-16]. We shall need to know the C_{PQ} coefficients to the first or to the second order. Their calculation will be shown later when discussing the connection of our approach with the gradient techniques.
So far we requantized only the H_A part of our V-E Hamiltonian (4.2). We can proceede in an analogous way for the H_B part of the V-E Hamiltonian (4.2). From (4.4) we have

$$H_B = \sum_r \hbar\omega_r \left(b_r^+ b_r + \frac{1}{2}\right) \tag{4.43}$$

Substituting the inverse transformations (4.7) and (4.8) into (4.43) we get

$$H_B = \sum_r \hbar\omega_r \left(\bar{b}_r^+ \bar{b}_r + \frac{1}{2}\right) + \sum_{PQr} \hbar\omega_r \left(\bar{b}_r^+ \bar{d}_{rPQ} + \bar{d}_{rQP} \bar{b}_r\right) \bar{a}_P^+ \bar{a}_Q + \sum_{PQRSr} \hbar\omega_r \bar{d}_{rQP} \bar{d}_{rRS} \bar{a}_P^+ \bar{a}_Q \bar{a}_R^+ \bar{a}_S \tag{4.44}$$

and

$$H'_B = H'_{B^\bullet} + H'_{B^0} + H'_{B'} + H'_{B''}, \tag{4.50}$$

where

$$H'_{B^\bullet} = \sum_{PQr} \hbar\,\omega_r \left(b_r^+ d_{rPQ} + d_{rQP} b_r\right) N\left[a_P^+ a_Q\right] \tag{4.51}$$

$$H'_{B^0} = \sum_{AIr} \hbar\,\omega_r\,(d_{rAI})^2 \tag{4.52}$$

$$H'_{B'} = \sum_{PQAIr} \hbar\, \omega_r \, (d_{rPA} d_{rQA} - d_{rPI} d_{rQI}) \; N\left[a_P^+ a_Q\right] \tag{4.53}$$

and

$$H'_{B''} = \sum_{PQRSr} \hbar\, \omega_r d_{rPS} d_{rQR} \, N\left[a_P^+ a_Q^+ a_S a_R\right] . \tag{4.54}$$

Note that due to the fact that the coefficients d_{rPQ} are given as an expansion also the perturbations (4.51) - (4.54) will be given as an expansion.
To distinguish the index r in the coefficients d_{rPQ} and therefore also in (4.51) - (4.54) from the one which originate from the Taylor expansion according to the normal coordinates $\left(\text{e.g. } C_{PQ}^{r_1 \ldots r_k}\right)$ we shall separate the former index r from the later by prime. Therefore we can introduce the following notation

$$m'^{r}_{PQ} = \hbar\, \omega_r d_{rPQ} \tag{4.55}$$

and

$$\mathbf{B}^{(n),r} = b_r^+ \, \mathbf{B}^{(n)} - \mathbf{B}^{(n)} \, b_r \, . \tag{4.56}$$

Which gives us this expression for the perturbation $H'_{B^\bullet}$.

$$\begin{aligned} H'_{B^\bullet} = \sum_{n=0}^{\infty} H_{B^\bullet_{n+1}} &= \sum_{n=0}^{\infty} \sum_{k=0}^{[n/2]} \sum_{PQr} \Big(\mathrm{m}_{PQ}^{(k,n-2k),r} \, . \, b_r^+ \, \mathbf{B}^{(n-2k)} \, + \\ &\quad + \mathrm{m}_{QP}^{(k,n-2k),r} \, . \, \mathbf{B}^{(n-2k)} \, b_r \Big) \; N\left[a_P^+ a_Q\right] = \\ &= \sum_{n=0}^{\infty} \sum_{k=0}^{[\frac{n}{2}]} \sum_{PQr} \mathrm{m}_{PQ}^{(k,n-2k),r} \, . \, \mathbf{B}^{(n-2k),r} \; N\left[a_P^+ a_Q\right] \end{aligned} \tag{4.57}$$

Further we omit the sign " - " on $\bar{d}_{rPQ}$ coefficients as well as on the $\bar{b}_r^+$ $\left(\bar{b}_r\right)$ and $\bar{a}_P^+ \left(\bar{a}_P\right)$ operators. Let us introduce the following functions of spinorbitals :

$$p(A) = 1; \quad p(I) = 0; \quad h(A) = 0; \quad h(I) = 1 \, . \tag{4.45}$$

Now we can apply the Wick theorem [4] to (4.44) and we get

$$a_P^+ a_Q = N\left[a_P^+ a_Q\right] + h(P)\, \delta_{PQ} \tag{4.46}$$

$$\begin{aligned} a_P^+ a_Q a_R^+ a_S = &- N\left[a_P^+ a_R^+ a_Q a_S\right] + N\left[a_R^+ a_S\right] h(P)\, \delta_{PQ} \, - \\ &- N\left[a_R^+ a_Q\right] h(P)\, \delta_{PS} + N\left[a_P^+ a_S\right] p(Q) \delta_{QR} \, + \\ &+ N\left[a_P^+ a_Q\right] h(R)\, \delta_{RS} + h(P)\, h(R)\, \delta_{PQ}\, \delta_{RS} \, + \\ &+ h(P) p(Q)\, \delta_{PS}\, \delta_{QR} \, . \end{aligned} \tag{4.47}$$

Using (3.25), (4.46) and (4.47) we get

$$H_B = \sum_r \hbar\omega_r \left(b_r^+ b_r + \frac{1}{2}\right) +$$
$$+ \sum_{PQr} \hbar\omega_r \left(b_r^+ d_{rPQ} + d_{rQP} b_r\right) N\left[a_P^+ a_Q\right] +$$
$$+ \sum_{AIr} \hbar\omega_r \left(d_{rAI}\right)^2 + \sum_{PQAIr} \hbar\omega_r \left(d_{rPA} d_{rQA} - d_{rPI} d_{rQI}\right) N\left[a_P^+ a_Q\right]$$
$$+ \sum_{PQRSr} \hbar\, \omega_r d_{rPS} d_{rQR} N\left[a_P^+ a_Q^+ a_S a_R\right] . \tag{4.48}$$

This is complete H_B Hamiltonian requantized with respect to our new quasiparticles given by the transformations (3.15) and (3.18). We can split it into the unperturbed part H_B^0 and the perturbations H_B'. Simplest way is to choose

$$H_B^0 = \sum_r \hbar\, \omega_r \left(b_r^+ b_r + \frac{1}{2}\right) \tag{4.49}$$

We can introduce the following diagrammatic representation. For the n-th order of H_B'. we have

$$\sum_{k=0}^{[n/2]} \sum_{PQr} \left(\mathbf{m}_{PQ}^{(k,n-2k),r} \cdot b_r^+ \mathbf{B}^{(n-2k)} + \mathbf{m}_{QP}^{(k,n-2k),r} \mathbf{B}^{(n-2k),b_r}\right) N\left[a_P^+ a_Q\right] \rightarrow \tag{4.58}$$

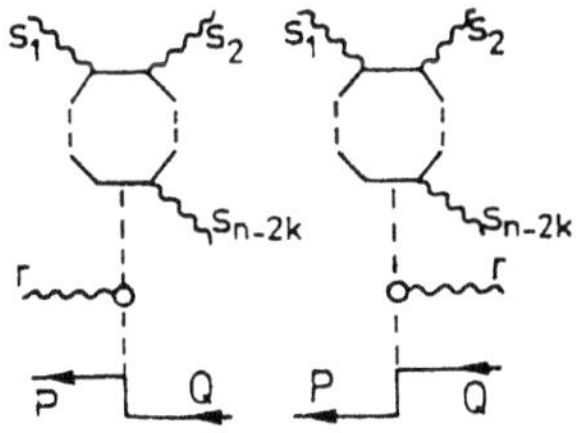

In order to simplify the notation in (4.53) - (4.54) where we hawe one - particle (fermion) and two-particle fermion operators we introduce the following notation

$$\tilde{f}'^{00}_{PQ} = \sum_{AIr} \hbar\omega_r (d_{rPA} d_{rQA} - d_{rPI} d_{rQI}) \tag{4.59}$$

$$\tilde{V}'^{00}_{PQRS} = 2 \sum_r \hbar\omega_r d_{rPS} d_{rQR} \tag{4.60}$$

and

$$\tilde{E}'^{00} = \sum_{AIr} \hbar\omega_r (d_{rAI})^2 \tag{4.61}$$

Now we can write the expansion for the perturbations $H'_{B^0}, H'_{B'}$ and $H'_{B''}$ in the form

$$H'_{B^0} = \sum_{n=0}^{\infty} H_{B^0_{n+2}} = \sum_{n=0}^{\infty} \sum_{k=0}^{[n/2]} \tilde{\mathbf{E}}^{(k,n-2k),00} \, . \, \mathbf{B}^{(n-2k)} \tag{4.62}$$

$$H'_{B'} = \sum_{n=0}^{\infty} H'_{B'_{n+2}} = \sum_{n=0}^{\infty} \sum_{k=0}^{[n/2]} \sum_{PQr} \tilde{\mathbf{f}}^{(k,n-2k),00} \, . \, \mathbf{B}^{(n-2k)} \, N \left[a^{+}_{P} a_{Q}\right] \tag{4.63}$$

$$H'_{B''} = \sum_{n=0}^{\infty} H'_{B''_{n+2}} = 12 \sum_{n=0}^{\infty} \sum_{k=0}^{[n/2]} \sum_{PQRS} \tilde{\mathbf{v}}^{(k,n-2k),00} \, . \, \mathbf{B}^{(n-2k)} \, . \, N \left[a^{+}_{P} a_{Q+} a_{S} a_{R}\right] \tag{4.64}$$

5 Non-adiabatic representation of the V-E Hamiltonian

For a great deal of quantum chemical calculation the adiabatic approximation is sufficient. The nuclear motion takes place on an effective potential surface that is the sum of the electronic energy and the nuclear repulsion as a function of the nuclear coordinates. As is well known, the Born-Oppenheimer approximation [(42)] is based on the fact that the spacing of electronic eigenvalues is generally large compared with the typical spacing associated with nuclear motion [(43)]. In other words this means that the vibrational quantum $\hbar\omega$ is much smaller than the energy difference $E_k - E_l$ between two electronic states. This is the case of non degeneracy of electronic states. When this condition is violated, the residual coupling via the nuclear kinetic energy operator causes transitions between the adiabatic electronic states. Usually only two or at most few electronic states are close in energy. In these cases it is necessary to use methods that abandom the Born-Oppenheimer approximation. In a recent paper Köppel et al. [(43)] discuss the concept of how, to handle non Born-Oppenheimer phenomena through the introduction of suitable "diabatic" states that may cross as a function of the internuclear distance, whereas the adiabatic electronic states are subjected to the non-crossing rule. In the previous chapter we have developed the method for the calculation of the energies of the first vibrational transitions is molecules in the adiabatic representation using the many-body perturbation theory. The corrections to the harmonic frequences which are calculated on the SCF level are due to the electron correlation, anharmonicity and non-adiabatic effects. The non-adiabatic corrections are for most molecules small. In the case of quasidegeneracy or degeneracy of electronic levels we cannot calculate the non-adiabatic corrections using the perturbation theory. This is the case when potential energy surfaces come close together. The vibrational energies in such cases are usually calculated using the concept of effective Hamiltonian and diabatic wave functions [43].

In this chapter we shall try to extend the technique of canonical transformations also for the case when the Born-Oppenheimer approximation fails. The basic idea will be the same as in the case of adiabatic representation where we started from the crude representation in which the V-E Hamiltonian was expressed through the standard electron and phonon operators. Then using the canonical transformations we arrived at the V-E Hamiltonian in which the new fermions corresponds to electrons which adiabaticaly follow the motion of nuclei. The new boson operators correspond to renormalized phonons where the renormalization brings the non-adiabatic corrections.

The transformation from the old set of electron and phonon operators to the new set of fermion and boson operators was described in chapter **4** and was gives as

$$\bar{a}_P = \sum_Q C_{PQ}(B) a_Q \tag{5.1}$$

$$\bar{a}_P^+ = \sum_Q C_{PQ}(B)^+ a_Q^+ \tag{5.2}$$

$$\bar{b} = b_r + \sum_{PQ} d_{rPQ}(B) a_P^+ a_Q \tag{5.3}$$

$$\bar{b}_r^+ = b_r^+ + \sum_{PQ} d_{rQP}(B)^+ a_P^+ a_Q \ . \tag{5.4}$$

In the case of non-adiabatic representation we shall try to generalize the transformations (5.1) - (5.4). The generalization can be done in a way that from operator coefficients $C_{PQ}(B)$ we arrive at general coefficients $C_{PQ}(b, b^+)$. Similar idea was studied by Wagner [44]. Let us introduce in analogy with normal coordinate operator B_r (2.30) the momentum operator $\tilde{B}_r$ as

$$\tilde{B}_r = b_r - b_r^+ \ . \tag{5.5}$$

Then we can generalize the equation (5.1) in the following way [36]

$$\bar{a}_P = \sum_Q C_{PQ}(B, \tilde{B}) a_Q \; . \tag{5.6}$$

We can understand the transformation in the following way. The new fermions represent the electron which move in dependence on coordinate and on momentum of nuclei. The electrons due to their finite mass do not follow adiabatically the motion of nuclei but their motion is phase shifted with respect to the motion of nuclei.
The general transformation (5.6) is complex and it brings rather complex conditions for the general coefficients $C_{PQ}(B, \tilde{B})$. Instead of using the general transformation (5.6) we shall try its simple version in which the general coefficients $C_{PQ}(B, \tilde{B})$ are expressed as a simple product of $C_{PQ}(B)$ and $\tilde{C}_{PQ}(\tilde{B})$ coefficients [36]. The coefficients $C_{PQ}(B)$ correspond to adiabatic transformation as described in previous chapter and the coefficients $\tilde{C}_{PQ}(\tilde{B})$ will correspond to the new transformation which we shall call non-adiabatic. The final non-adiabatic representation will be given by the combination of both transformations. The non-adiabatic transformation can be formulated analogously as the adiabatic transformation (4.5) - (4.8) except that the coefficients $C_{PQ}(B)$ and $d_{rPQ}(B)$ are substituted by $C_{PQ}(\tilde{B})$ coefficients which are functions of momentum operators $\tilde{B}_r$. The transformation from the old set of second quantized operators $\{a_P\}, \{a_P^+\}, \{b_r\}, \{b_r^+\}$ to the new set of operators $\{\bar{a}_P\}, \{\bar{a}_P^+\}, \{\bar{b}_r\}$, and $\{\bar{b}_r^+\}$ will have the form

$$\bar{a}_P = \sum_Q \tilde{C}_{PQ}(\tilde{B})\, a_Q \tag{5.7}$$

$$\bar{a}_P^+ = \sum_Q \tilde{C}_{PQ}(\tilde{B})^+\, a_Q^+ \tag{5.8}$$

$$\bar{b}_r = b_r + \sum_{PQ} \tilde{d}_{rPQ}(\tilde{B})\, a_P^+ a_Q \tag{5.9}$$

$$\bar{b}_r^+ = b_r^+ + \sum_{PQ} \tilde{d}_{rPQ}(\tilde{B})^+\, a_P^+ a_Q \; . \tag{5.10}$$

The coefficients $\tilde{C}_{PQ}$ and $\tilde{d}_{rPQ}$ are given as the Taylor expansion

$$\tilde{C}_{PQ}(\tilde{B}) = \sum_{k=0}^{\infty} \frac{1}{k\,!} \sum_{r_1 \dots r_k} \tilde{C}_{PQ}^{r_1 \dots r_k}\, \tilde{B}_{r_1} \dots \tilde{B}_{r_k} \tag{5.11}$$

$$\tilde{d}_{rPQ}(\tilde{B}) = \sum_{k=0} \frac{1}{k\,!} \sum_{s_1 \dots s_k} \tilde{d}_{rPQ}^{s_1 \dots s_k}\, \tilde{B}_{s_1} \dots \tilde{B}_{s_k} \; . \tag{5.12}$$

The new set of fermion operators (5.7) , (5.8) must fulfill the fermion anticommutation relations (3.10) and the new set of boson operators (5.9) , (5.10) must satisfy the boson commutation rule (3.20). This gives us some restrictive conditions on the coefficients $\tilde{C}_{PQ}$ and $\tilde{d}_{rPQ}$ (analogously as in the case of adiabatic transformation).

$$\sum_R \tilde{C}_{PR} \tilde{C}^{+}_{QR} = \delta_{PQ} \tag{5.13}$$

$$\tilde{d}_{rPQ} = \sum_R \tilde{C}^{+}_{RP} \left[b_r, \tilde{C}_{RQ}\right] \tag{5.14}$$

$$\tilde{d}_{rPQ} - \tilde{d}^{+}_{rPQ} = 0 . \tag{5.15}$$

For our purposes it will be sufficient to express the Taylor expansion of equations (5.13) - (5.15) to the second order of expansion.

$$\tilde{C}^{0}_{PQ} = \delta_{PQ} \tag{5.16}$$

$$\tilde{C}^{r}_{PQ} - \tilde{C}^{r*}_{QP} = 0 \tag{5.17}$$

$$\tilde{C}^{rs}_{PQ} + \tilde{C}^{rs*}_{QP} = \sum_R \left(\tilde{C}^{r}_{PR} \tilde{C}^{s*}_{QR} + \tilde{C}^{r}_{PR} \tilde{C}^{r*}_{QR}\right) \tag{5.18}$$

$$\tilde{d}^{0}_{rPQ} = - \tilde{C}^{r}_{PQ} \tag{5.19}$$

$$\tilde{d}^{s}_{rPQ} = - \tilde{C}^{rs}_{PQ} + \sum_R \tilde{C}^{s*}_{RP} \tilde{C}^{r}_{RQ} \cdot \tag{5.20}$$

Similarly as in the case of adiabatic transformation we have two invariants of the new non-adiabatic transformation. The number of fermion operator is the invariant of the transformation

$$\bar{N} = \sum_P \bar{a}^{+}_{P} \bar{a}_{P} = \sum_P a^{+}_{P} a_{P} = N \tag{5.21}$$

and the momentum operator $\tilde{B}_r$ is also the invariant of the transformation

$$\bar{\tilde{B}} = \tilde{B}_r . \tag{5.22}$$

In analogy with the adiabatic transformation it is easy to show the group property of the new non-adiabatic transformations. Therefore there exists the inverse transformation from the new set of second quantized operators to the old set of second quantized operators

$$a_P = \sum_Q \tilde{\bar{C}}_{PQ}(\tilde{\bar{B}})\ \bar{a}_Q \tag{5.23}$$

$$b_r = \bar{b}_r + \sum_{PQ} \tilde{\bar{d}}_{rPQ}(\tilde{\bar{B}})\ \bar{a}_P^+ \bar{a}_Q \tag{5.24}$$

and similarly for the creation operators a_P^+ and b_r^+. The coefficients $\tilde{\bar{C}}_{PQ}$ and $\tilde{\bar{d}}_{rPQ}$ are given through the coefficients C_{PQ} and d_{rPQ} in the following way

$$\tilde{\bar{C}}_{PQ}(\tilde{\bar{B}}) = \bar{C}_{PQ}(\tilde{B}) = \bar{C}_{QP}(\tilde{B})^+ \tag{5.25}$$

$$\tilde{\bar{d}}_{rPQ}(\tilde{\bar{B}}) = \tilde{\bar{d}}_{rPQ}(\tilde{B}) = -\sum_{RS} \tilde{C}_{PR}(\tilde{B})\ \tilde{d}_{rRS}(\tilde{B}) \cdot \tilde{C}_{QS}(\tilde{B})^+ \ . \tag{5.26}$$

Since further we shall work with three representations of V-E Hamiltonian namely with crude, adiabatic and non-adiabatic representation and with two transformations i.e. adiabatic and non-adiabatic transformation we introduce the new notation. We denote the second quantized operators in crude representation for electrons as $\{\bar{a}_P\}, \{\bar{a}_P^+\}$, for phonons as $\{\bar{b}_r\}, \{\bar{b}_r^+\}$, for coordinate operators as $\{\bar{B}_r\}$ and for momentum operators as $\{\tilde{\bar{B}}_r\}$.
The total V-E Hamiltomiam has in the crude representation the form (2.1)

$$\begin{aligned} H = {} & T_N(\tilde{\bar{B}}) + E_{NN}(\bar{B}) + \sum_{PQ} h_{PQ}(\bar{B})\ \bar{a}_P^+ \bar{a}_Q + \\ & + \frac{1}{2} \sum_{PQRS} v^0_{PQRS}\ \bar{a}_P^+ \bar{a}_Q^+ \bar{a}_S \bar{a}_R \ . \end{aligned} \tag{5.27}$$

The quantities $E_{NN}(\bar{B})$ and $h_{PQ}(\bar{B})$ are given through the Taylor expansion (2.12) and (2.13)

$$E_{NN}(\bar{B}) = \sum_{i=0}^{\infty} E^{(i)}_{NN}(\bar{B}) \tag{5.28}$$

and

$$h_{PQ}(\bar{B}) = h^0_{PQ} + \sum_{i=1}^{\infty} u^{(i)}_{PQ}(\bar{B}) \tag{5.29}$$

where

$$u_{PQ}(R) = < P \mid \sum_J \frac{-Z_J e^2}{\mid \bar{r} - \bar{R}_J \mid} \mid Q > . \tag{5.30}$$

After introducing the normal coordinates we are able to define the vibrational Hamiltonian

$$H_B = \sum_r \hbar\omega_r \left(\bar{b}_r^+ \bar{b}_r + \frac{1}{2} \right) . \tag{5.31}$$

The crucial step in the break down of the Born-Oppenheimer approximation is the quantization of vibrational Hamiltonian H_B. Let us discuss this problem in more details. The Hamiltonian H_B contains the kinetic energy and also the potential energy

$$H_B = E_{kin}(\tilde{\bar{B}}) + E_{pot}(\bar{B}) . \tag{5.32}$$

The potential energy is determined by the quadratic part of the nuclear energy $E_{NN}^{(2)}(\bar{B})$ as well as by some potential energy $V_N^{(2)}(\bar{B})$ which is also the quadratic function of coordinate operators $\bar{B}_r$ and which has the origin in the interaction of the electrons with the vibrating nuclei. Therefore we have

$$E_{pot}(\bar{B}) = E_{NN}^{(2)}(\bar{B}) + V_N^{(2)}(\bar{B}) . \tag{5.33}$$

In the case of kinetic energy term we could think that this is identical with the kinetic energy of the nuclei. This was true for the case of adiabatic approximation. In the case of break down of adiabatic approximation we have to have in mind the finite mass of electrons and therefore to introduce the more general kinetic energy term. Therefore we shall add to the kinetic energy of the nuclei $T_N(\tilde{\bar{B}})$ some another yet unknown term which we denote $W_N^{(2)}(\tilde{\bar{B}})$ and which will be the quadratic function of momentum operator

$$E_{kin}(\tilde{\bar{B}}) = T_N(\tilde{\bar{B}}) + W_N^{(2)}(\tilde{\bar{B}}) . \tag{5.34}$$

The total V-E Hamiltonian (5.27) can be divided into two parts

$$H = H_A + H_B \tag{5.35}$$

where H_B is the vibrational Hamiltonian and due to eqs. (5.33) and (5.34) has the from

$$H_B = T_N(\tilde{\bar{B}}) + W_N^{(2)}(\tilde{\bar{B}}) + E_{NN}^{(2)}(\bar{B}) + V_N^{(2)}(\bar{B}) . \tag{5.36}$$

The Hamiltonian H_A therefore is

$$H_A = E_{NN}(\bar{B}) + E_{NN}^{(2)}(\bar{B}) - \\ - V_N^{(2)}(\bar{B}) - W_N^{(2)}(\tilde{\bar{B}}) + \\ + \sum_{PQ} h_{PQ}(\bar{B})\, \bar{a}_P^+ \bar{a}_Q + \\ + \frac{1}{2} \sum_{PQRS} v_{PQRS}^0\, \bar{a}_P^+ \bar{a}_Q^+ \bar{a}_S \bar{a}_R \ . \tag{5.37}$$

Further we shall limit ourselves to such an approximation that our Hamiltonian (5.35) in non-adiabatic representation will be given up to the second order in boson indices. In individual orders of the Taylor expression we shall use the notation $H_{A_{\bar{i}(j,k)}}$ where $\bar{i}$ means the order of the Taylor expansion in the crude representation, j – means the power of coordinate operator $\bar{B}$ and k – means the power of momentum operator $\tilde{\bar{B}}$. Therefore we shall have

$$H_{A_{\bar{0}}} = E_{NN}^0 + \sum_{PQ} h_{PQ}^0\, \bar{a}_P^+ \bar{a}_Q + \\ + \frac{1}{2} \sum_{PQRS} v_{PQRS}^0\, \bar{a}_P^+ \bar{a}_Q^+ \bar{a}_S \bar{a}_R \tag{5.38}$$

$$H_{A_{\bar{1}(1,0)}} = \sum_r E_{NN}^r\, \bar{B}_r + \sum_{PQr} u_{PQ}^r\, \bar{B}_r\, \bar{a}_P^+ \bar{a}_Q \tag{5.39}$$

$$H_{A_{\bar{2}(2,0)}} = -\frac{1}{2} \sum_{rs} V_N^{rs}\, \bar{B}_r \bar{B}_s + \\ + \frac{1}{2} \sum_{PQrs} u_{PQ}^{rs}\, \bar{B}_r \bar{B}_s\, \bar{a}_P^+ \bar{a}_Q \tag{5.40}$$

$$H_{A_{\bar{2}(0,2)}} = -\frac{1}{2} \sum_{rs} W_N^{rs}\, \tilde{\bar{B}}_r \tilde{\bar{B}}_s \tag{5.41}$$

Further we shall try to determine the quantities V_N^{rs} and W_N^{rs} in order to get the vibrational frequences ω_r in the case of break down of the Born-Oppenheimer approximation.

Final results we shall get after passing to the non-adiabatic representation. We shall do no in two steps. First we pass from the crude representation to the adiabatic representation. Let us denote the fermion operators in adiabatic representation as $\{\bar{\bar{a}}_P\}, \{\bar{\bar{a}}_P^+\}$ the boson operators as $\{\bar{\bar{b}}_r\}, \{\bar{\bar{b}}_r^+\}$ the coordinate operators as $\{\bar{\bar{B}}_r\}$ and momentum as $\{\tilde{B}_r\}$. The transformations from crude representation to adiabatic representation will read

$$\bar{a}_P = \sum_Q C_{PQ}(\bar{\bar{B}})\ \bar{\bar{a}}_Q \tag{5.42}$$

$$\bar{a}_P^+ = \sum_Q C_{PQ}(\bar{\bar{B}})^+\ \bar{\bar{a}}_Q^+ \tag{5.43}$$

$$\bar{b}_r = \bar{\bar{b}}_r + \sum_{PQ}{}' d_{rPQ}\ (\bar{\bar{B}})\ \bar{\bar{a}}_P^+\ \bar{\bar{a}}_Q \tag{5.44}$$

$$\bar{b}_r^+ = \bar{\bar{b}}_r^+ - \sum_{PQ} d_{rPQ}\ (\bar{\bar{B}})\ \bar{\bar{a}}_P^+\ \bar{\bar{a}}_Q\ . \tag{5.45}$$

Than we pass from the adiabatic representation to the non-adiabatic by following transformation. Let us denote the fermion operators in non-adiabatic representation as $\{a_P\},\{a_P^+\}$ the operators as $\{b_r\},\{b_r^+\}$ the coordinate operator as $\{B_r\}$ and the momentum operator as $\{\tilde{B}_r\}$. The transformation from the adiabatic to non-adiabatic (momentum) representation will read

$$\bar{\bar{a}}_P = \sum_Q \tilde{C}_{PQ}\left(\tilde{B}\right)\ a_Q \tag{5.46}$$

$$\bar{\bar{a}}_P^+ = \sum_Q \tilde{C}_{PQ}\left(\tilde{B}\right)^+\ a_Q^+ \tag{5.47}$$

$$\bar{\bar{b}}_r = b_r + \sum_{PQ} \tilde{d}_{rPQ}\left(\tilde{B}\right)\ a_P^+\ a_Q \tag{5.48}$$

$$\bar{\bar{b}}_r^+ = b_r^+ + \sum_{PQ} \tilde{d}_{rPQ}\left(\tilde{B}\right)\ a_P^+\ a_Q\ . \tag{5.49}$$

This brings us the following expressions for the Hamiltomian

$$\begin{aligned} H_{A_0} &= E_{NN}^0 + \sum_{PQ} h_{PQ}^0\ a_P^+\ a_Q + \\ &+ \frac{1}{2} \sum_{PQRS} v_{PQRS}^0\ a_P^+\ a_Q^+\ a_S\ a_R \end{aligned} \tag{5.50}$$

$$\begin{aligned} H_{A_{1(1,0)}} &= \sum_r E_{NN}^r\ B_r + \\ &+ \sum_{PQr}\left[u_{PQ}^r + \sum_R \left(h_{PR}^0 C_{RQ}^r + h_{RQ}^0 C_{RP}^{r*}\right)\right] B_r a_P^+ a_Q + \\ &+ \sum_{PQRSTr}\left(v_{PQTS}^0 C_{TR}^r + v_{TQRS}^0 C_{TP}^{r*}\right) B_r a_P^+ a_Q^+ a_S a_R \end{aligned} \tag{5.51}$$

$$H_{A_{1(0,1)}} = \sum_{PQRr} \left(h^0_{PR} \widetilde{C}^r_{RQ} - h^0_{RQ} \widetilde{C}^{r*}_{RP} \right) \widetilde{B}_r \; a^+_P a_Q + \tag{5.52}$$

$$+ \sum_{PQRSTr} \left(v^0_{PQTS} \widetilde{C}^r_{TR} - v^0_{TQRS} \widetilde{C}^{r*}_{TP} \right) \widetilde{B}_r \; a^+_P a^+_Q a_S a_R$$

$$H_{A_{2(1,1)}} = \frac{1}{2} \sum_{PQRrs} \left\{ u^r_{PR} \widetilde{C}^s_{RQ} - u^r_{RQ} \widetilde{C}^{s*}_{RP} + \right.$$

$$+ \sum_S \left[\left(h^0_{PR} C^r_{RS} + h^0_{RS} C^{r*}_{RP} \right) \widetilde{C}^s_{SQ} - \right.$$

$$\left. \left. - \left(h^0_{SR} C^r_{RQ} + h^0_{RQ} C^{r*}_{RS} \right) \widetilde{C}^{s*}_{SP} \right] \right\} \cdot$$

$$\cdot \left(B_r \widetilde{B}_s + \widetilde{B}_s B_r \right) a^+_P a_Q + \tag{5.53}$$

$$+ \frac{1}{2} \sum_{PQRSTU} \left[\left(v^0_{PQTS} C^r_{TU} - v^0_{PQTU} C^r_{TS} \right) \widetilde{C}^s_{UR} + \right.$$

$$+ \left(v^0_{TURS} C^{r*}_{TQ} - v^0_{TQRS} C^{r*}_{TU} \right) \cdot$$

$$\left. \cdot \widetilde{C}^{s*}_{UP} + \left(v^0_{TQUS} - v^0_{TQSU} \right) \left(C^{r*}_{TP} \widetilde{C}^s_{UR} - C^r_{UR} \widetilde{C}^{s*}_{TP} \right) \right] \cdot$$

$$\cdot \left(B_r \widetilde{B}_s + \widetilde{B}_s B_r \right) a^+_P a^+_Q a_S a_R$$

$$H_{A_{2(2,0)}} = - \frac{1}{2} \sum_{rs} V^{rs}_N \; B_r \; B_s +$$

$$+ \sum_{PQrs} \left[\frac{1}{2} u^{rs}_{PQ} + \sum_R \left(\frac{1}{2} h^0_{PR} C^{rs}_{RQ} + \frac{1}{2} h^0_{RQ} C^{rs*}_{RP} + u^r_{PR} C^s_{RQ} + \right. \right.$$

$$\left. \left. + u^r_{RQ} C^{s*}_{RP} \right) + \sum_{RS} h^0_{RS} C^{r*}_{RP} C^s_{SQ} \right] B_r B_s a^+_P a_Q + \tag{5.54}$$

$$+ \frac{1}{2} \sum_{PQRSTrs} \left\{ v^0_{PQTS} C^{rs}_{TR} + v^0_{TQRS} C^{rs*}_{TP} + \right.$$

$$+ \sum_U \left[v^0_{PQTU} C^r_{TR} C^s_{US} + v^0_{TURS} C^{r*}_{TP} C^{s*}_{UQ} + \right.$$

$$\left. \left. + 2 \left(v^0_{TQUS} - v^0_{TQSU} \right) C^{r*}_{TP} \, C^s_{UR} \right] \right\} B_r B_s a^+_P a^+_Q a_S a_R$$

$$H_{A_{2(0,2)}} = - \frac{1}{2} \sum_{rs} W^{rs}_N \; \widetilde{B}_r \; \widetilde{B}_s + \sum_{PQRrs} \left(\frac{1}{2} h^0_{PR} \widetilde{C}^{rs}_{RQ} + \right.$$

$$+ \frac{1}{2} h^0_{RQ} \tilde{C}^{rs*}_{RP} - \sum_S h^0_{RS} \tilde{C}^{r*}_{RP} \tilde{C}^{s}_{SQ} \Big) \tilde{B}_r \tilde{B}_s a^+_P a_Q +$$

$$+ \frac{1}{2} \sum_{PQRSTrs} \Big\{ v^0_{PQTS} \tilde{C}^{rs}_{TR} + v^0_{TQRS} \tilde{C}^{rs*}_{TP} + \qquad (5.55)$$

$$+ \sum_U \Big[v^0_{PQTU} \tilde{C}^{r}_{TR} \tilde{C}^{s}_{US} + v^0_{TURS} \tilde{C}^{r*}_{TP} \tilde{C}^{s*}_{UQ} +$$

$$+ 2 \left(v^0_{TQSU} - v^0_{TQUS} \right) \tilde{C}^{r*}_{TP} \tilde{C}^{s}_{UR} \Big] \Big\} \tilde{B}_r \tilde{B}_s a^+_P a^+_Q a_S a_R$$

$$H_{A_{2(0,0)}} = 2 \sum_{PQr} E^r_{NN} \tilde{d}^0_{rPQ} a^+_P a_Q +$$

$$+ \sum_{PQRr} \Big\{ \Big[u^r_{PR} + \sum_S \left(h^0_{PS} C^r_{SR} + h^0_{SR} C^{r*}_{SP} \right) \Big] \tilde{d}^0_{rRQ} +$$

$$+ \Big[u^r_{RQ} + \sum_S \left(h^0_{RS} C^r_{SQ} + h^0_{SQ} C^{r*}_{SR} \right) \Big] \tilde{d}^0_{rPR} \Big\} a^+_P a_Q +$$

$$+ 2 \sum_{PQRSr} \Big[u^r_{PR} + \sum_T \left(h^0_{PT} C^r_{TR} + h^0_{TR} C^{r*}_{TP} \right) \Big] \tilde{d}^0_{rQS} a^+_P a^+_Q a_S a_R +$$

$$(5.56)$$

$$+ \sum_{PQRSTUr} \Big[\left(v^0_{PQTS} C^r_{TU} - v^0_{PQTU} C^r_{TS} \right) \tilde{d}^0_{rUR} +$$

$$+ \left(v^0_{TQRS} C^{r*}_{TU} - v^0_{TURS} C^{r*}_{TQ} \right) \tilde{d}^0_{rPU} +$$

$$+ \left(v^0_{TQUS} - v^0_{TQSU} \right) \left(C^{r*}_{TP} \tilde{d}^0_{rUR} + C^r_{UR} \tilde{d}^0_{rPT} \right) \Big] a^+_P a^+_Q a_S a_R +$$

$$+ 2 \sum_{PQRSTUVr} \left(v^0_{PQVT} C^r_{VS} + v^0_{VQST} C^{r*}_{VP} \right) \tilde{d}^0_{rRU} a^+_P a^+_Q a^+_R a_U a_T a_S$$

$$H_{B_0} = \sum_r \hbar\omega_r \left(b^+_r b_r + \frac{1}{2} \right) \qquad (5.57)$$

$$H_{B_{1(1,0)}} = \sum_{PQr} \hbar\omega_r \tilde{d}^0_{rPQ} B_r a^+_P a_Q \qquad (5.58)$$

$$H_{B_{1(0,1)}} = - \sum_{PQr} \hbar\omega_r d^0_{rPQ} \tilde{B}_r a^+_P a_Q \qquad (5.59)$$

$$H_{B_{2(1,1)}} = \frac{1}{2} \sum_{PQrs} \left(\hbar\omega_r \tilde{d}^s_{rPQ} - \hbar\omega_s d^r_{sPQ} \right) \cdot$$

$$(5.60)$$

$$\cdot \left(B_r \tilde{B}_s + \tilde{B}_s B_r \right) a^+_P a_Q$$

$$H_{B_{2(2,0)}} = 0 \qquad (5.61)$$

$$H_{B_{2(0,2)}} = - \sum_{PQRrs} \hbar\omega_r \left(d^0_{rPR} \, \overset{\sim s}{C}_{RQ} - \right.$$

$$\left. - d^0_{rRQ} \, \overset{\sim s*}{C}_{RP} \right) \tilde{B}_r \, \tilde{B}_s \, a^+_P \, a_Q \tag{5.62}$$

$$H_{B_{2(0,0)}} = \sum_{PQRr} \hbar\omega_r \left(\overset{\sim 0}{d}_{rPR} \overset{\sim 0}{d}_{rRQ} - d^0_{rPR} d^0_{rRQ} \right) a^+_P \, a_Q \, +$$

$$+ \sum_{PQRSr} \hbar\omega_r \left(\overset{\sim 0}{d}_{rPR} \overset{\sim 0}{d}_{rQS} - d^0_{rPR} d^0_{rQS} \right) a^+_P \, a^+_Q \, a_S \, a_R \tag{5.63}$$

To these equations we can apply the Wick theorem. For the fermions we use the indices P, Q, R, S, T, U, V. Further for occupied states we use indices A, B, C, D and for unoccupied I, J, K, L indices. The one-fermion terms we shall write in normal product form according to

$$\sum_{PQ} \lambda_{PQ} \, a^+_P \, a_Q = \sum_{PQ} \lambda_{PQ} \, N \left[a^+_P \, a_Q \right] + \lambda \, , \tag{5.64}$$

where

$$\lambda = \sum_I \lambda_{II} \, . \tag{5.65}$$

Analogously for two-fermion terms we have

$$\sum_{PQRS} \mu_{PQRS} \, a^+_P \, a^+_Q \, a_S \, a_R = \sum_{PQRS} \mu_{PQRS} \, N \left[a^+_P \, a^+_Q \, a_S \, a_R \right] +$$

$$+ \sum_{PQ} \mu_{PQ} \, N \left[a^+_P \, a_Q \right] + \mu \, , \tag{5.66}$$

where

$$\mu_{PQ} = \sum_I \left(\mu_{PIQI} + \mu_{IPIQ} - \mu_{PIIQ} - \mu_{IPQI} \right) \tag{5.67}$$

and

$$\mu = \sum_{IJ} \left(\mu_{IJIJ} - \mu_{IJJI} \right) . \tag{5.68}$$

Finally for the three-fermion terms

$$\sum_{PQRSTU} \nu_{PQRSTU} \, a^+_P \, a^+_Q \, a^+_R \, a_U \, a_T \, a_S = \sum_{PQRSTU} \nu_{PQRSTU}$$

$$N \left[a^+_P \, a^+_Q \, a^+_R \, a_U \, a_T \, a_S \right] + \sum_{PQRS} \nu_{PQRS} \, N \left[a^+_P \, a^+_Q \, a_S \, a_R \right] + \tag{5.69}$$

$$+ \sum_{PQ} \nu_{PQ} \, N \left[a^+_P \, a_Q \right] + \nu \, ,$$

where

$$\nu_{PQRS} = \sum_{I} (\nu_{PQIRSI} - \nu_{PQIRIS} + \nu_{PQIIRS} - \nu_{PIQRSI} + \qquad (5.70)$$

$$+ \nu_{PIQRIS} - \nu_{PIQIRS} + \nu_{IPQRSI} - \nu_{IPQRIS} + \nu_{IPQIRS})$$

$$\nu_{PQ} = \sum_{IJ} (\nu_{PIJQIJ} - \nu_{PIJQJI} - \nu_{PIJIQJ} + \nu_{PIJJQI} +$$

$$+ \nu_{PIJIJQ} - \nu_{PIJJIQ} - \nu_{IPJQIJ} + \nu_{IPJQJI} + \nu_{IPJIQJ} - \qquad (5.71)$$

$$- \nu_{IPJJQI} - \nu_{IPJIJQ} + \nu_{IPJJIQ} + \nu_{IJPQIJ} - \nu_{IJPQJI} -$$

$$- \nu_{IJPIQJ} + \nu_{IJPJQI} + \nu_{IJPIJQ} - \nu_{IJPJIQ}) ,$$

and

$$\nu = \sum_{IJK} (\nu_{IJKIJK} - \nu_{IJKIKJ} - \nu_{IJKJIK} + \qquad (5.72)$$

$$+ \nu_{IJKJKI} + \nu_{IJKKIJ} - \nu_{IJKKJI}) .$$

The individual terms of the non-adiabatic Hamiltonian which do not contain the fermion operators we denote as H^0, the terms which contain the normal product $N\left[a_P^+ a_Q\right]$ as H', the two-fermion terms which contain the normal product $N\left[a_P^+ a_Q^+ a_S a_R\right]$ as H'' and finally the three-fermion terms which contain the normal product $N\left[a_P^+ a_Q^+ a_R^+ a_U a_T a_S\right]$ as H'''.
Further we shall try to get the equations for the new unperturbed fermion energies and for the coefficients of the non-adiabatic transformation. We shall proceede in a similar manner as in the case of adiabatic representation where the new fermion energies were obtained by solving the $CPHF$ equations. Therefore analogously we want to diagonalize the one-fermion terms H' in the non-adiabatic representation, i.e. we want to have

$$\mathbf{H}' = \sum_{P} X_P \, N\left[a_P^+ a_P\right] . \qquad (5.73)$$

We shall try to diagonalize the eq. (5.73) to the first order of the Taylor expansion. In the zeroth order according to (5.50) we have

$$H_0' = H_{A_0'} = \sum_{PQ} \left[h_{PQ}^0 + \sum_{I} \left(v_{PIQI}^0 - v_{PIIQ}^0\right)\right] . \qquad (5.74)$$

$$. N\left[a_P^+ a_Q\right] = \sum_{P} \epsilon_P^0 \, N\left[a_P^+ a_P\right] ,$$

which is the diagonalization of the Hartree-Fock operator

$$f^0_{PQ} = h^0_{PQ} + \sum_I \left(v^0_{PIQI} - v^0_{PIIQ}\right) = \epsilon^0_P \, \delta_{PQ} \, . \tag{5.75}$$

The zeroth order is therefore the same as in the crude as well as in adiabatic representation. In the first order of the Taylor expansion we get according to (5.51), (5.52), (5.58) and (5.59)

$$\begin{aligned} H_{A'_{1(1,0)}} &= \sum_{PQr} \Big\{ u^r_{PQ} + \left(\epsilon^0_P - \epsilon^0_Q\right) C^r_{PQ} + \\ &+ \sum_{AI} \Big[\left(v^0_{PIQA} - v^0_{PIAQ}\right) C^r_{AI} - \\ &- \left(v^0_{PAQI} - v^0_{PAIQ}\right) C^r_{IA}\Big]\Big\} B_r \, N\left[a^+_P \, a_Q\right] , \end{aligned} \tag{5.76}$$

$$\begin{aligned} H_{A'_{1(0,1)}} &= \sum_{PQr} \Big\{ \left(\epsilon^0_P - \epsilon^0_Q\right) \tilde{C}^r_{PQ} + \\ &+ \sum_{AI} \Big[\left(v^0_{PIQA} - v^0_{PIAQ}\right) \tilde{C}^r_{AI} - \\ &- \left(v^0_{PAQI} - v^0_{PAIQ}\right) \tilde{C}^r_{IA}\Big]\Big\} \tilde{B}_r \, N\left[a^+_P \, a_Q\right] , \end{aligned} \tag{5.77}$$

$$H_{B'_{1(1,0)}} = - \sum_{PQr} \hbar \, \omega_r \, \tilde{C}^r_{PQ} \; B_r \, N\left[a^+_P \, a_Q\right] . \tag{5.78}$$

$$H_{B'_{1(0,1)}} = - \sum_{PQr} \hbar \, \omega_r \, C^r_{PQ} \; \tilde{B}_r \, N\left[a^+_P \, a_Q\right] . \tag{5.79}$$

If we diagonalize the terms $H'_{1(1,0)}$ and $H'_{1(0,1)}$ we get

$$H'_{1(1,0)} = \sum_{Pr} \epsilon^r_P \, B_r \, N\left[a^+_P \, a_P\right] \tag{5.80}$$

and

$$H'_{1(0,1)} = \sum_{Pr} \tilde{\epsilon}^r_P \, \tilde{B}_r \, N\left[a^+_P \, a_P\right] \tag{5.81}$$

which leads to the coupled equations for C^r_{PQ} and $\tilde{C}^r_{PQ}$ coefficients

$$\begin{aligned} & u^r_{PQ} + \left(\epsilon^0_P - \epsilon^0_Q\right) C^r_{PQ} + \\ &+ \sum_{AI} \Big[\left(v^0_{PIQA} - v^0_{PIAQ}\right) C^r_{AI} - \\ &- \left(v^0_{PAQI} - v^0_{PAIQ}\right) C^r_{IA}\Big] - \hbar \, \omega_r \, \tilde{C}^r_{PQ} = \epsilon^r_P \, \delta_{PQ} \end{aligned} \tag{5.82}$$

and

$$\begin{aligned} & \left(\epsilon^0_P - \epsilon^0_Q\right) \tilde{C}^r_{PQ} + \sum_{AI} \Big[\left(v^0_{PIQA} - v^0_{PIAQ}\right) \tilde{C}^r_{AI} - \\ &- \left(v^0_{PAQI} - v^0_{PAIQ}\right) \tilde{C}^r_{IA}\Big] - \hbar \, \omega_r \, C^r_{PQ} = \tilde{\epsilon}^r_P \, \delta_{PQ} \end{aligned} \tag{5.83}$$

The equations (5.82) and (5.83) differ from the analogous equations in the crude and in the adiabatic representation. For example in crude representation for which holds $C = 0$ and $\tilde{C} = 0$ we get the non-physical equation

$$u^{r}_{PQ} = \epsilon^{r}_{P}\, \delta_{PQ} \tag{5.84}$$

which of course cannot be diagonalized. In adiabatic representation for which holds $\tilde{C} = 0$ we get

$$\begin{aligned} u^{0}_{PQ} + \left(\epsilon^{0}_{P} - \epsilon^{0}_{Q}\right) C^{r}_{PQ} + \sum_{AI} \Big[\left(v^{0}_{PIQA} - v^{0}_{PIAQ}\right) C^{r}_{AI} - \\ - \left(v^{0}_{PAQI} - v^{0}_{PAIQ}\right) C^{r}_{IA}\Big] = \epsilon^{r}_{P}\, \delta_{PQ} \end{aligned} \tag{5.85}$$

and

$$-\hbar\, \omega_{r}\, C^{r}_{PQ} = \tilde{\epsilon}^{r}_{P}\, \delta_{PQ}\,. \tag{5.86}$$

Here we cannot diagonalize the equation (5.86). We can omit the equation (5.86) under the condition that

$$\hbar\, \omega_{r} \mid C^{r}_{PQ} \mid << u^{r}_{PQ} \sim \mid \epsilon^{0}_{P} - \epsilon^{0}_{Q} \mid \mid C^{r}_{PQ} \mid \tag{5.87}$$

which is equivalent to the condition

$$\hbar\, \omega_{r} << \mid \epsilon^{0}_{P} - \epsilon^{0}_{Q} \mid . \tag{5.88}$$

Condition (5.88) is nothing else than the condition for the validity of the adiabatic approximation. Therefore in the adiabatic representation we perform the diagonalization according to (5.85). When the condition (5.87) is not satisfied we have to perform the diagonalization according to equations (5.82) and (5.83). Solving equations (5.82) and (5.83) we obtain the non-diagonal adiabatic coefficients C^{r}_{PQ} and the non-diagonal coefficient $\tilde{C}^{r}_{PQ}$, $(P \neq Q)$. In the case $P = Q$ we can simply put

$$C^{r}_{PP} = 0 \tag{5.89}$$

and

$$\tilde{C}^{r}_{PP} = 0\,. \tag{5.90}$$

Further our strategy will be to continue in putting the V-E non-adiabatic Hamiltomian into the diagonal form in a similar manner we did for the adiabatic Hamiltomian. Doing this we get the expressions for W^{rs}_{N} and V^{rs}_{N}. The details of these derivations are described in our recent paper [45].

References

[1] P. Pulay, in : *Ab Initio Methods in Quantum Chemistry - II* (K. P. Lawley, ed), John Wiley & Sons Ltd. (1987).

[2] E. A. Salter, G. W. Trucks, and R. J. Bartlett, *J. Chem. Phys.* **90**, 1752 (1989).

[3] J. A. Pople, in : *Geometrical Derivatives of Energy Surfaces and Molecular Properties* (P. Jorgensen and J. Simons, Eds), p. 109, Reidel, Dordrecht (1986).

[4] N. H. March, W. H. Young and S. Sampanthar, *The Many - Body Problem in Quantum Mechanics*, Cambridge University Press (1967).

[5] J. Čížek, *J. Chem. Phys.* **45**, 4256 (1966).

[6] J. Čížek, *Adv. Chem. Phys* **14**, 35 (1969).

[7] J. Paldus and J. Čížek, *Adv. Quant. Chem.* **9**, 105 (1975).

[8] J. A. Pople, J. S. Binkley and R. Seeger, *Int. J. Quant. Chem.* Symp. **10**, 1 (1976).

[9] I. Hubač and P. Čársky, *Topics Curr. Chem.* **75**, 97 (1978).

[10] R. J. Bartlett, *Annu. Rev. Phys. Chem.* **32**, 359 (1981).

[11] S. Wilson, *Electron Correlation in Molecules*, Clarendon Press, Oxford (1984).

[12] K. Jankowski, in : *Methods in Computational Chemistry, Vol. 1, Electron Correlation in Atoms and Molecules* (S. Wilson, Ed.) p. 1. Plenum Press, New York (1987).

[13] M. Urban, J. Černušák, V. Kellö and J. Noga, in : *Methods in Computational Chemistry, Vol. 1, Electron correlation in Atoms and Molecules* (S. Wilson, Ed.) p. 117, Plenum Press, New York (1987).

[14] J. Gerratt and J. M. Mills, *J. Chem. Phys.* **49**, 1719 (1968).

[15] J. Gerratt and J. M. Mills, *J. Chem. Phys.* **49**, 1730 (1968).

[16] J. A. Pople, K. Raghavachari, H. B. Schlegel and J. S. Binkley, *Int. J. Quantum Chem.* Symp. **13**, 225 (1979).

[17] N. C. Handy, R. D. Amos, J. F. Graw, J. E. Rice and E. D. Simandiras, *Chem. Phys. Lett*, **120**, 151 (1985).

[18] R. J. Bartlett, in : *Geometrical Derivatives of Energy Surfaces and Molecular Properties* (P. Jorgensen and J. Simons, Eds.), p. 35, Reidel, Dordrecht (1986).

[19] E. A. Salter and R. J. Bartlett, *J. Chem. Phys.* **90**, 1967 (1989).

[20] P. Jorgensen and T. Helgaker, *J. Chem. Phys.* **89**, 1560 (1988).

[21] T. Helgaker, P. Jorgensen and N. C. Handy, *Theor. Chim. Acta* **76**, 227 (1989).

[22] C. Moller and M. S. Plesset, *Phys. Rev* **46**, 618 (1934).

[23] I. Hubač and M. Svrček *Int. J. Quant. Chem.* **23**, 403 (1988).

[24] I. Hubač, M. Svrček, E. A. Salter, C. Sosa and R. J. Bartlett, *Lecture Notes in Chemistry.* Vol. 52, p. 95, Springer, Berlin (1988).

[25] J. P. Blaizot and G. Ripka, *Quantum Theory of Finite Systems*, The MIT Press Cambridge, Massachusetts, London, England (1986).

[26] M. Hamermesh, *Group Theory and its Application to Physical Problems*, Addison - Wesley, Reading, Massachusetts, Palo Alto, London (1964).

[27] I. Hubač, V. Kvasnička and A. Holubec, *Chem. Phys. Lett.* **23**, 381 (1973).

[28] I. Hubač and M. Urban, *Theoret. Chim. Acta* **45**, 185 (1977).

[29] M. Svrček and I. Hubač, *Czech. J. Phys.* **41**, 556 (1991).

[30] R. J. Harrison, G. B. Fitzgerald, W. D. Laiding and R. J. Bartlett, *Chem. Phys. Lett.* **124**, 291 (1986).

[31] B. H. Brandow, *Rev. Mod. Phys.* **39**, 771 (1967).

[32] B. H. Brandow, *Int. J. Quant. Chem.* **15**, 207 (1979).

[33] J. Lindgren, *Int. J. Quant. Chem.* Symp. **12**, 33 (1978).

[34] V. Kvasnička, *Adv. Chem. Phys.* **36**, 345 (1977).

[35] M. Svrček and I. Hubač, *Int. J. Quant. Chem.* **31**, 625 (1987).

[36] M. Svrček, *Ph. D. Thesis, Faculty of Mathematics and Physics*, 1986, Bratislava, Czechoslovakia.

[37] T. H. Dunning, Jr, *J. Chem. Phys.* **53**, 2823 (1970).

[38] L. T. Redmon, G. D. Purvis III and R. J. Bartlett, *J. Am. Chem. Soc.* **101**, 2856 (1979).

[39] E. Davidson and D. Feller, *Chem. Phys. Lett.* **104**, 54 (1984).

[40] W. Kolos and L. Wolniewicz, *J. Chem. Phys.* **49**, 409 (1968).

[41] L. Wolniewicz, *J. Chem. Phys.* **78**, 6173 (1983).

[42] M. Born and R. Oppenheimer, *Ann. Physik.* (Leipzig **84**, 4357 (1927).

[43] H. Köppel, W. Domcke and L. S. Cederbaum, *Adv. Chem. Phys.* **57**, 59 (1984).

[44] M. Wagner, *Phys. Stat. Sol. (b)* **107**, 617 (1981).

[45] I. Hubač and M. Svrček, in : *Molecular Vibrations* , Plenum Press, to be published

RELATIVISTIC ELECTRONIC STRUCTURE THEORY

I.P.Grant

Department of Theoretical Chemistry §
University of Oxford

1. Introduction

The last decade has seen big changes in the status of relativistic electronic structure theory. Ten years ago, conferences were dominated by discussions on the validity of models of relativistic atomic and molecular structure, some of which raised serious doubts that it would ever be possible to devise procedures which would be recognized as theoretically well-founded. Today, we have arrived at a consensus on a hierarchy of approximations within quantum electrodynamics of atoms, ranging from mean-field one-body models through diagrammatic many-body perturbation theory and coupled-cluster all-order methods to the calculation of radiative corrections. Developments in the electronic structure of molecules lag somewhat behind those for atoms, but there are signs that we shall soon see rapid improvements.

My four lectures will survey the field as follows:

- Where do we look for relativistic effects?
- QED of atoms and molecules
- State of the art in relativistic atomic structure
- State of the art in relativistic molecular structure

The rest of this paper is divided into four sections summarizing each of the lectures along with a relatively small bibliography which will help those interested to get started.

2. Where do we look for relativistic effects?

It helps the novice to begin with a study of relativistic effects in one-electron atoms, the theory of which is in many standard textbooks; a good modern introduction from the viewpoint of a nuclear physicist is given by Greiner [1]. The hydrogenic case is easy to handle [2] as most of the quantities of interest - such as the electron density angular and radial distributions and their moments - are easy to write down analytically. The hydrogenic energy spectrum shows that some of the degeneracy of the nonrelativistic Kepler problem has been lifted; perturbation theory leads us to describe this as being due to the coupling of the intrinsic spin of the electron with its orbital motion. The existence of electron spin is also revealed by the behaviour of the atom in external electromagnetic fields.

---

§ Mathematical Institute, 24-9 St. Giles', Oxford, OX1 3LB, U.K.

Secondary effects [2] appear in many-electron systems. Electrons which have a substantial probability of being in regions with high potential energy, that is to say the neighbourhood of nuclei, would move on classical orbits intersecting those regions with a speed comparable to the speed of light *c*, and it is therefore the 'penetrating' orbitals that show the most characteristic dynamical effects of relativity. The corresponding orbitals, all characterized as having low angular momentum, are typically slightly contracted by comparison with their nonrelativistic equivalents. Orbitals with higher angular momentum have low density near the nuclei; the increased screening of the nucleus by the inner orbitals causes non-penetrating orbitals to expand. The effect of relativity in many-electron atoms is therefore a matter of subtle checks and balances, which need detailed study to understand properly.

Self-consistent field calculations on atoms give some idea of the way these competing influences operate. Both direct and induced changes due to relativity can be detected in the ordering of subshells, and trends in the predicted ionization potentials closely resemble the analogous trends in the Periodic Table. Relativistic effects are prominent in inner-shell processes - X-ray and Auger transitions - and in the spectra of highly-ionized high-Z atoms. The spectra of neutral species show more complicated effects which lie outside the scope of these lectures, as are the perturbations due to hyperfine interactions and weak parity-violating interactions.

The identification of relativistic effects in molecules is much more uncertain; whilst the relativistic corrections to the energy soon come to dominate corrections due to electron correlation as the atomic number increases, the nonlinear interaction between the two perturbing mechanisms is not very well understood. The theories are all rather more approximate than those for atoms, and it is not always obvious whether the effects that they predict are realistic or are just artifacts of the model. Pyykkö [3] has reviewed relativistic calculations on molecules, and Pitzer [4] and Pyykkö and Desclaux [5] have reviewed the chemical implications, some of which will be discussed in this lecture. A bibliography 1916-1985 has been compiled by Pyykkö [6] with over 3000 references.

3. QED of atoms and molecules

The majority of formulations of relativistic atomic structure theory take the bound state interaction picture of Furry [7] as their starting point. This picture depends on the use of a zero-order model in which the electrons move independently in some suitably chosen "mean-field" potential in the first instance. The states of the electron in this mean field - which in atomic physics consist of a discrete spectrum of bound states along with two continua, one of "positive energy" states which lie higher in energy, the other of "negative energy" states with energies which lie more than twice the rest-energy of the electron (approximately 511 keV) below the usual series limit - generate a representation of Fock space for the bare electron. The four-current, j^{μ}, of this quantized field is coupled to the four potential, A_{μ}, of the radiation field by the standard $j^{\mu}A_{\mu}$ coupling term, represented as a single vertex in Feynman diagram notation. The usual rules of many-body perturbation theory can then be invoked to compute whatever quantities are needed to describe states of the atom, in isolation as well as in interaction with the external world [8,9].

An effective relativistic interaction between electrons can be obtained by writing down the expected energy of the system for a state of the coupled fields with N electrons and no free photons. If the negative energy states are treated as inert - the "no-pair" approximation - the formalism closely resembles that of non-relativistic many-body theory, to which it indeed reduces in the non-relativistic limit, represented in mathematical terms as the limit in which $c \to \infty$. Expansion of the relativistic energy expression in powers of $1/c$ gives the nonrelativistic Pauli approximation in lowest order, and the terms of order $1/c^2$ give the relativistic perturbations - electron spin-orbit coupling, relativistic mass correction, Darwin correction, electron spin-spin interaction, electron spin-other-orbit and electron orbit-orbit interactions - described in standard quantum chemistry texts [10,11].

The relativistic energy formula can be interpreted [8] as the energy of a system of electrons interacting through an effective pair interaction which consists of the coulombic electron-electron repulsion modified by relativistic effects which depend upon the finite speed of light. The development of many-body theory within this framework is quite straightforward (see articles by Lindgren in [9, pp. 371-392] and [12, pp. 293-306] and by Quiney, Grant and Wilson [13, pp. 307-344]. Diagrammatic perturbation methods and all-order coupled-cluster methods have been developed in this way. Mean fields can be defined self-consistently by choosing them to cancel a suitable selection of second-order perturbation terms as in non-relativistic theory. For single-determinant wave-functions this leads to the Hartree-Fock model, and other possibilities such as *g*-Hartree potentials are available. It has also proved possible to include some contributions arising from virtual excitations involving negative energy states [13], though their effect is almost undetectable. Radiative corrections - the electron self-energy and vacuum polarization effects - which also involve interaction with negative energy states in part require special treatment to handle the renormalization of divergences, and this is currently under investigation.

Relativistic theories of molecular electronic structure within a fixed nucleus approximation pose no difficulties of principle, though technical difficulties of obtaining satisfactory numerical approximations to solutions of the Dirac equation have limited progress to self-consistent field calculations for simple diatomics [14]. This will be examined in my last lecture. A wide variety of more approximate computational schemes have been proposed. Dirac-Fock one-centre expansion calculations of simple hydrides were surveyed by Desclaux [15], and gave much qualitative information on relativistic effects on chemical properties across the Periodic Table. Many calculations have been done using *ab initio* relativistic effective core potentials for valence orbitals derived from atomic Hartree-Fock or Dirac-Fock calculations. Christiansen, Balasubramanian and Pitzer [16] have shown how to construct an effective spin-orbit operator within this formalism and to generate the necessary atomic and molecular orbitals. A very large number of such studies of molecules containing heavy elements have since appeared in the literature. Another approximate scheme, the self-consistent-field-$X\alpha$-Dirac-scattered-wave (DSW) molecular orbital method has been developed by Yang *et al.* and used extensively to study systems ranging from diatomic molecules to large metal complexes. A recent typical example of this sort of calculation, a study of bonding in metal carbonyls, has been published by Arratia-Perez and Yang [17].

The inclusion of nuclear motion within a fully relativistic theory remains a major challenge for practical calculations for atoms as well as for molecules. The problem of constructing a Hamiltonian formalism in classical relativistic mechanics which satisfies the requirements of Poincaré invariance, separability, world-line invariance and Einstein causality for N scalar particles interacting at a distance is non-trivial [18]. Sazdjian [19] has, for example, made some progress with deriving relativistic wave equations for two-body systems, but a corresponding development for even the simplest molecule is for the moment out of reach. The best that we can do in general is a hybrid treatment in which the nuclear motion is treated as a small non-relativistic perturbation. Fortunately, this is likely to be enough for most foreseeable applications.

4. State of the art in relativistic atomic structure

The first calculation of a relativistic self-consistent field for an atom in the Dirac-Hartree model was made in 1940 [20] using finite difference numerical methods similar to those still used today. Sophisticated packages are now available which are capable of modeling many-electron atoms in every row of the Periodic Table. The most widely used are due to Desclaux [21] and to Grant and his collaborators [22,23]. Both packages can handle open shell configurational states of considerable complexity using orbital wave-functions generated by finite difference methods. The most recent versions of the latter code [21] can do both relativistic multi-configurational self-consistent fields and CI calculations, including several hundred configurational states in the trial wave-function, and allowing for radiative corrections if required. Oscillator strengths for radiative transitions and other observable quantities can also be computed in this scheme. The wave-functions can also be used as starting points for calculations of electron scattering, photo-excitation and photo-ionization cross-sections.

Although finite difference methods are strongly established, they have drawbacks. It is well-known that MBPT calculations of atomic properties often require the calculation of contributions from continuum states as well as from bound states, and these are time-consuming to generate and use within a finite difference approximation scheme [25]. The problem is exacerbated in relativistic calculations because of the "negative energy" states, all of which are of continuum type. Moreover, finite difference methods are much less tractable for molecular calculations, where it is usually not possible to reduce the problem to solving a set of (possibly coupled) one dimensional differential equations. Numerical methods which exploit expansions in a set of functions which span as large a region as possible of the target Hilbert space are familiar to quantum chemists and in principle offer a way to bypass these difficulties. Many quantum chemists were convinced at one time that the so-called "variational collapse" problem was an almost impassable obstacle to basis set calculations with the Dirac Hamiltonian [27, 28]. There is now considerable evidence that one must work with "spinor" basis functions in which the components (two or four) are properly matched. They must satisfy the correct boundary conditions at the atomic nuclei so as to ensure an accurate representation of the wave function where it most matters - the regions of high potential where the velocities are relativistic [8, 28, 29, 30]. In our own work with atoms we have been able to identify several types: L-spinors, S-spinors and G-spinors, which are respectively the relativistic analogues of Coulomb Sturmian functions (linear combinations of Laguerre polynomials), on nonrelativistic STOs and on nonrelativistic spherical Gaussian functions. Recently B-spline representations of radial amplitudes have achieved a certain popularity [31, 32, 33]. Although they have great potential for a variety of applications in relativistic and non-relativistic atomic structure theory, there are a number of mathematical questions that restrict their use in electronic structure to self-consistent field and CI calculations. All of these computational schemes generate matrix representations of quantum mechanical operators, and the eigenvalue spectrum of the matrix of the Dirac Hamiltonian provides upper bounds to bound state eigenvalues as well as point eigenvalues in the region of the continuous spectrum of the original differential operator. Some of the eigenfunctions are good least squares approximations to bound state solutions of the target problem and the remainder represent square integrable wave-packets in the continuum. A strategy that demonstrates convergence of the theoretical predictions made with these approximate solutions by a systematic enlargement of the basis set is essential for confidence in the method. It is good that we have been able to show that the properties of our spinor basis sets are such that Rayleigh-Ritz approximations to eigenvalues and eigenfunctions of one-electron Dirac Hamiltonians, self-consistent field solutions, and expectations and transition matrices of most observables will converge to the correct limits [34]. It therefore makes sense to do MBPT calculations using the states generated in a finite basis, and even to attempt to calculate QED corrections this way.

5. State of the art in relativistic molecular structure

The first attempts to use basis set methods for atoms were made by Kim in 1967 [35], and the principles of his calculation can be extended to molecules very much as in the nonrelativistic case. Despite Kim's reservations in [35] about the future of the method, Roothaan-style formulations of the relativistic theory of molecular structure started to appear in journals [36]. Malli and Oreg [37] devised Dirac-Fock SCF equations for closed shell molecules in which each molecular spinor was expanded as a linear combination of atomic spinors. Matsuoka *et al.* [38] proposed a Dirac-Fock multi-configuration SCF theory for closed-shell molecules, and some calculations for diatomic and polyatomic molecules using spherical Gaussian functions were undertaken with this scheme [39]. At that time, the importance of using properly matched spinor basis functions had not been realized, so that the results can nowadays be regarded only as exploratory. The size of the basis set, roughly four times the dimension of the basis set for the corresponding nonrelativistic problem, increased the memory and processor requirements, emphasizing the need for a basis set design that minimized its dimensions and calling for an efficient program design. These considerations remain important today.

The current work on atoms demonstrates that it is important to choose basis sets that at least satisfy the "strict kinetic balance" criterion‡ to ensure the correct behaviour of the relativistic solutions in the nonrelativistic limit, to eliminate spurious solutions of the numerical problem and to maintain the formal symmetry between the positive and negative branches of the energy spectrum [8]. Kinetic balance provides a compact representation of the eigenfunctions, provided that the span of the exponent set is chosen properly. This focuses attention on the choice of exponents for the G-spinor sets that are most favoured for molecular calculations.

The commonest method of determining exponents uses some form of energy optimization [10, 41]. Each atom or molecule has to be done individually, and it is very often necessary to reoptimize the non-relativistic exponents in order to do the corresponding relativistic calculation. However, the use of a geometric sequence of exponents with a scale factor optimized on a nonrelativistic calculation has many attractive features which make it possible to construct highly efficient computer code [42]. These sets may be extended systematically without further optimization and may be transferred unchanged to a relativistic calculation. Many of the issues involved are discussed at considerable length in [44, Chapters 2, 3, 7 and 13].

Atomic calculations have revealed not only that it is necessary to use a relativistic one-body Hamiltonian to describe relativistic effects in heavy atoms properly but also that the introduction of magnetic and retardation terms into the electron-electron interaction operator generates sizable two-body relativity-correlation effects [29]. In addition to using basis sets of adequate realism to represent heavy atoms, molecular electronic structure calculations will in future have to incorporate the retarded electron-electron interaction, both at the self-consistent field level and in many-body calculations of correlation effects. This poses a major challenge, as today's codes - none of which can handle the retarded electron-electron interaction - can only be run on the most powerful computers available. Whilst we can still look forward to an increase in processing power and in memory, we should not rely on this alone to get us out of trouble. There is plenty to keep us busy for another 5 or 10 years.

References

[1] W.Greiner, "Relativistic Quantum Mechanics. Wave Equations", Springer-Verlag, Berlin (1987).

[2] I.P.Grant, *in* "Relativistic Effects in Atoms, Molecules and Solids", G.L.Malli, ed., p. 55, Plenum, New York (1983).

[3] P.Pyykkö, *Adv. Quantum Chem.* 11:353 (1978).

[4] K.S.Pitzer, *Acc. Chem. Res.* 12:271 (1979).

[5] P.Pyykkö and J.P.Desclaux, Acc. Chem. Res. 12:276 (1979).

[6] P.Pyykkö, "Relativistic Theory of Atoms and Molecules", (Lecture Notes in Chemistry **41**) Springer-Verlag, Berlin (1986).

[7] W.H.Furry, *Phys. Rev.* 81:115 (1951).

[8] I.P.Grant and H.M.Quiney, *in* "Advances in Atomic and Molecular Physics", Vol. 23, Sir David Bates, B.Bederson, eds., p.37, Academic Press, New York (1988).

‡ This means that the basis functions of large and small components are *individually* matched so that

$$\chi_{\rm small} = \sigma.{\rm p}\ \chi_{\rm large}$$

[9] "Relativistic, Quantum Electrodynamic, and Weak Interaction Effects in Atoms", (AIP Conference Proceedings 189), W.R.Johnson, P.J.Mohr and J.Sucher, eds., American Institute of Physics, New York (1989).
[10] R.McWeeny and B.T.Sutcliffe, "Methods of Molecular Quantum Mechanics", Chapter 8, Academic Press, New York (1969).
[11] R.McWeeny, "Methods of Molecular Quantum Mechanics", 2nd edition, Academic Press, London (1989).
[12] "Many-Body Methods in Quantum Chemistry", (Lecture Notes in Chemistry 52), U.Kaldor, ed., Springer-Verlag, Berlin (1989).
[13] H.M.Quiney, I.P.Grant and S.Wilson, *in* [12].
[14] S.Wilson, I.P.Grant and B.L.Gyorffy, eds., "The Effects of Relativity in Atoms, Molecules and the Solid State", Plenum Press, New York (1991).
[15] J.P.Desclaux, *in* "Relativistic Effects in Atoms, Molecules and Solids", G.L.Malli, ed., p. 213, Plenum Press, New York (1983)
[16] P.A.Christiansen, K.Balasubramanian and K.S.Pitzer, *J. Chem. Phys.* 76:5087 (1982).
[17] R.Arratia-Perez and C.Y.Yang, *J. Chem. Phys.* 83:4005 (1985).
[18] H.Sazdjian, *Ann. Phys. (N.Y.)* 136:136 (1981).
[19] H.Sazdjian, *Phys. Rev. D* 33:3401 (1986).
[20] A.O.Williams, *Phys. Rev.* 58:723 (1940).
[21] J.P.Desclaux, *Comput. Phys. Commun.* 9:31 (1975).
[22] I.P.Grant, B.J.McKenzie, P.H.Norrington, D.F.Mayers and N.C.Pyper, *Comput. Phys. Commun.* 21:207 (1980).
[23] B.J.McKenzie, I.P.Grant and P.H.Norrington, *Comput. Phys. Commun.* 21:233 (1980).
[24] K.G.Dyall, I.P.Grant, C.T.Johnson, F.A.Parpia and E.P.Plummer, *Comput. Phys. Commun.* 55:425 (1989).
[25] H.P.Kelly, *Phys. Rev.* 131:684 (1963).
[26] J.Sucher, *Int. J. Quantum Chem.* 25:3 (1984).
[27] W.Kutzelnigg, *Int. J. Quantum Chem.* 25:107 (1984).
[28] H.M.Quiney, I.P.Grant and S.Wilson, *J. Phys. B: At. Mol. Opt. Phys.* 22:L15 (1989).
[29] H.M.Quiney, I.P.Grant and S.Wilson, *J. Phys. B: At. Mol. Opt. Phys.* 23:L271 (1989).
[30] Y.Ishikawa, H.M.Quiney and G.L.Malli, *Phys. Rev. A* 43:3270 (1991).
[31] C.F.Fischer, *Int. J. Supercomputer Applics.* 1:5 (1991).
[32] W.R.Johnson, S.A.Blundell and J.Sapirstein, *Phys. Rev. A* 37:307 (1988).
[33] C.Bottcher and M.R.Strayer, *Ann. Phys. (N.Y.)* 175:64 (1987).
[34] I.P.Grant *in* [9, pages 235-253]
[35] Y.K.Kim, *Phys. Rev.* 17:154 (1967).
[36] G.Malli and J.Oreg, *Phys. Rev.* 63:830 (1975).
[37] G.Malli and J.Oreg, *Phys. Rev.* 68:529 (1979).
[38] O.Matsuoka, N.Suzuki, T.Aoyama and G.Malli, *J. Chem. Phys.* 73:1320 (1980).
[39] T.Aoyama, H.Yamakawa and O.Matsuoka, *J. Chem. Phys.* 73:1329 (1980).
[40] K.G.Dyall, K.Fægri, Jr., and P.R.Taylor, *in* [14].
[41] K.G.Dyall, P.R.Taylor, K.Fægri, Jr., and H.Partridge, *J. Chem. Phys.* (*in press*) (1991).
[42] H.M.Quiney *in* [43, pages 159-183, 185-199].
[43] R.G.Evans and S.Wilson, eds., "Supercomputational Science", Plenum Press, New York (1990).
[44] "Modern Techniques in Computational Chemistry: *MOTECC-90*™, E.Clementi, ed., ESCOM Science Publishers, Leiden (1990).

QUANTUM MECHANICAL SCATTERING THEORY FOR CHEMICAL REACTIONS

William H. Miller

Department of Chemistry, University of California, and
Chemical Sciences Division, Lawrence Berkeley Laboratory
Berkeley, California 94720

I. INTRODUCTION

Quantum mechanical reactive scattering theory provides the most complete description of a chemical reaction allowed by the basic laws of nature. Thus ever since the 1960's, when crossed molecular beam experiments opened the door to studying reactions at the most detailed state-to-state level, there has been intense interest and effort devoted to developing the theory to the practical stage that calculations can be carried out for real chemical reactions. These lectures review reactive scattering theory, in particular recent theoretical developments that have played a key role in the dramatic theoretical advances since ~1987. Two other recent reviews that I recommend are refs. 1 and 2.

II. WHY A BASIC SET VARIATIONAL APPROACH TO REACTIVE SCATTERING?

The theory of *reactive* scattering is more complicated than that for elastic and inelastic scattering[3a-c] because of *coordinates*, and different formulations of reactive scattering turn on how one deals with this coordinate problem. Figure 1 depicts the situation for the collinear A + BC → AB + C reaction. If one were treating only an inelastic scattering process (i.e. vibrational excitation),

$$A + BC(n) \rightarrow A + BC(n'), \tag{2.1}$$

then the standard Jacobi's coordinates (r_a, R_a) are the natural choice, and the coupled-channel expansion of the wavefunction has the form

$$\psi_{n_1} = \sum_n \phi_n(r_a) f_{n \leftarrow n_1}(R_a), \tag{2.2}$$

where $\{\phi_n\}$ are the (known) vibrational eigenfunctions for diatom BC and n_1 denotes the initial vibrational state. Substitution of this expansion into the Schrödinger equation leads to the standard coupled-channel equations for the unknown translational functions

Methods in Computational Molecular Physics, Edited by Stephen Wilson and
Geerd H.F. Diercksen, Plenum Press, New York, 1992

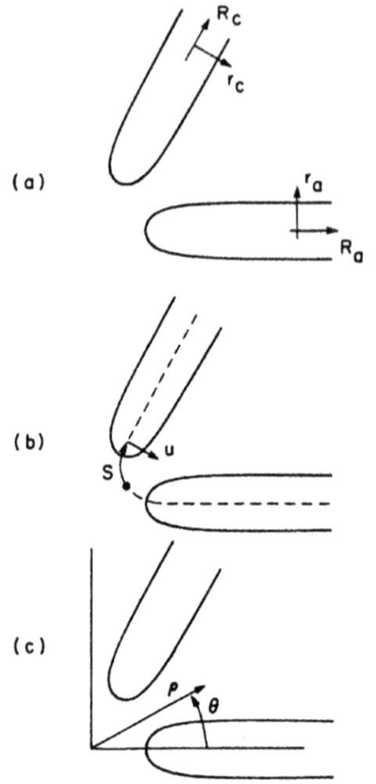

Figure 1 Schematic depiction of a collinear A + BC → AB + C potential energy surface and different ways of choosing coordinates.

(a) Jacobi coordinates for arrangement a(A + BC) and c(AB + C);

(b) reaction path ("natural collision") coordinates;

(c) hyperspherical (here simply polar) coordinates.

$$0 = \left(-\frac{\hbar^2}{2\mu}\frac{d^2}{dR_a^2} - E_n \right) f_{n\leftarrow n_1}(R_a) + \sum_{n'} V_{n,n'}(R_a) f_{n'\leftarrow n_1}(R_a), \tag{2.3a}$$

$$V_{n,n'}(R_a) = \int dr_a \phi_n(r_a)(V-v)\phi_{n'}(r_a); \tag{2.3b}$$

$E_n = E - \varepsilon_n$ is the translational energy for channel n, where E is the (fixed) total energy and ε_n the vibrational energy for state n, V is the total potential energy function, and v the potential for the isolated diatom BC.

The "coordinate problem" referred to above for reactive scattering is that the Jacobi coordinates (r_a, R_a) that are natural for describing the reactants A+BC are not appropriate for describing the products, AB+C. There are several ways to deal with this situation, but most of the recent progress in reactive scattering has been based on the formulation[4] in which the Jacobi coordinates for the various "arrangements" (i.e. A+BC, AB+C, AC+B) are all used simultaneously. For the collinear case of Figure 1, for example, the expansion for the wavefunction in this approach is

$$\psi_{\gamma_1 n_1} = \sum_n \phi_n^a(r_a) f_{an\leftarrow\gamma_1 n_1}(R_a) + \sum_n \phi_n^c(r_c) f_{cn\leftarrow\gamma_1 n_1}(R_c) \tag{2.4}$$

where γ = a(A+BC), b(B+AC), or c(C+AB) labels the arrangement of the atoms, and $\{\phi_n^a\}$ and $\{\phi_n^c\}$ are the vibrational eigenstates of diatoms BC and AB, respectively. Note that there are only two independent coordinates (degrees of freedom) in Eq. (2.4) for the collinear case shown in Figure 1; i.e., r_a and R_a are functions of r_c and R_c, or vice-versa (specifically, they are linear combinations of each other).

The philosophy of this approach is similar to that in quantum chemistry of using *multicenter* (LCAO = linear combination of atomic orbitals) expansions for molecular orbitals. For a diatomic molecule, for example, the molecular orbital $\chi(\mathbf{r})$ for an electron is expanded in basis functions utilizing the coordinates of the electron with respect to *both* nuclear centers,

$$\chi(r) = \sum_i a_i \phi_i^a(\mathbf{r}_a) + \sum_i b_i \phi_i^b(\mathbf{r}_b), \tag{2.5}$$

where $\mathbf{r}_a$ are the coordinates of the electron with respect to nucleus a and $\mathbf{r}_b$ those with respect to nucleus b. (Note that there are only three independent coordinates in Eq. (2.5), i.e., $\mathbf{r}_a$ is a function of $\mathbf{r}_b$, or vice-versa, specifically, $\mathbf{r}_a = \mathbf{r}_b + \mathbf{R}_b - \mathbf{R}_a$, where $\mathbf{R}_a$ and $\mathbf{R}_b$ are the coordinates of the nuclei.)

One may also think of this approach to reactive scattering as a generalization of the standard description of electron scattering.[5] Thus in Figure 1, consider the case that atom B = H^+ (a proton), and A and C are electrons, i.e. the collinear version of electron-hydrogen atom scattering. In this case $m_B \gg m_A, m_c$, and it is clear that the two translational coordinates coincide with the two interparticle coordinates, i.e. $R_c = r_a$ and $R_a = r_c$, and that the two sets of terms in Eq. (2.4) are related to one another simply by *exchange* of the electrons. (By symmetry, the two sets of terms are the same, with a + or - combination corresponding to the singlet and triplet case, respectively.) Thus even if one did not know that the (spatial) two-electron wavefunction should be symmetric or antisymmetric upon exchange of the two electrons, the fact that the electrons can actually interchange by virtue of the collision (a "chemical reaction") requires that the wavefunction include both sets of terms in Eq. (2.4). (The chemical reaction H+IH $\rightarrow$ HI+H is a molecular case very close to the e - H atom limit.) The general chemical reaction is more complicated than the electron scattering case because the finite mass of all atoms makes the relation between the various sets of Jacobi coordinates more complicated than simply exchanging them, but the basic idea is the same.

The expansion of the wavefunction in Eq. (2.4) is also essentially the same basic idea as the "resonating group model" (RGM) used in nuclear physics.[6] The different sets of Jacobi coordinates define different "groups" (or groupings) of atoms, and the fact that the wavefunction is a linear combination of these different terms allows for "resonance" (i.e. coupling, interaction) between them if there are non-zero matrix elements of the Hamiltonian ("resonance integrals") connecting them.

Equation 2.4 is thus a natural and efficient way to represent a reactive scattering wavefunction, but it introduces the complexity that the coupling between terms corresponding to different arrangements are nonlocal, exchange type interactions. The coupled-channel equations Eq. (2.3a) are thus generalized as follows:[4]

$$0 = \left(-\frac{\hbar^2}{2\mu_a}\frac{d^2}{dR_a^2} - E_n^a \right) f_{an\leftarrow\gamma_1 n_1}(R_a) + \sum_{n'} V^a_{n,n'}(R_a) f_{an'\leftarrow\gamma_1 n_1}(R_a)$$

$$\sum_{\substack{\gamma \\ \gamma\neq a}} \sum_{n'} \int dR_\gamma V_{an,\gamma n'}(R_a,R_\gamma) f_{\gamma n'\leftarrow\gamma_1 n_1}(R_\gamma). \tag{2.6}$$

The exchange interaction (the last term in the above equation), which couples states of different arrangements, is analogous to electron exchange interactions in quantum chemistry that arise from matrix elements in which the electron coordinates have been permuted (i.e.

exchanged). The coupled *integro*-differential equations for the translational functions, Eq. (2.6), are thus analogous to the Hartree-Fock equations of electronic structure theory, and as such they cannot be solved by finite difference algorithms.

Indeed, it is the presence of the exchange interaction in this formulation of reactive scattering that until recently has stymied this approach. Wolken and Karplus[7] made some early attempts using it, but these were not completely successful. It has ultimately become clear that the most satisfactory way of dealing with exchange is analogous to what quantum chemists do in the Hartree-Fock problem, namely to expand the unknown wavefunctions in a basis set and determine the expansion coefficients via a variational principle.

III. THE S-MATRIX VERSION OF THE KOHN VARIATIONAL PRINCIPLE

The Kohn variational principle[8,9] is essentially the Rayleigh-Ritz variational principle familiar from quantum eigenvalue problems, generalized to deal with scattering boundary conditions; i.e., the basic functional to be varied is $\langle\tilde{\psi}|H-E|\tilde{\psi}\rangle$, where $\tilde{\psi}$ is a trial function. We will employ S-matrix type boundary conditions[10] for the scattering wavefunctions and comment below on other possibilities. All of the relevant features are illustrated by simple s-wave potential scattering, so the methodology will first be described with regard to this problem, and the generalization to multichannel rearrangement scattering given at the end.

The Hamiltonian is of the standard form

$$H = \frac{-\hbar^2}{2\mu}\frac{d^2}{dr^2} + V(r), \tag{3.1}$$

where $V(r) \to 0$ as $r \to \infty$. The S-matrix version of the Kohn variational approximation to the S-matrix (at energy E) can be stated as[11]

$$S = \text{ext}[\tilde{S} + \frac{i}{\hbar} <\tilde{\psi}|H-E|\tilde{\psi}>], \tag{3.2}$$

where $\tilde{\psi}(r)$ is a trial wavefunction that is regular at $r = 0$ and has asymptotic form (as $r \to \infty$)

$$\tilde{\psi}(r) \sim -e^{-ikr}v^{-1/2} + e^{ikr}v^{-1/2}\,\tilde{S}, \tag{3.3}$$

where $v = \hbar k/\mu$ is the asymptotic velocity. (Note: The convention is used throughout this paper that the wavefunctions in the bra symbol $<|$ in bra-ket matrix element notation are <u>not</u> complex conjugated.) "ext" in Eq. (3.2) means that the quantity in square brackets is to be extremized by varying any parameters in $\tilde{\psi}(r)$. (Note that for a given trial function, Eq. (3.2) may also be viewed as the distorted wave Born approximation, where $\tilde{\psi}$ is the distorted wave.)

A linear variational form[11] is taken for the trial function $\tilde{\psi}(r)$,

$$\tilde{\psi}(r) = -u_0(r) + \sum_{t=1}^{N} u_t(r)c_t, \tag{3.4}$$

where $u_0(r)$ is a function that is regular at $r = 0$ and has the asymptotic form (as $r \to \infty$)

$$u_0(r) \sim e^{-ikr}v^{-1/2}. \tag{3.5}$$

A simple choice for $u_0(r)$ is

$$u_0(r) = f(r)e^{-ikr}v^{-1/2}, \tag{3.6}$$

where f(r) is a smooth cut-off function,

$$f(r) \to 0, r \to 0$$

$$f(r) \to 1, r \to \infty, \tag{3.7}$$

such as $f(r) = 1-e^{-\alpha r}$. The function $u_1(r)$ is

$$u_1(r) = u_0(r)^* \sim e^{ikr}\, v^{-1/2}, \tag{3.8}$$

and the basis functions $\{u_t(r)\}$, t = 2, ..., N are real, square-integrable functions. The coefficients $\{c_t\}$, t = 1, ..., N in Eq. (3.4) are the variational parameters in $\tilde{\psi}$.

With $\tilde{\psi}$ of Eq. (3.4) substituted into Eq. (3.2) and the coefficients $\{c_t\}$ varied to extremize it, one obtains the following expression for the S-matrix

$$S = \frac{i}{\hbar}(M_{0,0} - \mathbf{M}_0^T \cdot \mathbf{M}^{-1} \cdot \mathbf{M}_0), \tag{3.9}$$

where $M_{0,0}$ is a 1x1 "matrix", $\mathbf{M}_0$ a Nx1 matrix and $\mathbf{M}$ an NxN matrix,

$$M_{00} = \langle u_0|H-E|u_0\rangle \tag{3.10a}$$

$$(\mathbf{M}_0)_t = \langle u_t|H-E|u_0\rangle \tag{3.10b}$$

$$(\mathbf{M})_{t,t'} = \langle u_t|H-E|u_{t'}\rangle, \tag{3.10c}$$

for t,t' = 1, ..., N, and where "T" denotes matrix transpose. Note that all matrix elements involving the unbounded basis functions u_0 and u_1 exist because

$$\lim_{r\to\infty} (H-E)\begin{Bmatrix} u_0(r) \\ u_1(r) \end{Bmatrix} = 0. \tag{3.11}$$

At this point it is useful to compare the above procedure with the K-matrix version of the Kohn variational principle that has typically been used in the past.[12] The Kohn functional for K is

$$K[\tilde{\psi}] = \tilde{K} - \frac{2}{\hbar}\langle\tilde{\psi}|H-E|\tilde{\psi}\rangle, \tag{3.12a}$$

where here the (real) trial function $\tilde{\psi}$ has the asymptotic form

$$\tilde{\psi}(r) \sim \sin(kr)v^{-1/2} + \cos(kr)v^{-1/2}\tilde{K}. \tag{3.12b}$$

One then proceeds precisely as above, the difference being that now the "free" functions $u_0(r)$ are

$$u_0(r) = f(r)\sin(kr)v^{-1/2} \tag{3.13a}$$

$$u_1(r) = f(r)\cos(kr)v^{-1/2} \tag{3.13b}$$

(The cut-off function f(r) is actually not required in Eq. (3.13a) for u_0) The arithmetic is also similar, and the variational result obtain for the K-matrix is similar to Eq. (3.9)

$$K = \frac{-2}{\hbar}(M_{0,0} - \mathbf{M}_0^T \cdot \mathbf{M}^{-1} \cdot \mathbf{M}_0), \tag{3.14}$$

where the M-matrix elements are the same as those in Eq. (3.10) but here with the real functions u_0 and u_1 of Eq. (3.13). And now the problem appears: the matrix $H_{t,t'}$ in Eq. (3.10c) is real-symmetric in the present case, so that its eigenvalues are real. Every time that the energy E is equal to one of these eigenvalues, the matrix inverse $\mathbf{M}^{-1}$ in Eq. (3.14) is singular; this may happen as the energy E is varied, or at fixed E if non-linear parameters in the basis set $\{u_t\}$ are varied to cause one of the eigenvalues to pass through the value E. These "Kohn anomalies", or "false resonances", have been a plague of the Kohn variational principle ever since they were discovered by Schwartz in 1961.[12] In the limit of an infinitive basis they become infinitely narrow, and thus unobservable, but they have been a serious problem in practical calculations (which necessarily use finite basis sets).

In contrast, the S-matrix version[10,11] of the Kohn variational method has no anomalous singularities because the matrix of H in Eq. (3.10c) is complex-symmetric (because the function u_1 of Eq. (3.8) is complex and functions inside the bra symbol <| are not complex-conjugated). In fact, the condition that Eq. (3.9) is singular, namely

$$\det(\mathbf{M}) = \det[u_t|H-E|u_{t'}\rangle] = 0, \tag{3.15a}$$

t,t' = 1, ..., N, is the secular equation for eigenvalues of the Schrödinger equation

$$(H-E)\psi(r) = 0, \tag{3.15b}$$

with boundary condition (as $r \to \infty$)

$$\psi(r) \propto e^{ikr}; \tag{3.15c}$$

i.e., Eq. (3.15) is the expression that has been used before[13] for determining Siegert[14] eigenvalues, the complex energies that are the (physically correct) complex poles of the S-matrix which characterize the positions and widths of scattering resonances. Eq. (3.9) is thus singular only where it is supposed to be singular.

The S-matrix Kohn approach also allows one to identify a corresponding basis set approximation to matrix elements of the full outgoing wave Green's function $G^+(E) \equiv (E+i\varepsilon-H)^{-1}$. This is[10]

$$\langle a|G^+(E)|b\rangle = -\sum_{t,t'=1}^{N} \langle a|u_t\rangle(\mathbf{M}^{-1})_{t,t'}\langle u_{t'}|b\rangle, \tag{3.16}$$

where $\mathbf{M}$ is as above, Eq. (3.10c), and |a> and |b> are any square-integrable functions. Note that the complex-symmetric structure of the matrix $\mathbf{M}$ is the same as that in complex scaling/coordinate rotation theory,[15-18] and for the same reasons. If the functions |a> and |b> are real, then Eq. (3.16) leads to a useful way for calculating matrix elements of the microcanonical density operator,

$$\langle a|\delta(E-H)|b\rangle = -\pi^{-1}\mathrm{Im}\langle a|G^{+}(E)|b\rangle. \qquad (3.17)$$

In actual calculations for the S-matrix, Eq. (3.9), one does not wish to carry out numerical calculations with the complex symmetric matrix **M**. This can be avoided by the usual partitioning methods, so that Eq. (3.9) can be written in the equivalent form

$$S = \frac{i}{\hbar}(B - C \cdot B^{*-1} \cdot C), \qquad (3.18)$$

where B and C are the 1x1 "matrices"

$$B = M_{0,0} - \mathbf{M}_0^T \cdot \mathbf{M}^{-1} \cdot \mathbf{M}_0 \qquad (3.19a)$$

$$C = M_{1,0} - \mathbf{M}_0^{*T} \cdot \mathbf{M}^{-1} \cdot \mathbf{M}_0, \qquad (3.19b)$$

where $M_{0,0}$, $\mathbf{M}_0$, and **M** are as before, Eq. (3.10), except that t,t' = 2, ..., N (i.e., only the real basis functions), and

$$M_{1,0} = \langle u_0^*|H-E|u_0\rangle. \qquad (3.19c)$$

Here the matrix $(\mathbf{M})_{t,t'}$, t,t' = 2, ..., N is real and symmetric, and thus more easily dealt with. (One can readily verify that a value of E for which det(**M**) = 0 does <u>not</u> lead to a singularity in Eq. (3.18).

Finally, for general multichannel rearrangement scattering,[4,19] let (q_γ, r_γ) denote the internal coordinates and radial scattering (i.e., translational) coordinate for arrangement γ; $\{\phi_n^\gamma(q_\gamma)\}$ are the asymptotic channel eigenfunctions for the internal degrees of freedom. Eqs. (3.18) and (3.19) generalize as follows:

$$\mathbf{S} = \frac{i}{\hbar}(\mathbf{B} - \mathbf{C}^T \cdot \mathbf{B}^{*-1} \cdot \mathbf{C}), \qquad (3.20a)$$

where **S**, **B**, and **C** are "small" square matrices, the dimension of the number of open channels, e.g., $\mathbf{S} = [S_{n\gamma, n'\gamma'}]$, etc. **B** and **C** are given by

$$\mathbf{B} = \mathbf{M}_{0,0} - \mathbf{M}_0^T \cdot \mathbf{M}^{-1} \cdot \mathbf{M}_0 \qquad (3.20b)$$

$$\mathbf{C} = \mathbf{M}_{1,0} - \mathbf{M}_0^{*T} \cdot \mathbf{M}^{-1} \cdot \mathbf{M}_0, \qquad (3.20c)$$

where $\mathbf{M}_{0,0}$ and $\mathbf{M}_{1,0}$ are also "small" square matrices

$$(\mathbf{M}_{0,0})_{n\gamma,n'\gamma'} = \langle u_{0n}^\gamma \phi_n^\gamma|H-E|u_{0n'}^{\gamma'} \phi_{n'}^{\gamma'}\rangle \qquad (3.21a)$$

$$(\mathbf{M}_{1,0})_{n\gamma,n'\gamma'} = \langle u_{0n}^{\gamma *} \phi_n|H-E|u_{0n'}^{\gamma'} \phi_{n'}^{\gamma'}\rangle; \qquad (3.21b)$$

$u_{0n}^\gamma(r_\gamma)$ is a function regular at $r_\gamma = 0$ and with asymptotic form (as $r_\gamma \to \infty$),

$$u_{0n}^\gamma(r_\gamma) \sim e^{-ik_{n\gamma} r_\gamma}/v_{n\gamma}^{1/2}.$$

M is a "large" by "large" real symmetric matrix in the composite space, internal plus translational,

$$(\mathbf{M})_{tn\gamma,t'n'\gamma'} = \langle u^{\gamma}_{tn}\phi^{\gamma}_{n}|H-E|u^{\gamma'}_{t'n'}\phi^{\gamma'}_{n'}\rangle, \tag{3.21c}$$

where $\{u^{\gamma}_{tn}(r_{\gamma})\}$ is a square integrable basis (that need not depend on n - i.e. the same translational basis can be used for every channel). $\mathbf{M}_0$ is a "large" by "small" rectangular matrix

$$(\mathbf{M}_0)_{tn\gamma,n'\gamma'} = \langle u^{\gamma}_{tn}\phi^{\gamma}_{n}|H-E|u^{\gamma'}_{0n'}\phi^{\gamma'}_{n'}\rangle. \tag{3.21d}$$

Only open channels $\{n\gamma\}$ are included in the matrices $\mathbf{M}_{00}$, $\mathbf{M}_{1,0}$, and the "small" dimension of $\mathbf{M}_0$, while open and closed channels are required in the matrix $\mathbf{M}$ and the "large" dimension of $\mathbf{M}_0$.

Eqs. (3.20) - (3.21) thus express the S-matrix for reactive scattering in an extremely straight-forward manner: one chooses basis functions, computes matrix elements of the Hamiltonian, and then does a standard linear algebra calculation. Specifics related to A+BC → AB+C reaction in three-dimensional space - i.e. angular momentum coupling, elimination of the three Euler angles for overall rotation, conservation of total angular momentum, identical atom symmetry, etc. - are given in refs. 4 and 19.

IV. ENHANCEMENTS OF THE METHODOLOGY

Note that the generic form of the S-matrix Kohn trial function in the general multichannel case above is

$$\Psi_{\gamma_1 n_1} = -\Phi_{\gamma_1 n_1} + \sum_{\gamma,n} \Phi^{*}_{\gamma n}\, C_{\gamma n,\gamma_1 n_1} + \sum_{i} \chi_i\, C_{i,\gamma_1 n_1}, \tag{4.1}$$

where $\Phi_{\gamma n}(q_\gamma,r_\gamma)$ is an asymptotically incoming wave in channel γn,

$$\lim_{r_\gamma\to\infty} \Phi_{\gamma n}(q_\gamma,r_\gamma) = \phi^{\gamma}_{n}(q_\gamma)\, e^{-ik_{\gamma n}r_\gamma}/v_{\gamma n}^{1/2}, \tag{4.2}$$

$\Phi_{\gamma n}^{*}$ is the corresponding asymptotically outgoing wave, and $\{\chi_i\}$ is a L^2 (square integrable) basis that spans the interaction region. In Eqs. (3.20) - (3.21) above, the L^2 basis $\{\chi_i\}$ was chosen to be the direct product of the various channel eigenfunctions and a translational basis,

$$\chi_i = \phi^{\gamma}_{n}(q_\gamma)\, u_t(r_\gamma), \tag{4.3}$$

so that $i \equiv \gamma nt$, but this is not necessary. The $\{\chi_i\}$ basis can be chosen in a variety of ways, using a variety of coordinates. There is even some flexibility in the choice of the "free" wavefunctions $\{\Phi_{\gamma n}\}$, the only requirement being the asymptotic form in Eq. (4.2). Here we note several of these possibilities that have been tried so far.

a. Elimination of Exchange in the Free-Free and Bound-Free Matrix Elements

This was a simplification that was realized early on.[20] By proper choice of the cut-off function f(r) in the "free" functions u_0 and u_1 [cf. Eqs. (3.6) - (3.7)], one can insure that the matrix elements involving the free functions - $\mathbf{M}_{0,0}$, $\mathbf{M}_{1,0}$, and $\mathbf{M}_0$ in Eq. (3.21) - vanish between different arrangements; i.e.,

$$(\mathbf{M}_{0,0})_{n\gamma,n'\gamma'} = \delta_{\gamma,\gamma'}(\mathbf{M}^{\gamma}_{0,0})_{n,n'}, \tag{4.4}$$

and similarly for $\mathbf{M}_{1,0}$ and $\mathbf{M}_0$. This is quite important practically because it is these matrix elements which must be re-computed at each scattering energy E (because the "free" functions u_0 and u_1 depend on E), and furthermore, the exchange integrals are more difficult to evaluate numerically that the direct matrix elements (those with $\gamma = \gamma'$). With this simplification, the only non-zero exchange matrix elements are the bound-bound ones, Eq. (3.21c), but these are independent of the energy E, so that they can be computed once and then used to compute the S-matrix at many values of E.

b. Basis Set Contraction of the L^2 Basis

This is an obvious procedure borrowed from *ab initio* electronic structure theory. One initially evaluates the Hamiltonian in an L^2 basis of "simple", easy-to-use functions, and then transforms to a better basis, e.g., by diagonalizing some simpler reference problem. For the L^2 basis actually used for the calculation, one then chooses fewer of these good basis functions than the original number of the simple functions (cf. the transformation from many atomic orbitals to a fewer number of molecular orbitals). A quasi-adiabatic contraction[21] of the channel eigenfunctions has been especially easy to use, and in applications to the $H+H_2$ and $D+H_2$ reactions it approximately halves the number of basis functions necessary to achieve convergence. Since the computational time is proportioned to the cube of the size of the matrix **M** that is inverted in Eq. (3.20), this reduces the computational time by almost an order of magnitude.

Another important example of a contraction - or actually just a clever choice of basis - is to use the <u>body-fixed, or helicity representation</u> for the L^2 basis functions,[22] and the <u>space-fixed, or ℓ-representation</u> for the free functions. Thus the channel index n in Eq. (3.20) - (3.21) is $n \equiv (v,j,K)$ for the L^2 basis, where (v,j) are the vibrational and rotational quantum numbers for the diatom, and K is the projection quantum number for the diatom rotation with respect $\vec{r}_\gamma$ (the atom to diatom center of mass) as quantization axis; this is the body-fixed, or helicity representation. For the free functions in Eq. (3.20) - (3.21), the channel index is $n \equiv (v,j,\ell)$, where v and j are the same as before, and ℓ is the orbital angular momentum quantum number for the relative motion between the atom and the center of mass of the diatom; this is the space-fixed, or ℓ-representation. For the large separations between atom and diatom, the Hamiltonian trends to be more nearly diagonal in the ℓ-representation (because the centrifugal potential, $\ell^2/2\mu r^2$, is larger than the interaction potential for large r), but for small distances it is more nearly diagonal in the K-representation (because here the interaction is typically larger than the centrifugal potential). The transformation element $\langle k|\ell\rangle$ relating the ℓ- and k- representations is a Clebsch-Gordan coefficient.[4] Application to the $F+H_2 \rightarrow HF+H$ reaction has shown the enormous utility of this approach.[22]

Finally, an even more novel choice of L^2 basis is a point-wise basis, i.e., a pseudo-spectral, or discrete variable, or collocation representation.[23,24] This has the advantage of great flexibility, i.e., one has only to choose <u>points</u> that span the interaction region (of all arrangements). The other great advantage is that no integrals are required to obtain the matrix of the Hamiltonian in the L^2 representation. Initial applications have shown the feasibility of these approaches, and they are currently the object of intense research.

c. Distorted Free Functions

Another strategy to keep the L^2 basis as small as possible is to use better free functions,

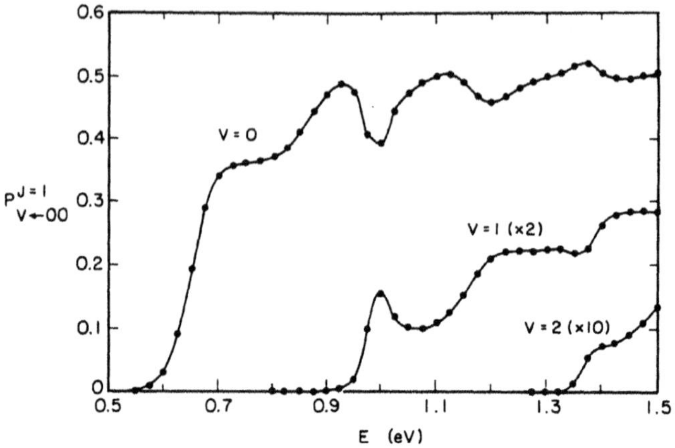

Figure 2 Reaction probability for $H + H_2(v = j = 0) \rightarrow H_2(v', \text{all } j') + H$, for total angular momentum $J = 1$, as a function of total energy E. The results for v' = 1 and v' = 2 have been multiplied by factors of 2 and 10, respectively, for convenience in displaying them.

i.e., functions u_0 and u_1 that describe the wavefunction accurately to smaller values of r_γ than the simple plane waves discussed above. For example, the function $\Phi_{\gamma n}$ of Eq. (4.1) and (4.2) can be chosen as

$$\Phi_{\gamma n}(q_\gamma, r_\gamma) = \sum_{n'} \phi^\gamma_{n'}(q_\gamma)\, f(r_\gamma)\, u^\gamma_{n'\leftarrow n}(r_\gamma)\ , \tag{4.5a}$$

where the translational function matrix $u^\gamma_{n'\leftarrow n}$ is the solution of the inelastic coupled channel Schrödinger equation for channel γ,

$$0 = \left(-\frac{\hbar^2}{2\mu_a}\frac{d^2}{dR_a^2} - E_{\gamma n'}\right) u^\gamma_{n'\leftarrow n}(r_\gamma) + \sum_{n''} V^\gamma_{n',n''}(r_\gamma) u^\gamma_{n''\leftarrow n}(r_\gamma)\ , \tag{4.5b}$$

with asymptotic boundary condition

$$\lim_{r_\gamma\to\infty} u^\gamma_{n'\leftarrow n}(r_\gamma) = \delta_{n',n}\, e^{-ik_{\gamma n} r_\gamma}\, v_{\gamma n}^{-1/2}\ ; \tag{4.5c}$$

this radial function is irregular as $r_\gamma \to 0$, but the cut-off function f(r) regularizes it. These "distorted" free functions[25] describe inelastic scattering separately in each arrangement, so that the L^2 basis can be restricted more specifically to the very short range exchange region (and thus be smaller). The free-free and bound-free matrix elements are also considerably simplified with these distorted free functions:[19]

$$(\mathbf{M}_{0,0})_{\gamma n,\gamma' n'} = \delta_{\gamma,\gamma'} \frac{\hbar^2}{2\mu_\gamma} \sum_{n''} \left\langle u^\gamma_{n''\leftarrow n} | f'^2 | u^\gamma_{n''\leftarrow n'} \right\rangle \tag{4.6a}$$

$$(\mathbf{M}_{1,0})_{\gamma n,\gamma' n'} = \delta_{\gamma,\gamma'} \left[\frac{i\hbar}{2}\delta_{n,n'} + \frac{\hbar^2}{2\mu_\gamma} \sum_{n''} \left\langle u^{\gamma *}_{n''\leftarrow n} | f'^2 | u^\gamma_{n''\leftarrow n'} \right\rangle\right] \tag{4.6b}$$

$$(M_0)_{t\gamma n,\gamma' n'} = -\delta_{\gamma,\gamma'} \frac{\hbar^2}{2\mu_\gamma} \left\langle u_{t\gamma n} | f'' + 2f' \frac{d}{dr_\gamma} | u^\gamma_{n\leftarrow n'} \right\rangle \tag{4.6c}$$

V. SOME APPLICATIONS

Accurate differential and integral cross sections have been calculated (in full three-dimensional space) with the methodology discussed above for the reactions $H + H_2(\text{para}) \rightarrow H_2(\text{ortho}) + H$,[26,27] $D + H_2 \rightarrow HD + H$,[19] and $F + H_2 \rightarrow HF + H$,[28] for a wide range of energies. Here I will review only the results for the $H + H_2$ reaction that has been of special interest because of experimental reports of resonance structure in the energy dependence of the integral cross section.

Figure 2 shows the reaction probability for $H+H_2(v = j = 0) \rightarrow H_2(v', \text{all odd } j') + H$ as a function of energy E, for various final vibrational states v', for one value of total angular momentum J. The resonance feature at $E \cong 1.0$ eV, due to a short-lived collision complex, is readily apparent. The integral cross section, however, is a sum of reaction probabilities over all values of J (which requires values up to ~20 for convergence in the present case),

$$\sigma_{v'j'\leftarrow vj}(E) = \pi[k_{vj}^2(2j+1)]^{-1} \sum_{J=0} (2J+1) \sum_{\ell,\ell'} |S^J_{v'j'\ell',vj\ell}(E)|^2, \tag{5.1}$$

and since the position of the resonance varies with J the resonance feature is averaged out (cf. inhomogeneous broadening) in the integral cross section.

Figure 3 shows the theoretical cross sections for final states v' = 1, j' = 1 and 3, compared to the experimental results of Nieh and Valentini.[29] Through the absolute magnitudes of the theoretical and experimental cross sections are in good agreement, the theoretical results do not show the resonance-like structure seen experimentally. This

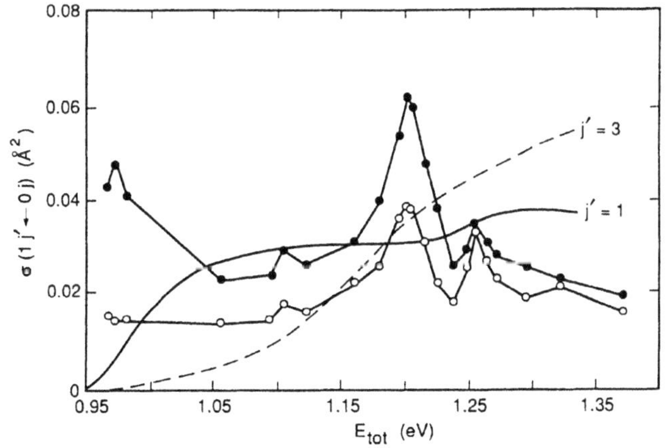

Figure 3 Integral cross section for the para → ortho reaction $H + H_2(v = 0,j) \rightarrow H_2(v' = 1,j') + H$ as a function of total energy. The solid and dashed lines are the theoretical results of ref. 26 for j' = 1 and 3, respectively; they both are the rotational ground state (j = 0). The solid and open circles are the corresponding experimental results of ref. 29 for which the initial rotational state is a Boltzmann distribution of para states (52% j = 0, 48% j = 2).

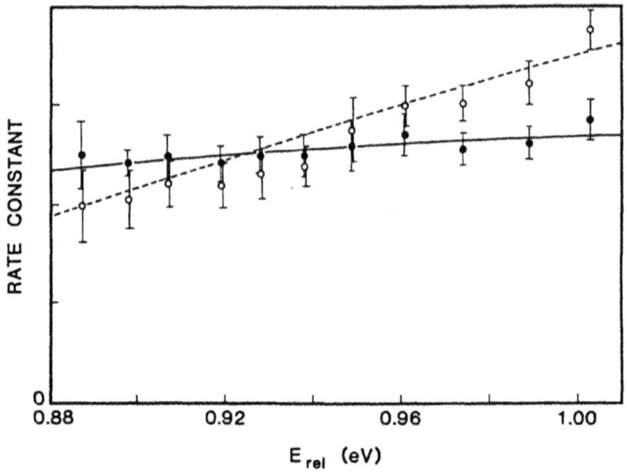

Figure 4 Same as Figure 3 except the points are the experimental results of ref. 30. (Precisely speaking, the experiments in both Figures 3 and 4 measure rate constants.) These experimental values have been normalized to the theoretical results (which have been appropriately averaged over the experimental energy resolution), though the j' = 1 and 3 values are absolute relative to one another; see ref. 30 for more details. Since E_{tot} of Figure 3 is related to E_{rel} here by (for j = 0) $E_{tot} \cong E_{rel} + 0.27$ eV, the energy scale here covers the range $E_{tot} \cong 1.15 - 1.28$ eV in Figure 3.

disagreement caused considerable concern until the very recent experiments[30] by Zare's group which agree well with the theoretical cross sections. These more recent experimental results are shown in Figure 4. Though over a narrower energy range, and only relative cross sections, they nevertheless cover the critical region where the structure was reported in the earlier experiments; they see no such structure and are in quantitative agreement with the theoretical results.

More recently, calculations have been carried out for the differential cross section over a range of energies,

$$\sigma_{v'j'K'\leftarrow vjK}(\theta,E) = \left|(2ik_{vj})^{-1} \sum_J (2J+1)\, d^J_{K'K}(\theta) S_{v'j'K',vjK}(J,E)\right|^2 . \tag{5.2}$$

Figure 5 shows these results[31] for the transitions v = j = 0 → v' = 0, 1, j' = 1, 3, 5. At low energies the cross section is strongly peaked in the backward direction (θ = 180°), but at the highest energies shown a peak begins to emerge in the forward scattering direction (θ = 0°). This is very typical for reactions with an activation barrier.

Of particular interest is a "ridge" along a line in the E-θ plane (seen most distinctly for final state v' = 0, j' = 1 in Figure 5). This feature in the cross section is attributed to the resonance relation $E = E_r(J)$, where $E_r(J)$ is the resonance energy as a function of total angular momentum J. I.e., the position of the resonance in Figure 2 for J = 1, E_r, is different for different values of J; $E_r(J)$ is essentially the rotational energy of the collision complex, and in fact the J-dependence is well described as

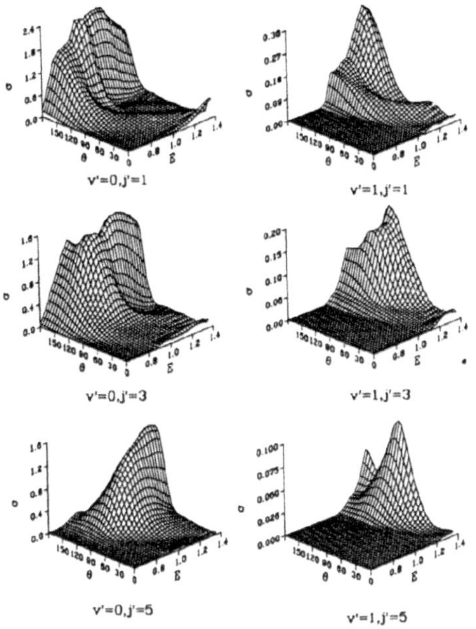

Figure 5 Three-dimensional plot of the differential cross sections $\sigma(\theta,E)$ as a function of θ and E, for the transitions $H + H_2(v = j = 0) \rightarrow H_2(v', j') + H$, for various final states (v', j').

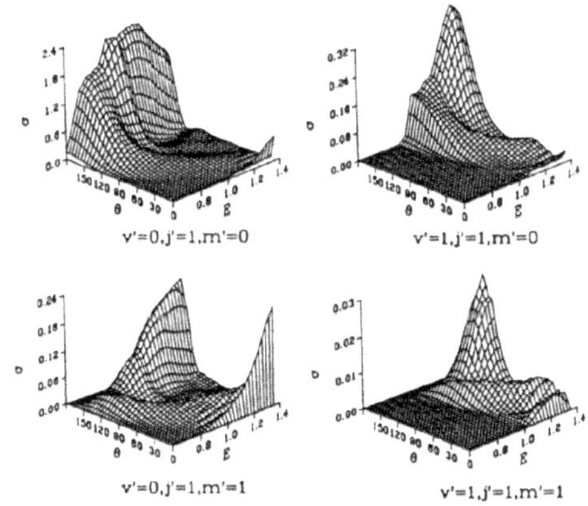

Figure 6 Three-dimensional plot of the completely state resolved differential cross section for $H + H_2(v = j = 0) \rightarrow H_2(v', j' m') + H$, as a function of θ and E. Note that the σ scale for m' = 1 is an order of magnitude smaller than that for m' = 0 final states.

$$E_r(J) = \text{constant} + BJ(J+1), \qquad (5.3)$$

where the rotation constant B~ 10-12 cm^{-1}, normal for an H•••H•••H geometry. Within a semiclassical analysis, furthermore, J is a function of θ by virtue of the stationary phase relation

$$\Theta(J) = \theta, \qquad (5.4)$$

where Θ(J) is the classical deflection function; i.e., by observing the differential cross section as a function of θ one to some extinct unfolds the sum over J that is inherent in the integral cross section (cf. Eq. 5.1). The ridge, or peak in the cross section along the line in the E-θ plane is thus identified as the resonance relation $E = E_r(J(\theta))$, with J(θ) determined by Eq. 5.4.

Finally, this is feature is seen even more clearly in the energy-dependent differential cross sections that are not summed over final helicity quantum number K', as shown in Figure 6 (here m ≡ K). These cross sections correspond to Eq. (5.2) with all possible initial and final quantum numbers specified and are thus the most complete description of the bimolecular reaction that is possible. The resonance ridge in the K'(≡ m') = 1 cross section for v' = 0, j' = 1 is stronger than that in the K'(≡ m') = 0 cross section, which is plausible because one expects most of the direct (i.e., non-resonant) scattering to appear in the ΔK = 0 channel (i.e., K' = 0 since K = 0 by virtue of the fact that j = 0).

VI. CONCLUDING REMARKS

One may thus say that the general prescription for carrying out quantum mechanical reactive scattering calculations is now quite clear but that one has just begun down the road of practical developments that can be made. Section IV describes some of these methodological developments - all of which can be characterized as how to choose basis functions more efficiently - but one expects to see many additional contributions. It very much reminds one of the early days of quantum chemistry, and it is clear already that many ideas developed in that field can usefully be brought to bear on quantum reactive scattering.

One major difference with quantum chemistry is that matrix elements of the Hamiltonian (i.e., integrals) will never be able to be evaluated analytically in the present case as they can be in electronic structure theory. (This is because the potential energy surface is a very complicated function of the nuclear coordinates, different for every system, and not simply Coulomb's law.) For reactions more complex than an atom-diatom system, integral evaluation will likely be a major bottleneck. This is one reason for interest in developing the pointwise representations noted in Section IVb. Some of you at this Advanced Study Institute, however, may find other ways to solve these problems. I wish you luck!

ACKNOWLEDGMENT

I am happy to acknowledge the significant contributions of my co-workers to much of the work surveyed here, including Mr. Scott Auerbach, Professor Shi-I Chu, Mr. Lionel Gaucher, Dr. Bernadette Jansen op de Haar, Dr. Andrew Peet, Professor Danny Yeager, Professor Weitao Yang, and especially Professor John Zhang. Research support by the Director, Office of Energy Research, Office of Basic Energy Sciences, Chemical Sciences Division of the US Department of Energy under Contract No. DE-AC03-76SF00098 and by the National Science Foundation, Grant CHE-8920690, is also gratefully acknowledged.

REFERENCES

1. D. E. Manolopoulos and D. C. Clary, Ann. Repts. C. (Roy. Soc. Chem.) **86**, 95 (1989).
2. W. H. Miller, Ann. Rev. Phys. Chem. **41**, 245 (1990).
3. (a) W. A. Lester, Jr., in Dynamics of Molecular Collisions, Part A, ed. W. H. Miller, p. 1, Plenum, New York, N.Y. (1976).
   (b) J. C. Light, in Atom-Molecule Collision Theory, ed. R. B. Bernstein, p. 239, Plenum, New York, N.Y. (1979).
   (c) D. Secrest, in Atom-Molecule Collision Theory, ed. R. B. Bernstein, p. 265, Plenum, New York, N.Y. (1979).
4. W. H. Miller, J. Chem. Phys. **50**, 407 (1969).
5. P. G. Burke and A. J. Taylor, Proc. Phys. Soc. London **88**, 549 (1966).
6. Y. C. Tang, M. LeMere, and D. R. Thompson, Phys. Rep. **47**, 167 (1978).
7. G. Wolken and Karplus, J. Chem. Phys. **60**, 351 (1974).
8. W. Kohn, Phys. Rev. **74**, 1763 (1948).
9. T.-Y. Wu and T. Ohmura, Quantum Theory of Scattering, pp. 57-68, Prentice-Hall, Englewood Cliffs, N.J. (1962).
10. (a) W. H. Miller and B. M. D. D. Jansen op de Haar, J. Chem. Phys. **865**, 6213 (1987).
    (b) M. Kamimura, Prog. Theor. Phys. Suppl. **62**, 236 (1977).
11. J. Z. H. Zhang, S.-I. Chu, and W. H. Miller, J. Chem. Phys. **88**, 6233 (1988).
12. (a) C. Schwartz Phys. Rev. **124** 1468 (1961); Ann. Phys. **10**, 36 (1961).
    (b) R. K. Nesbit, Variational Methods in Electron-Atom Scattering Theory, pp. 30-50, Plenum, New York, N.Y. (1980).
13. A. D. Isaacson, C. W. McCurdy, and W. H. Miller, Chem. Phys. **34**, 311 (1978).
14. A. J. F. Siegert, Phys. Rev. **56**, 750 (1939).
15. (a) J. Nuttall and H. L. Cohen, Phys. Rev. **188**, 1542 (1969).
    (b) F. A. McDonald and J. Nuttall, Phys. Rev. C **6**, 121 (1972).
    (c) R. T. Baumel, M. C. Crocker, and J. Nuttall, Phys. Rev. A **12**, 486 (1975).
16. (a) T. N. Rescigno and W. P. Reinhardt, Phys. Rev. A. **8**, 2828 (1973); **10**, 158 (1974).
    (b) B. R. Johnson and W. P. Reinhardt, *ibid.* **29**, 2933 (1984).
17. C. W. McCurdy and T. N. Rescigno, Phys. Rev. A **21**, 1499 (1980); **31**, 624 (1985).
18. For reviews, see
    (a) B. R. Junker, Adv. At. Mol. Phys. **18**, 207 (1982).
    (b) W. P. Reinhardt, Annu. Rev. Phys. Chem. **33**, 223 (1982).
    (c) Y. K. Ho, Phys. Rep. **99**, 1 (1983).
19. J. Z. H. Zhang and W. H. Miller, J. Chem. Phys. **91**, 1528 (1989).
20. J. Z. H. Zhang and W. H. Miller, J. Chem. Phys. **89**, 4454 (1988).
21. J. Z. H. Zhang and W. H. Miller, J. Phys. Chem. **94**, 7785 (1990).
22. J. Z. H. Zhang, J. Chem. Phys. **94**, 6047 (1991).
23. A. C. Peet and W. H. Miller, Chem. Phys. Lett. **149**, 257 (1988).
24. W. Yang, A. C. Peet, and W. H. Miller, J. Chem. Phys. **91**, 7537 (1989).
25. J. Z. H. Zhang, D. L. Yeager, and W. H. Miller, Chem. Phys. Lett. **173**, 489 (1990).
26. J. Z. H. Zhang and W. H. Miller, Chem. Phys. Lett. **153**, 465 (1988).
27. J. Z. H. Zhang and W. H. Miller, Chem. Phys. Lett. **159**, 130 (1989).
28. J. Z. H. Zhang, J. Chem. Phys., in press.
29. (a) J.-C. Nieh and J. J. Valentini, Phys. Rev. Lett. **60**, 519 (1988).
    (b) J.-C. Nieh and J. J. Valentini, J. Chem. Phys. **92**, 1083 (1990).
30. D. A. V. Kliner, D. E. Adelman, and R. N. Zare, J. Chem. Phys. **94**, 1069 (1991).
31. W. H. Miller and J. Z. H. Zhang, J. Phys. Chem. **95**, 12 (1991).

CLOSING REMARKS

G.H.F. Diercksen,
Max-Planck-Institut für Physik und Astrophysik,
Garching.

Before adjourning, we should be reminded that we are supported by society for pursuing our hobby - learning for no immediate return. We should be grateful for it and respond as

> *"men (and women) of learning [...], applying ourselves to seeking out the truth in the truth in the things as far as that has been permitted by God to human reason"*
>
> Nicolaus Copernicus
> *On the Revolutions, Book One*

It has been demonstrated at this Institute that sophisticated mathematical tools are ready. It is timely to apply them wisely in *"seeking out the truth"* in molecular physics.

Funds for basic research are limited, and will get more and more scarce in the future. We have to compete for them. Whatever the temptation may be, let us not get corrupted but stay honest and respect each other. What will count is not prestige, not power and not money but friends with an open and critical mind we can trust and rely on to provide the creative environment for *"seeking out the truth"*.

Finally, we should remember that our life is unique and limited - often much too short - while our work may, eventually, be carried on by others. Therefore, let us be wise enough to enjoy both life and work; at the right time, in the right way and in proper proportion.

It has been a pleasure to meet you all. I am sorry to see you leaving. Return home safely!

Faculty of the Advanced Study Institute

Geerd H.F. Diercksen,
Max-Planck-Institut für Astrophysik,
Karl-Schwarzschild-Str. 1,
8046 Garching bei München,
Germany.

Ian P. Grant
Department of Theoretical Chemistry,
University of Oxford,
5, South Parks Road,
Oxford OX1 3TG,
United Kingdom.
after 11 September, 1991:
Mathematical Institute,
University of Oxford,
24/29 St. Giles',
Oxford OX1 3LB,
United Kingdom.

Ivan Hubač
Department of Biophysics and Chemical Physics
Faculty of Mathematics and Physiks UK,
Mlynska dolina,
842 15 Bratislava,
Czechslovakia.

Per Jensen
Physikalisch-Chemisches Institut,
Justus-Liebug Universität Giessen,
Heinrich-Buff-Ring 58,
6300 Giessen,
Germany.

Poul Jørgensen
Chemistry Department,
Aarhus University,
Langelandsgade,
8000 Aarhus C,
Denmark.

Jacek Karwowski
Instytut Fizyki,
Uniwersytet Mikołaja Kopernika,
ul. Grudziqdzka 5/7,
87-100 Toruń,
Poland.

Roy McWeeny

Universita di Pisa,
Dipartimento di Chimica,
Via Risorgimento, 35,
86100 Pisa,
Italy.

William H. Miller

Department of Chemistry,
University of California,
Berkeley, CA 94720,
United States of America.

Jens N. Oddershede

Kemisk Institut,
Odense Universiet,
Campusvej 55,
5230 Odense M,
Denmark.

Josef Paldus

Department of Applied Mathematics,
Faculty of Mathematics,
University of Waterloo,
Waterloo, Ontario N2L 3G1,
Canada.

Victor R. Saunders

Science & Engineering Research Council,
Daresbury Laboratory,
Daresbury,
Warrington WA4 4AD,
United Kingdom.

Brian T. Sutcliffe

Department of Chemistry,
University of York,
Heslington,
York, YO1 5DD,
United Kingdom.

Stephen Wilson

Science & Engineering Research Council,
Rutherford Appleton Laboratory,
Chilton,
Oxfordshire OX11 0QX,
United Kingdom.

Participants at the Advanced Study Institute

Steven A. Alexander

Department of Physics,
University of Florida,
Gainesville, FL-32611,
United States of America.

Edoardo Aprá

Dipartmento di Chemica Inorganica,
Chimica Fisica e Chimica di Materiali,
Gruppo di Chimica Teorica,
Via P. Giuria 5,
10125 Torino,
Italy.

Per-Olof Åstrand

Department of Theoretical Chemistry,
University of Lund,
P.O. Box 124,
221 00 Lund,
Sweden.

Ferenc Bogar

Department of Theoretical Physics,
Jozsef Attila University,
Szeged,
Aradi vertanuk tere 1,
6720 Hungary.

Antonio C. Borin

Universidade de Sao Paulo,
Instituto de Quimica,
Lineu Prestes 748,
Sao Paulo,
Brazil CEP 05508

Robert Bukowski

Quantum Chemistry Laboratory,
Department of Chemistry,
University of Warsaw,
Pasteura 1,
02-093 Warsaw,
Poland.

Miroslaw Bylicki

Institute of Physics,
Nicholas Copernicus University,
Grudziadzka 5/7,
87 100 Toruń,
Poland.

Hazel Cox	C125 James College, University of York, Heslington, York YO1 5DD, United Kingdom.
François Culot	Laboratoire de Chimie Physique, CP 160, ULB, Avenue FD Roosevelt 50, 1050 Bruxelles, Belgium.
Michaël Deleuze	Laboratoire de Chimie Théorique Appliquée, F.U.N.B.P., Rue de Bruxelles 61, B5000 Namur, Belgium.
Matthias Ernzerhof	Institut für Physikalische und Theoretische Chemie, Universität Bonn, Wegelerstraße 12, 5300 Bonn 1, Germany.
Barbara Fehlner	TU-München, Theor. Chemie, Lichtenbergstr. 4, 8046 Garching bei München, Germany.
Berta Fernandez	Dpto. Quimica Fisica, Fac. Quimica Avda de las Ciencas, Universidad de Santiago de Compostela, 15702 Santiago de Compostela, Spain.
Graham Fletcher	Department of Physics, University of York, Heslington, York YO1 5DD, United Kingdom.
Norbert Flocke	Max-Planck-Institut für Astrophysik, Karl-Schwarzsschild Str. 1, 8046 Garching bei München, Germany.
Alessandro Fortunelli	Istituto di Chimica Quanistica del CNR, Via Risorgimento 35, 56100 Pisa, Italy.
Rudolf Friedmann	Department of Physical Chemistry, Martin Luther University, Halle-Wittenberg Mühlpforte 1, O-4010 Halle, Germany.

Stephan Fritzche	Department of Physics, University of Kassel, Heinrich-Plett-Str. 40, 3500 Kassel, Germany.
Alain D. Gorse	Laboratoire de Physicochimie Théorique, UA 503, Université de Bordeaux 1, 33405 Talence-Cedex, France.
Peter Graf	Max-Planck-Institut für Astrophysik, Karl-Schwarzsschild Str. 1, 8046 Garching bei München, Germany.
Steven Greiner	Allied-Signals R&T, 50 E. Algonquin Road, Des Plaines, Illinois 60019, United States of America.
Hülya Günay	Department of Chemistry, Middle East Technical University, 06531 Ankara Turkey
Richard Hargitai	Laboratory of Theoretical Chemistry, Eotvos University Budapest, 1518 Budapest 112, P.O. Box 32, Hungary.
Antonio J. Hernandez	Department of Chemistry, Simon Bolivar University, Apdo 89000, Caracas 1080 A Venezuela.
Hinne Hettema	Institute of Theoretical Chemistry, University of Nijmegen, Toernooiveld, 6525 ED Nijmegen, The Netherlands.
Wlodzimierz Jaskolski	Institute of Physics, Nicholas Copernicus University, Grudziadzka 5/7, 87 100 Toruń, Poland.
Tony Koures	Allied-Signals R&T, 50 E. Algonquin Road, Des Plaines, Illinois 60019, United States of America.

Milco Landtman
Valckenierstr. 65,
NL-1018 XE-Amsterdam,
The Netherlands.

Valentinas Lazauskas
Institute of Theoretical Physics and Astronomy
Groštanto 12,
232600 Vilnius,
Lithuania.

Ruth Le Sueur
Department of Physics and Astronomy,
University College London,
Gower Street,
London WC1E 6BT,
United Kingdom.

Gilson A. Rosa Lima
Dept. de Ciencia dos Materials e Mecanica,
IFUSP,
Universidade de Sao Paulo,
Caixa Postal 20516-01498,
Sao Paulo SP,
Brazil.

Algirdas Maknickas
Institute of Theoretical Physics and Astronomy
Groštanto 12,
232600 Vilnius,
Lithuania.

Massimo Malagoli
Dipartimento di Chimica,
Universita di Modena,
Via Campi 183,
41100 Modena,
Italy.

Laurence Malegat
Observatoire de Paris-Meudon,
Department Atomes et Molecules en Astrophysique,
5 Place Jules Janssen,
92195 Meudon Cedex,
France.

Piotr Malinowski
Institute of Physics,
Nicholas Copernicus University,
Grudziadzka 5/7,
87 100 Toruń,
Poland.

Christel Marian
Institute of Physical and Theoretical Chemistry,
University of Bonn,
Wegelerstr. 12,
5300 Bonn,
Germany.

Inmaculada Martin
Departamento de Química Fisica,
Facultad de Ciencias,
Universidad de Valladolid,
47005 Valladolid,
Spain.

Jozef Mášik
Faculty of Mathematics and Physics,
Comenius University,
842 15 Bratislava,
Czechoslovakia.

Jan von Milczewski
Max-Planck-Institut für Astrophysik,
Karl-Schwarzsschild Str. 1,
8046 Garching bei München,
Germany.

Robert Moszynski
Quantum Chemistry Laboratory,
Department of Chemistry,
University of Warsaw,
Pasteura 1,
02-093 Warsaw,
Poland.

Vladimiro Mujica
Escuela de Quimica,
Facultad de Ciencias,
Universidad Central de Venezuela,
Apartado 47102,
Caracas 1051,
Venezuela.

Ken Museth
Kemisk Institut,
Aarhus Universiteit,
Langelandsgade 140,
8000 Aarhus C,
Denmark.

Elke M.-L. Ohrendorf
Theoretische Chemie,
Physikalische-Chemisches Institut,
Universität Heidelberg,
Im Neuenheimer Feld 253,
6900 Heidelberg 1,
Germany.

Joakim Persson
Department of Theoretical Chemistry,
University of Lund,
P.O. Box 124,
221 00 Lund,
Sweden.

Josep Planelles
Secció Departamental de Química Física,
Collegi Universitari de Castelló,
Universitat de València,
Apartat 224,
12080 Castelló de la Plana,
Spain.

Wolfgang Quapp
Universität Leipzig,
Sektion Mathematik Z.307,
O-7010 Leipzig,
Germany.

Isabelle Rebatel
Laboratoire de Physique des Interactions,
Ioniques et Moléculaires,
Université de Provence,
Centre de St. Jérome - Case A61,
13397 Marseille Cedex 13,
France.

Maria del Mar Reguero de la Poza
Inst. Ciencia de Materiales CSIC,
C/Serrano 123,
28006-Madrid,
Spain.

Mary C. Salazar
Department of Chemistry,
Simon Bolivar University,
Apdo 89000,
Caracas 1080 A
Venezuela.

Markus Scheller
Physikalisch-Chemisches Institut,
Abt. für Theoret. Chemie,
INF 253,
6900 Heidelberg,
Germany.

Erich Schreiner
Max-Planck-Institut für Astrophysik,
Karl-Schwarzsschild Str. 1,
8046 Garching bei München,
Germany.

Fatma Sevin
Department of Chemistry,
Faculty of Engineering,
Hacettepe University,
06532 Beytepe,
Ankara,
Turkey.

Lidia Smentek-Mielczarek
Institute of Physics,
Nicholas Copernicus University,
Grudziadzka 5/7,
87 100 Toruń,
Poland.

Joaquim Jose Soares Neto
Departamento de Física,
Universidade de Brasilia,
CEP 70 910,
Brasilia-DF,
Brazil.

Jacek Styszynski	Katedra Fizyki, Universytet Szczeciński, Wielkopolska 15, 70 451 Szczegin, Poland.
Alessandra Tarantelli	Theoretische Chemie, Physikalische-Chemisches Institut, Im Neuenheimer Feld 253, 6900 Heidelberg, Germany.
Francisco Torrens	Department of Quimica-Fisica, Universitat de Valencia, Dr. Moliner-50, 46100-Burjassot, Spain.
Nathalie Vaeck	Van der Waals-Zeeman Laboratorium, University of Amsterdam, Valchenierstraat 65, 1018 XE-Amsterdam, The Netherlands.
Antoine D. Villesuzanne	Laboratoire de Physicochimie Théorique, UA 503, Université de Bordeaux 1, 33405 Talence-Cedex, France.
Luuk Visscher	Lab. voor Chemische Fysica, Nijenborgh 16, 9747 AG Groningen, The Netherlands.
Alex H. de Vries	Lab. voor Chemische Fysica, Nijenborgh 16, 9747 AG Groningen, The Netherlands.
Wolfgang Wenzel	Department of Physics, Ohio State University, 174 West 18th Avenue, Columbus, OH 43210, United States of America.
Graham Worth	Physical Chemistry Laboratory, South Parks Road, Oxford OX1 3QZ, United Kingdom

Index